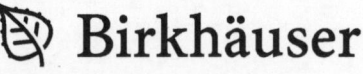
Birkhäuser

Geodynamics of the Latin American Pacific Margin

Edited by
William L. Bandy
Juanjo Dañobeitia
Carlos Mortera Gutiérrez
Yuri Taran
Rafael Bartolomé

Previously published in *Pure and Applied Geophysics*
(PAGEOPH), Volume 173, No. 10–11, 2016

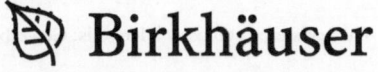

Editors

William L. Bandy
Instituto de Geofísica
Universidad Nacional
Autónoma de México
Ciudad Universitaria
Coyoacán, 04510, Mexico
DF, Mexico

Carlos Mortera Gutiérrez
Instituto de Geofísica
Universidad Nacional
Autónoma de México
04510, Mexico

Rafael Bartolomé
Instituto de Ciencias del
Mar-CSIC
P. Marítimo de la, Barceloneta 37-49
08003 Barcelona, Spain

Juanjo Dañobeitia
Unidad de Tecnología Marina, CSIC
Centro Mediterráneo de Investigaciones
Marinas y Ambientales Paseo
Marítimo de la Barceloneta 37–49
08003 Barcelona, Spain

Yuri Taran
Volcanology Department
Institute of Geophysics
UNAM, 3000 Av. Universidad
04510, Mexico D.F.
Mexico

The original version of the bookfrontmatter was revised: The spelling of the 3rd Editor's name and the affiliation of the 2nd Editor were corrected. The Erratum to the bookfrontmatter is available at http://dx.doi.org/10.1007/978-3-319-51529-8_24

ISBN 978-3-319-51528-1   ISBN 978-3-319-51529-8 (eBook)
DOI 10.1007/978-3-319-51529-8

Library of Congress Control Number: 2016963254

Mathematics Subject Classification (2010): 86-XX, 60-XX, 62-XX, 65-XX, 76-XX

Cover illustration: Based on Figure 6 of Bartolomé et al., this volume.

Cover design: deblik, Berlin

Printed on acid-free paper

This book is published under the trade name Birkhäuser (www.birkhauser-science.com)

The registered company is Springer International Publishing AG

The registered company address is: Gewerbestrasse 11, 6330 Cham, Switzerland

Contents

Pure Appl. Geophys. 173 (2016), 3213–3216
© 2016 Springer International Publishing
DOI 10.1007/s00024-016-1404-y

Introduction

WILLIAM L. BANDY,[1] JUANJO DAÑOBEITIA,[2] CARLOS MORTERA GUTIÉRREZ,[3] YURI TARAN,[4] and RAFAEL BARTOLOMÉ[5]

The Latin American Pacific margin from the Californias to the tip of Tierra del Fuego shares a common geodynamic heritage; the geologic sculptures of this heritage have been chiseled out in large part by the geodynamic processes associated with the subduction of oceanic lithosphere beneath the continental margins. Many aspects of these processes are addressed in the 22 articles comprising this volume, articles which cover a broad spectrum of scientific disciplines. These processes include: (1) the emplacement of magmas within, and subsequent deformation of, the overriding continental lithosphere (Longo et al.; Ochoa-Chávez et al.; Núñez-Cornú et al.; Neumann et al.), the magmas being produced by a variety of processes, such as the melting of the subducted lithosphere, decompression melting related to the tearing of both the subducting and overriding plates, and mantle flow through tears in the subducting plate; (2) the subduction of buoyant bathymetric features, such as seafloor spreading segments of the EPR off the Californias and Chile and the Cocos Ridge in central America and their effects on the overriding plate (Brandes et al.; Bourgois et al.); (3) plate motion reorganizations and related changes in the state of stress in the overriding continental crust (Mortera Gutiérrez et al.); (4) plate margin truncation such as that which occurred in southern Mexico due to the translation of the Chortís Block, a process that is most likely occurring at present along the Pacific coast of Baja California (Munguía, Mayer et al.; Munguía, González-Escobar et al.); (5) processes that generate the great megathrust earthquakes and tsunamis (Papadopoulos and Minadakis; Bartolome et al.; Dañobeitia et al.); (6) tectonic processes related to hydrocarbon generation and accumulation (Michaud et al.; Longo et al.); (7) processes that generate slow-slip events along the mega-thrust (Graham et al.); (8) processes that produce tearing of continental plates as is occurring in the Gulf of California (Dañobeitia et al.; Ortega et al.); (9) processes that produce forearc deformations, such as trench parallel transcurrent faults, translation of forearc slivers, and crustal uplift of the forearc area (Suárez et al.; Gaidzik et al.; Rousset et al.; Arzate et al.); (10) ridge-trench collisions, ophiolite emplacement and overthrusting of continental terrains (Arzate et al.; Bourgois et al.); and (11) mantle flow associated with, or induced by, the subducting lithosphere (Bernal-López et al.; Rosas et al.; Neumann et al.).

The following studies help to illustrate the diversity of the investigations that are being conducted along this margin, the results of which will be of great value in furthering our understanding of the complex array of geodynamic processes that have been sculpting the geologic landscape of the Latin American Pacific margin.

Bourgois et al. provide a review and synthesis of the geology and recent tectonic history of the Chile Triple junction, at which an active spreading center has been colliding with the subduction zone resulting in ophiolite emplacement, Adakite-like generation, and the development of a slab window beneath the South American continent. The tectonic history reveals the need for a westward relocation of some of

[1] Departamento de Geomagnetismo y Exploración, Instituto de Geofísica, Universidad Nacional Autónoma de México, Mexico DF, Mexico. E-mail: bandy@geofisica.unam.mx

[2] Unidad de Tecnología Marina-CSIC, P. Marítimo de la Barceloneta 37-49, 08003 Barcelona, Spain.

[3] Departamento de Sismología, Instituto de Geofísica, Universidad Nacional Autónoma de México, Mexico DF, Mexico.

[4] Departamento de Volcanología, Instituto de Geofísica, Universidad Nacional Autónoma de México, Mexico DF, Mexico.

[5] Instituto de Ciencias del Mar-CSIC, P. Marítimo de la Barceloneta 37-49, 08003 Barcelona, Spain.

the ridge segments as they near the trench, as has been observed elsewhere (e.g., off Mexico).

Papadopoulos and Minadakis investigate the foreshock activity associated with several great earthquakes occurring in the Chile subduction zone to verify and better understand previously noted 3-D precursory patterns. Using the great earthquake of 1 April 2014 as a reference event, they observe that similar foreshock 3-D patterns precede the great earthquakes of 27 February 2010 and 16 September 2015 within critical distances of about 170 and 50 km, respectively.

Longo et al. perform an aeromagnetic analysis of the subsurface structure and magma emplacement of the Auca Mahuida Volcano, Argentina and find that the magma emplacement is being controlled by the regional fault system, and has played a major role in the maturation and subsequent accumulation of oil below the volcano.

Michaud et al. investigate the spatial distribution of gas flares along the Ecuadorian margin using water column acoustic backscatter data. High-resolution seismic profiles show that most flares occur close to the surface expression of active faults, deformed areas, slope instabilities or diapiric structures.

Suárez et al. study the 10 April 2014 Nicaraguan earthquake. They find, in conjunction with the results of previous studies, that arc-parallel strains associated with the NW moving Nicaraguan forearc sliver, located between the volcanic arc and trench, are being accommodated both by block rotations onshore and by an NW translation of a smaller forearc sliver located in the offshore and near-shore areas.

Rosas et al. present the first 3-D, steady-state kinematic–dynamic thermal model for the Costa Rica–Nicaragua subduction zone. The models predict that the mega-thrust seismogenic zone decreases from about 100 km below Costa Rica to just a few kilometers below Nicaragua and also indicates that variations in slab dip induce an along-strike mantle flow in the mantle wedge. They conclude that 2-D models are not suitable for use in this area.

Brandes et al., using multichannel seismic reflection profiles, carry out 3-D kinematic retro-deformation modeling to analyze the spatial evolution of a bend in the South Limón fold-and-thrust belt, Costa Rica. They find that the bend can be modeled by a simple NNE-directed transport during a single deformational phase, and also indicates the need for the presence of a Trans-Isthmic fault system during this deformational phase.

Arzate et al. present a magnetotelluric profile oriented perpendicular to the trench in southern Mexico. They find evidence for a low angle thrust contact between the Oaxaca and Juárez terranes, with the older Oaxaca terrain overthrusting the younger Juárez terrane. They also find evidence that uplift in the Sierra Madre del Sur is facilitated by slab-dehydration driven bouyancy.

Bernal-López et al. employ shear wave splitting measurements to study the mantle flow under southern Mexico. They find that in the forearc, the fast axes are oriented NE–SW, whereas in the backarc, they are oriented N–S. The differences are proposed to be due to the entrainment of mantle flow under the subducting, subhorizontal slab in the forearc region and induced corner flow in the backarc mantle wedge.

Neumann et al. use scaled analog laboratory modeling to investigate the mantle flow beneath western Mexico near the gap between the Rivera and Cocos plates. They find a deep toroidal flow of asthenospheric mantle through the gap and a shallow counter-toroidal flow in the uppermost 100 km of the mantle wedge that draws mantle from the western Trans-Mexican volcanic Belt to the Jalisco block and then plunges into the deep mantle by the poloidal cell of the Cocos slab. They conclude that the model can explain the eruption of OIB lavas in the vicinity of Guadalajara.

Gaidzik et al. perform geomorphic, structural and fault kinematic analyses of the forearc area of Guerrero to determine the tectonic processes active in this area. They find evidence for active, sinistral transcurrent and normal faults oriented subparallel to the trench, consistent with GPS measurements, and the sense of oblique plate convergence in this area.

Graham et al. estimate the time-dependent slip distributions and Coulomb failure stress changes for six slow-slip events along Guerrero and Oaxaca using continous GPS data. They find evidence of slow slip on the mega-thrust everywhere between Oaxaca and Guerrero. In addition, slow slip reduces the slip deficit in the Guerrero Gap, whereas in Oaxaca, little or no slip is relieved by slow slip along the mega-thrust.

Rousset et al. use GPS data and morphology to explore links between variations in inter-slow-slip-event coupling along the southern Mexico subduction zone and the long-term topography of the coastal areas of Guerrero and Oaxaca. Their results favor a model in which frictional asperities partly control short-term inter-SSE coupling as measured by geodesy and in which those asperities persist through time.

Ochoa-Chávez et al. perform a P-wave tomographic analysis of the crustal and upper mantle structure of the Jalisco block and adjacent areas. They find that magma emplacement under the Colima Volcanic Complex is fracturing the crust, forming a well defined, classical, rift–rift–rift fracture pattern at mid crustal levels. No evidence is observed to support either a trenchward migration of the volcanic front or toroidal asthenospheric flow through the slab tears bounding the Jalisco Block to the NW and SE.

Núñez-Cornú et al. analyze seismic characteristics of explosions at Colima Volcano, Mexico, associated with the 2003–2005 eruption to determine characteristic features, propagation velocities, and origin times for both deep seismic sources and the associated explosions. The results suggest the presence of various magmatic pathways beneath the volcano.

Mortera Gutiérrez et al. determine the morphology, magnetic anomalies and shallow structure of the Bahía de Banderas, Mexico using multibeam bathymetry, marine total field magnetics and subbottom seismic reflection data. They find evidence that the stress field within the bay is presently extensional, oriented NNW–SSE, roughly parallel to the trench axis. Furthermore, they find no evidence for the previously proposed Bandera Fault, a regional fault proposed previously to extend westward from the bay to the Middle America Trench.

Bartolome et al. use multichannel seismic reflection data from TSUJAL project (2014) to obtain a better understanding of the complex interactions between Rivera and North American plates. They characterize the internal crustal structure of Rivera plate off Puerto Vallarta as a smooth dipping subduction of the Rivera plate beneath the North American plate, dominated by subduction–accretion

along the lower slope of the margin. They noted significant mass wasting of the continental slope and they concluded from the data that the region appears to be prone to generation of great earthquakes.

Dañobeitia et al. use multichannel and wide-angle data from TSUJAL project (2014) to obtain a better understanding of the complex interactions between Rivera and North American plates in the area of the Tres Marias Islands, located at the northern terminus of the Middle American Trench. These data show a crustal thickness of the oceanic slab of 6–7 km, and anomalous crustal velocity ($\leq$5.5 km/s) underneath Maria Magdalena Rise, located south of the islands, probably related to the initial phases of the Baja California Peninsula continental breakup. The Moho depth varies from 10 km west of TMI to >15 km east of the islands. The bathymetric escarpment located south of the islands is quite steep, resulting in numerous large slumps. These data also indicate compression west of the Islas Marias, suggested by the deformation of sedimentary wedges and elevated islands.

Ortega et al. investigate whether earthquake source mechanisms reflect important variations going from the Gulf of California to the East Pacific Rise. They find that the moment tensor solutions of the GC and EPR are similar; however, there is a clockwise rotation in the s_1 and s_3 directions for the GC compared to the EPR. They also found that the full moment tensor inversion best resolves complex faults, composed mainly of two double couples.

Munguía and Mayer et al. analyze the earthquake swarm of 2006 occurring within the Bahía Asunción, Mexico. They find that these earthquakes occurred on the coastline of the peninsula, east of the Tosco-Abreojos fault, indicating that the boundary between the Baja California and the Pacific plate is wider than previously thought.

Munguía and González-Escobar et al. analyze earthquake sequences located near San Carlos, Baja California Sur that occurred during three time periods. These events were found to be associated with the Santa Margarita fault, located about 60 km east of the Tosco-Abreojos fault system, and showed both normal and strike-slip components of fault motion. They conclude that the boundary between the Baja California and the Pacific plate is about 60 km wide in this area, and is a transtensional boundary.

González-Escobar et al. analyze the structure and stratigraphy of the Magdalena shelf located along the Pacific margin of Baja California using multichannel seismic reflection data. They imaged the old pre-Miocene forearc basin that is presently disrupted by the faults (Tosco-Abreojos to the west and the Santa Margarita and San Lázaro faults to the east) forming the transtensional boundary between the Baja California and the Pacific plate. The boundary is about 90 km wide in this area. They also observed a series of eastward deepening half-grabens containing a thick sediment infill.

Acknowledgments

We thank all the authors for their contributions to this special volume and Renata Dmowska for her help during the preparation of this volume and for handling the reviews of our contributions. We are most grateful to the many reviewers for their efforts to improve the quality of this special volume. These people include, in alphabetical order, José Miguel Azañón Hernández (U. Granada, Spain), Peter Bird (UCLA, USA), Thierry Calmus (UNAM, MEXICO), Eduardo Camacho (U. Panama, PANAMA), Paterno Castillo (UCSD, USA), Timothy Dixon (U. of South Florida, USA), Christian Escudero (U. Guadalajara, MEXICO), Donald Fisher (Penn State, USA), Victor Hugo Garduño Monroy (U. Michoacan, MEXICO), Antonio González (CICESE, MEXICO), Yoshihiro Ito (Kyoto U., JAPAN), Vladimir G. Kossobokov (IETP RAS, RUSSIA), Geoffroy Lamarche (NIWA, NEW ZEALAND), Dan Lizarralde (WHOI, USA), Maureen Long (Yale, USA), Héctor López Loera (IPICYT, MEXICO), John Makario Londoño (SGC, COLOMBIA), Jack Loveless (Smith College, USA), Anthony Lowry (Utah State, USA), Vlad C. Manea (UNAM, MEXICO), François Michaud (UPMC, FRANCE), Neil Mitchell (U. Manchester, UK), Hirata Naoshi (U. Tokyo, JAPAN), James Ni (New Mexico State U., USA), Jaime Humberto Ortega (U. Chile, CHILE), Javier Pacheco (UNA, COSTA RICA), Patricia Persaud (CalTech, USA), Cristina Pomposiello (INGEIS, ARGENTINA), Luis Quintanar Robles (UNAM, MEXICO), Charlotte Rowe (Los Alamos Nat. Lab., USA), Hartmut Seyfried (U. Stuttgart, GERMANY), Joann Stock (CalTech, USA), Gerardo Suárez (UNAM, MX), Patrick Taylor (NASA, USA), Gustavo Tolson (UNAM, MEXICO), Paul Umhoefer (Northern Arizona, USA), Gabriel Vargas (UCHILE, CHILE), Ikuko Wada (U. Minnesota, USA), F. Ramón Zúñiga (UNAM, MEXICO), and several anonymous reviewers who reviewed our contributions.

(Published online September 23, 2016)

Pure Appl. Geophys. 173 (2016), 3217–3246
© 2016 Springer International Publishing
DOI 10.1007/s00024-016-1317-9

▌Pure and Applied Geophysics

A Review on Forearc Ophiolite Obduction, Adakite-Like Generation, and Slab Window Development at the Chile Triple Junction Area: Uniformitarian Framework for Spreading-Ridge Subduction

JACQUES BOURGOIS,[1] YVES LAGABRIELLE,[2] HERVÉ MARTIN,[3] JÉRÔME DYMENT,[4] JOSE FRUTOS,[5] and
MARIA EUGENIA CISTERNAS[6]

Abstract—This paper aggregates the main basic data acquired along the Chile Triple Junction (CTJ) area (45°–48°S), where an active spreading center is presently subducting beneath the Andean continental margin. Updated sea-floor kinematics associated with a comprehensive review of geologic, geochemical, and geophysical data provide new constraints on the geodynamics of this puzzling area. We discuss: (1) the emplacement mode for the Pleistocene Taitao Ridge and the Pliocene Taitao Peninsula ophiolite bodies. (2) The occurrence of these ophiolitic complexes in association with five adakite-like plutonic and volcanic centers of similar ages at the same restricted locations. (3) The inferences from the co-occurrence of these sub-coeval rocks originating from the same subducting oceanic lithosphere evolving through drastically different temperature–pressure (*P–T*) path: low-grade greenschist facies overprint and amphibolite-eclogite transition, respectively. (4) The evidences that document ridge-jump events and associated microplate individualization during subduction of the SCR1 and SCR-1 segments: the Chonos and Cabo Elena microplates, respectively. The ridge-jump process associated with the occurrence of several closely spaced transform faults entering subduction is controlling slab fragmentation, ophiolite emplacement, and adakite-like production and location in the CTJ area. Kinematic inconsistencies in the development of the Patagonia slab window document an 11- km westward jump for the SCR-1 spreading segment at ~6.5-to-6.8 Ma. The SCR-1 spreading center is relocated beneath the North Patagonia Icefield (NPI). We argue

that the deep-seated difference in the dynamically sustained origin of the high reliefs of the North and South Patagonia Icefield (NPI and SPI) is asthenospheric convection and slab melting, respectively. The Chile Triple Junction area provides the basic constraints to define the basic signatures for spreading-ridge subduction beneath an Andean-type margin.

Key words: Chile Triple Junction, spreading-ridge subduction, forearc ophiolite, granite, adakite-like, patagonia slab window.

1. Introduction

The magmatic rocks, including mafic to ultramafic assemblages that emplaced along suture zones, typically display-varying geochemical affinities and metamorphic evolution. They were recently considered as resulting from oceanic spreading-ridge subduction. For instance, SCHARMAN et al. (2012) proposed that the high-grade metamorphism of the Chugach complex, Alaska, originated from the Kula-Farallon/Resurrection ridge subduction during Late Cretaceous–Paleocene, participating both in the growth and evolution of the local continental crust. Spreading-ridge subduction is a major geodynamic process that is considered to have dramatically affected the geology of both North and South American margins during the past 70 Myr (CHAN et al. 2012; McCRORY et al. 2009; SISSON and PAVLIS 1993; KAY et al. 1993; RAMOS and KAY 1992; DICKINSON and SNYDER 1979; ATWATER, 1970). In addition, spreading-ridge subduction was identified as efficiently accelerating the recycling of crustal material into the deep mantle (LIU et al. 2012a; STERN 2011; STERN and SCHOLL 2010; BOURGOIS et al. 1996).

---

[1] Institut des Sciences de la Terre Paris (iSTeP), Université Pierre et Marie Curie and Centre National de la Recherche Scientifique, Case 110, 4, place Jussieu, 75252 Paris Cedex 05, France. E-mail: jacques.bourgois@upmc.fr

[2] Observatoire des Sciences de l'Univers de Rennes, Geosciences Rennes, UMR 6118, B14B, 263 avenue du Général Leclerc, CS 74205, 35042 Rennes Cedex, France.

[3] Laboratoire Magmas et Volcans, Université Blaise Pascal, CNRS-IRD-OPGC, 5 rue Kessler, 63038 Clermont-Ferrand Cedex, France.

[4] Géosciences Marines, Institut de Physique du Globe (IPGP), 1, rue Jussieu, 75238 Paris Cedex 05, France.

[5] San Juan de Luz 4060, Depto. 703, Providencia, Santiago, Chile.

[6] Instituto de Geologia Economica Aplicada (GEA), Universidad de Concepcion, Concepción, Chile.

Alaskan-type mafic–ultramafic complexes, high-temperature metamorphic belts, and adakite emplacement are geological features considered as symptomatic of ridge–trench interaction for the evolution of the Paleozoic Central Asian Orogenic Belt (ZHANG et al. 2010, 2013; MAO et al. 2012; LIU et al. 2012a, b; TANG et al. 2012; YANG et al. 2012; CAI et al. 2010) and the melting of delaminated lower crust (CHEN et al. 2013; LING et al. 2013). These works were published in the wake of the conceptual model developed by COLE and STEWART (2009), which further clarified the North America continental margin magmatism, following Oligocene-to-Paleocene episodes of spreading-ridge subduction. Likewise, the Chile ridge subduction provides a natural living analog for the genesis of the Archaean continental crust through oceanic crust melting at depth (KOMIYA et al. 2015; MARTIN 1986, 1999; MARTIN et al. 2005; MOYEN and MARTIN 2012).

Because spreading-ridge subduction is a model for explaining past associations of different rock-units, the time has come to examine the detailed signatures and architecture from currently subducting active spreading centers and to constrain their evolution. Such an active situation exists at the modern Chile Triple Junction (CTJ) area (Fig. 1), where the Chile spreading center is entering the subduction zone beneath the South American plate (BOURGOIS et al. 2000; BEHRMANN et al. 1994; CANDE and LESLIE 1986). This triple junction is the only one worldwide, where the overriding plate is continental in character. Consequently, the CTJ appears to be a good analog for the past ridge-trench collision occurring along an Andean-type margin, a situation that likely happened during the subduction of the Tethys Ocean. The emplacement of ophiolite complexes (ANMA et al. 2009; SHIBUYA et al. 2007; BOURGOIS et al. 1993; MPODOZIS et al. 1985) and adakite-like rocks (BOURGOIS et al. 1996; STERN and KILIAN 1996) in association with magmatic products exhibiting highly variable geochemical characteristics (LAGABRIELLE et al. 2000; GUIVEL et al. 1999) is one of the more reliable signatures for spreading-ridge subduction.

In this paper, we use a multi-methodological approach based on kinematic, geologic, and magmatic signatures, to build a consistent and realistic scenario, accounting for the subduction of the Chile spreading-ridge during the past ∼6-to-7 Myr. Earlier ideas integrated with recent observations along the forearc area support a generic model for spreading-ridge subduction developed on a uniformitarian basis. One of the targets of this paper consists in providing basic constraints for identifying past spreading-ridge subduction along Andean-type forearcs, including when obscured by subsequent continental collision events, such as along the Great Tibetan Plateau (NIU et al. 2013). This paper also provides constraints on the Patagonia slab window development and location along with the potential effects on the back-arc magmatism in the Andean foreland.

2. Sea-Floor Kinematics

The South Chile Ridge (SCR) is the boundary between Nazca and Antarctica plates (Fig. 1), it consists of short segments (SCR1, SCR2, and SCR3) trending N160°E–165°E and separated by a series of parallel fracture zones (FZ). In the area surveyed during the CTJ cruise (BOURGOIS et al. 2000), these are the Darwin, Guamblin, and Guafo FZ north of the CTJ. To the south, the Taitao, Tres Montes, Esmeralda, and Madre de Dios FZ bound the already subducted ridge segments SCR0, SCR-1, and SCR-2.

Seafloor spreading along the SCR occurred at a rate of 70 mm year^{-1} (DeMETS et al. 1994) to 64 mm year^{-1} over the past 5 Myr (HERRON et al. 1981). Subsequently, TEBBENS et al. (1997) have constrained the average full rate to be 62 mm year^{-1} during the past 6 Myr. Magnetic data acquired during the CTJ cruise of the R/V L'Atalante (JB, Chief Scientist) allowed revising the spreading rates to be 60–66 mm year^{-1} during the past 3.5 Myr (anomalies 1–2A), whereas the older anomalies (3–5, i.e., ∼4 to 10 Ma) reveal a faster accretion rate of ∼80 mm year^{-1} (LAGABRIELLE et al. 2015). North of the Chile Triple Junction (BREITSPRECHER and THORKELSON 2009), the Nazca plate is being subducted beneath the South America plate at a rate of 75 mm year^{-1} in an N077°E direction during the past 3 Myr, 79 mm year^{-1} in an N079°E direction from 3 to 4 Ma, and 78 mm year^{-1} in an N078°E direction from 4 to 11 Ma. To the south, the Antarctic plate is subducting beneath the South America plate at a rate

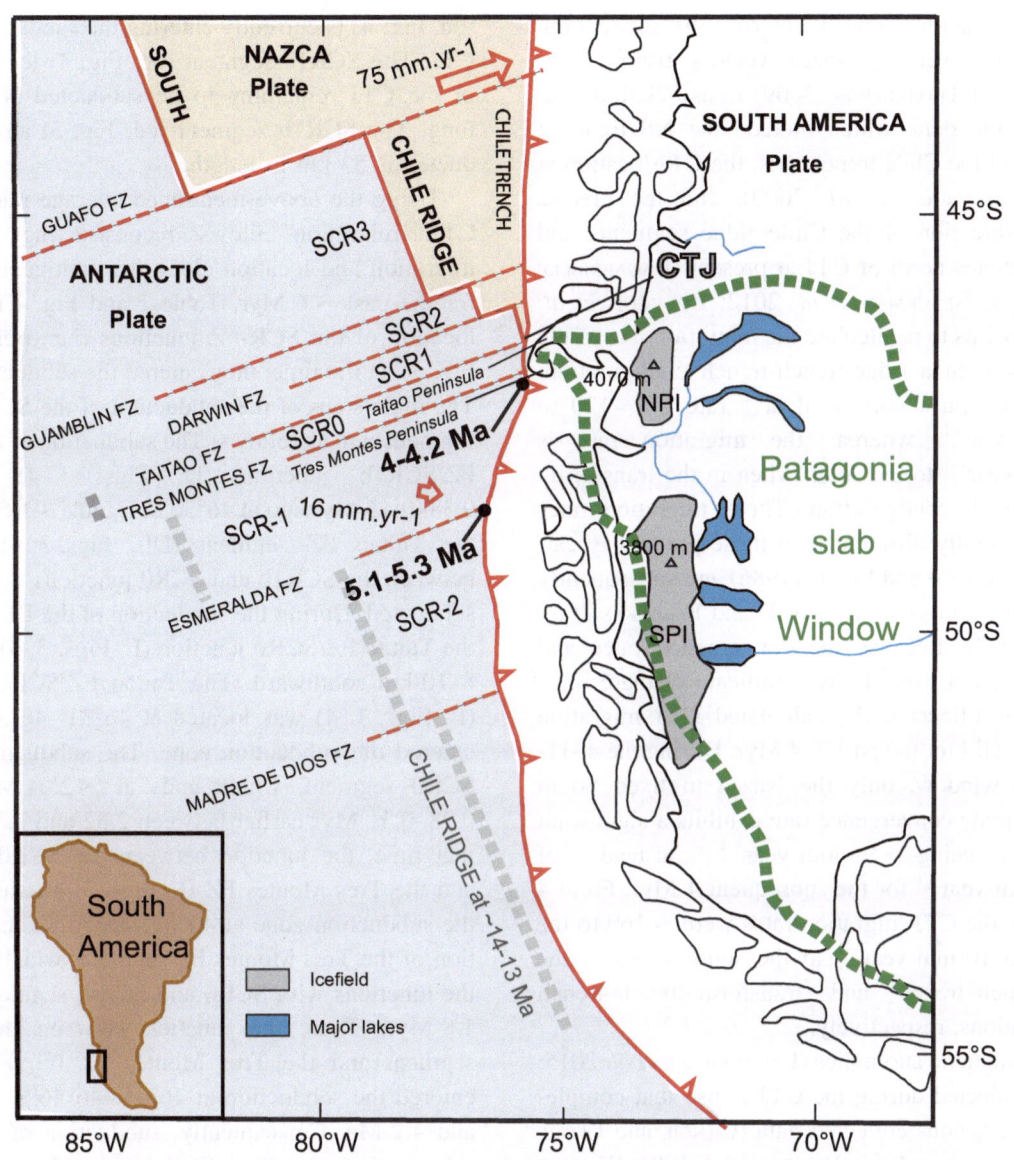

Figure 1

Spreading center at the Nazca-Antarctic plate boundary is subducting beneath the South America plate. Main oceanic features are shown, including the South Chile Ridge (SCR) and fracture zones (FZ). Convergence rates (direction shown by arrows) are those calculated by BREITSPRECHER and THORKELSON (2009). The *two bold black dots* and *number* show age of ridge/transform-trench intersection and its northward migration through time. Patagonia slab window geometry (*thick green dash line*) is from RUSSO *et al.* (2010a, b), BREITSPRECHER and THORKELSON (2009), and BOURGOIS and MICHAUD (2002). *Thick grey dash line* shows the position of the ridge crest relative to South America at 14–13 Ma (TEBBENS *et al.* 1997). *CTJ* Chile Triple Junction, *NPI* North Patagonia Icefield, *SPI* South Patagonia Icefield. *Inset* shows the general location

of 16 mm year^{-1} in an N100°E direction during the past 3 Myr, and 15 mm year^{-1} in an N101°E direction from 3 to 11 Ma.

Because the Chile ridge axis is ~10° to 12° oblique to the Chile trench axis, the actively spreading Chile ridge has been subducting beneath South America forming a generally northward moving ridge-trench-trench triple junction since ~14 Ma (LESLIE 1986; CANDE and LESLIE 1986; CANDE *et al.* 1982). At present, the SCR1 segment located north of the Taitao fracture zone is entering the subduction. The updated velocity space vectors from BREIT-SPRECHER and THORKELSON (2009) in association with the available data, which include the NS trending direction of the Chile trench axis, the CTJ location at 46°09′S (BOURGOIS *et al.* 2000), and the precise trending direction of the Chile ridge segments and Fracture zones north of CTJ at present (LAGABRIELLE *et al.* 2015; BLACKMAN *et al.* 2012; BOURGOIS *et al.* 2000) allow us to recalculate the migration rate of the CTJ. When, in a ridge-trench-trench configuration, the CTJ migrates northward at a rate of ∼120 to 130 mm year^{-1}, whereas the migration rate is 5–6 mm year^{-1} to the south when in the transform–trench–trench configuration. These migration rates are substantially different than those previously calculated by CANDE and LESLIE (1986) and subsequently assumed by GUIVEL *et al.* (2003) and BOURGOIS *et al.* (1996, 2000). Because the convergence rates and azimuth do not exhibit any significant change, since the Lower Pliocene, the calculated CTJ migration rates are valid for the past 3–4 Myr. During the 4–11-Ma time window, only the Nazca to fixed South America plate convergence rate exhibits a significant variation being 93 mm year^{-1} instead of 75–79 mm year^{-1} for the subsequent 4 Myr. From 4 to 11 Ma, the CTJ migration rates were ∼160 to the north and 10 mm year^{-1} to the south when in the ridge–trench–trench and transform–trench–trench configurations, respectively.

The magnetic anomalies (LAGABRIELLE *et al.* 2015; Fig. 2) collected during the CTJ cruise that complement the previous collected data (CANDIE and LESLIE 1986; CANDE *et al.* 1982; HERRON *et al.* 1981; WEISSEL *et al.* 1977) allow constraining the structure of the ocean floor between 45°20′ and 48°20′S. The length (Fig. 3) of the SCR segments is fairly well constrained being ∼44, ∼90, ∼39, and ∼181 km for SCR2, SCR1, SCR0, and SCR-1, respectively. Assuming that transform faults exhibit no significant change in length as they subduct (THORKELSON 1996; THORKELSON and TAYLOR 1989), the length of the ridge-to-ridge-transform segment is ∼60, ∼122, and ∼135 km along the Darwin, Taitao, and Tres Montes FZ, respectively. No data exist for constraining the length of the ridge-to-ridge-transform segment along the Esmeralda FZ. In addition, the SCR1 segment

(bd, Fig. 4) is currently entering the subduction at the CTJ. The SCR1a segment (bc, Fig. 4) located north of the CTJ remaining to be subducted is ∼37-km long. The SCR1b segment (cd, Fig. 5) already subducted is 53 km in length.

Using the above-mentioned average rates for the CTJ, migration allows reconstructing the CTJ migration and location along the continental margin for the past ∼6 Myr. Table 1 and Fig. 4 report the location of the SCR-FZ junctions (i.e., points D–G, Fig. 3); at the time, they entered the subduction zone. The main steps of the subduction of the SCR and FZ segments are as follows: The subduction of the Taitao FZ/SCR1b junction (D, Figs. 3, 4) occurred 0.38–0.42 Myr ago at 46°36′S. It took ∼1.62 Myr for the Taitao FZ segment (DE, Figs. 3, 4) located between the SCR1b and SCR0 junctions to be totally subducted. During the subduction of the DE segment, the Taitao FZ/SCR0 junction (E, Figs. 3, 4) migrated 8–10-km southward. The Taitao FZ/SCR0 junction (E, Figs. 3, 4) was located at 46°31′–46°32′S, as it entered the subduction zone. The subduction of the SCR0 segment, which ends at ∼2.04 Ma, began 0.28–0.31 Myr earlier between 2.32 and 2.35 Ma. At that time, the junction between the SCR0 segment and the Tres Montes FZ (F, Figs. 3, 4) was entering the subduction zone at 46°51′–46°52′S. The subduction of the Tres Montes FZ segment, which connects the junctions with SCR0 and SCR-1 segments lasted 1.8 Myr. Thus, the junction between the SCR-1 segment and the Tres Montes FZ (G, Figs. 3, 4) entered the subduction at 46°45′–46°46′S between 4 and 4.2 Ma. Consequently, subduction of the Chile ridge —i.e., segment SCR-1—ended shortly after 4–4.2 Ma in an area off the southern coastline of the Tres Montes Peninsula. Subduction of the SCR-1 segment is older than 4–4.2 Ma. The 181-km-long SCR-1 segment was entering the subduction zone between 5.1 and 5.3 Ma at ∼47°40′ to 47°50′S. Accepting that no major tectonic event, such as ridge jump and associated microplate evolution, has occurred, since the subduction of the SCR-1/Tres Montes FZ junction, subduction of ridge/transform junctions occurred as reported and portrayed in Fig. 4. In brief, underthrusting of ridge segments SCR-1, SCR0, and SCR1b beneath the Andean continental margin occurred between 4 and 5.3, 2.04 and

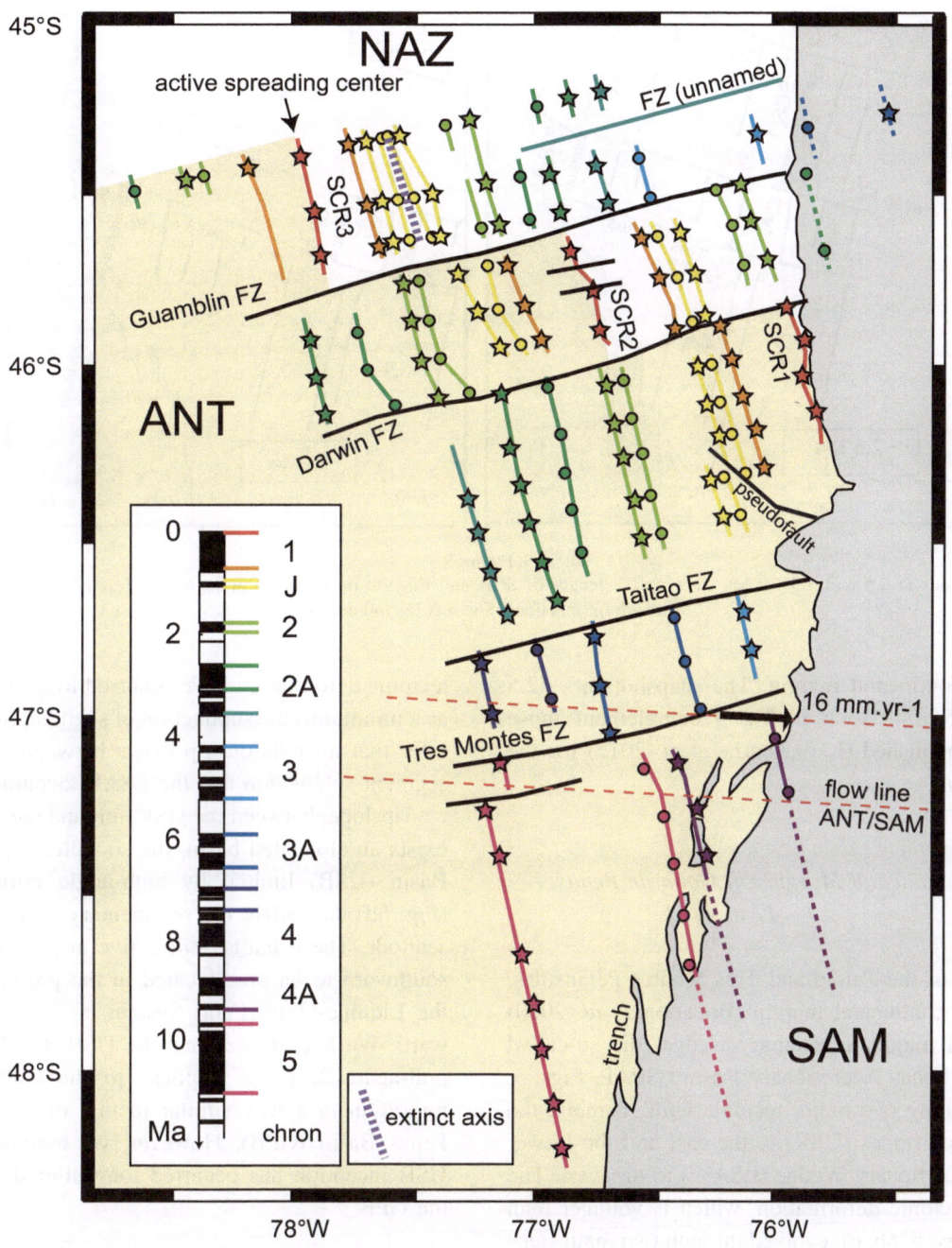

Figure 2
Magnetic picks, isochrons, and fracture zones deduced from the magnetic anomalies acquired during the CTJ cruise (from LAGABRIELLE *et al.*
2015). *Colored symbols* and *lines* represent the magnetic picks and isochrons, respectively, with *different colors* for different ages (see *inset*
for ages and names). *Stars* and *circles* mark the older and younger side, respectively, of normal polarity intervals. *Black lines* show the
location of fracture zones, which are also located from EM12 bathymetric data (BOURGOIS *et al.* 2000). *Dotted lines* denote isochrons projected
underneath the continental margin. *Thin dashed red lines* are the flow lines for the Antarctic/South American relative motion

2.35, and 0 and 0.4 Ma, respectively. Subduction of
the ridge-to-ridge segments along Taitao and Tres
Montes FZ occurred from 0.4 to 2.04 and from 2.35

to 4 Ma, respectively. These data allow reconstructing the evolution through time of the relationship
between the subducting oceanic plates and the

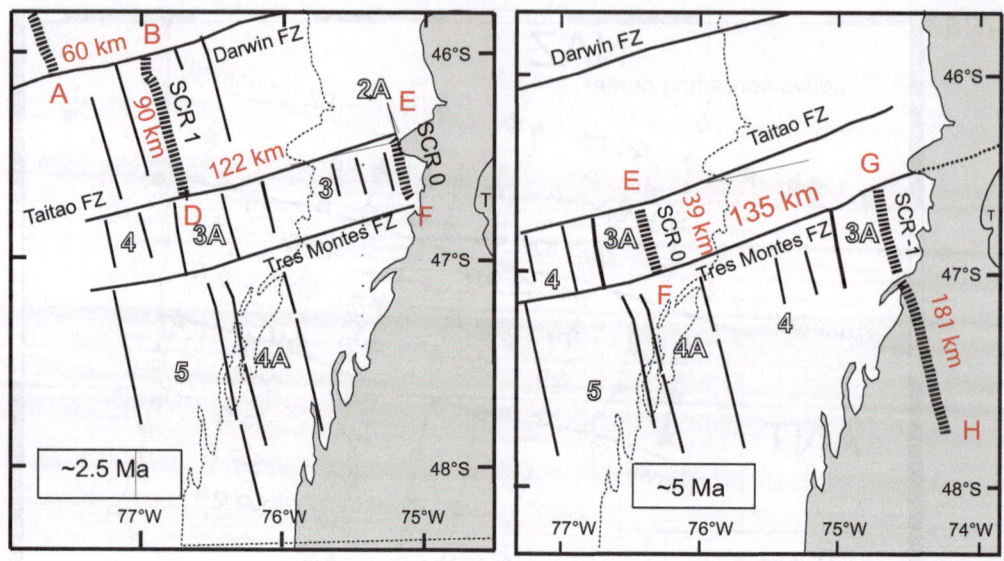

Figure 3

Reconstructions at 2.5 and 4.5–5.0 Ma showing the length of SCR and ridge-to-ridge-transform segments. *Letters A–H* refer to ridge-transform junctions. See text for more details

Andean continental margin. The snapshots at ∼2.5 and 4.5–5 Ma shown in Fig. 3 complement those recently published (LAGABRIELLE *et al.* 2015) for the past 3 Myr.

3. Continental Margin Off Golfo de Penas (∼47 to 49°s)

South of the Taitao and Tres Montes Peninsulas, the Chile continental margin (BOURGOIS *et al.* 2000) exhibits a major accretionary wedge, the so-called Golfo de Penas Accretionary Prism (GPAP, Figs. 5, 6a) exhibiting two major tectonic units, namely, the upper slope ridges (USR) to the east and the Lower Slope Accretionary Wedge (LSAW) to the west. The LSAW tectonic deformation, which is younger than 1 Ma (Figs. 5, 6b, c), evolved through two main steps during the last glacial-inter-glacial cycle. (1) The rapid increase of turbidite accumulation along the trench axis caused (2) the margin to switch from subduction-erosion or non-accretion to subduction-accretion after the inter-glacial optimum at ∼117 to 130 ka. During the subduction-accretion step, the USR unit acted as a backstop for the LSAW. The trench sediment accumulation and their subsequent

tectonic deformation have occurred long after (3 Myr as a minimum) the subduction of segment SCR-1. We infer that no relationship exists between the SCR-1 segment subduction and the LSAW accumulation.

Upslope, between the USR unit and the shelf area, exists an elongated basin, the so-called Upper Slope Basin (USB) limited by high-angle normal faults (Fig. 6d) that offset the sedimentary section and the seafloor. These faults, still active at present, branch southward to an area located in the prolongation of the Liquine-Ofqui Fault System, as it curves westward. We hypothesize that the USB developed as a pull-apart Basin in response to the Chiloe Block migration in a way similar to that of the Golfo de Penas Basin (GPB). However, we assume that the USB inception has occurred long after the onset of the GPB.

4. Golfo de Penas Area

From field, seismic reflection, and gravity data FORSYTHE and NELSON (1985a) have documented that the Liquine-Ofqui fault system (LOFS) controlled the evolution of the Golfo de Penas Basin (GPB), as a pull-apart basin located at the southern

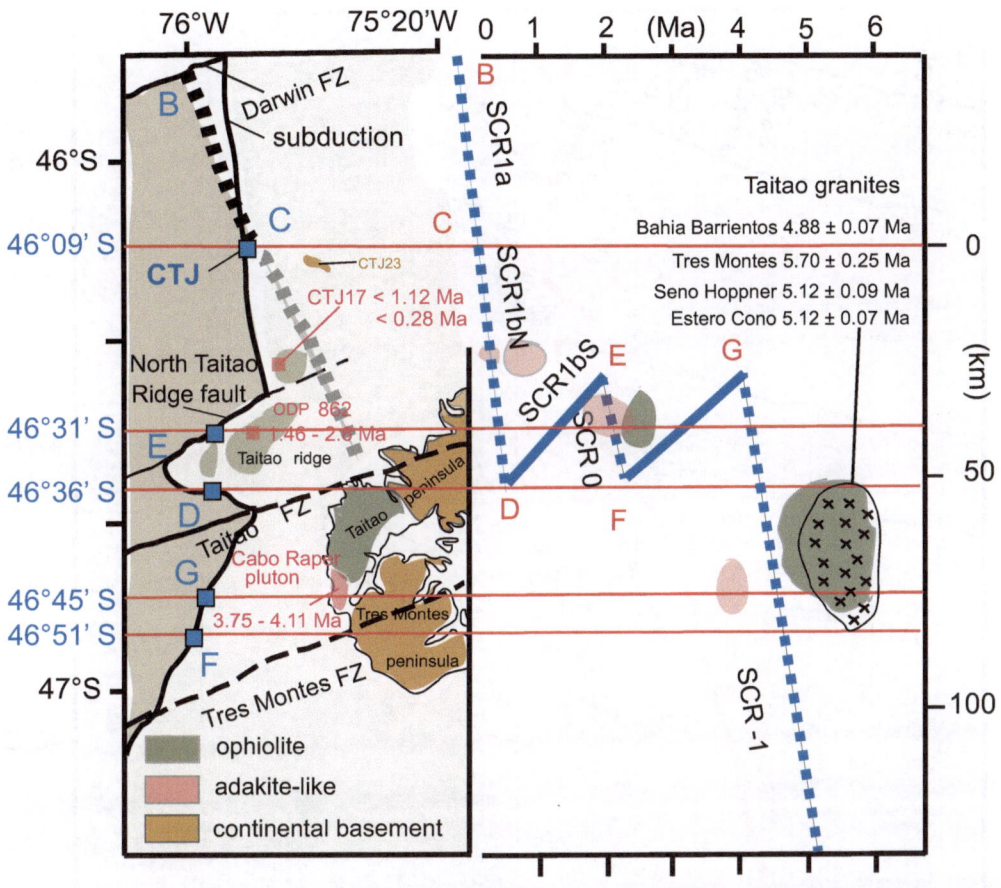

Figure 4

CTJ migration for the past 5–6 Myr considering that no geodynamic event, such as a ridge-jump event, has occurred. Frame on the *left* shows the location of ophiolite and adakite-like intrusive bodies and the CTJ migration through time. Frame on the *right* shows the latitudinal location versus time for (1) the subducted SCR segments and FZs and (2) the main ophiolite and adakite-like pieces. Note that no clear age correlation exists between ophiolite and adakite-like emplacement and the SCR segment and FZ subduction. The SCR1 segment is subdivided in sub-segments, including SCR1a, SCR1bN, and SCR1bS from north to south. *Letters E–G* are those shown in Fig. 3

termination of the Chiloe block (Fig. 5). They have identified two major normal faults bounding the deep GPB, the Chaicayan and the Seno Pulpo faults located to the NE and the SW, respectively. They have also argued that the Cenozoic sediment units (Fig. 7) exposed along the Taitao peninsula and the Golfo de Penas coastal area represents uplifted edging of the GPB. The Byron, Javier, and Chaicayan islands, and the Tres Montes peninsula exhibit Cenozoic sediments unconformably overlying the pre-Jurassic metamorphic basement of the Chile continental margin. At the base of the sequence (Byron Island), the Puerto Good sequence, which consists of terrigenous clastic beds, including siltstone, sandstone, and conglomerate, accumulated in an open marine environment. This sedimentary sequence, identified at two other sites along the Golfo de Penas shoreline (Puerto Barroso and north of Seno Hoppner, Fig. 8), is Middle Eocene in age. We infer that the GPB inception took place before the beginning of the subduction of the segment SCR-1, by about several tens of Myr. The activity of the right-lateral LOFS, which controls the tectonic evolution of the GPB, had to start during the Middle Eocene time. No relationship exists between the LOFS inception and evolution and the SCR subduction, at least during the Eocene and Lower Miocene times.

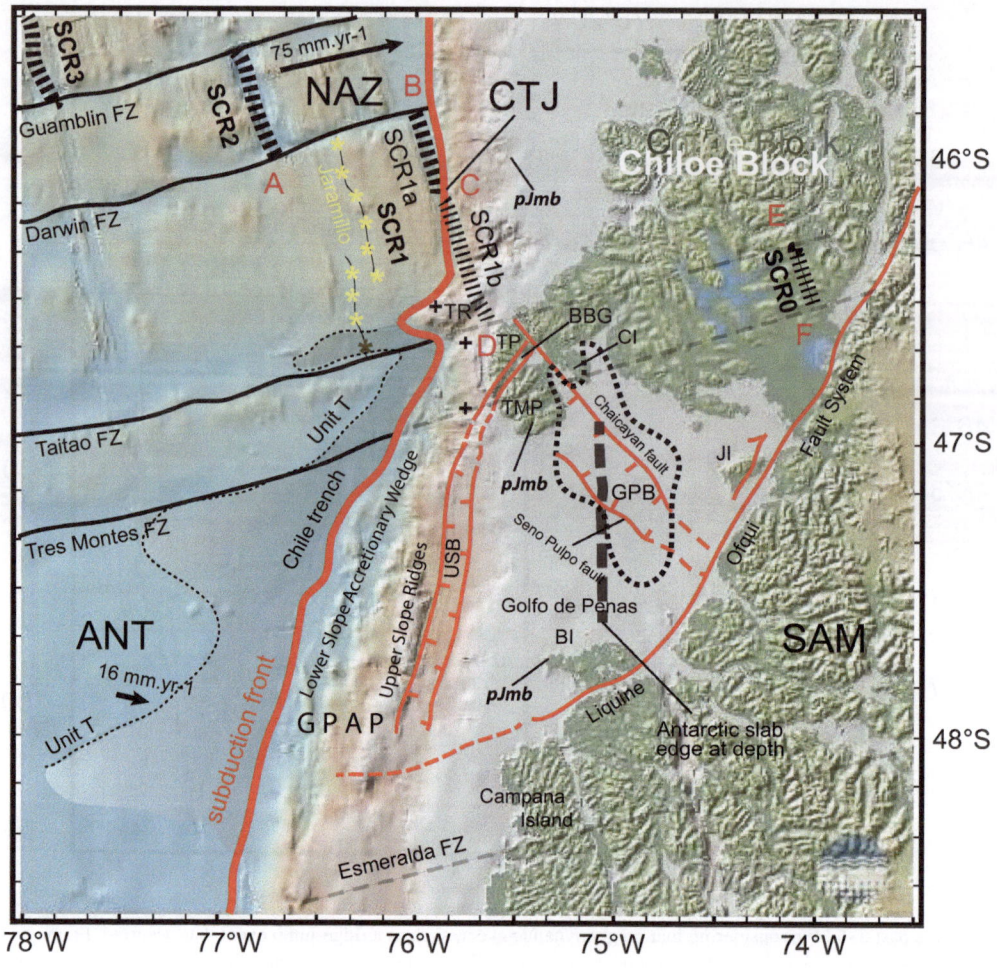

Figure 5
Geodynamic framework of the CTJ area. *Red letters A–F* show the SCR/FZ junctions (see Fig. 3). *Yellow stars* show the Jaramillo magnetic anomaly (∼1 Ma in age). The *thin black dot line* shows the westward extension of the so-called "Transparent Unit" (BOURGOIS *et al.* 2000). Note that the "Transparent Unit" (Unit T) is covering the Jaramillo magnetic anomaly. Unit T is younger than 1 Ma. The Golfo de Penas Basin (GPB) and the Upper Slope Basin (USB) developed as pull-apart basins along the right-lateral Liquiñe-Ofqui Fault system. Note that the Bahia Barrientos graben (BBG) is in the northward prolongation of the USB. *Black cross* shows the location of sites dredged during the MR-06-08 cruise (R/V *Mirai* dredge sites, ANMA and ORIHASHI 2013). *ANT* Antarctic plate, *BBG* Bahia Barrientos Graben, *BI* Byron Island, *CI* Chaicayan Islands, *CTJ* Chile Triple Junction, *FZ* fracture zone, *GPAP* Golfo de Penas accretionary prism, *GPB* Golfo de Penas Basin, *JI* Javier Island, *NAZ* Nazca plate, *pJmb* pre-Jurassic metamorphic basement, *SAM* South America plate, *SCR0 to SCR3* South Chile ridge segments (the same as at Fig. 1), *TMP* Tres Montes peninsula, *TP* Taitao peninsula, *TR* Taitao ridge, *USB* upper slope basin, *red letter* (the same as at Fig. 3), *thick dash line* Patagonia slab window limit at depth, *thick dotted line* limit of GPB, *thin dot line* seaward limit of Unit T

North of the Chaicayan fault (Fig. 8), the Chaicayan Group exposed along the Grosslet Island, the Hereford Island, and the San Pablo Fjord are unconformably (angular unconformity) overlying the metamorphic basement of the Chile continental margin. The Chaicayan Group consists of very coarse and poorly sorted conglomerate with material originating from the pre-Jurassic metamorphic basement. The

conglomerate grades upward to marine sediment, including moderately bedded mudstone, siltstone, sandstone, and conglomerate exhibiting shell debris. A very few volcaniclastic strata interbedded in the marine sequence were identified. The top the sequence consists of a thick cover of volcaniclastic material, associated with porphyry intrusion and volcanic edifices. The Chaicayan Group is gently dipping to the

Table 1

Location of the ridge-transform fault junctions and subduction through time

| Junction | Entering subduction | | Location at depth | | Time (Myr) |
SCR-FZ	Lat (S)	Age (Ma)	Lat (S)	Long (W)	subduction
C	CTJ	46°09′	0	46°09′	75°48′	0.0
D	SCR1/Taitao	46°36′	0.4	46°36′	75°42′	0.4 (SCR1b)
E	SCR0/Taitao	46°31′	2.0	46°17′	74°09′	1.6 (D–E)
F	SCR0/Tres Montes	46°51′	2.3	46°33′	74°01′	0.3 (SCR0)
G	SCR-1/Tres Montes	46°45′	4–4.2	46°17′	72°18′	1.8 (F–G)
H	SCR-1/Esmeralda		5.1–5.3	47°48′	71°38′	0.9–1.3 (SCR-1)

Points C to H are located at Figs. 3, 4, and 12. The "Time subduction" column on the right refers to the calculated time period for the ridge or Fracture zone segment (in brackets) to be subducted

north and to the east connecting to the SE into the Golfo de Penas as suggested from multichannel seismic profiles (MORDOJOVICH 1981). Foraminifer assemblages document a Late Miocene age.

Volcanic and hypovolcanic activity occurred in the Chaicayan island area. It includes (Fig. 8) the Pan de Azucar (4.44 ± 0.16 Ma; ANMA and ORIHASHI 2013), the Cupula San Pablo (0.8 ± 0.4 Ma, MPODOZIS *et al.* 1985) and associated volcanic ejecta (3.4–6.4 Ma; MPODOZIS *et al.* 1985), and the San Pedro fjord volcano (2.85 ± 0.16 Ma; ANMA and ORIHASHI 2013). The Pan de Azucar as well as the San Pedro fjord volcano are adakitic in composition (GUIVEL *et al.* 1999; LE MOIGNE *et al.* 1996). A recent study of the Pan de Azucar by ANMA and ORIHASHI (2013) has corroborated its adakitic affinity. In this area, north of the Chaicayan fault, an age gap of 0.7–2.6 Myr exists between the sediment accumulation of the Chaicayan Group during the Upper Miocene and the volcanic activity that began in the Lower Pliocene. Therefore, the thick cover of volcaniclastic material (informally named Pan de Azucar Complex, Fig. 7) that tops the Chaicayan sequence is likely associated with the volcanic activity starting in the Lower Pliocene as recorded in the Chile margin unit (CMU) in the Taitao peninsula (see below).

5. Taitao and Tres Montes Peninsulas

5.1. Chile Margin Unit (CMU)

Along the Seno Hoppner shoreline at 46°39′53″S–75°25′38″W, the CMU is unconformably overlying

the metamorphic basement rock of the Chile continental margin (BOURGOIS *et al.* 1992, 1993). There, the base of the CMU sequence consists of gravels and coarse grain sands generated by alteration and erosion of the underlying metamorphic basement. Above, the CMU (Figs. 7, 8) consists of volcanic, volcaniclastic, and sedimentary strata well exposed along the Bahia Barrientos. Coarse conglomerates associated with pyroclastic material are typical of this sequence. The volcanic rocks are *ejecta*, including bombs with twisted tail, pumice, pyroclastic flows, rhyolitic tuffs, and scoria exposed in poorly sorted volcanic agglomerates. Sequences of pillow–lavas and pillow–breccias, several tens of meters thick, can be closely associated with sediments. In conglomerates and gravels, the detrital fraction consists of well-rounded quartz pebbles. The other sediments, which include sandstone, siltstone, marl, and shale debris, were all deposited in a shallow water (lower to middle shelf) environment. They frequently show cross-lamination and slumping features, with occasional erosional channels, having a width of one up to a few tens of meters. The bedding is often deformed by impacts of lava blocks dropped within the unconsolidated sediment. No ophiolitic pebble (serpentinite or gabbro) has been found within the CMU detrital sediments. Based on calcareous nannofossil assemblages, BOURGOIS *et al.* (1993) documented an Early Pliocene age for the base of the sequence (northern Seno Hoppner shoreline at 46°40′47″S–75°27′38″W). However, upsection to the SSW, along the western shoreline of the Bahia Barrientos (46°44′41″S–75°31′16″W) sediment, has provided younger nannofossil assemblages of Late Pliocene to Pleistocene age (FORSYTHE

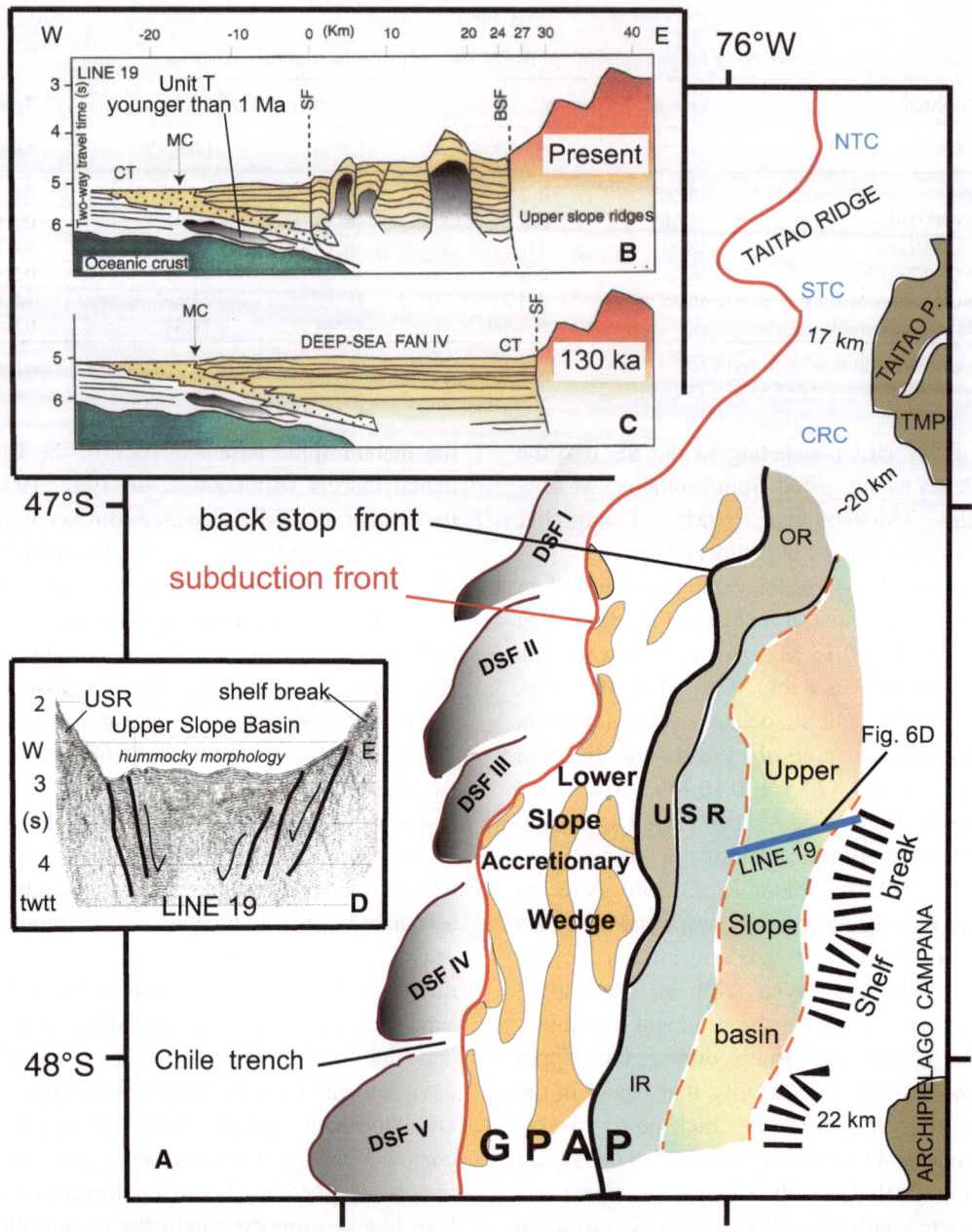

et al. 1985b). East of the Seno Hoppner, the Pan de Azucar Complex (see previous section) has recorded a major volcanic activity, which occurred as the CMU deposited. The CMU and associated volcanic products are likely correlating with the Pan de Azucar Complex, at least partly.

The Bahia Barrientos fjord is an NNE–SSW trending deep corridor, which exposes the CMU. To

the east (Tres Montes Peninsula), a major normal fault bounds the CMU from the metamorphic basement. There, the CMU bedding trending ~E–W is dipping 80° southward. Contrary to the main volcanic unit (MVU, see next section), the CMU did not undergo metamorphism.

The transtensional fault bordering the Bahia Barrientos graben (Fig. 8) to the east also bounds

◀Figure 6

Sketch map and profiles modified from BOURGOIS *et al.* (2000) showing the main morphostructural elements and evolution of the Golfo de Penas Accretionary Prism (GPAP). *A* GPAP exhibits two main units, the lower slope Accretionary Wedge downslope and the upper slope ridges (USR). USR shows two units: the outer (OR) and the inner (IR) ridges. *Grey color* show the extension of deep-sea fan (DSF) turbidite accumulated along the Chile trench. *B Line* drawing of the seismic line 19 shot during the CTJ cruises. *C* Reconstruction of the situation at ~130 ka during the Last Inter-Glacial optimum. The sedimentary material involved in the Lower Slope Accretionary Wedge is the eastward prolongation of the deep-sea fan IV, in which detrital accumulation is younger than the Unit T (i.e., Transparent Unit younger than 1 Ma, see Fig. 5). *D* The Upper Slope Basin along the seismic line 19. The hummocky morphology originated from ice-raft discharges. Normal fault systems bound the basin from the USR to the west and the shelf break slope to the east. *BSF* backstop front, *CRC* Cabo Raper Canyon, *CT* Chile Trench, *DSF I to V* deep-sea fan I to V, *GPAP* Golfo de Penas Accretionary Prism, *IR* inner ridge, *MC* Morning-ton Channel, *NTC* North Taitao Canyon, *OR* outer ridge, *SF* subduction front, *STC* South Taitao Canyon, *TMP* Tres Montes Peninsula, *USR* upper slope ridges

the continental basement from its Pliocene–Pleis-tocene sedimentary cover (the so-called CMU). This allows inference that the Chile continental basement

extends beneath the graben. To the west, a major fault bounds the CMU from the Taitao ophiolite. Because the CMU and the Pan de Azucar Complex are unconformably overlying the continental margin basement on one side and are metamorphism-free on the other side, we assume that the fault bounding the Bahia Barrientos corridor to the west is roughly following the Taitao ophiolite suture. At first, this fault worked as a transpressional or reverse fault (obduction of the ophiolite complex, see Sect. 7.2.1). This deep corridor bounded by two major normal faults is a graben that evolved until the Lower Pliocene. This latter is underlined by four intrusive bodies (Tres Montes, Bahia Barrientos, Seno Hoppner, and Estero Cono), whose ages range from 5 to 6 Ma. The Bahia Barrientos graben extends southwestward to the coastal line in the prolongation of the USB (Fig. 5). We infer that the right-lateral LOFS is likely controlling the most recent collapse (Upper Pliocene–Pleistocene) of the Bahia Barrientos graben as proposed for the USB (see above).

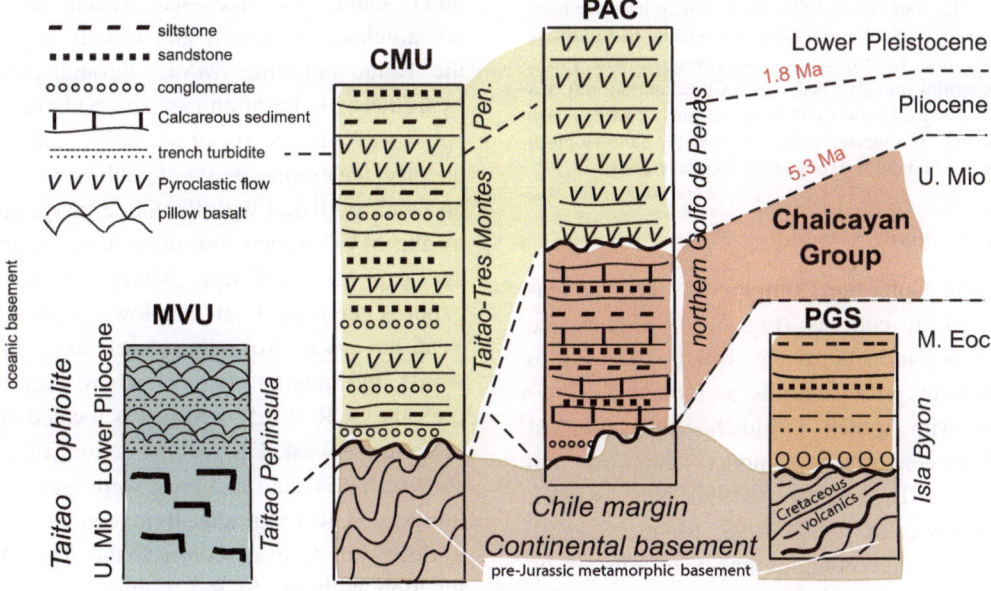

Figure 7

Synthetic columns of sedimentary units extending from the southern Golfo de Penas (47°50′S, Byron Island) to the northern Taitao Peninsula (46°30′S). The Puerto Good sequence, the Chaicayan Group, the Pan de Azucar complex, and the Chile margin unit accumulated in a shallow water environment, they unconformably overly the pre-Jurassic metamorphic basement of the Chile margin. As opposed the main volcanic unit, overlying the Taitao ophiolite accumulated in a deep-water environment (trench turbidite). *CMU* Chile margin unit, *MVU* main volcanic unit, *PAC* Pan de Azucar Complex, *PGS* Puerto Good sequence

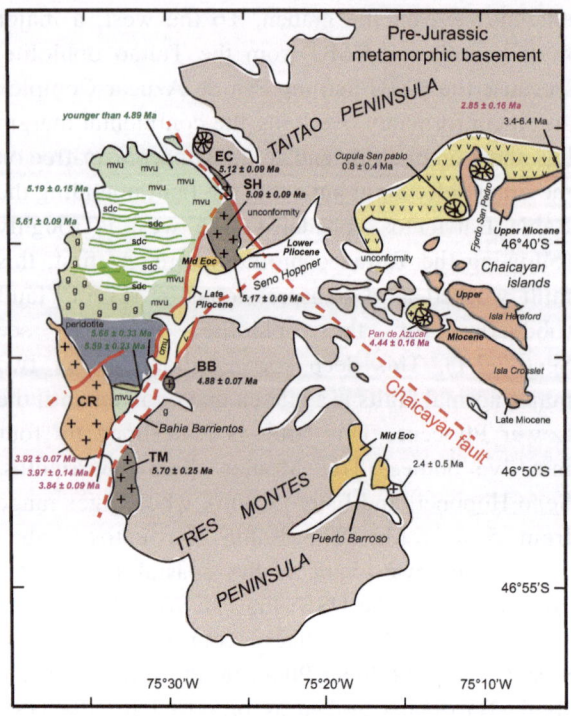

Figure 8
Geologic map of the Taitao and Tres Montes Peninsulas area.
SHRIMP U–Pb data from the Japan scientific community (ANMA
and ORIHASHI 2013; ANMA et al. 2006, 2009). Taitao intrusive body
(*black italic*); Taitao ophiolite (*green italic*); adakite-like intrusive
bodies (*purple italic*); K/Ar (*black regular*, MPODOZIS et al. 1985).
BB Bahia Barrientos intrusive body, *cmu* Chile margin unit, *CR*
Cabo Raper pluton, *EC* Estero Cono intrusive body, *g* gabbro, *mvu*
main volcanic unit, *sdc* sheeted dyke complex, *SH* Seno Hoppner
intrusive body, *TM* Tres Montes intrusive body

5.2. Taitao Ophiolite Complex

The Taitao Ophiolite Complex has been extensively studied. It consists (Fig. 7) of a complete
sequence of oceanic lithosphere, which, from top to
bottom, includes pillow basalts and massive flows
interbedded with trench turbidites (the so-called
MVU), a sheeted dyke complex, gabbros, and
ultramafic rocks (KAEDING et al. 1990; NELSON et al.
1993; BOURGOIS et al. 1992, 1993; MPODOZIS et al.
1985). This ophiolitic complex has been affected by
low-pressure metamorphism, the degree of which
increases with depth from the zeolite to the amphibolite facies (SHIBUYA et al. 2007; LE MOIGNE et al.
1993); the estimated temperatures rising from 230 °C
in the MVU up to 550 °C in the gabbros. Such
metamorphic features are characteristics of the

hydrothermal alteration typical of Mid-Ocean Ridge
environments. This leads to the conclusion that the
Taitao ophiolite has been generated and emplaced
close to the SCR axis (ANMA et al. 2006). At the base,
the ultramafic rocks underwent two episodes of
melting (SCHULTE et al. 2009), the most recent of
which having taken place, while the SCR was
magmatically active. The gabbros have normal
Mid-Ocean Ridge basalts (N-MORB) compositions,
while pillow basalts are enriched MORB (GUIVEL
et al. 1999; LE MOIGNE et al. 1996; LAGABRIELLE et al.
1994; KAEDING et al. 1990).

Table 2 and Fig. 9 report zircon U–Pb ages
(ANMA et al. 2006, 2009). The gabbro (Fig. 8 for
location) gives ages ranging from 5.5 to 6.0 Ma,
while the sheeted dyke complex is 5.19 ± 0.15 Ma
old. Similar ages (within the analytical error range)
were obtained from the gabbro zircon crystals, by the
fission-track method, which documents a very rapid
cooling from 750 to 800 °C in the zircon core, down
to ~ 320 °C in the annealed rim zone. The gabbros of
the Taitao ophiolite were located at shallow depth
($T < 320$ °C) at ~ 5.5 Ma (youngest age). On the
other hand, the deep-sea trench turbidites, the
volcaniclastic material, and basalts associated with
the Taitao ophiolite (MVU) through their common
hydrothermal alteration (see above) have provided an
age <4.89 Ma (ANMA et al. 2009, Table 2). Consequently, the emplacement of gabbros at shallow depth
and their collision with the Chile margin as recorded
by the MVU trench turbidites were separated by an
age gap of ~ 0.6 Myr. About 4.5 Ma, when the
gabbros emplaced at shallow depth, they were
~ 56 km away from the trench axis. In the other
words, the magma chamber where they crystallized
(i.e., the SCR at 5.5–6 Ma) was located significantly
more to the west. The reconstruction (Fig. 3) at 5 Ma
shows that Taitao gabbros were generated in the
northern SCR-1 segment. If no tectonic event, such as
a ridge jump, has occurred, the Tres Montes FZ
junction with the SCR-1 segment (area of point G
Fig. 3) occurred at 4–4.2 Ma. Consequently, the
Taitao gabbros (~ 5.7 Ma) and the Taitao sheeted
dyke complex (~ 5.2 Ma) would have completed
their development before the onset of the SCR-1
segment subduction at ~ 5.1 to 5.3 Ma.

Table 2

Compilation of SHRIMP U–Pb and fission-track ages on zircon and apatite

Author	Rock	Location	U–Pb	FT (z)	FT (a)	Age (Ma)	Range (Ma)
Hervé	Granite	Cabo Raper	x			3.97 ± 0.14	3.83–4.11
Hervé	Granite	Cabo Raper		x		3.5 ± 0.3	3.2–3.8
Hervé	Granite	Cabo Raper			x	2.9 ± 0.8	2.1–3.7
Hervé	Granite	Cabo Raper	x			3.84 ± 0.09	3.75–3.93
Hervé	Granite	Cabo Raper		x		4.1 ± 0.3	3.8–4.4
Hervé	Granite	Cabo Raper			x	1.9 ± 0.6	1.3–2.5
Anma 2	Granite	Cabo Raper	236			3.92 ± 0.07	3.85–3.99
Anma 2	Granite	Tres Montes	14			5.70 ± 0.25	5.45–5.95
Anma 2	Granite	Bahia Barr.	246			4.88 ± 0.07	4.81–4.95
Hervé	Granite	Bahia Barr.		x		3.5 ± 0.3	3.2–3.8
Hervé	Granite	Bahia Barr.			x	3.1 ± 1.3	1.8–4.4
Anma 2	Granite	Seno Hoep.	110			5.17 ± 0.09	5.08–5.26
Anma 2	Granite	Seno Hoep.	172			5.09 ± 0.09	5.00–5.18
Anma 2	Granite	Est. Cono	169			5.12 ± 0.09	5.03–5.21
Hervé	Granite	Est. Cono		x		3.5 ± 0.3	3.2–3.8
Anma 2	Ophiolite	MVU (sed)	338			Post 4.89	
Anma 1	Ophiolite	Taitao (sdc)	x			5.19 ± 0.15	5.04–5.34
Anma 1	Ophiolite	Taitao (gab)	x			5.66 ± 0.33	5.33–5.99
Anma 1	Ophiolite	Taitao (gab)		x		5.9 ± 0.4	5.5–6.3
Anma 1	Ophiolite	Taitao (gab)	x			5.61 ± 0.09	5.52–5.70
Anma 1	Ophiolite	Taitao (gab)		x		5.8 ± 0.2	5.6–6.0
Anma 2	Ophiolite	Taitao (gab)	107			5.59 ± 0.23	5.36–5.82

Number in the U–Pb column refers to numbering in original publications

5.3. Taitao Plutonic Bodies

Five granitic plutons (Cabo Raper, Tres Montes, Bahia Barrientos, Seno Hoppner, and Estero Cono) outcrop in the Taitao and Tres Montes peninsulas (Fig. 8); their U–Pb zircon ages range from 3.84 to 5.70 Ma (Table 2; ANMA et al. 2006, 2009; HERVÉ et al. 2003). Based on geochemical investigations, ANMA and ORIHASHI (2013) concluded that these granitoids were generated by the partial melting of the subducted oceanic basalts then transformed into garnet-free amphibolite at shallow depth (<30 km).

These inferences conflict with those of BOURGOIS et al. (1996) who have considered the Cabo Raper pluton, as having been generated by partial melting from subducted basalts transformed into garnet-bearing amphibolite or eclogite. Such conclusions are classically invoked for the genesis of adakitic magmas (DEFANT and DRUMMOND, 1990; MARTIN 1987; MARTIN et al. 2005; RAPP et al. 1999; SISSON et al. 2005). Geochemical modeling has led to the conclusion that (1) the probable source of the Cabo Raper magma is altered basalt, (2) garnet and hornblende must be residual phases, (3) the degree of partial melting is about 20 %, and (4) melting occurred at $P > 10$ kbar and $T > 650$ °C. Moreover, the Cabo Raper pluton, like the other four plutons, belongs to the "High-Silica Adakites" (HSA; MARTIN et al. 2005) that, contrary to "Low-Silica Adakites", have moderate-to-low Mg# (0.54) as well as low Ni and Cr contents (24 and 45 ppm, respectively), which precludes any significant interaction between the magma and mantle peridotite. Such evidence militates in favor of a lack of mantle wedge above the subducted plate, at the place where Cabo Raper adakites were generated. This conclusion is perfectly consistent with the fact that the Cabo Raper pluton is located at less than 17 km landward from the trench axis.

On the other hand, ZAMORA (2000) studied the composition of liquids experimentally generated by the partial melting of sample T19d, in a huge range of pressure and temperature conditions. T19d is a hydrothermally altered basalt belonging to the Bahia Barrientos ophiolite. Liquids with a chemical composition identical to that of Cabo Raper pluton are obtain for the degree of melting from 15 to 20 %, at

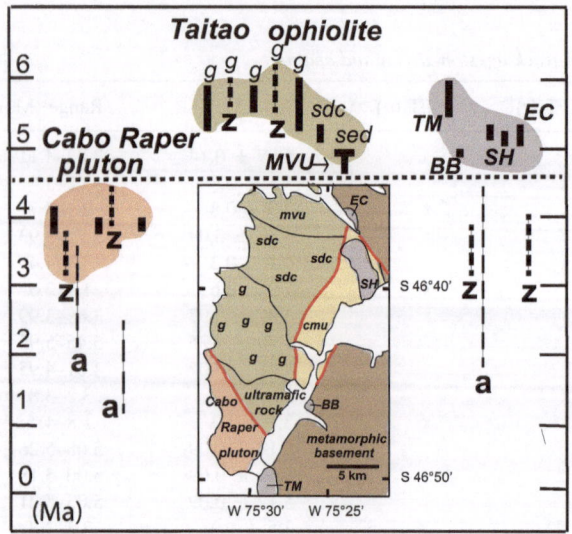

Figure 9

SHRIMP U–Pb (*thick black line*), zircon fission-track (*thick dash line*), apatite fission-track (*thin dash line*). Note that the Taitao ophiolite and the Taitao and Tres Montes intrusive bodies are significantly older than the adakite-like Cabo Raper pluton. *a* apatite, *BB* Bahia Barrientos intrusive body, *cmu* Chile margin unit, *EC* Estero Cono intrusive body, *g* gabbro, *mvu* main volcanic unit, *sdc* sheeted dyke complex, *SH* Seno Hoppner intrusive body, *TM* Tres Montes intrusive body, *z* zircon

pressure of 14 kbar and temperature of about 900 °C, the residue of melting consisting in an assemblage of Hornblende + Clinopyroxene + Plagioclase + Garnet, which is consistent with the geochemical modeling. These minimum P–T conditions constrain the source depth (CLOOS 1993) when the Cabo Raper pluton was generated and emplaced between 3.75 and 4.11 Ma.

Contrastingly, the Seno Hoppner granite does not show adakitic affinities and its composition is rather that of typical calc-alkaline series that GUIVEL et al. (1996) interpreted as resulting of partial melting of the forearc mantle wedge. On the other hand, ZAMORA (2000) proposed an alternative scenario, where the Seno Hoppner pluton could have been generated by melting of the subducted slab basalts outside of the garnet stability field, and followed by low degrees (10 %) of fractional crystallization of plagioclase. Finally, the Tres Montes intrusive body is geochemically very close to those of the Cabo Raper pluton, but older by about 1.25 Myr as a minimum.

The Tres Montes, Bahia Barrientos, Seno Hoppner, and Estero Cono granite stocks (Fig. 8; Tables 2, 3)

intrude the metamorphic basement of the continental wedge. These rocks and the CMU volcanism of similar age (∼5 to 6 Ma to Lower Pleistocene) do not display any adakitic affinity (ANMA and ORIHASHI 2013; KON et al. 2013; ANMA et al. 2009; LAGABRIELLE et al. 2000; GUIVEL et al. 1999), such that GUIVEL et al. (1999) consider them to be formed at shallow to moderate depths by partial melting of the subducted slab or from the SCR itself. In addition, these magmas also display a sedimentary signature, which leads to the conclusion that the MORB magmas formed in the subducting spreading center have been contaminated by the continental basement at limited sites (LAGABRIELLE et al. 1994).

Based on both their ages and their emplacement mode, the magmatic rocks can be divided into two groups (Table 2; Fig. 9). (1) The Cabo Raper pluton crystallized between 3.75 and 4.11 Ma (U–Pb zircon); ages are overlapped by the zircon fission-track ages. Its magma chamber emplaced into the forearc through a very short cooling event at a rate of several hundreds of °C Myr^{-1}. (2) The four others plutonic bodies emplaced about 0.7 Myr before, between 4.81 and 5.95 Ma (maximum time window); fission-track ages do not overlap crystallization ages, thus documenting a more steady process at the origin of their emplacement.

Today, the field relationship between the intrusions and their host rock is mainly tectonic, the original intruding features being only locally preserved. However, the Cabo Raper and the other four plutons exhibit only one kind of host rock, the ophiolite complex, and the metamorphic continental basement and its sedimentary cover, respectively. We infer that this situation reflects the original relationship with the host rock, stressing this way the difference between the two groups of intrusive bodies. The age and geochemical constraints associated with geological associations in the field allow us to consider different histories and mode of emplacement for the two groups of the Taitao intrusive bodies.

6. Continental Margin South of CTJ

The Chile continental margin extending from the CTJ (46°09′S) to the Taitao peninsula offshore area

Table 3

Compilation of published K/Ar and $^{40}Ar/^{39}Ar$ ages

Authors	Rock	Location	w	h	b	f	Age (Ma)	Range (Ma)
Mpodozis et al.	Granite	Cabo Raper		x			4.1 ± 2.4	1.7–6.5
Mpodozis et al.	Granite	Cabo Raper			x		3.6 ± 0.6	3–4.2
Mpodozis et al.	Granite	Cabo Raper		x			3.3 ± 0.3	3–3.6
Mpodozis et al.	Granite	Cabo Raper			x		3.4 ± 0.8	2.6–4.2
Guivel et al.	Granite	Cabo raper			x		5.1 ± 0.6	4.5–5.7
Guivel et al.	Granite	Cabo raper				x	4.5 ± 02	4.3–4.7
Guivel et al.	Granite	Cabo Raper			x		4.8 ± 0.3	4.5–5.1
Guivel et al.	Granite	Cabo raper				x	4.2 ± 0.1	4.1–4.3
Mpodozis et al.	Granite	Bahia Barr.			x		3.2 ± 1.2	2–4.4
Mpodozis et al.	Granite	Seno Hoep.			x		5.2 ± 0.3	4.9–5.5
Mpodozis et al.	Granite	Seno Hoep.			x		5.5 ± 0.4	5.1–5.9
Guivel et al.	Granite	Seno Hoep.			x		5.9 ± 0.5	5.4–6.4
Guivel et al.	Granite	Seno Hoep.				x	6.9 ± 1.0	5.9–7.9
Mpodozis et al.	Basalte	CMU	x				4.3 ± 1.2	3.1–5.5
Mpodozis et al.	Basalte	CMU	x				2.5 ± 1.3	1.2–3.8
Mpodozis et al.	Basalte	CMU	x				4.4 ± 0.6	3.8–5
Mpodozis et al.	Basalte	CMU	x				3.7 ± 0.6	3.1–4.3
Mpodozis et al.	Basalte	CMU	x				2.9 ± 0.8	2.1–3.7
Mpodozis et al.	Basalte	CMU	x				3 ± 1.4	1.6–4.4
Mpodozis et al.	Basalte	MVU	x				4.6 ± 1.0	3.6–5.6

W whole rock, h hornblende, b biotite, f feldspath

has been heavily studied during major campaigns conducted by the international scientific community. It includes geophysical surveys, dredge sampling, coring, and drilling during ODP Leg 141 (ANMA and ORIHASHI 2013; BOURGOIS et al. 2000; TEBBENS et al. 1997; BEHRMANN et al. 1994; CANDE et al. 1987; CANDE and LESLIE 1986).

6.1. Taitao Peninsula Segment

During the MR 08–06 cruise (ANMA and ORIHASHI 2013), the R/V *Mirai* dredged the continental margin at site D01 and D02 (Figs. 5, 10 for location) in an area off the Taitao and Tres Montes peninsulas. At site D01site, sediments associated with pyroclastic flows and lapilli were recovered in the seaward prolongation of the Bahia Barrientos graben. Both sediments and volcanics resemble those of CMU, which strengthens the assumption (proposed at Sect. 3) that the USB (~10-km south of D01 site) and Bahia Barrientos (~10 km northeast of D01 site) grabens are connected. Dredging at site D02 conducted off the Taitao ophiolite has sampled granitic rock, whose zircon crystals were dated at 4.06 ± 0.28

and 3.89 ± 0.18 Ma (samples G01 and G03, respectively). These ages are close to those of Cabo Raper granite (3.75–4.11 Ma). The D02 site location along a northwestward facing slope at 10–15 km off the Taitao ophiolite complex strongly suggests that the granite is outcropping nearby the dredged site. A prolongation of the Cabo Raper pluton likely exists at 10–15 km north of the Taitao peninsula.

6.2. Taitao Ridge Segment

EM12 bathymetry and imagery together with five dredges have documented that the Taitao ridge consists of accreted tectonic slices (BOURGOIS et al. 2000). Two thrusted sheets (R1 and R2, Fig. 10) of trench sediment have been stacked at the toe of a major thrust-fault that branches southward to the main subduction decollement. This major thrust-fault is bounding seaward the R3 tectonic unit that consists of a well-stratified pile of sediment covering an acoustic basement. Magmatic glass from basaltic-andesite pillow lavas dredged at site CTJ 11 (R2 or the base of the R3 tectonic unit) yields a $^{40}Ar/^{39}Ar$ age of 1.14 ± 0.11 Ma (GUIVEL et al. 2003). The

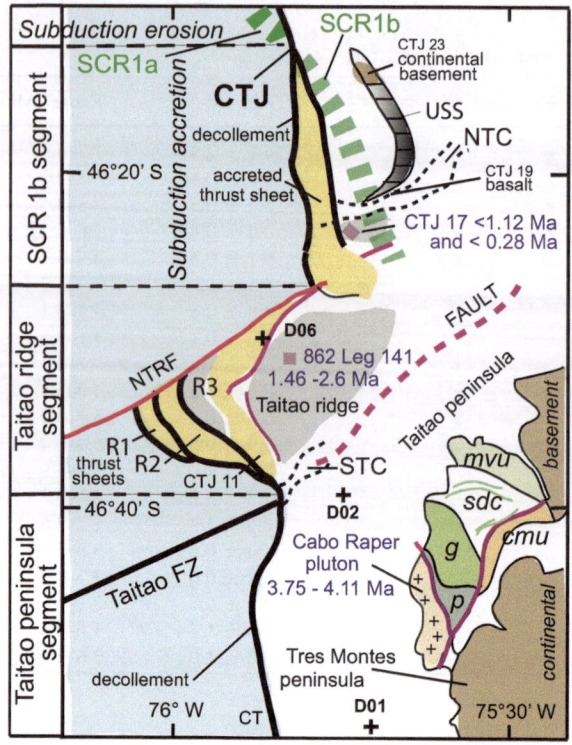

Figure 10

Sketch map showing the location of ophiolite bodies (Taitao ophiolite and Taitao Ridge ophiolite) and adakite-like magmatic elements together with tectonic features of the Chile continental margin extending from the CTJ to the north to the Taitao peninsula to the south. *Cmu* Chile margin unit, *CTJ* Chile Triple Junction, *CTJ 11, 17, 19, 23* sites dredged during the R/V *L'Atalante* cruise, *D01, D02, D06* Sites dredged during the R/V *Mirai* cruise, *g* gabbro, *mvu* main volcanic unit, *NTC* North Taitao canyon, *NTRF* North Taitao Ridge fault, *p* peridotite, *SCR1a* South Chile Ridge 1a sub-segment, *SCR1b* South Chile Ridge 1b sub-segment, *sdc* sheeted dyke complex, *STC* South Taitao canyon, *USS* upper slope scarp, *purple letter* adakite-like outcrop and age

association of the pillow basalt with the trench sediment allows inferring that the accretion of the R1 and R2 thrust sheets occurred after 1.25 Ma. To the north (Fig. 10), at dredge site D06, detrital zircon crystals from the R3 thick pile of sediment (Fig. 14 in BOURGOIS et al. 2000) gave (sample D06-S20) a weighted average age of 4.43 ± 0.76 Ma (ANMA and ORIHASHI 2013). These data allow inferring that the R3 thrust sheet—i.e., uplifted Antarctic plate basalt and overlying sediment—obduction occurred subsequently.

East of R3, ODP Leg 141 drilled the Taitao ridge at Site 862 (BEHRMANN et al. 1992, 1994); there, a thin

cover of sediment overlies a thick pile of volcanic material: tholeiitic basalts, andesites, dacites, and rhyolites. These samples (FORSYTHE et al. 1995a) show a wide diversity of compositions, which includes a bimodal suite of sub-alkaline basalts and dacites to rhyolites (KURNOSOV et al. 1995).

The trace element patterns and Pb isotopic compositions of the rhyolites indicate that their source consisted of a mixture of MORB and sediment. The rhyolites REE patterns are moderately fractionated ($[La/Yb]_N \leq 10$) with a concave shape at the heavy REE end (Fig. 5b in FORSYTHE et al. 1995a). Geochemical modeling based on trace element behavior indicates that they could be derived through 10–15 % melting of the above-mixed source, leaving a residue made up of plagioclase, hornblende, and garnet. Hornblende crystals extracted from rhyolites gave $^{40}Ar/^{39}Ar$ ages of 1.54 ± 0.08 and 2.2 ± 0.4 Ma (FORSYTHE et al. 1995b). By contrast with the Taitao peninsula, no clear adakitic affinity has been identified from the re-sampled rocks drilled at Sites 862 B and C (GUIVEL et al. 1999). Although younger, the Taitao ridge volcanics share some similarities with the Taitao peninsula ones; this includes a wide variety of chemical compositions associated with no evidence for continental crust melting or assimilation. Considering that the composition of the magmatic rocks sampled along the four incoming segments of the SCR (KLEIN and KARSTEN 1995; KARSTEN et al. 1996; SHERMAN et al. 1997) ranges from typical N-MORB to calc-alkaline dacite, the Taitao Ridge volcanic rocks can be grouped into two main clusters. (1) Basalt-to-calc-alkaline rock emplaced along the Chile spreading center (i.e., magmatic type 1 to 5 of GUIVEL et al. 1999) hereafter referred as "Taitao ridge ophiolite", and (2) adakitic andesites and dacites—i.e., magmatic type 6 of GUIVEL et al. (1999) and part of the Group 7 of FORSYTHE et al. (1995a)—which emplacement post-dates the obduction of the Taitao ridge ophiolite onto the continental wedge. This situation resembles that exposed along the Taitao peninsula, but shifted in time (i.e., younger).

The 2500-m high rectilinear cliff, which bounds the Taitao ridge to the north, is the trace of the North Taitao Ridge fault (NTRF, BOURGOIS et al. 2000). This fault, which cuts the R1-to-R3 thrust sheet, is at least partly, younger than 1.25 Ma. The sediments overlying the basement recorded the ~2000-m uplift of the Taitao ridge following the emplacement of the volcanics. The underthrust of the sheet R1 (Fig. 10) at the toe of the Taitao ridge accretionary wedge has occurred after 1.25 Ma. Obduction of the Taitao ridge ophiolite material located inward at ODP Site 862 occurred prior accretion of the R1 and R2 thrust sheets. Because the adakite-like material has provided ages ranging from 1.46 to 2.6 Ma, the obduction of the adjacent ocean floor could be older than 2.6 Ma.

6.3. CR1bN Segment

The continental margin located south of the CTJ has three main morphostructural features (Fig. 10). (1) The 1–1.5-km high Upper Slope Scarp (USS) with a seaward concave shape, it is associated with a detachment fault dipping seaward (BEHRMANN et al. 1994; BANGS et al. 1992). Along the USS, the metamorphic continental basement has been recovered at dredge site CTJ 23 during the CTJ cruise. (2) The E-W trending, 1250–2000-m deep North Taitao canyon is a deep window into the structure of the continental margin. The steep walls of this canyon were dredge at six sites (see BOURGOIS et al. 2000 for location). The adakitic dacites dredged at site CTJ 17 (Fig. 10) gave two $^{40}Ar/^{39}Ar$ ages at <1.12 and <0.28 Ma (GUIVEL et al. 2003). Subduction-related basaltic andesites were dredge along the North Taitao canyon. (3) At the toe of the continental margin, an accreted thrust sheet of trench sediment gradually widens southward from the CTJ, thus indicating the existence of relationships with the CTJ northward migration when in the trench–trench–ridge configuration.

7. Discussion

7.1. Slab Window Development

The development of a slab window in Patagonia has been identified as the result of the Chile ridge spreading center for a long time (BREITSPRECHER and THORKELSON 2009; BOURGOIS and MICHAU 2002;

GORRING et al. 1997; FORSYTHE et al. 1986). Subsequently, seismic tomography (RUSSO et al. 2010a, b) has allowed for a better definition of the boundaries of the slab-free area at depth. To the west, boundaries at 50- and 100-km depth are roughly trending N–S (white and green dash lines, Fig. 11), and cross the Golfo de Penas area at depth. To the north, the Nazca slab boundary roughly follows the Taitao and Tres Montes Transforms, while its along-ridge slab edge (Nazca-plate slab edge) remains poorly defined to the east.

Because (1) the age and location of the subduction of the SCR0 and SCR-1 segments are constrained as taking place at 2.0–2.3, and 4.0–5.3 Ma, respectively (Table 1), and (2) the Antarctic plate subduction rate remained constant at 16 mm year^{-1} for the past 11 Myr; the Antarctic slab edge location at depth can be confidently reconstructed (Fig. 11). Since the northern and the southern tip of the SCR-1 segment (points X and Y Fig. 11, respectively) have been entering the subduction zone, they have moved ~64 and ~80 km inward (to points X′ and Y′ Fig. 11, respectively). Because the tectonic evolution of the LSAW occurred long after the subduction of the SCR-1 segment, we have used the backstop front of the USR unit (grey dash line Fig. 11, BOURGOIS et al. 2000) as the base for the graphical reconstruction (black dash line Fig. 11). In the same way, the migration at depth of the Antarctic slab edge along the SCR0 segment is graphically reconstructed (black dash line Fig. 11); it is located ~37 km inward from the subduction front. The reconstructions (i.e., along SCR-1 and SCR0 segments) show that the Antarctic slab edge at depth (i.e., the western Patagonia slab window boundary) roughly fits that proposed from tomography by RUSSO et al. (2010a, b). However, these reconstructions predict that the Antarctic slab edge was roughly trending NS, drawing a >20° angle with the SCR spreading direction. This reflects not only the N10°E direction of the USR graphical base, but also the time-dependent migration of the slab window opening as controlled by the CTJ northward migration. It allows inferring that the USR clockwise rotation occurred after the SCR subduction. Consequently, we propose that this rotation is likely related to the evolution of the LOFS. If a ~20° anticlockwise rotation is applied, the reconstructed slab edge

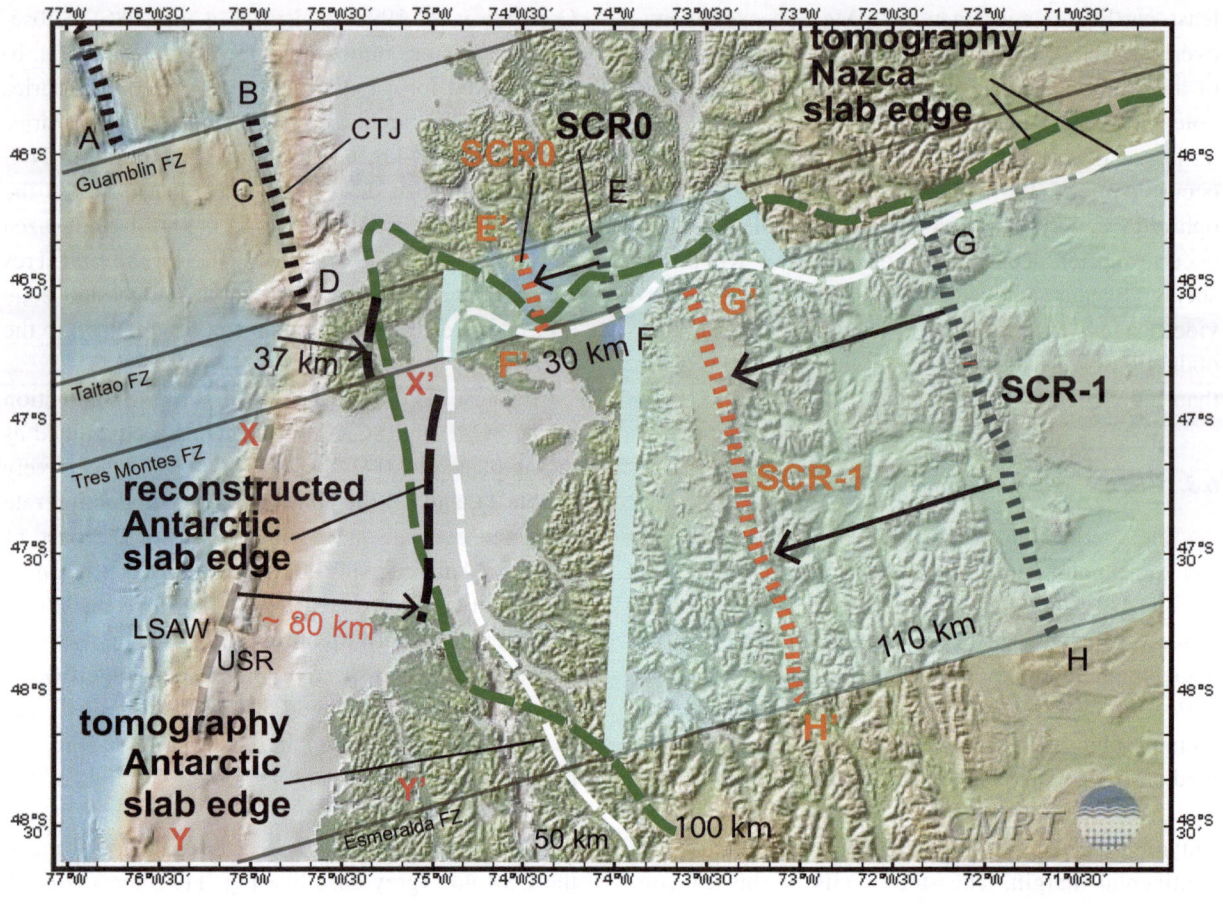

Figure 11

Patagonia slab window. *White* (50-km depth) and *green* (100-km depth) *dash lines* show the Patagonian slab window boundary defined by the P-wave tomography (Russo *et al.* 2010a, b). The boundary to the west follows the edge of the Antarctic slab. The boundary to the north is following the Taitao and Tres Montes FZ. *Black dash line* shows the Antarctic slab edge as reconstructed using the convergence rate between Antarctic and South America plate (*grey dash line*, base for reconstruction). This reconstruction fits the tomographic determination. *Black tight dotted line* (*EF* and *GH*) shows SCR0 and SCR-1 location at depth as reconstructed using length of structural elements (Fig. 3) and convergence rates. These spreading center locations allow reconstructing the slab edges (*pale green thick line*). Note that no fit exists between this reconstruction of the Antarctic slab edge and the tomographic determination. A 30- and 110-km westward shift (*orange tight dotted line*) for SCR0 and SCR-1, respectively, provides a proper fit between graphical reconstruction and tomography locations. (See text for more details)

matches almost perfectly the slab window edge as defined from P-wave anomalous travel times.

The convergence rates between the Nazca and South America plates (see Sect. 2) in association with the length of ridge-to-ridge segments along fracture zones (Fig. 3) allow reconstructing the eastern edge of the Patagonia slab window at depth. Assuming that no deformation of the fracture zones occurs during subduction, and that the half-spreading rate at depth remains similar to that documented from magnetic

anomalies at seafloor (Lagabrielle *et al.* 2015) enables us to locate the junctions between SCR and transforms at depth (i.e., points E–H, Fig. 11). These points can be used as a basis for reconstructing independently the edges of the slab window (thick light-blue-green line, Fig. 11). The reconstruction obtained from Nazca-plate convergence rate for the western boundaries along the SCR-1 and SCR0 segments shows a major discrepancy with both graphic and tomography reconstructions for the

Antarctic slab edge. The location of the SCR0 and SCR-1 spreading centers inferred from the Antarctica plate rate is 30 and 110 km (Fig. 11) west of those calculated from Nazca-plate rate, respectively. Because the graphic reconstruction from the Antarctic convergence rate fits properly the western slab-free boundary as localized from tomography, we infer that spreading center SCR-1 (i.e., H′G′ Figs. 11, 12) is suitably localized: it extends underneath the high relief occupied today by the North Patagonia Icefield (NPI).

However, the eastern edge of the SCR-1 slab-free segment (the Nazca slab edge) is located 110 km too far to the east when calculated using the Nazca-plate convergence rate. This can be

accounted for by the fact that the SCR-1 ridge that has entered the subduction from 5.1–5.3 to 4.0–4.2 Ma (Table 1) resulted from the westward ridge-jump of a ridge segment, previously entered in the subduction. The 110 km of missing space for matching the Nazca slab edge requires ~1.4 Ma of spreading-ridge activity at depth that is roughly the requisite time for the full subduction of the SCR-1 segment. We infer that the ridge-jump took place just before the onset of the SCR-1 segment subduction. It places the ridge jump at ~6.5 to 6.8 Ma. A similar situation is expected along the SCR0 segment that has experienced a ridge jump at 2.5–2.8 Ma before the onset of the SCR0 segment subduction.

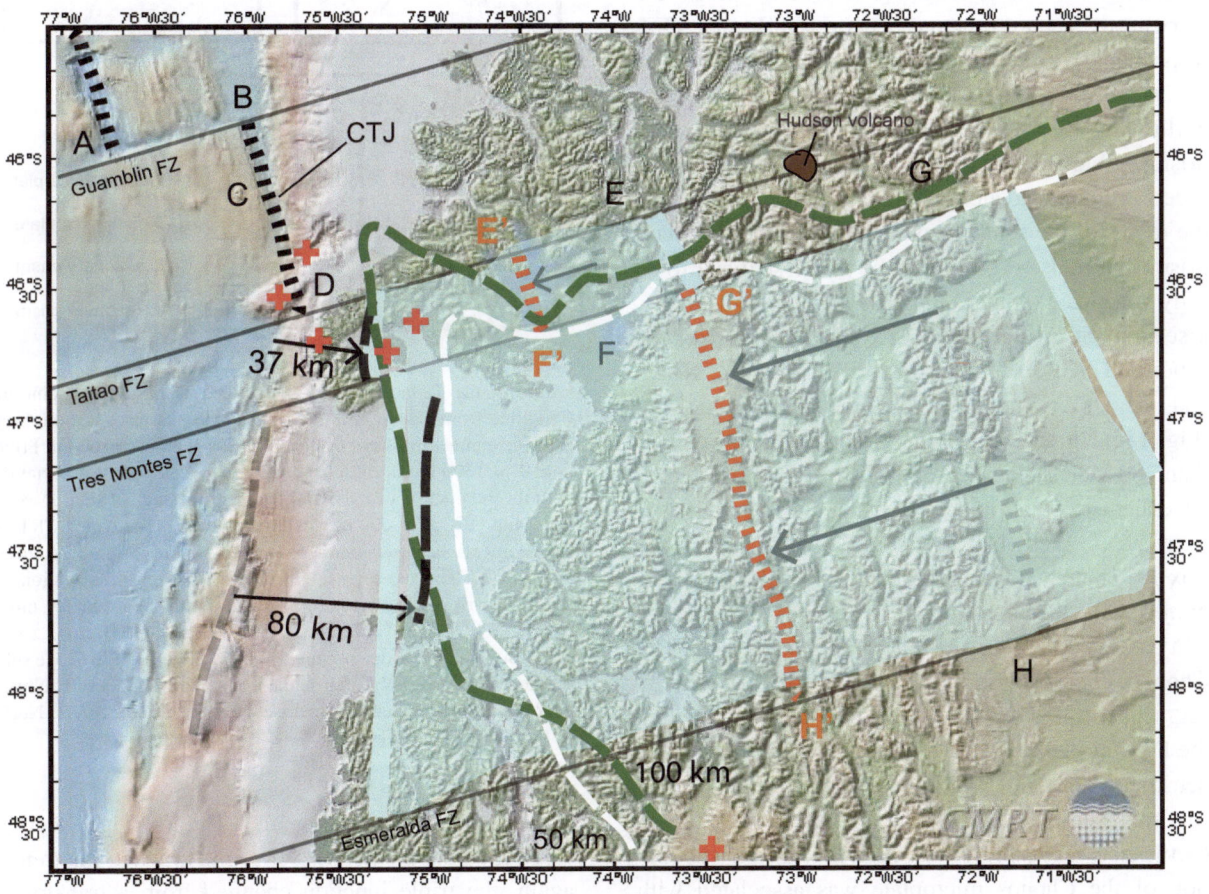

Figure 12

Patagonia slab window reconstruction (*pale green color*) using the westward shift (*orange tight dotted line*) for the SCR0 and SCR-1 spreading centers. Note the almost perfect fit between graphical reconstructions and tomographic determination (*green and white dash line*). *Red cross* location of adakite-like rock described in the text. *Red cross* at the *bottom* of the *figure* is located at the northern tip of the adakite Andean Austral Volcanic Zone (STERN and KILIAN 1996)

7.2. Ophiolite Obduction

7.2.1 Taitao Ridge Ophiolite

As exemplified along the SCR1 segment (Fig. 10), the tectonic regime of the Chile continental margin switches from subduction-erosion (north of CTJ, BEHRMANN *et al.* 1994) to subduction-accretion (south of CTJ, BOURGOIS *et al.* 2000) in relation with the CTJ northward migration. Because the Taitao ridge displays the tectonic pattern of an accretionary wedge (25–28 km from west to east) stacked against the Chile continental margin, BOURGOIS *et al.* (2000) concluded that the Taitao ridge was part of this accretionary wedge accumulated in the wake of the CTJ migration towards the north.

Along the SCR1 segment extending from the Darwin to Taitao Fracture Zones (DFZ and TFZ, respectively), a first ridge subduction event (Fig. 13a) began ~1300 ka ago. As the CTJ migrated toward the north (Fig. 13b), an accretionary prism (accretionary prism 1; Fig. 13) was rebuilt south of it. The thrust sheets accumulated along the Taitao Ridge transect are the relics of the southern part of this old accretionary prism 1. It took ~700 ka for the SCR1 segment to be fully subducted (point D to B in Fig. 3). Subsequently, a section (SCR1a and SCR1bN sub-segments) of the subducted SCR1 segment located between the NTRF and the DFZ jumped west of the subduction front (Fig. 13c), in an area of the Antarctic plate younger than 780 ka (Brunhes Chron). This ridge-jump created an ephemeral microplate (ROHR and FURLONG 1995; LUHR *et al.* 1985; DELONG *et al.* 1978; DELONG and FOX 1977), the Chonos microplate (Fig. 13c, BOURGOIS *et al.* 2000); the consequence of which is the transfer of a piece of the Antarctic plate to the Nazca plate. The Chonos microplate boundaries are the neo-SCR 1 segment (i.e., the SCR1a and SCR1bN sub-segments), the Darwin FZ, the Chile trench, and the NTRF (i.e., a neo-transform fault). During the time window needed for the neo-SCR1 segment to reach the trench (between ~200 and 600 ka, Fig. 13c, d), the subduction of the Chonos microplate was associated with frontal subduction-erosion (i.e., north of CTJ) that removed most of the previously accumulated accretionary prism 1. South of the NTRF, no ridge jump (SCR1bS sub-segment) occurred preventing

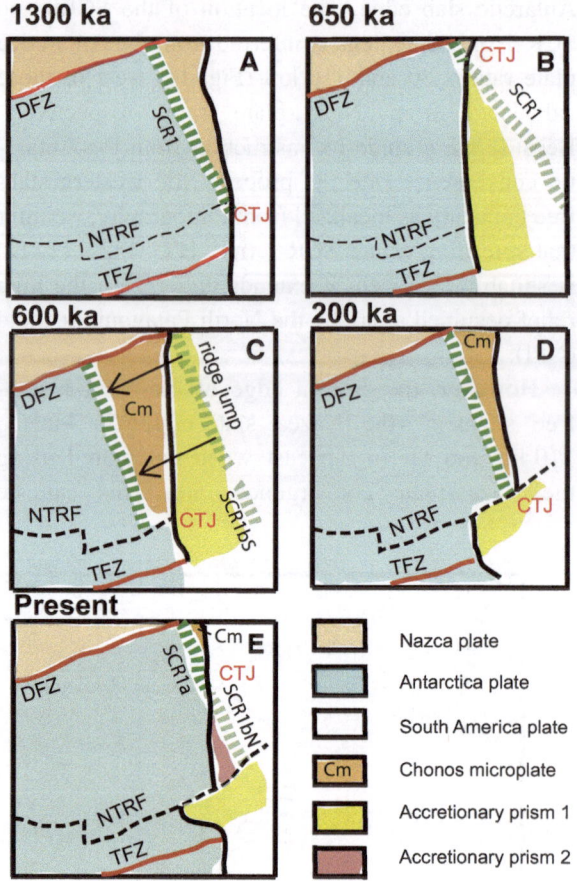

Figure 13
Origin of the Taitao ridge and associated Taitao Ridge ophiolite (modified from BOURGOIS *et al.* 2000). Two phases of accretionary prism accumulation have occurred during the past 1300 ka. From ~650 to 1300 ka, accretionary prism 1 developed in association with the subduction of the SCR1 segment located between Darwin and Taitao FZ. At ~600 ka ridge jump of the SCR1a and SCR1bN sub-segments occurred north of NTRF. No ridge jump occurred south of NTRF (sub-segment SCR1bS) allowing preservation of the Taitao ridge, while removed to the north. See text for more details. *CTJ* Chile Triple Junction, *DFZ* Darwin fracture zone, *SCR1* South Chile Ridge segment 1, *SCR1a* South Chile Ridge sub-segment 1a, *SCR1bN* South Chile Ridge sub-segment 1b North, *SCR1bS* South Chile Ridge sub-segment 1b South, *NTRF* North Taitao Ridge fault, *TFZ* Taitao fracture zone

subduction-erosion along the Taitao ridge section (Fig. 13c). As the neo-SCR1 segment reach the trench again, the triple junction changed from a transform-trench-trench to a ridge-trench-trench configuration (Fig. 13d, e). The section of the margin south of the CTJ (i.e., SCR1bN sub-segment) began developing an accretionary prism (accretionary prism 2, Fig. 13e); in the same time, the SCR1a sub-segment north of CTJ stayed in the subduction-erosion mode. We infer that the Taitao ridge is the remaining segment of an accretionary prism, which first exhibited a prolongation toward north of the NTRF, before being subsequently removed. The accretionary prism involving the Taitao Ridge ophiolite as a thrust sheet allows inducing its emplacement into the Chile continental margin (obduction) resulting from a related tectonic event. In other words, a transfer of a piece of the adjacent oceanic lithosphere and its incorporation into the Chile continental wedge has occurred during the tectonic development of the accretionary prism. The adakite-like magmatic rock drilled at Site 862 (Fig. 10) provides age constraints for the Taitao Ridge ophiolite, which could originate from an oceanic lithosphere older than 1.46–2.6 Ma. These ages slightly conflict with the inferred age for the Taitao Ridge ophiolite emplacement. We suspect a more complex emplacement mode for this ophiolite piece. The basalt recovered along the R1 and R2 thrust sheets constrains the age of the obduction that should be slightly older than 1.25 Ma (see Sect. 6.2). Therefore, the onset of the obduction matches the onset of the SCR1 subduction at 1.3 Ma (Fig. 13a). Because the subduction-accretion mode is following the ridge subduction, we suspect the Taitao Ridge ophiolite originated from the Antarctic plate lithosphere. The best candidate would be the wedge at the SCR1-Taitao junction.

7.2.2 Taitao Peninsula Ophiolite

To date, no direct evidence constrains the tectonic emplacement mode of the Taitao ophiolite into the Chile continental margin. Nevertheless, two lines of evidence provide insight into this puzzling issue. (1) The gabbros of the Taitao ophiolite cooled very rapidly from ~750 to 800 and ~320 °C in less than 200 ka, between ~5.5 and ~5.7 Ma. Consequently, these gabbros must have originated from a magma chamber located close to the SCR spreading center. (2) About 600 ka elapsed between the crystallization of the gabbros at shallow depth and their subsequent emplacement into the Chile continental margin, after 4.89 Ma as recorded by the MVU.

These evidences lead us to infer that the Taitao ophiolite emplacement has been carried out through two major tectono-magmatic and tectonic events. The first one took place in an area located west of the trench axis, the distance of which can be assessed using the convergence rate between the Nazca and South America plates at that time. The reconstruction illustrated in Fig. 14a shows the gabbros emplaced at shallow depth located at ~100 km from the trench axis at 5.5 Ma, at more than 120 km from their location today. The gabbro emplacement may have occurred through a lithospheric detachment associated with the evolution of an oceanic core complex (see the review by WHITNEY et al. 2013). As opposed to the typical geodynamic situation for such a detachment, commonly associated with slow-spreading mid-ocean ridges, the spreading rate was ~80 mm year^{-1} when the gabbro was emplaced. We suspect that its emplacement has been coeval to a ridge-jump event, allowing the local spreading rate to slow, the full rate being split between two different sites. The second tectonic event leading to the subsequent emplacement of the oceanic lithosphere (obduction) onto the Chile forearc occurred after 4.89 Ma. The reconstruction at 4.5 Ma (Fig. 14b) shows that the gabbros of the forthcoming Taitao ophiolite are still a component of the Nazca-plate lithosphere. At 4.5 Ma, they are located ~20 km west of the trench axis. The transfer of the oceanic lithosphere onto the South America continental margin has not yet occurred.

In Fig. 14, the geodynamic reconstructions are based on the fact that no major geodynamic event took place in the time window 4.5–5.5 Ma. Thus, they appear as being inconsistent with both the timing of emplacement and the location of the Taitao ophiolite. Accepting that the forthcoming Taitao ophiolites has collided the inner wall of the Chile trench at ~4.9 Ma or later, leads to locate the gabbro magma chamber at only ~60 km west of the trench axis (instead of ~100 km). We suggest (Fig. 15) that a ridge-jump of the SCR-1 segment has occurred. Subduction of the SCR-1 segment first occurred ~6 Ma ago (Fig. 15a, b). Subsequently, a westward ridge-jump (Fig. 15c) has placed the neo-SCR-1 at

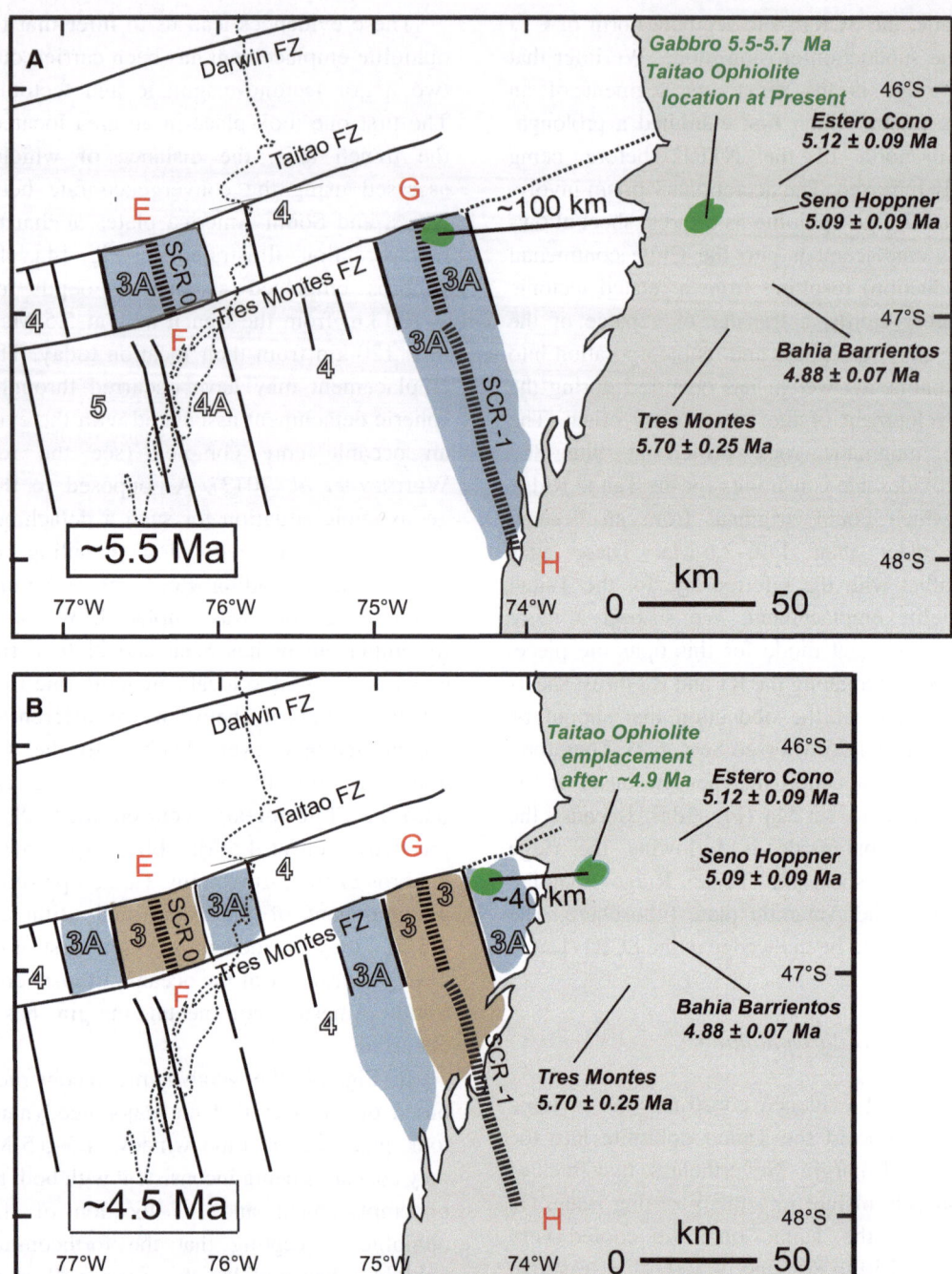

Figure 14

Geodynamic reconstructions at ~ 5.5 and 4.5 Ma considering that no major tectonic event, such as ridge-jump has occurred. **a** At 5.5 Ma: the SCR-1/Esmeralda FZ junction is close to the trench axis. At the latitude of the Taitao and Tres Montes peninsulas, the oceanic crust older than 5.5 Ma is several tens of km west of the trench axis. The emplacement of the Taitao Peninsula gabbro at shallow depth above 320 °C isotherm is completed. **b** At 4.5 Ma: the SCR-1 segment was at mid-term of subduction (Table 1). Subduction of the 4.5–6.3-Ma old oceanic crust (*brown* and *blue colors*) located at the latitude of the Taitao-Tres Montes peninsulas has not yet occurred. Emplacement of the Taitao and Tres Montes intrusive bodies (*black bold italic*) has already occurred. The obduction of the Taitao ophiolite is completed. The Cabo Raper pluton is not yet emplaced. *Cored letter* magnetic anomalies, *brown color* 4.5–5.5 Ma oceanic crust, *blue color* 5.5–6.3 Ma oceanic crust

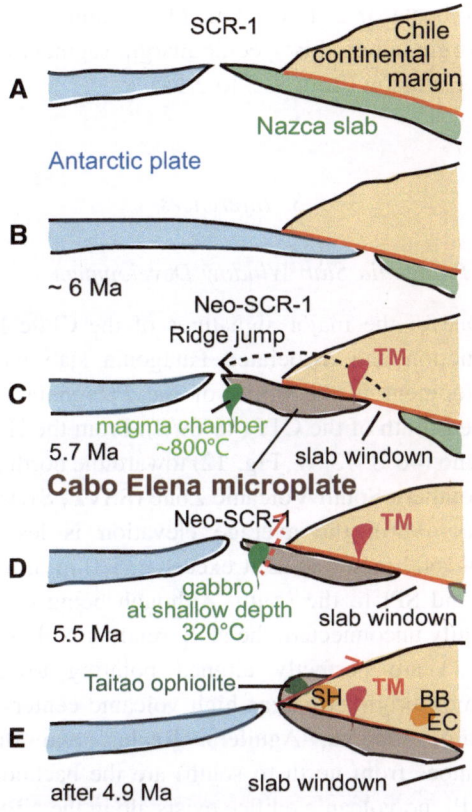

Figure 15
Taitao ophiolite emplaced through two tectonic events at 5.5-5.7 Ma (rapid gabbro uplift) and after 4.9 (obduction). **a, b** First subduction of the SCR-1 segment; **c** ridge jump of the SCR-1 spreading center, Cabo Elena microplate formation, gabbro magma chamber emplacement at depth close to the neo-SCR-1 segment; emplacement of the Tres Montes pluton; **d** Gabbro at shallow depth; **e** obduction of the Taitao ophiolite; the non-adakitic Seno Hoppner, Bahia Barrientos, and Estero Cono plutons intrusions stocks (ANMA and ORIHASHI 2013; KON et al. 2013; ANMA et al. 2009; LAGABRIELLE et al. 2000; GUIVEL et al. 1999) occurred between steps **d, e**

$\sim$60 km from the trench axis. A microplate, the Cabo Elena microplate, has been created. The Taitao gabbro magma chamber emplaced along the western side of the Cabo Elena microplate was close to the neo-SCR-1. The obduction of the Taitao ophiolite occurred through two tectonic phases: uplift of the gabbro (Fig. 15d) between 5.5 and 5.7 Ma, and thrusting of the oceanic lithosphere onto the adjacent continental margin (Fig. 15e), after $\sim$4.9 Ma.

7.3. Adakite-Like Generation

Today, adakitic volcanism is linked to the subduction of the Antarctic plate under the South America plate, all the volcanoes of the Austral Volcanic Zone of the Andes (AVZ) (South of Taitao) erupt such magmas (MARTIN 1987; SIGMARSSON et al. 1998; STERN et al. 1984; STERN and KILIAN 1996, among others). These latter are considered as generated by the melting of the slab basalts, in a hot subduction environment. Indeed, the age of the slab at depth, under the volcanic arc is typically <20 Ma, which induces high geothermal gradients along the Benioff plane. The Taitao Peninsula is located at the northern end of the AVZ. As mentioned above, adakite-like magmatic rocks were identified at five different sites (Fig. 12), including the Cabo Raper pluton (BOURGOIS et al. 1996) west of the Taitao peninsula, along the North Taitao canyon (GUIVEL et al. 2003), the Pan de Azucar, and the Fjordo San Pedro volcanic edifices (ANMA and ORIHASHI 2013; GUIVEL et al. 1999; LE MOIGNE et al. 1996), which are parts of the Pan de Azucar Complex (Fig. 7) unconformably overlying the metamorphic basement of the Chile continental margin. Moreover, rhyolites with an adakitic signature (see Sect. 6.2) were drilled at the ODP Site 862 (FORSYTHE et al. 1995a, b). As already proposed for most adakites worldwide (DEFANT and DRUMMOND 1990; MARTIN 1994, 1999, for review), BOURGOIS et al. (1996) identified the magma source for the Cabo Raper pluton as being a slab melt. This allows inducing that the slab has to be (1) buried down to the garnet-amphibolite facies conditions, and then (2) melted at $P > 10$ kbar and $T > 650$ °C. Thus, basalts were buried at $\sim$25-to-30-km depth for providing the magma source. Similar metamorphic and melting conditions were also documented at the four other sites listed above. In addition, all these adakitic rocks can be referred as "High-Silica Adakites" (HSA; MARTIN et al. 2005), which indicate that, contrarily to "Low-Silica Adakites", their interaction with mantle peridotite was extremely weak or even inexistent. Such evidence militates in favor of a lack of mantle wedge above the subducted plate, at the place where Cabo Raper adakites were generated.

BOURGOIS et al. (1996) have proposed the required P–T conditions being reached through the thickening of the continental wedge by the rapid accumulation of an accretionary prism on top the slab (Fig. 5 in BOURGOIS et al. 1996). We, therefore, propose the following evolution. At 4.9 Ma, as soon as the neo-SCR-1 spreading center has entered the subduction (Fig. 15E), the tectonic regime switched from sub-duction-erosion to subduction-accretion (see Sect. 6.3, BOURGOIS et al. 2000). An accretionary prism developed beneath the continental wedge, in front of the recently obducted Taitao ophiolite (see Fig. 4 in BOURGOIS et al. 1996). The Chile trench migrated oceanward ($\sim$23 km minimum), the thick-ening of the continental wedge allowed the basaltic slab to reach the pressure and temperature of the amphibolite-eclogite transition. It must also be con-sidered that the high temperature of the subducted basaltic crust could have been more easily reached, since the crust is young, i.e., it had not enough time for a significant cooling and dehydration. The Cabo Raper adakitic pluton has been generated by partial melting of this metamorphosed basalt, leaving a hornblende- and garnet-bearing residue. This event occurred at 3.83–4.11 Ma, then the magma very rapidly migrated upward, passing across the $\sim$800-to-320 °C isotherms in less than $\sim$0.55 Myr. We estimate that the heat flux climax able to initiate slab melting in the stability field of garnet took place about 4.38 Ma ago. This latter is the likely age for the transit of the neo-SCR-1 at depth able to produce the blowtorch effect (DELONG et al. 1979; KAEDING et al. 1990) beneath the Cabo Raper site. Just after the neo-SCR-1 transit at depth, the Tres Montes FZ initiated its southward migration beneath the Taitao peninsula. We propose that this drastic change in an area located beneath the Cabo Raper pluton has induced its rapid cooling at $\sim$3.8 Ma. The subduction-erosion mode began to work soon after $\sim$3.8 Ma in relation with the incoming SCR0 segment, in which subduction occurred at $\sim$2.3 Ma. The paleoreconstruction of the continental wedge at $\sim$4 Ma when compared to the modern configuration of the Taitao peninsula has allowed estimation of the crustal volume removed by subduction-erosion. BOURGOIS et al. (1996) have proposed a 23-km landward retreat of the continental wedge in association with a conservative subduction-erosion rate of 231–443 km^3 Myr^{-1} along each km of margin, along this specific margin segment during a time window from $\sim$2 to 3 Ma.

8. Inferences

8.1. Patagonia Slab Window Development

One of the major signatures of the Chile Ridge subduction and associated Patagonia slab window development is the uplift of the Patagonia Andes located south of the CTJ (46°09′S). From the Hudson volcano (46°S–73°W, Fig. 12) toward the north along the southern South Volcanic Zone (SSVZ, STERN and KILIAN 1996), the average elevation is less than 1500–1600 m, whereas it exceeds 3500 m along the NPI and SPI to the south. Although being morpho-logically unconnected, the N–S trending NPI and SPI (Fig. 1) are perfectly aligned pointing toward a common origin, whereas high volcanic centers (i.e., Lautaro, Viedma, Aguilera, Reclus, and Burney volcanoes from north to south) are the backbone of the SPI, no volcanic edifice exists along the NPI. The Upper Cretaceous Andean batholith is the backbone of the NPI, which matches the so-called volcanic gap (between the South and Austral Volcanic Zones). The SPI high relief follows the western boundary of the Patagonia slab window; it is the place of active adakite production related to the partial melting at the depth of the Antarctic slab edge (MARTIN 1987, SIGMARSSON et al. 1998; STERN et al. 1984; STERN and KILIAN 1996, among others). Instead of extending to the north, along the NPI, the Antarctic slab edge adakitic strip is bending westward toward the Golfo de Penas, the Taitao Peninsula, and the CTJ. Two points have to be noted: (1) the AVZ adakite strip does not follow the Antarctic slab edge bending at depth and (2) the adakitic volcano-plutonic centers of the CTJ area are older by about 3–4 Myr than the active AVZ. We suggest that the PT conditions for adakite production along the Taitao peninsula tran-sect have vanished, while being maintained southward along the AVZ. As opposed the SPI, the NPI shows its morphological backbone following the SCR-1 spreading segment at depth (G′H′, Fig. 12). We assume that the high relief of the NPI is dynamically sustained by upward asthenospheric convection along the SCR-1 spreading center at depth. The dynamic uplift of the NPI is also maintained and supplemented by glacial rebound (BOURGOIS *et al.* 2016).

8.2. Ophiolite Emplacement

Ophiolite emplacement is identified at two different sites, the Taitao Ridge and the Taitao Peninsula, both sites located south of CTJ. These two ophiolite pieces have different obduction ages: 1.25–2.60 and between 4.90 and 3.85 Ma (maximum limits), respectively. Regarding the Taitao Ridge ophiolite, evidences exist showing that emplacement was related to the construction of an accretionary prism at ~1.3 Ma and subsequently preserved to be removed owing to a ridge-jump event restricted to its northern part. Evidences document an emplacement of the Taitao Peninsula ophiolite through two tectonic events. The first one located at the SCR-1/ Tres Montes FZ junction area occurred away from the trench axis in relation with a westward ridge jump at ~5.7 Ma. The correlative slow-down of the spreading rate allows a lithospheric detachment to develop in association with an Ocean Core Complex that facilitated the gabbro emplacement. In addition, the ridge-jump produced slab fragmentation and individualization of an ephemeral microplate south of the Tres Montes fracture zone: the Cabo Elena microplate. Uplift of a piece of the oceanic lithosphere occurred in less than 200 ka. Ophiolite obduction has taken place subsequently. There is no direct evidence to determine if the USR (Fig. 6) is a southward prolongation of the Taitao Peninsula ophiolite. Thus, the extent of the ridge jump along the SCR segment is not known. The tectonic limit between the Taito Peninsula ophiolite and the Chile margin basement occurs along the major fault bounding the Bahia Barrientos graben to the west; we assume that this fault is inherited from the initial tectonic emplacement during obduction. As the faults bounding the Bahia Barrientos graben prolong the faults bounding the Upper Slope graben (Fig. 6), we propose that the USR is the likely offshore prolongation of the Taitao Peninsula ophiolite. If so, the ridge jump associated with the Taitao Peninsula ophiolite emplacement would extend to the southern tip of the USR, and ophiolite emplacement was not just restricted to areas characterized by being close to an FZ.

Ophiolite obduction is one of the main signatures of ridge subduction. Instead of exposing a complete sequence of an ophiolite piece, the Taitao Peninsula ophiolite is outcropping along a restricted location evidences that exist to infer the Taitao Ridge ophiolite having an extension similar to that of the Taitao Peninsula ophiolite. If the Taitao Peninsula ophiolite is spreading southward along the USR, the ophiolite unit could accordingly be more than 150-km long. The CTJ model could offer a new way to contemplate the emplacement mode for major ophiolite pieces.

8.3. Adakite Genesis

The genesis and emplacement of adakitic magmas appear to be a prominent feature related to the Chile Spreading center subduction, at least since ~6 Ma. To allow garnet stability in its residue of melting, the slab basalt must have been buried down to at least 30-km depth. These data compliment the relative contribution of the subducted slab, the mantle wedge, and the continental crust components in the generation of adakites as evaluated along the Austral Volcanic Zone (AVZ, STERN and KILIAN 1996; SIGMARSSON *et al.* 1998; Fig. 12). Subsequently, STERN (2011) has identified components originating from melting of subducted oceanic crust and sediments contributing to the adakite magma. Similarly, more to the north, in Ecuador, the adakites resulting of the subduction of the Carnegie ridge (oceanic plateau) had a significant sedimentary component in their source, in addition to subducted basalts (HIDALGO *et al.* 2007, 2012; SCHIANO *et al.* 2010). At the CTJ area, GUIVEL *et al.* (2003) considered that the sediments significantly contributed to the adakite source. These sediments came from the trench and were formed at the expense of the metamorphic continental basement and the Andean batholith. In other words, these sediments have an average composition similar to that of the continental crust. At the CTJ, the sediments have three potential sources: (1) the trench sediment itself, (2) the continental crust removed by subduction-erosion, and (3) the SCR basalt, which already carry a subduction signature (KLEIN and KARSTEN 1995). In other words, if both the CTJ and AVZ adakites have identical geochemical signatures, contrary to the AVZ, the CTJ adakites do not show any contribution from the mantle wedge (high Mg#, Ni and Cr contents; MARTIN et al. 2005; MARTIN and MOYEN 2002; PROUTEAU et al. 2001; RAPP et al. 1999).

As shown in Fig. 4, the subduction of spreading-ridges and transforms is closely correlated with the genesis of adakites. Refining the geodynamic history shows that ridge-jump was associated with the subduction of SCR-1, SCR0, and SCR1 segments. Consequently, the subducted slab was divided into small pieces (microplates) associated with the multiple slab window development (Fig. 15). For example, the Pan de Azucar emplacement likely correlates with the transit of the slab window that developed landward the Cabo Elena microplate. To account for the presence of two groups of dacites contemporaneously emplaced at the same site, GUIVEL et al. (2003) developed a model based on the vertical slab geometry. The latter can be further developed in the light of the data presented in the present paper.

Since (1) the adakitic rocks outcropping along the North Taitao canyon are located close to the landward prolongation of the NTRF, and (2) adakites and ridge-jump are temporally separated by a time gap of ~900 ka. Indeed, the oldest adakite sample (CTJ17–21, <1.12 Ma) could be emplaced in relation to the ridge-jump event (Fig. 13a–c) associated with the subduction of the SCR1 segment. In addition, Goss and KAY (2006) and Tagaki (2004) proposed that the adakite genesis could have been favored by the subduction-erosion processes, which appears as being a major agent for chemical, mechanical, and heat transfers at ridge-trench-trench triple junction, such as at CTJ. In addition to the subduction of a young and, consequently, warm oceanic slab, three additional factors may be contemplated to introduce complexity into the possible origin of adakites namely: (1) the occurrence of several transform faults close to each other entering the subduction that has been already recognized as a cause leading to unusual magmatism (GUIVEL et al. 2003; BOURGOIS et al. 1993, 1996; LE MOIGNE et al. 1996; LAGABRIELLE

et al. 1994; NELSON et al. 1993; KAEDING et al. 1990; MPODOZIS et al. 1985); (2) slab fragmentation associated with ridge-jump and microplate individualization that has been identified at several sites spread over time; and (3) dynamical processes, such as small-scale sublithospheric convection, producing melting and magmatism emplacement (KAISLANIEMI et al. 2014; SCHMERR 2012; BALLMER et al. 2007). Small-scale convection may accelerate upward magma transport and reintroduction of slab fragments beneath the SCR exhibiting a chemical diversity from typical N-type MORB to dacite with calc-alkaline affinities (KLEIN and KARSTEN 1995; KARSTEN et al. 1996; SHERMAN et al. 1997).

Acknowledgments

Jacques Bourgois, Maria Eugenia Cisternas, and Jose Frutos acknowledge funding from the ECOS program that allowed us to organize two field expeditions in the Taitao and Tres Montes peninsulas area. The first one, mainly funded by the Institut des Sciences de l'Univers (INSU), Centre National de la Recherche Scientifique (CNRS), France in 1992, was done using the Oxxean Tres fishing-boat (Puerto Montt). The second one has used a helicopter provided by a mining company thanks to one of us (Jose Frutos). In addition, we thank the Shipboard scientific party and the crew involved in the CTJ campaign of the R/V L'Atalante (JB, Chief Scientist). We thank William L. Bandy and an anonymous reviewer for thoughtful reviews that greatly improved the manuscript.

REFERENCES

ANMA, R., ORIHASHI, Y. (2013), Shallow-depth melt eduction due to ridge subduction: LA-ICPMS U-Pb igneous and detrital zircon ages from the Chile triple junction and the Taitao Peninsula, Chilean Patagonia, Geochemical Journal 47, 149–165.

ANMA, R., AMSTRONG, R., ORIHASHI, Y., IKE, S., SHIN, K-C., Kon, Y., KOMIYA, T., OTA, T., KAGASHIMA, S., SHIBUYA, T., YAMAMOTO, S., VELOSO, E.E., FANNING, M. HERVÉ, F. (2009), Are the Taitao granites formed due to subduction of the Chile ridge?, Lithos 113, 246–258.

ANMA, R., AMSTRONG, R., DANHARA, T., ORIHASHI, Y., IWANO, H. (2006), Zircon sensitive high mass-resolution ion probe U-Pb and fission-track ages for gabbros and sheeted dikes of the Taitao ophiolite, southern Chile, and their tectonic implications, Island Arc 15, 130–142.

ATWATER, T. (1970), *Implication of plate tectonics for the Cenozoic tectonic evolution of western North America*, Geological Society of America Bulletin *81*, 3513–3536.

BALLMER, M.D., VAN HUNEN, J., ITO, G., TACKLEY, P.J., and BIANCO, T.A. (2007), *Non-hotspot volcano chains originating from small-scale sublithospheric convection*, Geophys. Res. Lett. *34*, L23310, doi:10.1029/2007GL031636.

BANGS, N.L., CANDE, S., LEWIS, S.D., and MILLER, J.J. (1992), *Structural framework of the Chile Margin at the Chile ridge collision zone*, Proceedings of the Ocean Drilling Program, Initial Report *141*, 11–21.

BEHRMANN, J.H., LEWIS, S.D., CANDE, S.C., and ODP 141 Scientific Party (1994), *Tectonic and geology of spreading ridge subduction at the Chile triple junction. A synthesis of results from Leg 141 of the Ocean Drilling Program*, Geologische Rundschau *83*, 832–852.

BEHRMANN, J.H., LEWIS, S.D., and the Scientific Party, Proceedings of the Ocean Drilling Program. Initial Reports 141, 708 pp, Ocean Drilling Program, (College Station, Texas 1992).

BLACKMAN, D.K., APPLEGATE, B., GERMAN, C.R., THURBER, A.R., HENIG, A.S. (2012), *Axial morphology along the Southern Chile rise*, Marine Geology 315–318, 58–63.

BOURGOIS, J., CISTERNAS, M-E., BRAUCHER, R., BOURLÈS, D., FRUTOS, J. (2016), *Geomorphic records along the General Carrera (Chile)-Buenos Aires (Argentina) glacial lake (46–48°S), climate inferences and glacial rebound for the past 7–9 ka*, The Journal of Geology **(accepted)**.

BOURGOIS, J., and MICHAUD, F. (2002), *Comparison between the Chile and Mexico triple junction areas substantiates slab window development beneath northwestern Mexico during the past 12–10 Myr*, Earth and Planetary Science Letters *201*, 35–44.

BOURGOIS, J., GUIVEL, C., LAGABRIELLE, Y., CALMUS, T., BOULÈGUE, J., and DAUX, V. (2000), *Glacial-interglacial trench supply variation, spreading ridge subduction, and feedback controls on the Andean margin development at the Chile triple junction area (45–48°S)*, Journal of Geophysical Research *105*, 8355–8386.

BOURGOIS J., MARTIN H., LAGABRIELLE Y., LE MOIGNE J. and, FRUTOS JARA J. (1996), *Subduction- erosion related to spreading-ridge subduction: Taitao peninsula (Chile margin triple junction area)*, Geology 24, 723–726.

BOURGOIS, J., LAGABRIELLE, Y., LE MOIGNE, J., URBINA, O., JANIN, M. C., and BEUZART, P. (1993), *Preliminary results of a field study of the Taitao ophiolite (southern Chile): Implications for the evolution of the Chile triple junction*, Ofioliti *18*, 113–129.

BOURGOIS J., LAGABRIELLE Y., MAURY, R., LEMOIGNE, J., VIDAL P., CANTAGREL J.M., URBINA O. (1992), *Geology of the Taitao Peninsula (Chile Margin triple Junction area, 46°–47°S): Miocene to Pleistocene obduction of the Bahia Barrientos ophiolite*, EOS *73*, 592.

BREITSPRECHER, K., and THORKELSON, D.J. (2009), *Neogene kinematic history of Nazca-Antarctic-Phoenix slab windows beneath Patagonia and the Antarctic Peninsula*, Tectonophysics *464*, 10–20.

CAI, K., SUN, M., YUAN, C., ZHAO, G., XIAO, W., LONG, X., WU, F. (2010), *Geochronological and geochemical study of mafic dykes from the northwestern Chinese Altai: implications for petrogenesis and tectonic evolution*, Gondwana Research *18*, 638–652.

CANDE, S.C., LESLIE, R.B., PARRA, J.C., and HODBART, M. (1987), *Interaction between the Chile ridge and Chile trench: geophysical and geothermal evidences*, Journal of Geophysical Research, *92*, 495–520.

CANDE, S.C., and LESLIE, R.B. (1986), *Late Cenozoic tectonics of the southern Chile trench*, Journal of Geophysical Research *91*, 471–496.

CANDE, S.C., HERRON, E.M., HALL, B.R. (1982), *The early Cenozoic tectonic history of the southeast Pacific*, Earth and Planetary Science Letters *89*, 63–74.

CHAN, C.F., TEPPER, J.H., and NELSON, B.K. (2012), *Petrology of the Grays River volcanics, southwest Washington: Plume-influenced slab window magmatism in the Cascadia forearc*, Geological Society of America Bulletin *124*, 1324–1338, doi: 10.1130/B30576.1.

CHEN, J.L., WU, J.B., XU, J.F., DONG, Y.H., WANG, B.D., KANG, Z.Q. (2013), *Geochemistry of Eocene high-Mg# adakitic rocks in the northern Qiangtang terrane, central Tibet: implications for early uplift of the plateau*, Geological Society of America Bulletin *125*, 1800–1819, doi: 10.1130/B30755.1.

CLOOS, M. (1993), *Lithospheric buoyancy and collisional orogenesis: subduction of oceanic plateaus, continental margins, Island arcs, spreading ridges, and seamounts*, Geological Society of America Bulletin, *105*, 715–737.

COLE, R.B., STEWART, B.W. (2009), *Continental margin volcanism at sites of spreading ridge subduction: examples from southern Alaska and western California*, Tectonophysics *464*, 118–136.

DEFANT, M., and DRUMMOND, M.S. (1990), *Derivation of some modern arc magmas by melting of young subducted lithosphere*, Nature *347*, 662–665.

DELONG, S.E., SCHWARZ, W.M., ANDERSON, R.N. (1979), *Thermal effects of ridge subduction*, Earth and Planetary Science Letters *44*, 239–246.

DELONG, S.E., FOX, P.J., MCDOWELL, F.W. (1978), *Subduction of the Kula ridge at the Aleutian trench*, Geological Society of America Bulletin 89, 83–95.

DELONG, S.E., and FOX, P.J. (1977), Geological consequences of ridge subduction, In Island arc, deep sea trenches, and back-arc basins, Maurice Ewing Ser., vol. 1 (ed. TALWANI, M. and PITMAN III, W.C.) (AGU Washington D.C.) pp. 221–228.

DEMETS. C., GORDON, R. G., ARGUS, D. F., and STEIN, S. (1994), *Effect of recent revisions to the geomagnetic reversal time scale on estimate of current plate motions*, Geophysical Research Letters *21*, 2191–2194.

DICKINSON, W.R., SNYDER, W.S. (1979), *Geometry of subducted slabs related to San Andreas transform*, Journal of Geology 87, 609–627.

FORSYTHE, R.D., MEEN, J.K., BENDER, J., and ELTHON, D. (1995a), Geochemical data on volcanic rocks and glasses recovered from Site 86: implications for the origin of the Taitao ridge, Chile triple junction region, In Proceedings of the Ocean Drilling Program, Scientific Results 141 (eds. LEWIS, S.D., BEHRMANN, J.H., MUSGRAVE, R.J., and CANDE, S.C.) (College Station, TX, 1995) pp. 331–348.

FORSYTHE, R.D., DRAKE, R., and OLSSON, R.K. (1995b), Data Report: 40Ar/39Ar and additional paleontologic age constraints, Site 862, Taitao ridge, In Proceedings of the Ocean Drilling Program, Scientific Results 141 (eds. LEWIS, S.D., BEHRMANN, J.H., MUSGRAVE, R.J., and CANDE, S.C.) (College Station, TX, 1995) pp. 421–426.

FORSYTHE, R. D., NELSON, E. P., CARR, M. J., KAEDING, M. E., HERVÉ, M., MPODOZIS, C. M., SOFFIA, M. J., and HARAMBOUR, S. (1986), *Pliocene near trench magmatism in southern Chile: a possible manifestation of ridge collision*, Geology 14, 23–27.

FORSYTHE, R.D., and NELSON, E.P. (1985a), *Geological manifestations of ridge collision: evidence from the Golfo de Penas-Taitao Basin, southern Chile*, Tectonics 4, 477–495.

FORSYTHE, R.D., OLSON, R.K., HOSSON, C., NELSON, E.P. (1985b), *Stratigraphic and micropaleontologic observations from the Golfo de Penas-Taitao Basin, southern Chile*. Revista Geológica de Chile 25–26, 3–12.

GORRING, M.L., KAY, S.M., ZEITLER, P.K., RAMOS, V.A., RUBIOLO, D., FERNANDEZ, M.I., PANZA, J.L. (1997), *Neogene Patagonian plateau lavas: continental magmas associated with ridge collision at the Chile triple junction*, Tectonics 16, 1–17.

GOSS, A.R., and KAY, S.M. (2006), *Steep REE patterns and enriched Pb isotopes in southern Central American arc magmas: evidence for forearc subduction-erosion?*, Geochemistry, Geophysics, Geosystems 7, doi:10.1029/2005GC00116.

GUIVEL, C., LAGABRIELLE, Y., BOURGOIS, J., MARTIN, H., ARNAUD, N., FOURCADE, S., COTTEN, J., MAURY, R.C. (2003), *Shallow melting of oceanic crust during spreading ridge subduction: origin of near-trench Quaternary volcanism at the Chile Triple Junction*, Journal of Geophysical Research 108, 2345, doi:10.1029/2002JB002119,2003.

GUIVEL, C., LAGABRIELLE, Y., BOURGOIS, J., MAURY, R.C., FOURCADE, S., MARTIN, H., ARNAUD, N. (1999), *New geochemical constraints for the origin of ridge-subduction-related plutonic and volcanic suites from the Chile Triple Junction (Taitao Peninsula and Site 862, Leg ODP 141 on the Taitao Ridge)*, Tectonophysics 311, 83–111.

GUIVEL, C., LAGABRIELLE, Y., BOURGOIS, J., MAURY, R.C., MARTIN, H., ARNAUD, N., COTTEN, J. (1996), Magmatic responses to active spreading ridge subduction: multiple magma sources in the Taitao peninsula region (46°–47°S, Chile Triple Junction), Third ISAG, St Malo (France), 575–578.

HERRON, E.M., CANDE, S.G., HALL, B.R. (1981), *An active spreading center collides with a subduction zone: a geophysical survey of the Chile margin triple junction*, Memoir of the Geological Society of America 154, 683–701.

HERVÉ, F., FANNING, M.C., THOMSON, S.N., PANKHURST, R.J., ANMA, R., VELOSO, E.E., HERRERA, C. (2003), SHRIMP U-Pb and FT Pliocene ages of near-trench granites in Taitao peninsula, southern Chile, Short Paper-IV, South American Symposium on Isotope Geology, 190–193.

HIDALGO, S., GERBE, M.C., MARTIN, H., SAMANIEGO, P., and BOURDON, E. (2012), *Role of crustal and slab components in the Northern Volcanic Zone of the Andes (Ecuador) constrained by Sr-Nd-O isotopes*, Lithos 132–133, 180–192.

HIDALGO, S., MONZIER, M., MARTIN, H., CHAZOT, G., EISSEN, J.-P., and COTTEN, J. (2007), *Adakitic magmas in the Ecuadorian volcanic front: Petrogenesis of the Iliniza volcanic complex (Ecuador)*, Journal of Volcanology and Geothermal Research 159, 366–392.

KAEDING M., FORSYTHE R.D., and NELSON E. P. (1990), *Geochemistry of the Taitao ophiolite and near-trench intrusions from the Chile margin triple junction*, Journal of South American Earth Sciences 3, 161–177.

KAISLANIEMI, L., VAN HUNEN, J., ALLEN, M.B., NEILL, I. (2014), *Sublithospheric small-scale convection, a mechanism for collision zone magmatism*, Geology 42, 291–294.

KARSTEN, J. L., KLEIN, E. M., and SHERMAN, S. B. (1996), *Subduction zone geochemical characteristics in ocean ridge basalts from the southern Chile Ridge: implications of modern ridge subduction systems for the Archean*, Lithos 37, 143–161.

KAY, S.M., RAMOS, V.A., and MARQUEZ, M. (1993), *Evidence in Cerro Pampa volcanic rocks for slab-melting prior to ridge-trench collision in southern South America*, The journal of Geology 101, 703–714.

KLEIN, E.M., and KARSTEN, J.L. (1995), *Ocean ridge basalts with convergent margin geochemical affinities from the southern Chile ridge*, Nature 374, 52–57.

KON, Y., KOMIYA, T., ANMA, R., HIRATA, T., SHIBUYA, T., YAMAMOTO, S., and MARUYAMA, S. (2013), *Petrogenesis of the ridge subduction-related granitoids from the Taitao peninsula, Chile Triple Junction area*, Geochemical Journal 47, 167–183.

KOMIYA, T., YAMAMOTO, S., AOKI, S., SAWAKI, Y., ISHIKAWA, A., TASHIRO, T., KOSHIDA, K., SHIMOJO, M., AOKI, K., COLLERSON, K.D. (2015), *Geology of the Eoarchean, >3.95 Ga, Nulliak supracrustal rocks in the Saglek Block, northern Labrador, Canada: the oldest geological evidence for plate tectonics*, Tectonophysics, DOI: 10.1016/j.tecto.2015.05.003.

KURNOSOV, V., FORSYTHE, R., LINDSLEY-GRIFFIN, N., ZOLOTAREV, B., KASHINZEV, G., EROSHCHEV-SHAK, V., ARTAMONOV, A. and CHUDAEV, O. (1995), Comparison of the alteration and petrology of the Taitao ridge to the Taitao ophiolite, In Proceedings of the Ocean Drilling Program, Scientific Results 141 (eds. LEWIS, S.D., BEHRMANN, J.H., MUSGRAVE, R.J., and CANDE, S.C.) (College Station, TX, 1995) pp. 349–360.

LAGABRIELLE, Y., BOURGOIS, J., DYMENT, J., PELLETIER, B. (2015), *Lower plate deformation at the Chile Triple Junction from the paleomagnetic record (45°30' to 46°S)*, Tectonics 34, 1646–1660, doi:10.1002/2014TC003773.

LAGABRIELLE, Y., GUIVEL, C., MAURY, R., BOURGOIS, J., FOURCADE, S., MARTIN, H. (2000), *Magmatic-tectonic effects of high thermal regime at the site of active ridge subduction: the Chile triple junction model*, Tectonophysics 326, 255–268.

LAGABRIELLE Y., LE MOIGNE J., MAURY R.C., COTTEN J., BOURGOIS J. (1994), *Volcanic record of the subduction of an active spreading ridge, Taitao Peninsula (southern Chile)*, Geology 22, 515–518.

LE MOIGNE J., LAGABRIELLE Y., WHITECHURCH H., GIRARDEAU J., BOURGOIS J. and MAURY R. (1996), *Petrology and geochemistry of the ophiolitic and volcanic suites of the Taitao peninsula (Chile triple junction area)*, Journal of South American Earth Sciences 9, 43–58.

LE MOIGNE, J., LAGABRIELLE, Y., BOURGOIS, J., PALVADEAU, E. (1993), *Ophiolites en contexte de dorsale en subduction: nouvelles données sur la péninsule de Taitao (sud Chili)*, Comptes Rendus de l'Académie des Sciences de Paris 317, 403–410.

LESLIE, R.B. (1986), Cenozoic tectonics of southern Chile: triple junction migration, ridge subduction, and forearc evolution [Ph.D. thesis]: New York, University of Colombia, 276 p.

LING, M.-X., LI, Y., DING, X., TENG, F.-Z., YANG, X.-Y., FAN, W-M., XU, Y.G., SUN, W. (2013), *Destruction of the North China craton induced by ridge subduction*, The Journal of Geology 121, 197–213, DOI: 10.1086/669248.

LIU, W, LIU, X.J., XIAO, W.J. (2012a), *Massive granitoid production without massive continental crust growth in the Chinese Altay: insight into the source rock of granitoids using integrated zircon U-Pb age, Hf-Nd-Sr isotopes and geochemistry*, American Journal of Science 312, 629–684, DOI: 10.2475/06.2012.02.

LIU, Y., WANG, X., WANG, D., HE, D., ZONG, K., GAO, C., HU, Z., GONG, H. (2012b), *Triassic high-Mg adakitic andesites from Linxi, inner Mongolia: insight into the fate of the Paleo-Asian ocean crust and fossil slab-derived melt-peridotite interaction*, Chemical Geology, 328, 89–108.

LUHR, J.F., NELSON, S., ALLAN, J.F., and CARMICHAEL I.S.E. (1985), *Active rifting in south-western Mexico: manifestations of an incipient eastward spreading-ridge jump*, Geology *13*, 54–57.

MAO, Q., XIAO, W., FANG, T., WANG, J., HAN, C., SUN, M., YUAN, C. (2012), *Late Ordovician to Early Devonian adakites and Nb-enriched basalts in the Liuyuan area, Beishan, NW China: implications for early Paleozoic slab-melting and crustal growth in the southern Altaids*, Gondwana Research 22, 534–553.

MARTIN, H. (1999), *The adakitic magmas: modern analogues of Archaean granitoids*, Lithos 46, 411–429.

MARTIN, H. (1994), The Archean grey gneisses and the genesis of the continental crust, In archean crustal evolution (ed. CONDIE, X.E.) (Amsterdam, Netherlands, Elsevier), pp. 205–259.

MARTIN, H. (1987), *Archean and modern granitoids as indicators of changes in geodynamic processes*, Revista Brasileira de Geociência *17*, 360–365.

MARTIN, H. (1986), *Effect of steeper Archean geothermal gradient on geochemistry of subduction-zone magmas*, Geology *14*, 753–756.

MARTIN, H., SMITHIES, R.H., RAPP, R., MOYEN, J.-F. and CHAMPION, D. (2005), *An overview of adakite, tonalite-trondhjemite-granodiorite (TTG), and sanukitoid: relationships and some implications for crustal evolution*, Lithos, *79*, 1–24.

MARTIN, H. and MOYEN, J.-F. (2002), *Secular changes in TTG composition as markers of the progressive cooling of the Earth*, Geology 30, 319–322.

McCRORY, P.A.; WILSON, D.S., STANLEY, R.G. (2009), *Continuing evolution of the Pacific-Juan de Fuca-North America slab window system—A Trench-ridge-transform example from the Pacific rim*, Tectonophysics *464*, 30–42.

MORDOJOVICH, C. (1981), Sedimentary basins of Chilean Pacific offshore, In Energy Resources of the Pacific Region (ed. HALBOUTY, M.T.) (A.A.P.G. studies in Geology 12, Tulsa).

MPODOZIS, C.M., HERVÉ M.A., Nasi C., FORSYTHE R., and NELSON E. (1985), *El Magmatismo Plioceno de Península Tres Montes y su relación con la evolución del Punto Triple de Chile Austral*, Revista Geológica de Chile 25–26, 13–28.

MOYEN, J.-F. and MARTIN, H. (2012), *Forty years of TTG research*, Lithos *148*, 312–336.

NELSON, E., FORSYTHE, R., DIEMER. J., ALLEN, M., URBINA, O. (1993), *Taitao ophiolite: a ridge collision ophiolite in the forearc of southern Chile (46°S)*, Revista Geológica de Chile *20*, 137–166.

NIU, Y., ZHAO, Z., ZHU, D., MO, X. (2013), *Continental collision zones are primary sites for continental crust growth—A testable hypothesis*, Earth Science Reviews, doi:10.1016/j.earscirev.2013.09.004.

PROUTEAU, G., SCAILLET, B., PICHAVANT, M. and MAURY, R.C. (2001), *Evidence for mantle metasomatism by hydrous silicic melts derived from subducted oceanic crust*, Nature *410*, 197–200.

RAMOS, V.A., and KAY, S.M. (1992), Southern *Patagonian plateau basalts and deformation: backarc testimony of ridge collisions*, Tectonophysics *205*, 261–282, doi: 10.1016/0040-1951(92)90430-E.

RAPP, R.P., SHIMIZU, N., NORMAN, M.D. and APPLEGATE, G.S. (1999), *Reaction between slab-derived melts and peridotite in the mantle wedge: Experimental constraints at 3.8 GPa*, Chemical Geology *160*, 335–356.

ROHR, K.M.M., and FURLONG, K.P. (1995), *Ephemeral plate tectonics at the Queen Charlotte triple junction*, Geology *23*, 1035–1038.

RUSSO, R.M., VANDECAR, J.C., COMTE, D., MOCANU, V.I., GALLEGO, A., and MURDIE, R.E. (2010a), *Subduction of the Chile ridge: upper mantle structure and flow*, Geological Society of America Today 20, 4–10, doi: 10.1130/GSATG61A.1.

RUSSO, R.M., GALLEGO, A., COMTE, D., MOCANU, V.I., MURDIE, R.E., and VANDECAR J.C. (2010b), *Source-side shear wave splitting and upper mantle flow in the Chile Ridge subduction region*, Geology 38, 707–710, doi: 10.1130/G30920.1.

SCHARMAN, M.R., PAVLIS, T.L., RUPPERT, N. (2012), *Crustal stabilization through the processes of ridge subduction: examples from the Chugach metamorphic complex, southern Alaska*, Earth and Planetary Science Letters 329–330, 122–132.

SCHIANO, P., MONZIER, M., EISSEN, J-P., MARTIN, H. and KOGA, K. (2010), *Simple mixing as the major control of the evolution of volcanic suites in the Ecuadorian Andes*, Contribution to Mineralogy and Petrology *160*, 297–312.

SCHMERR, N. (2012), *The Gutenberg discontinuity: melt at the lithosphere-asthenosphere boundary*, Science *335*, 1480–1483.

SCHULTE, R.F., SCHILLING, M., ANMA, R., FARQUHAR, J., HORAN, M., KOMIYA, T., PICCOLI, P.M., PITCHER, L., WALKER, R. (2009), *Chemical and chronologic complexity in the convecting upper mantle: evidence from the Taitao ophiolite, southern Chile*, Geochimica et Cosmochismica Acta 73, 5793–5819.

SHERMAN, S. B., KARSTEN, J. L. and KLEIN, E. M. (1997), *Petrogenesis of axial lavas from the southern Chile ridge: major element constraints*, Journal of Geophysical Research *102*(B7), 14963–14990.

SHIBUYA, T., KOMIYA, T., ANMA, R., OTA, T., OMORI, S., KON, Y., YAMAMOTO, S., MARUYAMA, S. (2007), *Progressive metamorphism of the Taitao ophiolite; evidence for axial and off-axis hydrothermal alterations*, Lithos 98, 233–260.

SIGMARSSON, O., MARTIN, H., KNOWLES, J. (1998), *Melting of a subducting oceanic crust in Austral Andean lavas from U-series disequilibria*, Nature *394*, 566–569.

SISSON, T.W., RATAJESKI, K., HANKINS,W.B., GLAZNER, A.F. (2005), *Voluminous granitic magmas from common basaltic sources*, Contributions to Mineralogy and Petrology *148*, 635–661.

SISSON, V.B., PAVLIS, T.L. (1993), Geological *consequences of plate reorganization: an example from Eocene southern Alaska fore arc*, Geology *21*, 913–916.

STERN, C.R. (2011), *Subduction erosion: rates, mechanisms, and its role in arc magmatism and the evolution of the continental crust and mantle*, Gondwana Research 20, 284–308.

STERN, C.R., and SCHOLL, D.W. (2010), *Yin and yang of continental crust creation and destruction by plate tectonic processes*, International Geology Review 52, 1–31.

STERN, C.R., KILIAN, R. (1996), *Role of the subducted slab, mantle wedge and continental crust in the generation of adakites from the Andean Austral Volcanic Zone*, Contributions to Mineralogy and Petrology *123*, 263–281.

STERN, C.R., FUTA, K., MUEHLENBACHS, K. (1984), Isotopic and trace element data for orogenic andesites from the austral Andes, In Andean Magmatism, Chemical and Isotopic Constraints (eds. HARMON R.S., BARREIRO, B.A) (Shiva Geology Series. Nantwich), pp. 1–46.

TAKAGI, T. (2004), *Origin of magnetic- and ilmenite-series granitic rocks in the Japan arc*, American Journal of Science 304, 169–202.

TANG, G-J., WYMAN, D.A., WANG, Q., LI, J., LI, Z-X., ZHAO, Z-H., SIN, W-D. (2012), *Asthenosphere-lithosphere interaction triggered by a slab window during ridge subduction: trace element and Sr-Nd-Hf-Os isotopic evidence from Late Carboniferous tholeiites in the western Junggar area (NW China)*, Earth and Planetary Science Letters 329–330, 84–96.

TEBBENS, S.F., CANDE, S.C., KOVACS, L., PARRA, J.C., LaBRECQUE, J.L., and H. VERGARA (1997), *The Chile ridge: a tectonic Framework*, Journal of Geophysical Research *102*, 12,035–12059.

THORKELSON, D.J. (1996), *Subduction of diverging plates and the principles of slab window formation*, Tectonophysics *255*, 47–63.

THORKELSON, D.J., TAYLOR, R.P. (1989), *Cordilleran slab-windows*, Geology *17*, 833–838.

WEISSEL, J.K., HAYES, D.E., HERRON, E.M. (1977), *Plate tectonic synthesis: the displacements between Australia, New Zealand and Antarctica since Late Cretaceous*, Marine Geology *25*, 231–277.

WHITNEY, D.L., TEYSSIER, C., REY, P., and BUCK, R. (2013), *Continental and es*, Geological Society of Amererica Bulletin *125*, 273–298; doi: 10.1130/B30754.1.

YANG, G., LI, Y., GU, P., YANG, B., TONG, L., ZHANG, H. (2012), *Geochronological and geochemical study of the Darbut*

ophiolitic complex in the west Junggar (NW China): implications for petrogenesis and Tectonic evolution, Earth and Planetary Science Letters *21*, 1037–1049.

ZAMORA, D. (2000), Fusion de la croûte océanique subductée: approche expérimentale et géochimique, Unpublished thesis, Blaise Pascal University, Clermont-Ferrand, France, 314 pp.

ZHANG, C., MA, C., HOLTZ, F., KOEPKE, J., WOLFF, P. E., BERNDT, J. (2013), *Mineralogical and geochemical constraints on contribution of magma mixing and fractional crystallization to high-Mg adakite-like diorites in eastern Dabie orogeny, East China*, Lithos 172–173, 118–138.

ZHANG, Z., ZHAO, G., SANTOSH, M., WANG, J. DONG, X., and SHEN, K. (2010), *Late Cretaceous charnockite with adakitic affinities from the Gangdese batholith southeastern Tibet: evidence for Neo-Tethyan mid-ocean ridge subduction*, Gondwana Research *17*, 615–631.

(Received January 11, 2016, revised May 12, 2016, accepted May 12, 2016, Published online May 26, 2016)

Pure Appl. Geophys. 173 (2016), 3247–3271
© 2016 Springer International Publishing
DOI 10.1007/s00024-016-1337-5

Pure and Applied Geophysics

Foreshock Patterns Preceding Great Earthquakes in the Subduction Zone of Chile

G. A. Papadopoulos[1] and G. Minadakis[1]

Abstract—Foreshock activity is considered as one of the most promising precursory changes for the main shock prediction in the short term. Averaging over several foreshock sequences has shown that foreshocks are characterized by distinct 3D patterns: their epicenters move towards the main shock epicenter, event count accelerates, and b-value drops. However, these space–time-size patterns were verified so far only in a very few individual cases mainly due to inadequate seismicity catalogue data. We have investigated 3D foreshock patterns before the M_w 8.8 Maule in 27 February 2010, M_w 8.1 Iquique in 1 April 2014, and M_w 8.4 Illapel in 16 September 2015 great earthquakes in the Chile subduction zone. To avoid biased results, no a priori spatiotemporal definitions of foreshocks were inserted. The procedure was based on pattern recognition from statistically significant seismicity changes in the three domains. The pattern recognition in one domain was independent of the pattern recognition in another domain. We found and verified with two independent catalogue data sets (CSN, IPOC) that within a critical area of ca. 65 km from the main shock epicenter, the 2014 event was preceded by distinct foreshock 3D patterns. A nearly weak foreshock stage (20 January–14 March 2014) was followed by a main-strong stage (15 March–1 April 2014) highly significant in all domains, although foreshock activity slightly decreased in about the last 5 days. Seismic moment release also accelerated in the last stage due to the occurrence of a cluster of very strong foreshock events. Foreshock activity very likely occurred in the hanging-wall fault domain on the South American Plate overriding Nazca Plate. The 2014 foreshock activity was quite similar to the one preceding the 6 Apr. 2009 L' Aquila (Italy) M_w 6.3 earthquake associated with normal faulting. Using the 2014 earthquake as a reference event, we observed that similar foreshock 3D patterns preceded the 2010 and 2015 earthquakes within critical distances of about 170 and 50 km, respectively. However, the foreshock activities were only weak in both the cases likely because of poor catalogue completeness.

Key words: Foreshocks, pattern recognition, b value, earthquake prediction, subduction zone, Chile.

1. Introduction

The short-term foreshocks preceding the main shocks in a time frame varying from hours up to 5–6 months attracted interest since the 1960s (e.g., Mogi 1963a, b, 1985; Suyehiro et al. 1964; Papazachos 1975; Ishida and Kanamori 1978; Kagan and Knopoff 1978; Jones and Molnar 1979; Dodge et al. 1995). Since foreshocks may provide information which increases the probability for the occurrence of a future strong main shock in the short-term sense (e.g., Agnew and Jones 1991; Console et al. 1993; Michael 2012), they are generally considered as one of the most promising precursory phenomena (e.g., Wyss 1997; Vidale et al. 2001) regardless the various possible mechanisms that may drive foreshock generation (e.g., Helmstetter et al. 2003; Vidale and Shearer 2006; Peng 2007). For example, the short-term prediction of the large (M 7.2) earthquake of 4 February 1975 in Haicheng, China was based on several precursory phenomena particularly the short-term foreshocks (Raleigh et al. 1977; Scholz 1977).

A major problem, however, is that some main-shocks are preceded by foreshocks, while others do not; although in some regions, high foreshock rate was recognized, such as in the North Pacific, where large interplate earthquakes are preceded by fore-shock sequences at a rate of 70 % (Bouchon et al. 2013). On the other hand, if foreshocks are existent, then it is of great interest to examine how foreshock sequences are distributed in the space–time-size domains. Studies in several seismicity regimes showed that the short-term foreshock activity increases as the inverse of time (e.g., Papazachos 1975; Kagan and Knopoff 1978; Jones and Molnar 1979; Maeda 1999; Yamaoka et al. 1999; Papadopoulos et al. 2000, 2010). The acceleration of the fracturing process before the main shocks has also

[1] Institute of Geodynamics, National Observatory of Athens, Athens, Greece. E-mail: papadop@noa.gr

been supported by laboratory material fracture experiments (e.g., Mogi 1963a; Scholz 1968), numerical modeling in spring-block models (e.g. Hainzl et al. 1999) and analytical damage mechanics modeling (e.g. Main 2000). On the other hand, it was found that the foreshocks tend to move towards the main shock epicenter (Chen et al. 1999; Papadopoulos et al. 2010; Lippiello et al. 2012), which is an evidence that the foreshocks is an inherent property of the main shock nucleation process.

In the size domain, it has been supported that in foreshocks, the parameter b drops with respect to aftershocks and background seismicity (Mogi 1963a, b, 1985; Suyehiro and Sekiya 1972; Papazachos 1975; Jones and Molnar 1979; Main et al. 1989; Molchan et al. 1999; Papadopoulos et al. 2010; Chan et al. 2012; Kato et al. 2012; Nanjo et al. 2012) from geometrical point of view, parameter b, expresses the slope of the straight line in the magnitude–frequency or G–R relation (Ishimoto and Iida 1939; Gutenberg and Richter 1944):

$$\log N = a - bM \qquad (1)$$

where N is the cumulative number of events of magnitude equal to or larger than M, and a and b are the parameters determined by the data. In global scale seismicity, parameter b was estimated around unity (e.g., Frohlich and Davis 1993; Schorlemmer et al. 2005). From seismological point of view, b is an experimentally observed macroscopic parameter measuring the ratio of the small magnitude over the large magnitude events. In terms of geophysics, the b value depends on stress loading conditions (Mogi 1963a; Scholz 1968) as well as on the crustal heterogeneity (e.g., Abercrombie and Mori 1996). Therefore, b could be considered as a stress meter (Schorlemmer et al. 2005). High b value indicates low stress asymmetrically distributed. On the contrary, low b value is an evidence of concentrated high stress. This framework explains why high b value is usually associated with aftershocks or swarms of various origins, while foreshocks are characterized by low b value. Avlonitis and Papadopoulos (2014) showed that the low values of b experimentally observed in foreshock sequences can be modeled by a process of material softening in the seismogenic volume.

Due to the scarcity of data, the above 3D time–space-size foreshock patterns were found by either averaging over several foreshock sequences or analyzing synthetic earthquake catalogues. However, the co-existence of the three patterns in individual foreshock sequences has been verified only in a very limited number of cases so far. First, excellent example was the lethal earthquake ($M_w = 6.3$) of 6 April 2009 in L' Aquila, Italy, which was preceded by strong foreshock signal in the three domains of space, time, and size (Papadopoulos et al. 2010).

A point of controversy, however, is the definition of the short-term foreshocks (e.g., Yamashita 1998), and the recognition of foreshocks in seismic catalogues strongly depends on the definition adopted. For example, Wu et al. (2014) investigated for foreshocks preceding the 12 November 1999 Düzce earthquake (M_w 7.1) in the North Anatolian Fault within only ~65 h before and within 20 km around the mainshock. They found no foreshock activity. However, such spatiotemporal restrictions may exclude foreshock events occurring outside the pre-selected narrow space–time limits, thus leading to biased results. Pre-selected foreshock definitions may partly explain why only some main shocks are preceded by the short-term foreshocks. On the other hand, the incidence or not of foreshocks very likely depends on a variety of geophysical factors, such as the style of faulting, the focal depth, and the degree of small-scale crustal heterogeneity (e.g., Abercrombie and Mori 1996; Cheng and Wong 2016). The completeness of the catalogue (e.g., Yamaoka et al. 1999) is also an important factor. In seismic catalogues, particularly the ones produced by routine procedures, small foreshocks usually are overlooked and are not catalogued (Papadopoulos et al. 2006; Wu et al. 2014). Therefore, the role of microseismicity is important in recognizing foreshock sequences (Mignan 2014). To use foreshocks as a potential tool for the short-term earthquake prediction, it is of great importance to discriminate between foreshocks and other types of seismicity clusters, e.g., swarms and main shock-aftershock sequences, which again depends on the foreshock definition (e.g., Ogata et al. 1996).

Based on the results obtained so far, we may summarize as follows: (1) foreshocks are recognized only before some mainshocks and this perhaps is controlled not only by pre-selected foreshock definitions and catalogue completeness but also by geophysical factors and (2) when existent, foreshocks follow distinct patterns in space–time-size domains: epicenters move towards the main shock epicenter, the activity increases with the inverse of time, and the *b* value drops. As a consequence, we are not interested to investigate under which conditions a single event could be characterized as foreshock. On the contrary, our interest is focused on characterizing earthquake populations as foreshocks. In this paper, we have investigated space–time-size foreshock patterns preceding great earthquakes in the subduction zone of Chile and tested for their statistical significance. Emphasis was given to the 1 April 2014 great earthquake. Then, this event was used as a reference event to examine the great earthquakes of 27 February 2010 and 16 September 2015.

2. *Past Foreshock Studies in Chile*

In the subduction zone of Chile, foreshocks were recognized as early as 1960. As a matter of fact, the giant (M_w 9.5) earthquake of 22 May 1960, which was the largest ever instrumentally recorded, was preceded by 45 foreshocks in a time period of 33 h before the main shock, while 250 aftershocks were recorded in a 33 h time period after the main shock (Suyehiro 1966). Four foreshocks were bigger than magnitude 7.0, including a magnitude 7.9 on May 21 that caused severe damage in the Concepcion area.

The M_w 8.0 great earthquake that ruptured along the coast of central Chile on 3 March 1985 was preceded by intense foreshock activity, which started with an event of *mb* 4.7 on 21 February (Comte et al. 1986). The frequency of the foreshocks caused great alarm in Valparaiso, while in the next 11 days, the permanent central Chilean network recorded 360 earthquakes with coda magnitudes of >3.0. Comte et al. (1986) supported that the main shock began in the region of precursory cluster of the foreshock activity about 30 km offshore of Algarrobo. Those authors also provided evidence that similar precursory foreshock activity was noted before the great Valparaiso earthquake of 16 August 1906. The precursory activity presumably started on 18 June. Madariaga et al. (2010) reported on foreshock

activity that was observed in a small area from December 2009 to January 2010 before the great earthquake of 27 February 2010.

More recently, the rupture process of the 1 April 2014 Iquique great earthquake (M_w 8.1) and its possible initiation by the intense foreshock activity was examined by several authors, including Brodsky and Lay (2014), Yagi et al. (2014), Lay et al. (2014), Ruiz et al. (2014), Schurr et al. (2014), Kato and Nakagawa (2014), Bedford et al. (2015), Meng et al. (2015), Duputel et al. (2015).

Although there is a consensus that the 2014 earthquake was preceded by intense foreshock activity, different opinions were expressed as regard when the activity began and what was the extend of the foreshock area. For instance, Schurr et al. (2014) examined the foreshock evolution in the domains of space (epicentral locations), time (activity rate), and size (*b* value), and supported that foreshock activity was developed in at least the last 500 days or so before the great main shock, thus controlling the initiation of the main shock through gradual unlocking of the plate boundary. On the other hand, Ruiz et al. (2014) supported that the precursory foreshock activity started on 4 January 2014. As for the area covered by foreshocks, Schurr et al. (2014) suggested that it extended from 19.00 to 21.00 S (ca. 220 km) along the subduction zone strike, while according to Ruiz et al. (2014), the foreshocks delineated a region that spans ~150 km along the strike of the subduction zone.

Such discrepancies are due to that studies on the 2014 foreshock sequence focused rather to examine how foreshocks contributed to the stress accumulation near the epicenter of the great main shock than to investigate space–time-size foreshock patterns and their deviation from randomness through testing for statistical significance. Only variation of the activity rate was statistically tested by Schurr et al. (2014) who supported that they found significant changes during the last 250 days before the main shock.

The identification of foreshock patterns before individual main shocks occurring in different tectonic environments around the globe is of particular importance to advance research in earthquake prediction. The recognition of foreshock patterns may substantially help for not only better understand the precursory fracture process but also to establish computational tools useful for the real time, short-term hazard evaluation. Accordingly, Lay et al. (2014) questionned if it is viable to recognize a foreshock sequence like that in 2014 as a definite precursor to an imminent large event or not. Such a crucial question was examined some years earlier by Papadopoulos et al. (2010) who recognized three distinct seismicity stages preceding the L' Aquila (Italy) M_w 6.3 earthquake of 6 April 2009: background activity stage that lasted for several years, weak foreshock stage in the last 5 months or so, and strong foreshock stage in the last 10 days. Those authors supported that in a scheme of regular seismicity analysis and evaluation based on the daily seismicity monitoring and updating of the earthquake catalogue, the ongoing stage of weak foreshock activity would be detectable in about one or 2 months before the main shock. Furthermore, the detection of the strong foreshock signal would require at least 4 or 5 days after its onset on 27 March 2009. Then, the strong foreshock signal, being evident in space (dense concentration of epicenters in a very narrow area), in time (drastic increase of the daily number of events), and in size (drastic drop of b value) would be detectable a few days before the main shock occurrence.

In this paper, we investigated and tested for non-randomness 3D foreshock precursory patterns before great main shocks occurring in the subduction zone of Chile. The great earthquakes of 27 February 2010 (M_w 8.8), 1 April 2014 (M_w 8.1), and 16 September 2015 (M_w 8.4) were examined. The 14 November 2007 Tocopilla, northern Chile, M_w 7.7 earthquake was preceded by some seismic events within the last 6 months, and only three events within 42 h before the main shock could be considered as foreshocks (Schurr et al. 2012). However, this activity was not beyond randomness. The 15 October 1997 M_w 7.1 Punitaqui earthquake was preceded by the strong M_w 6.7 foreshock of 7 July 1997, but again, no significant foreshock activity was detected, likely because of the intraslab (normal faulting, ~ 70 km deep) nature of the main shock (Lay et al. 2014). For these reasons, the cases of 1997 and 2007 earthquakes were not further examined by us. However, the role of the high completeness magnitude cutoff in the earthquake catalogue of that period should not be ignored when considering possibilities for the recognition of significant foreshock activity.

3. Methodology and Data

3.1. Statistical Background and Computational Remarks

Several branching type models have been used to study spatiotemporal earthquake clustering. One of the most commonly used is the epidemic-type aftershock sequence (ETAS) (Ogata 1998) which has been tested for analyzing clustering features of foreshocks (Zhuang and Ogata 2006) and swarms (Hainzl and Ogata 2005). Here, we did not applied ETAS for the purpose to investigate foreshock patterns not only in space and time but also in the size domain. Although the so-called unified scaling law for earthquakes that describe changes in the domains of rate, energy, and space might be appropriate for such an investigation (e.g., Kosobokov and Nekrasova 2005; Nekrasova et al. 2011), we selected to test these domains independently. The reason is that we are interested to check if statistically significant changes in each one of the three domains start simultaneously or not. Changes in the space–time-size domains were investigated using as metrics the distance, D, of earthquake epicenters from the main shock epicenter, the seismicity (activity) rate, r, and the b value of the G–R relation.

In our analysis, we did not adopt pre-selected, narrow space–time windows for the foreshock investigation. We rather preferred to use wide enough spatiotemporal windows with the purpose to avoid missing foreshocks occurring at large space and time distances from the main shock focus. The windows were gradually narrowing until significant space–time seismicity changes, if any, were detected. We assumed that the combined significant increase of r and the significant drop of D and b indicate strong foreshock activity. In this sense, the definition of foreshocks activity is not based on a prior settings of space–time-size limits but on space–time-size patterns that have physical meaning as regard the foreshock generation. The significant change of only two out of three parameters may indicate weak foreshock activity.

The method of analysis employed in this study is part of the algorithm FORMA (FOReshock-Main shock-Aftershock) developed in-house for the statistical seismicity analysis (e.g., Papadopoulos et al. 2009). A first application of the algorithm was performed for the analysis of the foreshock and aftershock activity associated with the L'Aquila, Italy, strong (M_w 6.3) earthquake of 6 April 2009 and other seismic sequences (Papadopoulos et al. 2010). The algorithm FORMA permits to perform a long number of tests with the aim to find out the highest level of significance in the changes of the seismicity attributes D, r, and b. Let us assume that the seismicity is examined in the time interval T and that changes in one of the three attributes are investigated. Then, FORMA is set up to start investigation by comparing the mean values of the particular attribute in two arbitrarily selected sequential time intervals. The hypothesis is that the two time intervals represent background seismicity period, T_b, and foreshock activity, T_f, where $T_b + T_f = T$ (Fig. 1). The differences of the mean values of D and r in the catalogue segments of T_b and T_f are tested with the statistical z test. The differences of the mean values of b are tested with the Utsu test (Utsu 1992, 1966) which is based on a formulation for measuring the statistical significance of the difference in the b value between two earthquake samples. Specifically, the probability, P, that two samples may come from the same population is calculated by

$$P \approx \exp[-(dA/2) - 2] \qquad (2)$$

where N_b and N_f are the number of events in the two samples and

$$dA = -2N\ln N + 2N_b\ln[(N_b + N_f)(b_b/b_f)] + 2N_f\ln[N_b(b_f/b_b) + N_f] - 2 \qquad (3)$$

$$N = N_b + N_f \qquad (4)$$

Lower values of P (e.g., less than 0.05) indicate the higher statistically significant difference between two earthquake samples, implying that the two samples under study do not come from the same population.

To find out the optimum point of time, t_i (Fig. 1), where the seismicity change becomes significant, the time lengths of T_b and T_f are changed by removing one time unit (e.g., 1 day) or a certain number of events from T_b and adding it (or them) to T_f and inversely. In this way, a long number of T_b/T_f pairs are examined. The T_b/T_f pair for which the significance level exceeds a certain value (e.g., 0.95) for changes in D and r or it is less than a certain value (e.g., 0.05) for changes in b objectively determines the optimum point of time t_i. For the three seismicity attributes one, two or three different points of time t_i might be determined. This depends on the synchronization mode of changes in D, r, and b, that is whether or not the three changes appear simultaneously or not.

In our analysis, the length of time interval T is selected by considering some certain criteria. Namely, T should not be too long, thus securing that

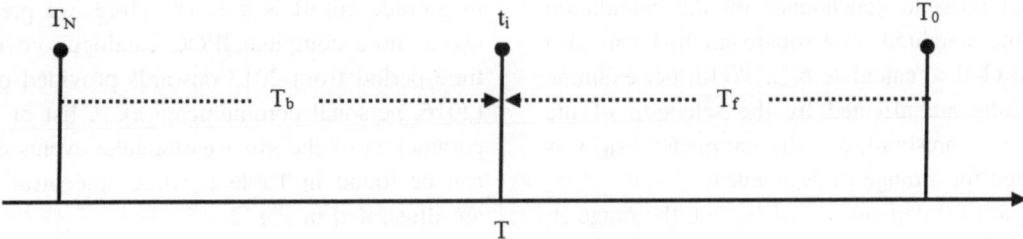

Figure 1
Schematic diagram for the identification of significant seismicity changes. A time period $T = T_0 - T_N$ is examined; T_0 = main shock origin time, T_N = start of the seismic catalogue segment selected to examine. The hypothesis is that the period of background seismicity, T_b, is followed by significant changes in one or more out of three seismicity attributes i (parameters D, r, and b) in the period, T_f, which is supposed to represent foreshock activity. Testing for significant changes, a long number of T_b/T_f pairs are examined. In practice, the time lengths of T_b and T_f are changed by removing one time unit (e.g., 1 day) or a certain number of events from T_b and adding it (or them) to T_f and inversely (see text for more details on the identification of the time point t_i as the onset of significant change of attribute i)

the completeness magnitude cutoff would be as low as possible. To avoid enhancement of the catalogue with abundant aftershocks of previous large events, T should start at least one year after the last large earthquake in the selected target area. Nevertheless, residual aftershock events may remain while smaller main shocks may also produce their own aftershocks. However, we did not decluster the earthquake catalogue with the purpose to keep it enhanced in smaller events, thus disfavoring the characterization of the activity rate in T_f as a significantly different one compared to the activity rate in T_b. Besides, it has been found that the choice of declustering algorithm and the variation in the related parameter values have the greatest impact on seismicity-rate-change calculations (Van Stiphout et al. 2011).

3.2. Control of the b-Value Variation

Apart from its dependence on various geophysical factors mentioned earlier, the calculation of parameter b is a quite sensitive procedure, since it directly depends on the computational techniques used. Namely, b depends on the magnitude cutoff in the earthquake sample, on the method of calculation, as well as on the range $\Delta = M_{max} - M_{min}$ inserted in the earthquake sample under examination. Calculations of both r and b were performed only for earthquake data sets being complete over a certain magnitude cutoff, M_c, determined from G–R diagrams. The standard method that we followed for the calculation of b, hereafter noted as b_{ML}, was the maximum likelihood approximation of Aki (1965). To control possible dependence on the calculation method, the weighted least-square method was also applied to G–R to calculate b_{GR}. To further examine if the results are affected by the selection of the completeness threshold, M_c, the parameter b_{ML} was recalculated for a range of M_c values.

As regard the dependence of b_{ML} on the range Δ, it was found that the b_{ML} variation is insignificant if the magnitude range in the sample exceeds 1.4, regardless the method used to estimate the parameter b (Papazachos 1974). Besides, it has been shown that the mean magnitude, M_m, in an earthquake sample is a function of the inverse of b (Utsu 1965; Lomnitz 1966; Hamilton 1967). Therefore, an equivalent way to examine the b-value variation is to look after the variation of M_m. To this aim, a conditional backward windowing technique was applied. More precisely, starting with a window of the last 100 events prior to the main shock (time T_0 in Fig. 1), both b_{ML} and M_m were calculated provided that the condition $\Delta > 1.4$ was fulfilled. Then, the process was repeated for the next 100 events sliding backwards with step of 1 event until the whole process reached the last 100 events from the beginning of the whole time frame under study (time T_N in Fig. 1). When the condition $\Delta > 1.4$ was not fulfilled, the calculation repeated by increasing the window with step of 1 event each time until the condition was fulfilled.

3.3. Data Sets

The seismicity data used in this study were taken from two different earthquake catalogues: (1) for the earthquake sequences of 2010, 2014, and 2015, we used the bulletins of the National Seismological Centre (CSN) of the University of Chile (http://www.sismologia.cl); (2) as regard the 2014 event, we also utilized the earthquake catalogue compiled by the Integrated Plate Boundary Observatory Chile (IPOC), which is a European-South American network of institutions and scientists (http://www.ipoc-network.org). The IPOC catalogue can be accessed via the web interface of the global seismological broadband network (GEOFON) operated by the German GeoForschungsZentrum (GFZ) (http://geofon.gfz-potsdam.de/). However, this IPOC catalogue was tested as for its completeness and found that the magnitude cutoff is 4.3. Therefore, we preferred to use a more complete IPOC catalogue covering the time period from 2013 onwards provided by Schurr (2016, personal communication). A list of the focal parameters of the great earthquake events examined can be found in Table 1, while epicentral locations are illustrated in Fig. 2.

4. Seismicity Analysis and Results

Results are presented first for the Iquique-Pisagua event of 1 April 2014, since the foreshock evidence published by Brodsky and Lay (2014), Yagi et al.

Table 1

Earthquake parameters of the three main shocks examined: Lat (S^0) and Lon (W^0) are epicentral latitude and longitude, respectively; M_w is moment-magnitude; h is focal depth

No.	Date and time	Lat	Lon	M_w	h (km)	Data source
1	27 Feb. 2010 06:34:08	36.2900	73.239	8.8	30.1	CSN
2	01 Apr. 2014 23:46:45	19.5720	70.908	8.1	38.9	CSN/IPOC
3	16- Sep. 2015 22:54:31	31.6370	71.741	8.4	23.3	CSN

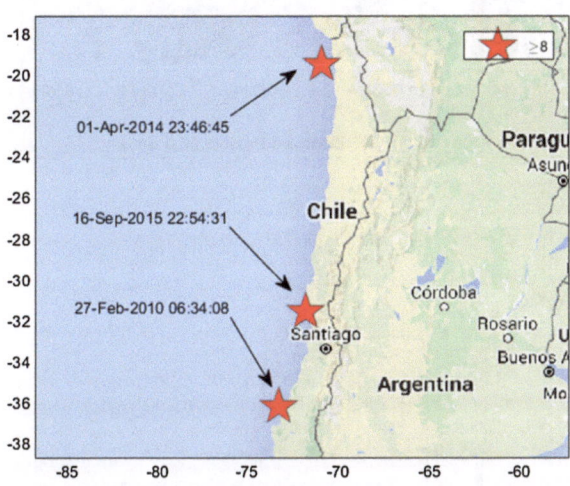

Figure 2
Epicenters of the three great earthquakes listed in Table 1

(2014), Lay et al. (2014), Schurr et al. (2014), and Bedford et al. (2015) was quite promising for the foreshock pattern recognition. Therefore, the fore-shock patterns that preceded the 2014 main shock could be used as a reference case for the analysis of the 2015 and 2010 sequences.

4.1. Earthquake of 1 April 2014—Analysis With the CSN Catalogue

To inspect the spatiotemporal distribution of seismicity before the main shock, a time period T extending from 1 January 2011 to the main shock occurrence was selected for examination. Allowing for spatial seismicity changes to occur in a large enough area, we initially chosen a circular area of radius $R = 200$ km around the main shock epicenter (Fig. 3 left). This area is larger than the dimension of the 2014 great earthquake fault rupture which has

been estimated with various waveform inversion techniques to not exceed about 160 km in length and 140 km in width (Yagi et al. 2014; Lay et al. 2014; Schurr et al. 2014).

It is evident that in the 18-day time period from 15 March 2014 up to the main shock occurrence, a dramatic drop of the average distance D occurred, which reflects a dense concentration of seismic epicenters within a critical target area of no more than 65 km away from the main shock epicenter (Fig. 3 left). At the same time, the number of events increased significantly during the 18-day time period preceding the main shock, although a slow increase already noted by the end of January 2014 (Fig. 4; Table 2). From 17 to 26 March 2014, a gradual shift of the foreshock activity towards the main shock epicenter noted (Fig. 3 right). In particular, in the time interval 22–26 March, the foreshocks concen-trated very close to the main shock epicenter at distances ranging from about 5 to 25 km. However, in the last 5 days or so before the great earthquake, the number of foreshock events decreased and distributed at random distances within the critical area of 65 km. Before 15 March 2014, a random variation of D was evident. It is noteworthy that two seismicity clusters appeared in the target area in August 2013 and in January–February 2014. Their possible foreshock nature is examined later.

4.2. Testing Foreshock Hypothesis

Based on the above observations, we put forward the hypothesis that in the target area, the mainshock was preceded by three sequential stages of seismicity characterized by statistically different features in their space, time, and size distributions: (1) back-ground seismicity (BGS; from the beginning of 2011

41

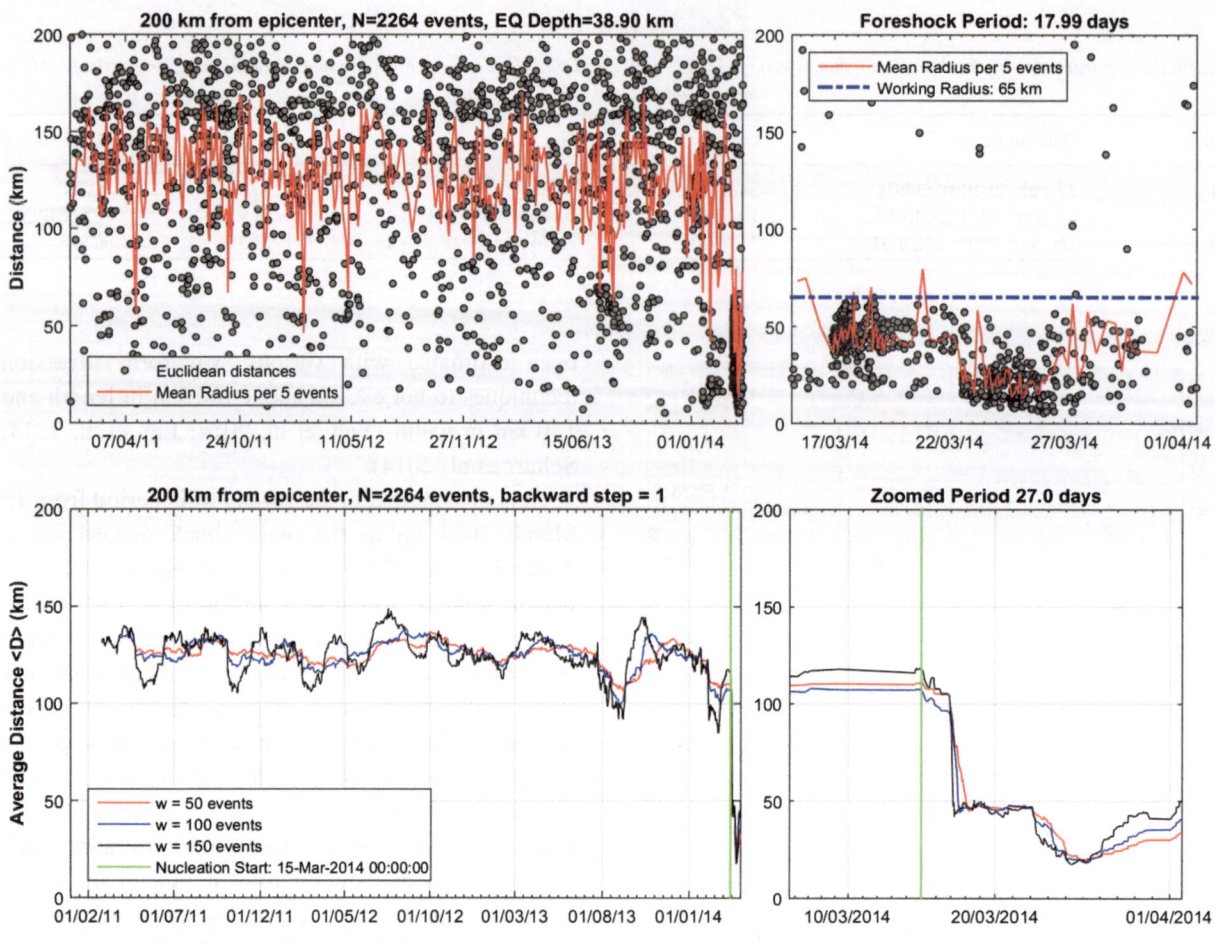

Figure 3
Top figure depicts the time variation of the average distance, D, of the earthquake epicenters (*solid circles*) from the main shock epicenter in the time period from 1 January 2011 to the origin time of the main shock of 1 April 2014. The dotted-blue line indicates critical distance of 65 km. Data taken from the CSN catalogue without magnitude cutoff. Distance was calculated for a non-overlapped window of 5 events. *Bottom figure* depicts the evolution of the average distance, D, using a backward sliding window of 50, 100, and 150 events, respectively, and sliding step of one event. The *green vertical line* indicates the estimated time stamp, t_i, where the foreshock activity starts

up to 19 January 2014; initially corresponding to the time interval T_b in Fig. 1), with the parameters D, r, and b varying randomly; (2) early weak foreshock activity (WFOR; from 20 January 2014 up to 14 March 2014); and (3) main-strong foreshock activity (FOR, T_f in Fig. 1) that preceded the main shock from 15 March 2014 up to the main shock occurrence. The hypothesis was tested on the basis of the statistical significance of the changes of the parameters D, r, and b (Table 2). From tests with the above three catalogue segments, it was found that the drop of distance D during the FOR period was highly significant as compared to that in the segments BGS, WFOR, and BGS+WFOR (cases 1, 2, and 4 in Table 2, respectively). In addition, the relative drop of D in WFOR was highly significant as compared to that in BGS (case 3 in Table 2).

The variations of both the activity rate, r and b_{ML}, were tested after analyzing the earthquake catalogue for data completeness. From the maximum curvature method in G–R diagrams, it was found that the magnitude completeness cutoff is $M_c = 2.6$ for the entire time interval from 2011 up to the main shock occurrence (Fig. 5 left). Taking events of magnitude equal to or larger than 2.6 and occurring in the target area, one may observe the highly significant increase

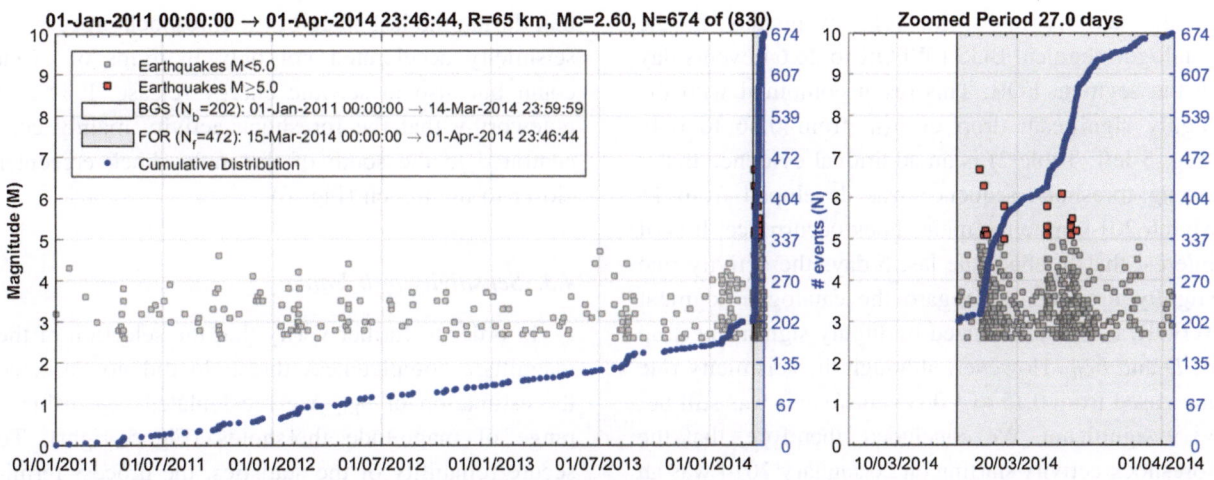

Figure 4

Time variation of the activity rate in the target area of radius 65 km for the time period from 1 January 2011 up to the main shock of 1 April 2014 (*left*). Grey and red squares indicate earthquakes of magnitude $M < 5$ and $M \geq 5$, respectively. BGS and FOR refer to the periods of background seismicity, T_b, and the foreshock activity, T_f. The catalogue data were taken for completeness magnitude threshold $M_c = 2.6$ from the CNS catalogue. *Vertical line* in the right-hand side shows the starting time of the main-strong foreshock activity at 15 March 2014. The cluster occurring before that date but after 20 January 2014 represents the early weak foreshock stage. *Right figure* depicts a zoomed version of 27 days period prior the occurrence of the main event: $T_{zoom} = T_f + \frac{T_f}{2}$

Table 2

List of the earthquake catalogue segments tested and results obtained

No.	Period	From	To	M_c	N	r	M_m	D	b_{GR}	b_{ML}	$P(b)$	$P(D)$	$P(r)$
1	BGS+ WFOR	01-Jan-2011 00:00:00	14-Mar-2014 23:59:59	2.6	202	0.17	3.21	40.0 ± 1.1	0.99 ± 0.16	0.66 ± 0.09	0.000332	99.82	99.99
	FOR	15-Mar-2014 00:00:00	01-Apr-2014 23:46:44	2.6	472	26.24	3.46	32.1 ± 0.6	0.68 ± 0.08	0.48 ± 0.04			
2	WFOR	20-Jan-2014 00:00:00	14-Mar-2014 23:59:59	2.6	37	0.69	3.37	40.0 ± 1.1	0.73 ± 0.18	0.53 ± 0.17	0.305476	99.51	83.40
	FOR	15-Mar-2014 00:00:00	01-Apr-2014 23:46:44	2.6	472	26.24	3.46	32.1 ± 0.6	0.68 ± 0.08	0.48 ± 0.04			
3	BGS	01-Jan-2011 00:00:00	19-Jan-2014 23:59:59	2.6	165	0.15	3.17	40.0 ± 1.1	0.97 ± 0.12	0.70 ± 0.11	0.112506	99.98	97.98
	WFOR	20-Jan-2014 00:00:00	14-Mar-2014 23:59:59	2.6	37	0.69	3.37	32.1 ± 0.6	0.73 ± 0.18	0.53 ± 0.17			
4	BGS	01-Jan-2011 00:00:00	19-Jan-2014 23:59:59	2.6	165	0.15	3.17	40.0 ± 1.1	0.97 ± 0.12	0.70 ± 0.11	0.000102	99.99	99.99
	FOR	15-Mar-2014 00:00:00	01-Apr-2014 23:46:44	2.6	472	26.24	3.46	32.1 ± 0.6	0.68 ± 0.08	0.48 ± 0.04			

Each catalogue segment is tested for changes in the parameters *D*, *r*, and *b* with respect to the catalogue segment preceding it in the list. For testing explanations, see in Sect. 2.1

The main event has been excluded from the analysis. Symbol $\pm$ indicates standard error

BGS background seismicity, *WFOR* week foreshock sequence, *FOR* strong foreshock sequence, M_c magnitude cutoff, *N* number of earthquake events, *r* seismicity rate (events/day), M_m mean magnitude, *D* average distance (km), b_{GR} weighted least-squares estimation of *b* value, b_{ML} maximum likelihood estimation of *b* value, *P(b)* probability according to formula 2 (Utsu test) for testing b_{ML} changes, *P(D)* probability according to *z* test for distance *D* testing, *P(r)* probability according to *z* test for rate *r* testing of the activity rate from 0.18 events/day in the catalogue segment BGS+WFOR to 26.64 events/day in the segment FOR. This result combined with the highly significant drop of b_{ML} from 0.66 to 0.48 (Fig. 5 left; Table 2) is an additional evidence that a strong foreshock sequence was developed from 15 March 2014 up to the main shock occurrence. It is of interest that in about the last 5 days the activity rate slighlty dropped. As regard the catalogue segment WFOR, it is characterized by highly significant drop of D and b_{ML}. However, although the seismicity rate increased from 0.15 to 1.06 events/day it was still not very significant. We concluded, therefore, that the foreshock activity starting on 20 January 2014 was an early-weak stage of the foreshock sequence which was followed by the main-strong foreshock stage developed from 15 March 2014 onwards.

Another important feature of the main-strong foreshock stage was unraveled by Fig. 4. Namely, during the 18-day main foreshock stage a cluster of strong events exceeding magnitude 5.0 occurred, while in the previous period extending to the beginning of 2011 the earthquake magnitudes did not exceeded 5.0. The strongest foreshock (M_w 6.7) occurred on 16 March 2014. This feature underlines

that during the last stage of the foreshock process the seismicity accelerated not only in terms of event count but also in seismic energy release. It is also noteworthy that the foreshock activity mainly concentrated to the south of the main shock epicenter closer to the trench (Fig. 6).

4.3. Sensitivity of b Value

In order to further verify that the selection of the magnitude completeness threshold did not affected the estimation of b_{ML} we recalculated b_{ML} using a range of magnitude thresholds (Fig. 5 right). To secure reliability of the statistics, the process terminated when a minimum number of 50 events was reached. The results obtained leave no doubt that the variation of b_{ML} in the foreshock activity remained constantly lower with respect to that in the background period, regardless the magnitude threshold applied. On the other hand, to control the possible dependence of the b value on the method of calculation, we calculated b_{GR} with the weighted least-squares method and found no different pattern. Figure 5 (left) depicts two G–R diagrams, one for the background seismicity period, T_b, and one for the

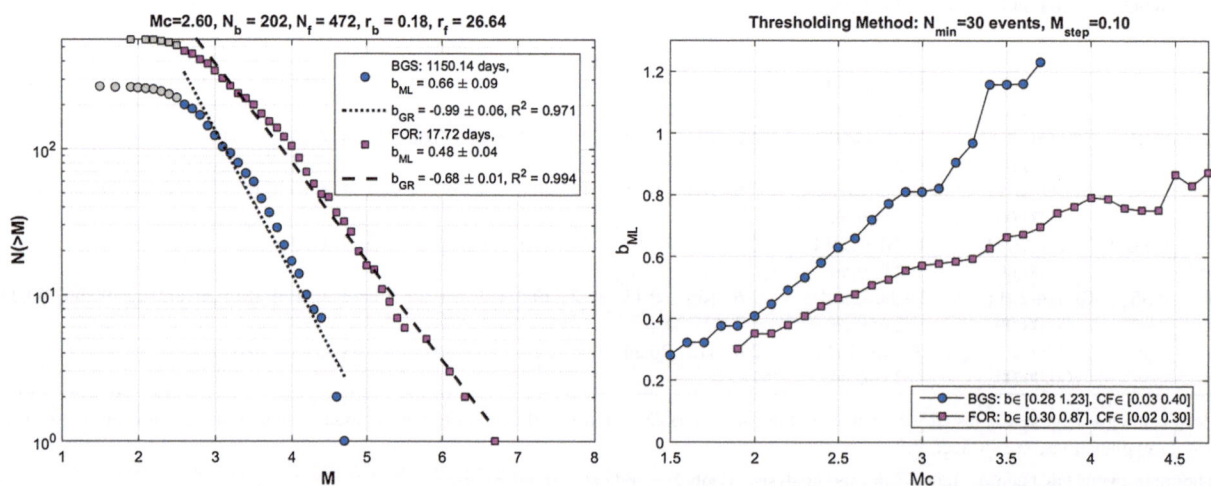

Figure 5

G–R diagrams for the background seismicity (BGS) period from 1 January 2011 to 14 March 2014 (N_b events, r_b seismicity rate) and for the 18-day foreshock (FOR) period (N_f events, seismicity rate r_f) calculated for magnitude threshold of $M_c = 2.60$ (left). b_{ML} and b_{GR} are b values calculated by the maximum likelihood approximation and the weighted least-square method, respectively. Variation of b_{ML} as a function of M_c calculated with magnitude step of M_{step} and minimum number of events of N_{min} (right). All calculations have been made for the target area of 65 km with the use of the CSN catalogue. The main shock was excluded from calculations

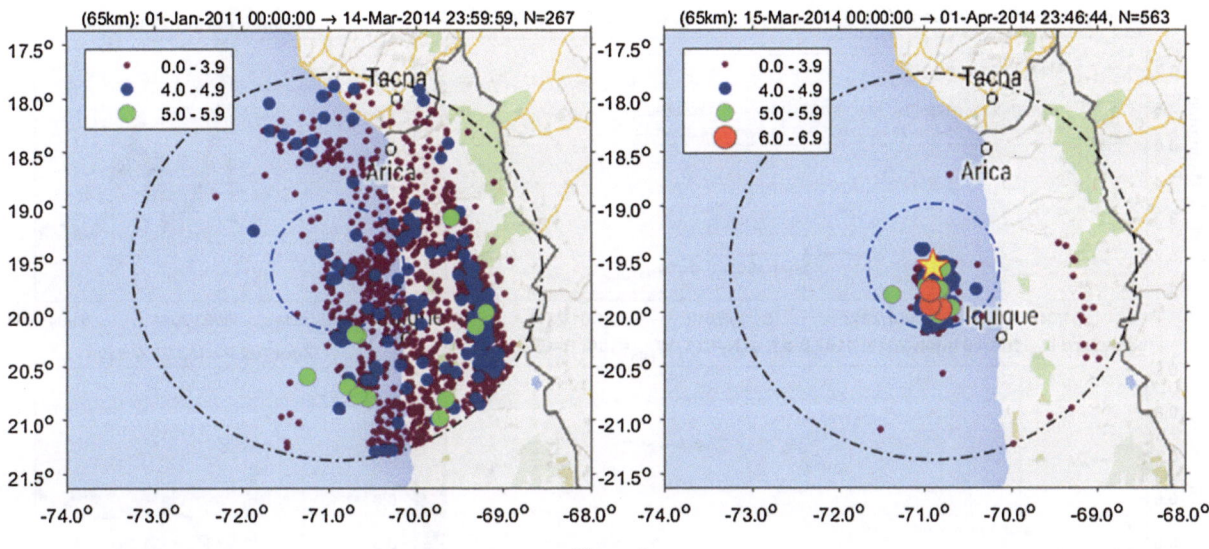

Figure 6

Epicentral distribution of the background seismicity for the period from 1 January 2011 up to 14 March 2014 (*left*), and of the foreshock sequence from 15 March 2014 up to the occurrence of the main shock (*right*). Data were taken from the CSN catalogue. Here, the black-dotted circle refers to a radius of approximately 200 km, while the blue one refers to the working radius of 65 km used for the analysis

18-day foreshock period, T_f. Here, background seismicity is meant by considering the catalogue segments BGS and WFOR combined together. It is observed that b_{GR} dropped very significantly (Table 2) from 0.99 in the BGS+WFOR period to 0.68 in FOR period.

The variations of the parameter b_{ML} within the target area were independently verified by the application of the conditional backward windowing technique (Fig. 7) described in Sect. 2.2. During the background period T_b, both parameters, b_{ML}, and mean magnitude, M_m, fluctuated randomly. In August 2013, a rather transient, insignificant change in b_{ML} occurred. From late January 2014, however, a gradual drop of b_{ML} and corresponding increase of M_m occurred until 14 March 2014. These changes, although not highly significant, may reflect initiation of instability during the early-weak foreshock period. However, in the 18-day time period T_f, the b_{ML} value dropped and the M_m increased dramatically. However, in about the last 5 days, the parameter b_{ML} slightly recovered.

The results obtained have verified the hypothesis that in the target area extending at critical distance of 65 km around the 1 April 2014 main shock epicenter, after the stage of a long-term stability of background

seismicity, the seismogenic system entered the stage of early weak foreshock activity by the end of January 2014. This stage lasted until 14 March 2014. From 15 March 2014 until the main shock occurrence, the main-strong foreshock stage preceded the great earthquake, although during the very last 5 days or so the intensity of the foreshock activity slightly decreased. It is of particular interest that the foreshock patterns preceding the 2014 Chile great earthquake are quite similar to the patterns found before the moderate (M_w 6.3) L'Aquila (Italy) earthquake of 6 April 2009 (Papadopoulos et al. 2010) (also in Fig. 2 by Daskalaki et al. 2016).

4.4. Earthquake of 1 April 2014—Analysis with the IPOC Catalogue

The results obtained with the CSN earthquake catalogue do not justify the suggestion of Shurr et al. (2014, their text and Figs. 2c, 3a, e) that the foreshock activity was developed at least 500 days before the great 2014 earthquake, that the cluster of July–August 2013 was foreshock activity, and that a gradual b_{ML}-value drop was evident at least 1000 days before the great earthquake, for that these seismicity changes did not pass non-randomness

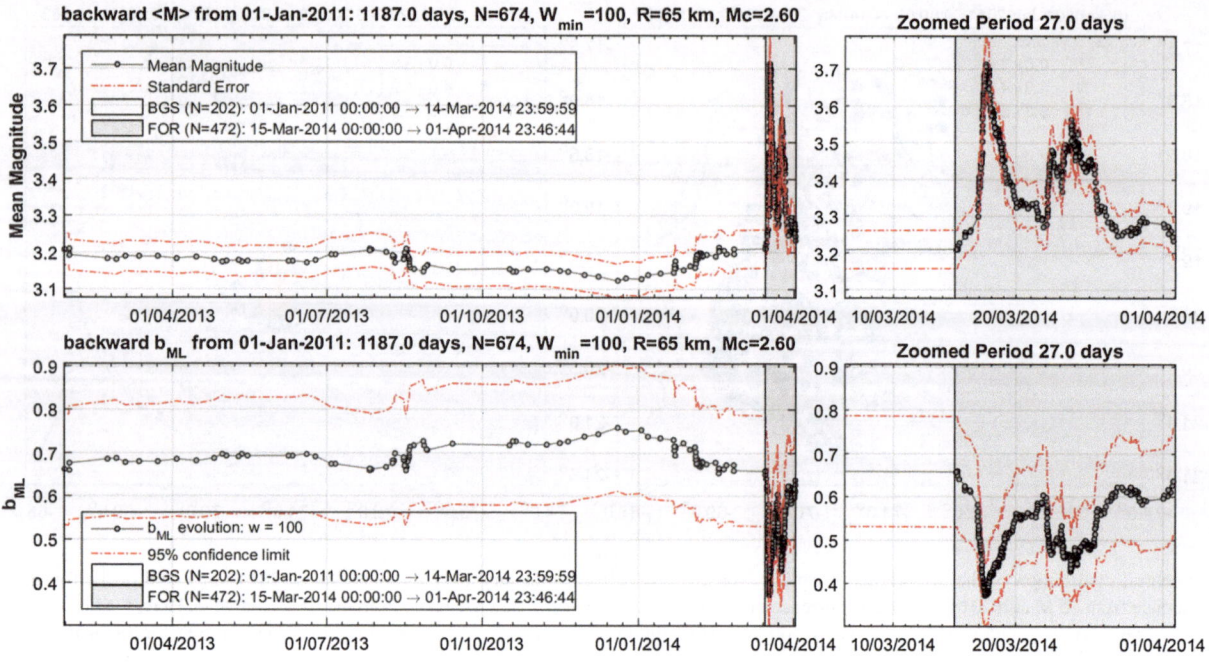

Figure 7

Conditional backward windowing technique, starting from the main shock occurrence, applied in the critical area of 65 km with the CSN catalogue and showing the time variation of the mean magnitude (*top chart*) and the b_{ML} (*bottom chart*). The procedure stopped at a point of time, where less than 100 events were remaining. Vertical bar in the right-hand side indicates the initiation of the strong foreshock stage on 15 March 2014. Red-dotted lines refer to the 95 % confidence level ($\pm 1.96\, b_{ML}/\sqrt{N}$, N = number of events). The main event has been excluded from the analysis. BGS and FOR as in Fig. 4. A zoomed version is also depicted in the right-hand side of both the charts, for a 27 days period:

$$T_{zoom} = T_f + \frac{T_f}{2}$$

tests. The analysis of Shurr et al. (2014) was based on their own IPOC catalogue starting from 2007. The seismicity changes that they interpreted as foreshock activity were observed in a rectangular area determined by the geographical coordinates from 19.00 to 21.00 S and from 70.00 to 71.50 W (Fig. 8) being of about 220 km in length and 170 km of width. This area is somehow different from the area with the initial radius of 200 km that we used to analyze CSN seismicity data. We showed, however, that the statistically significant foreshock features were only observed in a more narrow critical area of about 65 km around the main shock epicenter.

As a consequence, there is need to check if different seismicity catalogues produced different results. To this aim, the procedure described in Sect. 3.1 was repeated for the 1 April 2014 great earthquake using the IPOC earthquake catalogue which was available to us, as explained in Sect. 2.3. Our attention was focused on the rectangular area of

Shurr et al. (2014). We found completeness magnitude threshold comparable to that of the CSN catalogue ($M_c = 2.6$, Fig. 9 left). This is consistent with the $M_c = 2.6$ also found by Shurr et al. (2014) on their catalogue, although they preferred to analyze seismicity applying magnitude cutoff $M_c = 3.0$ to better securing completeness. For this reason, we initially performed our calculations with the IPOC catalogue with $M_c = 2.6$ and report results in the next lines. Recalculation by taking $M_c = 3.0$ produced very similar results leading to the same conclusions.

The plot of distances D against time (Fig. 10) clearly showed that increased activity was again densely concentrated only at a distance of about 65 km from the main shock epicenter in the last 18 days before the main shock. In addition, in the time interval 22–26 March, the foreshocks migrated very close to the main shock epicenter at distances from about 5 to 25 km, as shown by the CSN catalogue (Fig. 3 right). Again, in about the last 5 days before the

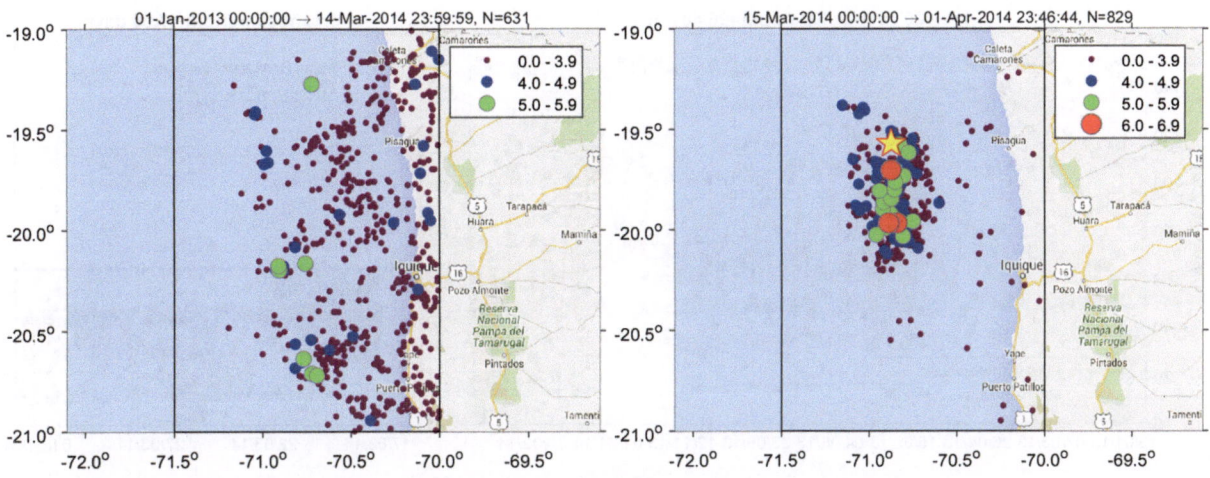

Figure 8

Seismicity in the area examined (*rectangular box*) from the beginning of the catalogue (1 January 2013) up to 14 March 2014 (*left*), and from the 15 March 2014 up to the main shock occurrence (*right*). The IPOC catalogue available to us was used

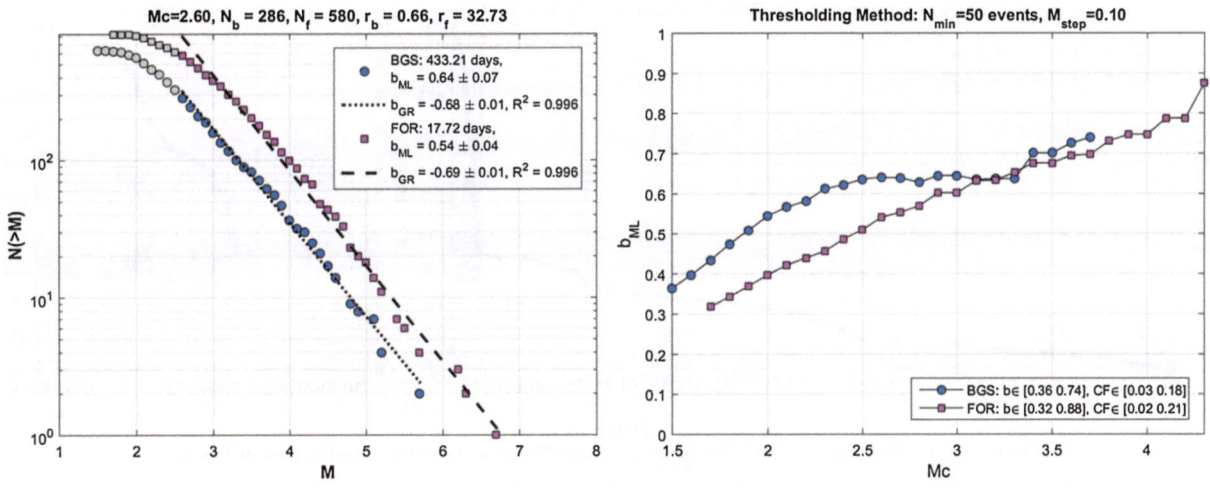

Figure 9

As in Fig. 5 for the background seismicity period from 1 January 2013 to 14 March 2014 and for the 18-day foreshock period (*left*). All calculations have been made with the IPOC catalogue available to us for the rectangular area mentioned in Sect. 3.4 and shown in Fig. 9

great main shock, the number of foreshock events slightly decreased and distributed at random distances within the critical area of 65 km. The b_{ML} value found for the main-strong foreshock activity of the last 18 days before the great earthquake remained lower than that found for the previous period regardless the magnitude cutoff used (Fig. 9 right).

Plotting the seismicity-rate variation (Fig. 11), we obtained the same pattern with that obtained from the CSN catalogue (Fig. 4). A difference is that the plot

in Fig. 11 is more enhanced in total event number as well as in the number of strong events ($M \geq 5$) occurring before the main-strong foreshock stage of the last 18 days. This difference is due to that the plot in Fig. 11 covers a larger area than the one in Fig. 4. However, the dramatic increase of the seismicity rate in the last 18 days was due to the event number increase within the critical area of 65 km.

Comparing the application of the conditional backward windowing technique (Fig. 12) within the

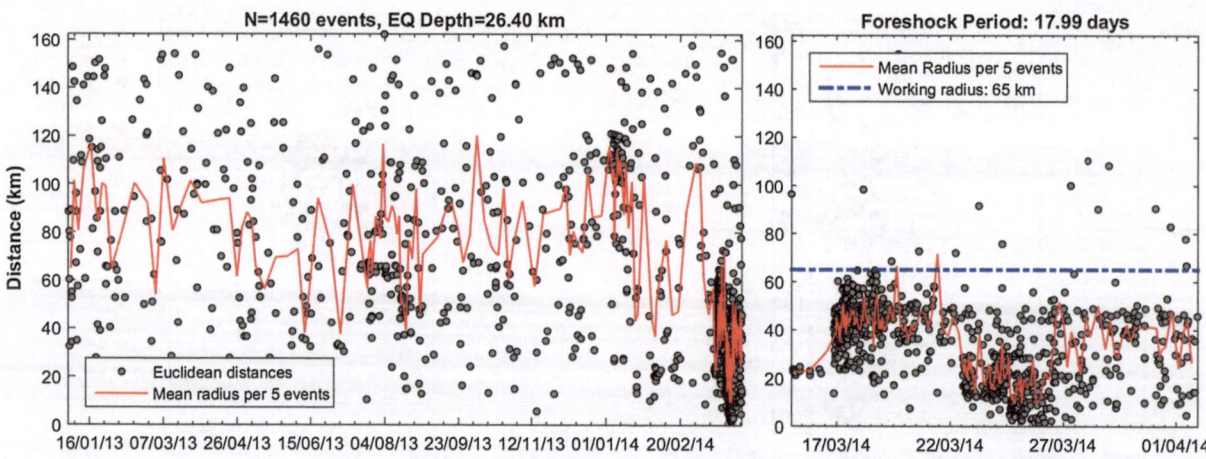

Figure 10
As in Fig. 3 for the time period from 1 January 2013 to the origin time of the main shock of 1 April 2014. Data taken from the IPOC catalogue available to us

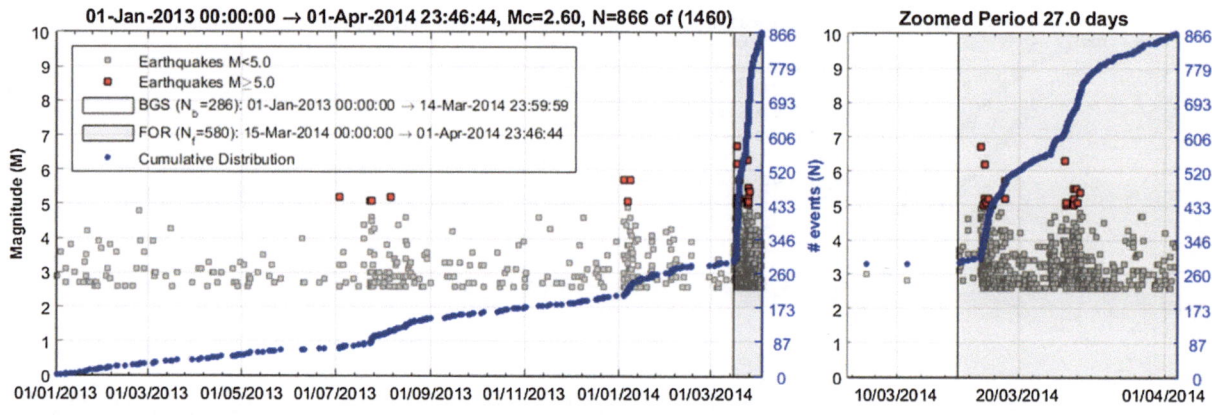

Figure 11
As in Fig. 4 with the use of the IPOC catalogue available to us for the rectangular area of Fig. 8

rectangular area of Fig. 8, with the one in the critical area of 65 km using the CSN catalogue (Fig. 7), we found that the plot in Fig. 12 smoothed out the transient drop of b_{ML} associated with the temporary cluster of July–August 2013. This happened also with the b_{ML} drop that started by the end of January 2014 that is with the beginning of the early weak foreshock stage. This is due to that in Fig. 12, b_{ML} was calculated for a large area, thus contaminating the seismicity of the critical 65-km area with the seismicity of surrounding seismic sources. This very likely explains also why the parameter b_{ML} although

in general dropped dramatically in Fig. 12, taking its lower value after the strong foreshock (M_w 6.7) of 16 March 2014, in some instances exceeded the value of b_{ML} prevailing before the 18-day period of strong foreshock activity. On the contrary, b_{ML} calculated with the CSN catalogue remained very significantly lower during that period (Fig. 7).

The analysis with the IPOC catalogue available to us showed that the foreshock precursory activity was again densely concentrated only within a critical distance of about 65 km from the main shock epicenter in the last 18 days before the main shock.

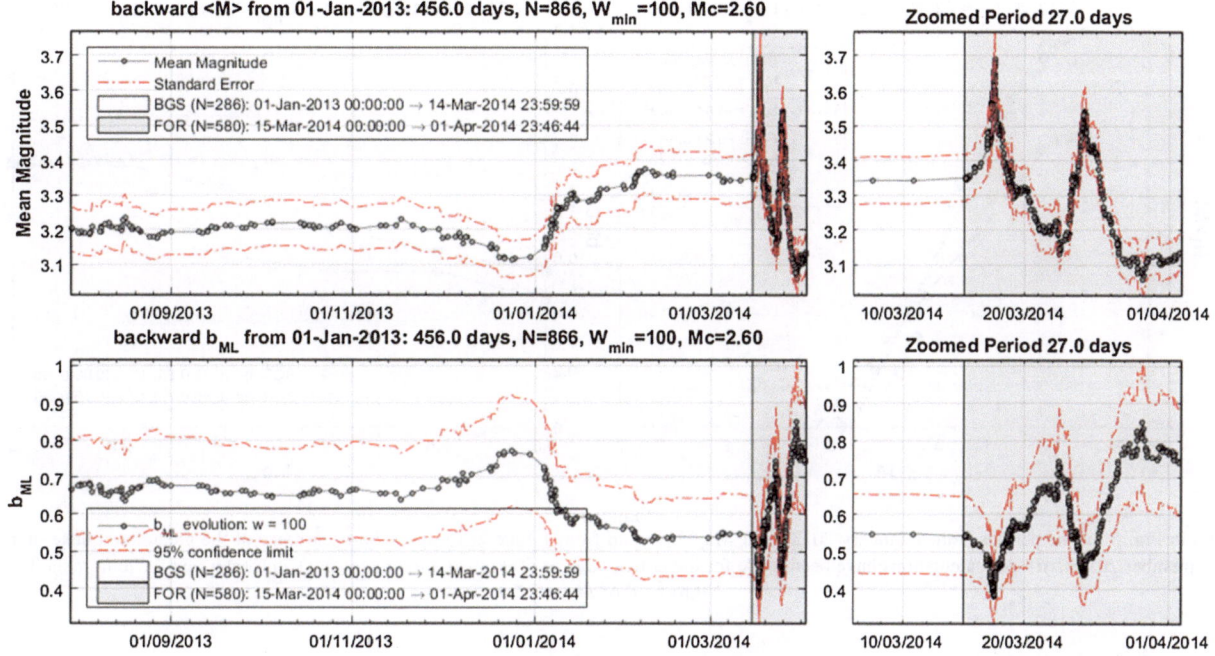

Figure 12
As in Fig. 7 with the use of the IPOC catalogue available to us for the rectangular area of Fig. 8

The dramatic increase of the seismicity rate in the last 18 days was due to the event number increase within the critical area of 65 km. The gradual decrease of b_{ML} became no evident in the long term but only by the end of January 2014 when the early weak foreshock stage initiated. The transient seismicity cluster of July–August 2013 did not show a significant deviation from background seismicity.

4.5. The 1 April 2014 Earthquake as a Reference Event

Summarizing the results obtained so far by analyzing both the CSN and IPOC catalogues, we may conclude that the great earthquake of 1 April 2014 was preceded by an early weak foreshock stage lasting for about 1.8 months and by a main-strong precursory foreshock stage that was evolved in the last 18 days before the main shock. Both the stages were developed in a critical area of about 65 km around the main shock epicenter mainly to the south of it. The strong foreshock stage was characterized by distinct, statistically very significant patterns in the space–time–size domains with respect to the background seismicity. Apart from the dense concentration of foreshocks around the main shock epicenter, the activity rate increased also dramatically, the b_{ML} value dropped, and the mean magnitude increased, while the level of seismic energy released also increased. Less intense was the foreshock activity in the early weak foreshock stage. In view of these highly significant seismicity changes, we considered the 1 April 2014 great earthquake as a reference event when examining the 2015 and 2010 large Chilean earthquakes. This consideration is also justified by that the 3D patterns revealed before the 2014 earthquake are quite similar to the ones that preceded the M_w 6.3 L'Aquila earthquake of 6 April 2009 (Papadopoulos et al. 2010; Daskalaki et al. 2016).

4.6. The case of 16 September 2015 Great Earthquake

The case of the great Chilean earthquake (M_w 8.4) of 16 September 2015 was examined by applying the

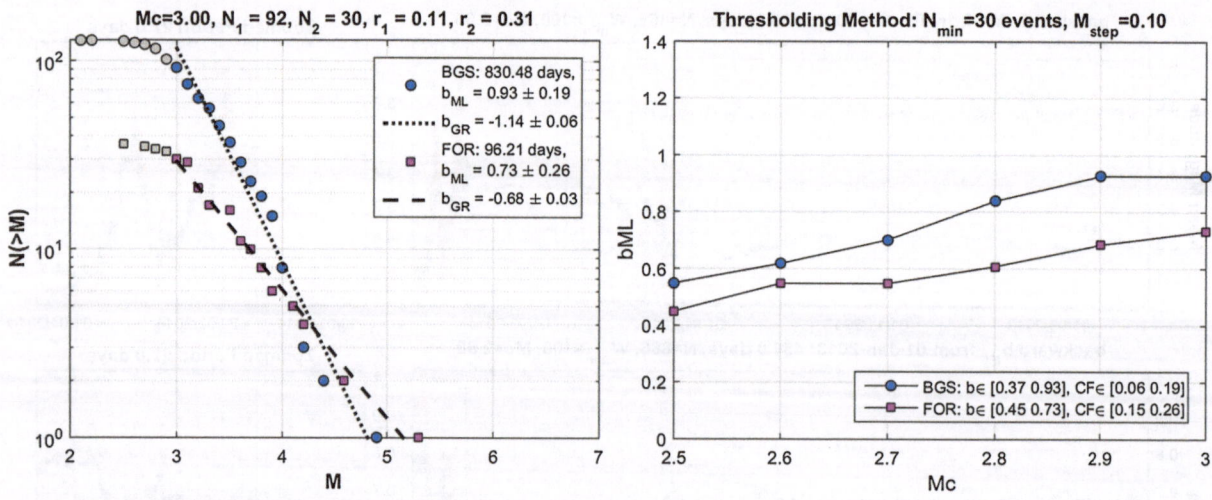

Figure 13

As in Fig. 5 for the periods from 1 January 2013 to 31 May 2015 and from 1 June 2015 up to the occurrence of the great earthquake of 16 September 2015 (*left*). All calculations have been made for the critical area of 50 km from the main shock epicenter and for joint magnitude threshold of $M_c = 3.0$

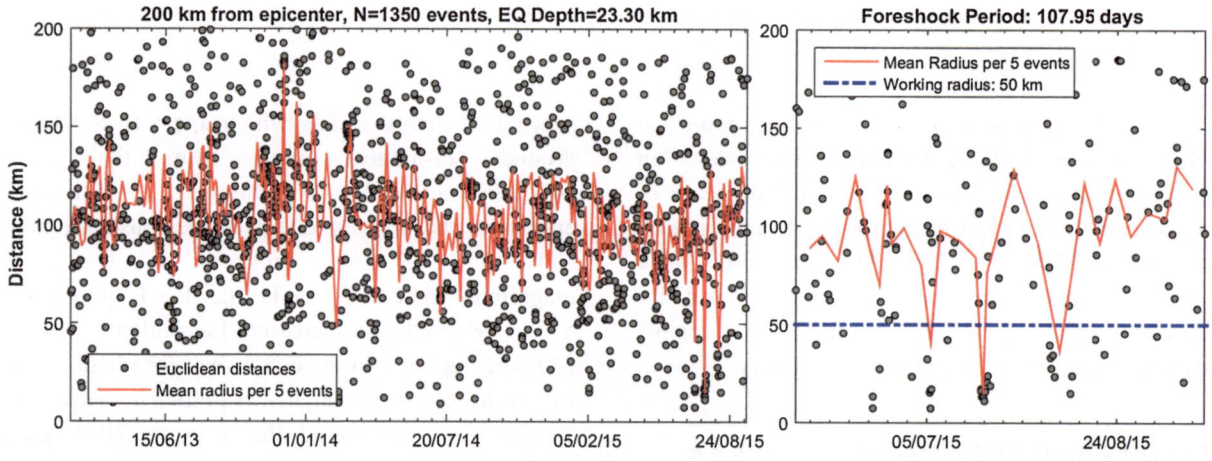

Figure 14

As in Fig. 3 for the time period from 1 January 2013 to the origin time of the main shock of 16 September 2015

procedure followed for the analysis of the 2014 event on the basis of the CSN seismicity catalogue. Unfortunately, the completeness magnitude cutoff was found large enough at $M_c = 3.0$ (Fig. 13) for an area extending up to 350 km around the great earthquake epicenter. The time variation of the mean distance, D, of the earthquake events from the main shock epicenter was tested for areas of several radii less than 350 km. A slightly increased concentration

of epicenters in an area of about 50 km was found (Fig. 14) from the beginning of June 2015 to 20 July 2015 when the activity concentrated at distances from 5 to 40 km. Spatially, this activity was not randomly distributed, as the background activity did (Fig. 15). On the contrary, it was concentrated along the plate boundary to the southwest and southeast from the main shock epicenter. Within the 50-km target area and for the time period from 1 June up to the main

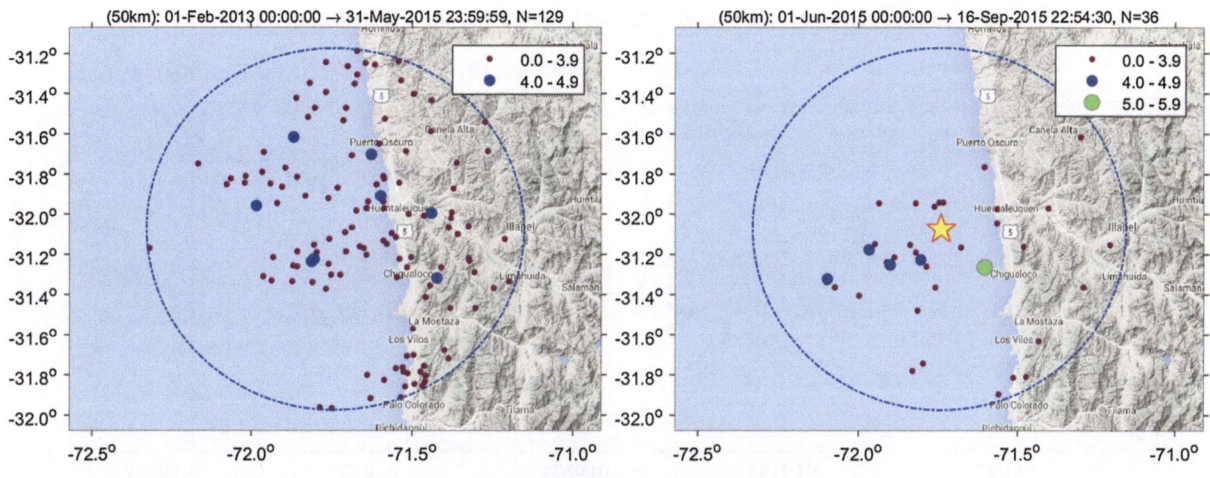

Figure 15

As in Fig. 6 for background seismicity of the period from 1 January 2013 up to 31 May 2015 (left) and of weak foreshock sequence from 1 June 2015 up to the 16 September 2015 main shock occurrence (right)

shock occurrence, the activity rate increased from 0.11 to 0.31 (Fig. 16), while b_{ML} dropped from 0.93 to 0.73 (Fig. 13 left). In about the last 10 days, the activity rate decreased. In the entire time period examined the largest earthquake magnitude of M 5.3 occurred on 12 August 2015 that is close in time with the main shock.

Our findings indicated that the 16 September 2015 great earthquake was preceded by foreshock activity for a period of about 3.5 months with foreshock patterns similar to the ones that preceded the 2014 event. The spatial cluster of seismicity around the main shock epicenter was highly significant, as shown by the difference in distance D (Table 3). However, we tested for the hypothesis that the time segments from 1 January 2013 to 31 May 2015 and from 1 June 2015 until the main shock of 16 September 2015 represented background and foreshock activities, respectively. The significance level for the changes of r and b_{ML} was found $P(r) = 88.30$ and $P(b_{ML}) = 0.19$, respectively (Figs. 13 left and 17, Table 3). Therefore, we suggested that only a weak foreshock stage was evident.

4.7. The Case of the 27 February 2010 Earthquake

The case of the great M_w 8.8 Chilean earthquake of 27 February 2010 was examined by analyzing the seismicity catalogue of CSN for the time period from 1 January 2007 up to occurrence of the great earthquake. We found magnitude completeness threshold of $M_c = 3.0$. Testing for critical areas of several radii, no significant spatial concentration of epicenters around the great earthquake epicenter was established. However, the occurrence of a small number of shocks occurring to the northeast and southeast from the main shock epicenter within a distance of about 170 km was observed from about 15 November 2009 up to the main shock origin time (Fig. 18). Therefore, further seismicity analysis was performed for these particular space and time windows.

The increase of the activity rate observed after 14 November 2009 (Fig. 18) was significant enough exceeding the 90 % level (Table 4). It is noteworthy that before the main shock, a cluster of three strong earthquakes occurred on 29 December 2009 ($M5.3$) and 21 January 2010 ($M5.6$ and $M5.2$). This feature reminds the increase of seismic moment release in the last days of the foreshock period before the 1 April 2014 earthquake and at lesser degree before the 16 September 2015 earthquake (Fig. 19).

In the time interval from 15 November 2009 up to the main shock occurrence, the b_{ML} value dropped in relation to the preceding period (Fig. 20). However, the change was insignificant (Table 4). On the other hand, spatially, the activity was not randomly distributed but concentrated around the main shock epicenter

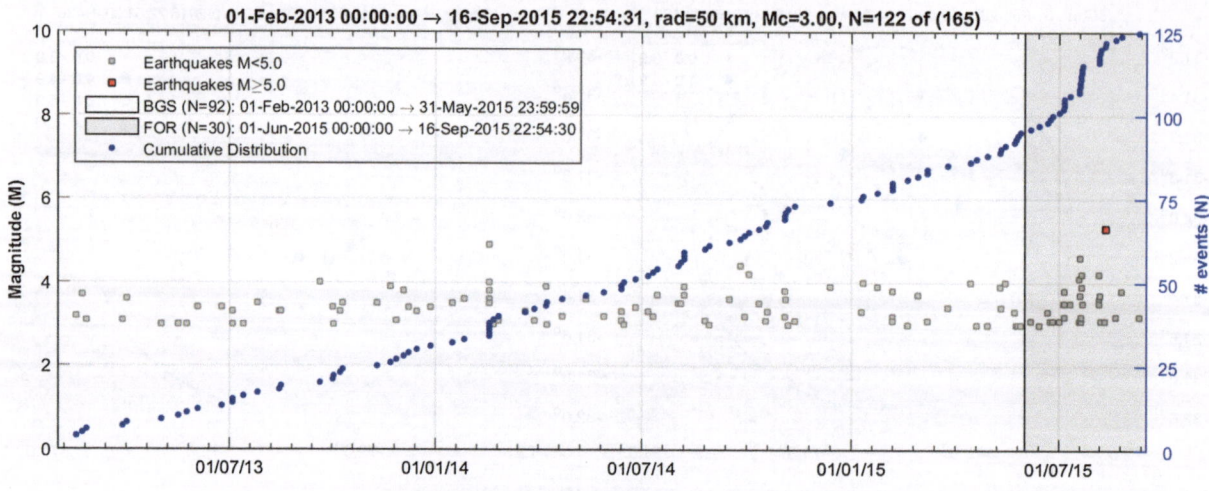

Figure 16

As in Fig. 4 for the 50-km target area from the main shock epicenter of 16 September 2015

Table 3

As in Table 2 for the 2015 great earthquake

Period	From	To	M_c	N	r	M_m	D	b_{GR}	b_{ML}	$P(b)$	$P(D)$	$P(r)$
BGS	01-Feb-2013 00:00:00	31-May-2015 23:59:59	3.0	92	0.11	3.42	35.5 ± 1.3	1.14 ± 0.12	0.93 ± 0.19	0.190847	99.99	88.30
FOR	01-Jun-2015 00:00:00	16-Sep-2015 22:54:30	3.0	30	0.28	3.54	27.4 ± 2.3	0.68 ± 0.06	0.73 ± 0.26			

particularly to the northeast and southeast of it (Table 4). The foreshock activity preceding the 2014 event also was selectively concentrated around the main shock epicenter, particularly to the south. Another similarity is the increase of the seismic moment release before both the 2010 and 2014 main shocks.

Taking into account the similarity in the seismicity patterns preceding the 2010 and the 2014 reference event as well as the highly significant change of D, and the significant change of r, we suggested that a weak foreshock activity perhaps preceded the 2010 great earthquake in an area of 170 km from the earthquake epicenter and in a time interval of about 104 days. That the drop of b_{ML} was insignificant and that the seismicity change was only significant could be attributed to the reduced number of events involved in the analysis due to large completeness level in the catalogue.

5. Discussion and Conclusions

Foreshocks attracted research interest aiming to the earthquake prediction in at least the last 50 years or so. In this context, seismicity data, laboratory experiments and simulation results showed that foreshock sequences are characterized by distinct precursory patterns in their 3D space–time–size distributions. Foreshocks move gradually towards the main shock epicenter, the event count accelerates and the b value of the G–R (magnitude–frequency) relation drops as the main shock approaches. However, these patterns have been mainly recognized either in aggregated data from several foreshocks or in synthetic seismicity catalogues. Only very few cases of individual foreshock sequences are known with the 3D precursory patterns being present at the same time. A characteristic example is the sequence

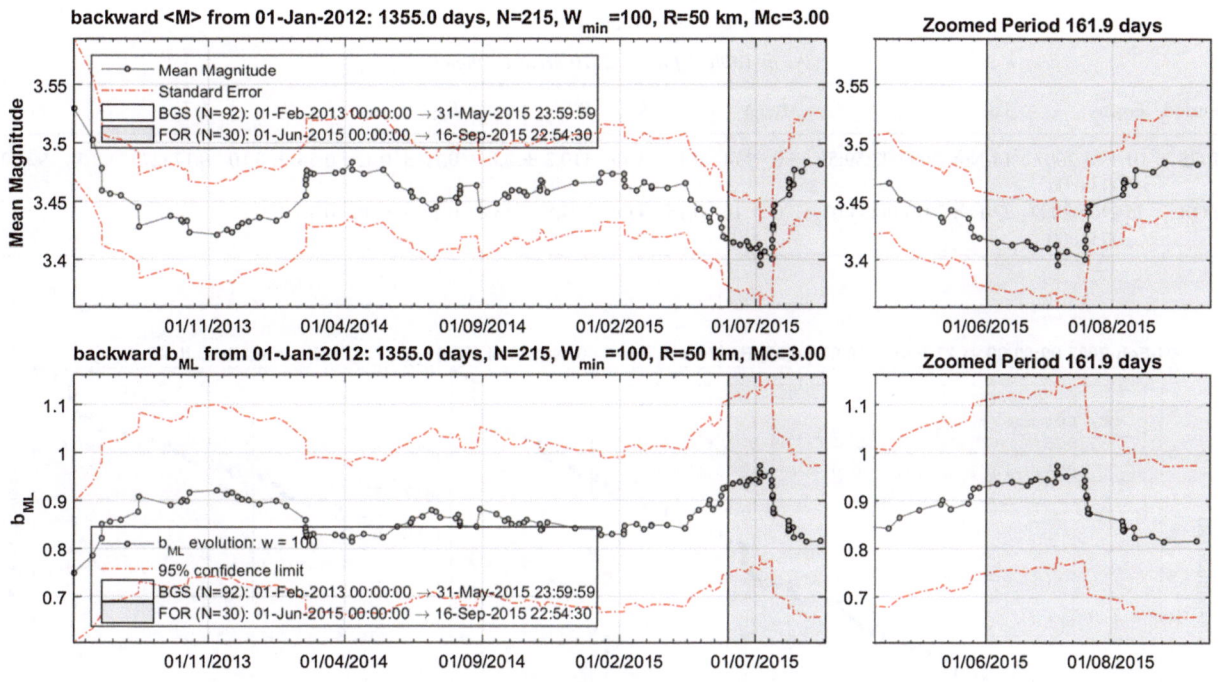

Figure 17

As in Fig. 7 for the 50-km target area from the main shock epicenter of 16 September 2015

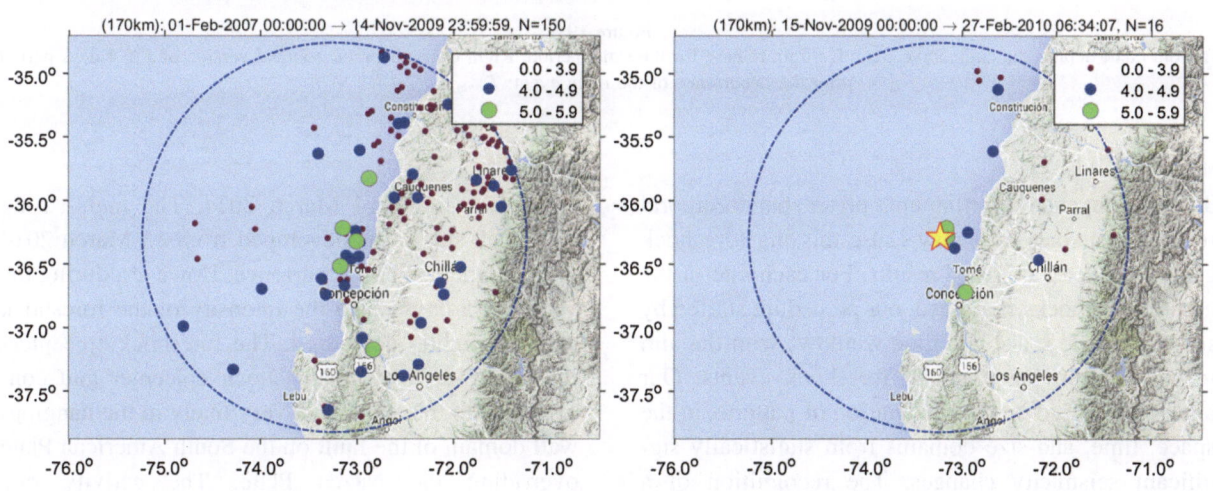

Figure 18

As in Fig. 6 for background seismicity of the period from 1 February 2007 up to 14 November 2009 (*left*) and of weak foreshock sequence from 15 November 2009 to 27 February 2010 main shock occurrence (*right*). Plots for the target area of 170 km form main shock epicenter

associated with the M_w 6.3 L'Aquila (Italy) earthquake of 6 April 2009 (Papadopoulos et al. 2010).

The study of more individual foreshock cases is of particular importance for verifying and better understand the features of the 3D precursory patterns. We

have examined three cases of relatively recent great earthquakes that ruptured the subduction zone of Chile, the Malue 27 February 2010, Iquique 1 April 2014 and Illapel 16 September 2015 earthquakes. The procedure followed for the foreshocks investigation

Table 4

As in Table 3 for the 2010 great earthquake

Period	From	To	M_c	N	r	M_m	D	b_{GR}	b_{ML}	$P(b)$	$P(D)$	$P(r)$
BGS	01-Feb-2007 00:00:00	14-Nov-2009 23:59:59	3.0	114	0.11	3.75	110.2 ± 3.8	0.75 ± 0.13	0.54 ± 0.10	0.173273	99.54	92.22
FOR	15-Nov-2009 00:00:00	27-Feb-2010 06:34:07	3.0	16	0.15	4.04	94.8 ± 13.8	0.42 ± 0.08	0.40 ± 0.19			

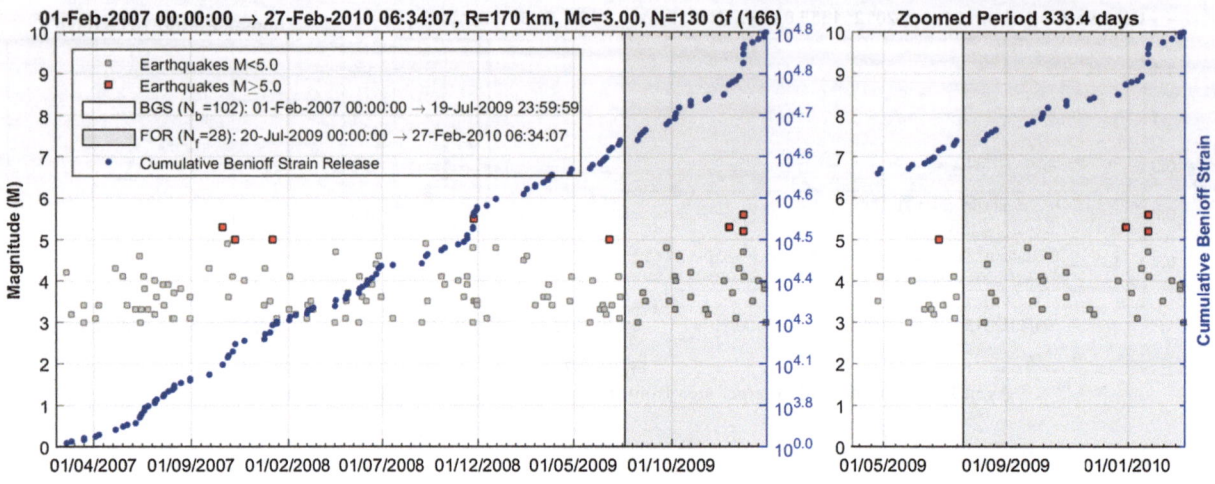

Figure 19

Bottom chart depicts the cumulative Benioff Strain release for the same period. Right charts depict a zoomed version of 156.4 days period prior the occurrence of the main event: $T_{zoom} = T_f + \frac{T_f}{2}$

does not depend on setting up a priori spatiotemporal restrictions, since they may cause missing foreshock events and lead to biased results. For each one of the great main shocks examined, our procedure started by selecting wide space and time windows with the aim to avoid removing potential foreshock events. Our analysis is based on the recognition of patterns in the space, time, and size domains from statistically significant seismicity changes. The recognition of a pattern in one domain is independent of the recognition of a pattern in another domain.

The foreshock sequence of the 1 April 2014 Iquique earthquake (M_w 8.1) was analyzed with the CSN and IPOC seismicity catalogues. The same results were found with both catalogues. Namely, the long-lasting stage of background seismicity that fluctuated randomly was followed by an early weak foreshock stage starting by the end of January 2014

and lasting until 14 March 2014. The main-strong foreshock stage was developed from 15 March 2014 until the main shock occurrence. However, during the very last 5 days or so the intensity of the foreshock activity slightly decreased. The foreshock epicenters moved towards the main shock epicenter and concentrated to the south of it very likely in the hanging-wall domain of the fault on the South American Plate overriding the Nazca Plate. The activity rate increased, the b-value dropped and the mean magnitude increased. All these seismicity changes were tested and found to be very significant in a critical area of 65 km from the main shock epicenter. Since the b-value calculation is quite sensitive depending on the catalogue completeness level and the computation techniques used, we performed calculations with both the maximum likelihood approximation and the weighted least-square method as well as for

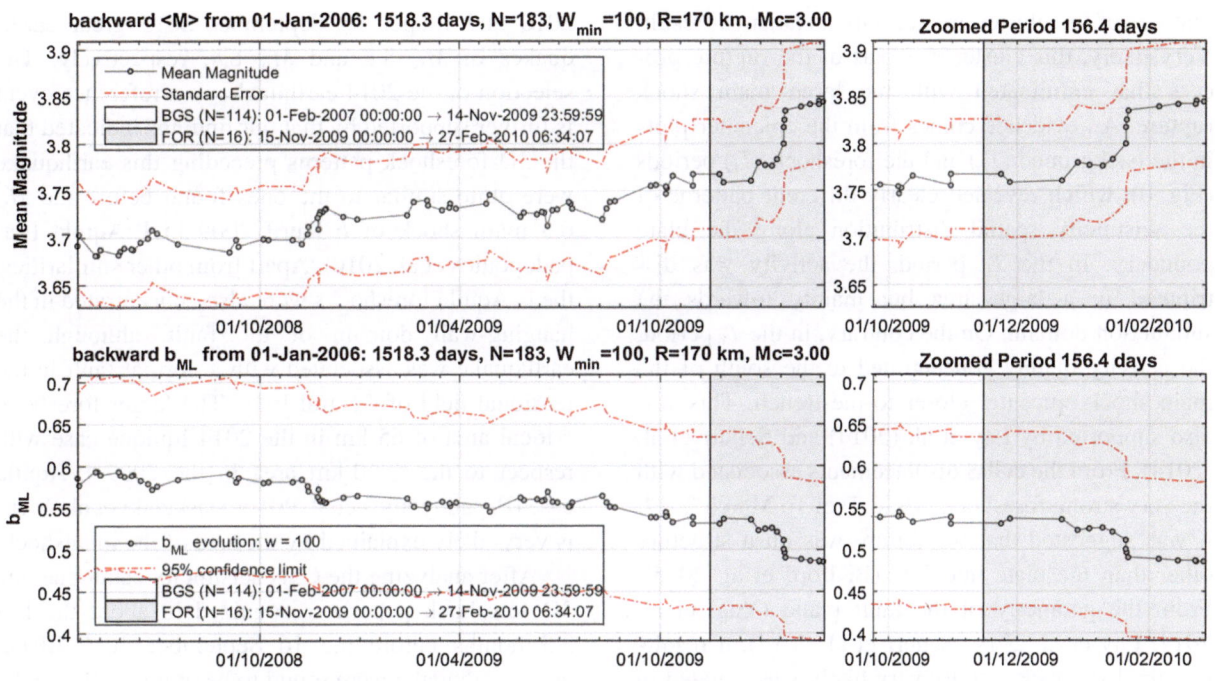

Figure 20
As in Fig. 7 for the 170-km target area from the main shock epicenter of 27 February 2010

several magnitude completeness thresholds. The results remained the same without changing the pattern recognized.

Our results are in part inconsistent to the ones reached by Schurr et al. (2014) who examined the same foreshock sequence based on the IPOC catalogue. Their claim that the foreshock activity was developed in at least the last 500 days or so before the great main shock in an area of about 220 km × 170 km was not verified by us. The deviation from randomness of the supposed foreshock activity distributions in space and size (*b* value) was not tested as for their statistical significance by Schurr et al. (2014). Only the seismicity-rate changes were tested by those authors who supported that they found significant changes during the last 250 days before the main shock. In addition, it remains doubtful if the transient seismicity cluster of July–August 2013 was part of the foreshock activity as Schurr et al. (2014) suggested. On the other hand, there is a general agreement that the intense foreshock activity in the last 18 days or so was concentrated to the south of the great earthquake and gradually migrated towards the main shock epicenter thus unlocking the plate boundary and controlling initiation of the 2014 great earthquake (Lay et al. 2014; Lay and Brodsky 2014; Schurr et al. 2014; Yagi et al. 2014 see also our results). Kato and Nakagawa (2014) suported that the foreshocks were associated with multiple slow-slip events before the main shock, a process which is similar to the sequence before the 2011 Tohoku-oki M9.0 earthquake.

Important feature of the 2014 sequence was also a cluster of very strong foreshock events that took place during the main foreshock stage, thus implying that the activity accelerated not only in terms of event count but also in seismic moment release. Of interest is that the foreshock activity decreased slightly in about the last 5 days.

The highly significant concentration of seismic epicenters around the main shock epicenter in the last 18 days and, at the same time, the acceleration pattern of the cluster leave no doubt that a very strong foreshock activity preceded the main shock of 1 April 2014. However, there is evidence that the space–time features of the cluster are not only a statistical result. Very likely, this cluster was part of the rupture process that culminated with the large main shock rupture. An evidence comes from the epicenter plots in the background (T_b) and the foreshock (T_f) periods (Fig. 6), which revealed clearly different patterns of the seismicity spatial distribution along the plate boundary. In the T_b period, the activity was distributed in a large area but mainly towards the subduction domain. On the contrary, in the T_f period, the activity mainly concentrated to the south of the main shock epicenter closer to the trench. This was also supported by Lay et al. (2014) and Schurr et al. (2014). From the cGPS displacements associated with the very strong foreshock (M_w 6.7) of 16 March 2014, it was suggested that the failure was on a structure other than the plate interface (Bedford et al. 2015). From the geometry of the fault plane (Yagi et al. 2014; Lay et al. 2014; Schurr et al. 2014), it results that the foreshock activity very likely was situated in the South American Plate overriding the subducting Nazca Plate that is on the hanging-wall of the fault.

The 3D foreshock pattern before the large 2014 Iquique-Pisagua earthquake was quite similar to the one that preceded the 6 April 2009 L' Aquila earthquake (M_w 6.3) in Central Italy, although the later was about two orders of magnitude less than the former one and occurred in a completely different seismotectonic regime that is in association with extensional field with normal faulting. In L'Aquila, after a nearly 4-month stage of weak foreshock activity, the strong foreshock stage was developed in the last 10 days before the main shock and became evident by dense epicenter concentration in the hanging-wall of the causative normal fault, increase of the seismicity rate and drop of the b value with respect to the background seismicity (Papadopoulos et al. 2010). All these changes were highly significant. In addition, the largest earthquake magnitudes were measured during the strong foreshock stage. These patterns were further analyzed by Daskalaki et al. (2016, their Fig. 2).

Since the 3D precursory foreshock patterns of the 2014 Iquique case are quite clear and highly significant from statistical point of view, we have used the 2014 great earthquake as a reference event in our foreshock investigation before the Malue 27 February

2010 and Illapel 16 September 2015 great earthquakes of M_w 8.8 and M_w 8.4, respectively. The selection of the 2014 earthquake as a reference event is further supported by that our findings indicated that the 3D foreshock patterns preceding this earthquake were quite similar to the ones found before the M_w 6.3 main shock of 6 April 2009 in L'Aquila (Papadopoulos et al. 2010). Apart from other similarities, the L'Aquila foreshocks were also concentrated in the hanging-wall domain of the fault, although the earthquake was associated with a normal fault in the tensional field of Central Italy. The larger foreshock critical area of 65 km in the 2014 Iquique case with respect to the 5–10 km area in the 2009 L'Aquila case (Papadopoulos et al. 2010; Daskalaki et al. 2016) is very likely explained by the size of the mainshock.

After analyzing the CSN seismicity catalogue, the 3D foreshock patterns recognized in about the last 3.5 months before the 16 September 2015 Illapel great earthquake were found to be quite similar to the ones that preceded the 2014 Iquique earthquake. Namely, foreshock concentration close to the main shock epicenter, increase of the activity rate, and b-value drop were observed in a critical area of about 50 km. However, although the change in epicentral distance D was highly significant parameters r and b did not changed significantly. Therefore, we concluded that it was only a weak foreshock sequence. Similar were also the 3D foreshock patterns found before the Malue 27 February 2010 great earthquake. In this case, however, the significance level in seismicity changes again varied from highly significant in D to significant in r and insignificant in b. We, therefore, again support that it was a weak foreshock activity, too. It is of importance to note that in both the 2010 and 2015 cases, the completeness magnitude cutoff was relatively high, $M_c = 3.0$. This means that the catalogue used for the analysis of the 2014 case with $M_c = 2.6$ is enriched in seismic events by a factor of 2.5. Therefore, one may not rule out that the recognition of foreshocks before the 2010 and 2015 earthquakes was disfavored by the catalogue completeness threshold. The problem of overlooking small foreshocks due to catalogue incompleteness has been already noted (e.g., Yamaoka et al. 1999; Papadopoulos et al. 2006). Therefore, estimations of the foreshock rate in several seismogenic zones of the globe should be carefully considered by taking into account the critical role of the seismicity catalogue completeness levels.

Acknowledgments

This is a contribution to the internal research project "EARTHWARN" of the Institute of Geodynamics, National Observatory of Athens, Greece. We are thankful to V. Kosobokov and to another anonymous reviewer, since their constructive comments helped to improve the original version of the paper. We also thank Dr. B. Schurr, GFZ, Potsdam (Germany), for providing us the IDOP earthquake catalogue.

REFERENCES

Abercrombie, R. E., & Mori, J. (1996). Occurrence patterns of foreshocks to large earthquakes in the western United States. *Nature, 381*, 303–307.

Agnew, D. C., & Jones, L. (1991). Prediction probabilities from foreshocks. *Journal Geophysical Research, 96*(B7), 11959–11971.

Aki, K. (1965). Maximum likelihood estimates of b in the formula logN = a-bM and its confidence limits. *Bull. Earth. Res. Inst. Univ. Tokyo, 43*, 237–239.

Avlonitis, M., & Papadopoulos, G. A. (2014). Foreshocks and b value: bridging macroscopic observations to source mechanical considerations. *Pure and Applied Geophysics.* doi:10.1007/s00024-014-0799-6.

Bedford, J., Moreno, M., Schurr, B., Bartsch, M., & Oncken, O. (2015). Investigating the final seismic swarm before the Iquique-Pisagua 2014 M_w 8.1 by comparison of continuous GPS and seismic foreshock data. *Geophysical Reseach Letters.* doi:10.1002/2015GL063953.

Bouchon, M., Durand, V., Marsan, D., Karabulut, H., & Schmittbuhl, J. (2013). The long precursory phase of most large interplate earthquakes. *Nature Geoscience, 6*(4), 299–302. doi:10.1038/ngeo1770.

Brodsky, E. E., & Lay, T. H. (2014). Recognizing foreshocks from the 1 April 2014 Chile earthquake. *Science, 344*, 700–702.

Chan, C.-H., Wu, Y.-M., Tseng, T.-L., Lin, T.-L., & Chen, C.-C. (2012). Spatial and temporal evolution of b-values before large earthquakes in Taiwan. *Tectonophysics.* doi:10.1016/j.tecto.2012.02.004.

Chen, Y., Liu, J. and Ge, H., 1999. Pattern Characteristics of Foreshock Sequences. Pageoph, 155, 2–4, 395–408.

Comte, D., Eisenberg, A., Lorca, E., Pardo, M., Ponce, L., Saragoni, R., et al. (1986). The 1985 Central Chile earthquake: a repeat of previous great earthquakes in the region? *Science, 233*, 449–453.

Console, R., Murru, M., & Alessandrini, B. (1993). Foreshock statistics and their possible relationship to earthquake prediction in the Italian region. *Bulletin of the Seismological Society of America, 83*, 1248–1263.

Daskalaki, E., Spiliotis, K., Siettos, C., Minadakis, G. and Papadopoulos G.A. (2016). Foreshocks and short-term hazard assessment to large earthquakes using complex networks: the case of the 2009 L'Aquila earthquake (submitted).

Dodge, D. A., Beroza, G. C., & Ellsworth, W. L. (1995). Foreshock sequence of the 1992 Landers, California, earthquake and its implications for earthquake nucleation. *Journal Geophysical Research, 100*(B6), 9865–9880.

Duputel, Z., Jiang, J., Jolivet, R., et al. (2015). The Iquique earthquake sequence of April 2014: Bayesian modeling accounting for prediction uncertainty. *Geophysical Reseach Letters.* doi:10.1002/2015GL065402.

Frohlich, C., & Davis, S. D. (1993). Teleseismic b values; or, much ado about 1.0. *Journal Geophysical Research, 98*(B1), 631–644.

Gutenberg, B., & Richter, C. (1944). Frequency of earthquakes in California. *Bulletin of the Seismological Society of America, 34*, 185–188.

Hainzl, S., & Ogata, Y. (2005). Detecting fluid signals in seismicity data through statistical earthquake modeling. *Journal Geophysical Research.* doi:10.1029/2004JB003247.

Hainzl, S., Zöller, G., & Kurths, J. (1999). Similar power laws for foreshock and aftershock sequences in a spring-block model for earthquakes. *Journal Geophysical Research, 104*, 7243–7253.

Hamilton, R. M. (1967). Mean magnitude of an earthquake sequence. *Bull. Seism. Soc. Am., 57*, 1115–1116.

Helmstetter, A., Sornette, S., & Grasso, J.-R. (2003). Main shocks are aftershocks of conditional foreshocks: how do foreshock statistical properties emerge from aftershock laws. *Journal Geophysical Research.* doi:10.1029/2002JB001991.

Ishida, M., & Kanamori, H. (1978). The foreshock activity of the 1971 San Fernando earthquake. California. *Bulletin of the Seismological Society of America 68*, 1265–1279.

Ishimoto, M., & Iida, K. (1939). Observations of earthquakes registered with the microseismograph constructed recently. *Bull. Earthq. Res. Inst. Tokyo Univ. 17*, 443–478.

Jones, L. M., & Molnar, P. (1979). Some characteristics of foreshocks and their possible relationship to earthquake prediction and premonitory slip on faults. *Journal Geophysical Research 84*, 3596–3608.

Kagan, Y., & Knopoff, L. (1978). Statistical study of the occurrence of shallow earthquakes. *Geophys. J. Roy. Astr. Soc. 55*, 67–86.

Kato, A., & Nakagawa, S. (2014). Multiple slow-slip events during a foreshock sequence of the 2014 Iquique, Chile M_w 8.1 earthquake. *Geophysical Research Letters.* doi:10.1002/2014GL061138.

Kato, A., Obara, K., Igarashi, T., Tsuruoka, H., Nakagawa, S., & Hirata, N. (2012). Propagation of slow slip leading up to the 2011 M_w 9.0 Tohoku-Oki earthquake. *Science 335*(6069), 705–708.

Kosobokov, V. G., & Nekrasova, A. K. (2005). Temporal variations in the parameters of the Unified Scaling Law for Earthquakes in the eastern part of Honshu Island (Japan). *Doklady Earth Sciences 405*, 1352–1356.

Lay, Th, Yue, H., Brodsky, E. E., & An, C. (2014). The 1 April 2014 Iquique, Chile, M_w 8.1 earthquake rupture sequence. *Geophysical Reseach Letters.* doi:10.1002/2014GL060238.

Lippiello, E., Marzocchi, W., de Arcangelis, L., & Godano, C. (2012). Spatial organization of foreshocks as a tool to forecast large earthquakes. *Scientific Reports.* doi:10.1038/srep00846.

Lomnitz, C. (1966). Magnitude stability in earthquake sequences. *Bulletin of the Seismological Society of America 56*, 247–249.

Madariaga, R., Métois, M., Vigny, Ch., & Campos, J. (2010). Central Chile finally breaks. *Science 328*, 181–182. doi:10.1126/science.1189197.

Maeda, K. (1999). Time distribution of immediate foreshocks obtained by a stacking method. *Pageoph 155*(2–4), 381–394.

Main, I. (2000). Apparent breaks in scaling in the earthquake cumulative frequency-magnitude distribution: fact or artifact? *Bulletin of the Seismological Society of America 90*(1), 86–97.

Main, I., Meredith, Ph G, & Jones, C. (1989). A reinterpretation of the precursory seismic *b*-value anomaly from fracture mechanics. *Geophysical Journal International 96*, 131–138.

Meng, L., Huang, H., Bürgmann, R., Ampuero, J.-P., & Strader, A. (2015). Dual megathrust slip behaviors of the 2014 Iquique earthquake sequence. *Earth and Planetary Science Letters 411*, 177–187.

Michael, A. (2012). Fundamental questions of earthquake statistics, source behavior, and the estimation of earthquake probabilities. *Bulletin of the Seismological Society of America*. doi:10.1785/0120090184.

Mignan, A. (2014). The debate on the prognostic value of earthquake foreshocks: a meta-analysis. *Scientific Reports*. doi:10.1038/srep04099.

Mogi, K. (1963a). The fracture of a semi-infinite body caused by an inner stress origin and its relation to the earthquake phenomena (second paper). Bull. Earthq. Res. Inst., Univ. Tokyo, 41, 595–614.

Mogi, K. (1963b). Some discussion on aftershocks, foreshocks and earthquake swarms – the fracture of a semi-infinite body caused by an inner stress origin and its relation to the earthquake phenomena (third paper). *Bulletin of the Earthquake Research Institute University of Tokyo 41*, 615–658.

Mogi, K. (1985). *Earthquake Prediction*. Tokyo: Academic Press. **355 pp**.

Molchan, G. M., Kronrod, T. L., & Nekrasona, A. K. (1999). Immediate foreshocks: time variation of the b-value. *Physics of the Earth and Planetary Interiors 111*, 229–240.

Nanjo, K. Z., Hirata, N., Obara, K., & Kasahara, K. (2012). Decade-scale decrease in b value prior to the M9-class 2011 Tohoku and 2004 Sumatra quakes. *Geophysical Reseach Letters*. doi:10.1029/2012GL052997.

Nekrasova, A., Kossobokov, V., Peresan, A., Aoudia, A., & Panza, G. F. (2011). A multiscale application of the unified scaling law for earthquakes in the Central Mediterranean Area and Alpine Region. *Pure and Applied Geophysics*. doi:10.1007/s00024-010-0163-4).

Ogata, Y. (1998). Space-time point-process models for earthquake occurrences. *Annals of the Institute of Statistical Mathematics 50*, 379–402.

Ogata, Y., Utsu, T., & Katsura, K. (1996). Statistical discrimination of foreshocks from other earthquake clusters. *Geophysical Journal International 127*, 17–30.

Papadopoulos, G. A., Charalampakis, M., Fokaefs, A., & Minadakis, G. (2010). Strong foreshock signal preceding the L'Aquila (Italy) earthquake (M_w 6.3) of 6 April 2009. *Natural Hazards & Earth System Science 10*, 19–24.

Papadopoulos, G. A., Drakatos, G., & Plessa, A. (2000). Foreshock activity as a precursor of strong earthquakes in Corinthos Gulf, Central Greece. *Physics and Chemistry of the Earth 25*, 239–245.

Papadopoulos, G. A., Latoussakis, I., Daskalaki, E., Diakogianni, G., Fokaefs, A., Kolligri, M., et al. (2006). The East Aegean Sea strong earthquake sequence of October–November 2005: lessons learned for earthquake prediction from foreshocks. *Natural Hazards and Earth Systems Sciences 6*, 895–901.

Papadopoulos, G.A., Minadakis, G. and Orfanogiannaki, K. (2009). The Prediction of the main shock from the algorithm FORMA: results from Greece and prospects for international testing. Seismol. Res. Lett., 80 (2), 375 (abstr.).

Papazachos, B. C. (1974). Dependence of the seismic parameter *b* on the magnitude range. *Pageoph 112*, 1059–1065.

Papazachos, B. C. (1975). Foreshocks and earthquake prediction. *Tectonophysics 28*, 213–226.

Peng, Z. G., et al. (2007). Seismicity rate immediately before and after main shock rupture from high-frequency waveforms in Japan. Journal of Geophysical Research-Solid Earth, 112(B3).

Raleigh, B., Benett, G., Craig, H., et al. (1977). Prediction of the Haicheng earthquake. *EOS Transactions, AGU 58*, 236–272.

Ruiz, S., Metois, M., Fuenzalida, A. et al. (2014). Intense foreshocks and a slow slip event preceded the 2014 Iquique M_w 8.1 earthquake, Science. doi:10.1126/science.1256074.

Scholz, C. H. (1968). Microfractures, aftershocks, and seismicity. *Bulletin of the Seismological Society of America 58*, 1117–1130.

Scholz, C. H. (1977). A physical interpretation of the Haicheng earthquake prediction. *Nature 267*, 121–124.

Schorlemmer, D., Wiemer, S., & Wyss, M. (2005). Variations in earthquake-size distribution across different stress regimes. *Nature 437*, 539–542.

Schurr, B., Asch, G., Hainzl, S., et al. (2014). Gradual unlocking of plate boundary controlled initiation of the 2014 Iquique earthquake. *Nature*. doi:10.1038/nature13681.

Schurr, B., Asch, G., Rosenau, M., Wang, R., Oncken, O., Barrientos, S., et al. (2012). The 2007 M7.7 Tocopilla northern Chile earthquake sequence: implications for along-strike and downdip rupture segmentation and megathrust frictional behavior. *Journal Geophysical Research 117*, B05305. doi:10.1029/2011JB009030.

Suyehiro, S. (1966). Difference between aftershocks and foreshocks in the relationship of magnitude to frequency of occurrence for the great Chilean earthquake of 1960. *Bulletin of the Seismological Society of America 56*, 185–200.

Suyehiro, S., Asada, T., & Ohtake, M. (1964). Foreshocks and aftershocks accompanying a perceptible earthquake in central Japan. *Meteor. Geophys. 15*, 71–88.

Suyehiro, S., & Sekiya, H. (1972). Foreshocks and earthquake prediction. *Tectonophysics 14*, 219–225.

Utsu, T. (1965). A method for determining the value of *b* in a formula logN = a-bM showing the magnitude–frequency relation for earthquakes. *Geophysical Bulletin Hokkaido University 13*, 99–103. (**in Japanese**).

Utsu, T. (1966). A statistical test of the difference in b-value between two earthquake groups. *J. Physics Earth 14*, 37–40.

Utsu, T. (1992). Representation and analysis of the earthquake size distribution: a historical review and some new approaches. *Pure and Applied Geophysics 155*, 509–535.

Van Stiphout, Th, Schorlemmer, D., & Wiemer, S. (2011). The effect of uncertainties on estimates of background seismicity rate. *Bulletin of the Seismological Society of America*. doi:10.1785/0120090143.

Vidale, J., Mori, J., & Houston, H. (2001). Something wicked this way comes: clues from foreshocks and earthquake nucleation. *EOS Transactions, AGU 82*, 68.

Vidale, J. E., & Shearer, P. M. (2006). A survey of 71 earthquake bursts across southern California: exploring the role of pore fluid pressure fluctuations and aseismic slip as drivers. *Journal Geophysical Research 111*, B05312. doi:10.1029/2005JB004034.

Wu, C., Meng, X., Peng, Z., & Ben-Zion, Y. (2014). Lack of spatiotemporal localization of foreshocks before the 1999 M_w 7.1 Düzce, Turkey, earthquake. *Bulletin of the Seismological Society of America, 104*, 560–566.

Wyss, M. (1997). Second round of evaluations of proposed earthquake precursors. *Pure and Applied Geophysics, 149*, 3–16.

Yagi, Y., Okuwaki, R., Enescu, B., et al. (2014). Rupture process of the 2014 Iquique Chile earthquake in relation with the foreshock activity. *Geophysical Reseach Letters*. doi:10.1002/2014GL060274.

Yamaoka, K., Ooida T. and Ueda Y. (1999). Detailed distribution of accelerating foreshocks before a M 5.1 earthquake in Japan. Pageoph, 155, 2–4, 335–353.

Yamashita, T. (1998). Simulation of seismicity due to fluid migration in a fault zone. *Geophysical Journal International, 132*, 674–686.

Zhuang, J., & Ogata, Y. (2006). Properties of the probability distribution associated with the largest event in an earthquake cluster and their implications to foreshocks. *Physical Review E.* doi:10.1103/PhysRevE.73.046134.

(Received February 3, 2016, revised June 17, 2016, accepted June 21, 2016, Published online July 4, 2016)

Pure Appl. Geophys. 173 (2016), 3273–3290
© 2015 Springer Basel
DOI 10.1007/s00024-015-1161-3

Pure and Applied Geophysics

Analysis of the Aeromagnetic Anomalies of the Auca Mahuida Volcano, Patagonia, Argentina

L. M. Longo,[1,2] R. De Ritis,[3] G. Ventura,[3,4] and M. Chiappini[3]

Abstract—We present the analysis of the subsurface structure of the Auca Mahuida volcano based on high-resolution aeromagnetic data integrated with the available geological information. Most of the detected magnetic anomalies have a dipolar structure opposite to that of the present geomagnetic field. According to the available geochronological data and paleomagnetic measurements, the source bodies of Auca Mahuida mainly emplaced in the Matuyama reverse polarity chron. The Reduction-to-the-Pole map confirms that the magnetization direction is mainly reverse with only few anomalies normally magnetized. Two opposite, coexisting polarities do not allow to fully remove the dipolar character of the field in the Reduction-to-the-Pole transformation. Therefore, we model the measured anomaly field by applying analytical techniques that are independent of the magnetization direction. The obtained anomaly strikes and source geometries indicate an emplacement of intrusive bodies controlled by the regional faults affecting the Auca Mahuida basement and the sedimentary successions of the Neuquén basin. Magma upraised along these faults and fractures feeding the volcanic activity and subsequently crystallized. The averaged power spectrum and Euler Deconvolution indicate source depths consistent with those of the intrusions recognized in wells. Borehole data highlight the widespread presence of intrusive bodies below the Auca Mahuida central crater and the peripheral sectors at depth of 2 km below sea level. These bodies have played a major role in the thermal maturation of hydrocarbons and in the subsequent accumulation of oil below the volcano. The obtained results shed light on the Auca Mahuida feeding system and on the intrusions geometry, also pointing out the effectiveness of the magnetic prospecting in the oil industry even in presence of strong remanent magnetization.

Key words: Magnetic anomalies, aeromagnetic data, oil fields and volcanoes, reverse polarity, Auca Mahuida volcano.

1. Introduction

Magnetic anomalies constitute a powerful tool to obtain information about the subsurface when magnetic susceptibility contrasts are present (Finn *et al.* 2001; Blanco-Montenegro *et al.* 2006); this is especially true when igneous rocks intrude sedimentary basins. From the magnetic point of view, sediments are mostly transparent (Reynolds 2011). Thus, structures such as magmatic dikes, sub-circular conduits, and sills emplaced within sedimentary sequences are likely sources of intense magnetic anomalies because of their high susceptibility contrast. Many studies of magnetic anomalies in volcanoes have demonstrated the usefulness of this type of data for the characterization of the subsurface (Blanco 2003; De Ritis *et al.* 2005).

The Auca Mahuida (hereafter AM) volcano is located in the north eastern sector of the Neuquén basin in Argentina (Fig. 1). A high-resolution aeromagnetic survey was conducted for YPF S.A. in the central AM area in 2001. The survey was designed to supplement the information provided by the seismic prospection, which does not cover the central crater area due to its rough topography.

Magnetic modeling has furnished a reliable image of the intrusive bodies inside the sedimentary rocks, their depth intervals and magnetization. Mostly, the sources of the magnetic anomalies emplaced in a reverse magnetic chron because they exhibit a dipolar structure opposite to that expected in the Southern hemisphere under a magnetizing Earth field with normal polarity. In fact, age determination of the AM samples show a time interval of 2.03–0.88 Ma (Bermúdez and Delpino 1998), which implies that volcanic activity concentrated during the reverse polarity Matuyama Chron. Therefore, the magnetic modeling technique adopted by us takes into account

[1] YPF, Gerencia de Geofísica, Buenos Aires, Argentina.
[2] Facultad de Ciencias Astronómicas y Geofísicas, Universidad Nacional de La Plata, La Plata, Argentina.
[3] Istituto Nazionale di Geofisica e Vulcanologia, Rome, Italy. E-mail: riccardo.deritis@ingv.it
[4] Istituto Ambiente Marino Costiero, Consiglio Nazionale delle Ricerche, Naples, Italy.

Figure 1
Top left Neuquén Basin position (*red line* inside the Argentina Republic map) together with the Auca Mahuida site within the Neuquén basin. *Top right* geological sketch map of the Payenia volcanic province. *Bottom* satellite image of the Auca Mahuida basaltic volcano complex

that the primary magnetization for the area is a strong reverse remanence. We computed the radially averaged power spectrum (SPECTOR and GRANT 1970) of the dataset, comparing the obtained depth to top of the magnetic sources with the intrusion depths measured in boreholes (VELA *et al.* 2006). Moreover, in order to obtain the 3D source geometries, we carried out the Euler Deconvolution (THOMPSON 1982), as well as inverse modeling (MACLEOD 2013). The obtained magnetic models were interpreted using the paleomagnetic measurements, borehole data and seismic profiles as constraints, shedding light on the inner structures of the central volcano. This is a valuable result since location, geometry, and volume of igneous bodies are parameters of primary importance for the detection and evaluation of possible oil reservoirs below AM.

2. Geological Background

AM volcano is the southernmost edifice of the Payenia retroarc volcanic province, which extends in the Andean foreland between 35°and 38° S latitude (KAY and RAMOS 2006; Fig. 1), and it is located about 400 km east of the outcropping front of the Andes. The volcanism started during Miocene and mainly developed between the Pliocene and the Quaternary, to the east of the roughly N–S striking Andes mountain front. The main igneous events occurred under extensional regimes that generated deep fractures, which are necessary for the uplift of the fluids from large depths. This extensional phase alternated with compressional events that built the thrust belt. The volcanism is absent to the south of the AM volcano, where the N100°E striking Cortaderas lineament, marks the southern boundary of the Miocene subduction (KAY and RAMOS 2006). This boundary is marked on the surface by N100°E to E–W striking faults affecting the Añelo depression (Fig. 1). AM volcanic field lies on a NNW-SSE to NW–SE striking anticline involving the complete sedimentary sequence of the Neuquén basin (ROSSELLO 2002; MOSQUERA and RAMOS 2006). This sequence is affected by NNW-SSE to NW–SE striking normal faults of the Entre Lomas half-graben (CRISTALLINI *et al.* 2006). The activity of these faults mainly developed from the

Late Jurassic to Upper Cretaceous periods, with minor activity in more recent times. The approximately 6 km thick stratigraphic sequence (Fig. 2) in the AM area includes a series of deposits with ages from the Upper Triassic to Lower Tertiary. These deposits were intruded by magma between 25 and 0.5 Myr ago (PÁNGARO *et al.* 2004). The basement is constituted by the Choiyoi group consisting of volcanoclastic Permo-Triassic rocks: pyroclastics, intrusive, and ignimbrites (SIGISMONDI and RAMOS 2008). These were followed, in the rift phase, by the volcanoclastic and epiclastic rocks of the Pre-Cuyo formation. AM edifice is located at 37°44'S, 68°55'W and reaches an altitude of 2258 m a.s.l. (Figure 1). AM consists of an E–W elongated lava plateau with monogenic vents and a summit polygenic cone (VENTURA *et al.* 2012). The lava flows cover a surface of about 2700 km². The thickness of the lavas is 500 m below the central crater and decreases to virtually 0 toward the periphery ROSSELLO 2002; LONGO *et al.* 2008).

The age of the AM rocks is between 2.03 ± 0.3 and 0.88 ± 0.3 Myr and the composition varies from within-plate basalts (WPB) to trachytes (KAY and RAMOS 2006; RAMOS and FOLGUERA 2010).

3. Data

3.1. Aeromagnetic Survey

A high-resolution aeromagnetic survey was conducted in the AM and Señal Cerro Bayo areas in 2001 by Carson Aerogravity company (CARSON AEREOGRAVITY 2001). Magnetic data were measured by a Geometrics high sensitivity Cesium Vapor magnetometer at a sampling rate of 1 Hz, at 2830 m a.s.l. The area was surveyed using profile and tie lines oriented N–S and E–W, respectively. Line spacing was 0.5 × 0.5 km above the AM central crater and 2 × 2 km in the surrounding areas. A magnetic base station was set up in the Rincon de los Sauces airport recording the diurnal variation. The field data were diurnally corrected using the base station, the IGRF2000 model was used in order to remove the main field component, and profile and tie lines leveled in order to get as precise as possible the Total Magnetic Intensity (TMI) field (Fig. 3).

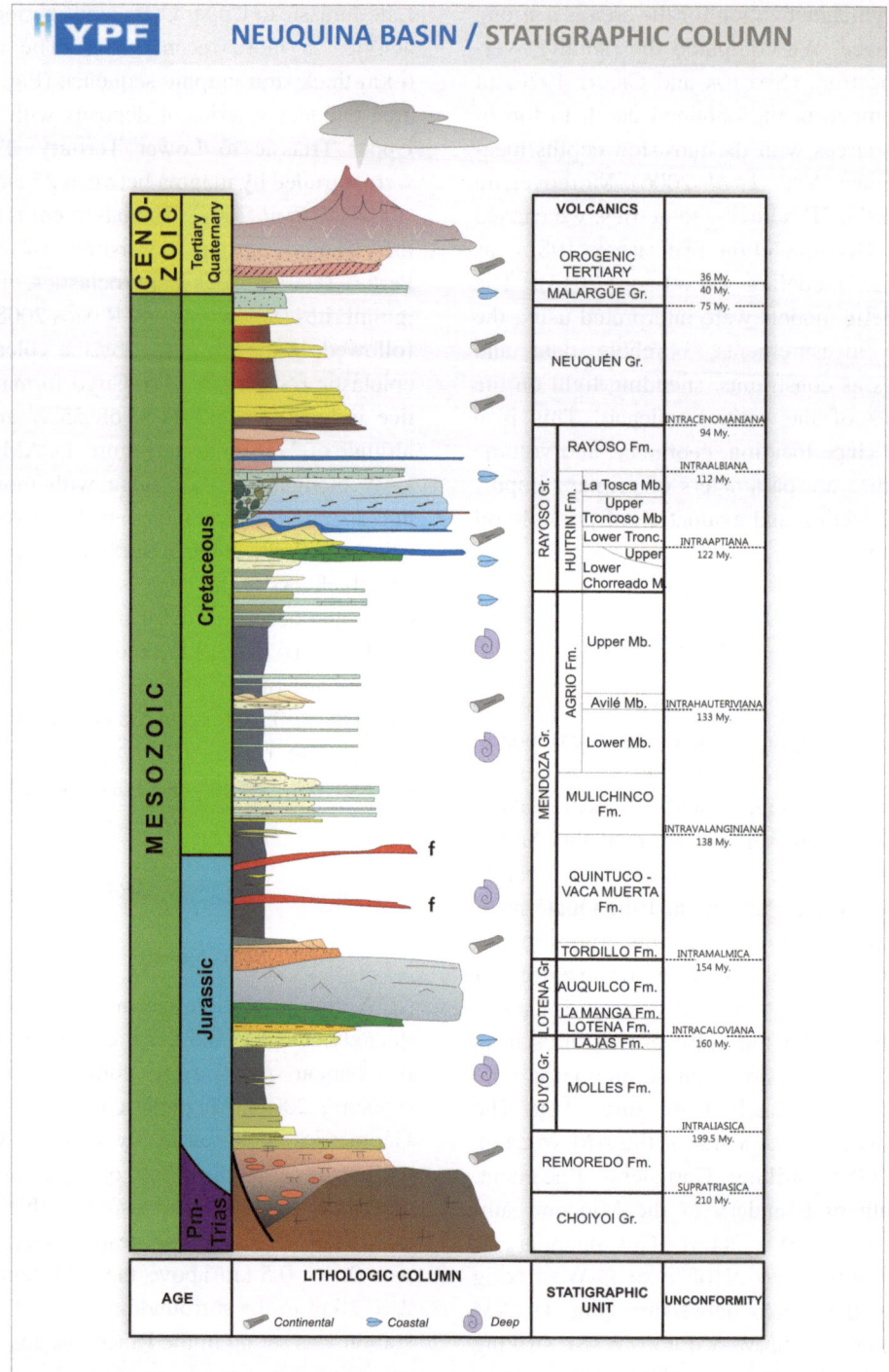

Figure 2
Stratigraphic column corresponding to the AM area. Adapted from BRISSON and VEIGA 1999

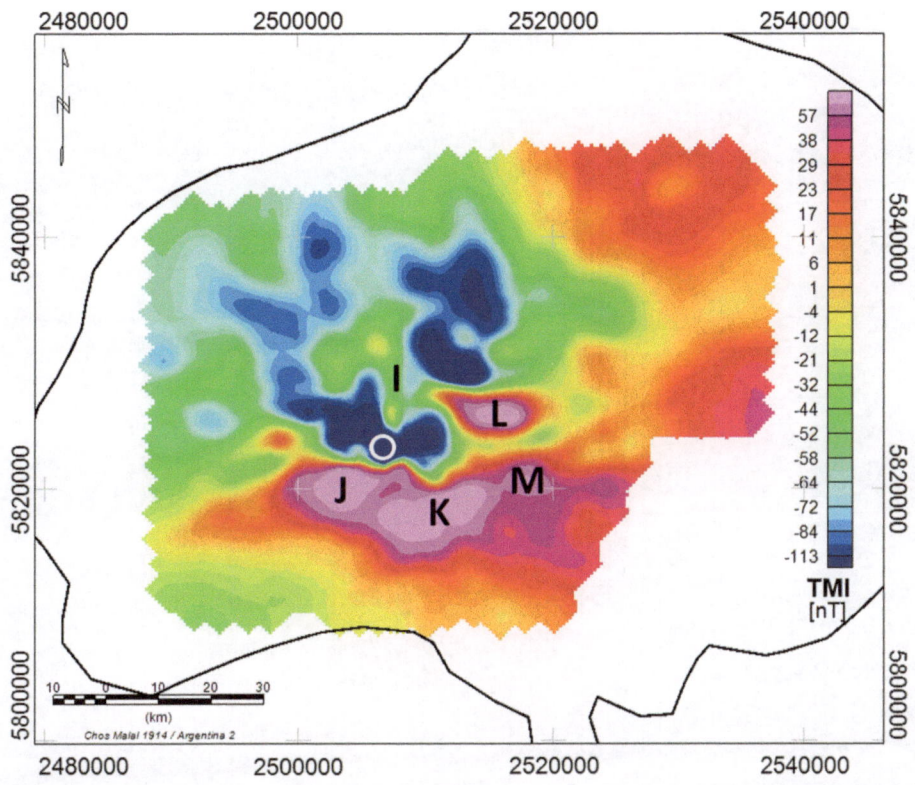

Figure 3

TMI anomaly map of the Auca Mahuida volcano. The *white circle* marks the central crater. Limit of the volcanic plateau is shown with a *black line*. The letters I, J, K, L, and M label specific anomalies discussed in the text. The aeromagnetic survey does not cover the entire plateau

3.2. *Magnetic Susceptibility Measurements and Rock Types*

Thirty-three samples were collected in the field in the AM central sector and flanks; some samples are from outcrops of the sedimentary upper Neuquén formation. The sample locations are shown in Fig. 4. The composition of the samples varies from basalts to trachytes. We measured the on-site susceptibility using a high sensitivity (1×10^{-7} SI) ZH - SM-30 magnetic susceptibility meter (Table 1; Fig. 4). The mass specific susceptibility was also determined (Table 1) in the laboratory with a Kappabridge system AGICO (KLY-2 model) on thirty samples. The measurements were carried out on the most representative lava flows, scorias and pumices of the AM area, as well as on siltstones and clays of the Neuquén formation. The measurements were carried out at the paleomagnetism laboratory of the Istituto Nazionale di Geofisica e Vulcanologia (INGV, Rome, Italy).

3.3. *Seismic and Well Data*

Seismic 2D lines and a 3D survey in time domain were available from YPF. The survey does not cover the whole AM area (Fig. 5a). Besides, the quality of the 2D lines is poor (Fig. 5 b) due to the significant thickness of surface basalts, which have a high acoustic impedance. Well data allow us to identify the formations of interest constraining the seismic section interpretation through synthetic seismograms, which provide seismic velocities. These data allow interpreting the shallower and intermediate formations, (e.g., Quintuco, Fig. 5c) at the borders of the volcanic plateau.

The depth reached by boreholes is up to 4 km below the ground level (Fig. 5), involving the

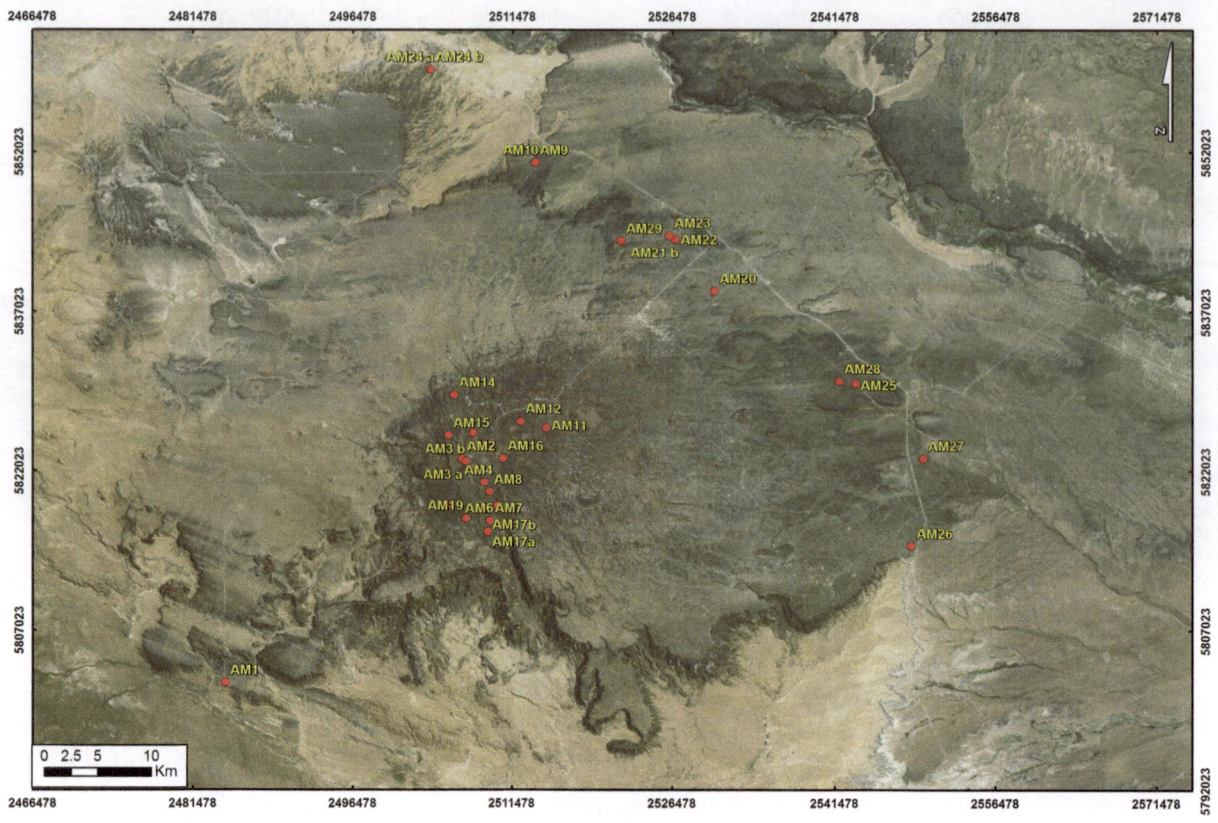

Figure 4
Location of paleomagnetic measurements on the AM complex

middle-upper parts of the sedimentary sequence. The available information includes the lithology, facies description, gamma ray, sonic and density logs (the last three are used to generate the synthetic seismograms). Moreover, samples of the sedimentary sequence and of the intrusions are available from cores of a few wells. These specimens are almost transparent and highly magnetic.

Subsurface temperature measurements were also available from the boreholes. The maximum temperature recorded at the bottom (~ 4 km b.s.l.) of 24 boreholes is about 180 °C, a value well below that of the Curie temperature. The depth variation of the equilibrium temperature (Fig. 6), after the usual corrections, agrees with the regional trend of the Neuquén basin (SIGISMONDI and RAMOS 2008). As a consequence, igneous rock bodies intruded into the sedimentary sequence can be magnetized and represent the source of magnetic anomalies.

4. Auca Mahuida Magnetic Anomaly Field

4.1. Topographic Effect of a Uniformly Magnetized Crustal Slab

In volcanic areas, shallow or outcropping sources (e.g., domes, cones) can be identified through their magnetic response. As a matter of fact, when a topographic magnetic effect is present, a direct correlation between topographic and magnetic features can be easily recognized (BLAKELY and GRAUCH 1983). Usually, this happens when a rough and very steep topography is present together with highly magnetized rocks. This is not the case of the AM volcano, which has a very low topographic gradient, and where such kind of correlation is not observable in the magnetic anomaly field. Nevertheless, in order to understand the possible wavelengths, amplitudes, and shape of the terrain magnetic effect, we

Table 1

Laboratory and field magnetic susceptibility measurements

Sample	k laboratory	k onsite measurementes	Lithology
AM1	0.00514	0.00837	Massive lava
AM2	0.02316	0.01366	Vesicular altered lava
AM3a	0.00346	0.02815	Massive lava
AM3b		0.02365	Massive lava
AM4	0.03302	0.01648	Massive lava
AM5	0.04029	0.02507	Massive lava
AM6	0.03839	0.01802	Massive lava
AM7	0.00506		Pumice
AM8	0.01283	0.00912	Massive lava
AM9	0.00667	0.00579	Vesicular lava close to a sedimentary dikes
AM10	0.00023	0.00009	Oxidade clay, volcaniclastic flow deposit
AM11	0.01834	0.01166	Vesicular lava close to a sedimentary dikes
AM12	0.02631	0.01400	Massive lava
AM13	0.02432	0.01933	Massive lava
AM14	0.01637	0.01314	Massive lava
AM15	0.02062	0.00911	Massive lava
AM16	0.01783	0.01077	Massive lava
AM17a	0.02071	0.01570	Massive lava
AM17b	0.01932	0.01456	Massive lava
AM18	0.02401	0.01559	Massive lava
AM19	0.02284	0.01036	Massive lava above the Neuquén sediments
AM20	0.01261	0.00752	Massive lava
AM21 a	0.01633	0.00256	Welded scoria
AM21 b		0.00089	Massive lava
AM22	0.02039		Vesicular lava
AM23	0.02535	0.01377	Massive lava
AM24 a	0.00017	0.00026	Red clays
AM24 b		0.00008	Red clay, Neuquen sedimentary sequence
AM25	0.00357	0.00289	Lava masiva
AM26	0.00797	0.00491	Vesicular lava
AM27	0.00984	0.00833	Massive lava
AM28	0.01658	0.00829	Massive lava
AM29	0.02569	0.00897	Massive lava

computed the signature of a crustal slab with a constant magnetization of 2 A/m (Fig. 7) placed in a reverse polarity field. This value has been chosen on the basis of the range of the susceptibilities we got with the inversion algorithm. In a more general sense, an estimation of the terrain magnetic effect might be useful, either to identify sources related to the topography or to enhance the magnetic anomaly of buried sources obscured by the terrain effect. The upper surface of the slab coincides with the high-resolution Digital Elevation Model provided by YPF S.A., and the lower one is a surface standing at 500 m below the topographic elevation and following the trend of the latter. The chosen area is a 125 × 125 km square grid, whose center corresponds to the AM central edifice, covering the whole AM volcanic plateau and a wide area of the surroundings.

4.2. Location and Depth of Magnetic Sources: The Euler Deconvolution

In the last decades, several techniques were developed for locating magnetic contacts, and one of the most widely used is the Euler Deconvolution. This technique was originally presented by THOMPSON (1982); it consists in solving the Euler's homogeneity equation for different windows of data covering the whole magnetic anomaly map. The horizontal position and depth of the simple equivalent magnetic sources are obtained. To solve the equations, it is

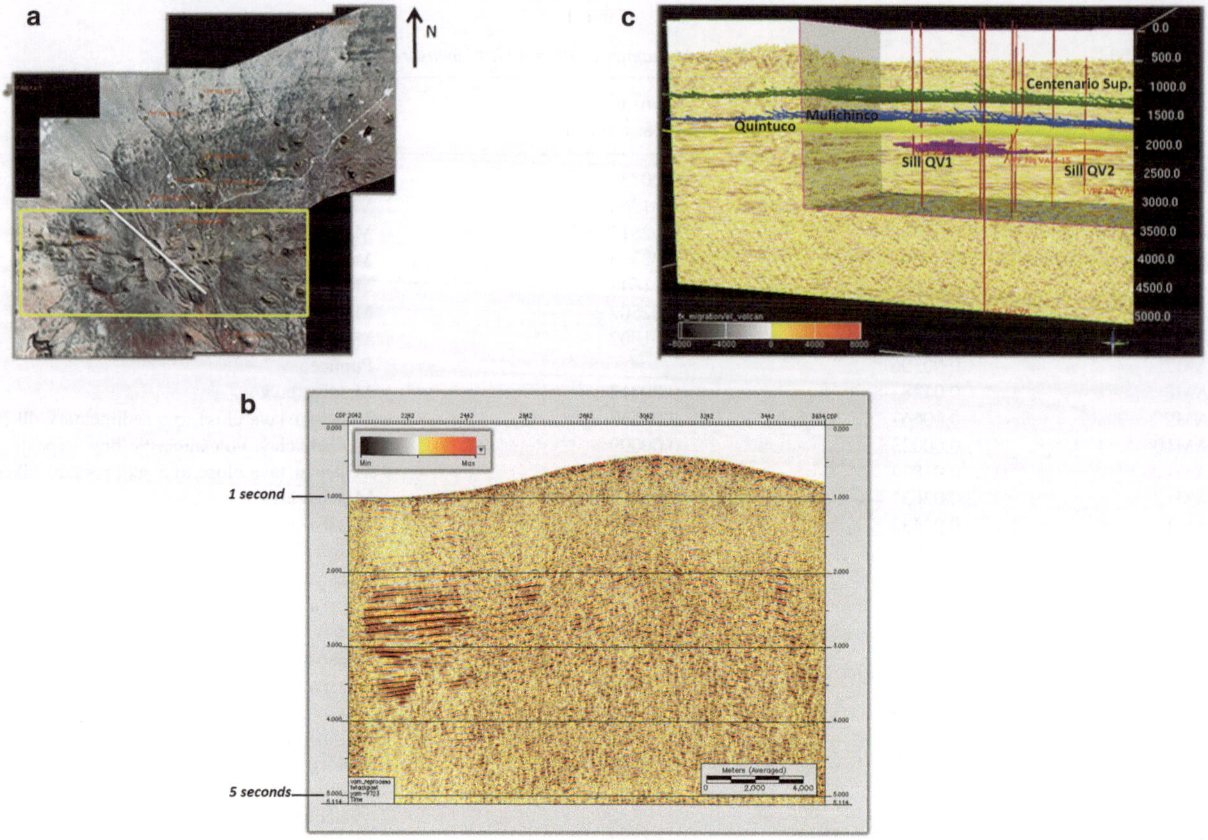

a

N

c

Centenario Sup.

Quintuco Mulichinco

Sill QV1   Sill QV2

b

1 second

5 seconds

Figure 5

a Ikonos satellite image of the AM volcanic plateau. The *yellow rectangle* indicates the area with no 3D seismic data. **b** 2D NW–SE two ways travel time seismic section (*white line* in **a**) showing the low signal-to-noise ratio in the central crater area. Each subdivision in the vertical scale is a time interval of 1000 ms (1 s), **c** Main seismic horizons interpreted around the central crater (Mulichinco and Centenario Superior). QV1 and QV2 are two sills intruded in Quintuco-Vaca Muerta. *Vertical lines* are borehole trajectories. Planes are sections of the seismic cube. *Vertical axis scale* is time in milliseconds

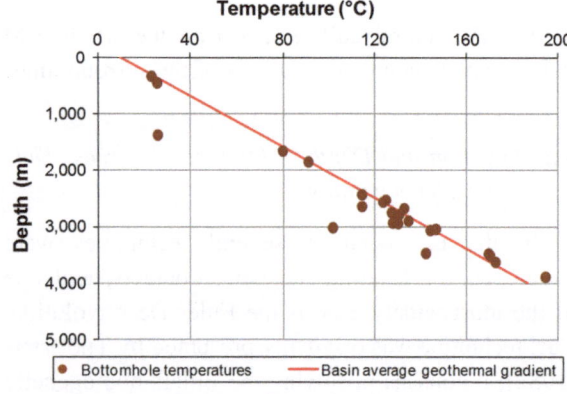

Figure 6
Subsurface temperature variation with depth. Points are temperature measurements in AM boreholes; the *solid line* corresponds to global average geothermal trend for the Neuquén Basin (SIGISMONDI and RAMOS 2008)

necessary to assign a parameter, the structural index (SI), which is characteristic of each magnetic source. This index represents the rate of attenuation of magnetic anomaly with distance to the source. SI values vary between 0 and 3 depending on the geometry of the anomalous body. This method works well for simple sources, such as a magnetic dipole, line of dipoles, monopole, etc. Real sources, however, are not point sources but extended bodies equivalent to ensembles of dipoles. REID *et al.* (1990), BARBOSA *et al.* (1999, 2000), MUSHAYANDEBVU *et al.* (2001) and HSU (2002), among others, also considered this method, with some variations. Model studies and theoretical work by REID *et al.* (1990) have led to the conclusion that the location of magnetic sources of varying strikes in presence of a non-vertical field

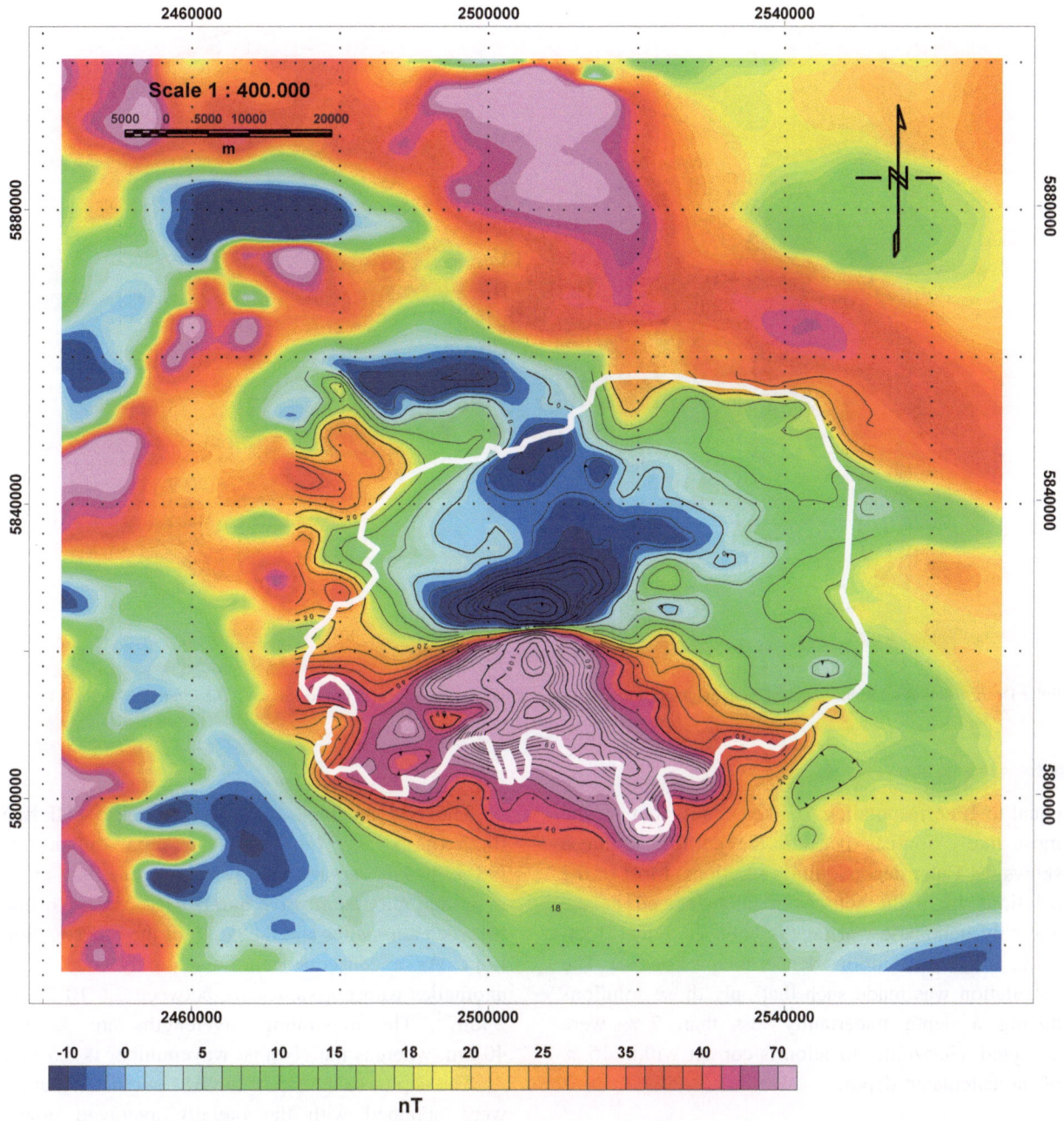

Figure 7
Computed topographic effect of a uniformly magnetized crustal slab placed in a reverse magnetizing field with $I = 57°$, $D = 142°$ and a constant magnetization of 2 A/m

can be accurately reproduced by Euler Deconvolution without applying Reduction-to-the-Pole. The method also yields useful results in the presence of remanence. No information on the dip of the sources is obtained; therefore, it must be estimated by other means, e.g., by inverse modeling.

The method was applied to the magnetic data of AM volcano using a 5 km window size and an SI

69

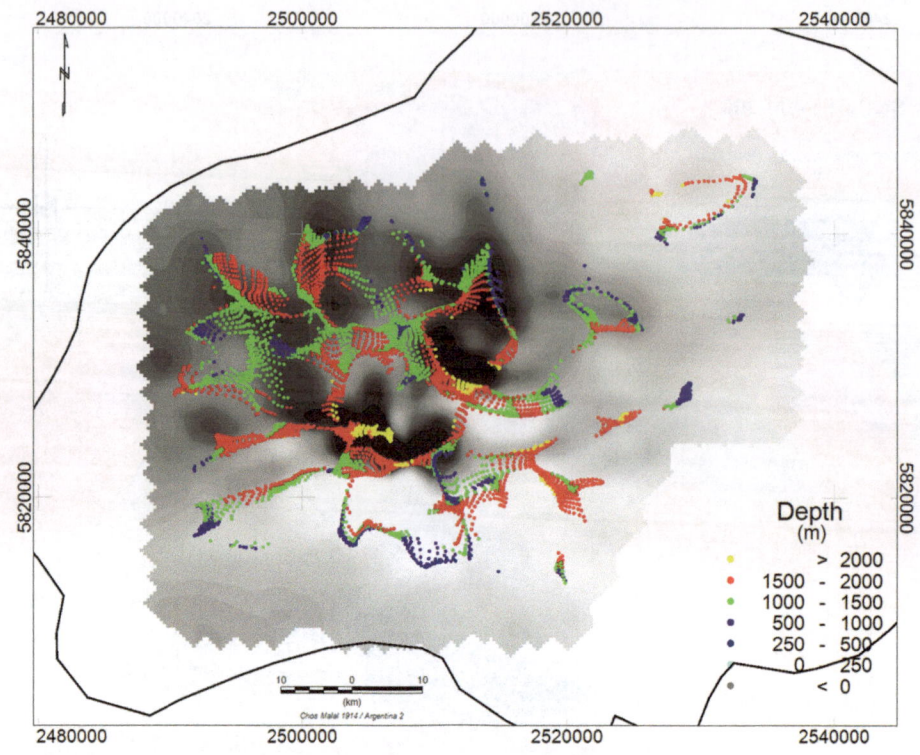

Figure 8
Euler Deconvolution solutions obtained from magnetic anomaly map with a window size of 5 km for SI = 1. The reference height is the sea level

equal to 1, corresponding to dikes or sills that are the most likely sources beneath the edifice. Figure 8 shows the Euler deconvolution solutions. Most of the solutions highlight sources mainly located at the anomaly boundaries, at depths between 1 and 2 km b.s.l., and other solutions deeper than 2 km b.s.l. The calculation was made such that only those solutions having a depth uncertainty less than 7 % were accepted. Horizontal location is correct within 15 % of the calculated depth.

4.3. Spectral Analysis of the Magnetic Anomalies

Fourier analysis represents a well-established procedure in potential field methods. It allows us to obtain information about the depth of the anomaly sources. Data spacing and the geographical extension of the survey poses some restrictions on the wavenumber and wavelength detectable by this analysis.

The AM magnetic anomaly field grid cell size (0.25 km x 0.25 km) and the area dimensions (35×40 km^2 approximately) determine the minimum and maximum wave number (k_{min}, k_{max}) of the recorded anomalies and the depth of the investigation limit. As a consequence, the anomaly map holds anomalies whose k values are between $2.6 \ 10^{-2}$ and 2 km^{-1}. The maximum wavelengths are 35 and 40 km, whereas the Nyquist wavenumber is 0.5 km.

The depths to the top of the AM magnetic sources were obtained with the radially averaged power spectrum technique. While the energy spectrum is a 2D function of the energy relative to wavenumber and direction, the radially averaged energy spectrum (RAPS) is a function of wavenumber alone, and it is calculated by averaging the energy for all directions for the same wavenumber. The Fourier transform of the magnetic field produced by a prismatic body has a broad spectrum whose peak location is a function of the depth to the prisms top and bottom surfaces, and whose magnitude is determined by the prisḿs magnetization. The peak wavenumber (k_{max}) can be determined by Eq. (1) (BLAKELY 1995):

$$k_{max} = \frac{\log(\frac{z_b}{z_t})}{z_b - z_t}, \qquad (1)$$

where z_t and z_b are the depths (measured from the flight elevation) to the top and bottom of the layer, respectively, k is the peak wavenumber in radians per unit of distance. When considering a grid large enough to include many sources, the log spectrum of the data can be interpreted to determine the statistical depth to the top. Therefore, depths are determined by measuring the slope of the energy power spectrum and dividing it by 4π (SPECTOR and GRANT 1970) as follows:

$$h = -s/4\pi, \qquad (2)$$

where h is the depth and s is the slope of the log (energy) spectrum.

Figure 9 shows depth estimates based on 5-point averages of the slope of the energy spectrum of the AM volcano, giving a maximum depth of the ensemble of magnetic sources at about 3.5 km (below flight elevation).

4.4. Reduction-to-the-Pole

Reduction-to-the-Pole (RTP) is one of the most widely used algorithms in magnetic data processing. It consists of generating the magnetic anomaly field that would be recorded, if the source bodies were placed at the earťhs magnetic poles. Therefore, the effects associated with the inclination of the field with latitude are removed and each dipolar anomaly is transformed into positive or negative counterparts located directly above the source facilitating its interpretation. The algorithm approximates the ambient magnetic field and the rock magnetization to constant average values (amplitudes and directions) for the entire study region. This is a reasonable assumption provided that the area of investigation has a limited geographical extension. An accurate RTP transformation in the presence of strong remanence requires both induction and remanent magnetization components. The AM volcanic activity developed during the Pliocene and Pleistocene times; in this

time span the Earťhs magnetic field reversed several times. In the previous sections, we argued that AM magnetic anomaly field shows prevailing reverse magnetized sources. Therefore, most of the magnetized sources emplaced in a reverse chron and the remanence is the dominant magnetization component of the AM rocks. Consequently, meaningful values for these parameters have to be selected in order to carry out the RTP transformation. Remanence values are chosen in agreement with previous paleomagnetic studies for the study area (Propiedades magnéticas de los basaltos del área del Auca Mahuida, Laboratorio de Paleomagnetismo 'Daniel Valencio', Universidad de Buenos Aires, unpublished data). Thus, we computed the Reduced-to-the-Pole field for two separate cases, one considering the source magnetization as due to induction alone (with angular values $I = -38°$ and $D = 3°$, Fig. 10a), and the other one as due to induction and remanence (Fig. 10b). As discussed above, the latter was considered to be acquired in an inverse chron with average inclination and declination of 38° and −177°, respectively.

4.5. Analytic Signal

The analytic signal (AS) method was introduced by NABIGHIAN (1972, 1984). It represents the energy envelope of the magnetic anomalies, and its amplitude $A(x, y)$ is a function of the orthogonal gradients of the anomalies (ROEST et al. 1992):

$$|A(x,y)| = \left[(\partial T/\partial x)^2 + (\partial T/\partial y)^2 + (\partial T/\partial z)^2\right]^{1/2}, \qquad (3)$$

where T is the total magnetic intensity anomaly. The AS is independent of the direction of magnetization and of the Earťhs magnetic field. This characteristic is particularly advantageous where strong remanence is present, or at low magnetic latitude. In fact, MACLEOD et al. (1993) proposed that AS is an alternative to Reduction-to-the-Pole for low latitudes. Independence of magnetization direction means that all bodies with the same geometry have the same analytic signal. Furthermore, as the peaks of analytic signal functions are symmetric and occur directly over the edges of wide bodies and over the centers of narrows bodies, the interpretation of analytic signal

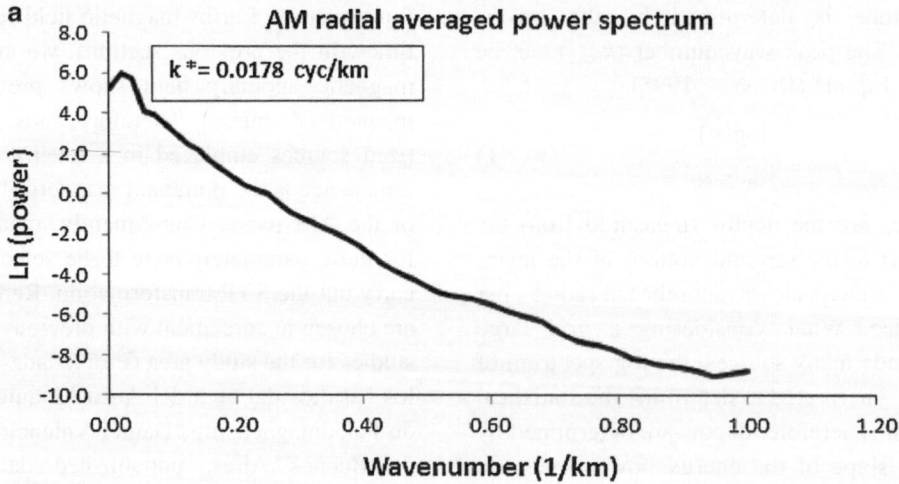

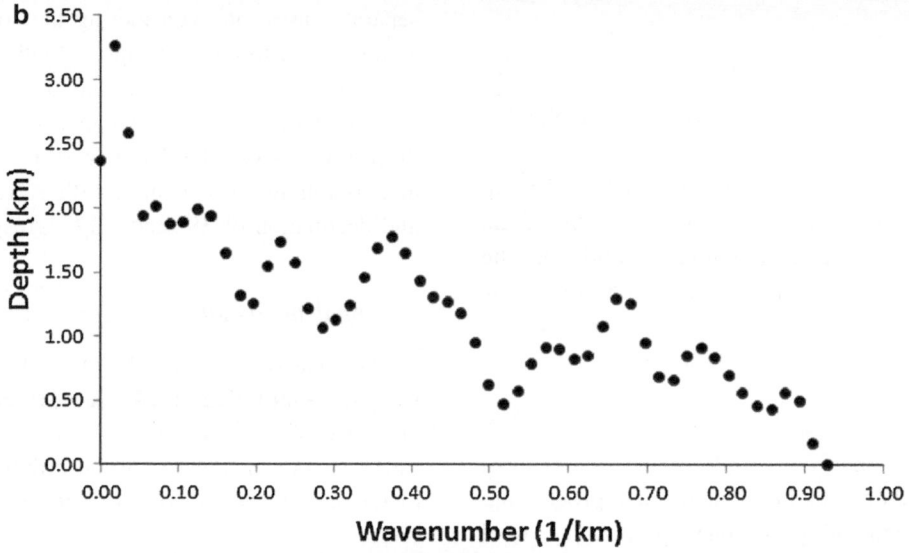

Figure 9

a Depth estimation of magnetic sources from radially averaged logarithmic energy spectrum of the AM anomaly field. The spectrum corresponds to ensembles of magnetic bodies with the same average depth; $k*$ is the wavenumber value for the maximum of the power spectrum. **b** Depth estimates from the slopes of the spectrum taken in 5-point intervals

maps should provide indications of magnetic source geometry (GUNN 1997). NABIGHIAN (1972) showed that the maxima of AS are located over the magnetization contrasts, so they can be used to indentify the main contrasts.

NABIGHIAN (1972) used the AS amplitude to estimate the depth to the magnetic contact. ROEST *et al.* (1992) and ATCHUTA RAO *et al.* (1981) used the anomaly width at half the amplitude to derive the depths. However, the measured amplitude is usually due to many overlapping anomalies and, even when the amplitude can be well determined for a single source, the resulting depth will be not well constrained if the correct model is not used. For the above mentioned reasons, we do not use this technique to estimate depths since the TMI anomaly map at AM volcano is produced by an ensemble of small anomalies.

In order to determine the main magnetization contrast of the TMI map, we calculate the AS

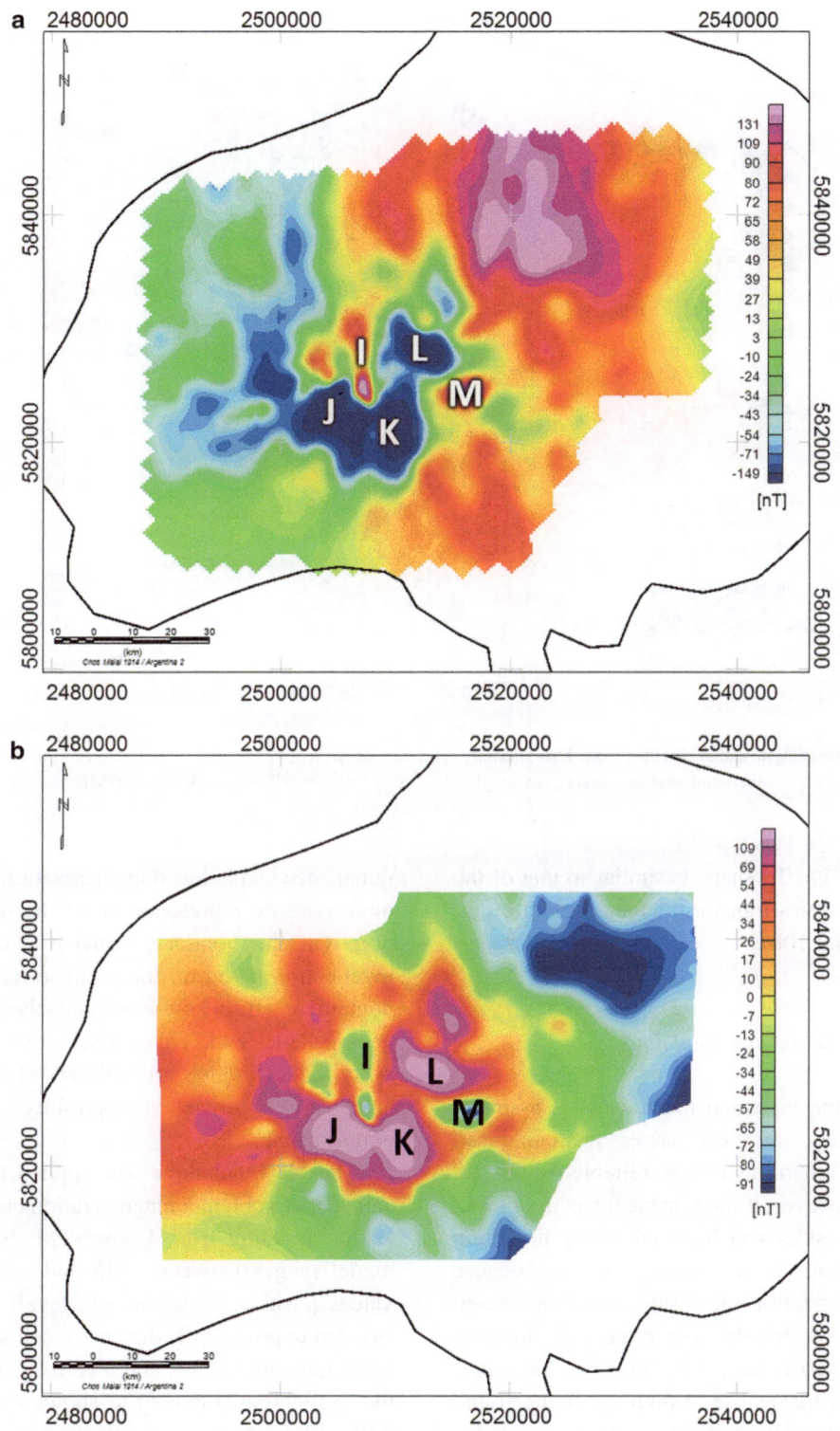

Figure 10

a RTP transform computed considering only the induction component ($I = -38°$ and $D = 3°$). **b** RPT transform computed with both induced and remanent magnetizations ($I = 38°$ and $D = -177°$). The anomalies are labeled with *black letters* (see Fig. 3 for comparison)

73

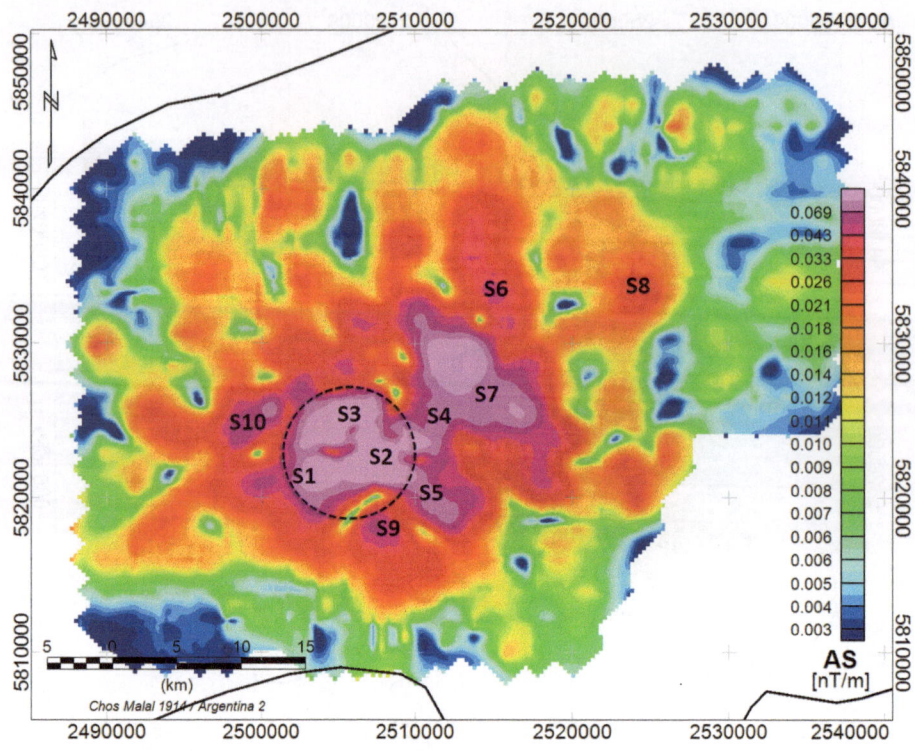

Figure 11
Amplitude map of the analytic signal obtained from TMI anomaly. The *dashed circle* indicates the central crater of the volcano. S1 to S10 are the main sources causing the amplitude peaks. Units of AS amplitude are [nT/M]

amplitude (Fig. 11). Its shape is similar to that of the Reduced-to-the Poles transform anomaly with remanence effect (Fig. 10b).

5. Inverse modeling

In most of the inversion methods, it is assumed that the magnetic response arises from induced magnetization. When this is true, reliable results can be obtained. However, if the remanence is important, more realistic results will be obtained by modeling the magnetization as a vector. This is because remanent magnetization can distort inversions based on the assumption that the source has only induced magnetization (MACLEOD 2013). The magnetic vector inversion (MVI) makes use of both the induced and remanent magnetization without a priori knowledge of the direction of the latter. MVI allows to change the direction of magnetization inside the model and thus it takes into account the combined effects of

remanence and induced magnetization. The result is a more realistic representation of the rock magnetization. On the contrary, conventional susceptibility inversion is based on the premise that the magnetic domains in all rocks orient themselves parallel to the geomagnetic field. This results in negative susceptibility values, which are artifacts of the calculation, since a negative susceptibility is physically unreasonable.

The MVI technique was applied to the AM volcano data to obtain a magnetization cube using VOXI Earth Modeling from GeosoftTM. Elevation in the model ranges between 2175 and −5038 m (with z values positive above the sea level). The grid constructed to process the data has a cell size of 500 m in the x and y directions, and a vertical size of 465 m in the z direction. Figure 12a shows the isosurface of 0.012 SI from the magnetic vector inversion (light blue shapes surrounding the central areas) together with the topography grid. The highest susceptibility values we got are about 0.13 SI; Fig. 12b is a plot of

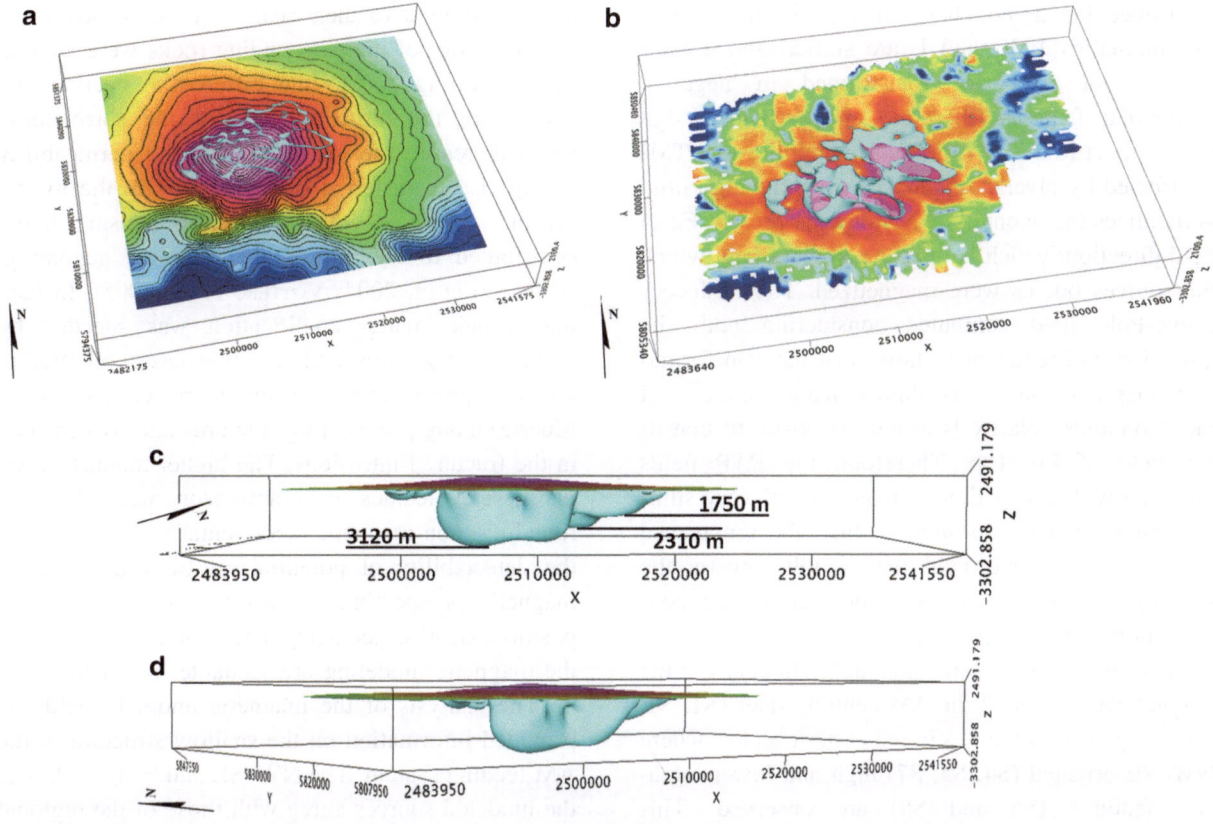

Figure 12

a 3D representation of the magnetic vector inversion results, the 0.012 SI isosurface roughly represents the magnetic anomalies source bodies (*light blue shapes* surrounding the central areas) along with the topography layer. **b** The 0.012 SI isosurface plot along with the analytic signal **c, d** Different views of the 0.012 SI isosurface near the central crater. The *color surface on top* is the topography. Z direction is positive above sea level

the analytic signal derived from the AM TMI anomaly data, together with the same 0.012 SI isosurface. It can be seen that there is a good correspondence between the source locations and the analytic signal map. A similar behavior was observed in other studies, and it was interpreted as having geological consistency (MacLeod 2013). Finally, panels (c) and (d) of Fig. 12 show two views of an E-W section of the 0.012 SI isosurface, where the depth extent of the bodies reaches about 3 km b.s. l.

6. Discussion and Conclusions

The magnetic anomaly map of the Auca Mahuida volcano (Fig. 3) shows an ensemble of sources with a prevailing reverse magnetization. The age of the AM rocks (Ramos and Folguera 2010) suggest that most of the sources emplaced during the reverse polarity Matuyama Chron. AM anomalies lie around the central crater with wavelengths between 1.5 and 5 km. These anomalies do not correlate with any topographic effect shown on Fig. 7, confirming their deep origin. The maximum subsurface temperatures, recorded at approximately 4 km b.s.l., are about 180 °C, a value lower than the Curie temperature, as previously reported. Thus, the igneous bodies intruded into the sedimentary sequence of the Neuquén basin are not thermally demagnetized.

The Reduced-to-the-Pole field computed considering only the induction component of the magnetization shows that the algorithm is not able to correctly transform the normally magnetized anomalies (such as the features I and M of Fig. 3 and

10a) since they do not change their positions. Instead, the anomalies J, K, and L are shifted above their source, but they are transformed in negative monopolar features (therefore changing their sign with respect to the TMI field). This implies that TMI is affected by inverse remanence since the algorithm verticalizes the anomaly axis along the present Earth field direction, which is opposite to that along which the sources bodies were magnetized. The Reduced-to-the-Pole field computed considering both the induction and remanence shows that the transformation does not remove the dipolar feature either, but the anomalies polarity is exactly opposite to that of the field of Fig. 10b. Therefore, the RTP fields obtained with and without remanence are very similar, but with opposite sign. In fact, the remanence relates to a reverse field that simply inverts the polarity and does not change considerably the configuration of the magnetic pattern.

The analytic signal maxima highlight peaks of the magnetization around the AM central crater (S1, S2, S3 in Fig. 11), as well as in the eastern sector, where NW–SE oriented (S4, S5, S7) high and lower intensity features (S6 and S8) are observed. This preferential direction coincides with the strike of the Entre Lomas fault system and with the axis of the anticline over which AM emplaced.

The magnetic field spectral analysis indentifies sources at shallow depth with a maximum of 3.5–4 km below flight elevation. Besides, most of the Euler Deconvolution solutions highlight depths around 1 km and 2 km b.s.l, and few others deeper than 2 km. These results are consistent with the depth of the bodies obtained using the MVI inversion modeling (Fig. 12) clustered in the central zone and elongated mainly in NW–SE direction. This orientation indicates that the shallow intrusive structures of the volcano are controlled by the NW–SE striking Entre Lomas faults (Late Jurassic to Upper Cretaceous periods), which affects the basement below the eastern sector of the volcanic edifice. The sub-volcanic bodies represent the crystallized dikes and sills mainly intruded in the Quintuco - Vaca Muerta and Mulichinco Formations, and the Cuyo Group. Their geometry is consistent with that inferred from seismic and well data (Fig. 5c). The magma upraised through previously existing faults

and contributed to their sealing; as a consequence, the properties of the surrounding rocks were altered. The fracturing due to the differential heating and cooling of the intrusions and of the surrounding rocks increased the local porosity and permeability giving rise to different rock qualities in the hydrocarbon reservoir. Moreover, volcanism likely contributed to the thermal hydrocarbon generation (PÁNGARO et al. 2004; VOTTERO et al. 2005). In fact, the organic matter coalification was speeded by higher temperatures and in some cases oil prematurely migrated from the source rocks (e.g., Vaca Muerta) along pre-existing fractures and was trapped in the fractured intrusions. The higher magnetization of these volcanics in comparison with the surrounding non-magnetic sedimentary layers allows the detectability of potential oil traps through the magnetic prospecting. As a consequence, intrusive position, depths, geometry, and volume provided by the magnetic modeling are valuable information.

The analysis of the magnetic anomaly field has provided information on the shallow structure of the AM feeding system. The NW–SE and E–W strikes of the modeled sources agree with those of the regional fault systems proving their key role in the control of the AM volcanism. These results point out the effectiveness of the magnetic prospecting methods in presence of strong remanent magnetization.

Acknowledgments

We are grateful to YPF S.A. Company for providing the data used in this study and the permission to publish. We also thank Telma Aisengart of Geosoft for the support to build the inverse model and Hernán Scuka of YPF for the valuable help to improve the quality of the figures. We also acknowledge the two anonymous reviewers for their efforts to improve the quality of the work and the Editor William L. Bandy for the handling of the manuscript.

REFERENCES

ATCHUTA RAO, D., RAM BABU, H.V., and SANKER NARAYAN, P.V., 1981, *Interpretation of magnetic anomalies due to dikes: the complex gradient method*. Geophysics, 46, 1572–1578.

BARBOSA V., SILVA J. and MEDIROS, W., 1999, *Stability analysis improvement of structural index estimation in Euler Deconvolution*. Geophysics, 64, 48–60.

BARBOSA V., SILVA J. and MEDIROS, W., 2000, *Making Euler Deconvolution applicable to small ground magnetic surveys*. Journal of Applied Geophysics, 43, 55–68.

BLAKELY, R.J., and GRAUCH, V.J.S., 1983. *Magnetic models of crystalline terrane: accounting for the effect of topography.* Geophysics, 48, 1551–1557.

BLANCO-MONTENEGRO I., l TORTA J. M., GARCÍA A., Araña V., 2003, *Analysis and modeling of the aeromagnetic anomalies of Gran Canaria (Canary Islands).* Earth and PlanetaryScienceLetters, 206, 601–616.

BLANCO-MONTENEGRO, I., De Ritis, R., & CHIAPPINI M., 2006. *Imaging and modeling the subsurface structure of volcanic calderas with high-resolution aeromagnetic data at Vulcano (Aeolian islands, Italy)*, B. Volcanol., 69 (6), p. 643–659, doi:10.1007/s00445-006-0100-7.

BERMÚDEZ, A., DELPINO, D., 1998, Estudio de testigos corona de rocas ígneas intrusivas reservorios de hidrocarburos y de las secuencias del Volcán Auca Mahuida. Repsol YPF. Unpublished report.

BLAKELY, R. J. Potential Theory in Gravity and Magnetic Applications (Cambridge University Press, New York, 1995)

BRISSON, I., and VEIGA, V., 1999, YPF internal report.

CARSON AEREOGRAVITY, 2001,VolcánAucaMahuida and Señal Cerro Bayo Exploration Lots (March 28, 2001 - June 10, 2001). For YPF S.A., Argentina. Data Processing Report.

CRISTALLINI, E.O., BOTTESI, G., GAVARRINO, A., RODRIGUEZ, L., TOMEZZOLI, R., COMERON, R., 2006, Synrift geometry of the Neuquén Basin in the northeastern Neuquén Province, Argentina, InEvolution of the Andean margin: a tectonic and magmatic view from the Andes to the Neuquén Basin (35 –39 422 latitude): Geological Society of America. eds. Kay, S. M. and Ramos, V. A.). Special Paper 407, pp. 147–161.

DE RITIS, R., BLANCO-MONTENEGRO, I., VENTURA, G., CHIAPPINI, M., 2005, Aeromagnetic data provide new insights on the volcanism and tectonics of Vulcano Island and offshore (Southern Tyrrhenian Sea, Italy), Geophysical Research Letters, 32, L15305, doi:10.1029/2005GL023465.

FINN, C., SISSON, T. W. & DESZCZ-PAN, M., 2001. *Aerogeophysical measurements of collapse-phrone hydrothermally altered zones at Mount Rainer volcano*, Nature, 409, 600–603.

GUNN P.J., 1997, *Quantitative methods for interpreting aeromagnetic data: a subjective review*. AGSO Journal of Australian Geology & Geophysics, 17(2), 105–113.

HSU, S., 2002, *Imaging magnetic sources using Euler's equation*. Geophysical Prospecting, 50, 15–25.

KAY, S.M., and RAMOS, V.A., 2006, Evolution of an Andean Margin: A tectonic and magmatic view from the Andes to the Neuquén Basin (35°–39°S): Geological Society of America, Special Papers, v. 407, 19–60, 10.1130/2006.2407(02).

LONGO, L.M., RAVAZZOLI, C.L.,CHIAPPINI, M., 2008, Proyecto de interpretación de datos aerogravimétricos y magnéticos en el Volcán Auca Mahuida. Cuarto encuentro ciéntifico del ICES (E-ICES4), Malargüe, Mendoza, Argentina.

MACLEOD, I.N, JONES, K., FAN DAI, D., 1993, 3-D analytic signal in the interpretation of total magnetic field data at low magnetic latitudes. Explor. Geophys., 24, 679-688.

MACLEOD, I.N., ELLIS, R.G., Magnetic Vector Inversion, a simple approach to challenge of varying direction of rock magnetization.23 rd International Geophysical conference and exhibition, 11–14 August 2013. ASEG-PESA 2013Melbourne, Australia.

MOSQUERA, A. and RAMOS, V.A., 2006, Intraplate deformation in the Neuquén Basin, In: KAY S.M. and RAMOS, V.A., eds. Evolution of an Andean margin: a tectonic and magmatic view from the Andes to the Neuquén Basin (35 –39 S latitude). Geological Society of America Special Paper, 407, 97–124.

MUSHAYANDEBVU, M.,VAN DRIEL, P., REID, A. and FAIRHEAD, J., 2001, *Magnetic source parameters of two-dimensional structures using extended Euler Deconvolution*. Geophysics, 66, 814-823.

NABIGHIAN, N.M., 1972, *The analytic signal of two-dimensional magnetic bodies with polygonal cross section: its properties and used for automated anomaly interpretation.* Geophysics 37, 507-517.

NABIGHIAN, N.M., 1984, *Toward a three-dimensional automatic interpretation of potential field data* via *generalized Hilbert transforms: fundamental relations*, Geophysics 49, 780-786.

PÁNGARO, F., VILLAR, H., VOTTERO, A., BOJARSKI, G., RODRÍGUEZ ARIAS, L., 2004, Eventos volcánicos y sistemas petroleros: El caso del Volcán Auca Mahuida, Cuenca Neuquina, Argentina. IX Congreso Latino americano de Geoquímica orgánica, México, 23-27.

RAMOS, V.A. and FOLGUERA, A., 2010, Payenia volcanic province in the Southern Andes: an appraisal of an exceptional Quaternary tectonic setting: Journal of Volcanology and Geothermal Research, v. 201, 53–64, doi:10.1016/j.jvolgeores.2010.09.008.

REID, A., ALLSOP, J. GRANSER, H., MILLET, A. and Somerton, I., 1990, *Magnetic interpretation in three dimensions using Euler Deconvolution.* Geophysics, 55, 80-91.

REYNOLDS J. M., An Introduction to Applied and Environmental Geophysics (John Wiley & Sons, 2011-2nd Edition)

ROEST, W.R VERHOEFF, J. PILKINGTON, M., 1992. *Magnetic interpretation using the 3-D analytic signal.* Geophysics 57, 116–125.

ROSSELLO, E.A., COBBOLD, P.R., DIRAISON, M., and ARNAUD. N., 2002, Aucamahuida (Neuquén basin, Argentina): A Quaternary shield volcano on a hydrocarbon-producing substrate, in 6th International Symposium on Andean Geodynamics (ISAG 2002). Extended Abstracts, Barcelona, Univ. De Barcelona: Instituto Geologico y Minero de España, 549–552.

SIGISMONDI, M. and RAMOS, V., 2008,El flujo de calor de la cuenca Neuquina, Argentina. VII Congreso Exploración y Desarrollo (Simposio de La Geofísica: Integradora del conocimiento del subsuelo).

SPECTOR, A. and GRANT, F.S., 1970, *Statistical models for interpreting aeromagnetic data*, Geophysics, 35, 293–302.

THOMPSON, D., 1982, EULDPH: *A new technique for making computer-assisted depth estimates form magnetic data.* Geophysics, 47, 31–37.

VELA R., SANCHO V., FASOLA M., 2006, Integración de datos geoquímicos en el desarrollo del yacimiento volcán Auca Mahuida. III Workshop de Geoquimica de Sistemas Petroleros, Quito, Ecuador, Abril 2006.

VENTURA, G., DE RITIS, R., LONGO, M., CHIAPPINI, M., 2012, *Terrain characterization and structural control of the Auca Mahuida volcanism (Neuquén Basin, Argentina)*: International Journal of Geographical Information Science, v. 27, 1469-1480, doi:10.1080/13658816.2012.741241.

VOTTERO, A., RODRÍGUEZ, L., VELA, R., 2005, Trampas de hidrocarburos en el centro este de la Cuenca Neuquina. VI Congreso de Exploración y Desarrollo de Hidrocarburos- Las trampas de hidrocarburos en las cuencas productivas de Argentina, 189–208.

(Received May 4, 2015, revised July 31, 2015, accepted August 1, 2015, Published online August 14, 2015)

Pure Appl. Geophys. 173 (2016), 3291–3303
© 2016 Springer International Publishing
DOI 10.1007/s00024-015-1230-7

Flare-Shaped Acoustic Anomalies in the Water Column Along the Ecuadorian Margin: Relationship with Active Tectonics and Gas Hydrates

Francois Michaud,[1,2] Jean-Noël Proust,[3] Alexandre Dano,[2] Jean-Yves Collot,[2,4] Grâce Daniella Guiyeligou,[2] María José Hernández Salazar,[5] Gueorgui Ratzov,[2] Carlos Martillo,[2,3,6,7] Hugo Pouderoux,[3] Laure Schenini,[2] Jean-Frederic Lebrun,[8] and Glenda Loayza[6,7]

Abstract—With hull-mounted multibeam echosounder data, we report for the first time along the active Ecuadorian margin, acoustic signatures of water column fluid emissions and seep-related structures on the seafloor. In total 17 flare-shaped acoustic anomalies were detected from the upper slope (1250 m) to the shelf break (140 m). Nearly half of the flare-shaped acoustic anomalies rise 200–500 m above the seafloor. The base of the flares is generally associated with high-reflectivity backscatter patches contrasting with the neighboring seafloor. We interpret these flares as caused by fluid escape in the water column, most likely gases. High-resolution seismic profiles show that most flares occur close to the surface expression of active faults, deformed areas, slope instabilities or diapiric structures. In two areas tectonic deformation disrupts a Bottom Simulating Reflector (BSR), suggesting that buried frozen gas hydrates are destabilized, thus supplying free gas emissions and related flares. This discovery is important as it opens the way to determine the nature and origin of the emitted fluids and their potential link with the hydrocarbon system of the forearc basins along the Ecuadorian margin.

Key words: Ecuador, Subduction, Continental margin, Multibeam, Backscatter, Fluid, Seepage, Acoustics.

1. Introduction

In the past decade, the increase in multibeam seabed and water column mapping shows that seepage activity at continental margins is a relatively widespread phenomena (Loncke *et al.* 2004; Dupré *et al.* 2010; 2014; Mascle *et al.* 2014). In seafloor seepage activity areas, vents are detected by images of plumes in the water column (Merewether *et al.* 1985; Hornafius *et al.* 1999; German *et al.* 1996; Greinert *et al.* 2006; Dupré *et al.* 2010). The basic principle for detecting plumes with echo sounders is the high backscattering of the pressure wave at the impedance contrasts between water and gas bubbles (Schneider Von Deimling *et al.* 2007). This creates flare-shaped backscatter features in echograms. Most acoustic anomalies recorded in the water column with multibeam echo sounders are caused by fluid escape at the seabed, most likely gases (Judd and Hovland 2007).

Along continental margins, submarine cold seeps are sites where fluids, either pore water or free gas, migrate upward from buried sedimentary horizons. They are also often inferred to be associated with gas hydrates in the underlying sediment, which can act as sources for methane, the dominant gas component at most seeps (Krabbenhöft *et al.* 2010; Römer *et al.* 2014). Hydrates are evidenced and mapped based on the distribution of Bottom Simulating Reflector (BSR), a characteristic seismic reflection caused by

[1] Univ. Pierre et Marie Curie, UPMC, CNRS, IRD, Observatoire de la Côte d'Azur, Géoazur UMR 7329, 250 rue Albert Einstein, 06560 Sophia Antipolis, Valbonne, France. E-mail: micho@geoazur.unice.fr

[2] Univ. Nice Sophia Antipolis, CNRS, IRD, Observatoire de la Côte d'Azur, Géoazur UMR 7329, 250 rue Albert Einstein, 06560 Sophia Antipolis, Valbonne, France.

[3] Géosciences Rennes, CNRS, Université de Rennes 1, Campus de Beaulieu, 35042 Rennes Cedex, France.

[4] Investigador Prometeo, Instituto Geofísico, Escuela Politécnica Nacional, Ladrón de Guevara E11-253, Aptdo.2759 Quito, Ecuador.

[5] Departamento de Geología, Escuela Politécnica Nacional, Ladrón de Guevara E11-53, Quito, Ecuador.

[6] Facultad de Ingenieria en Ciencias de la Tierra, Escuela Politécnica del Litoral, km 30.5 Vía Perimetral Campus "Gustavo Galindo", Guayaquil, Ecuador.

[7] Instituto Oceanográfica de la Armada de Ecuador (INOCAR), Avenida 25 de Julio vía Puerto Marítimo, Base Naval Sur, Guayaquil, Ecuador.

[8] Univ. des Antilles et de la Guyane, Campus de Fouillole, 97159 Pointe a' Pitre Cedex, France.

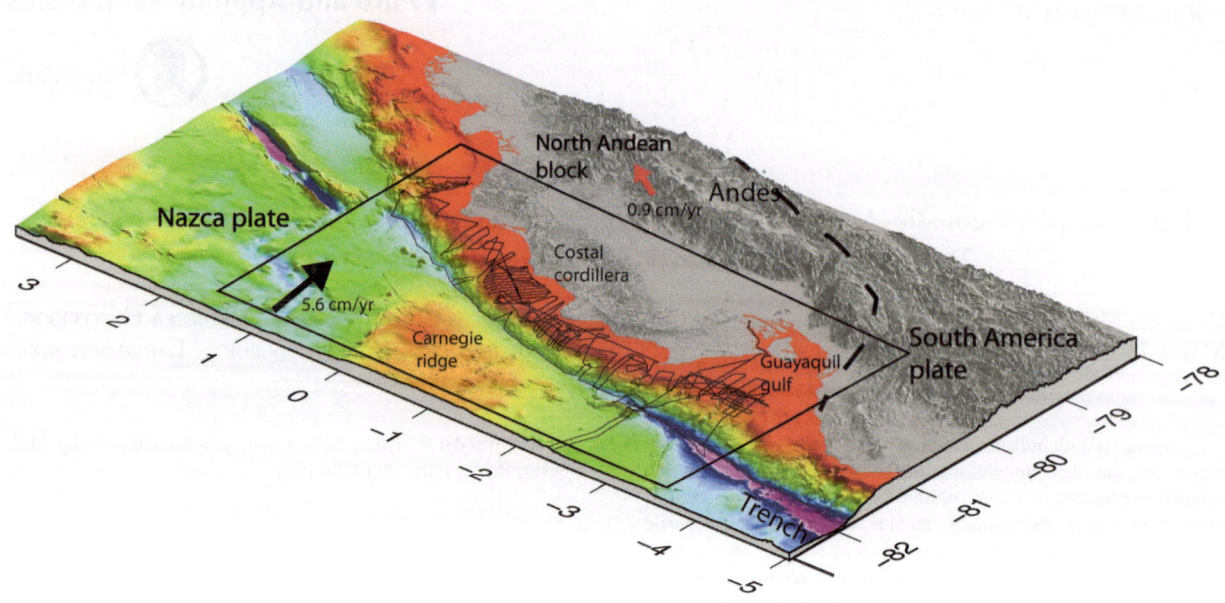

Figure 1
Geodynamic framework of the study area. The Nazca plate subducts beneath the South American continent at velocity of 5.6 cm/year (NOCQUET *et al.* 2009) and the North Andean block move to the North at a velocity of 0.9 cm/year (NOCQUET *et al.* 2014). The forearc area is characterized by the arrival in the trench of the Carnegie Ridge and the escape to the North of the North Andean block. Location of the Fig. 2 (*box*). *Thick black dashed line* = eastern boundary of the North Andean block from BOURGOIS (2013). *Thin black lines* = ATACAMES shiptracks

the strong impedance contrast between hydrate-containing sediment above and gas-filled pore space below.

Seafloor fluid flow occurs in a wide range of oceanographic environments and geological contexts (JUDD and HOVLAND 2007). Seepage may be associated with undercompaction sediments (JUDD and HOVLAND 2007), slope instabilities (PAULL *et al.* 1998; PRAEG *et al.* 2014) and seafloor deformation (DEYNOUX *et al.* 1990; DANO *et al.* 2014). Some seeps are associated with active faulting (GAY *et al.* 2007; GELI *et al.* 2008; DUPRÉ *et al.* 2015) with release of fluids along the faults. Such tectonically driven mechanism could feed hydrate-related vents and associated seepage (SUESS *et al.* 1999). Consequently, in many cases the detection of flare in the water column indicates active tectonic structure, or slope instabilities.

This paper provides a first, non-exhaustive, account of water-column acoustic flares along the active Ecuadorian margin and a discussion on the potential first-order relationships with tectonics and sedimentary features. This fluid system was

discovered by multibeam bathymetric data acquired in 2012 during the ATACAMES campaign (MICHAUD *et al.* 2013) conducted on the R/V *L'Atalante* (IFREMER).

2. Geological Setting

Along the Ecuadorian margin, the Nazca plate (Fig. 1) subducts eastwards at a 5.6 cm year^{-1} rate beneath the South America plate (NOCQUET *et al.* 2014). The evolution of the Ecuador margin is strongly influenced by the subduction of the Carnegie Ridge (MICHAUD *et al.* 2009; COLLOT *et al.* 2009) and the northward tectonic escape of the North Andean block at a 0.95 cm year^{-1} rate (NOCQUET *et al.* 2014) (Fig. 1). In the forearc, sediment supply from the Andes is deflected northwards and southwards by the uplift of the coastal cordillera (COLLOT *et al.* 2009). The uplift of the coastline and coastal range appears to have related to the subduction of the Carnegie Ridge (PEDOJA *et al.* 2006; GUTSCHER *et al.* 1999; PROUST *et al.* 2016) for, at least, the last 1.3 Ma

(GRAINDORGE *et al.* 2004) or 4–5 Ma (COLLOT *et al.* 2009). The subduction of the Carnegie Ride is synchronous with the acceleration of the North Andean block escape (WITT *et al.* 2006) and an increase of the subsidence rate in the Gulf of Guayaquil (DENIAUD *et al.* 1999) since the Early Pleistocene.

The usually smooth and linear continental margin slope becomes irregular south of 1°35′S (COLLOT *et al.* 2009), where the margin faces the incoming Carnegie Ridge. Several 1–2 km high, 10–15 km large subducting seamounts indent locally the lower margin slope (Fig. 2), (SAGE *et al.* 2006; MARCAILLOU *et al.* 2016). As shown along other convergent margins, morphologic embayment-related subducted-seamounts result from slope material removal by tunneling of underthusting seamounts (DOMINGUEZ *et al.* 1998; RANERO and VON HUENE 2000; MARCAILLOU *et al.* 2016).

Several main regional faults deform the margin seafloor. In the North of Ecuador, the trench sub-parallel Ancon fault that is locally associated with a crustal splay fault deforms sediments along the trench slope break (COLLOT *et al.* 2004, 2005, 2008); the onshore Galera fault system (EGUEZ *et al.* 2003) may extend seaward oblique to the shelf break and upper margin slope (MICHAUD *et al.* 2015), and, in Central Ecuador, the oblique to the margin Jama fault system extends offshore along a small upper slope sedimentary basin and is interpreted as a negative flower structure (Fig. 2), (COLLOT *et al.* 2004; MICHAUD *et al.* 2015).

Bottom Simulating Reflectors (BSR) are widely extended along the southern Colombian margin (MARCAILLOU *et al.* 2006). Along the Ecuadorian margin, multichannel seismic lines shot in the Ancon fault area (MARCAILLOU *et al.* 2006) (Fig. 2) show a BSR; and southward, until 0°00′N of latitude, BSR is also reported along the slope (MARCAILLOU *et al.* 2016).

3. Methods

Multibeam data were acquired using a hull-mounted Simrad EM710 and EM122 multibeam echosounder, yielding a Digital Terrain Models with a resolution of 25 m for bathymetry and 10 m for backscatter imagery. The data were processed using the seafloor mapping software CARAIBES (developed by IFREMER). Water column data were acquired along selected swaths, processed and visualized using Sonarscope software (developed by IFREMER). Sub-bottom profiles were acquired with a hull-mounted Chirp system (1.8–5.3 kHz). Seismic reflection data were recorded using a 72-channel digital streamer towed at 2 m of water depth (channel length 6.25 m). The source array towed at 2.1 m of water depth consisted of two ramps mounted with three 13/13 cubic-inch plus three 24/24 cubic-inch mini GIgun. Shots were fired at 140 bars every 25 m. Given the shot rate and the streamer configuration, this seismic reflection system ensures a ninefold stack. The seismic lines were processed onboard with the Seismic Unix (SU) software (Center of Wave Phenomena, Colorado School of Mines) for Band Pass Filtering, spherical divergence correction (water velocity)—NMO velocity analysis and correction, ninefold stack and constant velocity time migration (1490 m/s).

4. Results

We detected 17 plume sites along the Ecuadorian margin (Table 1, black dots in Fig. 2). The seeps concentrate in four main areas (Fig. 2). Two sites are located near the shelf break (site 3 and 15) whilst all others are seated on the upper slope, ranging from 140 to 1250 m in water depth.

To the north (Esmeraldas canyon area, Fig. 3) three plume sites were discovered on the western side of the Esmeraldas canyon. Sites 1 and 2 are very close together (3 km apart) and located around 1100–1200 m of water depth. The most spectacular is the 500 m high plume at site 1 (Fig. 3b). The seafloor backscatter image shows that plume sites 1 and 2 correspond to ~100 m large high-reflectivity sub-circular patches (Fig. 3c). Site 3 is located near the shelf break, just on the edge of a 5-km wide semi-circular slope re-entrant with irregular rim; this re-entrant does not breach the shelf break. The NS seismic profile shot across site 1 (Fig. 3d) shows deformed strata and lateral seismic facies changes. Plume site 1 locates above a transparent seismic

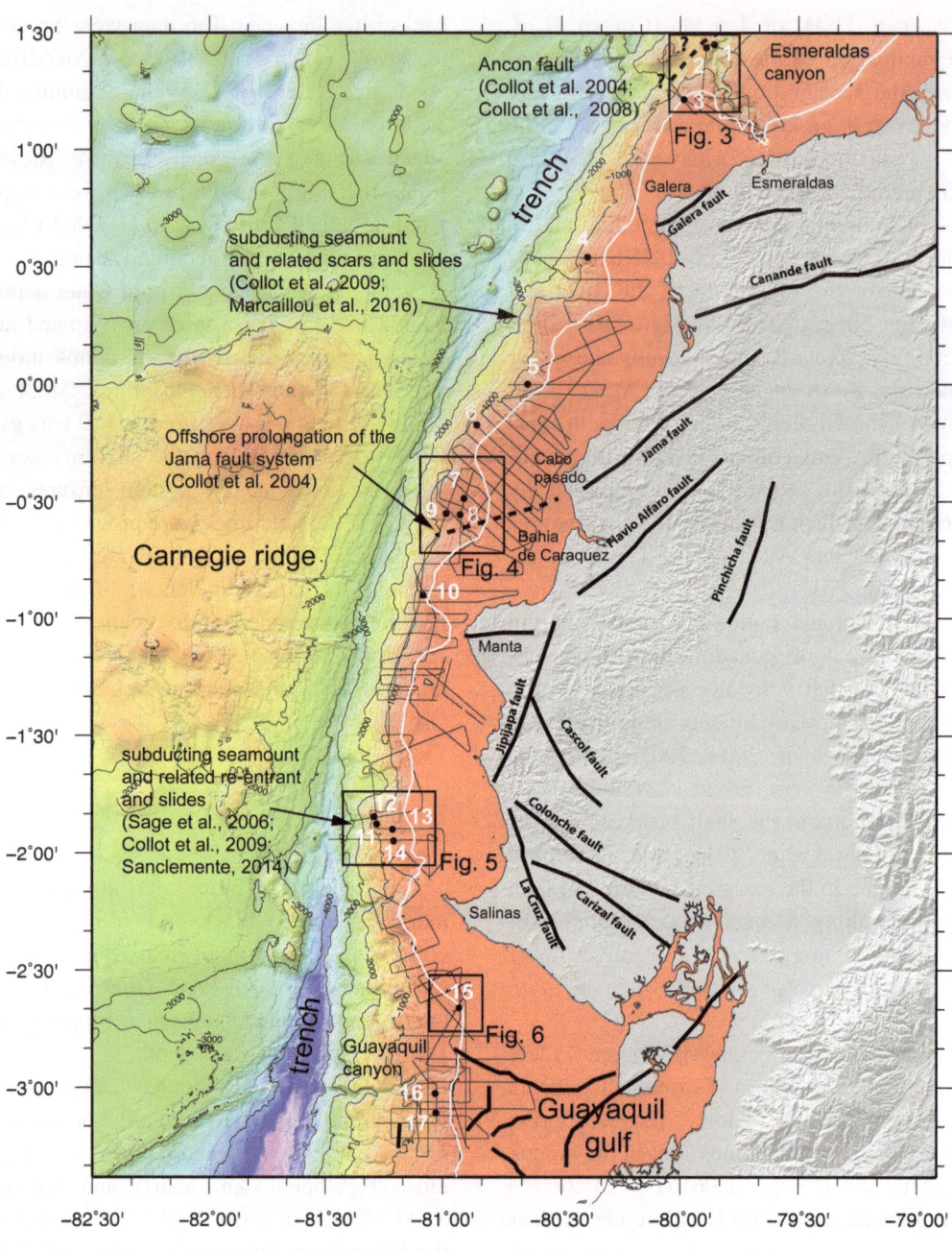

Figure 2
Bathymetric map (*black circles with number* = location of plume sites detected in this work). *Thick black lines* = principal faults on land from REYES and MICHAUD (2012); Offshore faults in the Gulf of Guayaquil are from WITT *et al.* (2006). *Thick dotted black line* = prolongation offshore of the Jama fault system inferred by COLLOT *et al.* (2004). *Fine white line* = 150 m isobath. *Fine black lines* = location of seismic profiles recorded during the ATACAMES cruise (MICHAUD *et al.* 2013).

facies, which underlines a diapir structure more and less rooted in a BSR. Southward of the diapir, the most recent sediments show important change of seismic facies from chaotic (b on Fig. 3d), to sub

parallel reflector facies dipping to the south (a on Fig. D) and to the north (c on Fig. D). This sediment fill is bounded to the South by a faulted monocline (Fig. 3d).

Table 1

Location and main characteristics of acoustic anomalies

Acoustic anomaly	Latitude (°)	Longitude (°)	Depth in meter	Plumes high in meter	Areas
1	1.4367	−79.8915	1225	500	Esmeraldas canyon area
2	1.4501	−79.8589	1100	300	
3	1.2153	−79.986	140	50	
4	0.5425	−80.3932	630	>100	Isolated anomaly on the upper slope
5	−0.1733	−80.8577	600	>100	
6	−0.001	−80.6443	675	150	
7	−0.5567	−80.928	475	100	Offshore prolongation of the Jama fault system area
8	−0.4885	−80.9118	575	>100	
9	−0.5522	−80.988	612	220	
10	−0.9006	−81.0826	600	>100	Isolated anomaly
11	−1.8406	−81.1973	790	300	Subducting seamount area
12	−1.8449	−81.2923	700	250	
13	−1.8769	−81.2817	660	300	
14	−1.8994	−81.2131	650	300	
15	−2.6609	−80.931	160	100	Gulf of Guayaquil area
16	−3.0264	−81.0317	550	150	
17	−3.1092	−81.0287	560	350	

Plume sites 4, 5, and 6 (Fig. 2) are isolated along ship tracks. They are located on the upper slope between 600 and 675 m of water depth (Table 1). They are more and less 100 m high. On the seismic profiles crossing these plume areas, sub-vertical acoustic anomalies are present and could be indicative of fluids flows from bottom levels. But the seismic profiles do not show any related specific tectonic or slope failure structures.

The offshore prolongation of the Jama fault system area (Fig. 4) shows three plume sites 7, 8, and 9, located at a water depth of about 500 m (Table 1; Figs. 2 and 4a). Plume 9 is about 200 m high (Fig. 4b) when plumes 7 and 8 are 100 m high (Fig. 4b). Plume sites 7 and 8 are located in a deformed sedimentary basin (unit c on Fig. 4c). Plume site 7 is detected in an area of the basin affected by sub vertical faults. Plume 8 is associated with a diapir expressed on the seafloor by a sub-circular, 15 m high and 500 m wide feature (Fig. 4d). Plume site 9 is located at the southern edge of this basin (Fig. 4e, d). At the site 9, unconformities are observed on the seismic profile (Fig. 4e); a chaotic unit (a on Fig. 4e) is topped by a unit with NE-inclined parallel reflectors (b on Fig. 4e). At this place the slope is affected by a semi-circular 2 km wide re-entrant oriented in a seaward direction that breaches a 100 m high linear scarp (Fig. 4d).

Plume site 10 (Table 1; Fig. 2) is isolated on the upper slope. The plume is less than 150 m high and is poorly developed in the water column.

Plume sites 11, 12, 13, and 14 (Table 1; Fig. 2) are grouped in an area that is approximately 20 km long ranging between 800 and 600 m of water depth. Each site shows a ∼300 m high flare. Plume sites 11 and 12 are located on the NS-trending crest of a 100-m bathymetric high (Fig. 5) on the slope between 600 and 1000 m of water depth. The seismic profile shows that this high is highly deformed (Fig. 5d) flanked by a continuous BSR partially disrupted at the top of the ridge. The seafloor backscatter image shows that plume sites 11 and 12 correspond to ∼100 m sub-circular high-reflectivity patches (Fig. 5c). Plume sites 13 and 14 are located higher on the slope bathed in 650 m of water depth but the seismic profiles do not show clear related structure.

Plume site 15 is located at the shelf break of the gulf of Guayaquil (Table 1; Figs. 2 and 6a). This site is remarkable because it exhibits several plumes (Fig. 6b). This site is in the center of a rough seafloor morphology (Fig. 6a and e) that defines an irregular 2 km large and 20 m high dome above the surrounding seafloor. The seafloor backscatter image (Fig. 6c) shows that this dome is associated with a set of high-reflectivity backscatter patches; the plume is at the center of an ellipsoid-shaped low-reflectivity patch. In

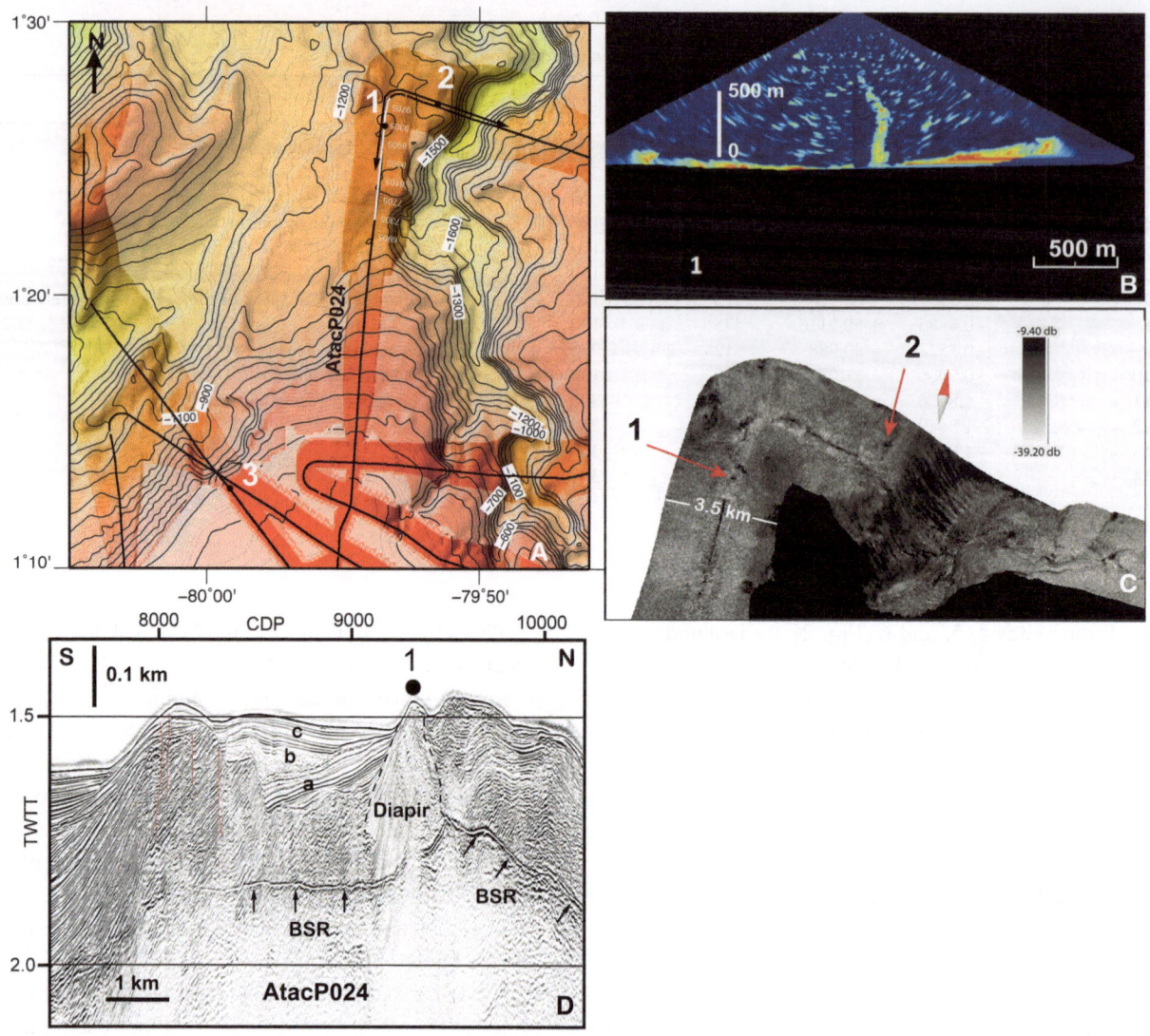

Figure 3

a Bathymetric map with curves every 20 m (location on Fig. 2) of the Canyon Esmeraldas area (grid cell 50 m). More intense color corresponds to multibeam data recorded during the campaign ATACAMES (Michaud *et al.* 2013). *Less intense color* = older multibeam data from Michaud *et al.* (2006); *less intense color* = area without multibeam; *thin black lines* = ship track during the ATACAMES cruise; *black circles with number* = location of plume sites; **b** *Flare-shaped backscatter* feature in echogram at site 1. **c** *Seabed backscatter* along the Atacames track (location on **a** = *fine black line* ending with arrows) **d** Seismic profile (location on **a** = *white line*)

detail (Fig. 6e), the dome exhibits NW–SE elongated 20 m high bathymetric ridges. On the seismic profile (Fig. 6d) the dome appears located above a large diapir outcropping at the seafloor. To the Northeast, below the platform, the NE-SW trending seismic profile exhibits well-stratified sedimentary unit (a on Fig. 6d) that overlay a deeper unit (b on Fig. 6d), which is highly deformed. These two units are affected by sub-

vertical faults. Next to the fault located furthest to the East (CDP 12500, Fig. 6d), the folded geometry of the reflectors of the upper unit indicates some contraction; meanwhile the vertical offset of the top of deeper unit indicates a normal component. The two plume sites 16 and 17 are located southward on the upper slope (site 16 and 17, Table 1; Fig. 2) of the Gulf of Guayaquil. These plumes are located south of the Guayaquil

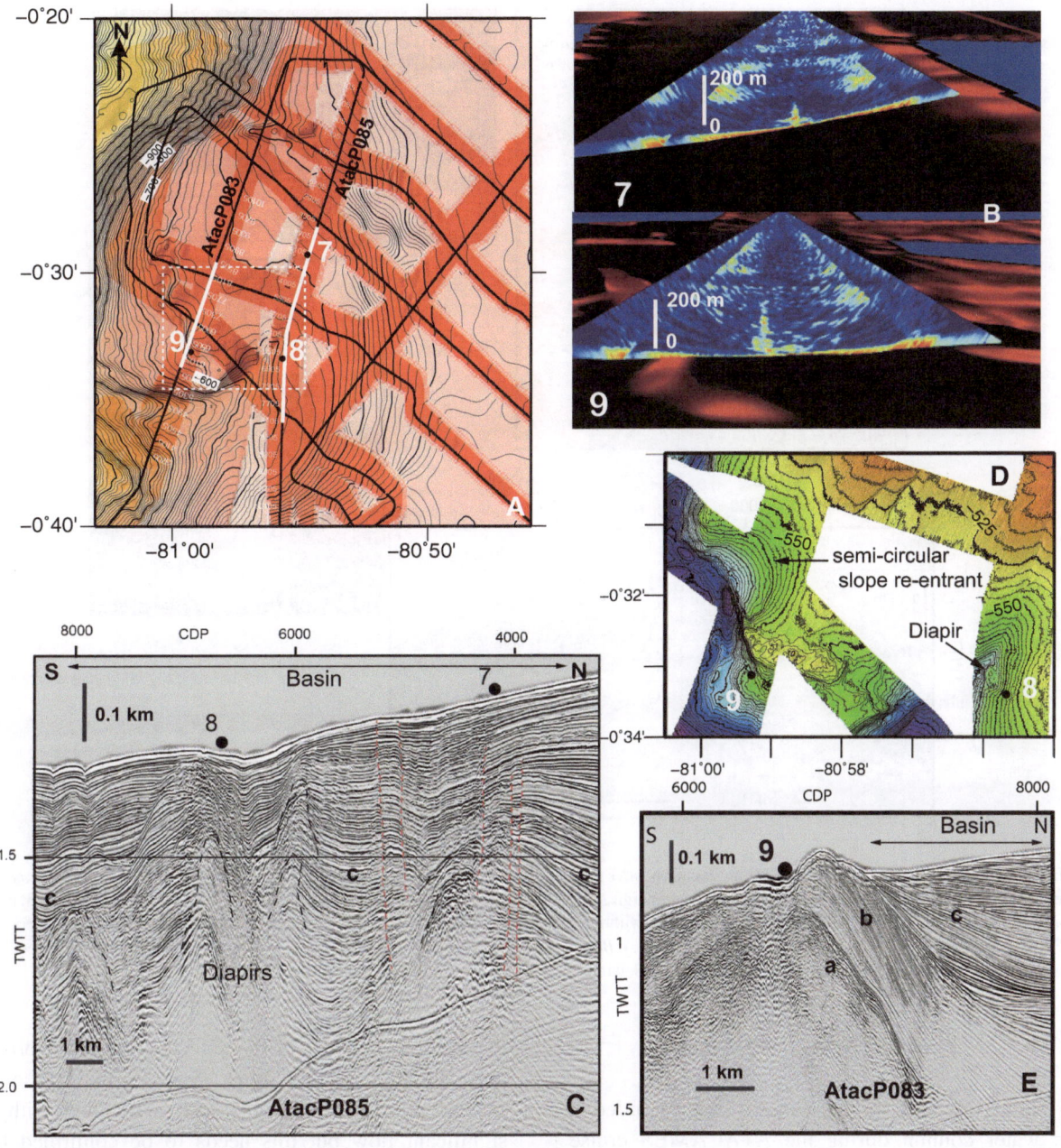

Figure 4

a Bathymetric map with *curves* every 20 m (location on Fig. 2) of the Jama area (grid cell 50 m). *More intense color* corresponds to multibeam data recorded during the campaign ATACAMES (MICHAUD *et al.* 2013). *Less intense color* = older multibeam data from MICHAUD *et al.* (2006); *less intense color* = area without multibeam. *Thin black lines* = ship track during the ATACAMES cruise. *Thin dotted white line* = location of bathymetric zoom of **d**. **b** *Flare-shaped backscatter* feature in echogram at site 7 and site 9. **c** Seismic profile in the basin (location **a** = *white line*). **d** Bathymetric zoom (*curves* every 5 m) of the site 8 and 9 area (grid cell 10 m). **e** Seismic profile crossing the basin southern boundary (location on **a** = *white line*)

canyon (Fig. 2), which is characterized by an irregular morphology and many landslides (LOAYZA *et al.* 2014). Plume 17 has a height of about 350 m. From the seabed, it remains in a vertical position for about 250 m, and then the top of the plume is slightly deflected.

85

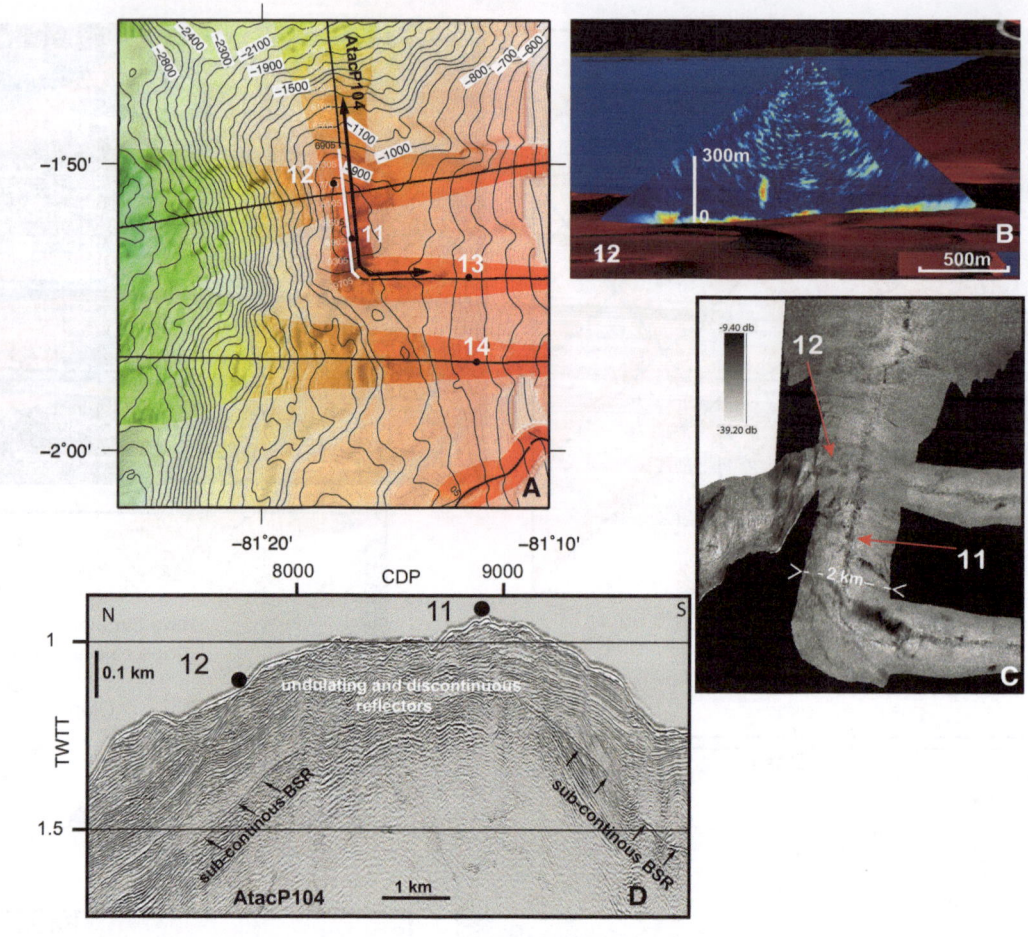

Figure 5

a Bathymetric map with *curves* every 20 m (location on Fig. 2) of the subducting seamount area (grid cell 50 m). *More intense color* corresponds to multibeam data recorded during the campaign ATACAMES (MICHAUD *et al.* 2013). *Less intense color* = older multibeam data from MICHAUD *et al.* (2006); *less intense color* = area without multibeam. *Thin black lines* = ship track during the ATACAMES cruise. **b** *Flare-shaped backscatter* feature in echogram at site 12. **c** *Backscatter* along the ATACAMES track (location on **a** = *fine black line* ending with *arrows*). **d** Seismic profile (location on **a** = *white line*)

5. *Potential Geological Controls*

Numerous and various echoes were recorded in the water column during the ATACAMES cruise. They show a high variability in size but they usually form subvertical elongated flares rooted at the seafloor. This geometry of the echoes is typical of gas emissions (OBZHIROV *et al.* 2004; DUPRÉ *et al.* 2014). We interpret the 17 detected acoustic anomalies as the presence of gas bubbles in the water column. Few of them are slightly inclined probably depending on the strength of the mid-water depth current effects. The plume site 1, which is more than 500 m high, is

the most spectacular of the detected sites with probably the highest flow rates. Site 15 that presents several plumes is probably also associated with a significant flow but this needs to be confirmed by in situ measurements,

5.1. *Seepage Distribution and Active Tectonics*

5.1.1 *Faults*

Some of the plume sites are clearly related to regional fault systems. Plume site 1 is located just above a highly deformed area (Fig. 3d). The Ancon fault,

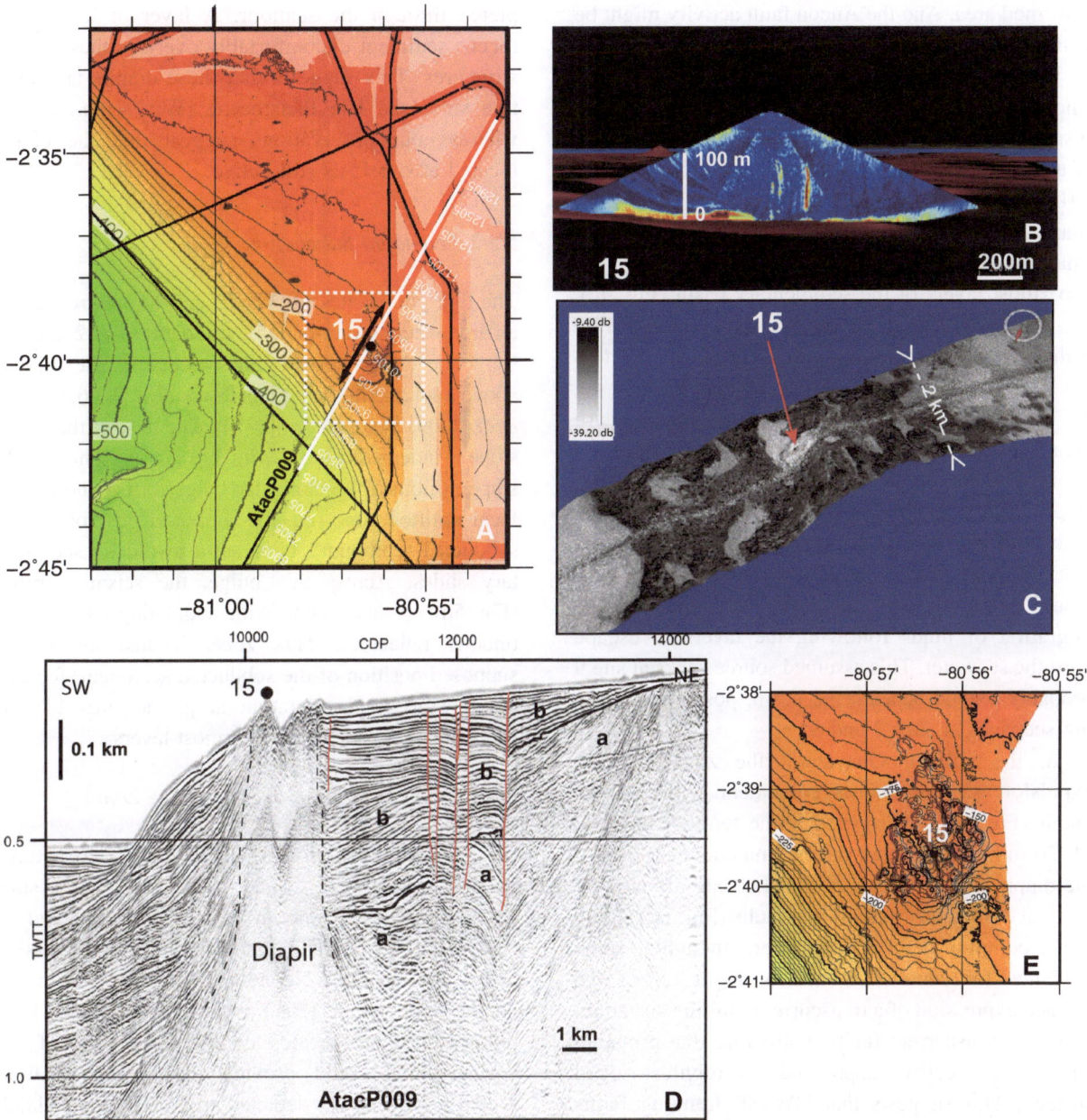

Figure 6

a Bathymetric map with *curves* every 20 m (location on Fig. 2) of the Guayaquil gulf area (grid cell 50 m). *More intense color* corresponds to multibeam data recorded during the campaign ATACAMES (MICHAUD *et al.* 2013). *Less intense color* = older multibeam data from MICHAUD *et al.* (2006) and from PAZMIÑO *et al.* 2010); *less intense color* = area without multibeam. *Thin red lines* = ship track during the ATACAMES cruise. **b** *Flare-shaped backscatter* feature in echogram at site 15. **c** *Backscatter* along the ATACAMES track (location **a** = *fine black line* ending with *arrows*). **d** Seismic profile (location on **a** = *white line*). *Red lines* = faults. **e** Bathymetric zoom (*curves* every 5 m, grid cell 10 m) of the diapir area (location on **a** = *white dotted line square*)

which is a major regional tectonic structure, is described a few kilometers to the north (COLLOT *et al.* 2004, 2008) but, so far, its southern tip has

remained unclear. We cannot clearly state concerning the origin and causes of the seepage at site 1. Nevertheless, this site is closely associated with a deformed area. And the Ancon fault activity might be at the origin of these observed deformations.

Plume sites 7, 8, and 9 are detected in the area where the offshore prolongation of the Jama fault system is suggested (COLLOT et al. 2004) at the edge of an upper slope sedimentary basin (MICHAUD et al. 2015). In contrast to the site 1 area, this area is not characterized by a significant deformation but by small faults rooted in diapirs cutting through the uppermost layers of the sedimentary basin (Fig. 4c). We infer that these small faults act as fluid migration pathways, which seeps on the seafloor at site 7. Site 8 lies on a diapir exposure on the sea bottom (Fig. 4c, d). Site 9 is located at the foot of linear scarp breached by a spoon-shaped reentrant (Fig. 4d, e) interpreted here as the scar of a slope failure. These slope instabilities might have exposed the source rocks and allowed the fluids to escape. Furthermore, this assumed source rocks are tilted and this geometry would perhaps have been able to facilitate the migration of fluids following the layers to escape until the seawater. This assumed source rocks at site 9 probably extend beneath the basin, possibly supplying seepages at sites 7 and 8.

In the gulf of Guayaquil, the zone of fluid expulsion 15 (Fig. 6a) is coincident with a large diapir (Fig. 6d), outcropping at the seafloor in NW–SE 20 m high elongated ridges. Immediately north of the diapir, the sediments of the shelf are affected by vertical dominantly extensional faults (Fig. 6d) showing a complex deformation pattern including some evidence for contraction, which might reflect the surface expression of a transcurrent faulting structure. It is this transcurrent faulting structure that probably shapes the nearby diapir and its topmost-ridged surface. This suggests that NW–SE trending faults parallel to the shelf break control the diapir and the associated plume 15. The root of the diapir is not identified in the seismic line (Fig. 6d), indicating that the under-compacted material below, from which it originates, is located deeper than 1 s TWTT below seafloor. Southward, several mud diapirs were reported along the N-S trending shelf break of the Guayaquil gulf (WITT et al. 2006). But on the contrary to the diapir at the site 15, none of them

pierce through the seafloor. A layer of late Pleistocene sediment uncomfortably overlays these diapir structures. This suggests that these diapirs are inactive since the late Pleistocene (WITT et al. 2006), whereas our observations suggest that the diapir located at site 15 is still active.

5.1.2 Seamount Subduction

The four plume sites 11, 12, 13, and 14 are grouped in an area where swath-bathymetric data (Figs. 2 and 5a) show the deformation of the surface of the margin above a subducting seamount (SAGE et al. 2006; COLLOT et al. 2009; SANCLEMENTE 2014). The base of the lower slope is indented (Fig. 2) with a re-entrant. The middle slope exhibits a N-S elongated bulge corresponding to a local uplifted zone (i.e. the supposed position of the subducted seamount), bounded seaward by sedimentary slides. Across this bulge, the seismic profile (Fig. 5d) reveals a zone with undulating and discontinuous reflectors. This zone, located above the supposed position of the subducted seamount, focuses the upward fluid migration at plume sites 11 and probably 12 through the uppermost layers.

5.1.3 Isolated Plumes and No-Plume Zones

No obvious tectonic or slope failure structures underline sites 4, 5, 6, and 10 (Fig. 2). These sites are located on a smooth regular slope bathed at 600 m of water depth. At the latitudes of sites 4, 5, and 6, a large thick Neogene sedimentary basin sits on the upper slope (HERNANDEZ et al. 2014). These plumes sites are located on the outer edge of the basin, which could provide the source of fluid seepage. No site was detected around La Plata Island, whereas it is considered as one of the more actively uplifting area of the continental shelf (PEDOJA et al. 2006; PROUST et al. 2016). Due to the acoustic sampling strategy during ATACAMES cruise, we cannot exclude that the seep area are not fully constrained. However, the reduced sediment thickness around La Plata Island (PROUST et al. 2016), might not provide enough fluids to supply seepage at the sea floor.

5.2. Seepage Distribution and BSR

Plume sites 1, 11, and 12 are located above local BSR segments (Figs. 3d and 5d). These BSR-related plume sites are the most spectacular of plume sites recognized along the margin. The first one exhibits the highest plume (500 m high, plume site 1), which may correspond to the highest fluid flow rate. The second one is the area affected by a subducting seamount, which shows the densest occurrence of seeps (four plume sites 11, 12, 13, and 14). Accumulation of gas hydrates corresponding to the BSR should inhibit any migration and associated seabed escapes (JUDD and HOVLAND 2007). Nevertheless, in these two areas, abundant evidences of seabed seepage correlate with deformed and discontinuous reflectors providing pathways for active seeps. We suggest that, active deformation probably disrupts the BSR, which then supplies the plumes. At site 1, close to the Ancon fault, the BSR is partly interrupted as indicated by the seismic profile (Fig. 3d). At site 11 above the subducting seamount (Fig. 5a), the BSR is continuous on both sides of the ridge, while directly below the ridge, it is highly disrupted (Fig. 5d).

6. Conclusions

This paper documents for the first time 17 acoustic flares, interpreted as originating from gas expulsions, on the seafloor of the Ecuadorian margin. Water column acoustic backscatter data acquired during the ATACAMES campaign in 2012 image these flares. From this first, non-exhaustive account of water-column acoustic flares, acoustic anomalies are mainly observed on the upper slope or at the shelf break. Whatever the fluid origin, the spatial distribution of the seeps suggests potential first-order relationship with tectonics and sedimentary features. At regional scale, most seeps follow the deformation structures seen in high-resolution seismic surveys. Further investigation involving full multibeam coverage and dedicated seismic data would help to refine the fluid migration pathways in the sedimentary column and their link to the regional hydrocarbon system.

Acknowledgments

We thank the crew of R/V L'Atalante and GE-NAVIR. This work was supported by Institut National des Sciences de l'Univers du Centre National de la Recherche Scientifique (INSU). Thanks to Institut de la Recherche et du Développement (IRD), the Laboratoire Mixte International "Séismes et Volcans"(LMI) and to the Instituto Oceanografico de la Armada (INOCAR). This work was partially sponsored by the Ministry of Higher Education, Science, Technology and Innovation of the Republic of Ecuador (SENESCYT). We thank the anonymous reviewers for their constructive comments, which helped us to improve the manuscript.

REFERENCES

BOURGOIS J (2013) *A review on tectonic record of strain buildup and stress release across the andean forearc along the gulf of guayaquil-tumbes basin (GGTB) near ecuador-peru border.* Intern Journ Geosc 4:618–635. doi:10.4236/ijg.2013.43057.

COLLOT J-Y, MARCAILLOU B, SAGE F, MICHAUD F, AGUDELO W, CHARVIS P, GRAINDORGE D, GUTSCHER MA, SPENCE G (2004) *Are rupture zone limits of great subduction earthquakes controlled by upper plate structures? Evidence form MCS data acquired accross the N-Ecuador-SW Colombia margin.* J Geophys Res 109:B11103. doi:10.1029/2004JB003060.

COLLOT J-Y, MIGEON, S, SPENCE, G, LEGONIDEC, Y, MARCAILLOU, B, SCHNEIDER, J-L, MICHAUD, F, ALVARADO, A, LEBRUN, J-F, SOSSON, M, PAZMIÑO, A, 2005. *Seafloor margin map helps in understanding subduction earthquakes.* EOS Transactions, American Geophysical Union, 86(46): 464–466.

COLLOT J-Y, AGUDELO W, RIBODETTI A, MARCAILLOU B (2008) *Origin of a crustal splay fault and its relation to the seismogenic zone and underplating at the erosional north Ecuador–south Colombia oceanic margin.* J Geophys Res 113:B12102. doi:10.1029/2008JB005691.

COLLOT J-Y, MICHAUD F, ALVARADO A, MARCAILLOU B, SOSSON M, RATZOV G, MIGEON S, CALAHORRANO A, PAZMIÑO A (2009) Visión general de la morfología submarina del margen convergente de Ecuador- Sur de Colombia: implicaciones sobre la transferencia de masa y la edad de la subducción de la Cordillera de Carnegie. In: COLLOT JY, SALLARES V, PAZMIÑO A (ed) Geologia y Geofisica Marina y Terestre del Ecuador. Publicacion CNDM-INOCAR-IRD, PSE001-09, Guayaquil, Ecuador, pp. 47–74.

DANO A, MIGEON S, PRAEG D, CERAMICOLA S, AUGUSTIN JM, KETZER JM, AUGUSTIN AH, DUCASSOU E, MASCLE J, (2014) Fluid Seepage in Relation to Seabed Deformation on the Central Nile Deep-Sea Fan, Part 1: Evidence from Sidescan Sonar Data S. KRASTEL et al. (eds.), *Submarine Mass Movements and Their Consequences*, Advances 129 in Natural and Technological Hazards Research 37, DOI 10.1007/978-3-319-00972-8 12:129–139 .

DENIAUD Y, BABY P, BASILE C, ORDOÑEZ M, MONTENEGRO G, MAS-CLE G (1999) *Ouverture et évolution tectono- sédimentaire du Golfe de Guayaquil: bassin d'avant- arc néogène et quaternaire du Sud des Andes équatoriennes.* C. R. Acad. Sci. Paris, *328*: 181–187.

DEYNOUX M, PROUST JN, DURAND J, MERINO E (1990) *Water transfer cylindrical structures in the Late Proterozoïc eolian sandstones in the Taoudeni Basin, West Africa.* Sedim. Geol., 66:227-242.

DOMINGUEZ S, LALLEMAND S, MALAVIEILLE J, VON HUENE R (1998) *Upper plate deformations associated with seamount subduction,* Tectonophysics, *293*: 207–224.

DUPRÉ S, WOODSIDE J, KLAUCKE I, MASCLE J, FOUCHER J-P (2010) *Widespread active seepage activity on the Nile Deep Sea Fan (offshore Egypt) revealed by high-definition geophysical imagery.* Marine Geology, *275*(1–4), 1–19.

DUPRÉ S, bERGER L, Le BOUFFANT N, SCALABRIN C, BOURILLET J-F (2014) *Fluid emissions at the Aquitaine Shelf (Bay of Biscay, France): A biogenic origin or the expression of hydrocarbon leakage?* Continental Shelf Research, *88*, 24–33.

DUPRÉ S, SCALABRIN C, GRALL C, AUGUSTIN J-M, HENRY P, SEN-GÖR AMC, GÖRÜR N, ÇAGATAY MN, GÉLI L (2015) *Tectonic and sedimentary controls on widespread gas emissions in the Sea of Marmara: Results from systematic, shipborne multibeam echo sounder water column imaging,* J. Geophys. Res. Solid Earth, *120*, doi:10.1002/2014JB011617.

EGUEZ A, ALVARADO A, YEPES H, MACHETTE M, COSTA C, DART R (2003) *Database and map of Quaternary faults and folds of Ecuador and its offshore regions.* US Geological Survey, Open File Rep, 03–289.

GAY A, LOPEZ M, BERNDT C, SÉRANNE M, (2007) *Geological controls on focused fluid flow associated with seafloor seeps in the Lower Congo Basin,* Marine Geology *244*; 68–92.

GELI L, HENRY P, ZITTER T, DUPRE S, TRYON M, CAGATAY M, De LEPINAY B, Le PICHON X, SENGOR A, GORUR N, NATALIN B, UCARKUS G, OEZEREN S, VOLKER D, GASPERINI L, BURNARD P, BOURLANGE S (2008). *Gas emissions and active tectonics within the submerged section of the North Anatolian Fault zone in the Sea of Marmara.* Earth and Planetary Science Letters, *274*(1–2): 34–39.

GERMAN CR, PARSON LM, MILLS RA (1996) Mid-ocean ridges and hydrothermal activity. In: Oceanography. An Illustrated Guide. C.P. SUMMERHAYES and S.A. THORPE, eds, p. 152–164.

GRAINDORGE D, CALAHORRANO A, CHARVIS P, COLLOT J-Y, BETHOUX N (2004) *Deep structures of the margin and the Carnegie Ridge, possible consequence on great earthquake recurrence interval.* Geoph Res Lett 31. doi:10.1029/2003GL018803.

GREINERT, J, ARTEMOV, Y, EGOROV, V, DE BATIST, M, MCGINNIS, D, (2006) *1300-m-high rising bubbles from mud volcanoes at 2080 m in the Black Sea: Hydroacoustic characteristics and temporal variability.* Earth and Planetary Science Letters *244*:1–15. doi:10.1016/j.epsl.2006.02.011.

GUTSCHER M-A, MALAVIEILLE J, LALLEMAND S, COLLOT J-Y (1999) *Tectonic segmentation of the North Andean margin: impact of the Carnegie ridge collision.* Earth and Planetary Science Letters *168*:255–270.

HERNÁNDEZ M-J, MICHAUD F., COLLOT J-Y, PROUST J-N, ORTEGA R., ALEMAN A-M (2014) The Neogene Forearc Basins of the Ecuadorian Shelf (1°N-2°20′S): Preliminary Interpretation of a Dense Grid of Mcs Data Abstract T11C-4578 presented at 2014 Fall Meeting, AGU, San Francisco, Calif., 15–19 Dec.

HORNAFIUS JS, QUIGLEY D, LUYENDYK BP (1999) *The world's most spectacular marine hydrocarbon seeps (Coal Oil Point, Santa Barbara Channel, California): Quantification of emissions.* Journal of Geophysical Research *104*. doi: 10.1029/1999JC900148.

JUDD AG, HOVLAND, M (2007) Seabed Fluid Flow. The Impact on Geology, Biology and the Marine Environment. Cambridge University Press, Cambridge.

KRABBENHÖFT A, NETZEBAND G, BIALAS J, PAPENBERG C (2010) *Episodic methane concentrations at seep sites on the upper slope Opouawe Bank, southern Hikurangi Margin, New Zealand,* Marine Geology, *272* (1/4):71–78. DOI 10.1016/j.margeo.2009.08.001.

LOAYZA G, PROUST J-N, MICHAUD F, COLLOT J-Y (2014), Evolution pléistocène du système de canyons du golfe de Guayaquil (Equateur). Contrôles paléo-climatique et tectonique, *14 éme ASF congrés*, Paris, page 248.

LONCKE L., MASCLE J., and FANIL SCIENTIFIC PARTIES, (2004) *Mud volcanoes, gas chimneys, pockmarks and mounds in the Nile deep-sea fan (Eastern Mediterranean): geophysical evidences,* Marine and Petroleum Geology *21*,6: 669–689.

MARCAILLOU, B, SPENCE G, COLLOT J-Y, WANG K (2006) *Thermal regime from bottom simulating reflectors along the north Ecuador–south Colombia margin: Relation to margin segmentation and great subduction earthquakes,* Journal Geophysisycal Research 111, B12407, doi:10.1029/2005JB004239.

MARCAILLOU B, COLLOT J-Y., RIBODETTI A, d'ACREMONT E, MAHAMAT AA, ALVARADO A (2016) *Seamount subduction at the North-Ecuadorian convergent margin: Effects on structures, inter-seismic coupling and seismogenesis,* Earth and Planetary Science Letters *433*, 1 January 2016, Pages 146–158.

MASCLE J, FLORE M, PRAEG D, BROSOLO L, CAMERA L, CERAMICOLA S, DUPRE S (2014) *Distribution and geological control of mud volcanoes and other fluid/free gas seepage features in the Mediterranean Sea and nearby Gulf of Cadiz.* Geo-marine Letters, *34*(2–3), 89–110. http://dx.doi.org/10.1007/s00367-014-0356-4.

MICHAUD F, COLLOT J-Y, ALVARADO A, Lopez E y el personal científico y tecnico del INOCAR, (2006) Republica del Ecuador, Batimetría y Relieve Continental, publicación IOA-CVM-02-Post. INOCAR, Guayaquil.

MICHAUD F, WITT C, ROYER J-Y (2009) Influence of the Carnegie ridge subduction on Ecuadorian geology: reality and fiction: In: KAY S, RAMOS V, and DICKINSON WR (eds) Backbone of the Americas: Shallow Subduction, Plateau Uplift and Ridge and terrane Collision. Geol Soc Am Memoir 204:217–228 doi:10.1130/2009.1204.10.

MICHAUD F, PROUST J-N, COLLOT J-Y, the ATACAMES *scientific team* (2013) Sediments distribution and tectonic faults (Ecuadorian shelf): preliminary results of the ATACAMES Cruise (2012), Abstract T23B-06 presented at *2013 Meeting of the Americas, AGU,* Cancun, Mexico, 14-17 May.

MICHAUD F, PROUST J-N, COLLOT J-Y, LEBRUN J-F, WITT C, RATZOV G, POUDEROUX H, MARTILLO C, HERNÁNDEZ M-J, LOAYZA G, PE-NAFIEL L, SCHENINI L, DANO A, GONZALEZ M, BARBA D, DE MIN L, PONCE ADAMS G, URRESTA A, CALDERON M, (2015) *Quaternary sedimentation and active faulting along the Ecuadorian shelf: preliminary results of the ATACAMES Cruise (2012),* Marine Geophysical Research, *36*, 1:81–98.

MEREWETHER R, OLSSON M-S, LONSDALE P (1985) *Acoustically detected hydrocarbon plumes rising from 2-km depths in*

Guaymas Basin, Gulf of California J. Geophys. Res., *90* (1985), pp. 3075–3085.

Nocquet J-M, Mothes P, Alvarado A (2009) Geodesy, geodynamics and earthquake cycle in Ecuador. Geology and marine and onland Geophysics of Ecuador: from the continental coast to the Galapagos Islands. In Collot JY, Sallares V, Pazmiño A (eds) Geologia y Geofisica Marina y Terrestre del Ecuador. Publicacion CNDM-INOCAR-IRD, PSE001-09, Guayaquil, Ecuador pp 83–94.

Nocquet J-M, Villegas-Lanza JC, Chlieh M, Mothes PA, Rolandone F, Jarrin P, Cisneros D, Alvarado A, Audin L, Bondoux F, Martin X, Font Y, Régnier M, Vallée M, Tran T, Beauval C, Maguiña Mendoza JM, Martinez W, Tavera H, Yepes H, (2014) *Motion of continental slivers and creeping subduction in the northern Andes*, Nature Geosciences, *7* (4) (2014), pp. 287–291. http://dx.doi.org/10.1038/ngeo2099

Obzhirov A, Shakirov R, Salyuk A, Suess E, Biebow N, Salomatin A (2004) *Relations between methane venting, geological structure and seismo-tectonics in the okhotsk sea*, Geo-marine letters, *24*:135–139.

Paull CK, Borowski WS, Rodrigues N M, (1998) ODP Leg 164 Shipboard Scientific Party, Marine Gas Hydrate Inventory: Preliminary Results of ODP Leg 164 and Implications for Gas Venting and Slumping Associated with the Blake Ridge Gas Hydrate Field. In: Henrient, J. P., Mienert, J., eds., Gas Hydrates: Relevance to World Margin Stability and Climate Change. *Geol. Soc. London Spec. Publ.*, 137: 153–160.

Pazmiño A, Zapata C, Michaud F, Martillo C, Loayza G (2010) Reporte Científico a Bordo del B.I.- 91 "Orion" Estudio Geológico del Margen de Plataforma Continental del Golfo de Guayaquil (GEMAC- 1), Diciembre 2010.

Pedoja K, Dumont JF, Lamothe M, Ortlieb L, Collot JY, Ghaleb B, Auclair M, Alvarez V, Labrousse B (2006) *Plio-quaternary uplift of the manta peninsula and La Plata Island and the subduction of the Carnegie Ridge, central coast of Ecuador*. J S Am Earth Sci 22:1–21.

Praeg D, Ketzer JM, Augustin AH, Migeon S, Ceramicola S, Dano A, Ducassou E, Dupre S, Mascle J, Rodrigues LF (2014) *Fluid seepage in relation to seabed deformation on the central Nile Deep-Sea Fan, part 2: evidence from multibeam and sidescan imagery.* Advances in Natural and Technological Hazards Research, *37,* 141–150.

Proust J-N, Martillo C, Michaud F, Collot J-Y, Dauteuil O (2016) *Subduction of seafloor asperities revealed by detailed stratigraphic analysis of the active margin shelf sediments of Central Ecuador.* Marine Geology (**in press**).

Ranero CR, Von Huene R (2000) *Subduction erosion along the Middle America convergent margin*: Nature, v. *404*, p. 335–357.

Reyes P, Michaud F (2012) Mapa Geologico de la Margen Costera Ecuatoriana (1500000) EPPetroEcuador-IRD (eds), Quito Ecuador.

Römer M, Sahling H, Pape Th, dos santos Ferreira C, Wenzhöfer F, Boetius A, Bohrmann G, (2014) *Methane fluxes and carbonate deposits at a cold seep area of the Central Nile Deep Sea Fan, Eastern Mediterranean Sea*, Marine Geology, *347*: 27–42.

Sanclemente E (2014) Seismic imaging of the structure of the central Ecuador convergent margin: relationship with the interseismic coupling variations, Earth Sciences. Thèse Université de Nice-Sophia Antipolis 2014.

Sage F, Collot J-Y , Ranero CR (2006) *Interplate patchiness and subduction-erosion mechanisms: Evidence from depth-migrated seismic images at the central Ecuador convergent margin*, Geology, v. *34*; no. 12; pp. 997–1000; doi: 10.1130/G22790A.1.

Schneider Von Deimling J, Brockhoff J, Greinert J (2007) *Flare imaging with multibeam systems: Data processing for bubble detection at seeps*, Geochem. Geophys. Geosyst., *8*, Q06004, doi:10.1029/2007GC001577.

Suess E, Torres Ł, M.E., Bohrmann G, Collier RW, Greinert J, Linke P, Rehder G, Trehu A, Wallmann K, Winckler G, Zuleger E, (1999) Gas hydrate destabilization: enhanced dewatering, benthic material turnover and large methane plumes at the Cascadia convergent margin, *Earth and Planetary Science Letters* 170 (1999) 1–15.

Witt C, Bourgois J, Michaud F, Ordoñez M, Jimenez N, Sosson M (2006) *Development of the Gulf of Guayaquil (Ecuador) as an effect of the North Andean Block tectonic escape since the lower Pleistocene*. Tectonics 25. doi:10.1029/2004TC001723.

(Received July 2, 2015, revised December 18, 2015, accepted December 20, 2015, Published online January 27, 2016)

Reprinted from the journal

Pure Appl. Geophys. 173 (2016), 3305–3315
© 2015 Springer Basel
DOI 10.1007/s00024-015-1201-z

Pure and Applied Geophysics

The 10 April 2014 Nicaraguan Crustal Earthquake: Evidence of Complex Deformation of the Central American Volcanic Arc

GERARDO SUÁREZ,[1] ANGÉLICA MUÑOZ,[2] ISAAC A. FARRAZ,[3] EMILIO TALAVERA,[2] VIRGINIA TENORIO,[2]
DAVID A. NOVELO-CASANOVA,[1] and ANTONIO SÁNCHEZ[4]

Abstract—On 10 April 2014, an M_w 6.1 earthquake struck central Nicaragua. The main event and the aftershocks were clearly recorded by the Nicaraguan national seismic network and other regional seismic stations. These crustal earthquakes were strongly felt in central Nicaragua but caused relatively little damage. This is in sharp contrast to the destructive effects of the 1972 earthquake in the capital city of Managua. The differences in damage stem from the fact that the 1972 earthquake occurred on a fault beneath the city; in contrast, the 2014 event lies offshore, under Lake Managua. The distribution of aftershocks of the 2014 event shows two clusters of seismic activity. In the northwestern part of Lake Managua, an alignment of aftershocks suggests a northwest to southeast striking fault, parallel to the volcanic arc. The source mechanism agrees with this right-lateral, strike-slip motion on a plane with the same orientation as the aftershock sequence. For an earthquake of this magnitude, seismic scaling relations between fault length and magnitude predict a sub-surface fault length of approximately 16 km. This length is in good agreement with the extent of the fault defined by the aftershock sequence. A second cluster of aftershocks beneath Apoyeque volcano occurred simultaneously, but spatially separated from the first. There is no clear alignment of the epicenters in this cluster. Nevertheless, the decay of the number of earthquakes beneath Apoyeque as a function of time shows the typical behavior of an aftershock sequence and not of a volcanic swarm. The northeast–southwest striking Tiscapa/Ciudad Jardín and Estadio faults that broke during the 1972 and 1931 Managua earthquakes are orthogonal to the fault where the 10 April earthquake occurred. These orthogonal faults in close geographic proximity show that Central Nicaragua is being deformed in a complex tectonic setting. The Nicaraguan forearc sliver, between the trench and the volcanic arc, moves to the northwest relative to the Caribbean plate at a rate of 14 mm/year. Part of the deformation is apparently accommodated by strain partitioning in the form of bookshelf faulting, on a system of orthogonal faults. The sinistral faults striking northeast–southwest rotate blocks of the Caribbean plate in a clockwise manner. The recent crustal earthquakes in central Nicaragua in 1931, 1972 and 2005 earthquakes took place on these left-lateral faults. The motion of the forearc sliver is also accommodated by a second set of right-lateral, strike-slip faults oriented parallel to the volcanic arc. Faults with this orientation and direction of motion are responsible for the 2014 and possibly the 1955 earthquakes. The presence of this geometry of orthogonal crustal faults highlights the seismic hazard posed by this complex faulting system, not only in the capital city of Managua, but also to the major Nicaraguan cities, which lie close to the volcanic arc.

Key words: Nicaraguan tectonics, crustal deformation, bookshelf faulting, Central American, seismicity.

1. Introduction

The seismicity in Central America is dominated by the subduction of the Cocos plate beneath the Caribbean plate. The majority of the earthquakes in the region occur along the subduction zone. In the past 450 years, several large subduction earthquakes ($M_w \geq 7.5$) have been reported along the Pacific coast of Central America (WHITE and HARLOW 1993; ROJAS *et al.* 1993). However, the catalog of historical Central American earthquakes shows the presence of relatively frequent crustal events along the volcanic arc, particularly in Nicaragua (WHITE and HARLOW 1993; WHITE 1991; AMBRASEYS and ADAMS 1996).

Although some have been highly destructive, the magnitude of these shallow earthquakes beneath the Nicaraguan volcanic arc has not exceeded $M \sim 6.5$ during historical and instrumental times (WHITE and HARLOW 1993; AMBRASEYS and ADAMS 1996). One of the more damaging of these crustal earthquakes occurred on 23 December 1972 (M_s 6.2). The fault that ruptured during the 1972 earthquake was mapped based on aftershocks located using a local network installed a few days after the earthquake occurred.

[1] Instituto de Geofísica, Universidad Nacional Autónoma de México, 04510 México City, Mexico. E-mail: gerardo@geofisica.unam.mx; gersua@yahoo.com
[2] Instituto Nacional de Estudios Territoriales (INETER), Managua, Nicaragua.
[3] Terracon Ingeniería S.A. de C.V., México City, Mexico.
[4] Trimble Navigation Ltd., Plano, TX, USA.

The rupture took place on a left-lateral, strike-slip fault oriented northeast to southwest, located beneath the city of Managua and extending into Lake Managua.

The earthquakes of 1972 and the recent event on 10 April 2014 occurred in close geographical proximity and are almost of identical magnitude (Fig. 1). In sharp contrast to the 1972 earthquake, the 10 April 2014 event caused relatively little damage. In this paper, we study the fault geometry and the seismic characteristics of the 10 April 2014 earthquake and compare it to the destructive 1972 event. Our results indicate that although both earthquakes are part of the same tectonic environment, they take place on faults perpendicular to one another. Both sets of faults accommodate the displacement of the Nicaraguan forearc with respect to the Caribbean plate, observed on the basis of geodetic and geological measurements (e.g., DeMets 2001; DeMets et al. 2007; Alvarado et al. 2011; La Femina et al. 2002, 2009).

2. History of Crustal Earthquakes in Central Nicaragua

Shortly after the occurrence of the 1972 earthquake, Algermissen et al. (1974) and Langer et al. (1974) deployed a network of portable seismographs to record the aftershocks. Their results show that the 1972 earthquake ruptured a left-lateral, strike-slip fault oriented northeast to southwest. The location of

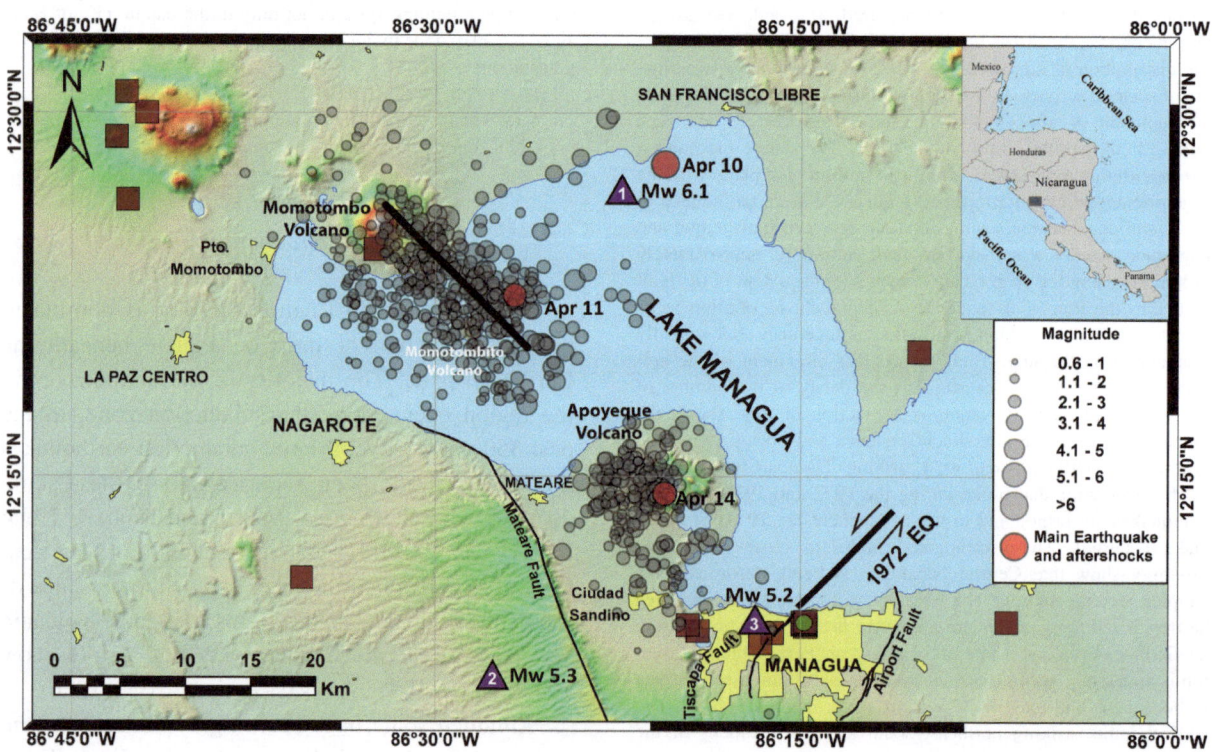

Figure 1

Epicenters of the aftershocks located in the first 20 days after the 10 April 2014 mainshock (*solid gray circles*) located based on arrival times read on the INETER seismic network. The fault length predicted from scaling relations of fault length versus the moment magnitude of the main event of approximately 16 km is shown as a *solid black line*. This fault length (*black line*) coincides with the extent of the cluster of aftershocks located beneath Lake Managua. The *red circles* are the epicentral locations of the main earthquake of 10 April and the two largest aftershocks of 11 and 14 April. *Triangles* indicate the locations of the mainshock and the two largest aftershocks from the USGS/NEIC (http://earthquake.usgs.gov/earthquakes). Seismic stations of the Nicaraguan seismic network used to locate the aftershocks are shown as *brown squares* and the *green dot* in central Managua shows the location of the strong motion instrument that recorded the main event. The left-lateral, strike-slip fault of the 1972 Managua earthquake, based on the aftershock distribution observed by Algermissen et al. (1974), is shown as *solid line*

the faults beneath the city and the poor quality of the local constructions were responsible for the extensive loss of life and material damage (CLUFF and CARVER 1973; WRIGHT and KRAMER 1973; BROWN et al. 1974; WYLLIE et al. 1974). It is estimated that during the 1972 earthquake 11,000 people were killed, 20,000 were injured and that the massive destruction rendered many constructions in Managua uninhabitable, leaving 250,000 people homeless (KATES et al. 1973; WRIGHT and KRAMER 1973; BROWN et al. 1974).

Based on reconnaissance studies conducted after the 1972 earthquake, LANGER et al. (1974) and WARD et al. (1974) suggest the earthquake occurred on the Tiscapa/Ciudad Jardín fault system (Fig. 1). The Tiscapa and the Ciudad Jardín faults are part of an active fault system oriented perpendicular to the subduction zone (McBIRNEY and WILLIAMS 1965; CARR 1976; WHITE 1991; WEINBERG 1992). STOIBER and CARR (1973) explain of these faults as a result of tears in the subducted Cocos plate. More recently, these faults have been explained as due to bookshelf tectonic deformation of the Caribbean plate, induced by the oblique subduction of the Cocos plate displacing the forearc to the northwest, relative to the rest of the Caribbean plate (LA FEMINA et al. 2002, 2009).

LEEDS (1974) and WHITE and HARLOW (1993) compiled catalogs of Nicaraguan earthquakes since 1520, which were strongly felt or caused damage in central Nicaragua, damaging cities like Managua, León, Momotombo, and others. These catalogs of shallow crustal earthquakes identify two events near the city of Managua: on 31 March 1931 and on 30 April 1955. The 1931 earthquake severely damaged the city of Managua (DURHAM 1931; SULTAN 1931). It is estimated that about 2500 people died and between 30,000 and 45,000 were left homeless. WHITE and HARLOW (1993) suggest a magnitude M_s 6.0, similar to that of the 1972 earthquake. COWAN et al. (2002) suggest that the 1931 earthquake occurred on the Estadio Fault. This fault beneath the Managua Graben is parallel and lies immediately to the west of the Tiscapa fault responsible for the 1972 event (Fig. 1).

WHITE and HARLOW (1993) report also a crustal earthquake on 30 April 1955 with an estimated magnitude M_s 6.0. However, the maximum intensity reported in Managua for this earthquake is MMI VI–VII (Modified Mercalli Intensity). No casualties are recorded and the number of people left homeless remains uncertain. This earthquake took place in northwestern Lake Managua; the location and damage pattern of the 1955 event is very similar to that of the 2014 earthquake. Thus, it appears that both the 1955 and the 2014 events occurred on the same fault system. More recently, another moderate crustal earthquake took place on 3 August 2005 (M_w 6.3) on a fault beneath Lake Nicaragua. This earthquake was well recorded by the permanent Nicaraguan seismic network and by temporary seismographs deployed at the time. The aftershock sequence and the focal mechanism define a left-lateral fault oriented northeast to southwest (FRENCH et al. 2010).

3. Aftershock Distribution and Fault Geometry of the 10 April 2014 Earthquake

3.1. Location of the Aftershocks: Mapping the Ruptured Fault

The 10 April 2014 earthquake was located by the National Earthquake Information Center of the United States Geological Survey (NEIC, USGS) in central Nicaragua. It was strongly felt in Managua and in the towns around Lake Managua. The focal mechanisms obtained by the Global Centroid Moment Tensor catalog and by the NEIC indicate a strike-slip event (http://www.globalcmt.org/; http://earthquake.usgs.gov/earthquakes). One of the nodal planes shows right-lateral strike-slip faulting, parallel to the volcanic arc. The other nodal plane is parallel to the fault orientation of the 1972 Managua earthquake.

The aftershocks of the 10 April 2014 event that occurred within the first 20 days after the mainshock were located using the stations of the Nicaraguan seismic network of the Instituto Nacional de Estudios Territoriales (INETER). This period was selected because after this time the number of aftershocks in the fault region decreased rapidly. The Nicaraguan network is based on short period, one or three component seismometers. The data are telemetered to a central facility located in INETER, in the city of Managua. Information from the national Nicaraguan network was complemented with data read from seismic stations in neighboring countries.

At the time of the occurrence of the earthquake, there were 13 operational seismic stations around Lake Managua. Thus, the permanent Nicaraguan seismic network provides excellent azimuthal coverage of the aftershock area. In fact, the closest station is at the foot of Momotombo Volcano and lies within the aftershock zone (Fig. 1). Approximately, 630 aftershocks were located in the sequence. The epicentral distribution of the aftershocks suggests two separate sources of seismic activity. One is aligned in a northwest to southeast direction and located beneath Lake Managua, to the southeast of the Momotombo volcano (Fig. 1). A second cluster of aftershocks took place beneath the Apoyeque volcano. It was initiated by the 14 April aftershock. The two clusters of aftershocks are distinctly separated (Fig. 1).

A selection of the best-located aftershocks was made, based on tests using different seismic velocity structures. The results show that the more stable and better-located epicenters are those with an azimuthal gap of less than 180°, a minimum distance to the closest station of 20 km (a distance similar to the maximum focal depth expected for a crustal fault), a residual of less than 0.3 s and a minimum of five reporting stations, of which at least one is an S-wave arrival. Similar criteria to evaluate the quality of hypocenters located using local networks have been proposed by other authors (e.g., CHATELAIN et al. 1980; KISSLING 1988; GOMBERG et al. 1990; BONDÁR et al. 2004). A total of 220 aftershocks passed these criteria. The distribution of the earthquakes selected defines a more clearly delineated trend of seismicity beneath Lake Managua oriented parallel to the volcanic arc. This alignment of aftershocks is interpreted as the trace of the fault ruptured during the April 10 event (Fig. 2).

The initial location of the 10 April mainshock lies about 15 km to the northeast of the main aftershock trend (Fig. 1). This original epicenter of the main event is not as accurate as that of the subsequent aftershocks, due to the lack of on-scale S waves on the local seismic records. To remedy this situation, the mainshock was relocated relative to the largest aftershock of April 11, using only the seismic stations

that were common to both events and fixing its hypocentral depth to 15 km. The relocated epicenter of the mainshock lies on the main trend of aftershocks reflecting the fault ruptured during the 10 April 2014 earthquake (Fig. 2).

The aftershock distribution shows that the 10 April 2014 event took place on a fault oriented northwest to southeast, parallel to the volcanic arc, and in a direction perpendicular to the faults mapped in the Managua graben where the 1931 and 1972 earthquakes took place (Fig. 2). The focal mechanism reported by the Global Centroid Moment Tensor catalog and by the NEIC indicates right-lateral strike-slip faulting parallel to the trend of aftershocks beneath Lake Managua (http://www.globalcmt.org/; http://earthquake.usgs.gov/earthquakes). One of the two largest aftershocks recorded took place on 11 April (M_w 5.3), the day after the main event (Table 1). The epicentral location of this aftershock is on the presumed fault and the source mechanism reported shows also right-lateral slip on a fault that is oriented parallel to the aftershock distribution (Fig. 2; Table 1).

Based on scaling relations of magnitude versus length of rupture of strike-slip earthquakes, the magnitude of the 10 April event (M_w 6.1) corresponds to a sub-surface fault length of approximately 16 km (e.g., WELLS and COPPERSMITH 1994; STIRLING et al. 2002; WESNOUSKY 2008). The extension of the aftershock lineament beneath Lake Managua corresponds well with this predicted fault length (Fig. 2). FRENCH et al. (2010) suggest a rupture length of 14–25 km for the 2005 Lake Nicaragua earthquake (M_w 6.3). LANGER et al. (1974), based on the extent of the aftershock zone of the 1972 earthquake, propose a fault length of 15–20 km. Thus, these three crustal events in central Nicaragua of approximately the same magnitude have similar rupture lengths inferred from their aftershock locations.

3.2. Aftershock Distribution Beneath the Apoyeque Volcano

The cluster of aftershocks beneath the Apoyeque volcano shows a clear spatial separation from the presumed rupture to the northwest (Figs. 1, 2). The seismicity beneath the Apoyeque volcano initiated

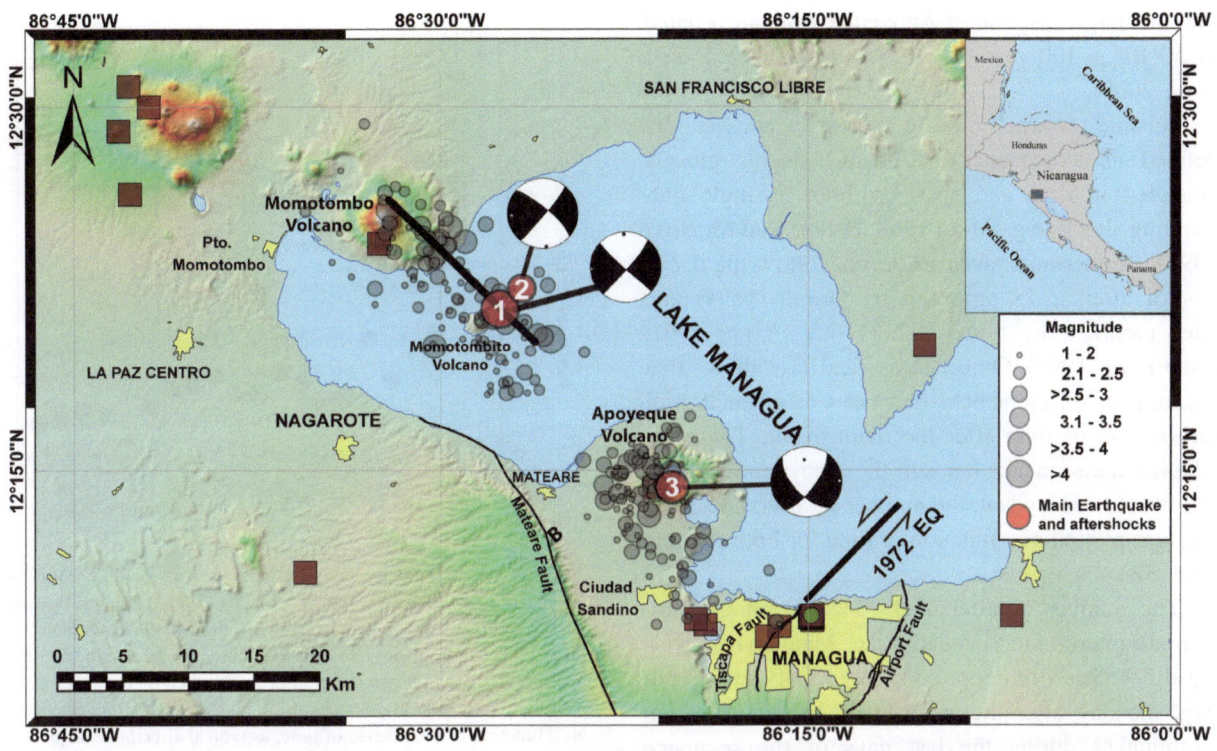

Figure 2

Epicenters of the selected aftershocks that occurred within the first 20 days after the 10 April 2014 mainshock (*solid circles*). The *red circles* labeled with *numbers* indicate the relocated epicenter of the main event of 10 April (1) and the two largest aftershocks of 11 April (2) and 14 April (3). The focal mechanisms of the mainshock and the two largest aftershocks are from the Centroid Moment Tensor catalog (DZIEWONSKI *et al.* 1981, http://www.globalcmt.org/). Other *symbols* are as in Fig. 1

Table 1

Source parameters of the 2014 crustal earthquakes in Central Nicaragua

Event date	Origin time[a]	M_w	Lat N	Long W	Strike	Dip	Rake	Depth (km)
10/Apr/2014[a]	23:27:45	6.1	12.4°	86.4°	306°	88°	177°	13
11/Apr/2014[b]	00:01:23	5.3	12.1°	86.5°	310°	77°	175°	11
14/Apr/2014[b]	05:07:03	5.2	12.2°	86.3°	138°	59°	−157°	10

[a] Focal parameters and depth from the National Earthquake Information Center, USGS

[b] Focal parameters from Global Centroid Moment Tensor Catalog (DZIEWONSKI *et al.* 1981)

after the 14 April event and took place simultaneously with the aftershocks beneath Lake Managua. One of the nodal planes of the focal mechanism of this event also shows right-lateral motion (Fig. 2; Table 1).

There is a question whether this activity is a volcanic swarm due to the reactivation of the volcanic plumbing system of Apoyeque by the earthquake of 10 April, or whether it reflects seismic activity on a fault, which is a continuation of the

ruptured fault to the northwest. Apoyeque volcano is within sight of the city of Managua, where personnel from INETER continuously monitor the volcano. Although there have been swarms of earthquakes reported beneath Apoyeque volcano in 2009 and 2012, neither INETER nor the Global Volcanism Program Weekly Reports report a reactivation of the volcano in the days following the mainshock (http://ineter.gob.ni/articulos/comunicaciones/comunicados/geofisica/comunicados.html; http://volcano.si.edu).

The last large eruption of Apoyeque volcano is dated as 50 BP ± 100 years (AVELLÁN *et al.* 2012; PARDO *et al.* 2009).

Seismic swarms due to volcanic activity are defined as a sequence of seismic events closely clustered in time and space without a single outstanding shock (e.g., MOGI 1963; BENOIT and MCNUTT 1996). Furthermore, swarms do not follow the decay law of aftershocks proposed by OMORI (1894) and later modified by other authors (e.g., UTSU 1961; UTSU *et al.* 1995). Omori's empirical law shows that the number of aftershocks decreases as a function of the inverse of time after the main event. To verify whether the sequence beneath the Apoyeque volcano follows this decay law, the number of aftershocks per day was plotted as a function of time for both clusters (Fig. 3).

The number of aftershocks located on the presumed ruptured fault beneath Lake Managua shows a rapid decrease from approximately ninety events per day, the day after the mainshock, to less than ten earthquakes during the last days of the sequence (Fig. 3). Beneath the Apoyeque volcano, a few aftershocks took place within the next 3 days of the main event. The 14 April aftershock triggers most of the seismicity observed beneath Apoyeque volcano (Fig. 3). The aftershocks under the volcano also follow a rapid decay in time and do not show the characteristic random behavior of volcanic swarms. Thus, the aftershocks beneath the Apoyeque reflect rupture on a tectonic fault rather than a volcanic tremor.

The aftershocks located in the vicinity of the Apoyeque volcano do not show a clear alignment supporting a particular fault orientation. Thus, it is not possible to categorically affirm from the aftershocks whether faulting took place on a northwest–southeast fault, like the main shock. Rupture may have taken place also on a left-lateral fault oriented northeast–southwest, parallel to the 1972 rupture.

4. Strong Motion Data and Reported Damage

The earthquake on 10 April and the two larger aftershocks caused relatively light damage and only minor effects were observed in the city of Nagarote

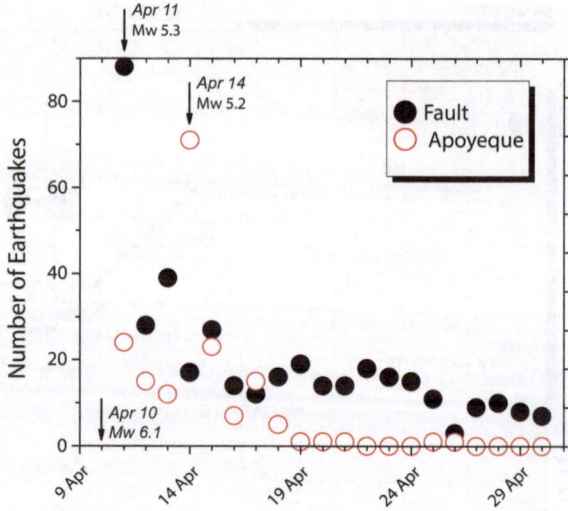

Figure 3

Number of aftershocks per day plotted versus time after the occurrence of the main event of 10 April. The aftershocks located on the fault beneath Lake Managua are shown as *solid circles* and those that took place beneath the Apoyeque volcano as *open circles*. Arrows indicate the date of occurrence of the main event and the two largest aftershocks. Both clusters of aftershocks decay as a function of the inverse of time, as typical aftershock sequences (*e.g.,* OMORI 1894)

(Figs. 1, 2). This is in sharp contrast to the 1931 and the 1972 earthquakes, which caused widespread damage in Managua. However, local newspapers report that some buildings damaged during the 1972 earthquake that were never repaired, such as the old Managua cathedral, suffered minor collapses due to the shaking induced by the 10 April event. The absence of damage observed not only in Managua but in the coastal towns of Ciudad Sandino, Mateare, Puerto Momotombo and Nagarote, for example, is due to the fact that the fault slip took place beneath Lake Managua, several kilometers away from the main cities and towns (Figs. 1, 2). Similarly, the 3 August 2005 earthquake took place beneath Lake Nicaragua (FRENCH *et al.* 2010). It was also felt strongly in the nearby region but caused very little damage.

The 1972 Managua earthquake was recorded by a strong motion instrument located in the ESSO refinery, to the southwest of the epicenter, and by seismoscopes installed in various parts of the capital (KNUDSON *et al.* 1974). The maximum acceleration observed during the earthquake was on the east–west component with values of 0.39 g. On the basis of the accelerogram recorded in the refinery and from the records obtained by the seismoscopes, KNUDSON *et al.* (1974) estimated that in the central part of the city of Managua accelerations exceeded 0.5 g.

As part of a project to instrument the city of Managua and the surrounding areas, INETER acquired several strong motion recorders. By chance, these authors completed the installation of the first accelerometer of this new array in the main offices of INETER, a few minutes before the earthquake of 10 April. The recorded accelerogram provides the only strong motion evidence in the vicinity of the earthquake (Fig. 4). The observed peak acceleration reached 0.045 g. This is an order of magnitude smaller than the one observed by KNUDSON *et al.* (1974) during the 1972 earthquake. The sharp differences in damage during the 1972 and the 2014 events in the city of Managua are due to this stark contrast of peak ground acceleration. This peak acceleration measured in Managua during the 2014 earthquake explains the observed intensities MMI V-VI.

5. Tectonic Significance of the 10 April Earthquake

Geodetic observations in Central America show that the Nicaraguan forearc, located between the subduction zone and the volcanic arc, moves to the northwest relative to the Caribbean plate (DeMETS 2001; DeMETS *et al.* 2007; CORREA-MORA *et al.* 2009; LA FEMINA *et al.* 2009; ALVARADO *et al.* 2011; KOBAYASHI *et al.* 2014). This type of partitioning in subduction margins, between motion perpendicular to the trench and oblique deformation of the overriding plate, has been observed in Alaska and Indonesia (e.g., FITCH 1972; JARRARD 1986; MCCAFFREY 1992; GEIST *et al.* 1988). In Nicaragua, the northwestward transport of the forearc relative to the Caribbean plate takes place at a rate of 1.4–1.6 cm year^{-1} (TURNER *et al.* 2007; DeMETS *et al.* 2007; CORREA-MORA *et al.* 2009; ALVARADO *et al.* 2011).

The presence of crustal earthquakes along the volcanic arc of central Nicaragua reflects the right-lateral motion of the forearc sliver relative to the Caribbean plate (Fig. 5). Although the strain rate is relatively low, due to the slow relative motion of the forearc, five earthquakes with magnitudes greater than M_w 6 have taken place in central Nicaragua during the last 85 years. The crustal earthquakes on the volcanic arc indicate that the relative motion of the forearc sliver is absorbed by strain partitioning on a set of orthogonal faults: dextral, strike-slip faults parallel to the volcanic arc, like the one responsible for the 2014 earthquake, and by left-lateral faults striking northeast–southwest as in the 1931, 1972 and 2005 events (Fig. 5). A similar scenario, where the deformation of the forearc to the oblique subduction is absorbed by rotation of blocks on orthogonal faults is observed in the Aleutians (GEIST *et al.* 1988).

To the northeast of the volcanic belt, the deformation of the Caribbean plate decreases rapidly (DeMETS *et al.* 2007; CORREA-MORA *et al.* 2009; LA FEMINA *et al.* 2009; ALVARADO *et al.* 2011; KOBAYASHI *et al.* 2014) and there is also no evidence of crustal seismicity (Fig. 5). LA FEMINA *et al.* (2002, 2009) suggested that bookshelf faulting accommodates the deformation of the Caribbean plate due to the motion of the forearc. In this scenario, sinistral slip on northeast southwest oriented faults absorbs the clockwise rotation of crustal blocks of the Caribbean plate induced by the drag of the forearc (Fig. 6a). The faults on which the 1931, 1972 and 2005 earthquakes took place appear to be the limits of some of these blocks (Fig. 6a). However, the presence of northwest to southeast right-lateral, strike-slip faults parallel to the volcanic belt shows that the deformation of the Caribbean plate is more complex. Slip of the forearc sliver relative to the Caribbean is also absorbed by right-lateral faults oriented parallel to the volcanic arc. The 2014 earthquake, and possibly the 1955 event, occurred on these faults. A simplified sketch of this fault geometry and style of deformation of the overriding Caribbean plate due to the relative motion of the forearc is shown in Fig. 6b. Thus, the rotation of small blocks in the forearc where strain partitioning is absorbed on perpendicular faults, as sketched on Fig. 6b, appears to be the manner in which the forearc absorbs the deformation induced by the oblique subduction of the Cocos plate.

The presence of these earthquakes of moderate magnitude, but occasionally of devastating consequences in the volcanic arc of Nicaragua, represents a major challenge to estimate seismic hazard in the

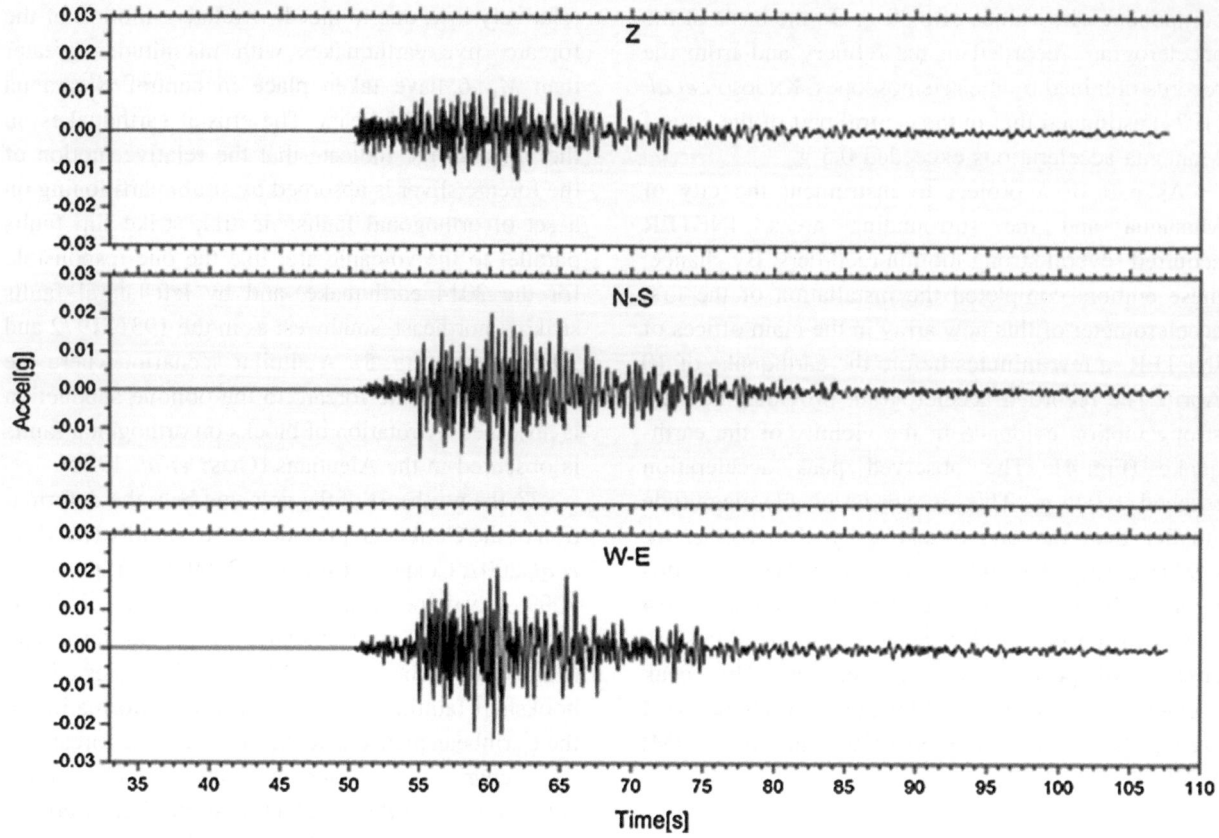

Figure 4
Acceleration records from the earthquake of 10 April 2014 obtained by the strong motion instrument installed minutes before the occurrence of this event in the main office of INETER. The peak acceleration on the two horizontal components reached 0.045 g

country. The most populated cities of Nicaragua are located on or near the volcanic arc, due to the more benign weather conditions and to the richness of the soil. Thus, these relatively moderate events represent a greater seismic hazard for the most important Nicaraguan cities, than the much larger subduction earthquakes. The presence of these damaging events, near the main cities of Nicaragua, stresses the need to understand better the seismicity and the complex tectonic structure and slip history of seismically active faults in Nicaragua and in Central America.

6. Summary and Conclusions

The aftershock distribution of the earthquake of 10 April 2014 suggests that the slip took place on a fault about 16 km long beneath Lake Managua. The

orientation of the aftershocks coincides with a fault striking northwest–southeast, along the direction of the volcanic arc. The aftershock distribution of the 10 April earthquake shows two distinct clusters of activity; one to the northwest, beneath Lake Managua, and a second one, distinctly separated from the first one, located in the vicinity of Apoyeque volcano.

The seismicity beneath Apoyeque volcano would appear to indicate that shaking induced by the mainshock reactivated the plumbing system beneath the volcano. However, the behavior of the sequence is typical of aftershocks decreasing in number as the inverse of time after the main event. Furthermore, the third largest aftershock on 14 April (M_w 5.2) shows a source mechanism which is almost identical to that of the main event and to the second largest aftershock on 11 April (Table 1; Figs. 2, 5). Thus, the aftershocks in the region of the Apoyeque volcano suggest

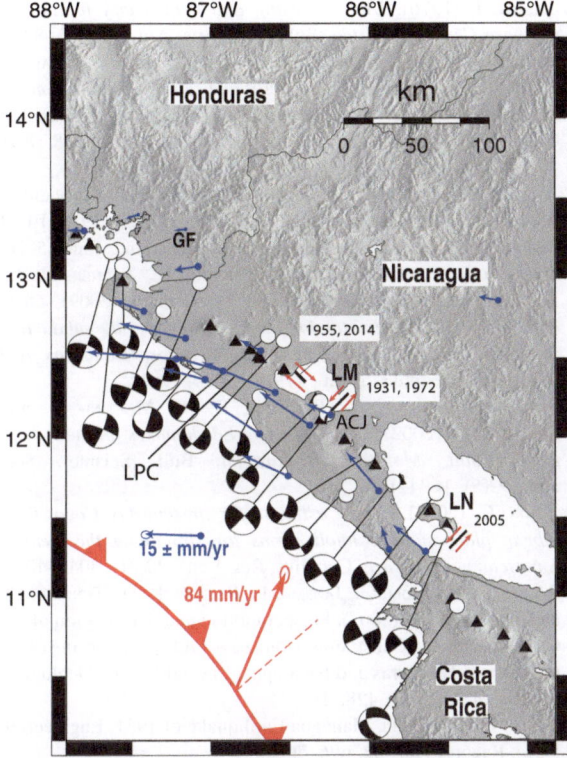

Figure 5

Focal mechanisms of crustal earthquakes reported in central Nicaragua; *solid* and *open quadrants* indicate compressional and dilational arrivals, respectively (http://www.globalcmt.org/). *Solid triangles* indicate active volcanoes. *Blue arrows* show the direction of motion of the forearc relative to the Caribbean plate from GPS data. *Red arrows* denote the slip on the faults ruptured during the 1931 and 1972, 1955 and 2014, and 2005 earthquakes. GF, LM and LN are Gulf of Fonseca, Lake Managua and Lake Nicaragua, respectively. *ACJ* represents the Aeropuerto-Ciudad Jardín fault system (modified from La Femina *et al.* 2002; Correa-Mora *et al.* 2009)

that stress transfer induced by the slip of the main-shock activated an extension or a sub-parallel fault to the one ruptured by the main event. This highlights the potential for the presence of larger earthquakes on these dextral faults if the segments were to coalesce and rupture in a single earthquake.

The dextral strike-slip motion of the focal mechanism of the main earthquake and the largest aftershocks indicate that these northwest–southeast faults are orthogonal to the seismically capable faults observed in the Managua basin. These left-lateral faults are oriented northeast–southwest. Slip on these faults caused the devastating earthquakes of 1931 and 1972. In contrast, the April 2014 earthquake took place on faults orthogonal to those

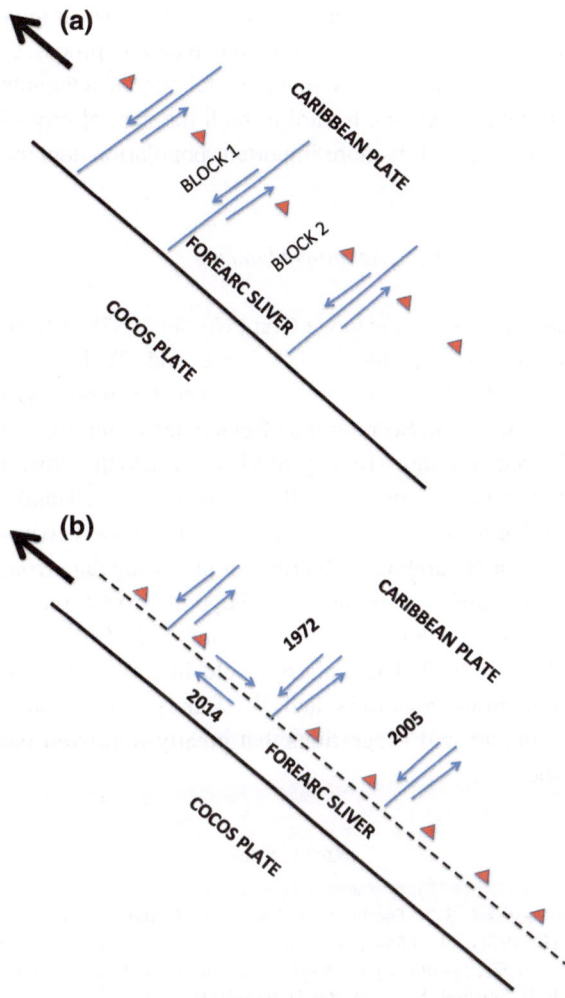

Figure 6

a Schematic representation of bookshelf faulting in Central Nicaragua where the relative motion of the forearc relative to the Caribbean plate is absorbed by the clockwise rotation of blocks delimited by left-lateral strike-slip faults. The *black arrow* indicates the motion of the forearc sliver relative to the Caribbean plate and the *triangles* represent the volcanic arc. **b** In this scenario, the tectonic deformation of the Caribbean plate due to the motion of the forearc sliver is accommodated by blocks rotating clockwise and by right-lateral faulting oriented parallel to the volcanic arc. In a schematic manner, this style of deformation on orthogonal faults reflects the one taking place in central Nicaragua. Other *symbols* as in Fig. 6a

mapped in the Managua basin. Based on the intensity reports, it appears that the 1955 earthquake took place on the same fault of the 2014 event.

The complexity of the geologic structure and the geometry of the seismically capable faults in the area highlight the need for detailed mapping of the faults in this region and a careful analysis of the seismicity in Nicaragua. Understanding the tectonic processes present in this area is fundamental for an adequate estimate of seismic hazard to both the capital city of Nicaragua and its more important population centers.

Acknowledgments

The authors acknowledge Grant Number 082821 from the Mexican Council of Science and Technology (CONACYT) for support in conducting this work and a grant from the Secretaría de Relaciones Exteriores de México (Foreign Ministry of Mexico), which allowed the authors to enjoy scientific exchange visits. Thanks are due to the Instituto Nacional de Estudios Territoriales of Nicaragua (INETER) for providing data from the national seismic network. Support from Jorge A. Castillo in relocating the mainshock of 10 April is acknowledged. The authors would like to thank two anonymous reviewers and W. Bandy for valuable comments and suggestions that greatly improved the paper.

REFERENCES

ALGERMISSEN, S. T., DEWEY, J. W., LANGER, C. J., and DILLINGER, W. H. (1974). *The Managua, Nicaragua, earthquake of December 23, 1972, Location, focal mechanism, and intensity distribution*, Bull. Seismol. Soc. Am., *64*(4), 993–1004.

ALVARADO, D., C., DEMETS, B. TIKOKK, D. HERNÁNDEZ, T.F. WAWRZYENIEC, C. PULLINGER, G. MATTIOLI, H.L. TURNER, M. RODRÍGUEZ, and F. CORREA-MORA. (2011). *Forearc motion and deformation between El Salvador and Nicaragua: GPS, seismic, structural and paleomagnetic observations*, Lithosphere, *3*, 3–21.

AMBRASEYS, N. N., and ADAMS, R. D. (1996). *Large-magnitude Central American earthquakes, 1898–1994*, Geophys. J. Int., *127*(3), 665–692.

AVELLÁN, D.R., MACÍAS, J.L., PARDO, N., SCOLAMACCHIA, T., and RODRIGUEZ, D. (2012). *Stratigraphy, geomorphology, geochemistry and hazard implications of the Nejapa Volcanic Field, western Managua, Nicaragua*, Journal of Volcanology and Geothermal Research, *213–214*: 51–71.

BENOIT, J. P., & MCNUTT, S. R. (1996). Global volcanic earthquake swarm database 1979–1989 (pp. 96–69). US Department of the Interior, US Gological Survey.

BONDÁR, I., MYERS, S. C., ENGDAHL, E. R., & BERGMAN, E. A. (2004). *Epicentre accuracy based on seismic network criteria*. Geophysical Journal International, *156*(3), 483–496.

BROWN, R. D., WARD, P. L., and PLAFKER, G. (1974). *Geologic and seismologic aspects of the Managua, Nicaragua, earthquakes of December 23, 1972*, Bull. Seismol. Soc. Am., *64*(4), 1031–1031.

CARR, M. J. (1976). *Underthrusting and Quaternary faulting in northern Central America*, Bull. Geol. Soc. Am., *87*, 825–829.

CHATELAIN, J. L., ROECKER, S. W., HATZFELD, D., & MOLNAR, P. (1980). *Microearthquake seismicity and fault plane solutions in the Hindu Kush region and their tectonic implications*, Journal of Geophysical Research: Solid Earth (1978–2012), *85*(B3), 1365–1387.

CLUFF, L. S. and CARVER, G. A. (1973). Geological observations, Managua, Nicaragua, earthquake of December 23, 1972, EERI Recon. Rept., Earthquake Eng. Res. Inst., Oakland, Calif, 5–20.

CORREA-MORA, F., DEMETS, C., ALVARADO, D., TURNER, H. L., MATTIOLI, G., HERNANDEZ, D., and TENORIO, C. (2009). *GPS-derived coupling estimates for the Central America subduction zone and volcanic arc faults: El Salvador, Honduras and Nicaragua*, Geophys. J. Int., *179*(3), 1279–1291.

COWAN, H., PRENTICE, C., PANTOSTI, D., DE MARTINI, P., and STRAUCH, W. (2002). *Late Holocene Earthquakes on the Aeropuerto Fault, Managua, Nicaragua*, Bull. Seismol. Soc. Am., *92*(5), 1694–1707.

DEMETS, C. (2001). *A new estimate for present-day Cocos-Caribbean plate motion: Implications for slip along the Central American volcanic arc*, Geophys. Res. Lett., *28*(21), 4043–4046.

DEMETS, C., MATTIOLI, G., JANSMA, P., ROGERS, R. D., TENORIO, C., and TURNER, H. L. (2007). Present motion and deformation of the Caribbean plate: Constraints from new GPS geodetic measurements from Honduras and Nicaragua, Special Papers, Geological Society of America, 428, 21.

DURHAM, H. W. (1931). Managua Earthquake of 1931, Engineering News Record, Apr. 22, 696–700.

DZIEWONSKI, A.M., CHOU T.A, and WOODHOUSE J.H (1981). *Determination of earthquake source parameters from waveform data for studies of global and regional seismicity*, J. Geophys. Res., *86*, 2825–2852, doi:10.1029/JB086iB04p02825.

FITCH, T. J. (1972). *Plate convergence, transcurrent faults, and internal deformation adjacent to southeast Asia and the western Pacific*, J. Geophys. Res., *77*(23), 4432–4460.

FRENCH, S. W., et al. Constraints on upper plate deformation in the Nicaraguan subduction zone from earthquake relocation and directivity analysis. Geochemistry, Geophysics, Geosystems *11*.3 (2010).

GEIST, E. L., J.R. CHILDS, and D.W. SCHOLL (1988). *The origin of Summit basins of the Aleutian Ridge: Implications for block rotation of an arc massif*, Tectonics, *7*(2), 327–341, doi:10.1029/TC007i002p00327.

GOMBERG, J. S., SHEDLOCK, K. M., & ROECKER, S. W. (1990). *The effect of S-wave arrival times on the accuracy of hypocenter estimation*, Bulletin of the Seismological Society of America, *80*(6A), 1605–1628.

JARRARD, R. D. (1986). *Terrane motion by strike-slip faulting of forearc slivers*, Geology, *14*(9), 780–783.

KISSLING, E. (1988). *Geotomography with local earthquake data*, Reviews of Geophysics, *26*(4), 659–698.

KATES, R. W., HAAS, J. E., AMARAL, D. J., OLSON, R. A., RAMOS, R., and OLSON, R. (1973). *Human impact of the Managua earthquake*, Science, *182*(4116), 981–990.

KOBAYASHI, D., LAFEMINA, P., GEIRSSON, H., CHICHACO, E., ABREGO, A. A., MORA, H., & CAMACHO, E. (2014). *Kinematics of the western Caribbean: Collision of the Cocos Ridge and upper plate deformation*. Geochemistry, Geophysics, Geosystems, *15*(5), 1671–1683.

KNUDSON, C. F., PEREZ, V., and MATTHIESEN, R. B. (1974). *Strong-motion instrumental records of the Managua earthquake of December 23, 1972*, Bull. Seismol. Soc. Am., *64*(4), 1049–1067.

LANGER, C. J., HOPPER, M. G., ALGERMISSEN, S. T., and DEWEY, J. W. (1974). *Aftershocks of the Managua, Nicaragua, earthquake of December 23, 1972*, Bull. Seismol. Soc. Am., *64*(4), 1005–1016.

LA FEMINA, P. C., DIXON, T. H., and STRAUCH, W. (2002). *Bookshelf faulting in Nicaragua*. Geology, *30*(8), 751–754.

LA FEMINA, P.C., DIXON, T.H., GOVERS, R., NORABUENA, E., TURNER, H., SABALLOS, A., MATTIOLI, G., PROTTI, M., and STRAUCH, W. (2009). *Forearc motion and Cocos Ridge collision in Central America*, Geochem. Geophys. Geosys., v. *10*, p. Q05S14, doi:10.1029/2008GC002181.

LEEDS, D. J. (1974). *Catalog of Nicaraguan earthquakes*, Bull. Seismol. Soc. Am., *64*(4), 1135–1158.

MCBIRNEY, A. R. and H. WILLIAMS (1965). *Volcanic history of Nicaragua*, Univ. Calif. Publ. Geol. Sci., *55*, 1–65.

MCCAFFREY, R. (1992). *Oblique plate convergence, slip vectors, and forearc deformation*, J. Geophys. Res., *97*(B6), 8905–8915.

MOGI, K. (1963). *Some discussions on aftershocks, foreshocks and earthquake swarms: the fracture of a semi-infinite body caused by an inner stress origin and its relation to the earthquake phenomena, 3*, Bull. Earthquake Res. Inst., Tokyo Univ. *41*. 615–658.

OMORI, F. (1894). *On the aftershocks of earthquakes*. Journal of the College of Science, Imperial University of Tokyo 7: 111–200.

PARDO, N., MACÍAS, J.L., GIORDANO, G., CIANFARRA, P., AVELLÁN, D.R., and BELLATRECCIA, F. (2009). *The ∼1245 yr BP Asososca maar eruption: The youngest event along the Nejapa-Miraflores volcanic fault, Western Managua, Nicaragua*, Journal of Volcanology and Geothermal Research, *184*: 292–312.

ROJAS, W., BUNGUM, H. and LINDHOLM, C. (1993). Historical and recent earthquakes in Central America, Revista Geológica de América Central, 16.

STIRLING, M. W., RHOADES, D., and BERRYMAN, K. R. (2002). *Comparison of scaling relations derived from data of the instrumental and preinstrumental eras*, Bull. Seismol. Soc. Am., *92*, 355–375.

STOIBER, R. and CARR, M. (1973). *Quaternary volcanic and tectonic segmentation of Central America*, Bulletin Volcanologique, *37*(3), 304–325.

SULTAN, D.I. (1931). *The Managua earthquake of 1931*, Military Engineer, *92*, 354–361.

TURNER, H. L., III, P. LAFEMINA, A., SABALLOS, G. S. MATTIOLI, P. E. JANSMA, and T. DIXON (2007). *Kinematics of the Nicaraguan forearc from GPS geodesy*, Geophys. Res. Lett., *34*, L02302, doi:10.1029/2006GL027586.

UTSU, T. (1961). *A statistical study of the occurrence of aftershocks*. Geophysical Magazine *30*: 521–605.

UTSU, T.; OGATA, Y.; MATSU'URA, R.S. (1995). *The centenary of the Omori formula for a decay law of aftershock activity*, Journal of Physics of the Earth *43*: 1–33.

WHITE, R. A. (1991). Tectonic implications of upper-crustal seismicity in Central America, in Neotectonics of North America: SLEMMONS, D. B., ENGDAHL, E. R., ZOBACK, M. D., and BLACKWELL, D. (Editors), Boulder, Colorado, Geological Society of America, pp. 323–338.

WHITE, R. A. and HARLOW, D. H. (1993). *Destructive upper-crustal earthquakes of Central America since 1900*, Bull. Seismol. Soc. Am., *83*(4), 1115–1142.

WARD, P. L., GIBBS, J., HARLOW, D., and ABURTO, A. (1974). *Aftershocks of the Managua, Nicaragua, earthquake and the tectonic significance of the Tiscapa fault*, Bull. Seismol. Soc. Am., *64*(4), 1017–1029.

WEINBERG, R. F. (1992). *Neotectonic development of western Nicaragua*, Tectonics, *11*(5), 1010–1017.

WELLS, D. L. and COPPERSMITH, K. J. (1994). *New empirical relationships among magnitude, rupture length, rupture width, rupture area, and surface displacement*, Bull. Seismol. Soc. Am., *84*(4), 974–1002.

WESNOUSKY, S.G. (2008). *Displacement and geometrical characteristics of earthquake surface ruptures: Issues and implications for seismic-hazard analysis and the process of earthquake rupture*, Bull. Seismol. Soc. Am., *98*(4), 1609–1632.

WRIGHT, R.N and KRAMER, S. (1973). Building performance in the 1972 Earthquake, National Bureau of Standards Technical Note 807, Washington D.C., 155 pp.

WYLLIE, L. A., WRIGHT, R. N., SOZEN, M. A., DEGENKOLB, H. J., STEINBRUGGE, K. V., and KRAMER, S. (1974). *Effects on structures of the Managua earthquake of December 23, 1972*, Bull. Seismol. Soc. Am., *64*(4), 1069–1133.

(Received July 3, 2015, revised October 22, 2015, accepted October 23, 2015, Published online November 7, 2015)

Pure Appl. Geophys. 173 (2016), 3317–3339
© 2015 Springer Basel
DOI 10.1007/s00024-015-1197-4

Pure and Applied Geophysics

Three-Dimensional Thermal Model of the Costa Rica-Nicaragua Subduction Zone

JUAN CARLOS ROSAS,[1] CLAIRE A. CURRIE,[1] and JIANGHENG HE[2]

Abstract—The thermal structure of a subduction zone controls many key processes, including subducting plate metamorphism and dehydration, the megathrust earthquake seismogenic zone and volcanic arc magmatism. Here, we present the first three-dimensional (3D), steady-state kinematic-dynamic thermal model for the Costa Rica-Nicaragua subduction zone. The model consists of the subducting Cocos plate, the overriding Caribbean Plate, and a viscous mantle wedge in which flow is driven by interactions with the downgoing slab. The Cocos plate geometry includes along-strike variations in slab dip, which induce along-strike flow in the mantle wedge. Along-strike flow occurs primarily below Costa Rica, with a maximum magnitude of 4 cm/year (~40 % of the convergence rate) for a mantle with a dislocation creep rheology; an isoviscous mantle has lower velocities. Along-margin flow causes temperatures variations of up to 80 °C in the subducting slab and mantle wedge at the volcanic arc and backarc. The 3D effects do not strongly alter the shallow (<35 km) thermal structure of the subduction zone. The models predict that the megathrust seismogenic zone width decreases from ~100 km below Costa Rica to just a few kilometers below Nicaragua; the narrow width in the north is due to hydrothermal cooling of the oceanic plate. These results are in good agreement with previous 2D models and with the rupture area of recent earthquakes. In the models, along-strike mantle flow is induced only by variations in slab dip, with flow directed toward the south where the dip angle is smallest. In contrast, geochemical and seismic observations suggest a northward flow of 6–19 cm/year. We do not observe this in our models, suggesting that northward flow may be driven by additional factors, such as slab rollback or proximity to a slab edge (slab window). Such high velocities may significantly affect the thermal structure, especially at the southern end of the subduction zone. In this area, 3D models that include slab rollback and a slab edge are needed to investigate the mantle structure and dynamics.

Key words: Subduction zones, thermal structure, geodynamics, numerical modeling, Middle America Trench.

1. Introduction

A subduction zone delineates the convergent boundary between tectonic plates, where oceanic lithosphere descends below a less dense oceanic or continental plate (STERN 2002). As an oceanic plate moves away from a mid-ocean ridge, it thickens and its density increases, eventually becoming negatively buoyant and sinking into the deep Earth (CLOOS 1993). It is generally accepted that this gravitational instability is the primary force driving the motion of the tectonic plates (FORSYTH and UYEDA 1975), and therefore it is not surprising that subduction zones are a topic of intense study. Among subduction zones, the Costa Rica-Nicaragua section of the Middle America Trench (MAT) has received particular attention in recent years. Figure 1 shows the configuration of the MAT in Central America, with the Cocos plate subducting beneath the Caribbean Plate. Subduction in this region is associated with the formation of the Central America Volcanic Arc (CARR et al. 2004). It is also responsible for earthquakes that occur within both the Cocos plate and the Caribbean plate, as well as megathrust earthquakes that occur on the inclined subduction interface. The most recent megathrust earthquake was the magnitude Mw 7.6 event that occurred on September 5th, 2012 below the Nicoya Peninsula of Costa Rica (PROTTI et al. 2014).

Despite the predominantly trench-perpendicular direction of plate convergence, there are numerous observations that suggest significant along-strike changes in the characteristics of this subduction zone. These include variations in the width of the megathrust seismogenic zone (NEWMAN et al. 2002; DESHON et al. 2006; SCHWARTZ and DESHON 2007; AUDET and SCHWARTZ 2013), variable amounts of hydrothermal circulation in the Cocos plate crust (FISHER et al. 2003; HUTNAK et al. 2008; HARRIS et al.

---

[1] Department of Physics, University of Alberta, 4-181 Centennial Centre for Interdisciplinary Science, Edmonton, AB, Canada. E-mail: jrosas@ualberta.ca

[2] Pacific Geoscience Centre, Geological Survey of Canada, 9860 West Saanich Road, Sydney, BC, Canada.

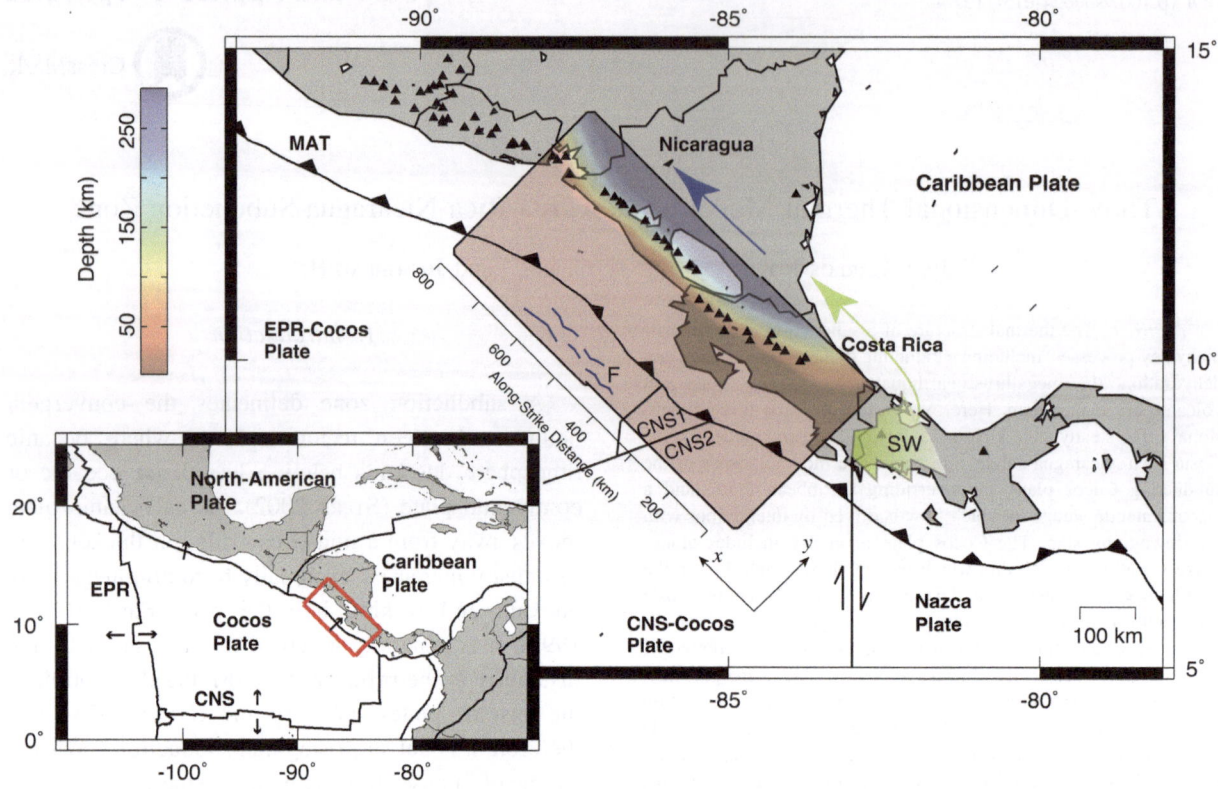

Figure 1

Map of the study area at the Costa Rica-Nicaragua subduction zone. *Colors* indicate the depth to the *top* of the subducted Cocos plate, as obtained from KYRIAKOPOULOS *et al.* (2015). *Scale* denotes along-strike distance (*x-axis*). 3D model extends from $x = 200$ km to $x = 700$ km, and from the trench landward until the slab reaches a depth of 300 km along the *y-axis*. *Dashed black lines* are the location of profiles A through D. Volcanoes are shown with *black triangles*. The Middle America Trench (MAT) runs approximately parallel to volcanic arc. Location of the Nicoya peninsula is shown. Plate boundaries between lithosphere generated at the East Pacific Rise (EPR) and lithosphere generated at the Cocos-Nazca Spreading Centre (CNS) are also shown (BARCKHAUSEN *et al.* 2001). Location of slab window (SW) is shown in *light green* (JOHNSTON and THORKELSON 1997), with *light green arrow* showing possible direction for oceanic-island basalt (OIB). B/La and Ba/La signal (*blue arrow*) increase in the northwest direction, indicating greater hydration in the mantle below Nicaragua. *Inset* shows study area (enclosed by *red box*) on a large-scale tectonic map. *Black arrows* denotes plate motion. Plate boundaries are from BIRD (2003)

2010a, b), and different levels of hydration for the descending slab and overlying mantle (ABERS *et al.* 2003; SYRACUSE *et al.* 2008; RYCHERT *et al.* 2008; Van AVENDONK *et al.* 2011; DINC *et al.* 2011). The geochemistry of arc magmas also varies along strike, with geochemical gradients in radiogenic isotope ratios, such as B/La and Ba/La, that are associated with fluid loss from the slab (PATINO *et al.* 2000; CARR *et al.* 2004). In addition, $^{208}Pb/^{204}Pb$ and $^{143}Nd/^{144}Nd$ ratios suggest that the magma source includes a component related to ocean-island basalt (OIB) from the Galapagos Hot Spot track (HERRSTROM *et al.* 1995; JOHNSTON and THORKELSON 1997; ABRATIS and WORNER 2001; HOERNLE *et al.* 2008). These data, together

with seismic anisotropy observations, have been used to argue for an along-strike flow of 6.3–19 cm/year within the mantle above the subducting Cocos plate in Costa Rica and Nicaragua (HOERNLE *et al.* 2008); flow is inferred to go from southeast to northwest (Fig. 1).

Many subduction zone processes, including the earthquake distribution, slab dehydration and arc volcanism, depend strongly on the thermal structure. A common approach to study the temperature distribution is to create two-dimensional (2D) steady-state numerical models of the subduction system (oceanic plate, overriding plate, and viscous mantle wedge) in which dynamically calculated mantle wedge flow is driven by a kinematically prescribed subducting plate (e.g., Peacock 1996). Three key factors control the thermal structure of a subduction zone: the temperature of the incoming oceanic plate (slab), the plate convergence rate, and the flow pattern of the overlying mantle wedge. The temperature of the slab is primarily a function of its age, with heat being transferred by conduction. In the mantle wedge, however, heat is transferred mainly by the motion of the viscous mantle. In the context of fluid dynamics, the downward motion of the descending oceanic plate drags the overlying mantle by viscous coupling and sets up a forced-convection flow pattern, i.e., corner flow (Batchelor 2000).

Several 2D thermal models have been developed for the Costa Rica-Nicaragua subduction zone. The objective of these models varies, and they can be roughly classified into two categories:

1. Models to study the relationship between hydrothermal circulation and shallow slab temperatures.
2. Models to study the role of viscous wedge flow in deep metamorphic and dehydration reactions within the slab.

The first category is reasonably well studied. These studies show that significant hydrothermal circulation must occur in the shallow (<1 km) oceanic crust, to explain the observed surface heat flow along the MAT in Central America (Fisher et al. 2003; Hutnak et al. 2008; Harris et al. 2010a). Hydrothermal circulation cools the oceanic crust and provides a relatively good match between crustal temperatures and distribution of seismicity (Harris and Wang 2002; Kummer and Spinelli 2008; Harris et al. 2010b; Cozzens and Spinelli 2012; Rotman and Spinelli 2013). For the second category, Peacock et al. (2005) demonstrated that a non-Newtonian mantle rheology is needed to sustain the high temperatures required for mantle melting and arc volcanism. This also results in an oceanic crust that is fully dehydrated and transformed to eclogite at depths of 70–100 km. However, they do not find significant variations in the thermal or flow structure along the strike of the subduction zone, which would be needed to fit the along-margin variations in arc geochemistry.

There are multiple factors that may explain the discrepancy between 2D models and the observed geochemistry. These include possible along-strike changes in the thermal state of the incoming oceanic plate (Rosas et al. 2015, submitted) or changes in the amount of water released into the mantle (Abers et al. 2003; Syracuse et al. 2008; Rychert et al. 2008; Van Avendonk et al. 2011; Dinc et al. 2011). An important one, however, is the intrinsic two-dimensionality of the corner flow model. It is known that geometrical factors such as along-strike variations in slab dip or trench curvature can induce along-strike mantle flow (Kneller and van Keken 2008; Bengtson and van Keken 2012; Wada et al. 2015). Such characteristics are present in Central America (Fig. 1). In this region, the Cocos plate changes its dip from approximately 70° in Nicaragua to 45° in Central Costa Rica at depths greater than 70 km. The change occurs over an along-strike distance of 200 km.

In this study, we present the first three-dimensional (3D) model of the thermal structure of the Costa Rica-Nicaragua subduction zone. Our approach is similar to previous 2D models, as we incorporate a subducting slab that drives mantle wedge flow by viscous coupling. However, we consider a three-dimensional slab with a non-Newtonian rheology for the mantle wedge, using the most up-to-date geometry for the Cocos plate. This is expected to induce significant lateral flow in the mantle wedge, therefore changing the overall temperature distribution relative to 2D models. Our objectives are to quantify these differences and provide a more detailed description of the thermal structure of this subduction zone.

2. Model Set Up

The numerical models use a kinematic-dynamic approach to model the steady-state thermal structure of the subduction zone (van Keken et al. 2008, 2002; Currie et al. 2004). The subducting oceanic plate and overriding continental plate have a fixed geometry and convergence rate. The mantle wedge has a viscous rheology and flows in response to the imposed subduction dynamics. The advantage of this approach over a completely dynamic subduction zone is that it allows for a much higher resolution of the thermal structure, especially in the mantle wedge corner region (BILLEN 2008). The disadvantage is that we neglect any dynamic factors, such as slab rollback or roll forward, which might induce 3D mantle flow as well.

To investigate 3D mantle wedge flow and the resulting temperature distribution of the Costa Rica-Nicaragua subduction zone, we use a three-dimensional finite-element mesh of the Costa Rica-Nicaragua subduction zone. The mantle wedge flow and thermal structure are computed using the equations of mass, momentum and energy. In that order, the equations are:

$$\nabla \cdot \mathbf{v} = 0, \tag{1}$$

$$\nabla P - \nabla \cdot \sigma = 0, \tag{2}$$

$$\nabla \cdot (k\nabla T) - \rho c_\mathrm{p}(\mathbf{v} \cdot \nabla T) + A = 0, \tag{3}$$

where $\mathbf{v}$ is the velocity, P is the dynamic pressure, σ is the deviatoric stress tensor, T is the temperature, k is the thermal conductivity, A is the rate of radiogenic heat production, ρ is the density, and c_p is the specific heat. In the calculations, the mantle is assumed to be a Boussinesq fluid with infinite Prandtl number, and flow in the mantle wedge is driven only by the subducting plate. The models use the finite-element code PGCtherm3D. This code is the three-dimensional version of PGCtherm2D, which has been previously benchmarked (van KEKEN et al. 2008) and used in other studies of different subduction zones (e.g., CURRIE et al. 2004; WADA et al. 2008; WADA and WANG 2009; WANG et al. 2015). PGCtherm3D was also used in a recent 3D modeling study of the northeast Japan subduction zone (WADA et al. 2015).

The three main units in the model are the overriding Caribbean plate, the subducting Cocos plate and the viscous mantle wedge; smaller units such as the sediment layer of the oceanic plate or the upper and lower continental crust are considered as part of these main units. The rigid overriding plate consists of a 35 km-thick crust and the top 5 km of upper mantle. This crustal thickness is consistent with the average Moho depth for Central America (MACKENZIE et al. 2008; MANEA et al. 2013). The geometry of the subducting plate is from KYRIAKOPOULOS et al. (2015). The subducting plate has a total thickness of

100 km, and it is assumed that this includes the oceanic sediments, crust and mantle lithosphere, as well as sub-lithospheric mantle that is entrained with the subducting lithosphere. The model domain extends from south Costa Rica to Nicaragua. We selected this region due to the strong along-margin variations in slab dip at depths 70 km, which changes from approximately 70° in Nicaragua to 45° in central Costa Rica. Further south, the Cocos plate is difficult to observe because of the lack of Wadati–Benioff seismicity and arc volcanism (PROTTI et al. 1995). The area may correspond to a slab window formed by subduction of the Cocos-Nazca spreading centre from late Miocene to late Pliocene (JOHNSTON and THORKELSON 1997; ABRATIS and WORNER 2001), as shown in Fig. 1. The x-axis in our models is approximately aligned with the trench, the y-axis is in the landward direction, and the z-axis is the depth. The origin of the model grid is at $-84°\mathrm{W}, 7°\mathrm{N}$, and therefore, the model domain extends from $x = 200$ to $x = 700$ km. The trench corresponds to the seaward boundary of the model. The location of the backarc (landward) boundary is taken to be where the top of the oceanic plate is at a depth of 300 km; this keeps our modeling domain in the upper mantle. As the slab dip varies along the strike of the subduction zone, the distance between the trench and backarc boundary is variable, ranging between 200 and 300 km from the wedge corner. Tests show that the location of the backarc boundary has only a minor effect on temperatures in the mantle wedge, as long as the distance from the wedge corner to the boundary does not change by more than 200 km along the strike.

The thermal parameters for each model material follow those used by HARRIS et al. (2010b) and are given in Table 1. The overriding plate has a velocity of 0 cm/year, and the subducting plate has an assigned convergence velocity of 9.1 cm/year, in agreement with the average velocity in our modeling area (DEMETS 2001). For the mantle wedge, we present models with either an isoviscous rheology (viscosity of 10^{21} Pa s) or a power-law (non-Newtonian) rheology. The latter is based on the flow law for dislocation creep of wet olivine (KARATO and WU 1993):

$$\eta = A(\dot{\epsilon})^{\left(\frac{1}{n}-1\right)}\exp\left(\frac{E}{nRT}\right) \tag{4}$$

Table 1

Radioactive heat generation (A), thermal conductivity (k), density (ρ) and heat capacity (c_p) for the subdomains in the numerical models

Subdomain	A (μW/m³)	k (W/mK)	ρ (g/cm³)	c_p (J/kg K)
Continental crust	0.2	2.9	3.3	1250
Mantle wedge	0.02	3.1	3.3	1250
Oceanic slab	0.2	2.9	3.3	1250

The thermal diffusivity (K) is given by $k/\rho c_p$

where $A = 28{,}968.6$ Pa s$^{1/n}$ is the pre-exponential factor, $\dot{\epsilon}$ is the strain rate, $E = 430$ kJ/mol is the activation energy, R is the universal gas constant, T is the temperature, and $n = 3$ is the power-law exponent.

The boundary conditions for the model are shown in Fig. 2. The top of the model is rigid with a temperature of 0 °C, and the bottom of the model corresponds to the base of the oceanic plate, which has a velocity of 9.1 cm/year and temperature of 1450 °C. Free-slip and insulating boundary conditions are used on the side boundaries at $x = 200$ km and $x = 700$ km, such that $v_x = 0$ and there is no

heat flow through these boundaries. The effect of the boundary conditions on the modeling results is discussed in Appendix. At the oceanic boundary, a fixed velocity of 9.1 cm/year is assigned. The temperatures along this boundary are based on the GDH1 oceanic plate cooling model (STEIN and STEIN 1992), using the age of the plate at each point along the trench. The complete oceanic boundary temperature can be seen in Fig. 2a. Three age domains are included. Lithosphere generated at the Cocos-Nazca Spreading Centre (CNS) is divided into two sections: CNS1 and CNS2 (von HUENE *et al.* 2000; BARCKHAUSEN *et al.* 2001). These segments have ages of 17.5–20, and

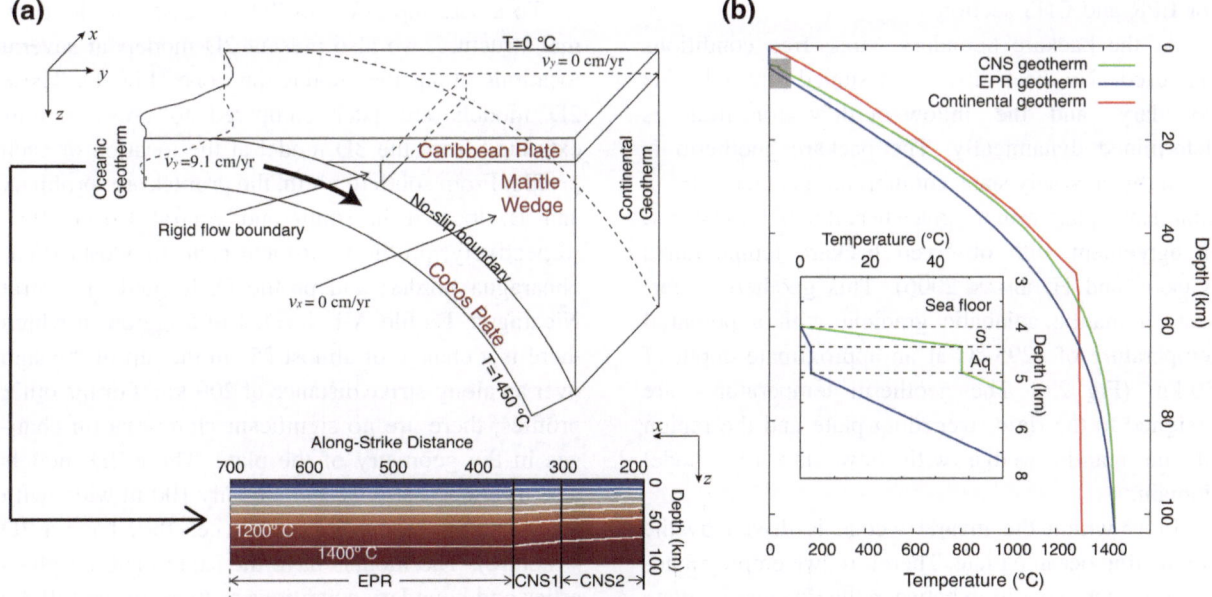

Figure 2

a Schematic diagram of the 3D model and boundary conditions. The 2D temperature boundary condition for the oceanic boundary is shown, with isotherms shown every 200 °C. Location of the boundary between CNS and EPR lithosphere is shown. **b** Typical geotherms for CNS (*green*) and EPR (*blue*) oceanic lithosphere. *Inset* shows the oceanic geotherms at shallow depths, with sediments (S) and aquifer (Aq) having thicknesses of 400 m and 500 m, respectively

20–24 Myear, respectively. An age of 24 Myear is used for lithosphere generated at the East Pacific Rise (EPR). The boundary between EPR and CNS2 lithosphere is found offshore the Nicoya peninsula (Fig. 1); in our coordinate system is located at approximately $x = 360$ km. At each point along the trench, the 1D oceanic geotherm incorporates the effects of sedimentation and sediment permeability, following LANGSETH and SILVER (1996), HARRIS and WANG (2002) and HARRIS et al. (2010b). In this area, temperatures at the uppermost oceanic crust (aquifer) may be affected by hydrothermal circulation. HARRIS et al. (2010b) showed that the offshore heat flow data are consistent with ventilated hydrothermal circulation for the EPR segment and insulated hydrothermal circulation for the CNS segment. In their models, hydrothermal circulation is allowed to continue to greater depths after subduction by introducing conductivity proxies in the crustal aquifer (DAVIS et al. 1997). Our geotherms use the thermal structure proposed by HARRIS et al. (2010b) for the oceanic boundary but do not include conductivity proxies for the aquifer. Our approach is thus similar to that employed by LANGSETH and SILVER (1996) and HARRIS and WANG (2002). Figure 2b shows typical geotherms for EPR and CNS sections.

At the backarc boundary, stress-free conditions are used. No heat flow is assumed through this boundary, and the inflow–outflow transition is determined dynamically. The backarc geotherm is given by a steady-state equilibrium geotherm for a continental plate with a surface heat flux of 90 mW/m², in agreement with observed backarc temperatures (CURRIE and HYNDMAN 2006). This geotherm intersects a mantle adiabatic gradient with a potential temperature of 1295 °C at an approximate depth of 50 km (Fig. 2b). The geotherm temperatures are assigned to the rigid overriding plate and the region of the mantle wedge with flow into the model domain.

Flow within the mantle wedge is driven by the subducting oceanic plate. Therefore, we employ a no-slip boundary condition between the downgoing plate and the lower boundary of the mantle wedge. The upper wedge boundary has a fixed velocity of 0 cm/year. For computational efficiency, a rigid vertical flow boundary is placed in the wedge. This is also

consistent with serpentinization of the forearc mantle, which is proposed to decouple the subducting slab from the mantle (MANEA et al. 2005, 2008; WADA et al. 2008). The boundary is located at the point at which the slab reaches a depth of 70 km, and it extends from the surface of the slab to the base of the upper plate. This location is consistent with observations that show that stagnation of the wedge corner is required to fit observations of a rapid landward increase in surface heat flow, from low values in the forearc to high values near the volcanic arc and into the backarc (WADA et al. 2008; WADA and WANG 2009). In the models, there are high gradients in pressure and temperature near the mantle wedge corner and along the surface of the slab, especially when a non-linear rheology is used (Eq. 4). Thus, our models use a variable element size in the direction of subduction, where elements are as small as a few meters in the mantle wedge tip and as large as ~ 10 km in the upper crust and deep slab. Along the strike of the subduction zone, the element width is 2 km. These sizes are based on benchmark tests (van KEKEN et al. 2008; BENGTSON and van KEKEN 2012). In total, we employ 1,138,800 cubic elements that give a total of 9,262,617 grid nodes.

To assess the effect of 3D processes on the thermal structure, we also present 2D models at several locations along the subduction zone (Fig. 1). These 2D models are later compared to cross-sections extracted from the 3D model at the location of each profile. From south to north, the profiles are: profile A and B, located in south and central Costa Rica, respectively; profile C, located near the Costa Rica-Nicaragua border; and profile D, located in central Nicaragua. Profile A is located in a region in which there is a change of almost 25° in the dip of the slab over an along-strike distance of 200 km. For the other profiles, there are no significant along-margin changes in the geometry of the plate. These 2D models also use PGCtherm3D, but are only 100 m wide, with no variations in slab geometry (i.e., they have a 2D structure). The models have the same material properties and boundary conditions as those in the full 3D model. Given that convergence of the Cocos plate in the MAT along Costa Rica-Nicaragua is mostly trench-perpendicular ($\sim 10°$ obliquity) (DEMETS 2001) and that the trench is almost straight in our study region, our 2D profiles are taken parallel to the convergence direction.

3. Results

Flow in the mantle wedge is a primary control factor on the thermal distribution of a subduction zone, especially beneath the volcanic arc. The nature of flow depends on the rheology of the mantle wedge (e.g., van KEKEN *et al.* 2002). Shear-wave splitting studies demonstrate that the dislocation creep is the dominant deformation mechanism in the mantle wedge (KNELLER and van KEKEN 2008; LONG and SILVER 2008; HOERNLE *et al.* 2008; SOTO *et al.* 2009). However, we also present isoviscous models to gain a basic understanding of the factors that control the direction and magnitude of mantle wedge flow.

It is illustrative to start with a brief discussion of 2D models. Figure 3a shows the basic structure of a 2D corner flow for an isoviscous mantle wedge for profile A. In a subduction zone, the downward motion of the slab drags the overlying mantle through viscous coupling (yellow ellipse). This creates a region of low pressure near the wedge corner (green circle) which induces flow that brings hot mantle from the backarc to replenish the lost material. The magnitude and size of this low-pressure area depends on the dip of the slab, with shallower dips tending to create lower pressures and a larger low-pressure region (TURCOTTE and SCHUBERT 2002; MANEA and GURNIS 2007). As a result of this flow pattern, the thermal structure of the subduction zone has two distinct regimes. The forearc is relatively cool, owing to conductive cooling by the underlying oceanic plate; the arc and backarc regions are heated by the mantle wedge flow.

For an isoviscous rheology, the constant viscosity of the mantle allows flow through the backarc boundary at any depth. However, a dislocation creep rheology is temperature dependent (Eq. 4) and thus flow is generally limited to temperatures greater than 1200 °C (van KEKEN *et al.* 2002; CURRIE *et al.* 2004). Figure 3b shows a 2D model that uses a dislocation

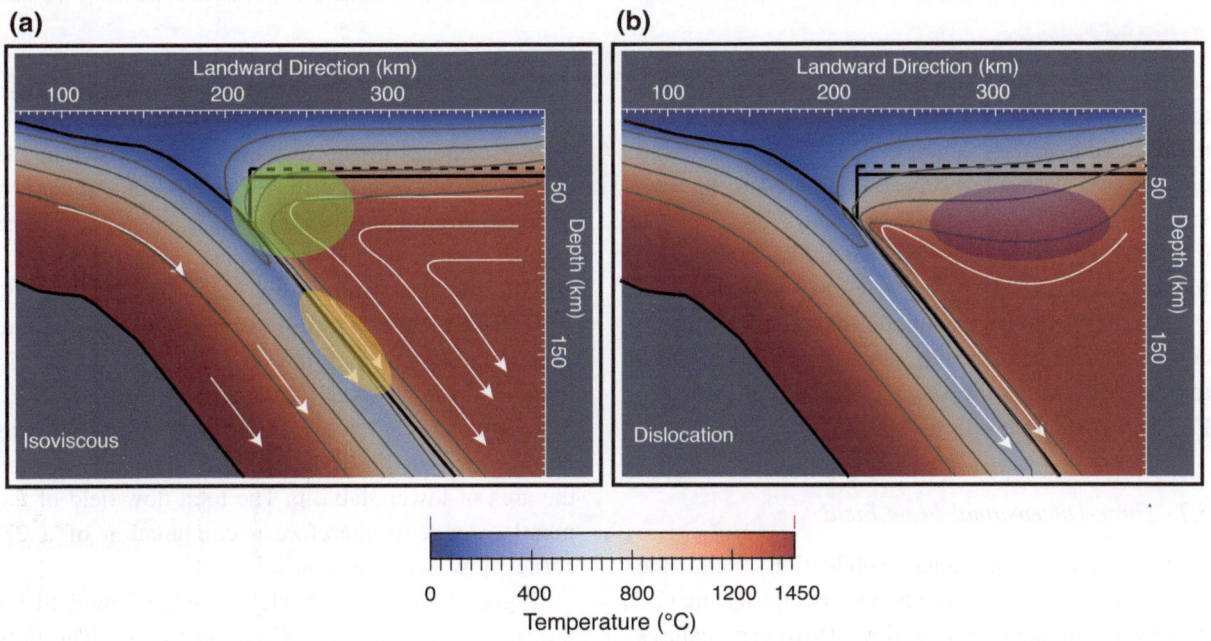

Figure 3

Two-dimensional (2D) corner flow models for **a** an isoviscous and **b** a dislocation creep wedge. Figures are for profile A. *Black dashed line* indicates the continental Moho. *White arrows* are streamlines. *Green circle* shows the low-pressure corner, which induces flow that brings hot mantle from the backarc. *Yellow ellipse* shows viscous coupling between slab and mantle wedge. *Purple ellipse* shows the stagnant lid for a dislocation creep rheology. Temperature contours are shown every 300 °C

111

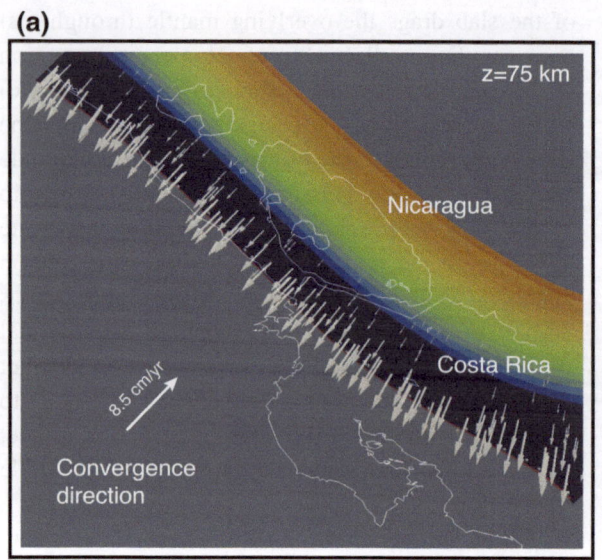
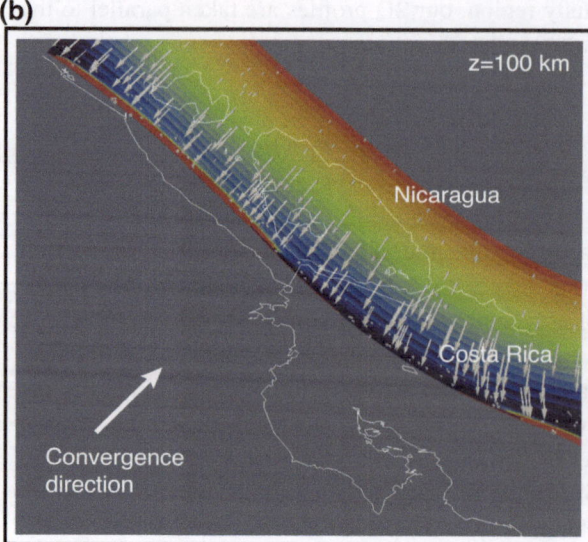

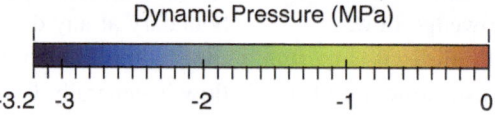

Figure 4

Dynamic pressure for the 3D model in horizontal planes located at depths of **a** 75 km and **b** 100 km. *Color scale* the magnitude in MPa. *Gray arrows* mantle wedge flow along each plane. Mantle wedge flow is mostly parallel (2D) to convergence direction below Nicaragua and northern Costa Rica. From mid-Nicoya to southern Costa Rica, mantle flow has an along-strike component (3D). Mantle flow departs from the 2D corner flow in response to changes in the low-pressure region (*blue*), which in turn are associated with changes in the dip of the slab

creep mantle wedge rheology. With this rheology, flow enters the model domain at a greater depth along the backarc boundary, and it is more strongly focused upward into the wedge corner, compared to the isoviscous model. As a result, the temperature in the mantle wedge corner is ∼100 to 200 °C higher than for the isoviscous case. However, the upper part of the mantle wedge is essentially stagnant, leading to cool temperatures in this area (∼800 °C). This region is known as the stagnant lid and is shown in Fig. 3b by a purple ellipse.

3.1. Three-Dimensional Flow Field

In a three-dimensional subduction zone, the subducting plate also drags the overlying mantle downward, inducing corner flow. However, features such as along-strike changes in slab dip, obliquity, or trench curvature will induce lateral pressure gradients, which can lead to along-strike flow (KNELLER and van KEKEN 2008; BENGTSON and van KEKEN

2012). For the case of Costa Rica-Nicaragua, along-strike variations in the dip of the Cocos plate (Fig. 1) induce changes in the dynamic pressure within the mantle wedge (Fig. 4). In response to this change, a component of along-strike flow is produced in the mantle wedge, with flow directed toward the southeast, from the region of high slab dip (70°) below Nicaragua to low slab dip (45°) below Costa Rica. This is consistent with the idea that a shallower dip induces lower pressures in the mantle wedge (e.g., TURCOTTE and SCHUBERT 2002; MANEA and GURNIS 2007), and thus the corner flow is deflected towards the area of lower slab dip. The total flow field of the mantle wedge is therefore a combination of a 2D corner flow and lateral flow.

Figure 5 shows the modeled wedge flow field for the isoviscous and dislocation creep cases. The flow field is visualized through streamlines that track the path of particles. The color scale denotes along-strike velocity (v_x), with positive values indicating north-westward flow. The dominant component of the flow

(a)　　　　　　　　　　　　　　　　　　　　**(b)**

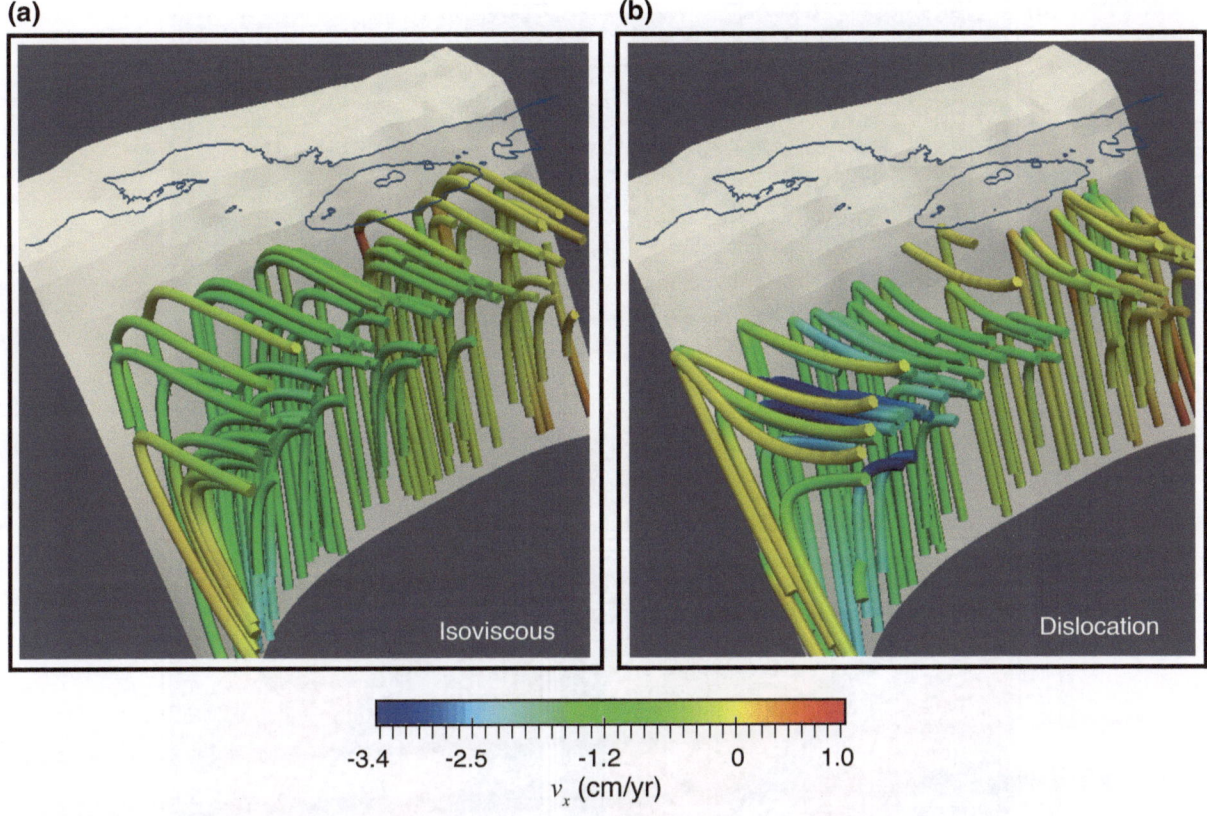

Figure 5

Three-dimensional flow field in the mantle wedge for **a** an isoviscous and **b** a dislocation creep rheology. *Color scale* denotes along-strike velocity, v_x, with *blue* denoting southeast flow and red denoting northwest flow. Maximum v_x for the isoviscous and dislocation creep models (not shown in figure) is -2.5 and -4.0 cm/year, respectively

field is the slab-driven corner flow, where particles are pulled toward the wedge corner and then descend with the subducting plate. As seen in the 2D models, a dislocation creep rheology leads to a strong focusing of flow toward the wedge corner (Fig. 3b), whereas an isoviscous wedge has subhorizontal flow in the upper part of the wedge (Fig. 3a). In both cases, the flow also exhibits an along-strike component. This is largest below Costa Rica in the southern part of the model domain, where there are along-strike changes in slab dip. In the isoviscous case (Fig. 5a), a maximum along-strike velocity of -2.5 cm/year is observed, whereas the dislocation creep rheology results in a maximum along-strike velocity of over -4 cm/year. This difference is caused by a lower mantle viscosity due to the higher temperatures in the dislocation creep model; flow is

concentrated in the central wedge region where temperatures are highest. The along-strike flow is minimal in the northern part of the model area, as there are only minor dip variations here.

It is important to mention that our current choice of boundary conditions could have an effect on the observed mantle flow. The $v_x = 0$ restriction imposed on the side boundaries affects the flow pattern, and as a result, our 3D model may underestimate the along-strike flow component. A detailed discussion of the effect of the boundary conditions in mantle flow is given in Appendix.

3.2. Three-Dimensional Thermal Structure

Our three-dimensional thermal model for Costa Rica-Nicaragua is shown in Fig. 6a for a dislocation

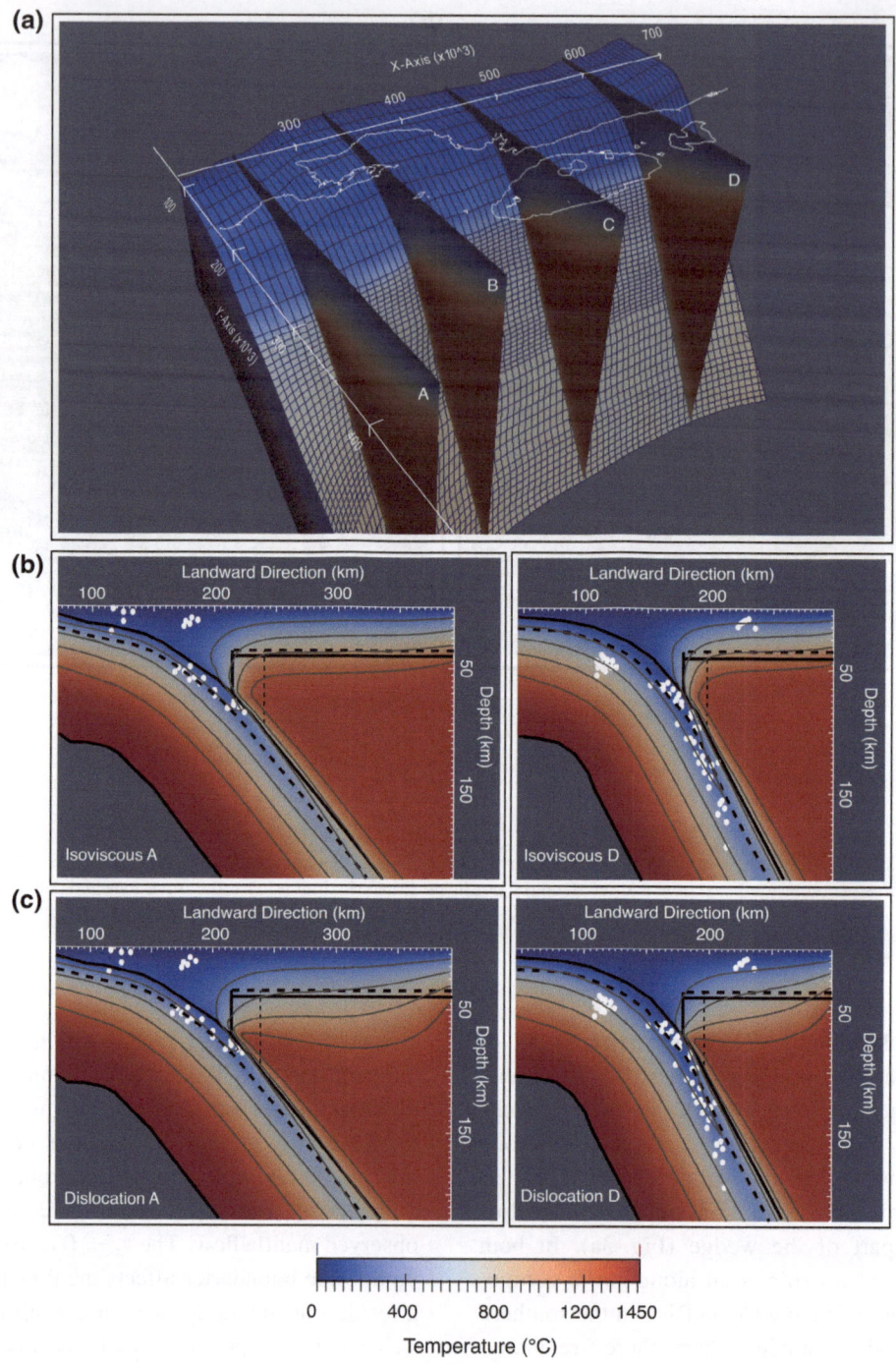

Figure 6
a Three-dimensional (3D) thermal model of the Costa Rica-Nicaragua subduction zone, with location of profiles A through D shown. The rheology for this model is dislocation creep. Cross-sections extracted from 3D model for profiles A and D are shown in two-dimensional view for **b** an isoviscous and **c** a dislocation creep rheology. Temperature contours are shown for every 300 °C. *White dots* are seismicity from the TUCAN (ABERS *et al.* 2004) and CRSEIZE (http://es.ucsc.edu/~hdeshon/crseize_homepage.html) deployments. *Thick dashed lines* represent the oceanic and continental Moho. *Thin dashed line* represents location of volcanic arc geotherm (Fig. 10a)

creep rheology. The location of profiles A through D is shown. The detailed model results are also shown for profiles A and D for an isoviscous (Fig. 6b) and dislocation creep rheology (Fig. 6c). The large-scale temperature field for profile A is similar to that seen in the 2D models (Fig. 3). If the wedge has an isoviscous rheology, isotherms are subhorizontal in the shallow backarc (Fig. 6b). In contrast, a dislocation creep rheology leads to a strong focusing of flow from depth into the wedge corner and a cool stagnant lid is created in the backarc (Fig. 6c).

To assess the effects of along-margin mantle flow, Figs. 7 and 8 show the difference between 3D and 2D mantle wedge temperatures and the along-strike mantle flow for an isoviscous and non-Newtonian rheology, respectively. For both cases, the largest temperature difference is observed in profile A. For profiles B, C and D, the magnitude of the difference decreases in the northwestward direction. This result is consistent with the location of strong along-margin slab curvature, mostly located in the southeast side of our domain area (Fig. 1), which creates pressure gradients that drive the flow in the southeastward direction (Fig. 4).

For an isoviscous rheology, the along-strike flow decreases temperatures in the 3D model by 10–30 °C with respect to the 2D model along profile A (Fig. 7). The along-strike component of the flow has a magnitude of $\sim$1.5 cm/year. For profile B, temperatures do not change appreciably in the mantle wedge, despite the stronger along-strike flow ($\sim$2 cm/year). The difference between profiles A and B can be understood by considering the motion of a single fluid element in the mantle. Given the higher along-strike velocities along profile B, a fluid element in this area would have less time to cool than a fluid element in the vicinity of profile A at a given distance from the backarc boundary. The result is that 3D flow cools down the mantle more efficiently for profile A than for profile B. For profiles C and D, there are no variations in mantle wedge temperatures between 2D and 3D models and almost no along-strike flow (<1 cm/year), indicating the mantle flow in this region is mostly 2D corner flow.

For a mantle with a dislocation creep rheology (Fig. 8), the temperature differences between 2D and

3D models are largest for profile A, but the effect of 3D flow is opposite to that in the isoviscous case. For this profile, the mantle wedge in the 3D model is up to 50 °C hotter than the 2D model, with the largest change in the uppermost mantle (stagnant lid). The along-strike flow has a maximum magnitude of $\sim$4 cm/year for profile A. The thermal changes decrease in the northwest direction. Profile B still shows a difference of 20 °C, with an along-strike flow of $\sim$2.5 cm/year. For profiles C and D, only minor changes in the mantle wedge temperatures are observed. In these models, the viscosity of the mantle wedge depends on both the temperature and stress (Eq. 4), which results in a more complex feedback between the flow field and thermal structure than for the isoviscous case. In general, however, the along-strike flow for the dislocation creep case has a relatively higher magnitude than in the isoviscous case. This rapid flow may limit the thickness of the stagnant lid, and as a result, the upper part of the mantle wedge will be somewhat warmer compared to a purely 2D model.

On all profiles, moderate changes in thermal structure are observed in a local region (<5 km wide) in the wedge corner near the stagnant wedge boundary (Figs. 8, 9). This region is characterized by strong pressure and thermal gradients that require high resolution to properly resolve for the temperature (van KEKEN et al. 2002). Thus, the temperatures in the wedge corner have some uncertainty that may arise from numerical artifacts.

4. Thermal Structure of the Costa Rica-Nicaragua Subduction Zone

The thermal structure of a subduction zone is a crucial control on key processes, including magmatism, slab metamorphism and dehydration, and earthquake distribution. Here, we examine the temperatures of the subducting slab and mantle wedge of our full 3D model with a dislocation creep rheology (Fig. 6a). Seismic anisotropy studies for Central America suggest flow by dislocation creep is the most appropriate rheology (HOERNLE et al. 2008).

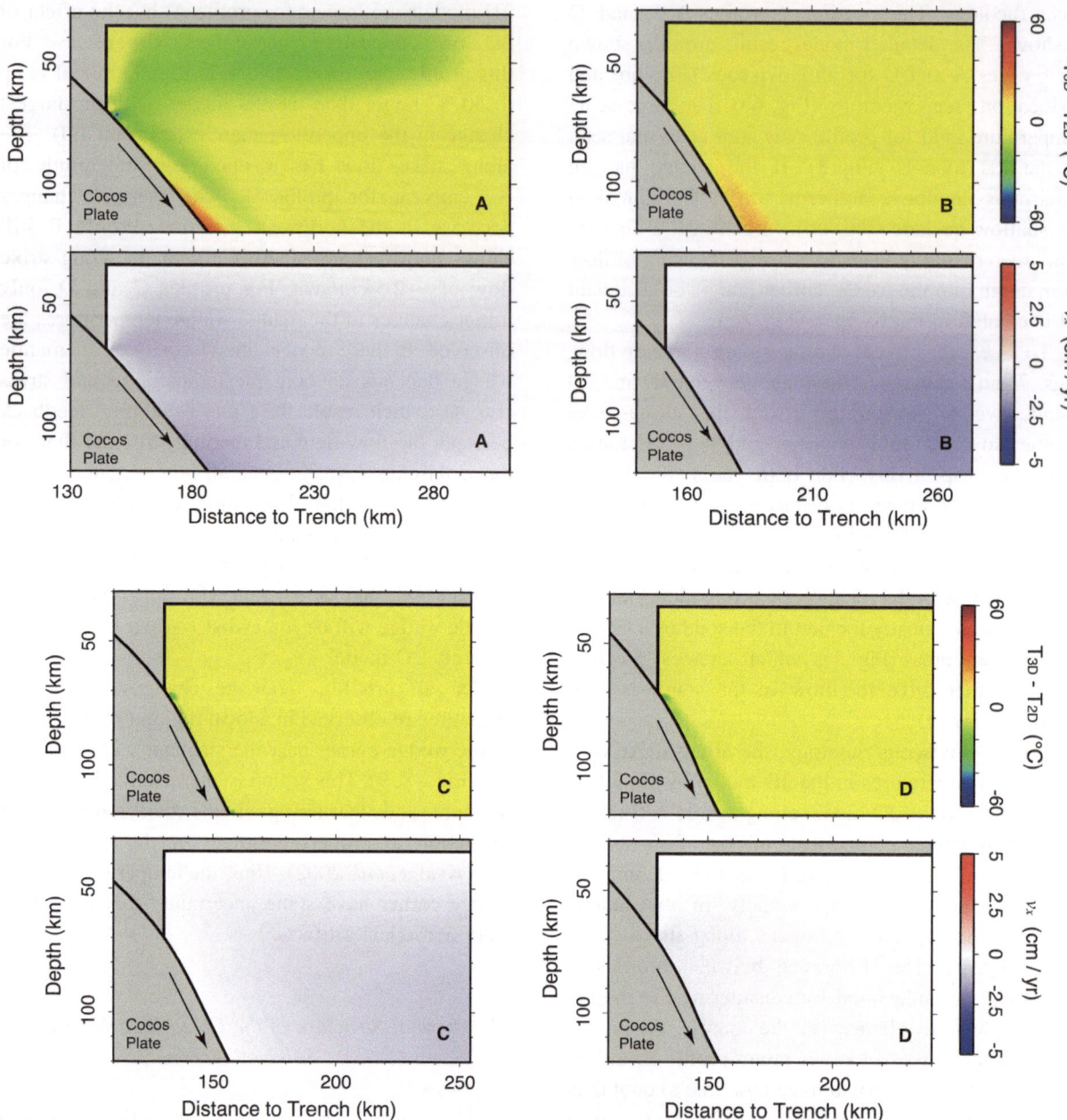

Figure 7
Temperature difference (*T3D–T2D*; *upper plot*) and along-strike velocity component (*v$_x$*; *lower plot*) for profiles A through D with an isoviscous (10^{21} Pa s) mantle wedge. *Black arrow* denotes subduction direction. *Negative sign* in the along-strike velocity indicates southeastward flow

4.1. Megathrust Earthquake Seismogenic Zone

We first address the implications of the 3D thermal model for megathrust earthquakes, which are earthquakes that occur on the subduction interface. The seismogenic zone corresponds to the part of

the interface that exhibits velocity-weakening behavior, and this may depend on interface temperatures (e.g., HYNDMAN *et al.* 1997). The updip limit is usually placed at temperatures of 100–150 °C, while the downdip limit is at either 350–450 °C or the

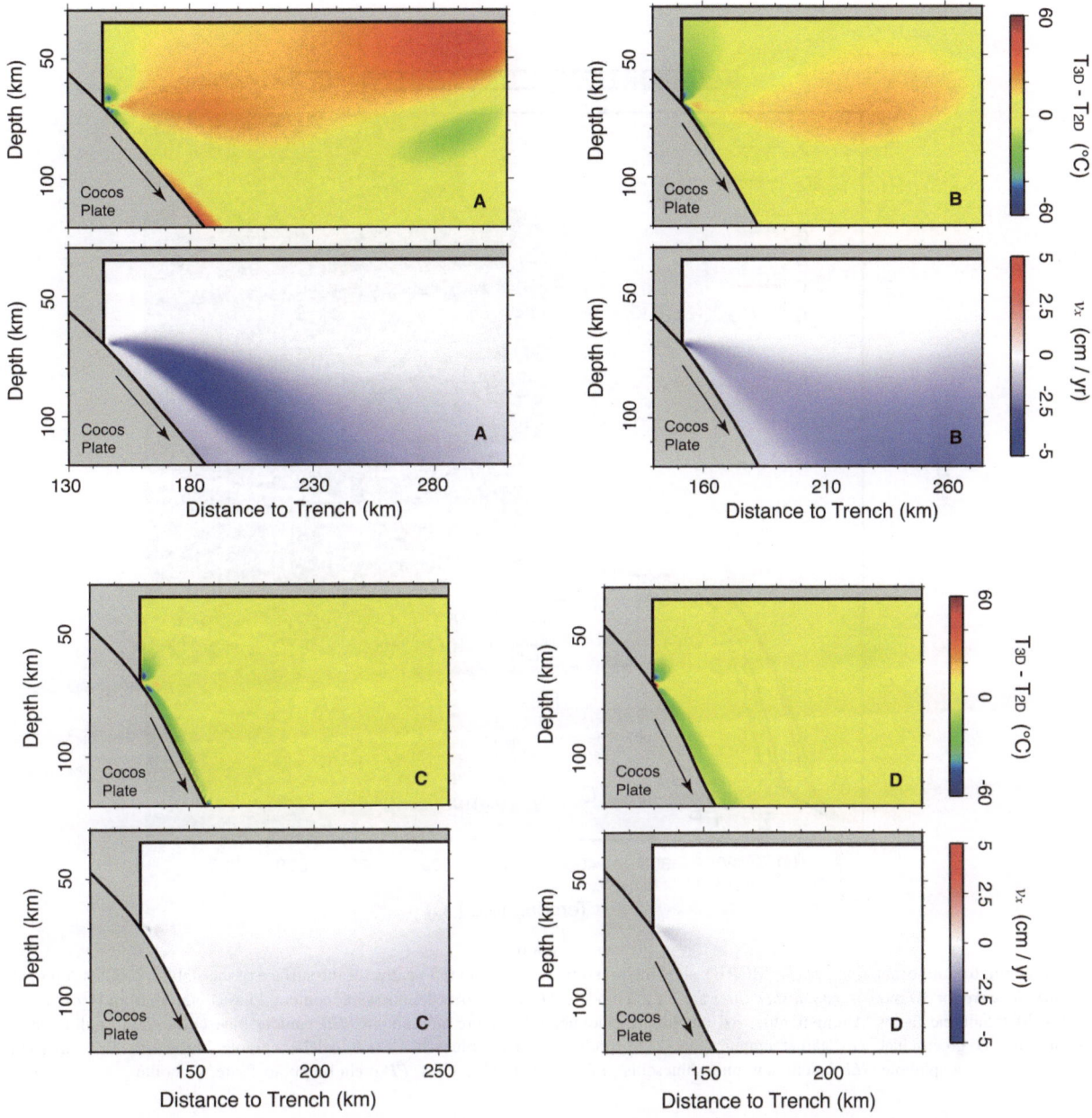

Figure 8
Same as in Fig. 8, but for a non-Newtonian mantle wedge that flows through dislocation creep (Eq. 4)

intersection between the interface and the upper plate Moho.

Figure 9 shows pressure–temperature (PT) paths for the surface of the Cocos plate for profiles A through D, for both the 3D (solid line) and 2D (dashed line) models with a dislocation creep mantle; note that pressure has been converted to depth using the material density. A temperature of 100 °C occurs at a depth of 15–20 km for the southern profiles (A and B), and at ~10 km depth for the northern profiles (C and D). For all profiles, a temperature of 350 °C occurs at depths greater than 60 km. The

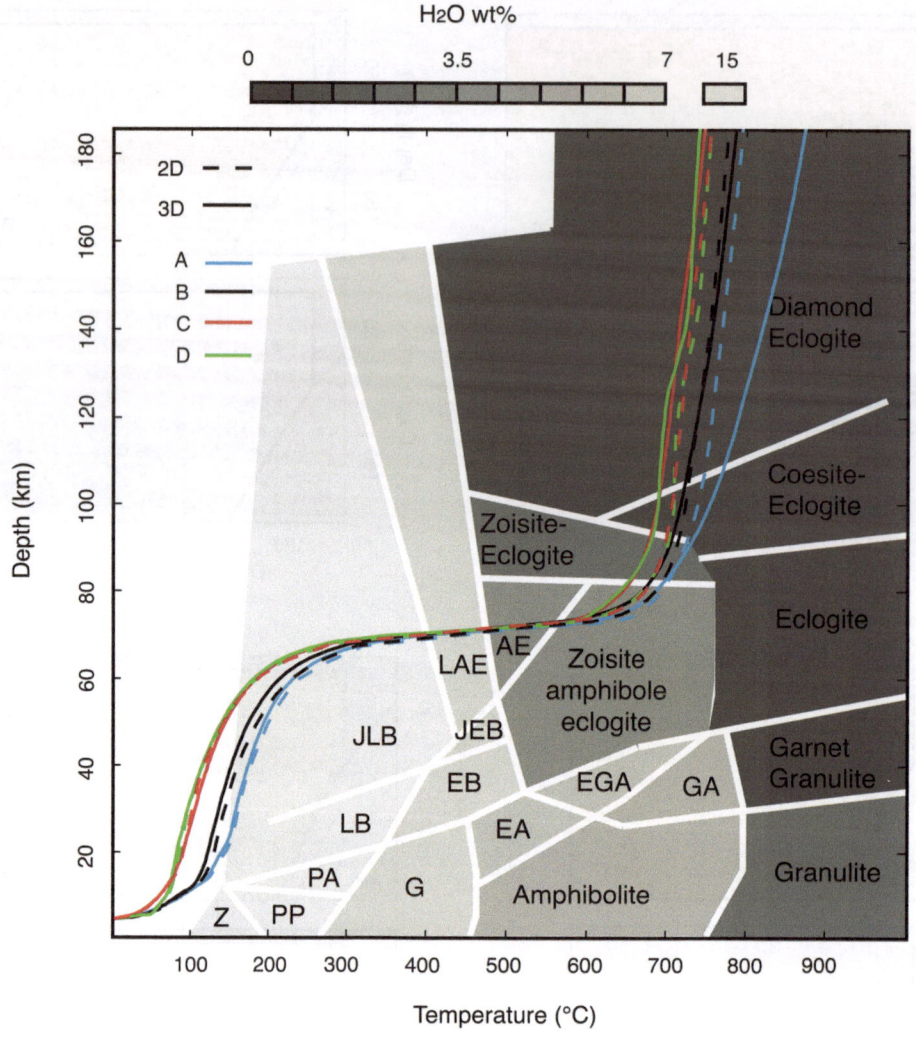

Figure 9

Phase diagram for mid-ocean ridge basalt (MORB), showing pressure–temperature (PT) paths for the surface of the slab for profiles A through D. *Solid lines* are for 3D models, and *dashed lines* are for 2D models. *Shades* of *gray* denote water content. Phase diagram from HACKER *et al.* (2003). Metamorphic facies: *A* amphibolite, *AE* amphibole eclogite, *EA* epidote amphibolite, *EB* epidote blueschist, *EGA* epidote garnet amphibolite, *G* greenschist, *GA* garnet amphibolite, *JLB* jadeite lawsonite blueschist, *JEB* jadeite epidote blueschist, *LAE* lawsonite amphibole, *JLB* jadeite lawsonite blueschist, *LB* lawsonite blueschist, *PP* prehnite pumpellyite, *Z* zeolite

Moho depth for the Caribbean plate is 35 km (MACKENZIE *et al.* 2008). Thus, it is likely that the downdip limit of the seismogenic zone corresponds to the Moho intersection and not the 350 °C isotherm. This is consistent with the conclusion of HARRIS *et al.* (2010b).

Figure 10 shows a map view of the location of the 100 °C isotherm on the subduction interface for our 3D model, as well as the location of the Moho intersection (corresponding to a slab depth of 35 km).

There is an abrupt seaward shift in the position of the 100 °C isotherm at the position of the Nicoya Peninsula. This point marks the change from oceanic lithosphere created at the East Pacific Rise (EPR) in the north and Cocos-Nazca Spreading Centre (CNS) in the south (Fig. 1). At the trench, the incoming EPR lithosphere is cooler than the CNS lithosphere (Fig. 2b), which translates to a cooler subduction interface and a more landward location of the 100 °C isotherm for the EPR segment. For the CNS segment,

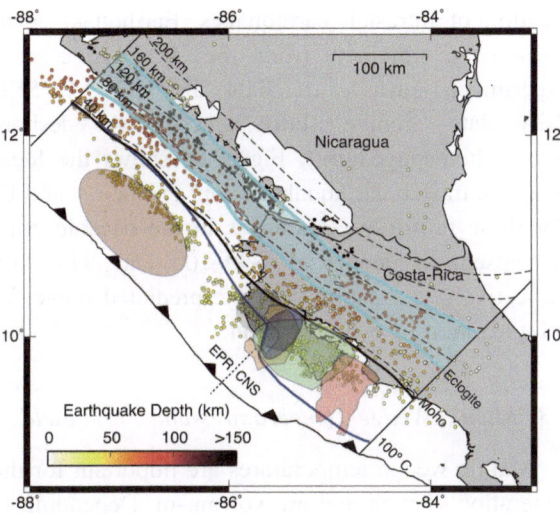

Figure 10

Map of the modeling area, with earthquake locations obtained by the TUCAN (ABERS *et al.* 2004) and CRSEIZE (http://es.ucsc.edu/ ~hdeshon/crseize_homepage.html) deployments. The 100 °C isotherm for the slab surface is shown in *dark blue*. The range over which the entire oceanic crust transforms to eclogite is shown in *light blue*. The intersection of the slab surface with the upper plate (Caribbean) Moho is shown in *black*. Slab depth contours are shown with dashed *black lines*. Approximate rupture areas for the 1900 (*black* Mw 7.2), 1950 (*green* Mw 7.7), 1978 (*blue* Mw 6.9), 1990 (*red* Mw 7.0), and 2012 (*orange* Mw 7.6) Nicoya megathrust earthquakes (YUE *et al.* 2013; PROTTI *et al.* 2014), as well as the 1992 (*brown* Mw 7.7) Nicaragua earthquake (KANAMORI and KIKUCHI 1993) are shown. The *dotted line* divides East Pacific Rise (EPR) and Cocos-Nazca Spreading Centre (CNS) lithospheres

the predicted seismogenic zone starts at 20–30 km from the trench and terminates at 70 km from the trench. In contrast, a seismogenic zone width of 20–30 km is predicted for the southeastern EPR segment, starting at 70–80 km from the trench. The width of the predicted seismogenic zone then decreases dramatically to a width of 10 km or less close to the border of Costa Rica with Nicaragua, and remains like that further northwest along the margin.

Figure 10 also shows rupture areas for several historical megathrust earthquakes in the Nicoya region (YUE *et al.* 2013; PROTTI *et al.* 2014). Each rupture area is roughly constrained by our model-predicted megathrust seismogenic zone. Of particular importance is the reduction in width of the predicted seismogenic zone in northern Nicoya, which correlates well with the reduction in rupture area for the 1900 (black), 1999 (blue) and 2012 (orange) Nicoya earthquakes. Further north, however, our results do

not fit the inferred rupture area of the 1992 Nicaragua earthquake (KANAMORI and KIKUCHI 1993; WANG *et al.* 2015). Our model predicts a cool subduction interface, such that the critical temperature for the updip limit of the seismogenic zone is not reached until a depth of ~ 35 km. This could indicate that our chosen geotherm for the north EPR section is too cold. We also note that the published depth of this earthquake is 45 km, which places it well below the subduction interface (KIKUCHI and KANAMORI 1995), indicating that this earthquake may not have occurred on the plate interface.

In general, the predicted megathrust seismogenic zone in our 3D models is in good agreement with that of previous 2D thermal models for this region (HARRIS *et al.* 2010b), as well as other studies of the megathrust seismogenic zone that rely on more direct methods, such as earthquake locations and GPS observations (NEWMAN *et al.* 2002; DESHON *et al.* 2006; SCHWARTZ and DESHON 2007). It should be noted that the temperatures of the shallow plate interface are primarily determined by the thermal structure of the oceanic plate at the trench, as well as its geometry and convergence rate; mantle wedge flow does not significantly affect the shallow interface temperatures. The good agreement between our 3D models and previous 2D models suggests that along-margin heat transport is negligible for the shallow subduction interface. As shown in Fig. 10, the transition in the location of the 100 °C isotherm between the EPR and CNS segments occurs over an along-strike width of less than 30 km, suggesting that 2D models are suitable for modeling interface temperatures for much of this subduction zone.

4.2. Slab Temperatures and Eclogitization of the Cocos Plate

The width of the megathrust seismogenic zone is controlled by the temperature along the interface between the oceanic plate and the overriding plate. For deeper sections within the plate, temperature also regulates the depth of release of water stored in the plate and the distribution of intraslab earthquakes (KIRBY *et al.* 1996). As the Cocos plate subducts, it is progressively being exposed to greater pressures and temperatures, causing it to undergo several metamorphic and dehydration reactions. The main components of the oceanic crust and mantle lithosphere are mid-ocean ridge basalt (MORB) and harzburgite, respectively (IRIFUNE 1993). For the case of the oceanic crust, the MORB eventually transforms to eclogite. This process dehydrates the slab and significantly increases its density. For the oceanic mantle lithosphere, hydration of harzburgite leads to serpentinization of the mantle. Serpentinite is usually stable until depths of 60–70 km for most subduction zones (SCHMIDT and POLI 2003). At larger depths, serpentinite starts to dehydrate. In general, water trapped in serpentinized harzburgite may be the most efficient transport mechanism of water to the deep mantle (RÜPKE et al. 2004). For this study, however, we only investigated the 3D temperature distribution within the oceanic crust. A full discussion of lithosphere mantle temperatures and dehydration is given in ROSAS et al. (2015, submitted).

Figure 9 shows the PT paths of the top of the Cocos plate superimposed on a phase diagram for MORB (HACKER et al. 2003). The PT paths show minimal variations between 3D and 2D models at shallow depths. For all 4 profiles, the surface of the slab goes through the jadeite lawsonite blueschist (JLB) and lawsonite amphibole eclogite (LAE) facies, before entering the amphibole eclogite (AE) facies at 70–75 km depth. At this point, the oceanic crust is almost dry, with less than 2 wt% H_2O. At greater depths, differences between 3D and 2D models are larger due to along-strike flow. For profile A, the difference between 3D and 2D models steadily increases with depth. At a depth of 180 km, the difference is 80–90 °C. For profiles B, C and D, the difference is much less (10–20 °C).

Figure 10 shows a map view of the predicted location of the transition to eclogite (light blue) in the oceanic crust for our 3D model, assuming no kinetic delay. The seaward boundary of the region represents the point at which the top of the slab transforms to eclogite (Fig. 9), whereas the landward boundary represents the point at which the entire oceanic crust (assumed to have a thickness of 7 km) is predicted to undergo complete eclogitization. The depth for complete eclogitization of the oceanic crust ranges between 120 and 160 km along our modeling area. The predicted phase change can be compared to the

location of intraslab earthquakes. Earthquake locations were obtained from the TUCAN seismic experiment (ABERS et al. 2004) and UCSC-CRSEIZEA data archive (http://es.ucsc.edu/~hdeshon/crseize_homepage.html). Figure 6c shows the location of intraslab earthquakes along Profiles A and D. For these profiles, earthquakes in the oceanic crust are absent at depths of >90–100 and >170 km, respectively, consistent with the predicted range for eclogitization of our 3D model.

4.3. Mantle Wedge Temperatures and Flow Field

Mantle wedge temperatures are important for the generation of melt and arc volcanism. Depending in the amount of water in the mantle wedge, temperatures of 1100–1300 °C are needed for melting (SCHMIDT and POLI 1998). Figure 11 shows the geotherms along a vertical line located at a point where the slab reaches a depth of 100 km, consistent with the global average location of the volcanic arc

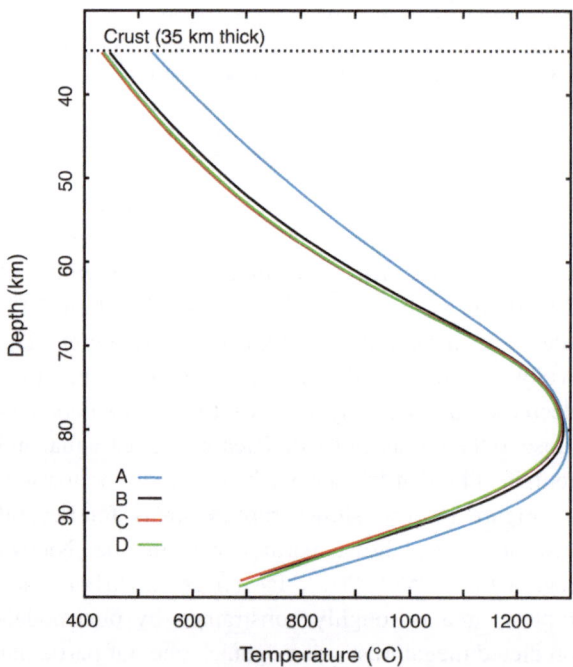

Figure 11

Volcanic arc geotherms for profiles A through D for our 3D thermal model. Geotherms were taken along a vertical line located at a point where the slab reaches a depth of 100 km (see Fig. 6c for profiles A and C). *Thin dashed line* shows the Moho of the continental crust

(SYRACUSE and ABERS 2006). For profile A, temperatures are up to 100 °C higher than for the other profiles. As shown in Fig. 8, 3D modeling predicts larger temperatures than 2D modeling (up to 40 °C) below the volcanic arc for this profile. It is unclear how this will affect melt production and arc volcanism, as melting also depends on the water content (SCHMIDT and POLI 1998). Assuming a similar water content for the mantle wedge along the strike, the higher temperatures observed below Costa Rica would suggest more melting in this region than for Nicaragua. However, seismic studies near the trench show that the incoming Cocos plate might carry large amounts of water along its Nicaraguan section, where significant fracturing occurs in the outer rise prior to subduction (ABERS et al. 2003; SYRACUSE et al. 2008; RYCHERT et al. 2008; Van AVENDONK et al. 2011; DINC et al. 2011). Thus, the Cocos plate may release more water into the wedge below Nicaragua than for Costa Rica, which can significantly affect the generation of melt.

As discussed in Sect. 3.1, an important result from our 3D models is that the observed along-strike changes in the slab geometry can induce along-strike mantle flow of up to 4 cm/year (Fig. 4b). In our models, significant along-strike flow toward the southeast is predicted below Costa Rica. This flow is driven by the southward decrease in dip angle of the subducting plate. In contrast, geochemical and seismic studies indicate a northward lateral flow, with a magnitude of 6–19 cm/year (HOERNLE et al. 2008). Our models show that such flow is not driven by along-strike changes in slab dip and trench curvature. If these observations are correct, an additional driving mechanism for along-strike flow must be considered. One possibility is mantle flow through the slab window just to the southeast of the study area (Fig. 1). Such flow may be enhanced by rollback of the Cocos plate as it subducts. Given that our moderate along-strike flow can increase temperatures by up to 40 °C in the mantle wedge (Figs. 8, 11) and by 80–90 °C along the slab at depths >180 km (Fig. 9) along profile A, we expect the 6–19 cm/year flow predicted by geochemical studies to have an even larger effect on mantle and slab temperatures. We are currently working on numerical models that

incorporate this flow to assess its effect in the thermal structure of the subduction zone.

5. Conclusions

Previous 3D thermal modeling studies of subduction zones investigated the relation between flow and anisotropy (KNELLER and van KEKEN 2008) or the effects of obliquity and trench curvature on slab surface temperatures (BENGTSON and van KEKEN 2012). However, these studies do not discuss the thermal structure of the wedge in detail. WADA et al. (2015) presented a 3D thermal model for the subduction zone of northeast Japan and found that the obliquity of the trench generates an along-strike flow component that results in a different thermal structure than for a 2D corner flow model. In our study, we present the first 3D model of the Costa Rica-Nicaragua subduction zone.

The key conclusions of this work are:

1. Velocity field for the mantle:
   Variations in the slab dip in the Costa Rica-Nicaragua subduction zone lead to along-strike mantle flow, with a maximum magnitude of 2.5 cm/year for an isoviscous wedge and 4 cm/year for a wedge that deforms through dislocation creep (approximately 40 % of the slab convergence rate). The predicted flow direction is toward the southeast, in the direction of decreasing slab dip. This is opposite to the flow direction inferred from geochemical and seismic observations (HOERNLE et al. 2008), which suggests that an additional mechanism to create along-margin flow is needed for this subduction zone. Just south of our modeling area, there appears to be a slab window that is created by the subduction of the Cocos-Nazca spreading centre (JOHNSTON and THORKELSON 1997; ABRATIS and WORNER 2001). Future 3D models should investigate how this may affect the mantle wedge flow field, especially below Costa Rica.

2. Thermal effects of 3D mantle flow:
   The thermal structure of the subduction zone depends on the rheology of the mantle wedge. In addition, along-strike flow can change the

temperature of the mantle wedge. In our 3D models, we find that along-strike flow of an isoviscous wedge results in cooler temperatures below the volcanic arc with respect to the corresponding 2D model. A dislocation creep rheology has the opposite effect: higher along-strike flow velocities lead to a hotter 3D model. The temperature differences between 2D and 3D are up to 50 °C and are largest below central Costa Rica (profile A), where along-strike flow has the highest magnitude. Given that dislocation creep is considered the primary mechanism for mantle wedge deformation and that dislocation creep has the opposite effect on temperatures than the isoviscous case, we conclude that 3D isoviscous are probably not well suited for studies of 3D subduction zone thermal structure.

3. Slab temperatures in Costa Rica-Nicaragua:
The temperatures of the shallow subducting plate (<35 km depth) are not significantly altered by 3D effects. Our predicted megathrust seismogenic zone correlates well with that obtained in other 2D modeling studies (HARRIS et al. 2010b) in the vicinity of the Nicoya peninsula. It also matches rupture areas of historical earthquakes in this area. For the Nicaragua section, our model suggests a very narrow seismogenic zone (<10 km wide). This does not fit the observed rupture area of the 1992 earthquake (Fig. 9), and this discrepancy needs to be explored in more detail. For the deeper parts of the slab, temperatures are affected by mantle wedge flow, resulting in difference of up to 80–90 °C between 2D and 3D modeling. This can affect factors such as eclogitization and dehydration of the subducting oceanic plate, as well as the stability of serpentinized mantle within the subducting mantle.

4. Applicability of two-dimensional models to the Costa Rica-Nicaragua Subduction Zone:
Our 3D models show that changes in slab dip lead to along-strike flow up to 4 cm/year, which results in temperature variations up to 50 °C for the subducting oceanic plate and mantle wedge compared to 2D models. This variation may have implications for metamorphism and dehydration of the deep slab and mantle wedge melting. In addition, if the significant along-strike flow

(6–19 cm/year) from geochemical and seismic studies (HOERNLE et al. 2008) is correct, it is reasonable to expect thermal changes that are much larger than those observed in our study. This indicates that 2D models are not suitable for modeling the thermal structure of the Costa Rica-Nicaragua subduction zone at mantle depths, as they are unable to capture all the complexity of mantle wedge dynamics.

6. Appendix: Effect of side boundary conditions on mantle wedge flow

In Sect. 2, the side boundaries of the model domain are described as free-slip, insulated boundaries. This no-flow condition ($v_x = 0$) forces the mantle to follow a 2D corner flow pattern near the side boundaries. In 3D models, along-strike flow can result from along-strike changes in the dip, trench curvature, or obliquity in convergence direction relative to the trench. Thus, it is clear that the side boundary conditions become important, and may adversely affect the model results if there are any of these geometrical factors near the boundaries, as the $v_x = 0$ restriction would act to inhibit any along-strike flow.

For our Costa Rica-Nicaragua model, profiles A and D are the closest ones to the southeast and northwest boundaries, respectively. For profile A, the dip at depths of ~ 40 km changes from $\sim 70°$ below northern Costa Rica to $\sim 45°$ in central and southern Costa Rica. The thermal changes between 3D and 2D models observed in Figs. 8 and 9 result from this change in slab dip. However, the distance of this profile to the side boundary is 50 km, which means that the effect of the no-flow boundary condition described above could affect the along-strike component of mantle flow and the resulting temperature distribution.

To assess the effects of the side boundary, we have tested an additional model in which the along-strike model width is increased by 250 km. Figure 12 shows the top view of the extended 3D model, with the original boundaries indicated. In the extended model, the southern boundary is 150 km south of the

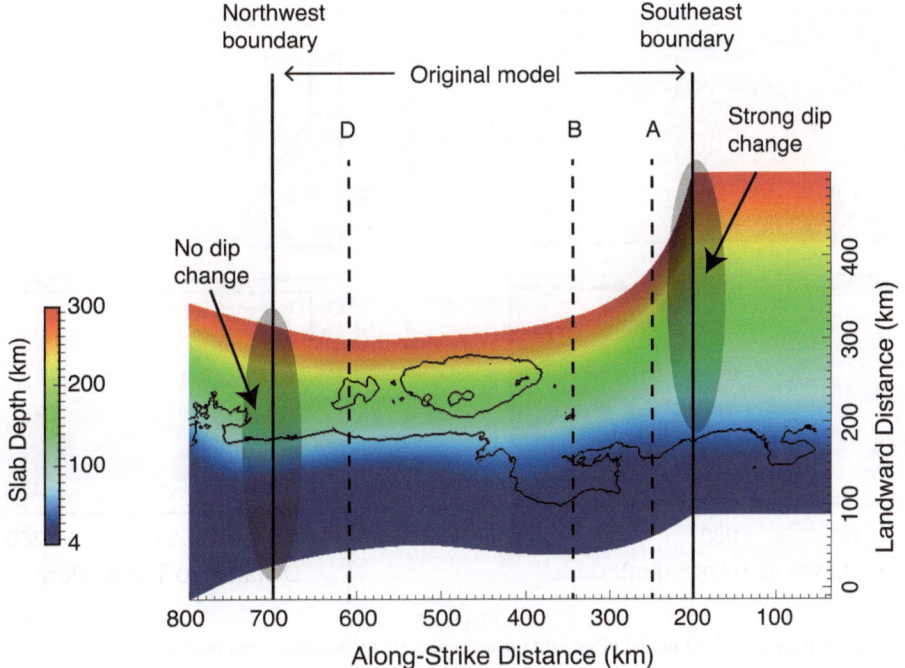

Figure 12

Slab geometry of the 3D extended model. *Color scale* denotes the depth of the slab. *Solid black lines* are the original model boundaries; *dashed lines* the location of the profiles *A*, *B* and *D*. Profile *C* is omitted, as it is located far away from the boundaries. In the extended model, the southeast boundary is located 150 km further south of that in the original model, assuming a constant slab geometry. For the northwest boundary, the extended model is 100 km further south of the original model boundary and the slab geometry is taken from KYRIAKOPOULOS *et al.* (2015). *Black circle* in the vicinity of the southern boundary shows the location of a strong along-strike dip gradient. For the northwest boundary, no significant dip gradient is observed

original model boundary. To assign the slab geometry in the new region, there are two possibilities. First, the slab geometry from KYRIAKOPOULOS *et al.* (2015) could be used (Fig. 1). However, the data show a significant decrease in dip south of the original model area, with a slab that is imaged to a maximum depth of ∼75 km at the southern limit of the KYRI-AKOPOULOS *et al.* (2015) study. This is a problem because of the maximum slab depth of 300 km imposed in our models. To solve this, we could extrapolate the slab geometry to a depth of 300 km, but this would result in a considerable shift of the backarc boundary in the landward direction, resulting in a highly distorted model geometry, which may lead to further numerical artifacts. We also note a complete absence of volcanoes (Fig. 1) and slab seismicity at depths >100 km (PROTTI *et al.* 1995) in this region, which suggest that the slab geometry should not be simply extrapolated to 300 km depth.

A second possibility is to take the geometry of the south boundary and apply it to the new section of the model. This would allow along-margin flow to pass through the original model boundary and therefore reduce the effect of the no-flow boundary condition on temperatures near the boundary. However, because the extended region has a constant dip, a strong along-strike gradient in dip is generated near the location of the original boundary (black circle in Fig. 12).

Figure 13 shows the temperature difference between the original 3D model and the extended 3D model, for a mantle wedge with an isoviscous rheology. The slab geometry of the extended region is the geometry of the south boundary, as discussed in the previous paragraph. The figure also shows the along-strike flow component for profiles A and B in the extended model; the along-strike flow for these profiles in the original model is shown in Fig. 7. For

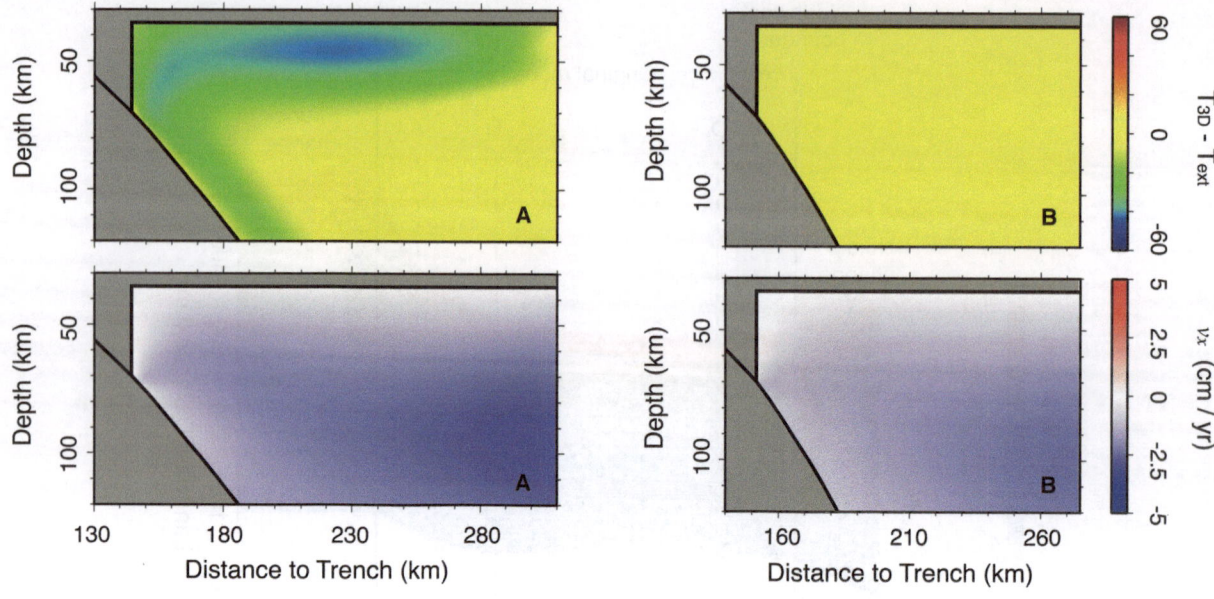

Figure 13
Thermal difference between the original 3D model (T_{3D}) and the extended 3D model (text), and the along-strike velocity component for the extended model, for profiles *A* and *B*

profile A, the extended model has that temperatures in the upper mantle wedge are ∼30 to 50 °C cooler than in the original models. This appears to be related to the higher magnitude of along-strike flow that is generated in the extended model (compare Figs. 13, 7). For profile B, there is very little difference between our original model and the extended model, which indicates that the boundary condition does not influence the thermal structure at this location.

From this, we conclude that the model results for Profile A in the original models may be affected by the side boundary. However, as the geometry from the extended model is not a real feature of the Cocos plate (is merely an extension of the geometry from the south boundary of the original model), the large temperature change observed in Fig. 13 is also not a feature that would be observed in reality. We thus prefer the original model over the extended model because the former is limited to real features of the slab geometry as provided by KYRIAKOPOULOS *et al.* (2015), although we acknowledge the effect of the side boundary is important.

At the north end of the modeling area, the original side boundary is moved 100 km to the north in the extended model, and the slab geometry in this section uses the geometry from KYRIAKOPOULOS *et al.* (2015)

(Fig. 12). In the extended geometry, profile D is located 190 km away from the boundary. The along-strike variations in dip near this profile are not as large as for profile A. Figure 14 shows there are negligible differences in the thermal structure and a comparison

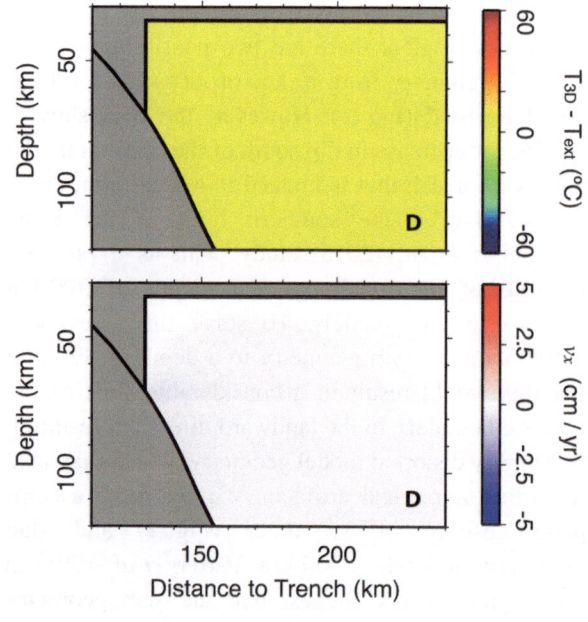

Figure 14
Same as Fig. 2a, but for profile *D*

of the along-strike velocity for this profile in Figs. 8 (original model) and 14 (extended model) shows no obvious difference. Therefore, we do not believe that there are any adverse effects from the side boundary at the north end of our original models.

Acknowledgments

We thank Christodoulos Kyriakopoulos for providing the updated geometry of the Cocos plate prior to publication, Felipe da Cruz Pimentel Moreira Santos for his technical support in creating the 2D meshes, Geoffrey Abers and Ellen Syracuse for providing earthquake locations shown in Figs. 4 and 11, and Robert Harris for discussions about hydrothermal circulation. We also thank Ikuko Wada and Vlad Manea for helpful and thoughtful reviews and William Bandy for his editorial handling of the manuscript. Research was funded by the Natural Sciences and Engineering Research Council of Canada (NSERC) and the National Council for Science and Technology of Mexico (CONACYT).

REFERENCES

ABERS, G. A., AUGER, L., SYRACUSE, E., PLANK, T., FISCHER, K. M., RYCHERT, C., WALKER, A., PROTTI, J., GONZALEZ SALAS, V., STRAUCH, W., and PEREZ, P. (2004), Imaging the Subduction Factory Beneath Central America: The TUCAN Broadband Seismic Experiment. AGU Fall Meeting Abstracts.

ABERS, G. A., PLANK, T., and HACKER, B. R. (2003), *The wet Nicaraguan slab*, Geophysical Research Letters. *30*, 1–4.

ABRATIS, M. and WORNER, G. (2001), *Ridge collision, slab-window formation, and the flux of Pacific asthenosphere into the Caribbean realm*, Geology. *29*, 127–130.

AUDET, P. and SCHWARTZ, S. Y. (2013), *Hydrologic control of forearc strength and seismicity in the Costa Rican subduction zone*, Nature Geoscience. *6*, 852–855.

BARCKHAUSEN, U., RANERO, C. R., HUENE, R., CANDE, S. C., and ROESER, H. A. (2001), *Revised tectonic boundaries in the Cocos plate off Costa Rica: implications for the segmentation of the convergent margin and for plate tectonic models*, Journal of Geophysical Research: Solid Earth. *106*, 19207–19220.

BATCHELOR, G. K. An introduction to fluid dynamics (Cambridge University, Cambridge, 2000).

BENGTSON, A. K. and van KEKEN, P. E. (2012), *Three-dimensional thermal structure of subduction zones: effects of obliquity and curvature*. Solid Earth. *3*, 365–373.

BILLEN, M. I. (2008), *Modeling the dynamics of subducting slabs*. Annual Review of Earth and Planetary Sciences. *36*, 325–356.

BIRD, P. (2003), *An updated digital model of plate boundaries*. *Geochemistry*, Geophysics, Geosystems. *4*. doi:10.1029/2001 GC000252.

CARR, M. J., FEIGENSON, M. D., PATINO, L. C., and WALKER, J. A. (2004) Volcanism and geochemistry in Central America: progress and problems, In Inside the Subduction Factory (ed. EILER. J.) (American Geophysical Union, Washignton D.C. 2004) pp. 153–174.

CLOOS, M. (1993), *Lithospheric buoyancy and collisional orogenesis: subduction of oceanic plateaus, continental margins, island arcs, spreading ridges, and seamounts*, Geological Society of America Bulletin. *105*, 715–737.

COZZENS, B. D. and SPINELLI, G. A. (2012), *A wider seismogenic zone at Cascadia due to fluid circulation in subducting oceanic crust*, Geology. *40*, 899–902.

CURRIE, C., WANG, K., HYNDMAN, R. D., and HE, J. (2004), *The thermal effects of steady-state slab-driven mantle flow above a subducting plate: the Cascadia subduction zone and backarc*, Earth and Planetary Science Letters. *223*, 35–48.

CURRIE, C. A. and HYNDMAN, R. D. (2006), *The thermal structure of subduction zone back arcs*. Journal of Geophysical Research. *111*. doi:10.1029/2005JB004024.

DAVIS, E. E., WANG, K., HE, J., CHAPMAN, D. S., VILLINGER, H., and ROSENBERGER, A. (1997), *An unequivocal case for high Nusselt number hydrothermal convection in sediment-buried igneous oceanic crust*, Earth and Planetary Science Letters. *146*, 137–150.

DEMETS, C. (2001), *A new estimate for present-day Cocos-Caribbean plate motion: implications for slip along the Central American volcanic arc*, Geophysical Research Letters. *28*, 4043–4046.

DESHON, H. R., SCHWARTZ, S. Y., NEWMAN, A. V., GONZALEZ, V., PROTTI, M., DORMAN, L. M., DIXON, T. H., SAMPSON, D. E., and FLUEH, E. R. (2006), *Seismogenic zone structure beneath the Nicoya peninsula, Costa Rica, from three-dimensional local earthquake P- and S-wave tomography*, Geophysical Journal International. *164*:109–124.

DINC, A. N., RABBEL, W., FLUEH, E. R., and TAYLOR, W. (2011). *Mantle wedge hydration in Nicaragua from local earthquake tomography*, Geophysical Journal International. *186*, 99–112.

FISHER, A. T., STEIN, C. A., HARRIS, R. N., WANG, K., SILVER, E. A., PFENDER, M., HUTNAK, M., CHERKAOUI, A., BODZIN, R., and VILLINGER, H. (2003), *Abrupt thermal transition reveals hydrothermal boundary and role of seamounts within the Cocos plate*, Geophysical Research Letters. *30*. doi:10.1029/2002GL016766.

FORSYTH, D. and UYEDA, S. (1975), *On the relative importance of the driving forces of plate motion*, Geophysical Journal of the Royal Astronomical Society. *43*, 163–200.

HACKER, B. R., ABERS, G. A., and PEACOCK, S. M. (2003), *Subduction factory 1. Theoretical mineralogy, densities, seismic wave speeds, and H_2O contents*, Journal of Geophysical Research. *108*. doi:10.1029/2001JB001127.

HARRIS, R. N., GREVEMEYER, I., RANERO, C. R., VILLINGER, H., BARCKHAUSEN, U., HENKE, T., MUELLER, C., and NEBEN, S. (2010a), *Thermal regime of the Costa Rican convergent margin: 1. Along-strike variations in heat flow from probe measurements and estimated from bottom-simulating reflectors*, Geochemistry, Geophysics, Geosystems. *11*. doi:10.1029/2010GC003272.

HARRIS, R. N., SPINELLI, G., RANERO, C., GREVEMEYER, I., VILLINGER, H., and BARCKHAUSEN, U. (2010b), *Thermal regime of the Costa Rican convergent margin: 2. Thermal models of the shallow Middle America subduction zone offshore Costa Rica*, Geochemistry, Geophysics, Geosystems. *11*. doi:10.1029/2010 GC003273.

HARRIS, R. N. and WANG, K. (2002), *Thermal models of the Middle America trench at the Nicoya peninsula*, Costa Rica, Geophysical Research Letters. *29*, 1–4.

HERRSTROM, E. A., REAGAN, M. K., and MORRIS, J. D. (1995), *Variations in lava composition associated with flow of asthenosphere beneath southern Central America*, Geology. *23*, 617–620.

HOERNLE, K., ABT, D. L., FISCHER, K. M., NICHOLS, H., HAUFF, F., ABERS, G. A., van den BOGAARD, P., HEYDOLPH, K., ALVARADO, G., PROTTI, M., and STRAUCH, W. (2008), *Arc-parallel flow in the mantle wedge beneath Costa Rica and Nicaragua*, Nature. *451*, 1094–1097.

HUTNAK, M., FISHER, A. T., HARRIS, R., STEIN, C., WANG, K., SPINELLI, G., SCHINDLER, M., VILLINGER, H., and SILVER, E. (2008), *Large heat and fluid fluxes driven through mid-plate outcrops on ocean crust*, Nature Geoscience. *1*, 611–614.

HYNDMAN, R.D., YAMANO, M., OLESKEVICH, D.A. (1997), *The seismogenic zone of subduction thrust faults*, Island Arc. *6*, 244–260.

IRIFUNE, T. (1993), *Phase transformations in the earth's mantle and subducting slabs: implications for their compositions, seismic velocity and density structures and dynamics*, Island Arc. *2*, 55–71.

JOHNSTON, S. T. and THORKELSON, D. J. (1997), *Cocos-Nazca slab window beneath Central America*, Earth and Planetary Science Letters. *146*, 465–474.

KANAMORI, H. and KIKUCHI, M. (1993), *The 1992 Nicaragua earthquake: a slow tsunami earthquake associated with subducted sediments*, Nature. *361*, 714–716.

KARATO, S.-i. and WU, P. (1993), *Rheology of the upper mantle: a synthesis*, Science. *260*, 771–778.

KIKUCHI, M. and KANAMORI, H. (1995), *Source characteristics of the 1992 Nicaragua tsunami earthquake inferred from teleseismic body waves*, Pure and Applied Geophysics. *144*, 441–453.

KIRBY, S., ENGDAHL, R.E., DENLINGER, R. (1996), Intermediate-depth intraslab earthquakes and arc volcanism as physical expressions of crustal and uppermost mantle metamorphism in subducting slabs, in Subduction Top to Bottom (ed. BEBOUT. G., SCHOLL. D., KIRBY. S., PLATT. J.) (American Geophysical Union, Washignton D.C 1996) pp. 195–214.

KNELLER, E. A. and van KEKEN, P. E. (2008), *Effect of three-dimensional slab geometry on deformation in the mantle wedge: implications for shear wave anisotropy*, Geochemistry, Geophysics, Geosystems. *9*. doi:10.1029/2007GC001677.

KYRIAKOPOULOS, C., NEWMAN, A.V., THOMAS, A.M., MOORE-DRISKELL, M., FARMER, G.T. (2015), *A new seismically constrained subduction interface model for Central America*, Journal of Geophysical Research. *120*, 5535–5548.

KUMMER, T. and SPINELLI, G. A. (2008), *Hydrothermal circulation in subducting crust reduces subduction zone temperatures*, Geology. 36, 91–94.

LANGSETH, M. G. and SILVER, E. A. (1996), *The Nicoya convergent margin—a region of exceptionally low heat flow*, Geophysical Research Letters. *23*, 891–894.

LONG, M. D. and SILVER, P. G. (2008), *The subduction zone flow field from seismic anisotropy: a global view*, Science. *319*, 315–8.

MACKENZIE, L. S., ABERS, G. A., FISCHER, K. M., SYRACUSE, E. M., PROTTI-QUESADA, J. M., GONZÁLEZ-SALAS, V., and STRAUCH, W. (2008), *Crustal structure along the southern Central American volcanic front*, Geochemistry, Geophysics, Geosystems. *9*. doi:10.1029/2008GC001991.

MANEA, VLAD C., MANEA, M., KOSTOGLODOV, V., SEWELL, G. (2005), *Thermo-mechanical model of the mantle wedge in Central Mexican subduction zone and a blob tracing approach for the magma transport*, Physics of the Earth and Planetary Interiors. *149*, 165–186.

MANEA, VLAD C. and GURNIS, M. (2007), *Subduction zone evolution and low viscosity wedges and channels*, Earth and Planetary Science Letters. *264*, 22–45.

MANEA, M. and MANEA, VLAD C. (2008), *On the origin of El Chichón volcano and subduction of Tehuantepec Ridge: a geodynamical perspective*, Journal of Volcanology and Geothermal Research. *175*: 459–471.

MANEA, VLAD C., MANEA, M., FERRARI, L. (2013), *A geodynamical perspective on the subduction of Cocos and Rivera plates beneath Mexico and Central America*, Tectonophysics. *609*, 56–81.

NEWMAN, A. V., SCHWARTZ, S. Y., GONZALEZ, V., DESHON, H. R., PROTTI, J. M., and DORMAN, L. M. (2002), *Along-strike variability in the seismogenic zone below Nicoya peninsula, Costa Rica*, Geophysical Research Letters. *29*, 1–4.

PATINO, L. C., MICHAEL, A., CARR, J., MARK, A., and FEIGENSON, D. (2000), *Local and regional variations in Central American arc lavas controlled by variations in subducted sediment input*, Contributions to Mineralogy and Petrology. 2000, 265–283.

PEACOCK, S., van KEKEN, P., HOLLOWAY, S., HACKER, B., ABERS, G. A., and FERGASON, R. (2005), *Thermal structure of the Costa Rica-Nicaragua subduction zone*, Physics of the Earth and Planetary Interiors. *149*, 187–200.

PEACOCK, S. M. (1996), Thermal and petrologic structure of subduction zones, in subduction top to bottom (ed. BEBOUT. G., SCHOLL. D., KIRBY. S., Platt. J.) (American Geophysical Union, Washington D.C 1996) pp. 119–133.

PROTTI, M., GUENDEL, F., McNALLY, K. (1995), *Correlation between the age of the subducting Cocos plate and the geometry of the Wadati–Benioff zone under Nicaragua and Costa Rica*, Geological Society of America Special Papers. *295*, 309–326.

PROTTI, M., GONZALEZ, V., NEWMAN, A. V., DIXON, T. H., SCHWARTZ, S. Y., MARSHALL, J. S., FENG, L., WALTER, J. I., MALSERVISI, R., and OWEN, S. E. (2014*), Nicoya earthquake rupture anticipated by geodetic measurement of the locked plate interface*, Nature Geoscience. *7*, 117–121.

ROSAS, J.C., CURRIE, C., HARRIS, R., He, J. (2015), *Effect of hydrothermal circulation on slab dehydration for the subduction zone of Costa Rica and Nicaragua*, Physics of the Earth and Planetary Interiors. Submitted.

ROTMAN, H. M. M. and SPINELLI, G. A. (2013), *Global analysis of the effect of fluid flow on subduction zone temperatures*, Geochemistry, Geophysics, Geosystems. *14*, 3268–3281.

RYCHERT, C. A., FISCHER, K. M., ABERS, G. A., PLANK, T., SYRACUSE, E., PROTTI, J. M., GONZALEZ, V., and STRAUCH, W. (2008), *Strong along-arc variations in attenuation in the mantle wedge beneath Costa Rica and Nicaragua*, Geochemistry, Geophysics, Geosystems. *9*. doi:10.1029/2008GC002040.

RÜPKE, L., MORGAN, J., HORTH, M., CONNOLLY, J. (2004), *Serpentine and the subduction zone water cycle*, Earth and Planetary Science Letters. *223*, 17–34.

SCHMIDT, M. W. and POLI, S. (1998), *Experimentally based water budgets for dehydrating slabs and consequences for arc magma generation*, Earth and Planetary Science Letters. *163*, 361–379.

SCHMIDT, M. W. and POLI, S. (2003), Generation of mobile components during subduction of Oceanic Crust, in Treatise on

Geochemistry (ed. RUDNICK, R., HOLLAND, H.D., TUREKIAN, K.K.) (Elsevier, 1996) pp. 567–591.

SCHWARTZ, S. Y. and DESHON, H. R. (2007), Distinct geodetic and seismic up-dip limits to the northern Costa Rica seismogenic zone: evidence for two mechanical transitions, in The Seismogenic Zone of Subduction Thrust Faults (ed. DIXON. T., MORE. J.) (Columbia University Press, New York 2007) pp. 576–599.

SOTO, G. L., NI, J. F., GRAND, S. P., SANDOVOL, E., VALENZUELA, R. W., SPEZIALE, M. G., GONZÁLEZ, J. M. G., and REYES, T. D. (2009), Mantle flow in the Rivera-Cocos subduction zone, Geophysical Journal International. 179, 1004–1012.

STEIN, C. and STEIN, S. (1992), A model for the global variation in oceanic depth and heat flow with lithospheric age, Nature. 359, 123–129.

STERN, R. J. (2002), Subduction zones, Reviews of Geophysics. 40, 1–38.

SYRACUSE, E. M., ABERS, G. A., FISCHER, K., MACKENZIE, L., RYCHERT, C., PROTTI, M., GONZALEZ, V., and STRAUCH, W. (2008), Seismic tomography and earthquake locations in the Nicaraguan and Costa Rican upper mantle. Geochemistry, Geophysics, Geosystems. 9. doi:10.1029/2008GC001963.

SYRACUSE, E. M. and ABERS, G. A (2006), Global compilation of variations in slab depth beneath arc volcanoes and implications, Geochemistry, Geophysics, Geosystems. 7. doi:10.1029/2005 GC001045.

TURCOTTE, D. L. and SCHUBERT, G., Geodynamics (Cambridge University Press, Cambridge, 2002).

VAN AVENDONK, H. J. A., HOLBROOK, W. S., LIZARRALDE, D., and DENYER, P. (2011), Structure and serpentinization of the subducting Cocos plate offshore Nicaragua and Costa Rica, Geochemistry, Geophysics, Geosystems. 12. doi:10.1029/2011 GC003592.

van KEKEN, P. E., CURRIE, C., KING, S. D., BEHN, M. D., CAGNIONCLE, A., HE, J., KATZ, R. F., LIN, S.-C., PARMENTIER, E. M., SPIEGELMAN, M., and WANG, K. (2008), A community benchmark for subduction zone modeling, Physics of the Earth and Planetary Interiors. 171, 187–197.

van KEKEN, P. E., KIEFER, B., and PEACOCK, S. M. (2002), High-resolution models of subduction zones: implications for mineral dehydration reactions and the transport of water into the deep mantle, Geochemistry, Geophysics, Geosystems. 3. doi:10.1029/ 2001GC000256.

von HUENE, R., RANERO, C.R., WEINREBE, W. (2000), Quaternary convergent margin tectonics of Costa Rica, segmentation of the Cocos plate, and Central American volcanism, Tectonics. 19-314-334.

WADA, I. and WANG, K. (2009), Common depth of slab-mantle decoupling: reconciling diversity and uniformity of subduction zones, Geochemistry, Geophysics, Geosystems. 10. doi:10.1029/ 2009GC002570.

WADA, I., WANG, K., HE, J., and HYNDMAN, R. D. (2008), Weakening of the subduction interface and its effects on surface heat flow, slab dehydration, and mantle wedge serpentinization, Journal of Geophysical Research. 113. doi:10.1029/2007JB005190.

WADA, I., J. HE, A. HASEGAWA, and J. NAKAJIMA (2015), Mantle wedge flow pattern and thermal structure in northeast Japan: effects of oblique subduction and 3-D slab geometry, Earth and Planetary Science Letters. 426, 76–88.

WANG, K., HE, J., SCHULZECK, F., HYNDMAN, R. D., and RIEDEL, M. (2015). Thermal condition of the 27 October 2012 Mw 7.8 Haida Gwaii subduction earthquake at the obliquely convergent Queen Charlotte margin, Bulletin of the Seismological Society of America. 115. doi:10.1785/0120140183.

YUE, H., LAY, T., SCHWARTZ, S. Y., RIVERA, L., PROTTI, M., DIXON, T. H., OWEN, S., and NEWMAN, A. V. (2013), The 5 September 2012 Nicoya, Costa Rica Mw 7.6 earthquake rupture process from joint inversion of high-rate GPS, strong-motion, and teleseismic P-wave data and its relationship to adjacent plate boundary interface properties, Journal of Geophysical Research. 118, 5453–5466.

(Received July 1, 2015, revised October 9, 2015, accepted October 13, 2015, Published online October 26, 2015)

Pure Appl. Geophys. 173 (2016), 3341–3356
© 2016 Springer International Publishing
DOI 10.1007/s00024-016-1263-6

Pure and Applied Geophysics

Kinematic 3-D Retro-Modeling of an Orogenic Bend in the South Limón Fold-and-Thrust Belt, Eastern Costa Rica: Prediction of the Incremental Internal Strain Distribution

CHRISTIAN BRANDES,[1] DAVID C. TANNER,[2] and JUTTA WINSEMANN[1]

Abstract—The South Limón fold-and-thrust belt, in the back-arc area of southern Costa Rica, is characterized by a 90° curvature of the strike of the thrust planes and is therefore a natural laboratory for the analysis of curved orogens. The analysis of curved fold-and-thrust belts is a challenge because of the varying structural orientations within the belt. Based on seismic reflection lines, we created a 3-D subsurface model containing three major thrust faults and three stratigraphic horizons. 3-D kinematic retro-deformation modeling was carried out to analyze the spatial evolution of the fold-and-thrust belt. The maximum amount of displacement on each of the faults is (from hinterland to foreland); thrust 1: 800 m; thrust 2: 600 m; thrust 3: 250 m. The model was restored sequentially to its pre-deformational state. The strain history of the stratigraphic horizons in the model was calculated at every step. This shows that the internal strain pattern has an abrupt change at the orogenic bend. Contractional strain occurs in the forelimbs of the hanging-wall anticlines, while a zone of dilative strain spreads from the anticline crests to the backlimbs. The modeling shows that a NNE-directed transport direction best explains the structural evolution of the bend. This would require a left-lateral strike-slip zone in the North to compensate for the movement and thereby decoupling the South Limón fold-and-thrust belt from northern Costa Rica. Therefore, our modeling supports the presence of the Trans-Isthmic fault system, at least during the Plio-Pleistocene.

Key words: Fold-and-thrust belt, kinematic modeling, active margin, Central America, Costa Rica, Cocos Ridge.

1. Introduction

Southern Central America is a complex Late Mesozoic/Cenozoic island-arc system. The Costa Rican part of the island arc can be subdivided into a northern and a southern arc segment (SEYFRIED *et al.*

1991). The southern Costa Rican arc segment is influenced by flat subduction of the Cocos Ridge (CORRIGAN *et al.* 1990; GARDNER *et al.* 2013), which results in deformation and uplift on the upper-plate forearc (MORELL *et al.* 2008, 2012) (Fig. 1). Early work on the bivergent southern Costa Rican land-bridge was carried out by GREB *et al.* (1996). The structural evolution of the forearc fold-and-thrust belt (Fila Costeña) has been comprehensively studied during the last decade (MENDE 2001; FISHER *et al.* 2004; SITCHLER *et al.* 2007; MORELL *et al.* 2008, 2012, 2013; GARDNER *et al.* 2013). The onshore and offshore back-arc fold-and-thrust belt (South Limón belt) was analyzed by MENDE (2001) and BRANDES *et al.* (2007a, b, c; 2008); the latter focuses on basin dynamics (burial history and temperature evolution) and the architecture of the fold-and-thrust belt. An important structural feature in the area is the bend of the South Limón fold-and-thrust belt to the north-west, close to the city of Puerto Limón, where the strike of the thrusts changes through 90° (Fig. 2). This bend was first illustrated by PROTTI and SCHWARTZ (1994). BRANDES *et al.* (2007c) analyzed the sediment distribution on the offshore part of the South Limón fold-and-thrust belt and suggested that most of the deformation took place during the Pleistocene. Initial movements could have started already in the Late Pliocene (BRANDES *et al.* 2007c), but evidence is limited.

Kinematic evolution of fold-and-thrust belts is commonly analyzed in 2-D cross sections. The construction of 2-D balanced sections has evolved into a key technique in the analysis of contractional tectonics (e.g., DAHLSTROM 1969, 1990; COOPER and TRAYNER 1986; DEPAOR 1988; WU *et al.* 2005; TANNER *et al.* 2011). However, plane strain is typically

[1] Institut für Geologie, Leibniz Universität Hannover, Callinstraße, 30167 Hannover, Germany. E-mail: brandes@geowi.uni-hannover.de
[2] Leibniz Institute for Applied Geophysics (LIAG), Stilleweg 2, 30655 Hannover, Germany.

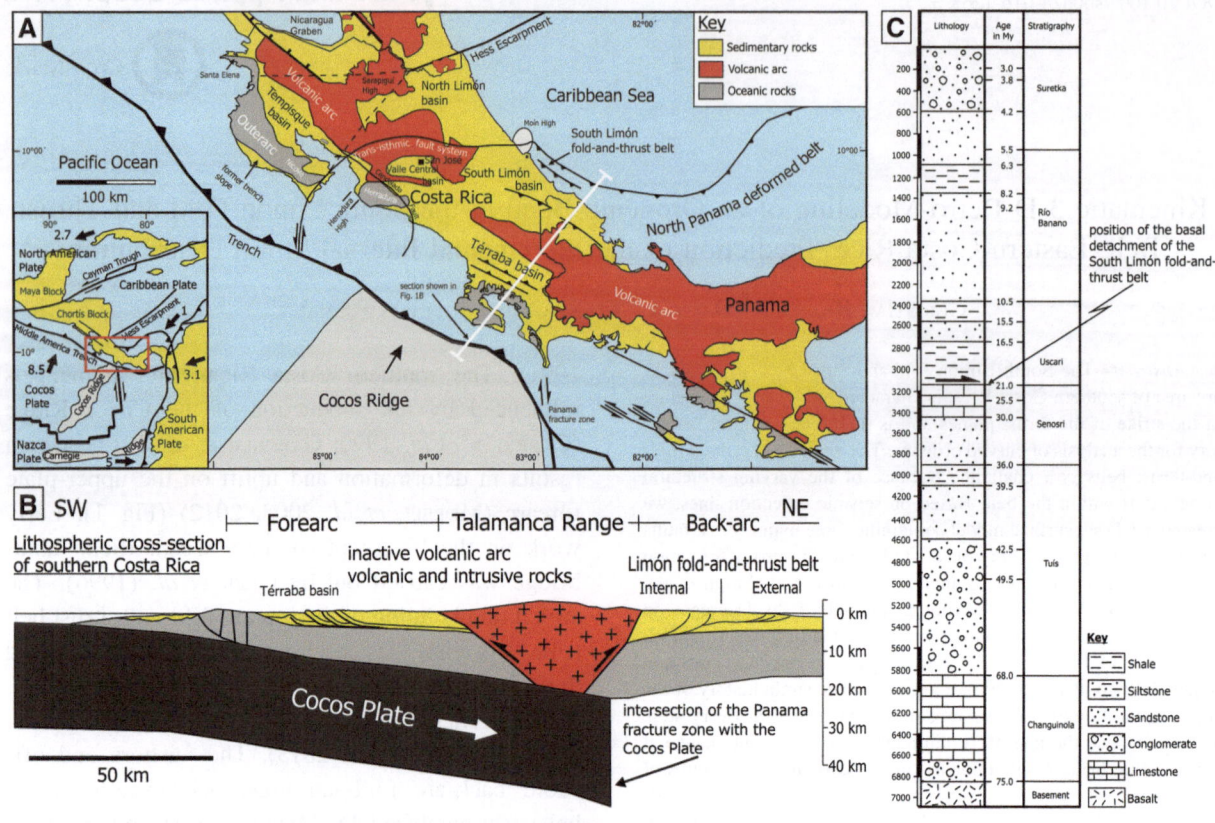

Figure 1

a Tectonic map of Costa Rica. The South Limón fold-and-thrust belt is located behind the island arc and extends along the southeastern Caribbean coast (modified after BARBOZA *et al.* 1997; FERNANDEZ *et al.* 1997). The plate tectonic map of the Caribbean region (insert on the *left side*) is modified after ROSS and SCOTESE (1988), DONNELLY (1989), MESCHEDE and FRISCH (1998), and DEMETS (2001). *Red box* shows location of detailed study area. Numbers are modern, absolute plate vectors (in cm year^{-1}). **b** Crustal-scale cross section of southern Costa Rica (modified after FISHER *et al.* 2004, MORELL 2016) that shows the bivergent structure of the active margin. **c** Lithologic log of the South Limón Basin (modified after CAMPOS 2001 and BRANDES *et al.* 2008). The rheological contrast between the platform carbonates of the Senosri Formation and the overlying shales of the Uscari Formation defines the depth of the detachment of the South Limón fold-and-thrust belt

assumed. The analysis of fold-and-thrust belts in map view (e.g., AFFOLTER and GRATIER 2004) and 3-D (e.g., TANNER *et al.* 2003, SALA *et al.* 2014) also underwent considerable progress in the last decades. A 3-D approach integrates lateral geometric variations obtained from map and depth views into the cross section-based structural restoration (TANNER *et al.* 2003). According to HINDLE and BURKHARD (1999), only 3-D restoration of a curved fold-and-thrust belt can provide reliable information on true displacement and the amount of internal strain.

This study presents the 3-D structural analysis of the north-western bend of the South Limón fold-and-thrust belt, based on a 3-D model built from intersecting 2-D seismic reflection lines (Fig. 2). Our aim is to understand the kinematic evolution of the structure.

1.1. Salients, Oroclines and Syntaxes

Curved fold-and-thrust belts were recognized first by SUESS (1908) and ARGAND (1924), and have been studied worldwide. In the context of curved fold-and-thrust belts, a specific terminology has been developed. The term salient is used to describe orogens that are convex to the foreland (MACEDO and MARSHAK 1999; Fig. 3a). Salient development is controlled by the pre-deformation geometry of the sedimentary basin or the interaction of the fold-and-thrust belt with an indenter, strike-slip faults, or other

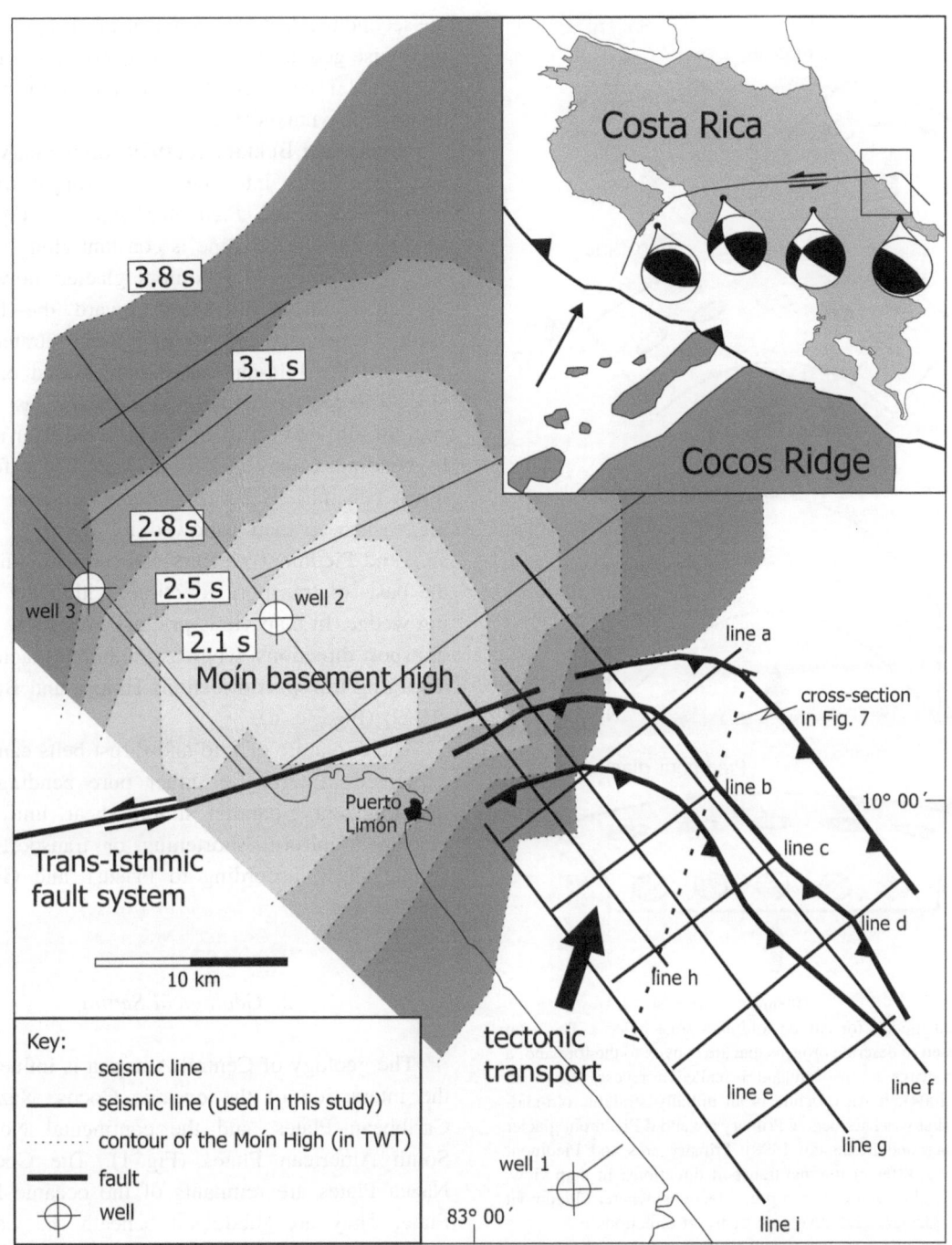

Figure 2

Seismic lines and well locations in the eastern coastal area of the Costa Rican back-arc. The *bold lines* indicate the locations of the seismic sections used in this study. The position and geometry of the Moín basement high is shown as isolines in two-way travel time (modified after BRANDES *et al.* 2007a, b, c, 2009). The trace of the cross section in Fig. 7 is shown. The South Limón fold-and-thrust belt is closely related to the Moín basement high and the Trans-Isthmic fault system. Focal mechanisms of MARSHALL and FISHER (2000) imply strike-slip movements along the Trans-Isthmic fault system that transform into reverse movements in the area of the South Limón fold-and-thrust belt

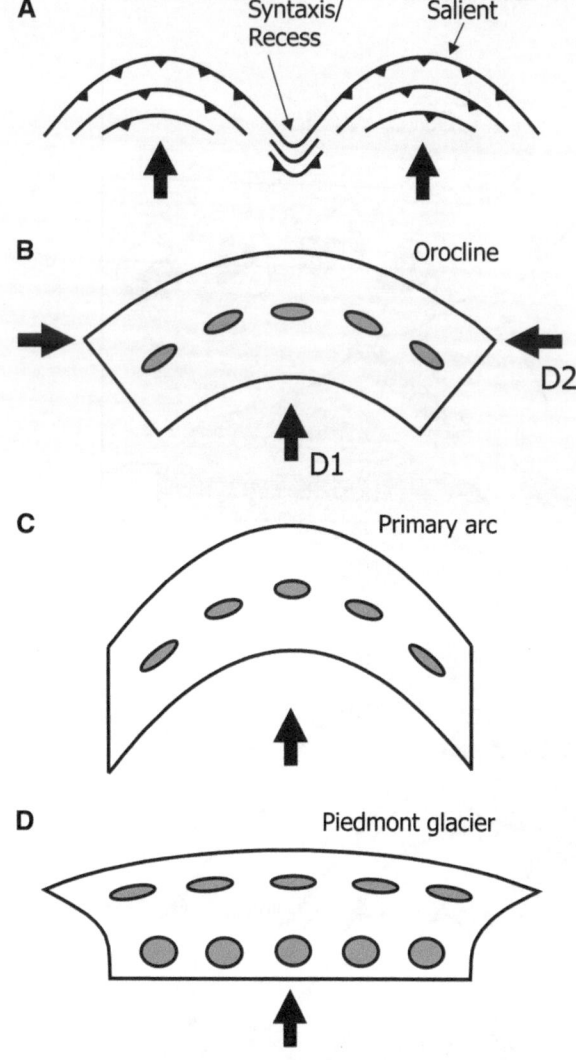

the second bends it. A syntaxis is an abrupt bend in an otherwise geometrically straight orogen (GATES *et al.* 2004). MARSHAK (2004) used the term recess to describe a syntaxis (Fig. 3a).

HINDLE and BURKHARD (1999) divide curved fold-and-thrust belts into the three groups: *oroclines*, *primary arcs* and *Piedmont glaciers* (Fig. 3b–d). Strain within an orocline is constant along strike, as it is for a so-called Piedmont glacier; however, in the latter, strain increases toward the foreland. Within a primary arc, strain is concentrated at the edges of the fold-and-thrust belt. Based on CAREY (1955, 1958), oroclines are defined as initially straight fold-and-thrust belts that were later modified by bending. CAREY (1958) assumes that a fold-and-thrust belt initially forms under pure shear shortening, which is then followed by bending. Primary arcs and Piedmont glaciers can be distinguished on the basis of transport directions within the deforming wedge. In map view, primary arcs show uniform transport directions, whereas Piedmont glaciers have diverging transport directions HINDLE and BURKHARD (1999) (Fig. 3c, d).

The curvature of fold-and-thrust belts can be also explained in terms of either pure bending, radial thrusting, curve-parallel simple shear, uniform displacement/uniform shortening or transport-parallel simple shear, according to FERRILL and GROSHONG (1993).

Figure 3
Conceptual models for curved fold-and-thrust belts. **a** The term salient is used to describe orogens that are convex to the foreland; a segment concave to the foreland is called a recess (based on MARSHAK 2004). **b** An orocline is an initially straight fold-and-thrust belt that was later bent. **c** Primary arc and **d** Piedmont glacier (after HINDLE and BURKHARD 1999). Primary arcs and Piedmont glaciers have different internal transport directions. In map view, primary arcs show uniform transport directions, whereas Piedmont glaciers have diverging transport directions

2. Geological Setting

The geology of Central America is influenced by the interaction of the oceanic Cocos, Nazca, and Caribbean Plates, and the continental North and South American Plates (Fig. 1). The Cocos and Nazca Plates are remnants of the oceanic Farallon Plate. They are subducted beneath the Caribbean Plate along the NW–SE trending Central America trench. The present-day subduction velocity off Costa Rica, relative to the Caribbean Plate, is 8.5 cm year^{-1} (DEMETS 2001). The geological evolution of southern Costa Rica has been studied by many authors. Studies were carried out on many different topics, such as the arc-related sedimentary basins (CORRIGAN *et al.* 1990; ASTRORGA *et al.* 1991;

obstacles (MARSHAK 2004). A segment that is concave to the foreland is called a recess (VAN DER PLUIJM and MARSHAK 2004) (Fig. 3a). The transformation of an initially straight orogen front into a curved one is called oroclinal bending (CAREY 1955, 1958). Oroclinal bending usually involves two deformation phases: the first creates the fold-and-thrust belt and

SEYFRIED et al. 1991; AMANN 1993; BOWLAND 1993; VON EYNATTEN et al. 1993; CAMPOS 2001; BRANDES et al. 2007a, b, c) and offshore geophysics (RANERO and VON HUENE 2000; BARCKHAUSEN et al. 2003) but detailed knowledge of the back-arc area and the related fold-and-thrust belts is still limited.

Southern Central America is characterized by the northeastward subduction of the Cocos Plate. An important feature of the Cocos Plate is the NE–SW-trending, aseismic Cocos Ridge, which is interpreted as the trace of a hot spot (e.g., WALTHER 2003). The entire Cocos Ridge is more than 1000 km long, 250–500 km broad, and rises ca. 2 km above the ocean floor. The timing of the onset of ridge subduction is still in debate. ABRATIS and WÖRNER (2001) assume that the Cocos Ridge has been subducting since 8 Ma, whereas other studies propose a much younger history, depending on the method used: 3.6 Ma (COLLINS et al. 1995) and 3–2 Ma (MACMILLAN et al. 2004). The studies of MORELL et al. (2012), GARDNER et al. (2013), and LONSDALE and KLITGORD (1978) support the assumption that the front edge of the ridge arrived at the trench later than 3 Ma. In the most recent modeling study, ZEU-MANN and HAMPEL (2015) derive an onset of Cocos Ridge subduction at ∼2 Ma.

The Limón back-arc basin belongs to the southern Central American arc-trench system and is situated beneath the present-day coastal plain and continental shelf of eastern Costa Rica (Fig. 3). Its northern boundary is the Hess Escarpment. To the west and to the south the basin is fronted by the volcanic arc. The eastward extent is defined by the 200 m bathymetric contour line of the Caribbean Sea and by the extent of the Limón fold-and-thrust belt, respectively (Fig. 3). The fill of the Limón Basin consists of Upper Cretaceous to Pleistocene deep marine to continental volcaniclastic rocks (SHEEHAN and PENFIELD 1990; COATES et al. 1992; AMANN 1993; BOTTAZZI et al. 1994; FERNANDEZ et al. 1994; MCNEILL et al. 2000; MENDE 2001; CAMPOS 2001; COATES et al. 2003) (Fig. 4). Deposition of shallow-water carbonates occurred during Late Cretaceous, Eocene and Oligocene times on local structural highs. The Limón Basin can be subdivided into a northern and a southern sub-basin, separated by the E-W trending Trans-Isthmic Fault System.

2.1. Geology and Stratigraphy of the Limón Back-Arc Basin

The South Limón Basin contains approx. 6–10 km-thick sedimentary rocks that were deformed by NE-directed folding and thrusting during the Late Neogene (BOTTAZZI et al. 1994; MENDE 2001). The resulting fold-and-thrust belt is called the South Limón fold-and-thrust belt in this study. Stratigraphic information for the South Limón Basin is based on onshore outcrops and well data. The oldest sediments consist of ∼1280 m-thick pelagic limestones and intercalated volcaniclastic rocks of Late Campanian to Maastrichtian age (Changuinola Formation, Fig. 1). The Changuinola Formation is overlain by up to 3000 m-thick Paleocene to Lower Eocene, coarse-grained, volcaniclastic turbidites, debris flow deposits, lava flows and tuffs of the Tuís Formation, which represent a prograding deepwater apron system (MENDE 2001). An early contractional deformation phase during Eocene to Oligocene times resulted in the uplift of the Moín High (BRANDES et al. 2009) and the formation of significant tectonic and topographic relief, as implied by the coeval deposition of 150-200 m-thick shallow-water limestones of the Las Animas Formation on local structural highs (AMANN 1993; MENDE 2001), and of 700–900 m-thick hemipelagic mudstones, calcareous turbidites, and carbonate debris flow deposits of the Senosri Formation in adjacent basin areas (MENDE 2001). During the Late Oligocene a basin-wide unconformity formed, probably caused by uplift of the island arc in combination with a major sea-level fall (SEYFRIED et al. 1991; AMANN 1993; KRAWINKEL et al. 2000). Subsequently, carbonate ramps built up on top of the uplifted areas. These carbonate ramps are overlain by up to 2000 m-thick shallow-water volcaniclastic sediments of the Upper Oligocene to Upper Miocene Uscari Formation, interpreted as delta-influenced shelf deposits (AMANN 1993; MENDE 2001). The Uscari Formation is overlain by the shallow-water limestones and volcaniclastic rocks of the 400–1800 m-thick Río Banano Formation (AMANN 1993; BOTTAZZI et al. 1994; MENDE 2001). The South Limón fold-and-thrust belt developed in the Neogene. On the fold belt, wedge-top basins formed (BRANDES et al. 2007a), which were filled by shallow marine and

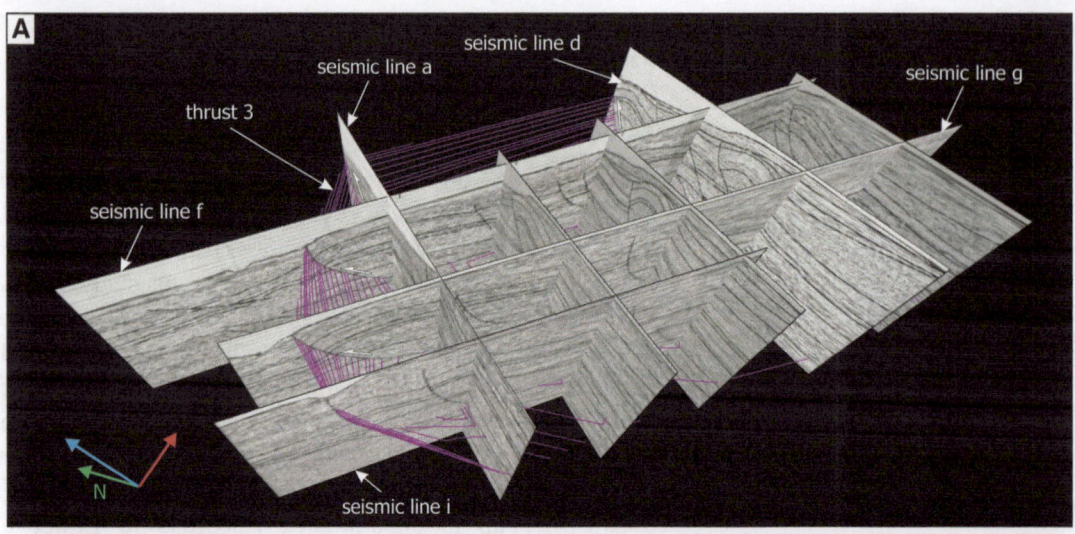

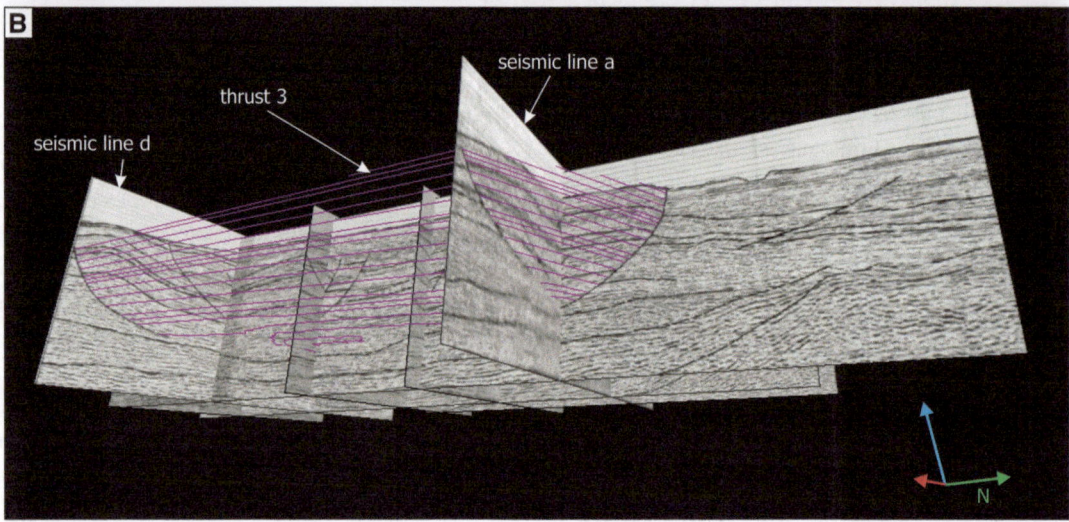

Figure 4
Construction of the 3-D model using the 2-D seismic reflection lines. All thrust traces from the individual seismic sections form a prominent bend in the fold-and-thrust belt. The thrusts were linearly splined to form 3-D fault surfaces. The fault surface of thrust 3 is shown as a set of isolines

continental deposits of the Plio-Pleistocene Suretka Formation (AMANN 1993; BOTTAZZI *et al.* 1994; MENDE 2001).

3. *Database and Methods*

This study is based on nine two-dimensional seismic reflection lines that were acquired during onshore and offshore campaigns in the 1970s and 1980s and made available by the Costa Rican Ministry of Environment and Energy (MINAE) (Fig. 2). Stratigraphic and lithological information for the seismic interpretation was derived from well 1 located close to the NE–SW-trending seismic sections and well 2 on the top of the Moín High (Fig. 2). The latter well penetrates Pleistocene to Miocene sandstones, shales, and limestones. To correlate the major reflectors on the five NE–SW-trending sections (lines a–e), four NW–SE-trending cross-lines (lines f–i) were used. The interpretation of the seismic sections has been presented by BRANDES *et al.* (2007c).

3.1. Model Building

We constructed a 3-D subsurface model in the software package 3DMove™ by placing the seismic sections in their correct geographical positions (Fig. 4a, b) and interpolating traces of stratigraphic horizons and their cutoffs at faults. The model contains three major listric thrust faults, which we name, from the trailing- to the leading-edge, thrusts 1, 2 and 3, respectively. In Fig. 4a, b we show, based on thrust 3, how the thrust traces from the individual seismic sections form the prominent bend in the fold-and-thrust belt. The thrusts were first linearly splined to form 3-D fault planes and then the traces of three stratigraphic horizons (base Pleistocene and two intra Miocene reflectors) were taken from the interpretation and meshed into 3-D surfaces with a spline algorithm. The limited coverage of seismic data caused a small triangular hole (approx. 2 km edge length) in the model (Fig. 5a1 and a4). However, this did not negatively affect the modeling. The resulting 3-D structural model was consistent and robust for all further analysis.

All faults detach at one level, at ca. 5 km depth below sea level. The detachment depth is controlled by a lithological contrast between the underlying platform carbonates of the Senosri Formation and the incompetent shales of the Uscari Formation BRANDES et al. (2007b, c). Above all thrust faults, hanging-wall anticlines were developed (Fig. 5b). These structures can be classified as fault-propagation folds (SUPPE and MEDWEDEFF 1990; BRANDES and TANNER 2014), based on: (a) the offset along the thrusts decreases up-dip of the fault and (b) the position of the anticlines near to the fault tip. They are not fault-bend folds, because the slight bend in, e.g., thrust 1 (Fig. 5b) is minor and could not cause the observed anticline amplitude by pure fault-bend folding.

The structural restoration was performed with 3DMove™, using the fault-parallel flow algorithm (ZIESCH et al. 2014). This method constructs flow paths for all particles of the hanging wall. The flow paths are at all times parallel to the fault plane in the transport direction. Fault-parallel flow is a suitable approach for thrusts with cutoff angles below 30° (ZIESCH et al. 2014). With the use of the strain toolbox in Move™, we recorded the dilatation strain history of the stratigraphic horizons in the model during the sequential restoration.

4. Structural Analysis

In map view the traces of the thrust planes in the model are convex to the foreland. In the east the thrusts strike NW–SE and curve to strike NE–SW in the west, i.e. there is a 90° bend in the strike of the thrust planes (Figs. 2, 4, 5a, 6a, b). The seismic lines show that there is displacement on all of the thrusts, as shown by displacements of the seismic reflectors, irrespective of their orientation. Figure 6b shows the along-strike heave on all three thrusts in map view. This illustrates that the maximum heave on all three faults is in the middle of the bend of the fold-and-thrust belt, implying that it is not a lateral/frontal ramp system, but rather a true array of curved faults. Correlating the maximum heave from thrust to thrust gives the vector of the main transport direction (azimuth N018°), which we chose as the transport direction to retro-deform the model.

4.1. Structural Restoration

Based on the match of stratigraphic cutoffs (base Upper Miocene), we first determined the maximum amount of displacement on each of the faults (thrust 1: 800 m; thrust 2: 600 m; thrust 3: 250 m; see Fig. 7). We assumed foreland propagation of the deformation front, i.e. thrust 1 moved first and was followed by thrusts 2 and 3; i.e., each successive thrust was moved by a smaller amount than the previous one.

In 3D-Move™ the fault-parallel flow algorithm can be implemented in 3-D, albeit with a constant transport direction. The structural restoration involved first restoring the hanging wall of fault 3, then that of fault 2, and finally that of fault 1. In each case the footwall and hanging-wall cutoffs of the base Miocene were matched. Throughout the restoration process, we tracked the strain evolution of all hanging-wall objects.

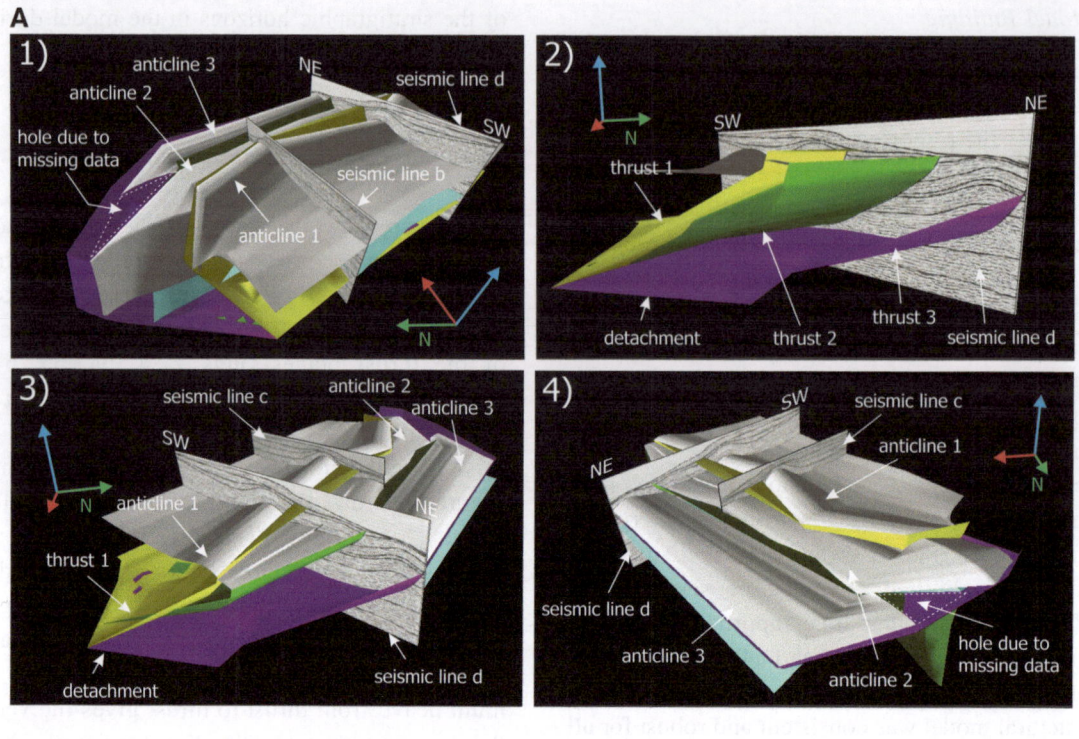

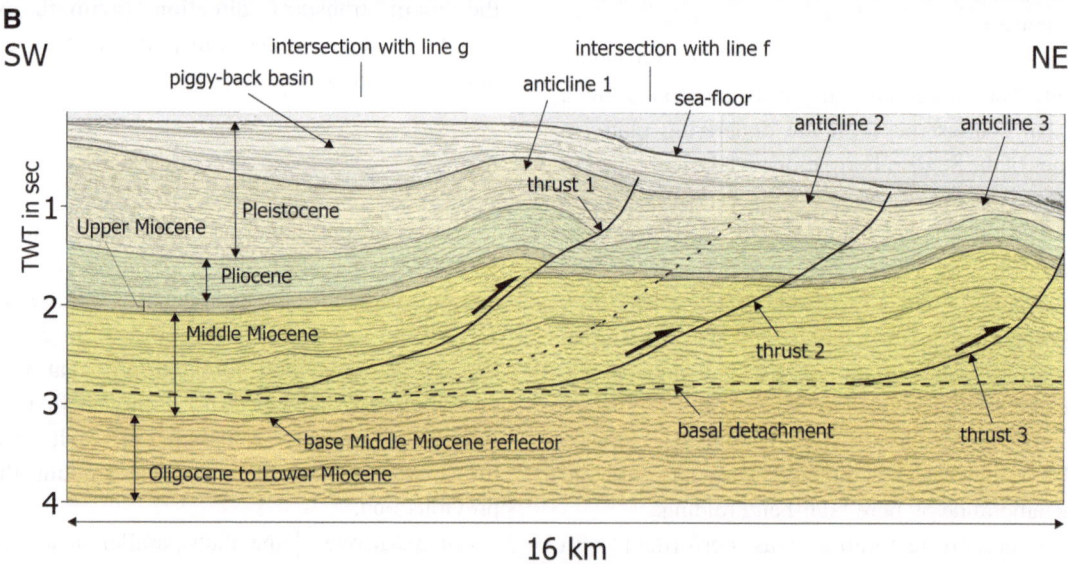

Figure 5

a The bend of South Limón fold-and-thrust belt as 3-D subsurface model in 3D Move™. The structural model contains three thrusts and their related hanging-wall anticlines. The 2-D traces of three stratigraphic horizons (base Pleistocene and two intra Miocene reflectors) were taken from the interpreted seismic sections and meshed into 3-D surfaces using a spline algorithm. **a**1, **a**4 Due to missing data there is a triangular hole in the model. **b** Interpreted seismic section of the South Limón fold-and-trust belt. The basal detachment lies at the base of the Middle Miocene deposits, controlled by a lithological change from limestones to overlying shales

4.2. Strain Modeling

3-D kinematic modeling can also provide information on the strain distribution within a model. The strain is displayed as an attribute on a stratigraphic surface, in this case the base of the Pleistocene and coded according to color (Fig. 8). We will first

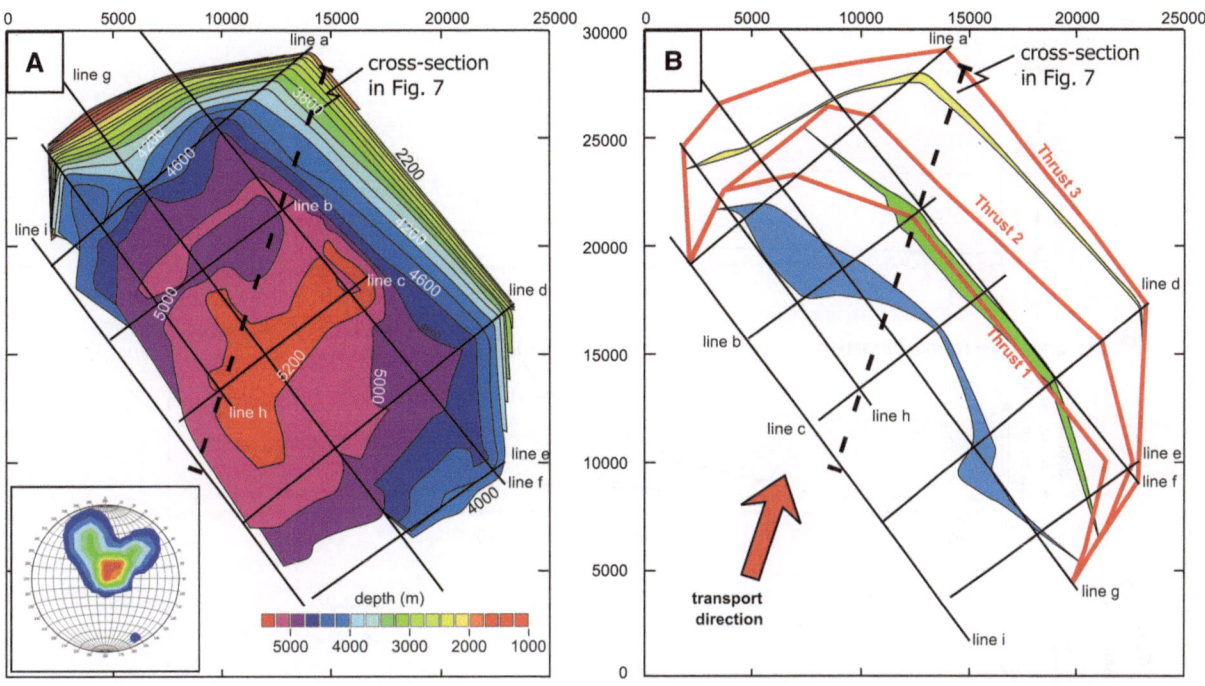

Figure 6

a Isoline map of the basal detachment surface. The detachment has a maximum depth of 5200 m. **b** Heave map for all three thrusts. This map shows that the maximum heave is in the middle of the bend of the fold-and-thrust belt, implying that it is not a lateral-frontal ramp system, but rather a true array of curved faults. Correlating the maximum heave from thrust to thrust gives the vector of the main transport direction (azimuth N018°)

describe the strain history in terms of retro-deformation, in which the youngest fault was retro-deformed first and the oldest fault was retro-deformed last. We invert this sequence to discuss the model in terms of forward deformation. During the retro-deformation, the amount of strain measured is the inverse of that produced during forward deformation. In addition, if the different phases of a strain path are reversed in the correct order (because strain is not commutative), the magnitude of finite strain within any fault block is identical after either retro- or forward deformation.

4.3. Structural Restoration

4.3.1 Thrust 1

Restoration of the leading-edge hanging-wall anticline results in a strain pattern nearly strike-parallel to thrust 1, irrespective of the thrust strike, with an abrupt change at the orogenic bend. Contractional strain is

manifested on the forelimb of the hanging-wall anticline, while a zone of dilative strain spreads from the anticline crest to the backlimb (Fig. 8a). Strain magnitudes are 5 % and greater. Not only is the hanging-wall of the active thrust deformed, but it also affects the hanging walls of thrusts 1 and 2. This can be seen especially in the region of the bend, where strain is characterized mainly by extension. A linear ENE-WSW trending zone of dilative strain is the most pronounced element. There are also subordinate, WNW-ESE striking, dilative-strain zones (Fig. 8a).

4.3.2 Thrust 2

During the subsequent retro-deformation step, the strain pattern remains the same as before. However, due to the fact that the spacing between the thrusts is smaller, the strain is further distributed over the second and third hanging-wall anticlines. The dilative strain in hanging-wall anticline 1 follows the trend of

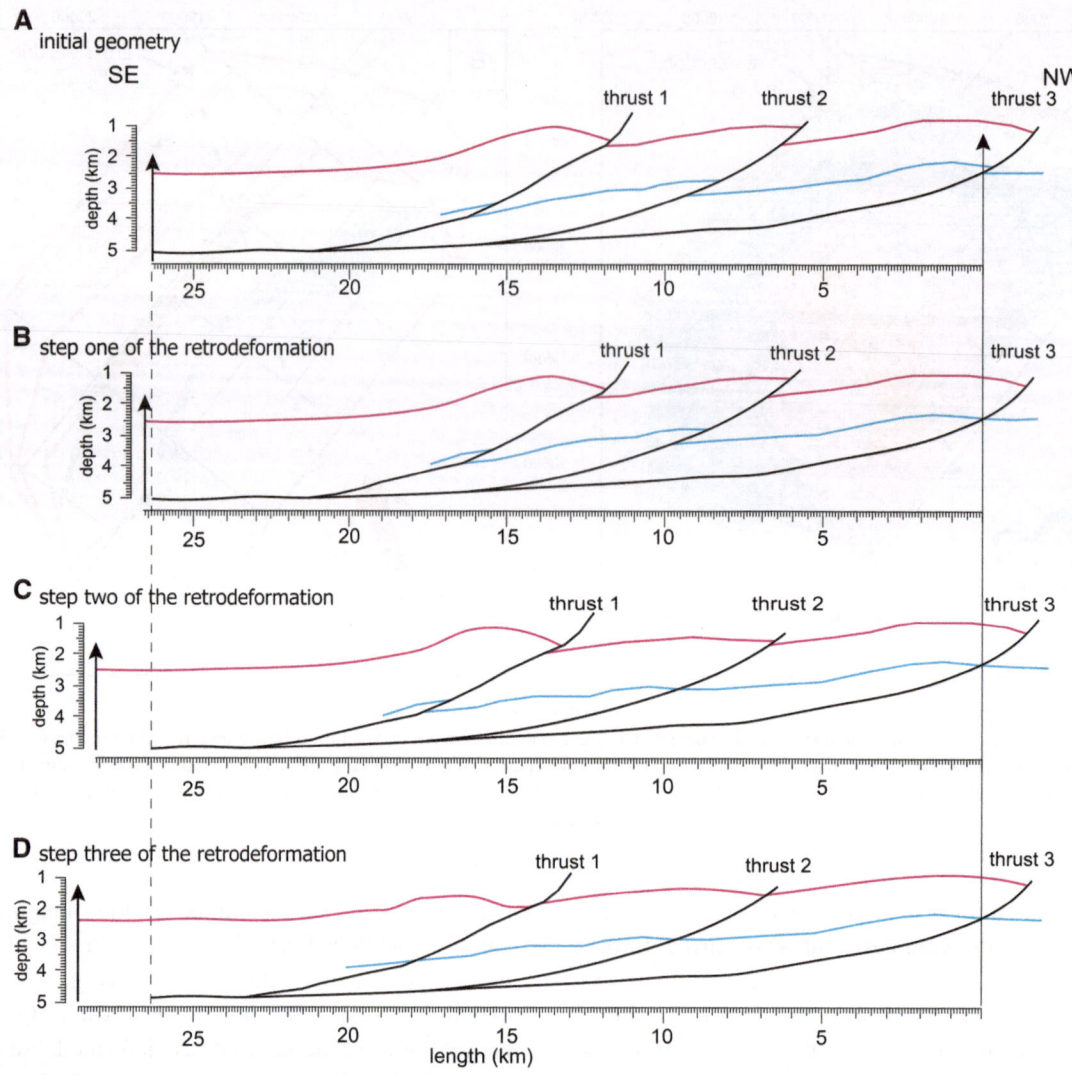

Figure 7

Structural restoration. A cross section (parallel to the transport direction), extracted from the 3-D model (see Fig. 2 for location), showing the present-day situation (**a**) and the three retro-deformation steps (**b–d**). The structural restoration involved first restoring the hanging wall of fault 3, then that of fault 2, and finally that of fault 1. In each case, the footwall and hanging wall cutoffs of the base of the Miocene were matched. Based on the structural restoration we estimate a horizontal shortening of 9 %, see Table 1 for shortening derived from the individual retro-deformation steps

Table 1

Shortening of the cross section shown in Fig. 7

	Length (in km)	Shortening (in km)	Shortening (in %)
Present-day	26.3		
Step 1	26.6	0.3	1.14
Step 2	28.1	1.8	6.84
Step 3	28.7	2.4	9.13

Step one after restoration of thrust 3, *step two* after restoration of thrust 2, *step three* after restoration of thrust 1

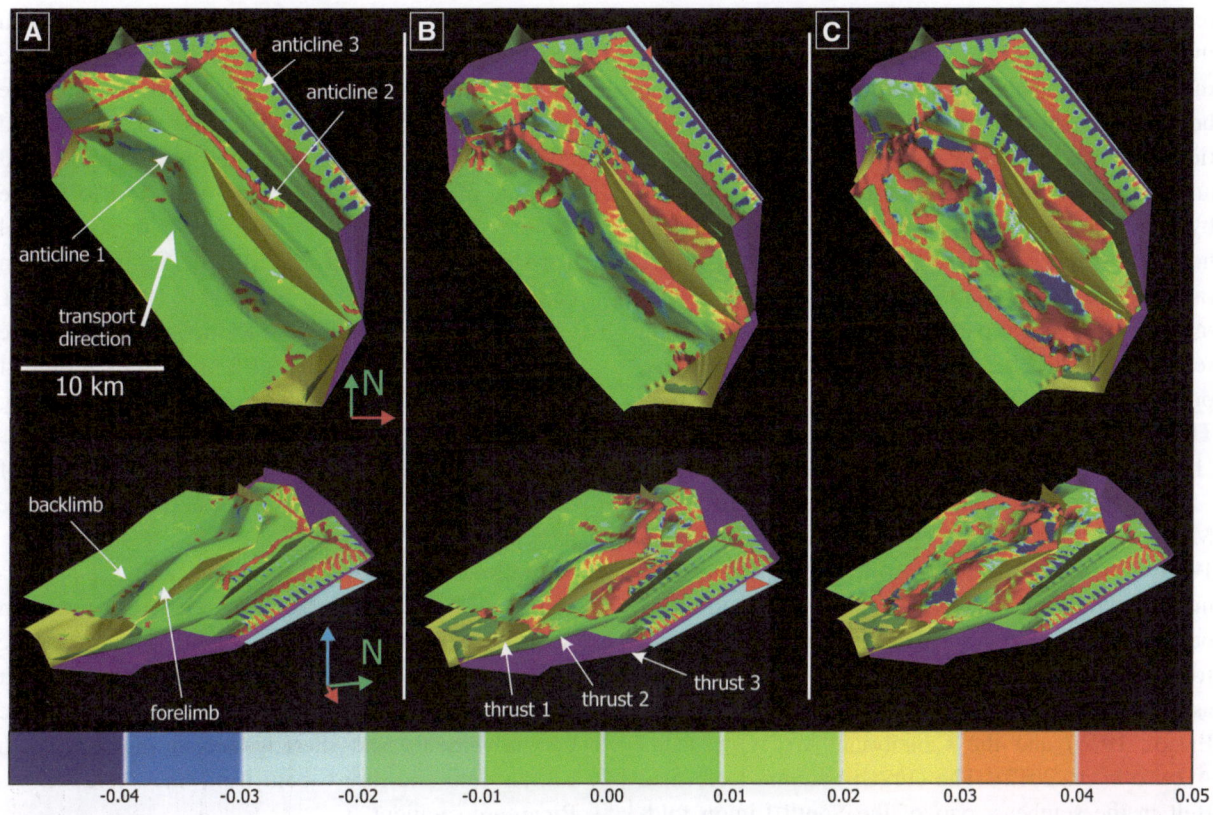

Figure 8
Strain distribution within the South Limón fold-and-thrust belt. **a** Situation after the restoration of thrust 3. **b** Situation after the restoration of thrust 2. **c** Situation after the restoration of thrust 1. The strain is displayed as a *color-coded attribute* on the base Pleistocene horizon, with dilative strain in *red* and contractional strain in *blue*. *Green colors* indicate low to zero strain. Strain values are dimensionless. The strain history is shown in terms of retro-deformation, in which the youngest fault was retro-deformed first and the oldest fault was retro-deformed last. The strain modeling shows contractional deformation concentrated on the forelimbs of the anticlines, whereas dilative deformation characterizes the anticlinal crests and backlimbs. As a result of the close thrust spacing strain not only affects the hanging-wall anticline of the active thrust, but also the hanging walls above it. The strain distribution in the bend area is similar to the general pattern in the hanging-wall anticlines and follows the same trend as the orogenic bend, i.e. from NW–SW in the west to SW–NE in the east

the hinge line (Fig. 8b). Strong modifications of the strain distribution occur in the bend. Previous ENE-WSW striking zones of dilative strain are now overprinted by dilative N–S trending strain zones.

4.3.3 Thrust 3

In this step, the strain pattern in hanging-wall anticline 3 drastically changes and increases. Zones of dilative strain parallel to the former hinge line are dissected and strongly replaced by anastomosing zones of contractional strain, as indicated by the blue colors (Fig. 8c). This affects the whole backlimb of hanging wall 3 up to 10 km from the front of thrust 1.

Strain zones in the bend are more curvilinear and thicker than in Fig. 8b.

5. Discussion

All seismic lines, irrespective of their orientation to the bend, show that there is displacement on all the thrusts. Following this observation, the structural modeling was carried out with a tectonic transport direction to the NNE that bisects the two major strike directions of the thrusts. The retro-deformation produced reasonable geometries once the fault slip was restored. With this one direction we produce a

139

reasonable restoration of the 3-D model, which indicated the validity of our method. This implies that the assumption of a NNE-directed transport in the bend area is suitable to explain the structural evolution, i.e., it is a primary arc (in the terms of HINDLE and BURKHARD 1999). This interpretation is supported by the heave map (Fig. 6b); for example, in that heave on thrust 3 is nearly constant parallel to strike (around the bend of the fold-and-thrust belt). Our modeling shows this part of the offshore South Limón fold-and-thrust belt, with the bend to the northwest, probably evolved in one single deformation phase. Therefore, it is not an orocline in the sense of CAREY (1955, 1958).

Analyzing salients (GRAY and STAMTAKOS 1997; MUKUL and MITRA 1998) and syntaxes (BUTLER et al. 1989; GATES et al. 2004) is an important step toward understanding the geodynamics of fold-and-thrust belts (POBLET and LISLE 2011). In many cases curved fold-and-thrust belts are regarded as oroclines, such as the Patagonian-Fuegian Andes (DALZIEL and ELLIOT 1973) and the Cantabrian Arc (GUTIÉRREZ-ALONSO et al. 2012). The curved Panama Deformed Belt in the southeast part of the South Limón fold-and-thrust belt has been considered an orocline with three main deformation stages (MONTES et al. 2012). MARSHAK (2004) showed that curvatures of fold-and-thrust belts can be caused by the interaction of the propagating belt with subsurface obstacles. Based on this model, BRANDES et al. (2007c, 2009) proposed that bending of the thrust planes was the result of the interaction of the northwestward-propagating fold-and-thrust belt with the Moín basement high in the subsurface (Figs. 1, 2).

The South Limón fold-and-thrust belt has been interpreted as the western prolongation of the North Panama deformed belt (SILVER et al. 1990; GOES et al. 1993). The latter was caused by the collision of Panama with South America around 25–23 Ma (FARRIS et al. 2011). Using a basin modeling study, BRANDES et al. (2008) showed that the subsidence in the South Limón basin increased at around 23 Ma, which may indicate the onset of fold-and-thrust belt formation in the back-arc. This fits the data of FARRIS et al. (2011).

A transition from marine to continental deposits occurred at the end of the Late Miocene. During the Plio-Pleistocene wedge-top basins were filled with coarse alluvial sediments (Suretka Formation). These deposits contain granodiorite clasts derived from the rising Talamanca Range (AMANN 1993). Young and ongoing deformation is indicated by strongly tilted deposits of the Suretka Formation in the Río Cerere (AMANN 1993), the recent opening of the Bocas del Toro Bay (GREB et al. 1996) and the 1991 Limón earthquake (PROTTI and SCHWARTZ 1994). MORELL (2016) postulates that Cocos Ridge subduction is the first-order driver for the South Limón fold-and-thrust belt, based on the supposition that shortening decreases eastwards. However, we speculate that this decrease in the shortening could also be a function of the bend in the fold-and-thrust belt that we describe here.

Following MARSHAK (2004), the curvature of fold-and-thrust belts can be also the result of an interaction with strike-slip faults. A NNE-directed tectonic transport of the South Limón fold-and-thrust belt requires a left-lateral strike-slip zone to compensate the movements and thereby decoupling the South Limón fold-and-thrust belt from the North Costa Rican arc segment.

For the South Limón fold-and-thrust belt transport-parallel simple shear can be ruled out, because the offset along the thrusts is mainly constant even along strike of the bend. Based on the model of FERRILL and GROSHONG (1993), uniform displacement/uniform shortening (NNE directed in the bend and NE directed in the main part of the fold belt) would be therefore a suitable kinematic classification for the South Limón fold-and-thrust belt.

5.1. Implications for the Geology of Costa Rica

There is a left-lateral fault zone in central Costa Rica; the so-called Trans-Isthmic fault system that has been analyzed over the last 25 years by studies that argue for (SEYFRIED et al. 1991; KRAWINKEL and SEYFRIED 1994; MARSHALL and FISHER 2000) and against (FERNÁNDEZ ARCE 1996), both its presence and kinematic timing. Some authors have interpreted the Trans-Isthmic fault system as an old structure that was already present during the early stages of island-arc evolution (SEYFRIED et al. 1991).

From the seismic lines it is evident that there is displacement on all of the thrusts, irrespective of their orientation. To allow movements in the structural modeling that were evenly distributed on all thrusts, we believe a transport direction that bisects the two major strike directions of the thrusts is required, and therefore we chose an azimuth of N018° for the thrust transport vector. This transport direction is almost identical to the plate vector of the Cocos Plate, as shown by NAIF et al. (2013; Fig. 2). The results show that our modeling is able to correctly restore the 3-D geometry of the South Limón fold-and-thrust with a single NNE-directed tectonic transport. Such a transport requires a left-lateral strike-slip zone that decouples the fold-and-thrust belt from the North Costa Rican arc segment and indicates that the Trans-Isthmic fault system must be present at this time. The results of our modeling are consistent with the model of FAN et al. (1993), where the South Limón fold-and-thrust belt interacts with a diffuse east–west trending, left-lateral strike-slip zone in the area of Puerto Limón and fits fault plane solutions from MARSHALL and FISHER (2000; Fig. 2) that imply that the back-arc thrusting is compensated inland along the Trans-Isthmic fault system by strike-slip motion.

Further evidence for the presence of the Trans-Isthmic fault system can be found in the work of PINDELL and KENNAN (2009) where lateral escape tectonics in Panama is postulated that also require an east–west trending fault zone in Costa Rica.

5.2. Implications for Subseismic-Scale Deformation Structures

In many fold-and-thrust belts, subseismic-scale deformation structures are manifested as fracture sets. The curvature of anticlines is a proxy for fracture density (LISLE 1994; FISHER and WILKERSON 2000). Therefore, the analysis of strain distribution over time (Fig. 8) has the potential to predict the fracture distribution within the analyzed part of the South Limón fold-and-thrust belt. In general, this modeling shows contractive strain of the forelimbs and dilative strain on the backlimbs of the hanging-wall anticlines. The contraction in a narrow strip on the leading edge of each hanging-wall anticline partially changes into dilative strain on the anticlinal crests.

The final step (i.e. after the restoration of the whole model) shows that both contractional and extensional strain zones exist as a tiled pattern (Fig. 8c). This is a result of the curvilinear nature of the hinge of anticline 1. Note that although strain in anticlines 2 and 3 is limited to the limbs, the strain of anticline 1 transgresses the hinge line, which might be caused by pronounced listric geometry of the thrust 1. We suggest that mean strains above 4 % are likely to cause permanent brittle deformation (VAN DER PLUIJM and MARSHAK 2004; LOHR et al. 2008). Furthermore, we propose that the strain distribution can be interpreted as a direct proxy for fracture intensity (LISLE 1999; LOHR et al. 2008).

6. Conclusions

The evolution of the bend in the South Limón fold-and-thrust belt can be modeled with a simple NNE-directed transport, implying one deformation phase, i.e., the bend is not an orocline, but rather a primary arc. The pronounced bend is most likely an effect of the interaction of the fold belt with the Trans-Isthmic fault system and the Moín basement high in the subsurface. The strain modeling shows contractional deformation concentrated on the fore-limbs of the anticlines, whereas dilative deformation focuses on the anticlinal crests and backlimbs. Due to the close thrust spacing, strain not only affects the respective hanging-wall anticline of the active thrust, but also the ones above it. The strain distribution in the bend area is similar to the strike of the hanging-wall anticlines and follows the trend of the bend.

Acknowledgments

We are grateful to W. Bandy for the opportunity to contribute to this volume and his editorial comments. We thank H. Seyfried and two anonymous reviewers for thoughtful and encouraging remarks. We would like to thank the Costa Rican Ministry of Environment and Energy (MINAE) for providing the seismic data. Midland Valley is gratefully thanked for an academic license for Move™.

REFERENCES

ABRATIS, M., and WÖRNER, G. (2001), *Ridge collision, slab-window formation, and the flux of Pacific asthenosphere into the Caribbean realm*, Geology 29, 127–130. doi:10.1130/0091-7613(2001)029<0127:RCSWFA>2.0.CO;2.

AFFOLTER, T., and GRATIER, J.-P. (2004), *Map view retrodeformation of an arcuate fold-and thrust belt: The Jura case*, Journal of Geophysical Research 108, B03404. doi:10.1029/2002JB002270.

AMANN, H. (1993), *Randmarine und terrestrische Ablagerungsräume des neogenen Inselbogensystems in Costa Rica (Mittelamerika)*, Profil 4, 161 pp.

ARGAND, E. (1924), *La tectonique de l'Asie*, 13th International Geological Congress Conf. Proceedings, Brussels, pp. 171.

ASTORGA, A., FERNANDEZ, J.A., BARBOZA, G., CAMPOS, L., OBANDO, J., AGUILAR, A., and OBANDO, L.G. (1991), *Cuencas sedimentarias de Costa Rica: evolucion geodinamica y potencial de hidrocarburos*, Revista Geologica de América Central 13, 25–59.

BARBOZA, G., FERNÁNDEZ, A., BARRIENTOS, J., and BOTTAZZI, G. (1997), *Costa Rica: Petroleum geology of the Caribbean margin*, Leading Edge 16, 1787–1794.

BARCKHAUSEN, U., RANERO, C.R., VON HUENE, R., CANDE, S.C., and ROESER, H.A. (2003), *Revised tectonic boundaries in the Cocos Plate off Costa Rica: Implications for the segmentation of the convergent margin and for plate tectonic models*, Journal of Geophysical Research 106, 19207–19220. doi:10.1029/2001JB000238.

BOTTAZZI, G., FERNANDEZ, A. and BARBOZA, G. (1994), *Sedimentología e historiatectono-sedimentaria de la cuenca Limón Sur. In* Seyfried H. & Hellmann W. (eds.) Geology of an Evolving Island Arc, The Isthmus of Southern Nicaragua, Costa Rica and Western Panamá, Profil 7, pp. 351–389.

BOWLAND, C.L. (1993), *Depositional history of the western Colombian Basin, Caribbean Sea, revealed by seismic stratigraphy*. GSA Bulletin 105, 1321–1345. doi:10.1130/0016-7606(1993)105<1321:DHOTWC>2.3.CO;2.

BRANDES, C., and TANNER, D.C. (2014), *Fault-related folding: a review of kinematic models and their application*, Earth Science Reviews 138, 352–370. doi:10.1016/j.earscirev.2014.06.008.

BRANDES, C., ASTORGA, A., and WINSEMANN, J. (2009), *The Moín High, East Costa Rica: Seamount, laccolith or contractional structure?* Journal of South American Earth Sciences 28, 1–13. doi:10.1016/j.jsames.2009.02.005.

BRANDES, C., ASTORGA, A., LITTKE, R., and WINSEMANN, J. (2008), *Basin modelling of the Limón Back-arc Basin (Costa Rica): burial history and temperature evolution of an island-arc related basin system*, Basin Research 20, 119–142. doi:10.1111/j.1365-2117.2007.00345.x.

BRANDES, C., ASTORGA, A., BACK, S., LITTKE, R., and WINSEMANN, J. (2007), *Fault controls on sediment distribution patterns, Limón Basin, Costa Rica*, Journal of Petroleum Geology 30, 25–40. doi:10.1111/j.1747-5457.2007.00025.x.

BRANDES, C., ASTORGA, A., BACK, S., LITTKE, R., and WINSEMANN, J. (2007), *Deformation style and basin-fill architecture of the offshore Limón Back-arc basin*, Marine and Petroleum Geology 24, 277–287. doi:10.1016/j.marpetgeo.2007.03.002.

BRANDES, C., ASTORGA, A., BLISNIUK, P., LITTKE, R., and WINSEMANN, J. (2007), Anatomy of anticlines, piggy-back basins and growth strata: a case study from the Limón Fold-and-thrust belt, Costa Rica. In: NICHOLS, G., WILLIAMS, E., and PAOLA, C. (eds)

Sedimentary Processes, Environments and Basins: A Tribute to Peter Friend, IAS Special Publication 38, pp. 91–110, Blackwell Science, Oxford. doi:10.1002/9781444304411.ch5.

BUTLER, R.W.H., PRIOR, D.J., and KNIPE, R.J. (1989), *Neotectonics of the Nanga Parbat Syntaxis, Pakistan, and crustal stacking in the northwest Himalayas*, Earth and Planetary Science Letters 94, 329–343. doi:10.1016/0012-821X(89)90150-7.

CAMPOS L. (2001), *Geology and basins history of middle Costa Rica: an intraoceanic island arc in the convergence between the Caribbean and the central pacific plates*, Tübinger Geowissenschaftliche Arbeiten, Reihe A 62, 138 pp.

CAREY, S.W. (1958), A tectonic approach to continental drift. In: CAREY, S.W. (ed) Continental Dift: A symposium, Tasmania, Hobart, pp. 177–355.

CAREY, S.W. (1955), *The orocline concept in geotectonics*, Proceedings Royal Society Tasmania 89, 255–288.

COATES, A.G., AUBRY, M-P., BERGGREN, W.A., COLLINS, L.S. and KUNK, M. (2003), *Early Neogene history of the Central American arc from Bocas del Toro, western Panama*. GSA Bulletin 115, 271–287.

COATES, A.G., JACKSON, J.B.C., COLLINS, L.S., CRONIN T.M., DOWSETT, H.J., BYBELL, L.M., JUNG, P. and OBANDO, J.A. (1992), *Closure of the Isthmus of Panama: The near-shore marine record of Costa Rica and western Panama*. GSA Bulletin 104, 814–828.

COOPER, M.A., and TRAYNER, P.M. (1986), *Thrust-surface geometry: Implications for thrust-belt evolution and section-balancing techniques*, Journal of Structural Geology 8, 305–312. doi:10.1016/0191-8141(86)90051-9.

COLLINS, L.S., COATES, A.G., JACKSON, J.B.C., and OBANDO, J.A. (1995), Timing and rates of emergence of the Limón and Bocas del Toro basins: Caribbean effects of Cocos Ridge subduction? In: MANN, P. (ed) Geologic and Tectonic Development of the Caribbean Plate Boundary in Southern Central America, Geological Society of America Special Paper 295, pp. 263–289. doi:10.1130/SPE295-p263.

CORRIGAN, J., MANN, P., and INGLE, JR, J.C. (1990), *Forearc response to subduction of the Cocos Ridge, Panama-Costa Rica*, GSA Bulletin 102, 628–652. doi:10.1130/SPE295-p263.

DAHLSTROM, C.D.A. (1969), *Balanced cross sections*, Canadian Journal of Earth Sciences 6, 743–757. doi:10.1139/e69-069.

DAHLSTROM, C.D.A. (1990), *Geometric constraints derived from the law of conservation of volume and applied to evolutionary models for detachment folding*, AAPG Bulletin 74, 336–344.

DALZIEL, I.W.D. and ELLIOT, D.H. (1973), The Scotia Arc and Antarctic margin. In: NAIRN, A.E.M., and STEHLI, F.G. (eds) The Ocean Basins and Margins, vol. 1: The South Atlantic. New York, Plenum Press, 171–245. doi:10.1007/978-1-4684-3030-1_5.

DEMETS, C. (2001), *A new estimate for present-day Cocos-Caribbean plate motion: Implications for slip along the Central American volcanic arc*, Geophysical Research Letters 28, 4043–4046. doi:10.1029/2001GL013518.

DEPAOR, D.G. (1988), *Balanced section in thrust belts part 1: construction*, AAPG Bulletin 72, 73–90. doi:10.1306/703C81CD-1707-11D7-8645000102C1865D.

DONNELLY, T.W. (1989), Geologic history of the Caribbean and Central America, In: BALLY, A.W. and PALMER A.R. (eds) The Geology of North America—An overview, Geological Society of America Special Paper A, pp. 299–321.

FAN, G., BECK, S.L., and WALLACE, T.C. (1993), *The seismic source parameters of the 1991 Costa Rica aftershock sequence:*

evidence for a transcurrent plate boundary, Journal of Geophysical Research *98* (B9), 15759–15778. doi:10.1029/93JB01557.

FARRIS, D.W., JARAMILLO, C., BAYONA, G., RESTREPO-MORENO, S.A., MONTES, C., CARDONA, A., MORA, A., SPEAKMAN, R.J., GLASCOCK, M.D., and VALENCIA, V. (2011), *Fracturing of the Panamanian Isthmus during initial collision with South America*, Geology *39*, 1007–1010. doi:10.1130/G32237.1.

FERRILL, D.A., and GROSHONG, R.H. (1993), *Kinematic model for the curvature of the northern Subalpine Chain, France*, Journal of Structural Geology *15*, 523–541. doi:10.1016/0191-8141(93)90146-2.

FERNANDEZ ARCE, M. (1996), *Evaluacion del hipotetico sistema de falla transcurrente este-oeste de Costa Rica*, Rev. Geol. Amér. Central. *19/20*, 57–74. doi:10.15517/rgac.v0i19-20.8626.

FERNANDEZ, J., ALVARO, A., GUILLERMO, B., BOTTAZZI, G., CAMPOS, L., OBANDO, J., TEJERA, R., ARRIETA, L., BARRIENTOS, J., BUSTOS, I., ESCALANTE, G., PIZARRO, D., VALERÍN, E., ASTORGA, A., BOLANOS, X., CALVO, C., LAURITO, C., ROJAS, J., and VALERIO, A. (1997), Mapa Geológico de Costa Rica, Ministerio del Ambiente y Energía, Costa Rica.

FERNANDEZ, J.A., BOTTAZZI, G., BARBOZA, G. and ASTORGA A. (1994), *Tectónica y estratigrafia de la Cuenca Limón Sur*. Rev Revista Geologica de América Central, Vol. Terremoto de Limón, 15–28.

FISHER, D.M., GARDNER, T.W., SAK, P.B., SANCHEZ, J.D., MURPHY, K., and VANNUCCHI, P. (2004), *Active thrusting in the inner forearc of an erosive convergent margin, Pacific coast, Costa Rica*, Tectonics *23*, TC2007, doi:10.1029/2002TC001464.

FISCHER, M.P. and WILKERSON, M.S. (2000), *Predicting the orientation of joints from fold shape: results of pseudo-three-dimensional modeling and curvature analysis*, Geology *28*, 15–18. doi:10.1130/0091-7613(2000)28<15:PTOOJF>2.0.CO;2.

GARDNER, T.W., FISHER, D.M., MORELL, K.D., and CUPPER, M.L. (2013), *Upper-plate deformation in response to flat slab subduction inboard of the aseismic Cocos Ridge, Osa Peninsula, Costa Rica*, Lithosphere *5*, 247–264. doi:10.1130/L251.1.

GATES, A.E., VALENTINO, D.W., CHIARENZELLI, J.R., SOLAR, G.S., and HAMILTON, M.A. (2004), *Exhumed Himalayan-type syntaxis in the Grenville orogen, northeastern Laurentia*, Journal of Geodynamics *37*, 337–359. doi:10.1016/j.jog.2004.02.011.

GOES S.D.B., VELASCO A.A., SCHWARTZ S.Y., and LAY T. (1993), *The April 22, 1991, Valle de la Estrella, Costa Rica (Mw = 7.7) earthquake and its tectonic implications: a broadband seismic study*, Journal of Geophysical Research *98*, B5, 8127–8142. doi:10.1029/93JB00019.

GRAY, M.B., and STAMATAKOS, J. (1997), *New model for evolution of fold and thrust belt curvature based on integrated structural and paleomagnetic results from the Pennsylvania salient*, Geology *25*, 1067–1070. doi:10.1130/0091-7613(1997)025<1067:NMFEOF>2.3.CO;2.

GREB, L., SARIC, B., SEYFRIED, H., BROSZONN, T., BRAUCH, S., GUGAU, G., WILTSCHKO, C., and LEINFELDER, R. (1996), *Ökologie und Sedimentologie eines rezenten Rampen-systems an der Karibikküste von Panamá*, Profil *10*, 1–168.

GUTIÉRREZ-ALONSO, G., JOHNSTON, S.T., WEIL, A.B., PASTOR-GALÁN, D., and FERNÁNDEZ-SUÁREZ, J. (2012), *Buckling an orogen: The Cantabrian Orocline*, GSA Today *22*, 4–8. doi:10.1130/GSATG141A.1.

HINDLE, D., and BURKHARD, M. (1999), *Strain displacement and rotation associated with the formation of curvature in fold belts;*

the example of the Jura arc, Journal of Structural Geology *21*, 1089–1101. doi:10.1016/S0191-8141(99)00021-8.

KRAWINKEL, H., SEYFRIED, H., CALVO, C. and ASTORGA, A. (2000), *Origin and inversion of sedimentary basins in southern Central America*. Zeitschrift für Angewandte Geologie SH *1*, 71–77.

KRAWINKEL, J., and SEYFRIED, H. (1994), A review of plate-tectonic processes involved in the formation of the southwestern edge of the Caribbean Plate, In: SEYFRIED, H., and HELLMANN, W. (eds) Geology of an Evolving Island Arc, The Isthmus of Southern Nicaragua, Costa Rica and Western Panamá, Profil *7*, 47–61.

LISLE, R. (1994), *Detection of abnormal strains in structures using Gaussian curvature analysis*, American Association of Petroleum Geologists Bulletin *78*, 1811–1819.

LISLE, R. (1999), *Predicting patterns of strain from three-dimensional fold geometries: neutral surface folds and forced folds*, Geological Society London, Special Publications *169*(1), 213–221. doi:10.1144/GSL.SP.2000.169.01.16.

LONSDALE, P., and KLITGORD, K. (1978), *Structure and tectonic history of the eastern Panama Basin*, GSA Bulletin *89*, 981–999.

LOHR, T., KRAWCZYK, C.M., ONCKEN, O., TANNER, D.C., SAMIEE, R., ENDRES, H., THIERER, P., TRAPPE, H., BACHMANN, R., and KUKLA, P.A. (2008), *Prediction of subseismic faults and fractures: Integration of three-dimensional seismic data, three-dimensional retro-deformation, and well data on an example of deformation around an inverted fault*, AAPG Bulletin *92/4*, 473–485. doi:10.1306/11260707046.

MACEDO, J., and MARSHAK, S. (1999), *Controls on the geometry of fold-and-thrust belt salient*, GSA Bulletin *111*, 1808–1822. doi:10.1130/0016-7606(1999)111<1808:COTGOF>2.3.CO;2.

MACMILLAN I., GANS P.B., and ALVARADO G. (2004), *Middle Miocene to present tectonic history of the southern Central American Volcanic Arc*, Tectonophysics *392*, 325–348. doi:10.1016/j.tecto.2004.04.014.

MARSHAK, S. (2004), Salients, recesses, arcs, oroclines, and syntaxes – a review of ideas concerning the formation of map-view curves in fold-thrust belts. In: MCCLAY, K.R. (ed) Thrust Tectonics and Hydrocarbon Systems, American Association of Petroleum Geologists, Memoir *82*, pp. 131–156.

MARSHALL, J.S., FISHER, D.M., and GARDNER T.W. (2000), *Central Costa Rica deformed belt: kinematics of diffuse faulting across the western Panama block*, Tectonics *19*, 468–492. doi:10.1029/1999TC001136.

MCNEILL, D.F., COATES, A.G., BUDD, A.F. AND BORNE, P.F. (2000), *Integrated paleontologic and paleomagnetic stratigraphy of the Upper Neogene deposits around Limon, Costa Rica: A coastal emergence record of the Central American Isthmus*. GSA Bulletin *112*, 963–981.

MENDE A. (2001), *Sedimente und Architektur der Forearc- und Backarc-Becken von Südost-Costa Rica und Nordwest-Panamá*, Profil *19*, 130 pp.

MESCHEDE, M., and FRISCH, W. (1998), *A plate tectonic model for the Mesozoic and Early Cenozoic history of the Caribbean Plate*, Tectonophysics *296*, 269–291. doi:10.1016/S0040-1951(98)00157-7.

MONTES, C., BAYONA, G., CARDONA, A., BUCHS, D.M., SILVA, C.A., MORÓN, S., HOYOS, N., RAMÍREZ, D.A., JARAMILLO, C.A., and VALENCIA, V. (2012), *Arc-continent collision and orocline formation: Closing of the Central American seaway*, Journal of Geophysical Research *117*, B04105. doi:10.1029/2011JB008959.

MORELL, K.D. (2016), Seamount, ridge and transform subduction in southern Central America. Tectonics 35, 10.1002/2015TC003950.

MORELL, K.D., GARDNER, T.W., FISHER, D.M., IDLEMAN, B., and ZELLNER, H. (2013), Active thrusting, landscape evolution and late Pleistocene sector collapse of Barú Volcano above the Cocos-Nazca slab tear, southern Central America, GSA Bulletin 125, 1301–1318. doi:10.1130/B30771.1.

MORELL, K.D., KIRBY, E., FISHER, D., and VAN SOEST, M. (2012), Geomorphic and exhumational response of the Central American volcanic arc to Cocos Ridge subduction, Journal of Geophysical Research 117, B04409, doi:10.1029/2011JB008969.

MORELL, K.D., FISHER, D.M., and GARDNER, T.W. (2008), Inner forearc response to subduction of the Panama fracture zone, southern Central America, Earth and Planetary Science Letters 265, 82–95. doi:10.1016/j.epsl.2007.09.039.

MUKUL, M., and MITRA, G. (1998), Finite strain and strain variation analysis in the Sheeprock Thrust Sheet: an internal thrust sheet in the Provo salient of the Sevier Fold-and-Thrust belt, Central Utah, Journal of Structural Geology 20, 385–405. doi:10.1016/S0191-8141(97)00087-4.

NAIF, S., KEY, K., CONSTABLE, S., and EVANS, R.L. (2013), Melt-rich channel observed at the lithosphere–asthenosphere boundary, Nature 495, 356–359. doi:10.1038/nature11939.

PINDELL, J.L., and KENNAN, L. (2009), Tectonic evolution of the Gulf of Mexico, Caribbean and northern South America in the mantle reference frame: an update, In: JAMES, K.H., LORENTE, M.A., and PINDELL, J.L. (eds) The Origin and Evolution of the Caribbean Plate, Geological Society, London, Special Publications 328, pp. 1–55. doi:10.1144/SP328.1.

POBLET, J., and LISLE, R.J. (eds), (2011), Kinematic evolution and structural styles of fold- and-thrust belts, Special Publication of the Geological Society of London 349. doi:10.1144/SP349.4.

PROTTI, M., and SCHWARTZ, S.Y. (1994), Mechanics of back arc deformation in Costa Rica: Evidence from an aftershock study of the April 22, 1991, Valle de la Estrella, Costa Rica, earthquake ($M_w = 7.7$), Tectonics 13, 1093–1107. doi:10.1029/94TC01319.

RANERO, C.R., and VON HUENE, R. (2000), Subduction erosion along the Middle America convergent margin, Nature 404, 748–752. doi:10.1038/35008046.

ROSS M.I., and SCOTESE C.R. (1988), A hierachical tectonic model of the Gulf of Mexico and the Caribbean region, Tectonophysics 155, 139–168. doi:10.1016/0040-1951(88)90263-6.

SALA, P., PFIFFNER, O.A., and FREHNER, M. (2014), The Alpstein in three dimensions: fold-and-thrust belt visualization in the Helvetic zone, eastern Switzerland, Swiss Journal of Geoscience 107, 177–195. doi:10.1007/s00015-014-0168-6.

SEYFRIED, H., ASTORGA, A., AMANN, H., CALVO, C., KOLB, W., SCHMIDT, H., and WINSEMANN, J. (1991), Anatomy of an evolving island arc: tectonic and eustatic control in the south Central American forearc area. In: MACDONALD, D.I.M. (ed.) Sedimentation, Tectonics and Eustacy: Sea-level changes at Active Margins, International Association of Sedimentologists, Special Publication 12, pp. 273–292. doi:10.1002/9781444303896.ch13.

SHEEHAN, C.A., PENFIELD, G.T., and MORALES E. (1990), Costa Rica geologic basins lure wildcatters. Oil Gas Journal Apr. 30, 74–79.

SILVER E.A., REED D.L., TAGUDIN J.E., and HEIL D.J. (1990), Implications of the north and south Panama thrust belts for the origin of the Panama orocline, Tectonics 9, 261–281. doi:10.1029/TC009i002p00261.

SITCHLER, J.C., FISHER, D.M., GARDNER, T.W., and PROTTI, M. (2007), Constraints on inner forearc deformation from balanced cross sections, Fila Costena thrust belt, Costa Rica, Tectonics 26, TC6012. doi:10.1029/2006TC001949.

SUESS, E. (1908), Das Antlitz der Erde. Erster Band, Wien, 778 pp.

SUPPE J., and MEDWEDEFF D.A. (1990), Geometry and kinematics of fault-propagation folding, Eclogae Geologicae Helvetiae 83(3), 409–454.

TANNER, D.C., BEHRMANN, J.H., and DRESMANN, H. (2003), Three-dimensional retro-deformation of the Lechtal Nappe, Northern Calcareous Alps, Journal of Structural Geology 25, 737–748. doi:10.1016/S0191-8141(02)00057-3.

TANNER, D.C., BENSE, F.A., and ERTL, G. (2011), Kinematic retro-modelling of a cross-section through the Western Irish Namurian Basin. In: POBLET, J., and LISLE, R.J. (eds) Kinematic evolution and structural styles of fold- and-thrust belts, Special Publication of the Geological Society of London 349, 61–76. doi:10.1144/SP349.4.

VAN DER PLUIJM, B.A., and MARSHAK, S. (2004), Earth Structure. Norton and Company, 656 pp.

VON EYNATTEN, H., SCHMIDT, H., and WINSEMANN, J. (1993), Plio-Pleistocene outer arc basins in southern Central America, In: FROSTICK, L., and STEELE, R. (eds) Sedimentation and Tectonics, IAS Special Publication 20, pp. 399–414. doi:10.1002/9781444304053.ch21.

WALTHER, C.H.E. (2003), The crustal structure of the Cocos Ridge of Costa Rica, Journal of Geophysical Research 108, 1–21. doi:10.1029/2001JB000888.

WU, S., YU, Z., ZHANG, R., HAN, W., and ZOU, D. (2005), Mesozoic-Cenozoic tectonic evolution of the Zhuanghai area, Bohai-Bay Basin, east China: the application of balanced cross-sections, Journal of Geophysics and Engineering 2, 158–168. doi:10.1088/1742-2132/2/2/011.

ZEUMANN, S., and HAMPEL, A. (2015), Deformation of erosive and accretive forearcs during subduction of migrating and non-migrating aseismic ridges: Results from 3D finite-element models and application to the Central American, Peruvian and Ryukyu margins, Tectonics 34, 1769–1791. doi:10.1002/2015TC003867.

ZIESCH, J., TANNER, D.C., and KRAWCZYK, C.M. (2014), Strain associated with the fault-parallel flow algorithm during kinematic fault displacement, Mathematical Geoscience 46(1), 59–73. doi:10.1007/s11004-013-9464-3.

(Received May 4, 2015, accepted February 26, 2016, Published online March 17, 2016)

Pure Appl. Geophys. 173 (2016), 3357–3371
© 2016 Springer International Publishing
DOI 10.1007/s00024-016-1295-y

Low Angle Contact Between the Oaxaca and Juárez Terranes Deduced From Magnetotelluric Data

JORGE A. ARZATE-FLORES,[1] ROBERTO MOLINA-GARZA,[1] FERNANDO CORBO-CAMARGO,[1] and VÍCTOR MÁRQUEZ-RAMÍREZ[1]

Abstract—We present the electrical resistivity model along a profile perpendicular to the Middle America trench in southern Mexico that reveals previously unrecognized tectonic features at upper to mid-crustal depths. Our results support the hypotheses that the upper crust of the Oaxaca terrane is a residual ∼20 km thick crust composed by an ∼10 km thick faulted crustal upper layer and an ∼10 km thick hydrated and/or mineralized layer. Oaxaca basement overthrust the younger Juárez (or Cuicateco) terrane. The electrical resistivity model supports the interpretation of a slab subducting at a low angle below Oaxaca. Uplift in the Oaxaca region appears to be related to fault reactivation induced by low angle subduction. In the Juárez terrane, isostatic forces may contribute to uplift because it is largely uncompensated. In the Sierra Madre del Sur, closer to the coast, uplift is facilitated by slab-dehydration driven buoyancy. Both gravity and resistivity models are consistent with a thinned upper crust in the northeast end of the profile.

Key words: Magnetotellurics, Subduction, Oaxaca terrane, Juarez terrane, Uplift.

1. Introduction

Southern Mexico is characterized by crustal discontinuities that separate contrasting basement units (ORTEGA-GUTIÉRREZ 1981). Older terranes were assembled between late Paleozoic and Cretaceous time, and its geology has been influenced by convergent margin tectonics and magmatism for most of the Cenozoic. The present plate tectonic setting is dominated by active subduction along the Pacific margin. Because of complex plate interactions, including the evolution of the Caribbean-Cocos-North America triple junction, the geometry of

subduction is variable along the continental margin (e.g., PARDO and SUAREZ 1995).

Arc magmatism linked to modern subduction occurs along the Trans-Mexican Volcanic Belt (TMVB), in Los Tuxtlas, and in the Chiapas region (Fig. 1). But magmatism is sparse, and large segments of the trench lack any manifestation of magmatism (MANEA *et al.* 2013). The subduction angle changes gradually from steep in the western TMVB to nearly flat in the eastern TMVB and under Oaxaca, and then dipping at a high angle under Chiapas and Central America (e.g., MANEA and MANEA 2010). As a result of truncation of the continental margin in the Cenozoic, magmatism migrated landward from the trench (MORÁN-ZENTENO *et al.* 1996; MOLINA GARZA *et al.* 2015). Evidence of truncation of the continental margin includes the lack of a Paleogene forearc, the eastward younging of the ages of Cenozoic plutons along the coast, and trench parallel shear zones such as Tierra Colorada, Chacalapa, and Juchatengo.

The relationships between subduction, tectonics and magmatism have been explored in other areas of the continental margin (e.g., FERRARI *et al.* 2012), but not in the Oaxaca region. Unlike most regions of flat-slab subduction Oaxaca is characterized by extensional tectonics in the overriding plate since Oligocene time (NIETO-SAMANIEGO *et al.* 2006), forming the Neogene basins of central Oaxaca (the Valles Centrales, VC on Fig. 1). Subduction has been also linked to impressive uplift along the continental margin in the Sierra Madre del Sur (DUCEA *et al.* 2004) around the contact zone between the Xolapa-Oaxaca terranes, and further inland in Sierra Juárez (CENTENO GARCÍA 1988) (Fig. 1). The Sierra Madre del Sur province north of Puerto Escondido is a

¹ Centro de Geociencias, Universidad Nacional Autónoma de México, Blvd. Juriquilla #3000, C.P. 76230 Juriquilla, QRO, Mexico. E-mail: arzatej@geociencias.unam.mx

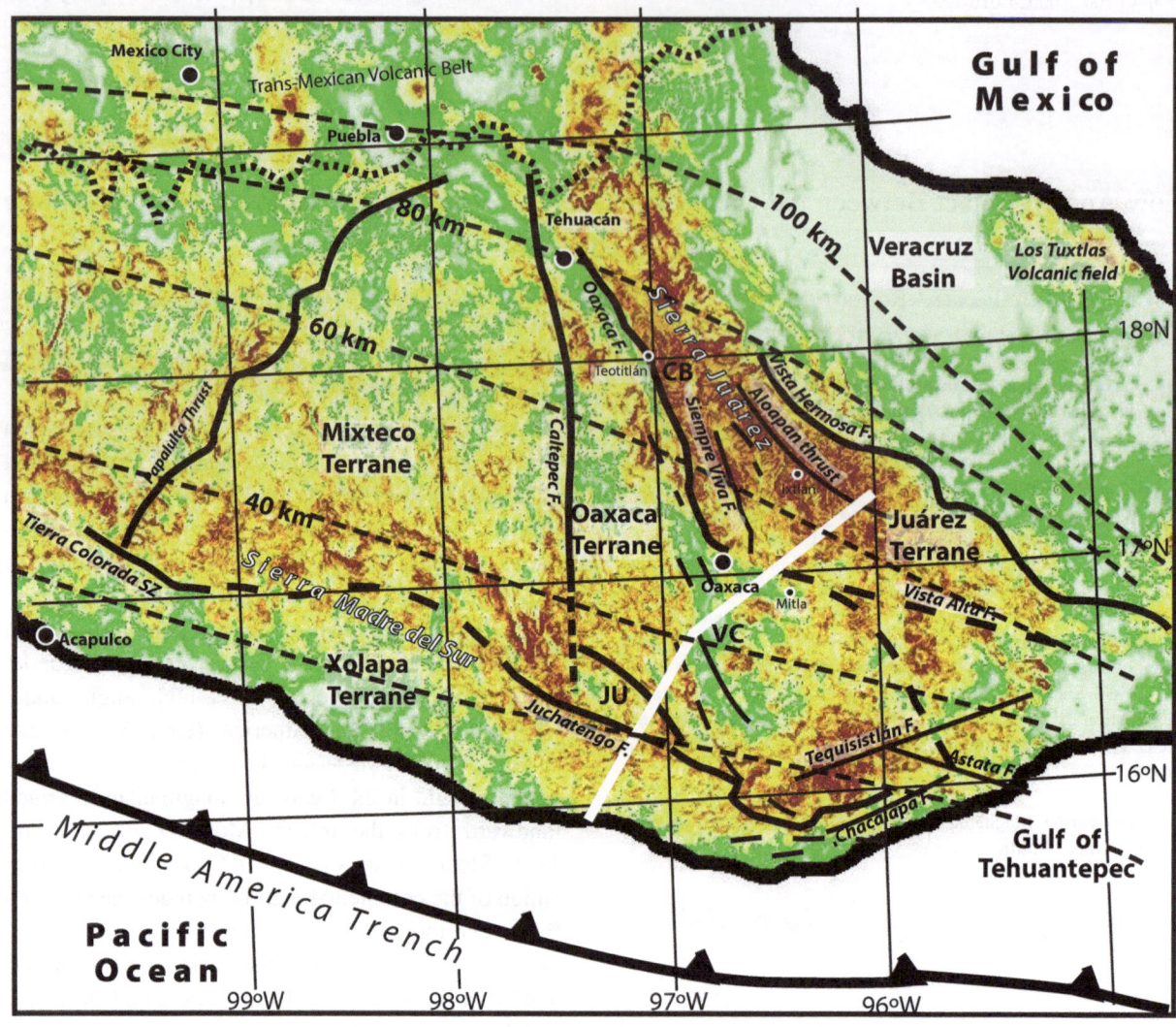

Figure 1
Relative elevation image of southwestern Mexico showing main regional structures and terrane limits. *Labeled dotted lines* represent depth to the subducting interface from PARDO and SUAREZ (1995). *White continuous white line* shows the trace of the interpreted magnetotelluric profile. *VC* Valles Centrales, *JU* Juchatengo terrane, *CB* Cuicateco Basin

mountainous region with exposures of crystalline rocks reaching elevations in excess of 3000 m.

Based on the construction of electric conductivity and gravity profiles, herein we show that the Oaxaca terrane is thrusted over the Cuicateco basin (CB on Fig. 1) of the Juárez terrane. The Juárez terrane must be largely isostatically uncompensated because its crust is relatively thin (e.g., BARBOZA GUDIÑO 1996), which may be a driver of uplift of Sierra Juárez. Dynamic topography may drive, in contrast, documented subsidence of the Veracruz basin in the east (Fig. 1). A large contrast in electric properties of the upper crust is reported here across the Aloapan thrust within the Juárez terrane. Herein we show that some terrane boundaries lack distinct contrasts in electric properties, and thus seem to have mid to lower crust of similar characteristics. The Juchatengo block (JU in Fig. 1) in southern Oaxaca is, however, much more resistive and of higher density than neighboring terranes. This supports models interpreting the Juchatengo terrane as a suture.

2. Tectonic Setting

2.1. Terranes in Southern Mexico

Continental crustal elements in southern Mexico include the Oaxaca terrane, with its Precambrian basement, and the Paleozoic Mixteco terrane (Fig. 1). The Mixteco and Oaxaca terranes were juxtaposed along the right-lateral Caltepec fault in the Middle Permian (ELÍAS HERRERA and ORTEGA GUTIÉRREZ 2002). Precambrian basement rocks of the Oaxaca terrane are mostly graphite bearing paragneiss, mafic orthogneiss, anorthosite, and migmatite. This basement rocks are generally of high magnetic susceptibility and conductive. They are intruded by late Paleozoic plutons, and are overlain by Paleozoic through Cretaceous strata. Cenozoic rocks in the Oaxaca terrane include Paleogene continental siliciclastic and volcanic rocks deposited in extensional basins (DÁVALOS-ÁLVAREZ et al. 2007).

The Oaxaca terrane is limited to the east by the Cenozoic Oaxaca fault system, which follows the trend of the Sierra Juárez Mylonite Belt (SJMB), and separates the Oaxaca and Juárez (or Cuicateco) terranes. Sierra Juárez is a prominent physiographic feature, with widespread exposures of pre-Jurassic schist and serpentinite-gabbro associations, Jurassic redbeds, latest Jurassic granitoids and migmatites, and a thick sequence of Lower Cretaceous continentally derived turbidites (ÁNGELES MORENO 2006). The turbidites were deposited in the Cuicateco basin. The sedimentary rocks include carbonates and are poor conductors. The contacts between different units that compose the Juárez terrane are tectonic.

Two of the most conspicuous features of the Juárez terrane are the Sierra Juárez (SJ) itself, an elevated plateau deeply incised by east flowing rivers, and the Oaxaca fault bordering SJ on the west. Sierra Juárez is a nearly north-south trending mountain range rising from central Oaxaca at an average elevation of 1500 m, up to a maximum height of 3200 m (Fig. 1). The range shows an imposing relief of about 2500 meters between the highest elevations and the river valleys, and drops rapidly to near sea level on its eastern margin. To the east of SJ lies the Veracruz basin, characterized by a thick Cretaceous to Neogene succession of sedimentary rocks. Besides the astounding relief, two features in SJ are remarkable: the presence of active normal faults in the uplifted region and the fact that the direction of extension is nearly parallel to the direction of plate convergence.

The basement of the Juárez terrane includes the Mazateco Complex (CARFANTÁN 1986; ÁNGELES MORENO 2006). The Lower Cretaceous siliciclastic rocks of the Juárez terrane have been mapped as the Chivillas Formation in the northern portion of the Cuicateco basin, where the maximum deposition age is mid-Aptian (MENDOZA-ROSALES et al. 2009). The Juárez terrane is limited to the east by the Vista Hermosa fault. The fault represents the boundary with the Veracruz basin of the Maya terrane, and it is a thrust that places pre-Jurassic schist and ultramafic rocks over Jurassic redbeds in the Valle Nacional area.

The Xolapa terrane extends along coastal Mexico, from west of Acapulco to the Tehuantepec isthmus (SOLARI et al. 2007). Xolapa consists primarily of amphibolite facies gneiss and migmatite, with protoliths of Jurassic and Early Cretaceous age, intruded by a suite of eastward younging calc-alkaline plutons (TALAVERA-MENDOZA et al. 2013; SCHAAF et al. 1995). It has been interpreted as the roots of a Jurassic-Early Cretaceous continental arc. Left lateral strike-slip faults separate Xolapa from the Mixteca and Oaxaca terranes, including the Tierra Colorada and Chacalapa shear zones (Fig. 1). Exposure of mid-crustal rocks in the Xolapa terrane has been linked to Oligocene–Miocene uplift of the region (DUCEA et al. 2004).

The Juchatengo block (labeled JU on Fig. 1) is a Paleozoic tectonic element of oceanic affinity enclosed in a relatively small area bounded by the Xolapa terrane to the south, the Mixteca terrane to the west, and the Oaxaca terrane to the east. It is characterized by pre-Early Permian mafic rocks with MORB chemistry intruded by late Paleozoic granitoides (GRAJALES NISHIMURA et al. 1999). A back-arc or a rift tectonic setting have been proposed; however, its relative position between the Oaxaca and Mixteco terranes has been interpreted as part of a suture (ELÍAS HERRERA and ORTEGA GUTIÉRREZ 2002).

2.2. The Oaxaca Fault

The Cenozoic Oaxaca fault (Fig. 1) is a brittle, N10°W trending, west dipping, extensional system with a length of ~250 km and a structural relief of about >3 km (DÁVALOS-ÁLVAREZ et al. 2007). It is characterized by a steep fault escarpment of about 1500 m (a mountain front). ALANIZ–ÁLVAREZ et al. (1994) have suggested that the Sierra Juarez Milonitic Belt (SJMB) is a NS trending, ~10 km wide, structural complex extending for 130 km parallel to the Oaxaca fault. Outcrops of the mylonite belt are, however, restricted to the vicinity of Oaxaca City. To the north of Oaxaca City, in the Teotitlán region, the trace of the fault exposes latest Jurassic-earliest Cretaceous migmatites and granitoides (ÁNGELES MORENO 2006). Further north, in the Tehuacán region, the Oaxaca fault zone juxtaposes Cenozoic and Lower Cretaceous clastic rocks, and there are no outcrops of mylonite (DÁVALOS-ÁLVAREZ et al. 2007). To the south of Oaxaca City, NW of the city of Mitla (Fig. 1), the trace of the mylonite belt zone is covered by Miocene ignimbrites. This portion of the fault is mapped as the Vista Alta fault, and there are outcrops of redbeds of inferred Jurassic age with intense deformation developing pencil cleavage. ALANIZ–ÁLVAREZ et al. (1994) described the fault as a site of Cenozoic rejuvenation of an older continuous structure, but the significance of the SJMB is not fully understood.

The SJMB is characterized by greenschist facies mylonite with a protolith that includes granitoids, mafic and ultramafic rocks, and granulitic gneiss (BARBOZA GUDIÑO 1996). The mylonite is characterized by a foliation that dips about 30° to the west, and kinematic indicators suggest right-lateral motion (ALANIZ-ÁLVAREZ et al. 1995). The SJMB was last active as a strike-slip fault during Middle Jurassic time (ALANIZ-ÁLVAREZ et al. 1995). The low angle foliation of the SJMB may not be a primary feature, because there are vertically dipping Jurassic clastic rocks along the fault trace northeast of Oaxaca City. The redbeds contain clasts of mylonite; the mylonite may thus have been tilted. The extensional Oaxaca fault system may be active and seismogenic (DÁVALOS-ÁLVAREZ et al. 2007). The western side of the fault is a series of half-grabens filled by Cenozoic

clastic rocks and isolated outcrops of Miocene volcanic rocks (DÁVALOS-ÁLVAREZ et al. 2007). These authors established four pulses of Cenozoic activity in the fault, initiating in the Eocene. Activity continued into the Pleistocene with deposition of the Coyoltepec conglomerate and late Holocene alluvial fans.

3. The Magnetotelluric Data Set

3.1. Geoelectric Strike

We have combined two magnetotelluric data sets from previous studies (ARZATE et al. 1995; CORBO 2013) to complete a 200 km long profile across the Xolapa-Oaxaca and Oaxaca-Juarez terrane boundaries (Fig. 2). We calculated electric strike directions for the 12 broadband magnetotelluric stations comprising the profile. The average electric strike azimuth along the MT transect defines two sectors with different strike directions; the SW part has an average azimuth of 17° whereas the NE half of the profile it is −3°. Figure 2 shows strike directions given by the major axis of the tensor phase ellipses (CALDWELL et al. 2004). In the upper right of the figure we show the rose diagram of the individual estimation of the electric strike direction for every sounding. The average values are indicated with blue and magenta lines for the NE and SW profile segments, respectively. In a similar way, we plotted the average azimuths of the induction vectors for all the frequency range (10^3–10^{-3} Hz) and every MT sounding (lower right of the figure). The induction vector azimuth is by definition perpendicular to the electric strike direction, which is to be expected for a meaningful 2D modeling of the data set. However, about 10° deviations are observed for both profiles suggesting a regional 3D induction source.

3.2. 2D Inversion of MT Data

Inversion of the profile data was done after rotating the southern and northern soundings to the respective average electric strike directions, i.e., 17° and −3°, respectively. Then, the non-linear conjugate gradients algorithm (RODI and MACKIE 2001) was

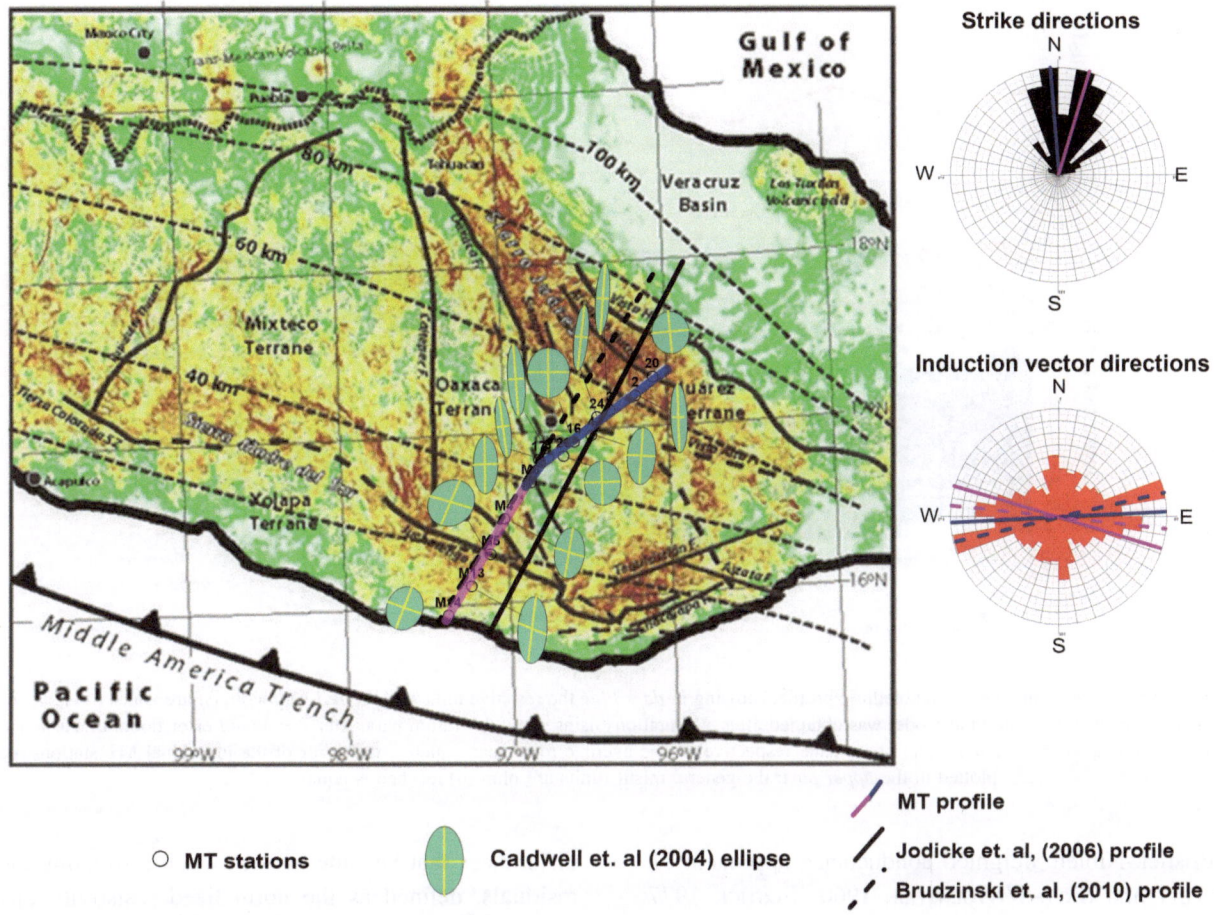

Figure 2

Location of the composite MT profile (*magenta* and *blue lines*) and average electric strike directions (*upper right of figure*) for the frequency range 320–10–3 Hz, that define two profile segments; the SW segment (*magenta*) has an average strike of 17°, whereas for the NE segment (in *blue*) is nearly NS (−3°). The major axis of the tensor phase ellipticity shows variations of strike due to crustal structure along the profile; major axis points to the regional strike locally. The azimuth of the induction vectors (*lower right, dotted lines*) indicates deviations of about 10° of theoretical 2D response in both profiles. The location of Jödicke (2006) MT profile is shown as a *continuous black line*, located approximately 100 km southeast of our profile. Brudzinski *et al.* (2010) seismicity cross section, which coincides exactly with the southern segment of our profile, is shown with an *orange line*

used to invert the impedance of both polarization modes TE and TM. Vertical data was generally noisy and we preferred to exclude it from the inversion process. We used a tested regularization parameter, and 189 rows by 46 columns mesh filled with homogeneous 100 O-m cells as the initial model. We inverted for the smoothed resistivity and phase curves assigning fixed 10 and 5 % error floors to the apparent resistivity and phase data, respectively. Several runs were done varying the number of iterations and regularization parameter τ with the result of basically the same electric structure but

increasing model smoothness as we increase the regularization parameter and number of iterations separately. Figure 3 shows the resulting resistivity image after 50 iterations and $\tau = 4$ that returns a misfit rms error of 3.6 %. The resistivity range of the obtained image lies between 1 and 10,000 O-m. The upper part of Fig. 3 shows a plot of the rms misfits for the individual MT stations. Larger errors are observed to concentrate around station 03 particularly at frequencies below 0.1 Hz. Vertical gray lines in the resistivity image represent the depth of investigation at every MT sounding estimated using

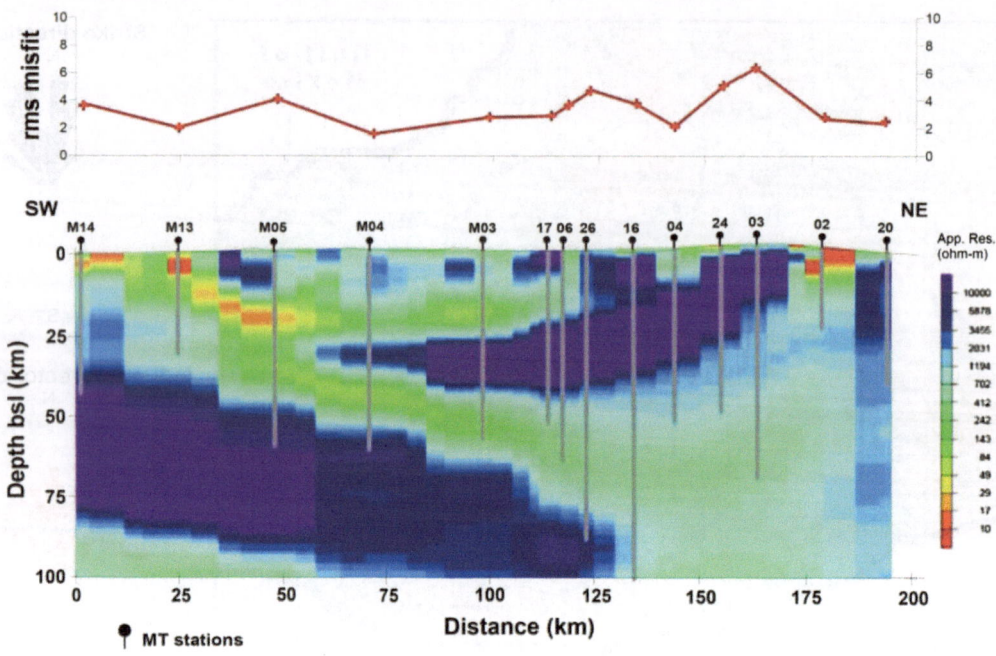

Figure 3

Resistivity model along the magnetotelluric profile, showing in *dark blue* the resistive units and the *red-yellow-green* are zones of enhanced electric conductivity. The final model was obtained after 50 iterations using a regularization parameter $t = 4$, and error floors of 10 % and 5 % for the apparent resistivity and phase data, respectively. The average root mean square (rms) error of the individual MT stations are plotted in the *upper part*; the general misfit (units are ohm-m) reached is equal to 3.6 %

Bostick-Niblett weighted conductance approach (NI-BLETT and SAYN-WITTGENSTEIN 1960; BOSTICK 1977; JONES 1983). The estimation of the depth of investigation can be performed using any of the modes of polarization (or invariants). We used the TE mode, which underestimates the values of penetration of EM fields relative to those determined using the TM polarization mode. This provides conservative estimates, which we prefer.

The pseudo-sections of both, the TE and TM modes are shown in Fig. 4. The observed (ρ_{obs}) and the model resistivity response (ρ_{resp}) as a function of frequency for both polarization modes are plotted in Fig. 4a, b, while the observed and calculated phases (ϕ_{obs} and ϕ_{resp}) are plotted in the lower part of the Fig. 4c, d, respectively. As observed, the obtained resistivity structure recovered from the model resembles fairly well the original resistivity data, which is also true for the phase structure except for a limited range of frequencies (>10 Hz) at stations 17, 06, and 26. Isolated frequencies at other sites, particularly, at stations 24 and 03, also show high estimated phase

differences that become evident when estimating the residuals, defined as the normalized resistivity difference between the observed and calculated data. As expected from the amplitude and phase pseudo-sections, the larger residuals and larger rms misfits are concentrated around sites 03 and 24, which is particularly true for the TE mode of polarization although either modes seem to be affected by EM distortion in this zone.

4. Interpretation of the Resistivity Model

4.1. Hydrated and Mineralized Upper Continental Crust

Mineralized fluids release and melt production are characteristic subduction related processes, which can be recognized by their associated anomalous electrical conductivity (e.g., JONES 1983). Many magnetotelluric studies have been conducted at the active continental margin along the west coast of the

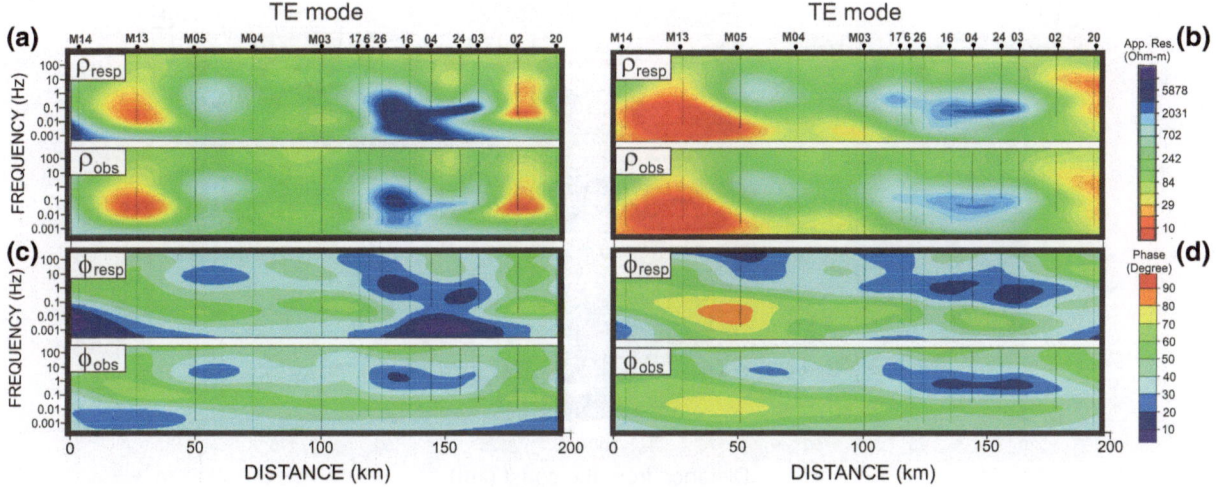

Figure 4

The pseudo-sections of both, the TE and TM modes are shown in Fig. 4. Comparison of the observed (ρ_{obs}) and model resistivity response (ρ_{resp}) as a function of frequency for the TE and TM polarization modes (**a** and **b**, respectively). The corresponding observed and calculated phases (ϕ_{obs} and ϕ_{resp}) are also shown (**c** and **d**, respectively)

American continent (e.g., KURTZ *et al*. 1986, 1990; WANNAMAKER *et al*. 1989; ARZATE *et al*. 1995; JÖDICKE 2006; CORBO 2013; SCHWARZ and KRÜGER 1997; BRASSE *et al*. 2002, 2008) that reveal a variety of conductivity anomalies in diverse subduction regimes associated with fluids, metamorphism, and partial melt. WANNAMAKER *et al*. (1989) first interpreted a landward dipping conductor in the forearc region of Juan de Fuca subducting plate in Oregon as directly imaging the interconnected water content of subducted sediments, and thus the imaging of the top of the subducting plate where the sediments lie. The existence of important quantities of water at crustal depths was later supported by experimental petrological studies (e.g., PEACOCK 1990; TATSUMI and EGGINS 1990; SCHMIDT and POLI 1998) that emphasize the importance of dehydration reactions in the downgoing oceanic slab.

JÖDICKE (2006) related enhanced conductivity zones to specific metamorphic processes in the southern Mexico subduction zone, in particular along a transect parallel to ours, ~100 km to the east. Subduction related metamorphism consists of mineralized water (brines), stemming either from water filled open porosity and fractures, or from water bound chemically to the lower oceanic crust. The main metamorphic reactions associated with subduction, are dependent on particular PTV conditions, and may be also influenced by earlier sea floor metamorphism and strong hydrothermal alteration during the early period of crustal formation at the East Pacific Rise. Such alteration results in the zonation of hydrated minerals ranging in a descending order in the crust from zeolite facies to prehnite-pumpellyite facies to greenschist facies (*ibid*.).

The interpreted conductivity structure of the MT section is shown in Fig. 5. The red dotted lines separate the different tectonic elements according to electrical conductivity contrasts. The main conductors are associated to at least three geodynamic processes, and are labeled with capital letters for their discussion. Isothermals in the section are those estimated by JÖDICKE (2006), and a black continuous line shows the top of the oceanic plate derived from refraction data (SPRANGER 1994) along a profile located 200 km east of the MT profile. Black dots represent hypocenters of seismicity, occurring from 1974 to 2013 within a 20 km band centered on the MT profile, of magnitude $M > 4.5$ obtained from the SSN catalogue. Arrows pointing downwards indicate the tectonostratigraphic limits of the Xolapa-Oaxaca terranes (black arrow) and the Oaxaca-Juarez terranes (white arrow) according to ORTEGA-GUTIÉRREZ (1981). MT stations are marked with inverted red triangles.

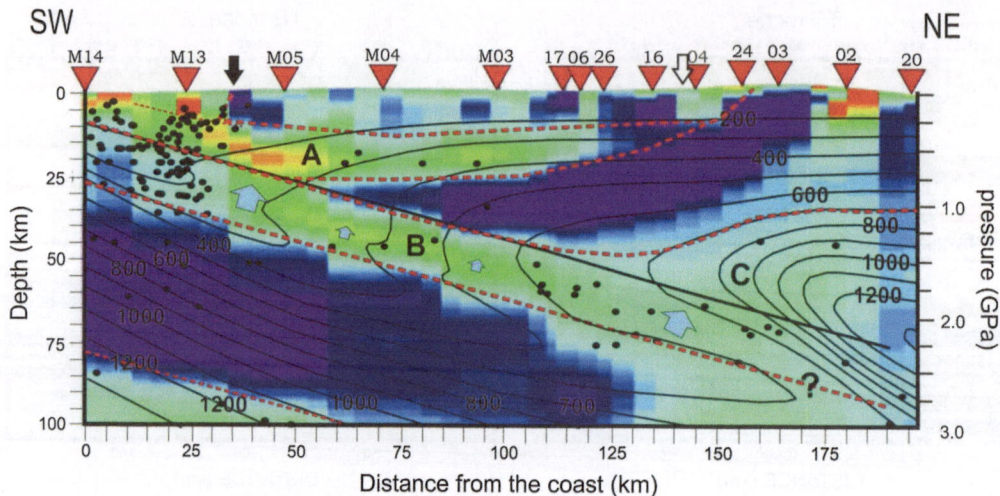

Figure 5

Interpreted conductivity structure of the MT section. The *red dashed lines* separate anomalous conductivity regions. These conductive zones are associated to at least three geodynamic processes leveled with *capitals letters*. Isothermals are those estimated by JÖDICKE (2006), and *black continuous line* shows the top of the oceanic plate derived from refraction data (SPRANGER 1994) along a profile located ~100 km eastward of this MT profile. *Black dots* represent hypocenters of seismicity within a 20 km band of magnitude $M < 4.5$ of the SSN catalogue from 1974 to 2013. *Arrows* pointing downwards indicate the tectonostratigraphic limit of the Xolapa-Oaxaca terrain (*black arrow*) and the Oaxaca-Juarez terrane (*white arrow*) according to CAMPA and CONEY (1983). *Blue arrows* indicate zones of major dehydration within the oceanic conductive upper slab (*B*). MT stations are marked with *inverted red triangles*

The anomalous conductivity zone closest to the coast (labeled A in Fig. 5) located at depths above 25 km is assumed to originate from water expelled from gradually closing pores and fractures on top of the bending oceanic slab that occur at a short distance from the coast margin. Another source for the water budget at similar depths is the dehydration and breakdown of clay minerals occurring at temperatures of 150–200 °C (PEACOCK 1990; RÜPKE *et al.* 2002).

A feasible explanation for the subhorizontal extension to the NE of this conductivity anomaly is that the large amount of fluids (20–30 %) released by the oceanic crust at the bending of the slab (SCHMIDT and POLI 1998) can migrate sub-horizontally through shear structures or secondary porosity (fractures) associated with the low angle convergence regime in this region. Fluids could remain for long periods within the anisotropic layered structure, trapped below impermeable silica layer that precipitates at ~200 °C (HYNDMAN and SHEARER 1989) preventing the fluids to migrate upwards. With a large stock of trapped fluids important volumes of ores (Fe-, Cu-, Zn-sulfide or oxide rich ore deposits, JÖDICKE 2006) can react and precipitate from the abundant mineralized fluids available. Large epithermal disseminated ore

impregnations along interconnected fracture zones may also help to explain the observed anomalous conductor. Also, disseminated graphite may contribute to the overall high conductivity, particularly in the Oaxaca terrane basement.

4.2. Dipping Angle of the Subducting Cocos Plate

Less conductive than the continental crust near the coast, the 20 km thick ~16° dipping conductor (B) observed in the conductivity model on top of the resistive oceanic upper mantle is interpreted as a continuous dehydration zone confined above by a resistive lower continental crust. The gradual transformation of zeolites to smectites and micas close to the convergent front evolves until the blueschist facies is reached, probably not farther than 50 km from the coast and at depths below 30–40 km. The blueschist reaction is reported to have an important water potential increasing with temperature, as high as 12 wt % (SCHMIDT and POLI 1998), which may explain the observed dipping extended conductive zone. The confining environment of such a conductive zone (a high density-low permeability lower continental crust on top) and a high density-low permeability eclogite shear zone underneath prevents fluid migration to shallow depths keeping the fluids trapped in the subduction interface. The fluids released by the oceanic slab and those released adiabatically at greater depths remain trapped for long periods of time on top of the subducted slab, and may be exerting a net buoyancy force against the whole continental crust in the area.

The angle of the interface between the Cocos and North American plates from refraction data (SPRANGER 1994), shown as a black continuous straight line in Fig. 5, is consistent with the dip angle of the enhanced conductivity zone. It fits quite well with the contrasting resistivity zones (conductive-resistive-conducting) of the MT image despite being estimated along a parallel profile approximately 80 km toward the east. This implies that regionally under the Oaxaca terrane, the subduction interface lies just above the conductive dipping layer. The Xolapa-Oaxaca terrane boundary interface with the oceanic crust below the Xolapa terrane coincides with that deduced by JÖDICKE (2006). However, the contradicting results, showing an isolated conductivity zone at ~ 120 km from the coast above the refraction interface, could be attributed to a static shift problem; particularly given the lower period range (10^{-1}–10^{-5} Hz) of the data acquired with the old Metronix instruments used in their work. The broader band MT equipment used in the present study (10^{3}–10^{-4} Hz) allows for better resolution of the resistive continental crust at shallow depths, and other important upper crustal structures.

4.3. The Gravity Model

To constrain the resistivity model, we modeled the free-air gravity anomaly data along a coincident profile, the data were extracted from the Geosat and ERS-1 satellite altimetry database (SANDWELL 1997; SANDWELL and SMITH 2009). These data were then reduced to simple Bouguer anomalies using a background density value of 2.67 g/cm^{3}. Despite the relatively high error in the data (± 3 mgal or more) and the low lateral resolution (SANDWELL 1997; SANDWELL and SMITH 2009; YALE et al. 1998; GREEN et al. 1998) the regional gravity model is regarded as a useful resource to test and compare with the resistivity model structure.

The crustal density model (Fig. 6b) was constructed assuming that the electric contrasts obtained from the MT inversion coincide with lithological interfaces. Figure 6b shows the results of the direct modeling of the simple Bouguer anomaly using a typical 2D Talwani (TALWANI and LANDISMAN 1959) type algorithm. In general, the Bouguer anomaly data can be fitted fairly well taking the reference values of 3.1 and 3.0 g/cm^{3} used by previous authors (e.g., BLAKELY et al. 2005; BANDY et al. 1999) for the upper oceanic mantle (OM) and subducting oceanic lithosphere (OL), respectively. We use an average value of 2.80 g/cm^{3} for the ~ 20 km thick altered oceanic crust (AOC). The SW sector of the upper crust (Ocf) is regarded as more intensely fractured with an average density of 2.7 g/cm^{3}. A thinned continental crust of assumed density of 2.6 g/cm^{3} lies at the convergence front that suddenly stops below the accretionary prism of variable density (density = 2.4–2.55 g/cm^{3}). The combined density units have been grouped as the continental convergence front, labeled in the figure as AC. Both the faulted oceanic crust (Ocf) and the fractured continental block and accretionary sediments (AC) appear to be confined by a regional fault system, not only defined electrically but also seismically, which can be match fairly well by the gravity model. Such a density contrast coincides with the tectonostratigraphic limit between the Xolapa and the Oaxaca terranes.

The upper continental crust of the Oaxaca terrane was divided in two, approximately horizontal, layers; the lower layer (UC) having a constant density value of 2.65 g/cm^{3} ORTEGA-GUTIÉRREZ et al. (2008), whereas the overlying more heterogeneous unit (OC) having a variable density structure that is necessary to fit the data. Although the lateral resolution of the MT data is insufficient to resolve variations of the uppermost crustal unit, the high frequency content in the gravity profile and scattered shallow resistivity variations within this unit supports the idea of a highly variable structural and lithological nature of the Oaxaca terrane down to maximum depths of ~ 10 km.

Upper crustal densities within the Juarez terrane were set to values of 2.8 and 2.9 g/cm^{3} for the near surface and deeper portion, respectively. The lower continental crust (LC) density under the Juarez terrane,

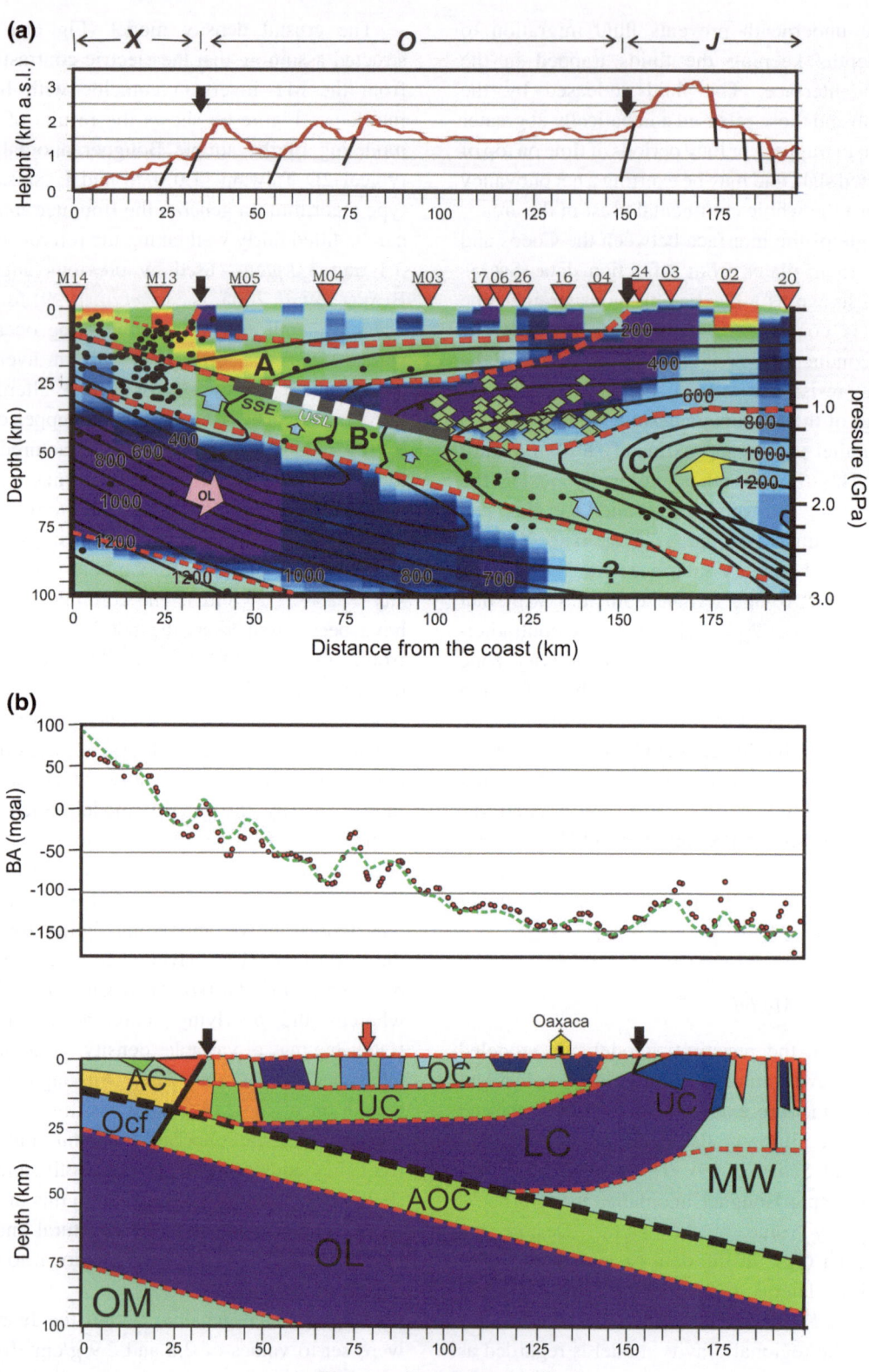

◄Figure 6

a *Bottom* Resistivity image and correlated seismicity associated to diverse tectonic processes across the SW–NE profile. Bending of the plate at the subduction front and dewatering reactions above the oceanic slab (*B* and *C*) define the main seismicity swarms observed. A third zone of aligned seismicity correlating with the horizontal conductor A may be also of regional dynamic significance. Non-volcanic tremors (NVT) registered by BRUDZINSKI *et al.* (2010) lie within and below the resistive lower Juárez continental crust (*green squares*) and are interpreted as being associated to dewatering related seismicity. Slow slip events (SSE) and an ultra-slow velocity layer (USL) interpreted by BRUDZINSKI *et al.* (2010) along a coincident profile (*gray and white dashed thick line*) define well a decoupled sector of the subducting slab at the contrasting resistivity interface. A highly hydrated slab reduces drastically frictional forces promoting aseismic subduction. The *yellow arrow* suggests a net isostatic component that induces active uplifting of the *upper* crust. *Top* Topographic profile. Approximate vertical exaggeration of 1:100. **b** *Top* Plot of the Satellite gravity data and gravity values calculated from a crustal density model (*bottom*). Model nomenclature is as per Table 1. The density model was based upon the electric interfaces from the MT conductivity model, and thus mirrors the electric structure along the studied profile. See text for explanation of the nomenclature

which wedges toward the trench (and is regarded to be in contact with the hydrated oceanic crust), was set at an average value of 2.9 g/cm³ (e.g., ORTEGA-GUTIÉRREZ *et al.* 2008). At the mantle wedge (MW), in the northeasternmost part of the section, the broad conductivity anomaly cannot distinguish any particular structure in the gravity; one exception is the subtle variations above the plate interface up to the surface where a high conductivity anomaly is unveiled. Although it appears as a single point anomaly, gravity anomaly in this sector reveals a highly variable density structure at the NE flank of Sierra de Juarez terrane. The limited number of MT stations in this part of the profile

prevents a more detailed description of the resistivity structure. However, the gravity data can be fitted using an average density value of 2.9 g/cm³, similar to that of the hydrated oceanic crust, but requires the incorporation graben-like structures of low density (2.4 g/cm³) in the upper Juarez crust that are regarded as associated with the Oloapan fault system (Fig. 1). Table 1 shows the summary of density values used for the gravity model.

4.4. The Oaxaca-Juarez Contact Zone

According to CAMPA and CONEY (1983), the Oaxaca-Juarez contact zone at the surface lies across the projected trace of the Oaxaca fault system (black arrow at 150 km distance from the coast in Fig. 6); however, the suture zone according to the geophysical data appears to dip southwards and it is affected by a steep normal fault producing large resistivity contrasts in the uppermost crust. The Oaxaca fault system, is projected between MT stations 04 and 24, and is identified by a shallow and pronounced lateral resistivity contrast.

Previous refraction and reflection studies across the Oaxaca fault (VALDES *et al.* 1986; SPRANGER 1994; GEOLIMEX 1999) did not confirm its continuation at depth as expected. Supporting this result, the regional MT study of JÖDICKE (2006) did not detect an expected large electric contrast across the Oaxaca Fault at crustal depths along a transect passing south Oaxaca city. By analyzing the electrical impedance along a regional transect, JORDING *et al.* (2000) concluded that the major structural change at depth in this region is

Table 1

Resume of density values used for the gravity model taken from ORTEGA-GUTIÉRREZ et al. (2008) and BLAKELY et al. (2005)

AC	Altered crust	2.7–2.4	Assumed
OC	Oaxacan crust	Variable (see text)	ORTEGA-GUTIÉRREZ *et al.* (2008)
Ocf	Oceanic crust (faulted)	2.70	Assumed
UC	Upper crust (below OT)	2.65	ORTEGA-GUTIÉRREZ *et al.* (2008)
UC	Upper crust (Juarez T.)	2.8–2.9	ORTEGA-GUTIÉRREZ *et al.* (2008)
LC	Lower crust	2.90	ORTEGA-GUTIÉRREZ *et al.* (2008)
AOC	Altered oceanic crust	2.80	Assumed
MW	Mantle wedge	2.90	BLAKELY *et al.* (2005)
OL	Oceanic lithosphere	3.00	Assumed
OM	Oceanic mantle	3.10	BLAKELY *et al.* (2005)

For text for further details

displaced northeastwards of the Mitla Valley, about 30 km to the east from the generally accepted Oaxaca-Juarez terrane boundary. A more detailed MT study of the region (Corbo 2013) concluded that the Oaxaca fault system does not appear to penetrate deep into the crust but remains relatively shallow (<5 km). Other results also suggest that the contact between the Oaxaca and Juarez terranes at crustal depths occurs along a SW dipping interface south of Oaxaca City (Campos-Enriquez et al. 2013).

According to our results, the Juárez terrane underlies the Oaxaca terrane along a low angle SW dipping contact zone of contrasting conductivities and densities, the contact finally becomes subhorizontal. Our results suggest that the Oaxaca terrane is the combination of two ∼10–12 km thick layers of differing electrical structure (Fig. 6). The upper layer is more heterogeneous electrically than the lower, which is a layer of enhanced conductivity (<50 Ohm-m). Geological mapping of the Oaxaca terrane supports the existence of a highly faulted upper crust. Nieto-Samaniego et al. (2006) also report widespread Cenozoic normal faults. Ortega Gutiérrez et al. (1990) have reported evidences of over-thrusting of the Oaxaca terrane over the Juárez terrane. They reported a low angle structural complex (the SJMB) dipping gently to the west but also that the rocks above this structure are older than the rocks below it. In our regional MT model the highly resistive (>5 k Ohm) Juárez terrane is under-thrusting the Oaxaca terrane unit along the gently dipping contact. We speculate that the observed anomalous conductivity of the lower layer of the Oaxaca terrane is related to the existence of large amounts of laminar mineralization, improving electrical current flow and promoting shear displacements induced by the presently active convergence process. The largely mafic nature of orthogneisses of the Oaxaca complex and the presence of graphite in paragneisses may also contribute to the high conductivity.

4.5. Metamorphic Reactions and Fluid Migration

The deeper conductive region along the subducting plate, which is fairly well constrained and labeled with the letter C in the MT model, appears to be a broad zone with a gradual increase of conductivity below the 700 °C isotherm. This conductive zone lies between the 700 °C isotherm and the top of the subducting slab. This conductor can be associated to a higher P–T metamorphic facies ($P \sim 1.5$ GPa, $T > 700$ °C,), occurring at distances >150 km away from the subduction front. According to the metamorphic facies path and thermal model proposed by Jödicke (2006) for the subducting Cocos plate in the Xolapa-Oaxaca-Juárez composite terrane, the pressure–temperature conditions of this reaction corresponds to the eclogite metamorphic facies which may release up to ∼1 wt % of water to the lower continental crust (Hyndman and Shearer 1989). This diffuse enhanced conductivity zone is interpreted as partly due to the upward migration of released water bounded to the hydrated and mineralized oceanic slab at ∼50 km of depth. Important seismicity in the depth range of 50–75 km occurs, almost one every second year, in a narrow band of 20 km projected hypocenters. At these depths and between 100 and 175 km distance from the coast, the fluids bounded to the oceanic slab appear to undergo a massive decompression stimulated by a more permeable and lighter lower crust. Seismicity in this region may be indicating rapid fluid migration through cracks to lower crustal depths. Upcoming fluid from the dehydration of the oceanic slab in combination with upcoming hydrated material from the ductile–brittle mantle wedge, could promote the upward migration of mixed mineralized fluids towards lower crustal depths.

4.6. Comparison of seismicity and anomalous conductivity

The seismicity data set plotted on the MT section corresponds to a period of 39 years registered at the National Seismological Service (SSN) catalogue from 1974 to 2013. The data hypocenters are from events that occurred within a band of 20 km centered at the MT profile trace. Most of the seismicity along the MT profile occurs at the convergence front, within the first 25 km of the coast (Fig. 6) as a result of stresses produced by plate bending and relative plate motions. Seismicity extends to depths deeper than 50 km in this distance range. The black arrow in Fig. 6 shows the Xolapa terrane surface limit, that also seems being marking a clear limit for bending slab seismicity.

Away from the convergence front, the observed seismicity occurs notably within the enhanced conductivity zones marked in Fig. 6 with letters A, B and C. Of singular interest is the seismicity occurring at a depth of ~ 20 km within the lower part of the Oaxaca terrane (conductor A), between 50 and 100 km from the coast. Given the close association of these events to the resistivity boundary marking the Oaxaca terrane, we hypothesize that the seismicity in this zone is triggered by slow slip displacements along this fault or perhaps along layered structures related to this major contact which bounds the Oaxaca and Juarez terranes. This proposal could be corroborated by analyzing the focal mechanism of the earthquakes occurring in this sector, but this is beyond the scope of the present study.

Projecting the location of two slow slip episodes (SSE) registered by Brudzinski et al. (2010) onto the magnetotelluric model profile (Fig. 6) it is observed that SSEs coincide with a sharp conductivity interface within an approximated depth range of 25–40 km. The ultra-slow velocity layer (USL) interpreted by these authors is consistent with the conductivity interface of an impermeable (resistive) and a hydrated (conductive) rheology, which would allow for stable sliding displacements. However, their non-volcanic tremors (NVT) hypocenters of July 2006 plotted onto the same MT profile occur also along a conductivity interface rather than within an anomalous conductivity zone, as when compared with Jödicke (2006) conductivity model. Differences can be due to the ~ 80 km offset of Jödicke (2006) profile with respect to our coincident MT model with Brudzinski et al. (2010) seismicity profile. Such discrepancy can also be attributed to the lower bandwidth of the data set (10^{-1}–10^{-5} Hz) used to compute their conductivity model, making thus more difficult to account for static shift problems.

A deeper seismicity zone within the dipping conductor at depths below 50 km and distances >100 km from the coast are spatially related to the NVT hypocenters (green diamonds in Fig. 6). Non-volcanic tremors registered by Brudzinski et al. (2010) have been proposed to be due to the dewatering of the oceanic crust at depths around 60 km (e.g., Manea and Manea 2010). The correlation of seismicity and anomalous conductivity in the temperature range of 600–700 °C show that dewatering processes in the Oaxaca region start at a depth of 75 km, and precludes NVTs that are trigger above depths of ~ 40 km as a consequence of fluid migration.

5. Conclusions

Broad band magnetotelluric data (10^3–10^{-3} Hz) provided good vertical resolution of upper crustal structures down to depths below 75 km. Rotated to the strike transfer functions were inverted using a minimal structure non-linear conjugate gradients algorithm, yielding a repeatable and stable model. The zones of enhanced conductivity correlates well with seismicity and define at least three subduction related processes. The closest to the trench anomalous conductivity zone (A) suggest a highly fractured crust saturated with seawater and mineralized fluids some of which could be underplating the Oaxaca terrain. Large amounts of seawater-saturated sediments at the convergence front may contribute as well to the anomalous conductivity in the zone. With a large stock of trapped fluids, important volumes of ores can react and precipitate from the abundant mineralized fluids available. Large epithermal disseminated ore impregnations along interconnected fracture zones may also contribute to explain the observed anomalous conductor. The anomalous conductivity dipping layer is associated with a highly hydrated low angle subducting oceanic slab (B), while the broad anomalous conductor (C) is related to the combination of migrating dehydration fluids from the eclogite metamorphic facies above the plate (depth approximately 60 km) and the fluids from the converging hydrated mantle wedge. Upcoming fluid from the dehydration of the oceanic slab in combination with upcoming hydrated material from the ductile–brittle mantle wedge, could promote the upward migration of mixed mineralized fluids towards lower crustal depths, and hence uplifting.

We conclude that the slow slip events (SSE) and the ultra-slow velocity layer (USL) interpreted by Brudzinski et al. (2010) occur at the electrical interface between the subducting slab and the resistive lower crust. A highly hydrated oceanic slab under the impermeable and confined lower continental crust would reduce drastically frictional forces promoting aseismic subduction. We also find a good correlation of the anomalous conductivity zones with the observed seismicity implying a connection with fluids displacement (natural fracking) in most of the cases. According to our results the non-volcanic tremors (NVT) occur within a broad interface between the resistive Juarez lower crust and dehydrated conductive oceanic slab at depths above 40 km, preceded by the dewatering of the subducting plate from depths of $\sim$75 km and above.

Both gravity and resistivity models are consistent with a thinned upper crust in the northeast end of the profile. We hypothesize that the documented uplift of this area results from buoyancy due to a hydrated and lighter lower crust and mantle wedge. The seismicity occurring in this zone is allegedly due to upwards fluid migration through nearly vertical paths along the weakened and thinned lower continental crust under the Juarez terrane.

References

ALANIZ–ÁLVAREZ, S.A., NIETO-SAMANIEGO, A.F., and ORTEGA-GUTIÉRREZ, F. (1994), *Structural evolution of the Sierra Juárez mylonitic complex, state of Oaxaca, Mexico*, Revista Mexicana de Ciencias Geológicas *1*, 147–156.

ALANIZ-ALVAREZ, S.A., VAN DER HEYDEN, P., NIETO SAMANIEGO, A.F., and ORTEGA-GUTIÉRREZ, F. (1995), *Radiometric and kinematic evidence for Middle Jurassic strike-slip faulting in southern Mexico related to the opening of the Gulf of Mexico*, Geology *24* (5), 443–446.

ÁNGELES MORENO, E., (2006) Petrografía, geología estructural y geocronología del borde noroccidental del terreno Cuicateco, sierra Mazateca, estado de Oaxaca, México, M.Sc. Thesis, Universidad Nacional Autónoma de México (Ciudad de México, 2006)

ARZATE J., MARESCHAL M., and LIVELYBROOKS D. (1995), *Electrical image of the subducting Cocos plate from magnetotelluric observations*, Geology *23* (8), 703–706.

BANDY, W., KOSTOGLODOV V., HURTADO-DÍAZ A., and MENA M. (1999), *Structure of the southern Jalisco subduction zone, Mexico, as inferred from gravity and seismicity*, Geofís. Internacional *38*, 127–136.

BARBOZA GUDIÑO, R. (1996), *Contribución a la geología de la Sierra de Juárez en el sur de México*, Zentralblatt für Geologie und Paläontologie, Teil *1* (1994 H. 7/8), 991–1005.

BLAKELY R., BROCHER T. and WELLS R. (2005), *Subduction-zone magnetic anomalies and implications for hydrated forearc mantle*, Geology *33* (6), 445–448.

BOSTICK, F.X. (1977), *A simple almost exact method of magnetotelluric analysis. In: Ward, S. (ed.), Workshop of Electrical Methods in Geothermal Exploration*, Univ. of Utha, Res. Inst., U.S. Geol. Surv., contract N° 14080001-8-359.

BRASSE, H., LEZAETA, P., RATH, V., SCHWALENBERG, K., SOYER, W. and HAAK, V. (2002), *The Bolivian Altiplano conductivity anomaly*, J. Geophys. Res. *107*, 3 (1–14).

BRASSE, H., KAPINOS, G., MÜTSCHARD, L., ALVARADO, G.E., WORZEWSKI, T., and JEGEN, M., (2008), *Deep electrical resistivity structure of northwestern Costa Rica*. Geophys. Res. Lett. 36 (2), L02310 (1–5).

BRUDZINSKI, M., HINOJOSA PRIETO, H., SCHLANSER, K. M., CABRAL CANO, E., ARCINIEGA CEBALLOS, A., DÍAZ MOLINA, O., and DEMETS, C. (2010), *Nonvolcanic tremor along the Oaxaca segment of the Middle America subduction zone*, Journal of Geophys. Res. *115*, B00A23 (1–15).

CALDWELL, T. G., BIBBY, H. M., and BROWN, C. (2004), *The Magnetotelluric Phase Tensor*, Geophys. Journal Int. *158*, 457–469.

CAMPA, M.F., and CONEY, P.J. (1983), *Tectono-stratigraphic terranes and mineral resource distributions in Mexico*, Canadian Journal of Earth Sciences *20*, 1040–1051.

CAMPOS-ENRIQUEZ, J.O., CORBO-CAMARGO, F., ARZATE-FLORES J., KEPPIE J.D., ARANGO-GALVÁN, C., UNSWORTH, M., and BELMONTE-JIMÉNEZ S.I. (2013), *The buried southern continuation of the Oaxaca-Juarez terrane boundary and Oaxaca Fault, southern Mexico: magnetotelluric constraints*, J. South Am. Earth Sci. *43*, 62–73.

CARFANTÁN, J.C., (1986) Du systeme cordilleran nord-américain au domaine Caraibe - Étude géologique du Mexique meridional. PhD. Thesis. Chambérym Université de Savoie, (Savoie, France, 1986).

CENTENO GARCÍA, E., (1988) Evolución estructural de la falla de Oaxaca durante el Cenozoico. M.S. Thesis, Universidad Nacional Autónoma de México, (Ciudad de México, México, 1988).

CORBO, F. (2013), Estudio de la subducción y su relación con la presencia de fluidos a partir de sondeos magnetotelúricos en el Bloque de Jalisco y Oaxaca. PhD. Thesis. Centro de Geociencias, Universidad Nacional Autónoma de México, (Ciudad de México, México).

DÁVALOS-ÁLVAREZ, O.G., NIETO SAMANIEGO, A.F., ALANIZ ÁLVAREZ, S.A., MARTÍNEZ HERNÁNDEZ, E. and RAMÍREZ ARRIAGA, E. (2007), *Estratigrafía cenozoica de la región de Tehuacán y su relación con el sector norte de la falla de Oaxaca*, Revista Mexicana de Ciencias Geológicas *24*, 197–215.

DUCEA, M. N., GEHRELS, G. E., SHOEMAKER, S., RUIZ, J., and VALENCIA, V. A. (2004), *Geologic evolution of the Xolapa Complex, southern Mexico: Evidence from U-Pb zircon geochronology*, GSA Bulletin *116* (7-8), 1016–1025.

ELÍAS HERRERA, M., and ORTEGA GUTIÉRREZ, F. (2002), *Caltepec fault zone: An Early Permian dextral transpressional boundary between the Proterozoic Oaxacan and Paleozoic Acatlán complexes, southern Mexico, and regional tectonic implications*, Tectonics *21*, 4 (1–18).

FERRARI, L., OROZCO-ESQUIVEL, T., MANEA, V., and MANEA, M. (2012), *The dynamic history of the Trans-Mexican Volcanic Belt and the Mexico subduction zone*, Tectonophysics *522–523*, 122–149.

GRAJALES NISHIMURA, J.M., CENTENO-GARCIAA, E., KEPPIEA, J.D., and DOSTAL, J. (1999), *Geochemistry of Paleozoic basalts from the Juchatengo complex of southern Mexico: tectonic implications*, Journal of South American Earth Sciences *12*, 537–544.

GREEN, C. M., FAIRHEAD J. D., and MAUS D. S. (1998), *Satellite-derived gravity: Where we are and what's next*, The Leading Edge *17*, 77–79.

HYNDMAN, R.D., and SHEARER, P.M. (1989), *Water in the lower continental crust: modelling magnetotelluric and seismic reflection results*, Geophysical Journal International *98* (2), 343–365.

JÖDICKE, H., JORDING A., FERRARI L., ARZATE J., MEZGER K., and RÜPKE L. (2006), *Fluid release from the subducted Cocos plate and partial melting of the crust deduced from magnetotelluric studies in southern Mexico: Implications for the generation of volcanism and subduction dynamics*, J. Geophys Res. *111*, B08102 (1–22).

JONES A. G. (1983), *On the equivalence of the Niblett and Bostick transformations in the magnetotelluric method*, J. Geophys Prosp. *14*, 72–73.

JORDING A., FERRARI L., ARZATE J., and JODICKE H. (2000), *Crustal variations and terrane boundaries in southern Mexico as imaged by magnetotelluric transfer functions*, Tectonophysics *327*, 1–13.

KURTZ R.D., DELAURIER J.M., and GUPTA, J.C. (1986), *A magnetotelluric sounding across Vancouver Island detects the subducting Juan de Fuca plate*, Nature *321*, 596–599.

KURTZ, R.D., DELAURIE, J.M., and GUPTA, J.C. (1990), *The electrical conductivity distribution beneath Vancouver Island: A region of active plate subduction*, J. Geophys. Res. *95*, 10929–10946.

MANEA, M., and MANEA, V.C. (2010), *Curie Point Depth Estimates and Correlation with Subduction in Mexico*, Pure and Applied Geophysics *168* (8), 1489–1499.

MANEA, V.C., MANEA, M., and FERRARI, L. (2013), *A geodynamical perspective on the subduction of Cocos and Rivera plates beneath Mexico and Central America*, Tectonophysics *609*, 56–81.

MENDOZA-ROSALES C.C., CENTENO-GARCÍA E., SILVA-ROMO G., CAMPOS-MADRIGAL E., and BERNAL J.P. (2009), *Barremian rift-related turbidites and alkaline volcanism in southern Mexico and their role in the opening of the Gulf of Mexico*, Earth and Planetary Science Letters *295*, 419–434.

MOLINA GARZA, R.S., GEISSMAN, J.W., WAWTZYNIEC, T.F., PEÑA ALONSO, T.A., IRIONDO, A., WEBER, B., and ARANDA-GARCÍA, J.J. (2015), *Geology of the coastal Chiapas (Mexico) Miocene plutons and the Tonalá shear zone: Syntectonic emplacement and rapid exhumation during sinistral transpression*, Lithosphere *7* (8), 257–274.

MORÁN-ZENTENO, D. J., CORONA-CHÁVEZ, P., and TOLSON, G. (1996), *Uplift and subduction erosion in southwestern Mexico since the Oligocene: pluton geobarometry constraints*, Earth and Planetary Science Letters *141*, 51–65.

NIBLETT E. R., and SAYN-WITTGENSTEIN C. (1960), *Variation of electrical conductivity with depth by the magneto-telluric method*, Geophysics *25* (5), 998–1008.

NIETO-SAMANIEGO A.F., ALANIZ-ALVAREZ S.A., SILVA-ROMO G., EGUIZA-CASTRO M.H., and MENDOZA-ROSALES C.C. (2006), *Latest Cretaceous to Miocene deformation events in the eastern Sierra Madre del Sur, Mexico, inferred from the geometry and age of major structures*, GSA bulletin *118* (1–2), 238–252.

ORTEGA-GUTIÉRREZ F. (1981), *Metamorphic belts of southern Mexico and their tectonic significance*, Geofísica Internacional *20*, 177–202.

ORTEGA GUTIÉRREZ, F., MITRE-SALAZAR, L.M., ROLDÁN-QUINTANA, J., SÁNCHEZ-RUBIO, G., and DE LA FUENTE, M. (1990), *North American Continent-Ocean Transect Program, Transect H-3 – Acapulco Trench to the Gulf of Mexico across southern Mexico*, Geological Society of America, Decade of North American Geology Program, scale 1:500000, (México).

ORTEGA-GUTIÉRREZ F., ELIAS HERRERA M., and ELIZONDO DÁVALOS M. G. (2008), *On the nature and role of the lower crust in the volcanic front of the Trans-Mexican Volcanic Belt and its fore-arc region, southern and central México*, Revista Mexicana de Ciencias Geológicas *25* (002), 346–364.

PARDO M., and SUAREZ G. (1995), *Shape of the subducted Rivera and Cocos plates in southernMexico: Seismic and tectonic implications*, J. Geophys Res. 100, *12*,357–12,373.

PEACOCK S.M. (1990), *Fluid Processes in Subduction Zones*, Science *248*, 329–337.

RODI W., and MACKIE R. (2001), *Nonlinear conjugate gradients algorithm for 2-D magnetotelluric inversion*, Geophysics *66* (1), 174–187.

SANDWELL, D.T., and SMITH W. H. (1997), *Marine gravity anomaly from Geosat and ERS 1 satellite altimetry*, J. Geophys. Res. *102*, 10039–10054.

SANDWELL D. T. and SMITH W. H. (2009), *Global marine gravity from retracked Geosat and ERS-1 altimetry: Ridge segmentation versus spreading rate*, J. Geophys Res. *114*, B01411 (1–18).

SCHAAF, P., MORÁN-ZENTENO, D., HERNÁNDEZ-BERNAL, M., SOLÍS-PICHARDO, G., TOLSON, G., and KÖHLER, H. (1995), *Paleogene continental margin truncation in southwestern Mexico, Geochronological evidence*: Tectonics *14*, 1339–1350.

SCHMIDT M. W., and POLI S. (1998), *Experimentally based water budgets for dehydrating slabs and consequences for arc magma generation*, Earth and Planetary Sc. Letters *163*, 361–379.

SCHWARZ G., and KRÜGER D. (1997), *Resistivity cross section through the Southern Central Andean Crust as inferred from 2-D modelling of magnetotelluric and geomagnetic deep sounding measurements*, J. Geophys Res. *102* (B6), 11957–11978.

SOLARI, L.A., TORRES DE LEÓN, R., HERNÁNDEZ PINEDA, G., SOLÉ, J., SOLÍS-PICHARDO, G., and HERNÁNDEZ-TREVIÑO, T. (2007), *Tectonic significance of Cretaceous–Tertiary magmatic and structural evolution of the northern margin of the Xolapa Complex, Tierra Colorada area, southern Mexico*, Geological Society of America Bulletin *119*, 1265–1279.

SPRANGER, W.J. (1994), M. GEOLIMEX: Eine erste Geotraverse durch Sudmexiko; Auswertung des refraktions seismischen Profils, PhD. Thesis, Christian-Albrechts University, (Kiel, Germany).

TALAVERA-MENDOZA, O., RUIZ, J., CORONA-CHAVEZ, P., GEHRELS, G.E., SARMIENTO-VILLAGRANA, A., GARCÍA-DÍAZ, J.L., and SALGADO-SOUTO, S.A. (2013), *Origin and provenance of basement metasedimentary rocks from the Xolapa Complex: New constraints on the Chortis–southern Mexico connection*, Earth and Planetary Science Letters *369*, 188–199.

TALWANI, M., J. L. WORZEL., and LANDISMAN, M. (1959), *Rapid gravity computations for two-dimensional bodies with applications to the Mendocino submarine fracture zone*, J. Geophys. *64*, 49–59.

TATSUMI Y., and EGGINS S. (1990), Subduction Zone Magmatism (Oxford, England).

VALDES, C.M., MOONEY, W. D., SINGH, S. K., MEYER, R. P., LOMNITZ, C., LUETGERT, J. H., HELSLEY, C. E., LEWIS, B. T. R., and MENA, M. (1986), *Crustal structure of Oaxaca, Mexico, from seismic refraction measurements*, Bulletin of the Seismological Society of America 76, 547–563.

WANNAMAKER P.E., BOOKER J.R., JONES A.G., CHAVE A.D., FILLOUX J.H., WAFF H.S. and LAW L.K. (1989), *Resistivity cross-section through the Juan de Fuca subduction system and its tectonic implications*, J. Geophys Res. *94* (14), 127–144.

YALE M. M., SANDWELL D.T. and HERRING A.T. (1998), *What are the limitations of satellite altimetry?*, The Leading Edge *17*, 73–73.

(Received September 4, 2015, accepted April 8, 2016, Published online April 28, 2016)

Pure Appl. Geophys. 173 (2016), 3373–3393
© 2015 Springer International Publishing
DOI 10.1007/s00024-015-1214-7

❚ Pure and Applied Geophysics

Seismic Anisotropy and Mantle Flow Driven by the Cocos Slab Under Southern Mexico

LESLIE A. BERNAL-LÓPEZ,[1] BERENICE R. GARIBALDI,[2] GERARDO LEÓN SOTO,[3] RAÚL W. VALENZUELA,[2] and
CHRISTIAN R. ESCUDERO[1]

Abstract—Shear wave splitting measurements were made using SKS and SKKS waves recorded by the Meso-American Subduction Experiment, which was deployed in southern Mexico starting at the coast of the Pacific Ocean and running north toward the Gulf of Mexico. In this segment of the Middle America Trench the oceanic Cocos plate subducts under the continental North American plate. The active volcanic arc is located at the southern end of the Trans-Mexican Volcanic Belt. Unlike most subduction zones, however, the volcanic arc is not subparallel to the trench. In the fore-arc, between the trench and the Trans-Mexican Volcanic Belt, the Cocos slab subducts subhorizontally. Beneath the volcanic belt, however, the slab dives steeply into the mantle. A marked difference in the orientation of the fast polarization directions is observed between the fore-arc and the back-arc. In the fore-arc the fast axes determined using *SKS* phases are oriented NE–SW, in the same direction as the relative motion between the Cocos and North American plates, and are approximately perpendicular to the trench. Physical conditions in the subslab mantle are consistent with the existence of A-type olivine and consequently entrained mantle flow is inferred. Strong coupling between the slab and the surrounding mantle is observed. In the back-arc SKS fast polarization directions are oriented N–S and are perpendicular to the strike of the slab. Given the high temperatures in the mantle wedge tip, the development of A-type, or similar, olivine fabric throughout the mantle wedge is expected. The orientation of the fast axes is consistent with corner flow in the mantle wedge.

Key words: Shear wave splitting, upper mantle anisotropy, mantle flow, Mexico, Middle America Trench, flat slab subduction.

Electronic supplementary material The online version of this article (doi:10.1007/s00024-015-1214-7) contains supplementary material, which is available to authorized users.

[1] Centro de Sismología y Volcanología de Occidente, Universidad de Guadalajara, Puerto Vallarta, Jal., Mexico.

[2] Departamento de Sismología, Instituto de Geofísica, Universidad Nacional Autónoma de México, Circuito de la Investigación S/N, Cd. Universitaria, Del. Coyoacán, 04510 Mexico, DF, Mexico. E-mail: raul@ollin.geofisica.unam.mx

[3] Instituto de Investigaciones en Ciencias de la Tierra, Universidad Michoacana de San Nicolás de Hidalgo, Morelia, Mich., Mexico.

1. Introduction

Southern Mexico, where the Cocos plate subducts beneath the North American plate (Fig. 1), represents a natural laboratory for studying subduction geodynamic processes. The region presents special characteristics such as flat subduction (PARDO and SUÁREZ 1995; PÉREZ-CAMPOS *et al.* 2008; HUSKER and DAVIS 2009; KIM *et al.* 2010), non-volcanic tremors (PAYERO *et al.* 2008), slow slip events (RADIGUET *et al.* 2012; SONG and KIM 2012a), and an active volcanic arc which does not run parallel to the offshore trench (GILL 1981; SUAREZ and SINGH 1986; FERRARI *et al.* 2012). Fragmentation of the Farallon plate (∼23 Ma) gave rise to the Cocos plate to the north and the Nazca plate, which subducts under the South American plate (ATWATER and STOCK 1991). The northern edge of the Cocos plate is adjacent to the small Rivera plate that detached from the Cocos plate and has moved independently for the last ∼10 Ma (DEMETS and TRAYLEN 2000). The subduction angle of the Cocos plate in southern Mexico varies along the Middle America Trench (MAT). In the northwest, near the Rivera plate, the Cocos slab dips at ∼50°; toward the southeast, beneath Guerrero state, the slab undergoes flat subduction; farther southeast, near the Tehuantepec Ridge, the subduction angle becomes normal again, dipping at ∼30° (PARDO and SUÁREZ 1995). The Cocos slab dips shallowly to the north at 15° for 80 km from the Guerrero coast and then it extends subhorizontally for about 200 km to the Trans-Mexican Volcanic Belt (TMVB), where it subducts steeply (∼75°) down to a depth of 500 km (PÉREZ-CAMPOS *et al.* 2008; HUSKER and DAVIS 2009; KIM *et al.* 2010). The flat subduction beneath Guerrero has been associated with the atypical location of the TMVB which is not parallel to the trench. Early

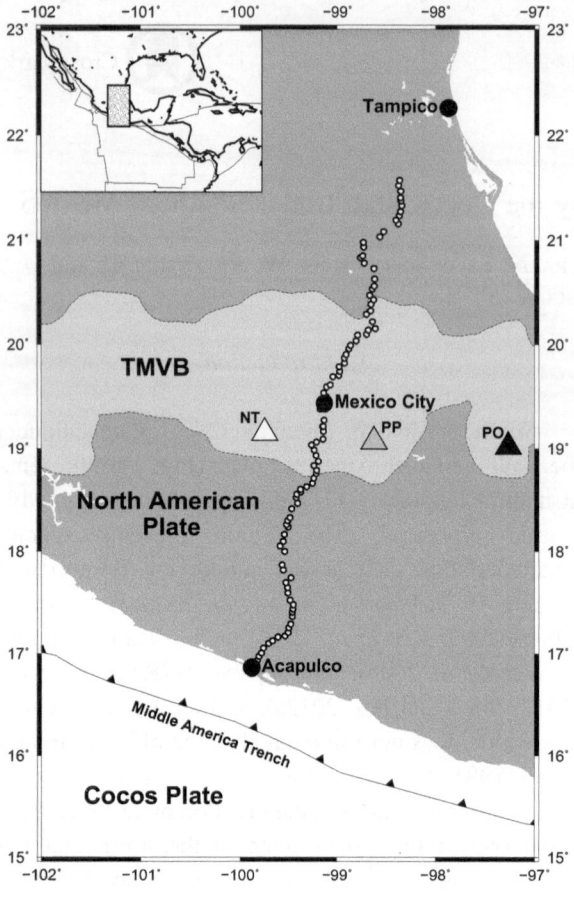

Figure 1

Tectonic map of the area of study. *Light gray* indicates the location of the Trans-Mexican Volcanic Belt (TMVB). *Small open circles* stand for the locations of the seismic stations in the Meso American Subduction Experiment (MASE). The Cocos plate subducts at the Middle America Trench under North America. *Triangles* show the positions of active andesitic volcanoes near the area of the study; *white triangle* (NT) stands for Nevado de Toluca, *gray triangle* (PP) is Popocatépetl, and *black triangle* (PO) shows the location of Pico de Orizaba. Black dots are main cities. *Inset* shows location of the study area in Mexico and relative to North America

on, the volcanic arc started migrating away from the trench 16 Ma (FERRARI *et al.* 2012), reached its maximum extent northward ~10 Ma, and then began its migration back toward the trench.

Many questions remain open in our understanding of the subduction recycling machine, in particular those related with upper mantle dynamics. The role of mantle flow in the slab morphology (or the other way around), the generation and transport of melt, the mantle material dragged by the subducting slab, the mixing rates of upper and lower mantle material, and

the hydration rate and viscosity of the mantle wedge are some of the processes that are still not well understood (LONG 2013).

Seismic anisotropy studies based on shear wave splitting have been incorporated as a standard tool for understanding upper mantle dynamics. Since there is a direct link between the anisotropic parameters and mantle deformation, seismic anisotropy results may shed light in our understanding of the present geodynamic processes. Two parameters are needed to quantify anisotropy, the fast polarization direction and the delay time between the fast and slow waves (e.g. SILVER and CHAN 1991). Shear wave splitting measurements in the upper mantle have commonly been interpreted in terms of the Lattice Preferred Orientation (LPO) of olivine minerals. For olivine, the main component of the upper mantle, the seismic anisotropy of S waves varies between 3.3 and 17.4 % with a mean value of 9.4 % (ISMAÏL and MAINPRICE 1998). Experimental results suggest that under relatively low stresses, high temperature, and low water content, A-type LPO fabric is developed (KARATO *et al.* 2008). Under these circumstances it is possible to assume hexagonal anisotropy with a horizontal axis of symmetry where the fast polarization orientation is nearly parallel to the horizontal asthenospheric flow or to the long axis of the finite strain ellipse (LONG 2013). The presence of water may affect the olivine type of fabric developed under stresses. As shown by JUNG *et al.* (2006) high water content, relatively low temperatures, and high stresses result in the development of B-type fabric for olivine minerals leading to fast polarization orientations nearly perpendicular to the mantle flow.

Core phases such as SKS and SKKS are naturally polarized in the radial direction after they leave the liquid outer core. For these phases, the recorded anisotropy lies anywhere in the path between the core-mantle boundary and the receiver. The shear wave splitting technique has excellent lateral resolution (e.g., SILVER and KANESHIMA 1993), but it lacks direct vertical resolution, consequently additional information is required for constraining the depth of anisotropy. In most studies, however, anisotropy is interpreted to reside in the uppermost upper mantle (SILVER 1996; SAVAGE 1999). In a subduction system, anisotropy may be found in the subslab region, the subducting slab itself, the mantle wedge, and the overriding plate. In addition to the subduction system, there is evidence that in some cases the anisotropic parameters could have another contribution associated to the D" layer at the bottom of the lower mantle (e.g., LONG 2009). Using stations in California, northwestern Mexico, and Geoscope station UNM in Mexico City, LONG (2009) found anisotropy in a region of D" beneath the eastern Pacific Ocean.

Shear wave splitting measurements in subduction zones have been reported worldwide. Combining different phases and techniques, it has been possible, in many cases, to infer the anisotropy in the different parts within a subduction system. In the subslab region most subduction zones exhibit trench-parallel fast polarization orientations (LONG and SILVER 2008, 2009) such as South America (RUSSO and SILVER 1994), Northern Tonga (FOLEY and LONG 2011), Tonga-Kermadec (LONG and SILVER 2008), Hikurangi (AUDOINE et al. 2004), Sumatra-Indonesia (LONG and SILVER 2008; HAMMOND et al. 2010), Sangihe (DI LEO et al. 2012a, b), New Hebrides (KIRÁLY et al. 2012), Izu-Bonin (WIRTH and LONG 2010), Marianas (WOOKEY et al. 2005), Ryukyu (LONG and VAN DER HILST 2005, 2006), Kamchatka (PEYTON et al. 2001; LEVIN et al. 2004), Aleutians (LONG and SILVER 2008), Caribbean (PIÑERO-FELICANGELI and KENDALL 2008), Calabria (BACCHESCHI et al. 2007), Central America (ABT et al. 2010), and Scotia (LYNNER and LONG 2013). A few subduction zones show subslab anisotropy with a pattern of fast polarization perpendicular to the trench (LONG and SILVER 2008, 2009) as in Cascadia (CURRIE et al. 2004; LONG and SILVER 2008; RUSSO 2009; EAKIN et al. 2010), western Mexico (VAN BENTHEM 2005; STUBAILO and DAVIS 2007, 2012a, b, 2015; BERNAL-DÍAZ et al. 2008; LEÓN SOTO et al. 2009; ROJO-GARIBALDI 2011; PONCE-CORTÉS 2012; VAN BENTHEM et al. 2013; BERNAL-LÓPEZ 2015; STUBAILO 2015) and Alaska (PERTTU et al. 2014). Most of the previous studies made measurements of upper mantle anisotropy beneath the seismic station using records of teleseismic SKS and other core-transmitted phases. More recently, a number of studies have determined source-side, subslab anisotropy using direct teleseismic S phases after accounting for anisotropy under the receiver (LYNNER and LONG 2013,

2014a, b). LYNNER and LONG (2014a) concluded that trench-perpendicular fast splitting directions in the subslab mantle are more common than previously thought (LONG and SILVER 2008, 2009).

LONG and SILVER (2008) and LONG and WIRTH (2013) reviewed the nature of seismic anisotropy within the mantle wedge in subduction zones around the world and pointed out that splitting patterns in the mantle wedge are more variable than in the subslab mantle. Many subduction zones exhibit a transition from trench-parallel fast directions close to the trench (LONG and SILVER 2008) to trench-perpendicular farther away (e.g., Ryukyu, Marianas, and Tonga), while others exhibit the opposite pattern (e.g., Kamchatka). In the particular case of Mexico, LEÓN SOTO and VALENZUELA (2013) used S waves from local, Cocos intraslab earthquakes to quantify anisotropy in the mantle wedge in the region of the Isthmus of Tehuantepec, southeast of the area in the present study. Northeast of the 100 km isodepth contour they found a clear pattern of trench-perpendicular fast polarization directions, which is consistent with the existence of A-type olivine and 2-D corner flow. In the region southwest of the 100 km isodepth contour, where the slab is shallower, the anisotropy pattern is less clear given that fast axes are oriented in different directions. For these shallower earthquakes, the path through the mantle wedge is shorter, and thus the anisotropy contributions from the continental crust and also from the slab itself become more significant (LEÓN SOTO and VALENZUELA 2013). It should be noted that there is no active andesitic volcanism in the Isthmus of Tehuantepec even as the Cocos slab reaches depths in excess of 110 km (RODRÍGUEZ-PÉREZ 2007; CASTRO-ARTOLA 2010), a depth where this type of volcanism is common in subduction zones worldwide (TATSUMI and EGGINS 1995; SYRACUSE and ABERS 2006). This observation might be an indication that no dehydration fluids from the slab are present in the mantle wedge. Teleseismic measurements using SKS waves in the same region reveal trench-perpendicular fast polarization directions and are consistent with both corner flow in the mantle wedge, and entrained flow beneath the slab (BERNAL-DÍAZ et al. 2008; PONCE-CORTÉS 2012; VAN BENTHEM et al. 2013).

Previous studies of seismic anisotropy where the Cocos slab undergoes flat subduction have relied mostly on data from a permanent but sparse network and have established that the fast polarization direction is trench-perpendicular (VAN BENTHEM 2005; PONCE-CORTÉS 2012; VAN BENTHEM et al. 2013). Under the TMVB only a couple of permanent stations are available and the fast axes there are oriented N–S (PONCE-CORTÉS 2012; VAN BENTHEM et al. 2013). In the present article data from a temporary but dense array, the Meso-American Subduction Experiment (MASE 2007; PÉREZ-CAMPOS et al. 2008), were used to improve coverage of these regions, most especially under the TMVB and north of it. This has led to a better understanding of upper mantle flow and its relationship to the subducted Cocos plate, both south and north of the TMVB. Permanent and temporary networks are complementary and combined provide a more thorough picture of anisotropy throughout the region. Broadly speaking, the permanent network covers a large area sparsely, whereas the temporary deployment spans a smaller area with denser coverage.

2. Data and Method

The data set used in this study was provided by the Meso-American Subduction Experiment (MASE) and was made up of 100 broadband stations in a dense linear array (MASE 2007; PÉREZ-CAMPOS et al. 2008). The instruments were deployed along a nearly North–South line starting at the coast of the Pacific Ocean and heading north toward the Gulf of Mexico for a distance of about 600 km (Fig. 1). Data were recorded between January 2005 and June 2007. A total of 34 teleseismic earthquakes with epicentral distances greater than 85° and magnitudes larger than or equal to 6.0 provided useful records (Table S1 and Fig. 2). The geographic distribution of the sources determined the range of the back azimuths, ϕ_b, used. Except for one event in Mozambique ($\phi_b \approx 100°$) and one in the South Sandwich Islands ($\phi_b \approx 140°$), observations were made in the back azimuth range from 240° to 340° (Fig. 2). A bandpass filter between 0.03 and 1 Hz was used to reduce noise. Core phases (SKS and SKKS) were cut in windows beginning

approximately 10 s before the phase arrival and included the full waveform (Fig. 3). Core phases with a signal-to-noise ratio (SNR) greater than 3 were kept for further analysis.

Shear wave splitting is the result of an S wave traveling through a region of transverse anisotropy. A shear wave propagating in a direction perpendicular to the symmetry axis splits into two orthogonal phases traveling with different velocities, i.e. fast and slow waves. The anisotropic parameters that can be inferred from seismic recordings are the fast polarization direction, ϕ, and the lag time (or delay time) between the fast and slow phases, δt.

The anisotropic parameters for each event-station pair were measured using the method of SILVER and CHAN (1991). It was assumed that the symmetry axis is horizontal and that a single layer of transverse anisotropy is present under the station. The limitations of this assumption will be discussed in the following section. The Silver and Chan method is a geometrical procedure for recovering the incident wave before it enters the anisotropic region. A grid search was used to minimize the energy in the tangential component by trial and error. The trial angles ranged from −90° to 90° and the delay times from 0 to 4.5 s. The anisotropic parameters are proposed as the grid point of minimum corrected tangential energy. Figures 3 and 4 show an example of the method for station MAZA where four clearly split measurements were made. Error bounds were determined using an inverse F test in the manner of SILVER and CHAN (1991), choosing the 95 % confidence region. The bounding values of the confidence region along the parameter axes correspond to 2σ marginal uncertainties for ϕ and δt. Likewise, one half of the bounding values of the confidence region along the parameter axes represent the 1σ (one standard deviation) uncertainties. Individual measurements for each station were averaged with the stacking method (Fig. 4) of WOLFE and SILVER (1998).

3. Results

Shear wave splitting measurements for SKS and SKKS phases were determined for 757 source–receiver pairs (Table S2). A total of 339 phases

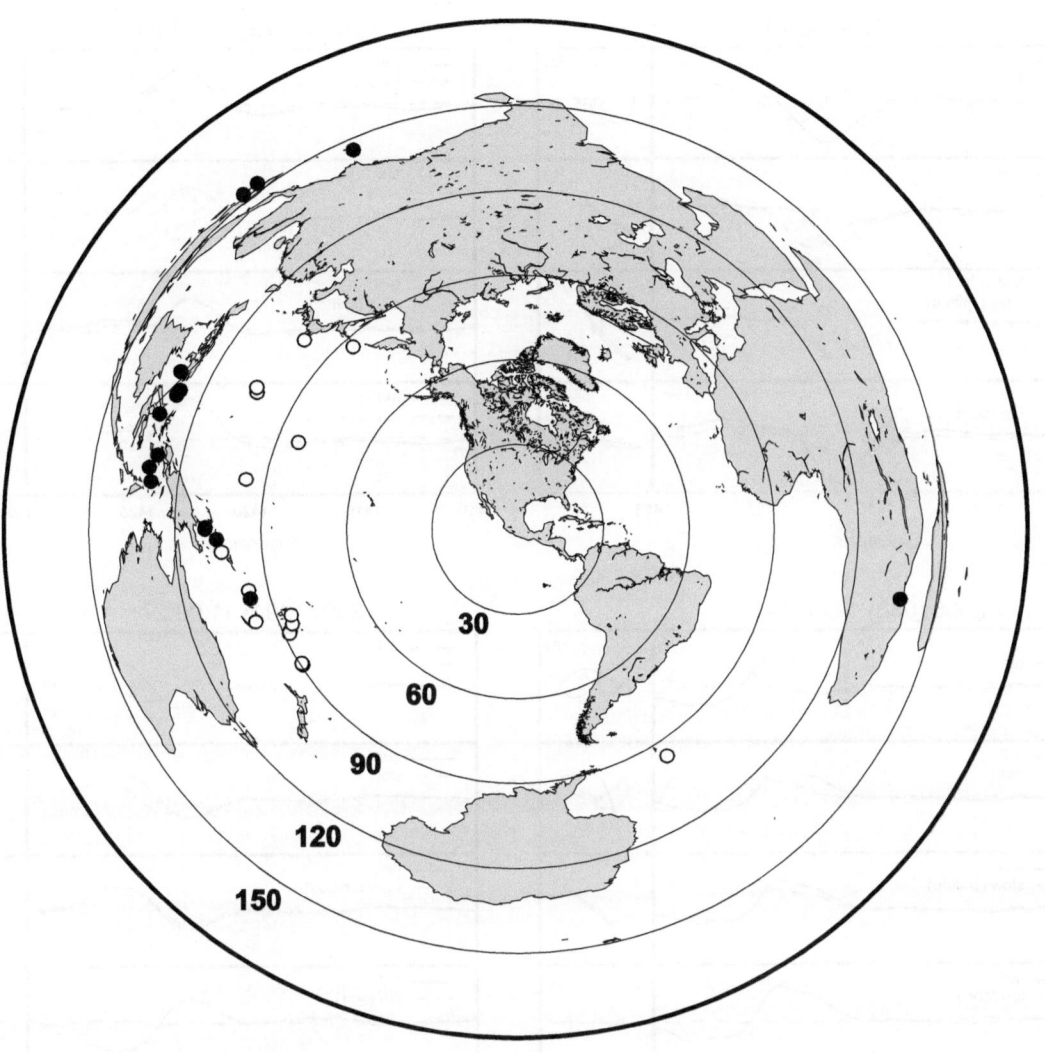

Figure 2

Teleseismic events used in this study. *Open (closed) circles* represent the epicenter of events used to make SKS (SKKS) measurements

registered resolvable splitting. Out of these, 221 measurements were made with SKS waves and 118 with SKKS. Null measurements numbered 418. Null measurements have an ambiguous interpretation. If the shear wave is propagating with a polarization in the orientation of the fast or slow axis before entering into the anisotropic layer, the phase does not show any splitting (SILVER and CHAN 1991). Resolution of the technique is also an issue because delay times smaller than ∼0.4 s are hard to resolve, leading to a null measurement (SILVER and CHAN 1991; SILVER 1996). Another possibility is a multilayer setting,

where the relative orientations of the fast polarization axes result in cancellation of the overall splitting.

Stacked results are listed in Table S3 and shown in Fig. 5. Shear wave splitting results (PONCE-CORTÉS 2012; VAN BENTHEM *et al.* 2013) of nearby stations belonging to the Mexican Seismology Bureau (Servicio Sismológico Nacional, SSN) are also shown (Fig. 5). Fast polarization orientations are shown as black bars with lengths proportional to the averaged delay time. Table S3 lists the station-averaged measurements for SKS and SKKS phases separately, and these are also shown in Fig. 5a and b, respectively.

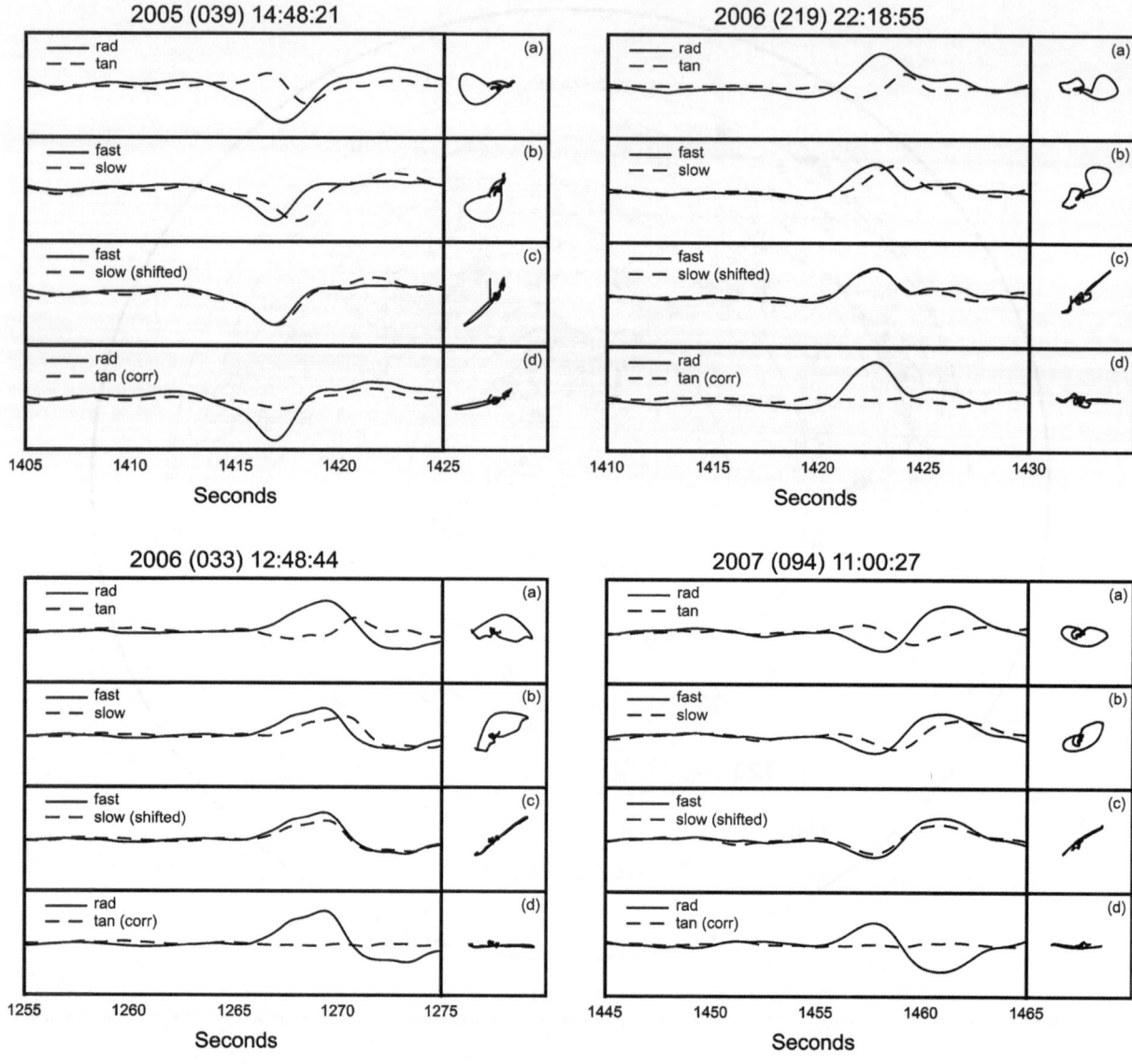

Figure 3
Example of shear wave splitting determination at station MAZA using individual measurements from four different earthquakes. *Particle motion* is shown to the right of each seismogram. For each event, **a** shows the radial and tangential components recorded by the station, **b** the horizontal components rotated to the fast and slow axes, **c** the fast component and the slow component shifted by the measured delay time, and panel **d** the radial and tangential components after correcting for anisotropy

Data from SKS waves reveal a clear change in the orientation of the fast polarization directions between the southern and northern parts of the array (Fig. 5a). The transition between these two domains takes place between the 110 and 150 km isodepth contours (Fig. 5a). Fast polarization directions determined using SKS phases in the southern group (fore-arc) show a trench-perpendicular pattern oriented NE–SW

(N37°E ± 9°) and delay times of 1.13 ± 0.24 s (Fig. 5a), which is the arithmetic average of the stacked values at each southern station. On the other hand, SKS measurements for stations located in the TMVB and northwards (back-arc) show fast polarization directions oriented N–S (N1°E ± 12°), with an average delay time of 1.03 ± 0.39 s. Measurements using SKKS phases show fast polarization

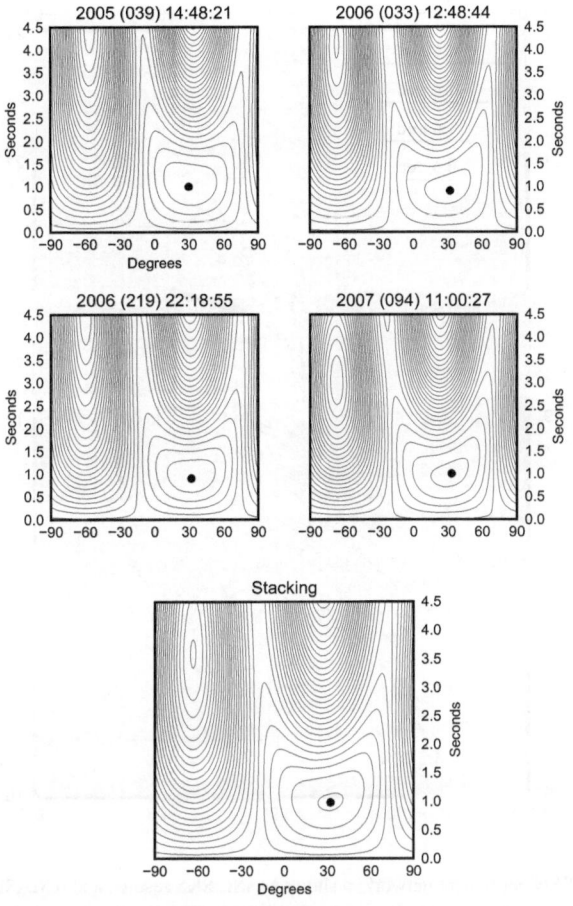

Figure 4

Energy plots for the shear wave splitting measurements recorded at station MAZA (seismograms in Fig. 3). Anisotropic parameters for event 2005 (039) 14:48:21 were estimated as ($\phi = 30° \pm 12°$, $\delta t = 1.0 \pm 0.2$ s), for event 2006 (033) 12:48:44 as ($\phi = 32° \pm 16°$, $\delta t = 0.9 \pm 0.3$ s), for event 2006 (219) 22:18:55 as ($\phi = 32° \pm 15°$, $\delta t = 0.9 \pm 0.3$ s), and for event 2007 (094) 11:00:27 as ($\phi = 33° \pm 11°$, $\delta t = 1.0 \pm 0.2$ s). Stacked results show an average of ($\phi = 32° \pm 5°$, $\delta t = 0.98 \pm 0.09$ s)

directions consistently oriented NE–SW along the whole profile (Fig. 5b). For the sake of comparison with SKS data, anisotropy parameters for SKKS observations were calculated separately for the southern and northern ends of the array (Table 1). Fast polarization directions determined using SKKS phases in the fore-arc average N52°E $\pm$ 8° and the average delay time is 1.51 ± 0.30 s. In the back-arc, the SKKS anisotropy parameters are ϕ = N46°E $\pm$ 18° and $\delta t = 1.28 \pm 0.54$ s.

To show the variation of the anisotropy parameters, ϕ and δt, along the array, these were plotted as a

function of latitude (Fig. 6). The change in the orientation of the SKS fast axes between the fore-arc and the back-arc is clear in Fig. 6a. The standard deviations, σ_ϕ and $\sigma_{\delta t}$, of the measurements are larger in the back-arc than in the fore-arc (Fig. 6a and c; Table 1). Two reasons can account for this observation. First, stations at the northern end of the MASE array ran for a shorter period of time than those in the south, so fewer measurements were possible. Most importantly, however, the fast polarization direction for the northern segment is oriented N–S, which makes it difficult to measure the anisotropy parameters given that the back azimuth of most of the events is in the range $245° \leq \phi_b \leq 280°$ (Fig. 2). When the fast axis, ϕ, is oriented (SAVAGE 1999) parallel ($\phi = \phi_b \pm 15°$) or perpendicular to the event's back azimuth, ϕ_b, the shear wave splitting method will return a null measurement (SILVER and CHAN 1991). Indeed, for northern MASE, 238 of the 336 entries reported in Table S2 are null measurements. The fast polarization axes measured using SKKS phases are consistently oriented NE–SW along the entire array (Fig. 6b). Observed delay times do not show a significant dependence with latitude, but are in general larger for SKKS than for SKS phases (Fig. 6c and d; Table 1).

Another way to visualize the different orientation of the fast polarization direction for SKS observations between the fore-arc and the back-arc is by plotting the anisotropy parameters as a function of back azimuth. Figure S1 shows individual measurements of (ϕ, δt) with their corresponding 1σ uncertainties for various events at stations TICO and PASU (from Table S2). Because of their back azimuths, SKS measurements plot on the left side ($248° \leq \phi_b \leq 271°$) whereas SKKS observations are shown to the right ($270° \leq \phi_b \leq 294°$). The horizontal lines represent the average (stacked) parameters for the station for both SKS and SKKS phases (from Table S3). Station TICO is representative of measurements in the southern segment of the MASE array (Fig. S1a and S1b). The average parameters for TICO determined from SKS phases are (47°, 1.19 s) and correspond to the lower horizontal line in Figures S1a and S1b. They are shown as solid lines in the range of back azimuths where SKS phases were recorded (left hand side) and as dashed

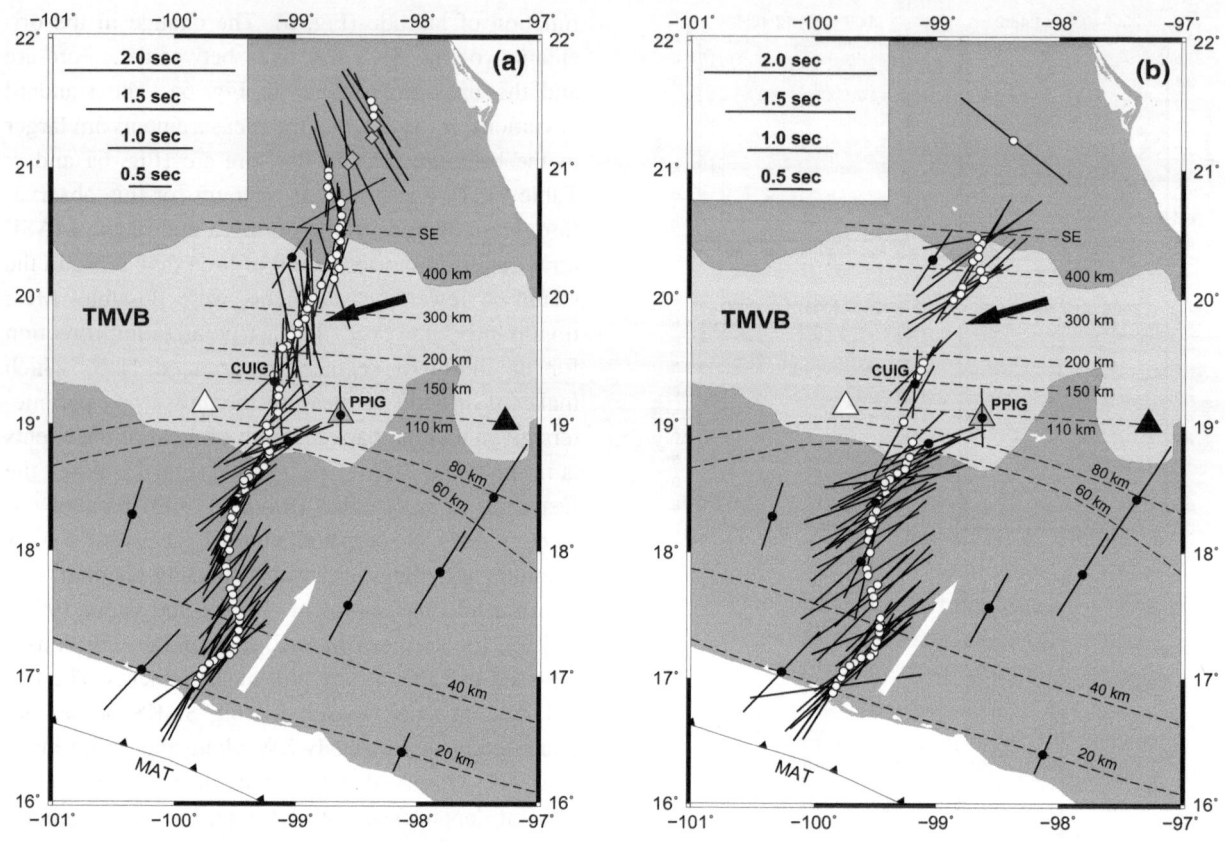

Figure 5
Stacked results for the MASE experiment and nearby stations of Mexico's SSN permanent network. **a** shows MASE SKS results, and **b** MASE SKKS results. *Solid bars* indicate fast polarization orientations at each station that clearly recorded splitting. The length of each bar corresponds to the delay time according to the legend. *White circles* stand for MASE stations (this study), and *black circles* for SSN ones (PONCE-CORTÉS 2012; VAN BENTHEM *et al.* 2013). A few *gray diamonds* at the northern end of the array are null measurements showing the fast polarization direction consistent with splitting observations at nearby stations; however, fast axes perpendicular to the ones shown, or no resolvable anisotropy ($\delta t \approx 0$ s), are also permitted by the data. *Open arrow* indicates the velocity of convergence of the Cocos plate with respect to North America at ~6.3 cm/year and N32°E (DEMETS *et al.* 2010). Solid arrow represents the absolute motion of the North American plate at ~4.0 cm/year and N254°E (GRIPP and GORDON 2002). The region in *light gray* represents the Trans-Mexican Volcanic Belt (TMVB) and the *triangles* stand for andesitic volcanoes according to caption of Fig. 1. Slab isodepth contours from 20 to 80 km are from FERRARI *et al.* (2012) and references therein. Determination of contours at 110 km and deeper is discussed in the text. The *slab edge* (SE) is the maximum depth of the steep slab at 500 km

Table 1

Comparison of averaged SKS and SKKS splitting parameters in southern and northern MASE

Phase	Regions			
	Southern MASE		Northern MASE	
	$\varphi \pm \sigma_\varphi$	$\delta t \pm \sigma_{\delta t}$	$\varphi \pm \sigma_\varphi$	$\delta t \pm \sigma_{\delta t}$
SKS	N37°E ± 9°	1.13 ± 0.24 s	N1°E ± 12°	1.03 ± 0.39 s
SKKS	N52°E ± 8°	1.51 ± 0.30 s	N46°E ± 18°	1.28 ± 0.54 s

lines where SKKS phases were available (right-hand side). The SKKS average at the same station is (49°, 1.25 s) and in this case the horizontal line is solid on the right-hand side and dashed on the left. For this station the anisotropy parameters show no dependence on the back azimuth. The SKS average is consistent with the SKKS individual measurements, and the other way around too, the SKKS average is consistent with the SKS individual measurements (Fig. S1a and S1b). Within the limited range of back azimuths available, this observation shows that the assumption of a single anisotropic layer stated previously is reasonable where the Cocos slab is subhorizontal. Station PASU is located in the back-arc, i.e. northern segment of the MASE array (Fig. S1c and S1d). The different orientation of the fast axes between the fore-arc and the back-arc for SKS observations is evident from Figures S1a and S1c. At TICO the individual SKS measurements are oriented NE–SW and span the range $35° \leq \phi \leq 51°$ (Fig. S1a), whereas at PASU the SKS fast axes are oriented N–S and their range is $-7° \leq \phi \leq 18°$ (Fig. S1c). At PASU, the SKKS fast axes are oriented NE–SW and fall within the range $41° \leq \phi \leq 66°$ (Fig. S1c). In contrast with TICO, the average SKS fast axis at PASU does not fit the individual SKKS measurements, and likewise, the average SKKS fast axis at PASU does not fit the individual SKS measurements (Fig. S1c). This behavior suggests that the orientations of the fast polarization directions depend on the back azimuth for stations in the northern segment of the MASE array.

Figure 7 shows the SKS waveforms and contour plots for the event of January 2nd, 2006 at stations ZACA and PASU, which are representative of stations in the fore-arc and back-arc, respectively. Differences between the records at these two stations are robust. Specifically, the fast SKS wave at PASU has a small amplitude compared to the fast wave at ZACA (Fig. 7). The contour plots are also different. The fast axis at ZACA, in the fore-arc, is $\phi = 27° \pm 17°$, whereas $\phi = -7° \pm 6°$ at PASU (Fig. 7).

It is interesting to consider the reasons for the different orientations of the fast axes determined from SKS and SKKS observations at the northern end of the array. SKS measurements were made predominantly in the back azimuth range from 245° to 260° (e.g., Fiji, Vanuatu, and Loyalty islands), whereas SKKS data are from back azimuths between 275° and 295° (e.g., New Britain, Banda Sea, and Celebes Sea); see Fig. 2 and S1. Within the limited range of back azimuths available, the orientation of the fast polarization directions appears to depend on the back azimuth of the sources (Fig. S1c). Previous studies have modeled back azimuthally dependent observations of anisotropy using two or more separate, horizontal anisotropic layers (SILVER and SAVAGE 1994; ÖZALAYBEY and SAVAGE 1994, 1995). Due to the complex structure of subduction zones, as many as four different layers may contribute to the observed anisotropy: the subslab mantle, the slab, the mantle wedge, and the overlying plate, and it is consequently difficult to tease out the individual contributions from each layer (LONG and SILVER 2008). In this light, our assumption of a single anisotropic layer where the Cocos slab dips steeply may be simplistic. Given the limited range of back azimuths in our sources, we are hesitant to model more than one anisotropic layer. As it will be discussed in the following section, we believe that the one-layer model proposed for the back-arc is plausible and makes sense based on physical arguments.

Other explanations may also be possible for the discrepancy in the orientation of the fast axes determined using SKS and SKKS phases in the back-arc. For instance, the geometry of the steeply dipping slab (PÉREZ-CAMPOS et al. 2008; HUSKER and DAVIS 2009; KIM et al. 2010) under the TMVB may affect the path of the upgoing waves and consequently the splitting measurements. Depending on the back azimuth, the paths to stations in one end of the array would go through the steep slab while rays reaching stations at the other end would avoid the steep slab altogether. The work of STUBAILO (2015) is similar to the one in the present study, but the analysis methods are somewhat different. He also measured SKS and SKKS splitting using the MASE data set. His results, however, show a consistent NE–SW (N30°E $\leq \phi \leq$ N65°E) orientation for the fast axes along the entire MASE array, in agreement with the SKKS-only observations in the present study. Lastly, comparison with one other study is warranted. LONG (2009) measured shear wave splitting using data from

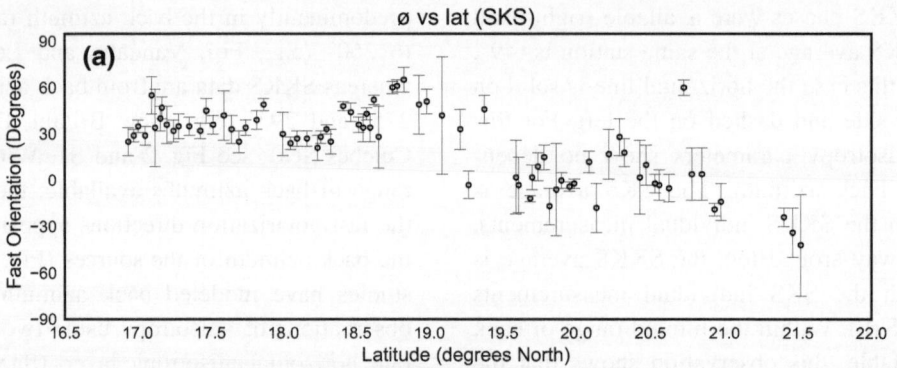

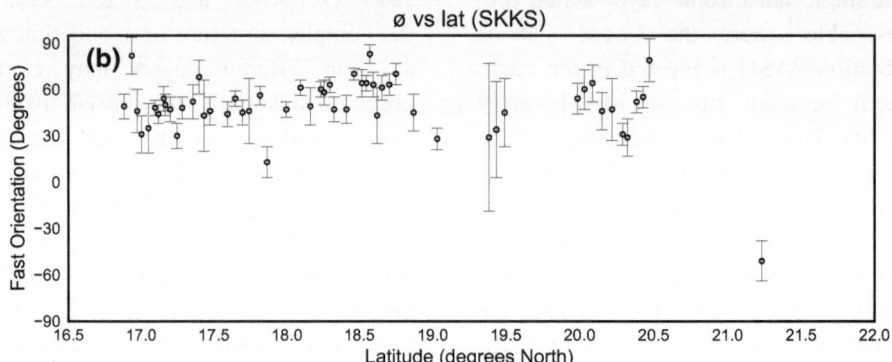

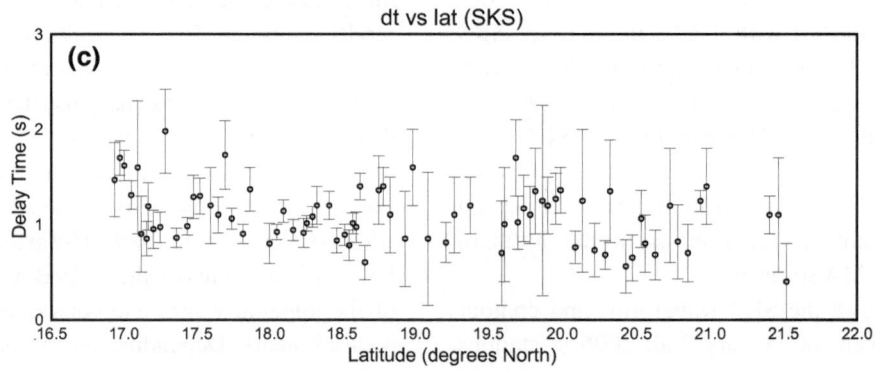

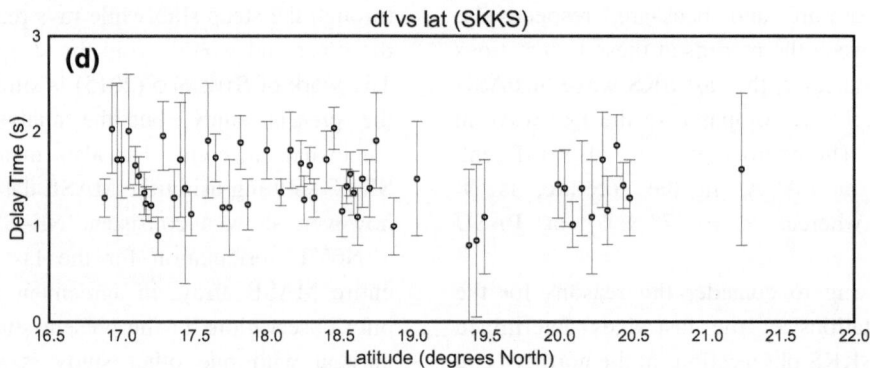

Figure 6
Anisotropy parameters, ϕ and δt, as a function of station latitude, including *error bars* for one standard deviation. **a** and **c** correspond to SKS measurements, while **b** and **d** are for SKKS phases. **c** and **d** show that the delay time stays fairly constant throughout the array. **a** Shows that the orientation of the fast polarization direction measured using SKS waves changes from NE–SW in the south of the array to N–S in the north. On the other hand, fast polarization directions determined from SKKS observations are oriented NE–SW throughout the entire array **b**

stations in California, the NARS-Baja California array in northwestern Mexico, and Geoscope station UNM. In some cases, she found significant differences between the SKS and SKKS anisotropy measurements obtained using the same event-station pair. LONG (2009) concluded that most of the SKKS paths sampled anisotropy in the D'' region of the lowermost mantle. Most of her SKS measurements

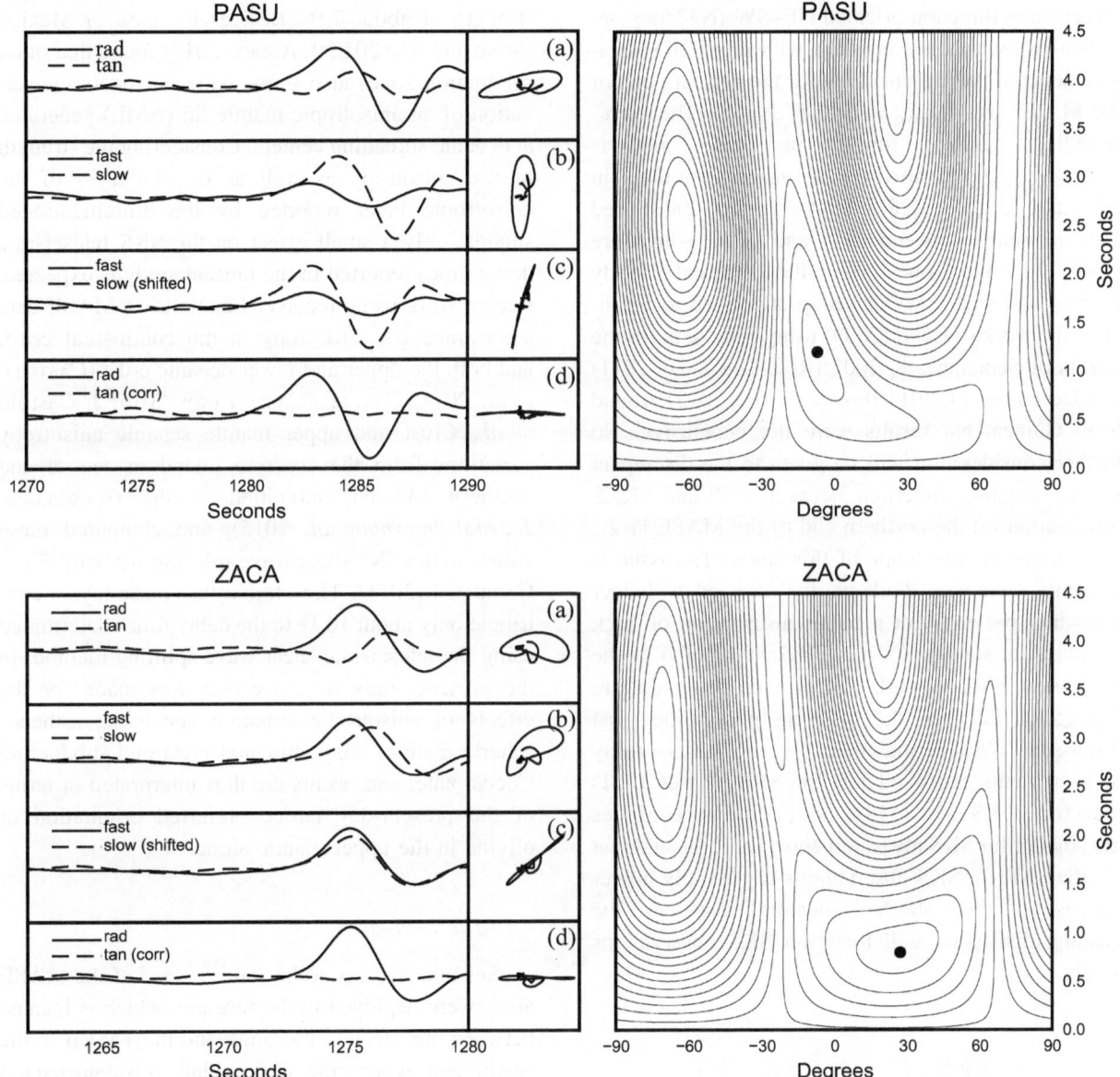

Figure 7
Comparison of SKS waveforms and contour plots for stations ZACA (fore-arc) and PASU (back-arc) using the event of January 2nd, 2006. The different orientation of the fast axes is clearly resolved. At station ZACA $\phi = 27° \pm 17°$, whereas at PASU $\phi = -7° \pm 6°$. Also notice that the fast wave at PASU is smaller than at ZACA. The location of the stations along the array is shown in Fig. 8

for the same event-station pair, however, returned null values. Her results from station UNM are especially relevant for the present study. UNM is co-located with SSN station CUIG and is located within the region of N–S oriented fast axes determined from SKS waves in the present study (Fig. 5a). The fast polarization direction at CUIG is also oriented N–S (PONCE-CORTÉS 2012). LONG (2009) reported 6 measurements using SKKS phases at UNM with the fast polarization direction oriented NE–SW (N42°E $\leq \phi \leq$ N68°E), which is consistent with the SKKS measurements of the present study for the northern end of the MASE array (ϕ_{avg} = N46°E $\pm$ 18°). The corresponding delay times in her analysis ($1.5 \leq \delta t \leq 3.1$ s), however, are larger than those in this study ($\delta t_{avg} = 1.28 \pm 0.54$ s). LONG (2009) used back azimuths between 271° and 277°, which are close to the ones in the present study ($275° \leq \phi_b \leq 295°$). Our group attempted to reproduce the work of LONG (2009) using SSN data for the events of September 8, 2002 (ROJO-GARIBALDI 2011) and December 14, 2011 (PONCE-CORTÉS 2012) around New Guinea, but results were not conclusive. No further consideration will be given to the discrepant fast polarization directions between SKS and SKKS observations at the northern end of the MASE array, as it is outside the scope of this study. The issue is certainly complex. Perhaps the use of a longer recording period, with a better distribution of back azimuths, at stations CUIG, PPIG, and DHIG of the permanent SSN network will provide better data to distinguish between the different possibilities just discussed. Considering that (1) more anisotropy measurements are available for SKS phases (221) than for SKKS ones (118), that (2) the uncertainties are smaller for SKS measurements (Table 1), and that (3) the SNR for SKS observations is generally higher because of their shorter epicentral distances, the ensuing discussion will focus on SKS observations solely.

4. Implications for Mantle Flow

In addition to the upper mantle anisotropy studies described in the introduction, other researchers have looked at the fine structure of anisotropy in the flat

slab segment of the MAT. SONG and KIM (2012b) found that the upper oceanic crust of the subducting Cocos slab is an ultraslow-velocity layer (USL). It is 3 to 5 km thick and it shows anisotropy larger than 5 %. SONG and KIM (2012a) discovered that the topmost 2 to 6 km of the Cocos subducted oceanic mantle is an anisotropic high-velocity lid (HVL). This layer has the fast polarization axes oriented in the direction of plate convergence and an anisotropic strength of about 7 %. Besides the case of Mexico (SONG and KIM 2012a), AUDET (2013) found that other subduction zones also show evidence for the preservation of an anisotropic mantle lid (AML) generated at oceanic spreading centers. Considering the strength of the anisotropy as well as the thickness of the anisotropic layer reported by the aforementioned studies, only a small effect on the SKS teleseismic delay times reported in the present study is expected. Recent work using receiver functions on MASE data determined the anisotropy in the continental crust, and both the upper and lower oceanic crust (CASTILLO et al. 2014; CASTILLO-CASTELLANOS 2015; J. Castillo et al., Crust and upper mantle seismic anisotropy variations from the coast to inland in central and southern Mexico, submitted to the *Geophysical Journal International*, 2015), and compared these values to the SKS shear wave splitting study by ROJO-GARIBALDI (2011). They found that these layers contribute only about 10 % to the delay times determined using the teleseismic shear wave splitting method. In the present study no correction was made for the effects of anisotropic structure above the asthenospheric mantle, i.e. continental crust and subducting Cocos plate, and results are thus interpreted in terms of the present-day lattice preferred orientation of olivine in the upper mantle alone.

4.1. The Fore-arc

Seismic stations at the southern end of the MASE array were deployed in the fore-arc, which is located between the MAT to the south and the TMVB to the north, and where the Cocos slab is subhorizontal (PARDO and SUÁREZ 1995; PÉREZ-CAMPOS et al. 2008; HUSKER and DAVIS 2009). Figure 5a shows the stacked SKS anisotropy parameters at each station. The arithmetic average of MASE stations in this segment is $(\phi, \delta t) = $ (N37°E ± 9°, 1.13 ± 0.24 s). Shear wave splitting measurements using data from Mexico's SSN permanent network (PONCE-CORTÉS 2012; VAN BENTHEM et al. 2013) are also shown in Fig. 5a. The SSN network is sparser than the MASE experiment but it covers a larger area. Overall, the pattern of shear wave splitting determined from MASE and SSN data is consistent throughout this region of flat slab subduction (Fig. 5a). Based on model PVEL (DEMETS et al. 2010), the relative velocity between the Cocos and North American plates at the MAT south of MASE is ~6.3 cm/year and the Cocos plate moves in the direction of ~N32°E (Fig. 5a). The observed fast axes are thus oriented in the direction of the relative plate motion (RPM) between the Cocos and North American plates, and approximately perpendicular to the MAT. Physical conditions beneath the subhorizontal slab are expected to be low stress, low water content, and relatively high temperature and are consistent with the existence of a A-type olivine (JUNG et al. 2006; LONG and SILVER 2008). The asthenosphere under the oceanic flat slab in Mexico is at temperatures above 1000 °C (MANEA et al. 2006; MANEA and MANEA 2011). Consequently, the orientation of the fast axes is indicative of entrained mantle flow beneath the flat slab, as represented in cross section and schematically in Figs. 8 and 9, respectively. Strong coupling between the slab and the underlying mantle is inferred. It should be mentioned that little or no mantle wedge exists above the subducted Cocos plate because of the subhorizontal geometry of the slab. PÉREZ-CAMPOS et al. (2008) imaged a thin (~10 km) low-velocity zone (LVZ) between the lower continental crust and the slab, which is likely altered oceanic crust or a mantle wedge remnant.

STUBAILO et al. (2012) studied the velocity structure of the region around the MASE deployment using Rayleigh waves. In their Fig. 13, STUBAILO et al. (2012) show phase velocity maps for waves of different periods in the range from 16 to 100 s. In a further step, STUBAILO et al. (2012) inverted the phase velocity maps and obtained a three-dimensional model for shear wave velocity and anisotropy (their Fig. 15). Their model is parameterized into three layers: the continental crust, mantle lithosphere, and asthenosphere down to a depth of 200 km. Broadly

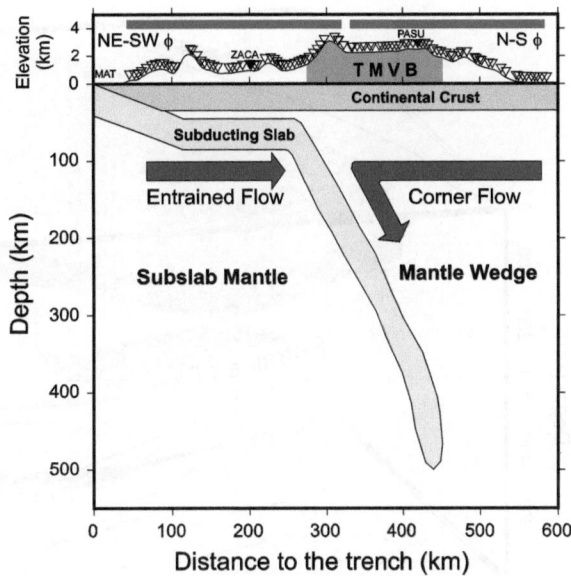

Figure 8
Cross section along the MASE array showing the geometry of the Cocos slab. Fast axes determined from SKS phases are oriented NE–SW in the fore-arc, and N–S in the back-arc. In the south, where the slab is subhorizontal, mantle flow is entrained under the slab. To the north, the steeply dipping slab drives corner flow in the mantle wedge. The region of fast polarization directions oriented N–S extends to the north beyond the northern end of the Trans-Mexican Volcanic Belt, which is also coincident with the edge of the slab. This cross section does not show the different orientation between the Middle America Trench to the south and the strike of the slab under the TMVB

speaking, in the fore-arc their two models show anisotropy with the fast axes oriented trench-perpendicular in the west but the fast axes become trench-parallel toward the east (their Fig. 13 middle and bottom; their Fig. 15b and c). The details of the transition from trench-perpendicular to trench-parallel fast polarization directions, however, are different between their two models. In their Fig. 15b and c, the fast directions become trench-perpendicular where the Orozco Fracture Zone (OFZ) is expected to extend landward under the North American plate. STUBAILO et al. (2012) interpret these results as toroidal flow into the mantle wedge coming in from the two edges that limit the segment of the Cocos slab located between the OFZ to the east and the Rivera slab to the west. We wish to point out some similarities between the anisotropy patterns in Fig. 13, middle and bottom, of STUBAILO et al. (2012) and our own results. The trench-perpendicular

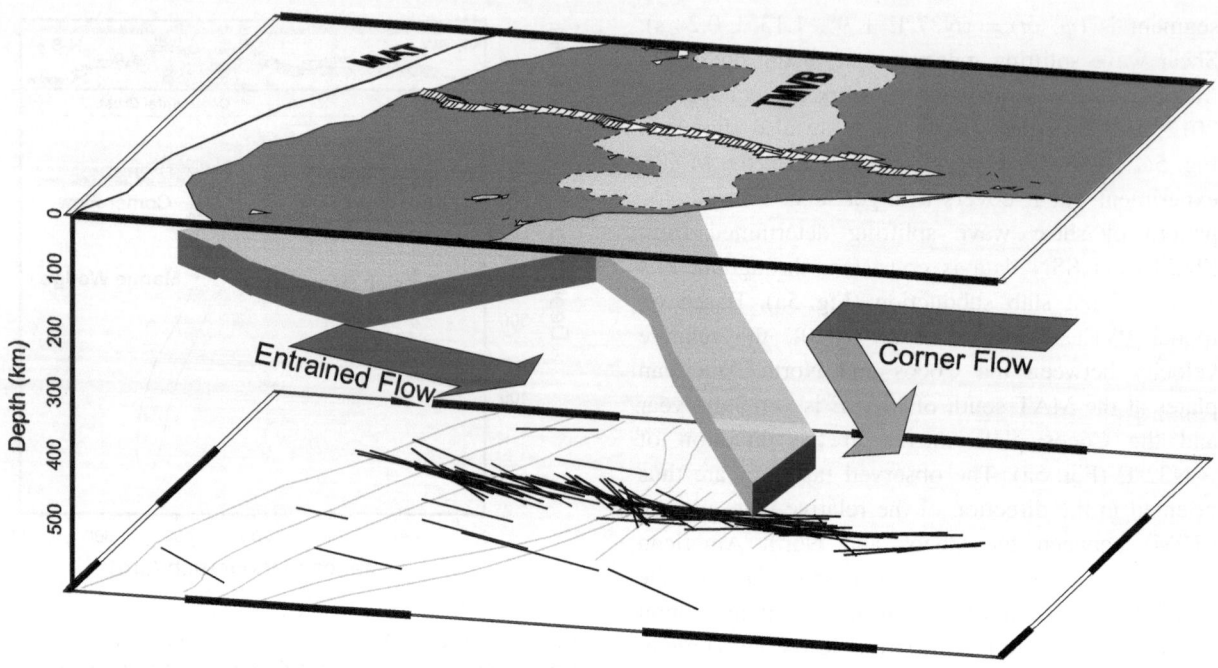

Figure 9

Cartoon illustrating a perspective view of the SKS shear wave splitting results. The perspective is 75° clockwise from north at an elevation of 15°. It only shows a section of the subducting slab corresponding to reconstruction from tomographic results from the isodepth contours (projected at the *bottom plane* as *gray lines*) shown in Fig. 5. Shear wave splitting results are projected at the *bottom plane* as *black lines*. The change in the orientation of the fast polarization directions between the fore-arc and the back-arc can be clearly seen. The strike of the Middle America Trench to the south is not subparallel to the volcanic arc at the southern end of the TMVB and neither is it subparallel to the strike of the steeply dipping slab. In the fore-arc, the subhorizontal slab drives entrained flow, whereas in the back-arc the steeply dipping slab produces corner flow

fast axes observed to the west in the fore-arc and even slightly east of the MASE profile, Fig. 13 bottom of STUBAILO *et al.* (2012), are oriented nearly in the same direction as the SKS shear wave splitting fast axes reported in this study (Fig. 5a) from both MASE and SSN data. Farther east, however, the fast polarization directions reported by STUBAILO *et al.* (2012) become trench-parallel while SSN (PONCE-CORTÉS 2012; VAN BENTHEM *et al.* 2013) SKS fast axes remain oriented trench-perpendicular (Fig. 5a). Figure 13, middle and bottom, of STUBAILO *et al.* (2012) shows that in the back-arc, the Rayleigh wave fast anisotropy axes are clearly rotated to a direction approximately N–S, or slightly west of north, which is different from the fore-arc and also consistent with SKS results in this study (Fig. 5a). Figure 13 bottom of STUBAILO *et al.* (2012) shows data for Rayleigh waves at a period of 85 s. The sensitivity kernel, Fig. 14 of STUBAILO *et al.* (2012), at a period of 85 s peaks at a depth of ~120 km thus sampling the asthenosphere

preferentially, which is also consistent with the results in the present study. The base of the oceanic lithosphere under the flat slab segment, including both the continental crust and the LVZ between the continental crust and the slab, is found at a depth of ~80 km based on thermal models (MANEA *et al.* 2006; HUSKER and DAVIS 2009; MANEA and MANEA 2011) and seismic observations (PÉREZ-CAMPOS *et al.* 2008; HUSKER and DAVIS 2009). As the flat slab (rigid part, or lithosphere) moves in the direction of relative plate motion between Cocos and North America, it drives entrained flow in the asthenosphere below.

Previous studies of SKS anisotropy in the Mexican subduction zone have also found trench-perpendicular fast polarization directions. In the region west of the present study, trench-perpendicular fast axes were found under the subducted Rivera plate (LEÓN SOTO *et al.* 2009; VAN BENTHEM *et al.* 2013), whereas in the region to the east, trench-perpendicular fast polarization directions were also reported beneath the Cocos slab (BERNAL-DÍAZ et al. 2008; PONCE-CORTÉS 2012; VAN BENTHEM et al. 2013). Recent source-side shear wave splitting observations also show a consistent pattern of plate motion parallel, i.e. nearly trench-perpendicular, fast directions in the subslab mantle for the Mexican subduction zone (LYNNER and LONG, 2014a). In the Isthmus of Tehuantepec, to the east of the present study, shear wave splitting measurements using local intraslab earthquakes found trench-perpendicular fast axes indicative of corner flow where the Cocos slab reaches depths greater than 100 km (LEÓN SOTO and VALENZUELA 2013). Another region where trench-perpendicular, subslab ϕ have been extensively documented is the Cascadia subduction zone (CURRIE et al. 2004; LONG and SILVER 2008; RUSSO 2009; EAKIN et al. 2010). Both the Mexican and Cascadia trenches share in common the subduction of young and hot remnants of the larger and now extinct Farallon plate (ATWATER and STOCK 1991).

Thorough compilations of subslab anisotropy, mostly using SKS data, showed that ϕ is trench-parallel in most subduction zones around the world (LONG and SILVER 2008, 2009; LONG 2013). LONG and SILVER (2008) explained these observations as the result of 3D return flow driven by trench migration, combined with a barrier to entrained flow beneath the slab. In this view, a thin decoupling zone, produced through shear strain, between the downgoing slab and the subslab mantle is needed to allow return flow (PHIPPS MORGAN et al. 2007; LONG and SILVER 2008, 2009). To explain the observed trench-perpendicular ϕ in Mexico and Cascadia, LONG and SILVER (2009) suggested that, because of the age and temperature of the slab, the amount of strain is not yet high enough for the shear mechanism to reach steady state and decoupling has not yet occurred. Farther southeast along the MAT from the region in this study, east of the projected landward extension of the Tehuantepec Ridge (MANEA and MANEA 2008) and near the Guatemala border, observed delay times are small and indicative of little horizontal mantle flow (PONCE-CORTÉS 2012). Farther east still, under Nicaragua and Costa Rica, ABT et al. (2010) used teleseismic SKS and local S observations and found trench-parallel flow beneath the subducted Cocos plate and in the mantle wedge. Source-side splitting measurements,

however, found plate motion parallel subslab fast directions under Nicaragua (LYNNER and LONG 2014a). LYNNER and LONG (2014a) proposed that the difference between their own results and those of ABT et al. (2010) arises because the two studies sample different volumes of the subslab mantle. For comparison, the age of the oceanic Cocos plate subducting at the MAT in the present study is roughly between 12 and 16 Myr (KANJORSKI 2003), while at offshore Nicaragua and Costa Rica it is between 18 and 24 Myr (ABT et al. 2009). In this case, the age difference between Guerrero and Nicaragua/Costa Rica seems too small to explain the difference between trench-perpendicular fast axes in Guerrero and trench-parallel fast axes in Nicaragua and Costa Rica. In the case of Nicaragua and Costa Rica, the mantle flows toward the west-northwest in the mantle wedge driven from a slab window located to the southeast, whereas subslab flow might be driven toward the east-southeast by along-arc gradients in slab rollback as well as regional trench and plate motions (ABT et al. 2010). The trench-perpendicular fast polarization directions and entrained subslab mantle flow observed in the present study under Guerrero are consistent with the absence of a thin decoupling layer between the downgoing slab and the subslab mantle, as proposed by LONG and SILVER (2009).

Other models have been advanced to explain the difference between trench-parallel and trench-perpendicular subslab fast axes observed worldwide. SONG and KAWAKATSU (2012) used a model whereby the oceanic asthenosphere is made up of horizontal melt-rich layers or sheets embedded in a meltless mantle (KAWAKATSU et al. 2009), also called the "millefeuille" model (LONG 2013). In this scenario, the oceanic asthenosphere is endowed with orthorhombic anisotropy, and consequently the orientation of the fast polarization directions is controlled by the dip of the subducting plate (SONG and KAWAKATSU 2012). In the following discussion SKS waves, with nearly vertical incidence angles of about 10° to 15°, are considered. For steep subduction zones, i.e. slab dip greater than 30° to 40°, the fast splitting direction is predominantly parallel or sub-parallel to the trench and it does not vary with ray back azimuth (SONG and KAWAKATSU 2012). For shallow subduction zones (slab dip between 10° and 20°), however, the fast splitting direction is predominantly normal or sub-normal to the trench when incident waves travel on paths that point away from the slab dip (i.e., the up-dip direction). For a subhorizontal slab (dip less than 10°) the SKS fast splitting direction is always normal or sub-normal to the trench (SONG and KAWAKATSU 2012). Given that in the area of the present study the slab is subhorizontal, the observed trench-perpendicular fast polarization directions and entrained mantle flow are consistent with the orthorhombic anisotropy model of SONG and KAWAKATSU (2012).

Recently, PACZKOWSKI et al. (2014a, b) conducted numerical modeling of slab-driven mantle flow. PACZKOWSKI et al. (2014a) accounted for the relative contributions of trench migration and global mantle flow to compute the background velocity field. PACZKOWSKI et al. (2014b) calculated the distribution of finite strain. In most cases PACZKOWSKI et al. (2014a) found that for long, steep slabs the ambient background mantle flow is deflected toward the edge of the slab, in general agreement with observed trench-parallel subslab fast polarization directions. On the other hand, short slabs that do not penetrate into the lower mantle predominantly deflect ambient background mantle flow beneath the bottom of the slab and are usually consistent with trench-perpendicular subslab fast axes. In the case of Mexico, the trench-perpendicular fast polarization directions observed under the subducted Rivera plate (LEÓN SOTO et al. 2009) match well with the predicted model results for a short slab and trench-perpendicular background mantle flow (PACZKOWSKI et al. 2014a). In the region of the present study, the slab in the fore-arc does not penetrate into the lower mantle, but the trench-perpendicular fast axes cannot be directly compared to the work by PACZKOWSKI et al. (2014a) because of the subhorizontal geometry of the slab. The smallest slab dip angle in the numerical models of PACZKOWSKI et al. (2014a) is 30°.

In their thorough reviews of anisotropy in subduction zones, LONG and SILVER (2008, 2009) pointed out that most studies based on core-transmitted phases, such as SKS, found predominantly trench-parallel subslab fast polarization directions, with only a few trench-perpendicular exceptions. More recently, however, source-side shear wave splitting measurements (e.g., LYNNER and LONG 2014a, b) have led to the conclusion that subslab trench-perpendicular fast axes occur more frequently than previously thought (LONG and SILVER 2008, 2009). Furthermore, trench-perpendicular fast polarization directions have been correlated with younger oceanic lithosphere whereas older lithosphere has been associated to trench-parallel fast axes (LYNNER and LONG 2014a, b), with the transition marked by a slab age of ∼95 Ma. Under younger lithosphere strong coupling with a thick, ∼200 km, mantle layer would produce entrained flow in A-type (or similar) olivine fabric (LYNNER and LONG 2014a, b). On the other hand, older lithosphere would be poorly coupled to the underlying mantle, possibly by a thin and weak layer, and thus return flow could be driven by trench migration. In the Mexican subduction zone young lithosphere is being subducted, and so observed trench-perpendicular fast anisotropy axes (BERNAL-DÍAZ et al. 2008; LEÓN SOTO et al. 2009; PONCE-CORTÉS 2012; VAN BENTHEM et al. 2013), including the present study, are consistent with the model of LYNNER and LONG (2014a, b).

4.2. The Back-arc

The northern segment of the MASE array spanned the back-arc, including the TMVB and the region north of it. The currently active volcanic arc is located at the southern end of the TMVB (Figs. 1, 5a), where the Cocos slab dips steeply to the north (PÉREZ-CAMPOS et al. 2008; HUSKER and DAVIS 2009; KIM et al. 2010). Likewise, the orientation of the SKS fast polarization direction changes from NE–SW under southern MASE to N–S in the back-arc (Fig. 5a). The arithmetic average of MASE stations for the northern segment is $(\phi, \delta t) = (\text{N1}°\text{E} \pm 12°, 1.03 \pm 0.39 \text{ s})$. A few SSN stations are located on or near the northern end of the MASE line. Shear wave splitting measurements at stations CUIG and PPIG have their fast axes oriented N–S (PONCE-CORTÉS 2012) and are consistent with northern MASE (Fig. 5a). The change in the orientation of the fast axes from NE–SW in the fore-arc to N–S in the back-arc is caused by the particular geometry of the Mexican subduction zone, given that the TMVB is not subparallel to the offshore trench (GILL 1981; SUAREZ and SINGH 1986; FERRARI et al. 2012), as shown in Figs. 1 and 5a. The measured fast polarization directions in the back-arc are perpendicular to the strike of the steeply dipping Cocos slab, as defined by the isodepth contours under the TMVB (Fig. 5a). Determining the slab geometry at depth using seismicity data alone has not been possible because hypocenters end at a depth of ~80 km (PARDO and SUÁREZ 1995; MANEA et al. 2006; HUSKER and DAVIS 2009; PACHECO and SINGH 2010). Receiver function and tomography studies (PÉREZ-CAMPOS et al. 2008; HUSKER and DAVIS 2009; KIM et al. 2010) have unequivocally established the dip of the Cocos slab at its intersection with the MASE profile down to a depth of ~500 km. Because of the linear nature of the MASE array, however, it is difficult to determine the strike of the slab using receiver functions and tomography. To constrain the 110 km isodepth contour we used the relationship between slab depth and location of active andesitic volcanoes (TATSUMI and EGGINS 1995; SYRACUSE and ABERS 2006). These volcanoes are from west to east (MACÍAS 2005): Nevado de Toluca, Popocatépetl, and Pico de Orizaba (Figs. 1 and 5a). Determination of the deeper contours (from 150 to 500 km) is fraught with greater uncertainty. We took the depth of the slab at its intersection with the MASE array (HUSKER and DAVIS 2009; FERRARI et al. 2012) and drew lines parallel to the well constrained 110 km isodepth contour. Additionally, the E–W trend of the TMVB is consistent with this assumption (Fig. 5a). B-type olivine develops under low temperature, high water content, and high stress conditions and is often present in the mantle wedge tip (KNELLER et al. 2005; JUNG et al. 2006; LONG 2013). In the region of the present study, however, high temperatures, in excess of 900 °C, exist throughout the mantle wedge and dehydration of the slab occurs down to depths of 150 km (MANEA and MANEA 2011). Therefore, given that the transition from B- to C-type fabric occurs around 700–800 °C at mantle wedge stresses (KNELLER et al. 2005), C-type olivine is expected in the mantle wedge tip. Physical conditions in the mantle wedge core usually are low stress, low water content, and relatively high temperature and so the existence of A-type olivine is expected (KNELLER et al. 2005; JUNG et al. 2006; LONG

and SILVER 2008). For both A- and C-type olivine, the fast polarization directions should align in the direction of mantle flow. Consequently, the N–S fast polarization directions observed in the back-arc in the present study are the result of slab strike-perpendicular, 2-D corner flow in the mantle wedge as represented in Figs. 8 and 9. This is indicative of strong coupling between the downgoing slab and the asthenospheric mantle around it. It is also interesting that the N–S pattern of the anisotropic fast axes is consistent for stations both above the steeply dipping slab and farther north where no slab has been seismically imaged (Figs. 5a and 8). This interpretation is also consistent with the Rayleigh wave anisotropy results of STUBAILO et al. (2012), their Fig. 13 bottom and their Fig. 15c, which sample the asthenosphere and were explained by wedge flow.

5. Conclusions

Shear wave splitting measurements were made using SKS and SKKS phases recorded by an array deployed in southern Mexico. The roughly linear profile ran from the coast of the Pacific Ocean to the north, through the Trans-Mexican Volcanic Belt, and farther north close to the Gulf of Mexico. The oceanic Cocos plate subducts under the continental North American plate at the Middle America Trench, but the TMVB is not subparallel to the MAT. In the fore-arc the Cocos slab undergoes flat subduction; however, it dips steeply under the TMVB. The observed anisotropy pattern is controlled by the slab geometry. For a subset of data using only SKS observations, the fast polarization directions are oriented NE–SW under the fore-arc, in the same direction as the relative motion between the Cocos and North American plates, and nearly perpendicular to the MAT. Physical conditions in the subslab mantle are low stress, low water content, and relatively high temperature which are conducive to the development of A-type olivine fabric. It is thus inferred that mantle flow is entrained under the subhorizontal slab and also that the slab and the mantle are strongly coupled. The results from this study were compared against several of the models that have been proposed to explain subslab observations of seismic anisotropy. Our observations are consistent with the model by LONG and SILVER (2008, 2009) because the fast axes are trench-normal where the subducting slab is young and hot and no decoupling between the slab and the underlying mantle is expected. The results from the present study are consistent with the model by SONG and KAWAKATSU (2012) because the slab is subhorizontal and the fast polarization directions align trench-perpendicular. The results from our study are generally consistent with the numerical models by PACZKOWSKI et al. (2014a, b) in the sense that the subhorizontal slab does not penetrate into the lower mantle and the subslab fast polarization directions are trench-perpendicular. Unfortunately, given that PACZKOWSKI et al. (2014a, b) did not model a flat slab, a direct comparison with our study area is not possible. Lastly, LYNNER and LONG (2014a, b) noticed that under young lithosphere strong coupling of the slab with a thick mantle layer drives entrained flow under A-type olivine conditions. Our results are consistent with their model.

In the back-arc, where the Cocos slab subducts steeply, SKS fast axes are oriented N–S. This direction is perpendicular to the strike of the slab. In the mantle wedge core low stress, low water content, and relatively high temperatures are expected and are consistent with the development of A-type olivine. In the mantle wedge tip high temperatures and a hydrated mantle are consistent with the existence of C-type olivine fabric. For both, A- and C-type olivine, seismic fast axes align in the direction of mantle flow and so corner flow in the mantle wedge is inferred.

Acknowledgments

We are thankful to Manuel Velásquez for computer support; Rob Clayton, Vlad Manea, and Marina Manea for discussions and suggestions. We are also thankful to Xyoli Pérez-Campos, Rob Clayton, Arturo Iglesias, Shri Krishna Singh, Paul Davis, and Allen Husker for access to the MASE data; and also to all the volunteers who contributed their time for field work. We are thankful to Karen Fischer for providing the computer code used in the early stages of this project to measure the splitting parameters. The suggestions made by two anonymous reviewers greatly enriched the manuscript. One of us (GLS) received a postdoctoral fellowship from Mexico's Consejo Nacional de Ciencia y Tecnología for work at Centro de Sismología y Volcanología de Occidente, Universidad de Guadalajara. This work was funded by Universidad Nacional Autónoma de México through Programa de Apoyo a Proyectos de Investigación e Innovación Tecnológica, PAPIIT grant IN112814. The MASE experiment was supported by the Tectonics Observatory at the California Institute of Technology and by the Center for Embedded Network Sensors (CENS) at the University of California Los Angeles. The MASE experiment was funded by the Gordon and Betty Moore Foundation. The maps and figures in this study were made using the Generic Mapping Tools package (WESSEL and SMITH 1998).

REFERENCES

ABT, D.L., FISCHER, K.M., ABERS, G.A., STRAUCH, W., PROTTI, J.M., and GONZÁLEZ, V. (2009), *Shear wave anisotropy beneath Nicaragua and Costa Rica: Implications for flow in the mantle wedge*, Geochem. Geophys. Geosyst. *10*, Q05S15, doi:10.1029/2009GC002375.

ABT, D.L., FISCHER, K.M., ABERS, G.A., PROTTI, M., GONZÁLEZ, V., and STRAUCH, W. (2010), *Constraints on upper mantle anisotropy surrounding the Cocos slab from SK(K)S splitting*, J. Geophys. Res. *115*, B06316, doi:10.1029/2009JB006710.

ATWATER, T. and STOCK, J. (1991). *Pacific North America plate tectonics of the Neogene southwestern United States: An update*, Int Geol Rev *40*, 375 – 402.

AUDET, P. (2013), *Seismic anisotropy of subducting oceanic uppermost mantle from fossil spreading*, Geophys. Res. Lett. *40*, 173-177, doi:10.1029/2012GL054328.

AUDOINE, E., SAVAGE, M.K. and GLEDHILL K. (2004). *Anisotropic structure under a back arc spreading region, the Taupo volcanic zone, New Zealand*, J Geophys Res *109*. doi:10.1029/2003JB02932.

BACCHESCHI, P., MARGHERITI, L., and STECKLER, M.S. (2007). *Seismic anisotropy reveals focused mantle flow around the Calabrian slab (Southern Italy)*, Geophys Res Lett *34*, L05302.

BERNAL-DÍAZ, A., VALENZUELA-WONG, R., PÉREZ-CAMPOS, X., IGLESIAS, A., and CLAYTON, R.W. (2008), *Anisotropía de la onda SKS en el manto superior debajo del arreglo VEOX (abstract), Geos Boletín Informativo de la UGM 28 (2), 199-200.

BERNAL-LÓPEZ, L.A. (2015), *Anisotropía sísmica y flujo del manto producidos por la placa de Cocos subducida en el sur de México*, M. Sc. thesis, 65 pp., Centro de Sismología y Volcanología de Occidente, Universidad de Guadalajara, Puerto Vallarta, Jal., Mexico.

CASTILLO-CASTELLANOS, J.A. (2015), *Variaciones de la anisotropía sísmica en la corteza y manto superior en el centro-sur de*

México, M.Sc. thesis, 147 pp., Instituto de Geofísica, Universidad Nacional Autónoma de México, Mexico City, Mexico.

CASTILLO, J.A., PÉREZ-CAMPOS, X., HUSKER, A.L., and VALENZUELA-WONG, R. (2014), *Crust and mantle anisotropy variations from the coast to inland in central and southern Mexico*, Abstract DI33A-4303 presented at 2014 Fall Meeting, AGU, San Francisco, CA, 15-19 December.

CASTRO-ARTOLA, O.A. (2010), *Caracterización de la geometría de la zona Benioff con una red densa de banda ancha en el Istmo de Tehuantepec*, B.Sc. thesis, 65 pp., Facultad de Ingeniería, Universidad Nacional Autónoma de México, Mexico City, Mexico.

CURRIE, C.A., CASSIDY, J.F., HYNDMAN, R. and BOSTOCK, M.G. (2004). *Shear wave anisotropy beneath the Cascadia subduction zone and western North American craton*, Geophys J Int *157*, 341-353.

DEMETS, C., and TRAYLEN, S. (2000), *Motion of the Rivera plate since 10 Ma relative to the Pacific and North American plates and the mantle*, Tectonophysics *318*, 119-159.

DEMETS, C., GORDON, R.G., and ARGUS, D.F. (2010), *Geologically current plate motions*, Geophys. J. Int. *181*, 1-80.

DI LEO, J.F., WOOKEY, J., HAMMOND, J.O.S., KENDALL, J.-M., KANESHIMA, S., INOUE, H., YAMASHINA, T., and HARJADI, P. (2012a). *Deformation and mantle flow beneath the Sangihe subduction zone from seismic anisotropy*, Phys Earth Planet Inter *194-195*, 38-54. doi:10.1016/j.pepi.2012.01.008.

DI LEO, J.F., WOOKEY, J., HAMMOND, J.O.S., KENDALL, J.-M., KANESHIMA, S., INOUE, H., YAMASHINA, T., and HARJADI P. (2012b). *Mantle flow in regions of complex tectonics: Insights from Indonesia*, Geochem Geophys Geosyst *13*, Q12008. doi:10.1029/2012GC004417.

EAKIN, C.M., OBREBSKI, M., ALLEN, R.M., BOYARKO, D.C., BRUDZINSKI, M.R., and PORRITT, R.(2010), *Seismic anisotropy beneath Cascadia and the Mendocino triple junction: Interaction of the subducting slab with mantle flow*, Earth Planet. Sci. Lett. *297*, 627-632.

FERRARI, L., OROZCO-ESQUIVEL, T., MANEA, T. and MANEA, V.C. (2012). *The dynamic history of the Trans-Mexican Volcanic Belt and the Mexico subduction zone*, Tectonophysics *522-523*, 122 – 149.

FOLEY, B., and LONG, M.D. (2011), *Upper and mid-mantle anisotropy beneath the Tonga slab*, Geophys. Res. Lett.*38*, L02303, doi:10.1029/2010GL046021.

GILL, J.B., *Orogenic andesites and plate tectonics, "Minerals and rocks, vol. 16"* (Springer, Berlin 1981).

GRIPP, A.E., and GORDON, R.G. (2002), *Young tracks of hotspots and current plate velocities*, Geophys. J. Int. *150*, 321-361.

HAMMOND, J.O.S., WOOKEY, J., KANESHIMA, S., INOUE, H., YAMASHINA, T. and HARJADI, P. (2010). *Systematic variation in anisotropy beneath the mantle wedge in the Java-Sumatra subduction system from shear wave splitting*, Phys Earth Planet Inter *178*, 189-201.

HUSKER, A. and DAVIS, P.M. (2009), *Tomography and thermal state of the Cocos plate subduction beneath Mexico City*, J. Geophys. Res. *114*, B04306, doi:10.1029/2008JB006039.

ISMAÏL, W.B., and MAINPRICE, D. (1998), *An olivine fabric database: An overview of upper mantle fabrics and seismic anisotropy*, Tectonophysics *296*, 145-157, doi:10.1016/S0040-1951(98)00141-3.

JUNG, H., KATAYAMA, I., JIANG, Z., HIRAGA, T., and KARATO, S. (2006). *Effect of water and stress on the lattice preferred orientation (LPO) of olivine*, Tectonophysics *421*, 1-22.

KANJORSKI, M.N. (2003), *Cocos plate structure along the Middle America subduction zone off Oaxaca and Guerrero, Mexico: Influence of subducting plate morphology on tectonics and seismicity*, Ph.D. thesis, University of California, San Diego, CA, USA.

KARATO, S.-i., JUNG, H., KATAYAMA, I. and SKEMER, P. (2008). *Geodynamic significance of seismic anisotropy of the upper mantle: New insights from laboratories studies*, Annu. Rev. Earth Planet. Sci. *36*, 59-95.

KAWAKATSU, H., KUMAR, P., TAKEI, Y., SHINOHARA, M., KANAZAWA, T., ARAKI, E., and SUYEHIRO, K. (2009), *Seismic evidence for sharp lithosphere-asthenosphere boundaries of oceanic plates*, Science *324*, 499-502, doi:10.1126/science.1169499.

KIM, Y., CLAYTON, R.W., and JACKSON, J.M. (2010), *Geometry and seismic properties of the subducting Cocos plate in central Mexico*, J. Geophys. Res. *115*, B06310, doi:10.1029/2009JB006942.

KIRÁLY, E., BIANCHI, I., and BOKELMANN, G. (2012). *Seismic anisotropy in the south western Pacific region from shear wave splitting*, Geophys Res Lett *39*, L05302.

KNELLER, E.A, VAN KEKEN, P.E., KARATO, S.-i., and PARK, J. (2005), *B-type olivine fabric in the mantle wedge: Insights from high-resolution non-Newtonian subduction zone models*, Earth Planet. Sci. Lett. *237*, 781-797, doi:10.1016/j.epsl.2005.06.049.

LEÓN SOTO, G., and VALENZUELA, R.W. (2013), *Corner flow in the Isthmus of Tehuantepec, Mexico inferred from anisotropy measurements using local intraslab earthquakes*, Geophys. J. Int. *195*, 1230-1238, doi:10.1093/gji/ggt291.

LEÓN SOTO, G., NI, J.F., GRAND, S.P., SANDVOL, E., VALENZUELA, R.W., GUZMÁN SPEZIALE, M., GÓMEZ GONZÁLEZ, J.M., and DOMÍNGUEZ REYES, T. (2009), *Mantle flow in the Rivera-Cocos subduction zone*, Geophys. J. Int. *179*, 1004-1012, doi:10.1111/j.1365-246X.2009.04352x.

LEVIN, V., DROZNIN, D., PARK, J. and GORDEEV, E. (2004). *Detailed mapping of seismic anisotropy with local shear waves in southeastern Kamchatka*, Geophys J Int *158*, 1009-1023.

LONG, M.D. (2009). *Complex anisotropy in D" beneath the eastern Pacific from SKS-SKKS splitting discrepancies*, Earth Planet Sci. Lett. *283*, 181-189.

LONG, M.D. (2013). *Constraints on subduction geodynamics from seismic anisotropy*, Rev Geophys *51*, 76-112.

LONG, M.D. and SILVER, P.G. (2008). *The subduction zone flow field from seismic anisotropy: A global view*, Science *319*, 315-318.

LONG, M.D., and SILVER, P.G. (2009), *Mantle flow in subduction systems: The subslab flow field and implications for mantle dynamics*, J. Geophys. Res. *114*, B10312, doi:10.1029/2008JB006200.

LONG, M.D. and VAN DER HILST, R.D. (2005). *Upper mantle anisotropy beneath Japan from shear wave splitting*, Phys Earth Planet Inter *151*, 206-222.

LONG, M.D. and VAN DER HILST, R.D. (2006). *Shear wave splitting from local events beneath the Ryukyu arc: Trench parallel anisotropy in the mantle wedge*, Phys Earth Planet Inter *155*, 300-312.

LONG, M.D., and WIRTH, E.A. (2013), *Mantle flow in subduction systems: The mantle wedge flow field and implications for wedge processes*, J. Geophys. Res. Solid Earth *118*, 583–606, doi:10.1002/jgrb.50063.

LYNNER, C., and LONG, M.D. (2013), *Sub-slab seismic anisotropy and mantle flow beneath the Caribbean and Scotia subduction zones: Effects of slab morphology and kinematics*, Earth Planet. Sci. Lett. *361*, 367-378, doi:10.1016/j.epsl.2012.11.007.

LYNNER, C., and LONG, M.D. (2014a), *Sub-slab anisotropy beneath the Sumatra and circum-Pacific subduction zones from source-side shear wave splitting observations*, Geochem. Geophys. Geosyst. *15*, 2262–2281, doi:10.1002/2014GC005239.

LYNNER, C., and LONG, M.D. (2014b), *Testing models of sub-slab anisotropy using a global compilation of source-side shear wave splitting data*, J. Geophys. Res. Solid Earth *119*, 7226–7244, doi:10.1002/2014JB010983.

MACÍAS, J.L. (2005), *Geología e historia eruptiva de algunos de los grandes volcanes activos de México*, Bol. Soc. Geol. Mex. *LVII*, 379-424.

MANEA, M., and MANEA, V.C. (2008), *On the origin of El Chichón volcano and subduction of Tehuantepec Ridge: A geodynamical perspective*, J. Volcanol. Geoth. Res. *175*, 459-471, doi:10.1016/j.volgeores.2008.02.028.

MANEA, V.C., and MANEA, M. (2011), *Flat-slab thermal structure and evolution beneath central Mexico*, Pure Appl. Geophys. *168*, 1475-1487, doi:10.1007/s00024-010-0207-9.

MANEA, V., MANEA, M., KOSTOGLODOV, V., and SEWELL, G. (2006), *Intraslab seismicity and thermal stress in the subducted Cocos plate beneath central Mexico*, Tectonophysics *420*, 389-408, doi:10.1016/j.tecto.2006.03.029.

MASE (2007), *Meso America subduction experiment, Caltech, dataset, Pasadena, CA, USA*, doi:10.7909/C3RN35SP.

ÖZALAYBEY, S., and SAVAGE, M.K. (1994), *Double-layer anisotropy resolved from S phases*, Geophys. J. Int. *117*, 653-664.

ÖZALAYBEY, S., and SAVAGE, M.K. (1995), *Shear-wave splitting beneath western United States in relation to plate tectonics*, J. Geophys. Res. *100*, 18,135-18,149.

PACHECO, J.F., and SINGH, S.K. (2010), *Seismicity and state of stress in Guerrero segment of the Mexican subduction zone*, J. Geophys. Res. *115*, B01303, doi:10.1029/2009JB006453.

PACZKOWSKI, K., MONTÉSI, L.G.J., LONG, M.D., and THISSEN, C.J. (2014a), *Three-dimensional flow in the subslab mantle*, Geochem. Geophys. Geosyst. *15*, 3989-4008, doi:10.1002/2014GC005441.

PACZKOWSKI, K., THISSEN, C.J., LONG, M.D., and MONTÉSI, L.G.J. (2014b), *Deflection of mantle flow beneath subducting slabs and the origin of subslab anisotropy*, Geophys. Res. Lett. *41*, 6734-6742, doi:10.1002/2014GL060914.

PARDO, G., and SUÁREZ, M. (1995). *Shape of the subducted Rivera and Cocos plates in southern Mexico: Seismic and tectonic implications*, J Geophys Res *100*, 12,3357 – 12,373.

PAYERO, J.S., KOSTOGLODOV, V., SHAPIRO, N., MIKUMO, T., IGLESIAS, A., PÉREZ-CAMPOS, X. and CLAYTON, R.W. (2008). *Nonvolcanic tremor observed in the Mexican subduction zone*, J Geophys Res *35*, L07305.

PEYTON, V., LEVIN, V., PARK, J., BRANDON, M., LEES, J., GORDEEV, E., and OZEROV, A. (2001). *Mantle flow at a slab edge: Seismic anisotropy in the Kamchatka region*, Geophys Res Lett *28*, 379-382.

PÉREZ-CAMPOS, X., KIM, Y., HUSKER, A., DAVIS, P., CLAYTON, R., IGLESIAS, A., PACHECO, J.F., SINGH, S.K., MANEA, V.C. and GURNIS, M. (2008). *Horizontal subduction and truncation of the Cocos plate beneath Central Mexico*, Geophys Res Lett *35*, L18303.

PERTTU, A., CHRISTENSEN, D., ABERS, G., and SONG, X. (2014). *Insights into the mantle structure and flow beneath Alaska based on a decade of observations of shear wave splitting*, J Geophys Res *119*, 8366-8377.

PHIPPS MORGAN, J., HASENCLEVER, J., HORT, M., RÜPKE, L., and PARMENTIER, E.M. (2007), *On subducting slab entrainment of buoyant asthenosphere*, Terra Nova *19*, 167-173, doi:10.1111/j.1365-3121.2007.00737.x.

PIÑERO-FELICANGELI, L., and KENDALL, J.-M. (2008). *Sub-slab mantle flow parallel to the Caribbean plate boundaries: Inferences from SKS splitting*, Tectonophysics *462*, 22-34.

PONCE-CORTÉS, J.G. (2012), *Medición de la anisotropía de las ondas SKS en el manto superior, debajo de las estaciones permanentes del Servicio Sismológico Nacional instaladas a partir del año 2005*, B.Sc. thesis, 79 pp., Facultad de Ingeniería, Universidad Nacional Autónoma de México, Mexico City, Mexico.

RADIGUET, M., COTTON, F., VERGNOLLE, M., CAMPILLO, M., WALPERSDORF, A., COTTE, N., and KOSTOGLODOV, V. (2012). *Slow slip events and strain accumulation in the Guerrero gap, Mexico*, J Geophys Res *117*, B04305.

RODRÍGUEZ-PÉREZ, Q. (2007), *Estructura tridimensional de velocidades para el sureste de México, mediante el análisis de trazado de rayos sísmicos de sismos regionales*, M.Sc. thesis, 83 pp., Instituto de Geofísica, Universidad Nacional Autónoma de México, Mexico City, Mexico.

ROJO-GARIBALDI, B. (2011), *Anisotropía de las ondas SKS en el manto superior debajo de un arreglo sísmico entre Guerrero y Veracruz*, B. Sc. thesis, 84 pp., Facultad de Ciencias, Universidad Nacional Autónoma de México, Mexico City, Mexico.

RUSSO, R.M. (2009). *Subducted oceanic asthenosphere and upper mantle flow beneath the Juan de Fuca slab*, Lithosphere *1*, 195-205.

RUSSO, R.M., and SILVER, P.G. (1994). *Trench-parallel flow beneath the Nazca plate from seismic anisotropy*, Science *263*, 1105-1111.

SAVAGE, M.K. (1999), *Seismic anisotropy and mantle deformation: What have we learned from shear wave splitting?*, Rev. Geophys. *37*, 65-106.

SILVER, P.G. (1996), *Seismic anisotropy beneath the continents: Probing the depths of Geology*, Annu. Rev. Earth Planet. Sci. *24*, 385-432.

SILVER, P.G., and CHAN, W.W. (1991). *Shear wave splitting and subcontinental mantle deformation*, J Geophys R *96*, 16429 – 16454.

SILVER, P.G., and KANESHIMA, S. (1993), *Constraints on mantle anisotropy beneath Precambrian North America from a transportable teleseismic experiment*, Geophys. Res. Lett. *20*, 1127-1130.

SILVER, P.G., and SAVAGE, M.K. (1994), *The interpretation of shear-wave splitting parameters in the presence of two anisotropic layers*, Geophys. J. Int. *119*, 949-963.

SONG, T.-R.A., and KAWAKATSU, H. (2012), *Subduction of oceanic asthenosphere: Evidence from sub-slab seismic anisotropy*, Geophys. Res. Lett. *39*, L17301, doi:10.1029/2012GL052639.

SONG, T.-R.A., and KIM, Y. (2012a), *Anisotropic uppermost mantle in young subducted slab underplating central Mexico*, Nat. Geosci. *5*, 55-59, doi:10.1038/ngeo1342.

SONG, T.-R.A., and KIM, Y. (2012b), *Localized seismic anisotropy associated with long-term slow-slip events beneath southern Mexico*, Geophys. Res. Lett. *39*, L09308, doi:10.1029/2012GL051324.

STUBAILO, I. (2015), *Seismic anisotropy below Mexico and its implications for mantle dynamics*, Ph.D. thesis, 119 pp., University of California, Los Angeles, CA, USA.

STUBAILO, I., and DAVIS, P. (2007), *Shear wave splitting measurements and interpretation beneath Acapulco-Tampico transect in*

Mexico, Eos Trans. AGU *88* (52), Fall Meet. Suppl. Abstract T51B-0539.

STUBAILO, I., and DAVIS, P.M. (2012a), *Anisotropy of the Mexico subduction zone based on shear-wave splitting analysis (abstract)*, Seism. Res. Lett. *83* (2), 379.

STUBAILO, I., and DAVIS, P.M. (2012b), *Anisotropy of the Mexico subduction zone based on shear-wave splitting and higher modes analysis*, Abstract T11A-2538 presented at 2012 Fall Meeting, AGU, San Francisco, CA, 3-7 December.

STUBAILO, I., and DAVIS, P.M. (2015), *The surface wave, shear wave splitting, and higher mode seismic anisotropy comparison of the Mexican subduction zone (abstract)*, Seism. Res. Lett. *86* (2B), 677.

STUBAILO, I., BEGHEIN, C., and DAVIS, P.M. (2012), *Structure and anisotropy of the Mexico subduction zone based on Rayleigh-wave analysis and implications for the geometry of the Trans-Mexican Volcanic Belt*, J. Geophys. Res. *117*, B05303, doi:10.1029/2011JB008631.

SUAREZ, G., and SINGH, S.K. (1986), *Tectonic interpretation of the Trans-Mexican Volcanic BeltDiscussion*, Tectonophysics *127*, 155-158.

SYRACUSE, E.M., and ABERS, G.A. (2006), *Global compilation of variations in slab depth beneath arc volcanoes and implications*, Geochem. Geophys. Geosyst. *7*, Q05017, doi:10.1029/2005GC001045.

TATSUMI, W., and EGGINS, S., *Subduction Zone Magmatism, "Frontiers in Earth Science"* (Blackwell Science, Cambridge, MA, USA1995).

VAN BENTHEM, S.A.C.(2005), *Anisotropy and flow in the upper-mantle under Mexico*, M. Sc. thesis, 41 pp., Utrecht University, Utrecht, The Netherlands.

VAN BENTHEM, S.A.C., VALENZUELA, R.W., and PONCE, G.J. (2013), *Measurements of shear wave anisotropy from a permanent network in southern Mexico*, Geofís. Int. *52*, 385–402, doi:10.1016/S0016-7169(13)71485-5.

WESSEL, P., and SMITH, W.H.F. (1998), *New, improved version of Generic Mapping Tools released*, Eos Trans. AGU *79*, 579.

WIRTH, E., and LONG, M.D. (2010). *Frequency-dependent shear wave splitting beneath the Japan and Izu-Bonin subduction zones*, Phys Earth Planet Inter *181*, 141-154. doi:10.1016/j.pepi.2010.05.006.

WOLFE, C.J., and SILVER, P.G. (1998). *Seismic anisotropy of oceanic upper mantle: Shear wave splitting methodologies and observations*, J Geophys R *103*, 749-771.

WOOKEY, J., KENDALL, J.-M., and RUMPKER, G. (2005). *Lowermost mantle anisotropy beneath the north Pacific from differential S-ScS splitting*, Geophys J Int *161*, 829-838.

(Received August 30, 2015, revised November 17, 2015, accepted November 22, 2015, Published online December 21, 2015)

Pure Appl. Geophys. 173 (2016), 3395–3417
© 2015 Springer International Publishing
DOI 10.1007/s00024-015-1218-3

Pure and Applied Geophysics

Toroidal, Counter-Toroidal, and Upwelling Flow in the Mantle Wedge of the Rivera and Cocos Plates: Implications for IOB Geochemistry in the Trans-Mexican Volcanic Belt

Florian Neumann,[1] Alberto Vásquez-Serrano,[2] Gustavo Tolson,[3] Raquel Negrete-Aranda,[4] and Juan Contreras[4]

Abstract—We carried out analog laboratory modeling at a scale 1:4,000,000 and computer rendering of the flow patterns in a simulated western Middle American subduction zone. The scaled model consists of a transparent tank filled with corn syrup and housing two conveyor belts made of polyethylene strips. One of the strips dips 60° and moves at a velocity of 30 mm/min simulating the Rivera plate. The other one dips 45°, moves at 90 mm/min simulating the subduction of the Cocos plate. Our scaled subduction zone also includes a gap between the simulated slabs analogous to a tear recently observed in shear wave tomography studies. An acrylic plate 3 mm thick floats on the syrup in grazing contact with the polyethylene strips and simulates the overriding North America plate. Our experiments reveal a deep toroidal flow of asthenospheric mantle through the Cocos–Rivera separation. The flow is driven by a pressure gradient associated with the down-dip differential-motion of the slabs. Similarly, low pressure generated by the fast-moving Cocos plate creates a shallow counter-toroidal flow in the uppermost 100 km of the mantle wedge. The flow draws mantle beneath the western Trans-Mexican Volcanic Belt to the Jalisco block, then plunges into the deep mantle by the descending poloidal cell of the Cocos slab. Moreover, our model suggests a hydraulic jump causes an ~250 km asthenosphere upwelling around the area where intra-arc extensional systems converge in western Mexico. The upwelling eventually merges with the shallow counter-toroidal flow describing a motion in 3D space similar to an Archimedes' screw. Our results indicate the differential motion between subducting slabs drives mixing in the mantle wedge of the Rivera plate and allows the slab to steepen and retreat. Model results are in good agreement with seismic anisotropy studies and the geochemistry of lavas erupted in the Jalisco block. The model can explain the eruption of OIB lavas in the vicinity of the City of Guadalajara in western Mexico, and the south shoulder in the central part of the Tepic-Zacoalco fault system.

[1] CICESE, Ensenada, BC, Mexico. E-mail: fneumann@cicese.edu.mx
[2] Instituto de Geologia, UNAM, Mexico DF, Mexico. E-mail: alberto-vasquez@ciencias.unam.mx
[3] Departamento de Geologia, Instituto de Geologia, UNAM, Mexico DF, Mexico. E-mail: tolson@unam.mx
[4] Departamento de Geologia, CICESE, Ensenada, BC, Mexico. E-mail: rnegrte@cicese.mx; juanc@cicese.mx

Key words: Subduction zone, Trans-Mexican volcanic belt, TMVB, slab breakoff.

1. Introduction

In this paper we detail results of laboratory simulations of mantle flow beneath the arc and forearc of the Middle America Subduction Zone (MASZ), western Mexico (Fig. 1). With the help of this model we thoroughly investigate a proposed connection between a tear in the subducted slab imaged by tomographic models and the occurrence of magmatic rocks with an oceanic island basalt (OIB) affinity in the western Trans Mexican Volcanic Belt (TMVB) (Fig. 2; Ferrari 2004; Yang et al. 2009; Soto et al. 2009; Suhardja 2013). Indeed SKS splitting measurements around the MASZ, which convey information about shallow mantle streams, hint that enriched asthenosphere entered through the tear by means of toroidal flow (Fig. 3, Soto et al. 2009). This resulted, according to other authors, in mantle mixing and the development of compositional heterogeneities in the mantle wedge of the Rivera-North America subduction zone (Fig. 4; Ferrari 2004; Yang et al. 2009; Soto et al. 2009). Hence, it appears that the creation of the tear has the potential to unveil the elusive origin of the OIB observed along the axis of the Tepic-Zacoalco rift (Fig. 1). We will see, with the aid of our new analog simulations, that this model reconciles quite well the plate kinematics, mantle flow, and petrologic observations.

We further use our analog model to examine the driving force behind the postulated toroidal flow. As illustrated in Fig. 5, current conceptual models call

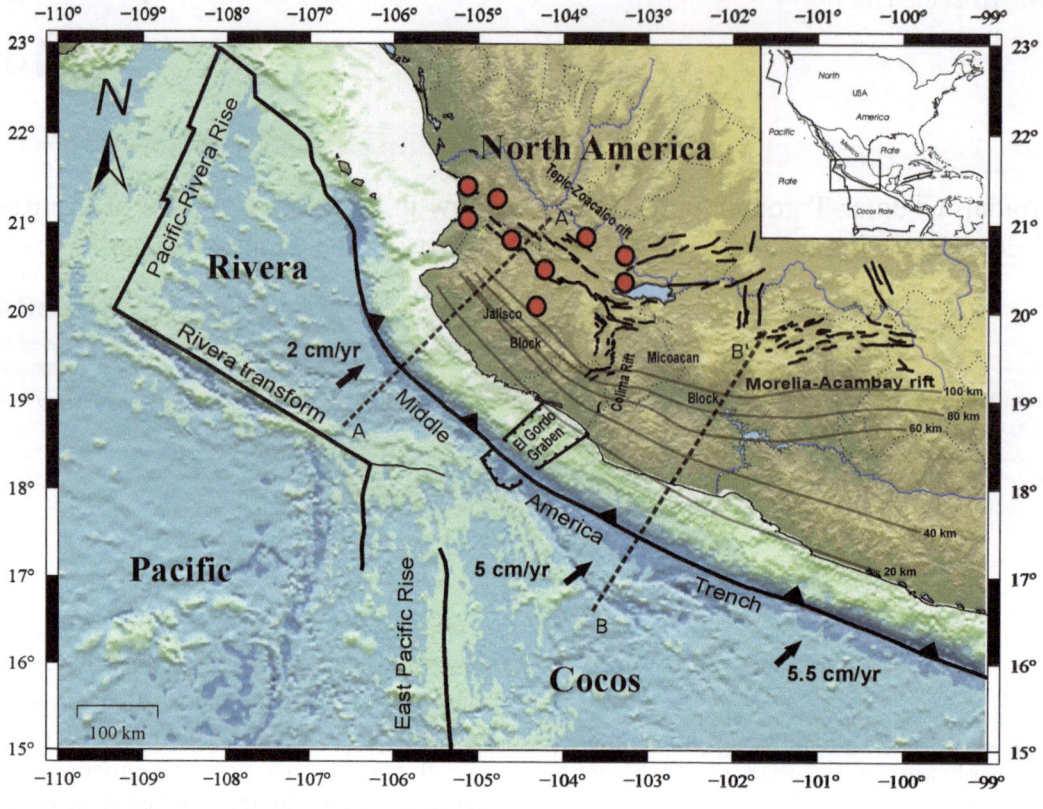

Figure 1

Tectonic configuration and major fault zones in southwestern Mexico. This is a region of complex geodynamics in which the young Rivera and Cocos plates subduct under the North America plate at different rates and angles. As explained in the main text, this causes flexure of the upper plate resulting in intra-arc extension forming a series of poorly linked fault systems. The Colima rift splits the forearc of the western Middle America subduction zone into two fault-bounded terrains denominated the Jalisco and Michoacán blocks. *Contours* indicate the depth in km of the Wadati-Benioff zone (PARDO and SUAREZ 1995). *Arrows* show direction and relative convergence rate of the Rivera and Cocos plate with respect to stable North America (from DEMETS and WILSON 1997). *Red circles* indicate the location of oceanic island basalt flows

for ~50 km seaward retreat of the Rivera slab during the late Miocene. Slab rollback, according to the model, induced the formation of two toroidal flow cells which transported asthenosphere mantle around the northern lateral slab edge and the gash in the south (FERRARI 2004; SOTO et al. 2009). Analog experiments validate this view since they have shown that fluid further from the slab edge moves around the plate in a downward sweeping arc; this material is later drawn into the wedge and towards the plate along a slight upward circulation flow (KINCAID and GRIFFITHS 2003; SCHELLART 2004). Thus, it is clear enough that lateral flow was allowed around the northern slab edge during rollback, which may also have provided a mechanism for transporting geochemically distinct mantle from the ocean side of the

slab into the wedge and producing anomalous melting patterns. It is not clear, however, how this happened in the southern edge where the gap is only 100 km wide; the problem is that fluid motion in the vicinity of retreating plate edges is dominated by shear flow (Fig. 5). As can be appreciated in this figure, the parcels of fluid in this flow regime move horizontally around the edges and then downward owing to viscous drag against the downgoing plate (KINCAID and GRIFFITHS 2003). Thus, the combined edge effects of both slabs would have hindered thermal or thermochemical ascent of asthenospheric mantle (e.g., TURNER and HAWKESWORTH 1998). Moreover, decompression melting in the mantle wedge requires an upward component of motion but rollback subduction results in flow trajectories that tend to be horizontal (KINCAID and GRIFFITHS 2003; SCHELLART 2004). A further problem in the hypothesis set forward by the previous authors (FERRARI 2004; YANG et al. 2009; SOTO et al. 2009) is that at the extremely low retreat rate of ~5 mm/year the toroidal flow must have been too weak to cause a significant advection of mantle from the ocean side of the slab.

The model used here makes use of materials that approximate reasonably well the rheology of the mantle. Moreover, boundary conditions in the experiments are consistent with the plate kinematics and tomographic images. Computer rendering of pathlines described by embedded particles in our experimental method bring to light complex toroidal, shallow counter-toroidal, and upwelling mantle flow patterns in the mantle wedge of the MASZ. The issues that will be addressed by this paper are the following: (1) how the tear and edgewise differential motion between the subducting slabs perturb the flow of the northern MASZ, (2) how they affect mantle mixing and OIB geochemistry, and (3) did rollback of the Rivera plate play a role in mantle mixing as has been proposed in the literature?

2. Geodynamics of the Northern Middle America Subduction Zone

Southwestern Mexico is a region of complex intra-arc deformation and oceanic lithosphere microplate activity (Fig. 1; STOCK and LEE 1994; LONSDALE 2005; MANN 2007; DEMETS and TRAYLEN 2000). In the following paragraphs we summarize what is presently known about the region's geodynamics since the Miocene. To begin with, deformation in the continental interior is characterized by a series of active, poorly linked fault systems arranged parallel to the axis of the volcanic arc causing limited intra-arc extension in an overall N–S direction (Fig. 1, SUTER et al. 1995, 2001; FERRARI and ROSAS-ELGUERA 1999). Even though individual fault segments seldom exceed 30 km in length and hundreds of meters of vertical displacement, these fault systems are often referred to as rifts, namely, the Tepic-Zacoalco rift, the Morelia-Acambay rift, and the normal-to-trench Colima rift (Fig. 1). These shear zones, in turn, bound two tectonic terrains: the Jalisco

and Michoacán blocks, lodged in the forearc of the Rivera-North America and Cocos-North America subduction zones, respectively.

The previously discussed field relations together with geochemical observations discussed below, have led some authors to propose the incipient rifting of the Jalisco and Michoacán blocks from the Mexican mainland since Pliocene times (e.g., LUHR et al. 1985; ALLAN 1986; MARQUEZ et al. 1999; VERMA 2002, 2009). However, as discussed by CONTRERAS (2013), the notion of western Mexico undergoing lithospheric extension conflicts with the state of stress of the highly coupled MASZ. The idea is inconsistent with the motion of North America, which has been advancing against the Rivera and Cocos plates at a rate of 8 cm/yr during the last ~10 Ma (DEMETS and TRAYLEN 2000). Modeling studies and seismic observations worldwide demonstrate that upper plate advance induces compression in the over-riding plate (e.g., VAN HUNEN et al. 2002, 2004; HEURET et al. 2007, SCHOLZ and CAMPOS 1995, 2012). In contrast, important clues about the origin of the extensional phase come from the careful analysis of the timing of faulting: fault activity appears to be intimately connected with periods of abrupt changes in convergence rate (FERRARI and ROSAS-ELGUERA 1999; DEMETS and TRAYLEN 2000; SUTER et al. 2001). CONTRERAS (2013) showed using a thermo-mechanical model that the periods of acceleration in convergence rate caused a state of unbalanced forces in the mantle wedge that, in turn, resulted in the bending of the overriding plate and the consequent brittle failure of the arc.

On the other hand, the Rivera and Cocos plates subduct at different angles and rates underneath the western edge of North America (Fig. 1; LUHR et al. 1985; PARDO and SUAREZ 1995; DEMETS and WILSON 1997; DEMETS and TRAYLEN 2000; YANG et al. 2009). The Rivera plate dips at an angle of ~60° under the forearc of the MASZ and moves at a rate of ~2 cm/year toward North America relative to the Pacific Plate. In contrast, the Cocos plate plunges into the mantle at an angle of ~45° and moves at a rate of ~5.5 cm/year near El Gordo graben, offshore Colima (Fig. 1). The differential motion of ~2.5 cm/year between the Rivera and Cocos plates is distributed along a north-trending shear zone ~50-km

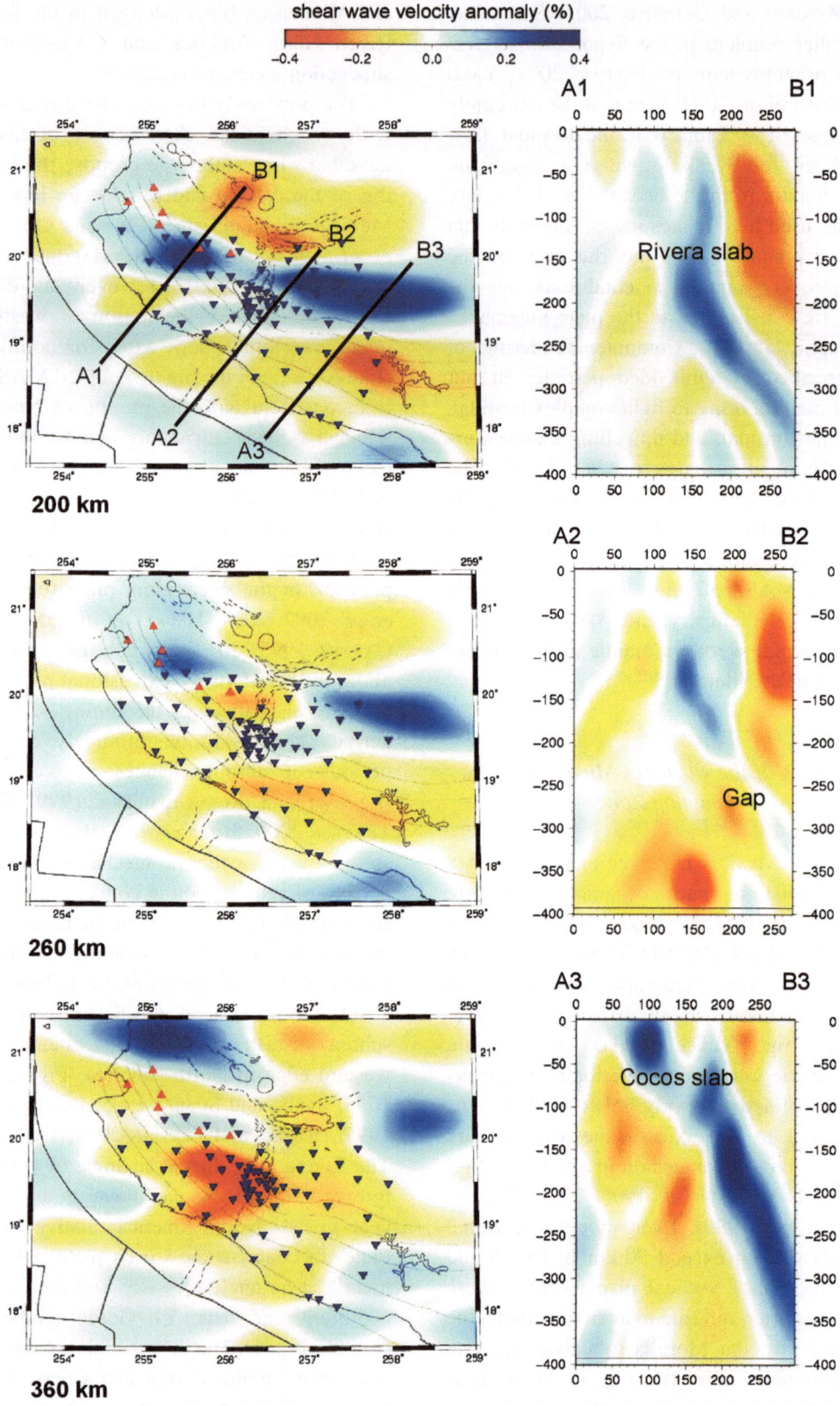

shear wave velocity anomaly (%)

Figure 2

Shear-wave velocity model of the mantle structure beneath the Jalisco and Michoacán blocks (SUHARDJA 2013). The *left column* displays velocity maps for different mantle depths. The *right column* corresponds to velocity profiles across the western Middle America subduction zone. Notice the oceanic slab looses continuity for depths in excess of ~250 km under the Colima graben, which appears to be a gap or gash between the slabs

wide connecting the Manzanillo spreading segment with El Gordo graben (DEMETS and WILSON 1997).

Recent results from the MARS and CODEX seismic experiments suggest the tangential velocity discontinuity between the plates correlates with a narrow zone of prominent low shear wave velocity located at a depth of ~200–300 km under the

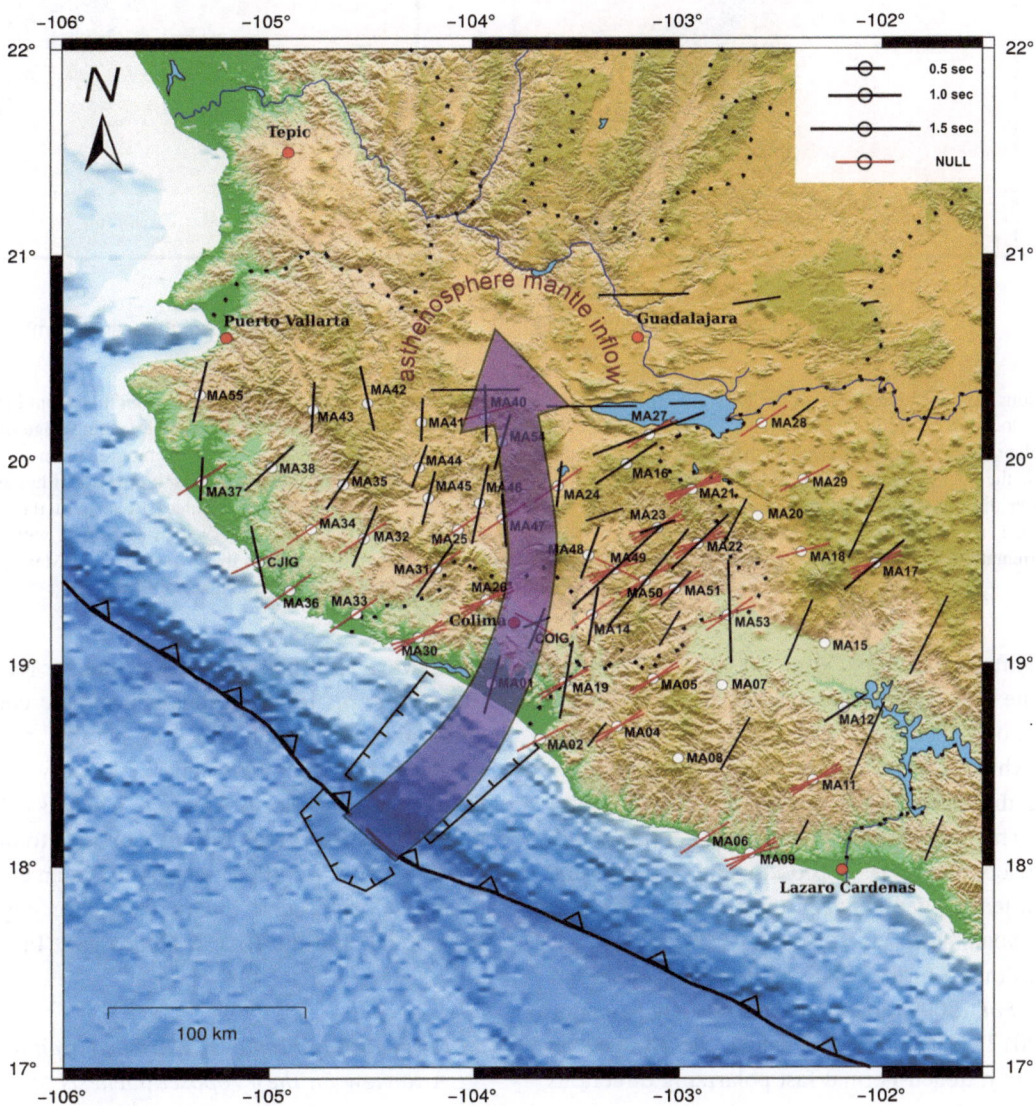

Figure 3

Results of SKS and SKKS shear wave splitting orientations obtained from the MARS experiment (SOTO *et al.* 2009; STUBAILO *et al.* 2012). *Four-letter labels* are station codes. *Black bars* indicate weighted average of fast polarization direction. *Bar length* is proportional to time delay; the longer the time delay, the more pronounced the anisotropy of the shallow mantle under the station. *Red bars* represents null measurements. Observe splitting directions under the forearc of the Michoacán block are trench-normal indicating arc corner flow is strongly controlled by the sinking slab. In contrast, splitting directions under the forearc of the Jalisco block are trench-oblique, suggesting an inflow of oceanic mantle through the gap between the slab

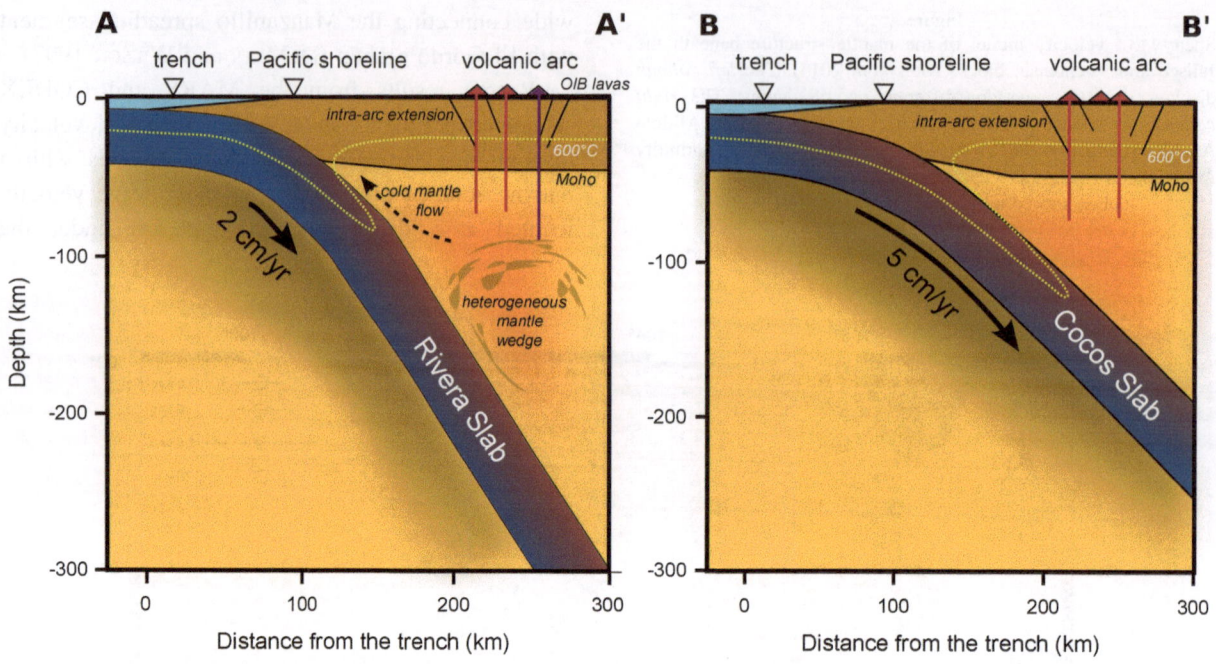

Figure 4

Cross sections of the mantle wedge of the Rivera-North America and Cocos-North America subduction zones (modified from FERRARI *et al.* 2012). The location of the profiles is shown in Fig. 1. Notice the thinning of the continental crust of near the leading edge of the North America plate, which has been recently imaged by magnetotelluric surveys (CORBO-CAMARGO *et al.* 2013) Tomographic models shown in Fig. 2 show the mantle wedge of the Middle America subduction has negative velocity anomalies suggesting it consists of hot and buoyant material. On the other hand, the presence of oceanic island basalts along the axis of the Tepic-Zoacalco rift suggest the presence of compositional heterogeneities in the mantle wedge of the Middle America subduction zone. This theory is in agreement with shear wave splitting orientations shown in Fig. 3. In contrast, the homogenous composition of magmatic materials in the central Trans-Mexican Volcanic Belt suggest a relatively compositionally uniform mantle wedge under the Michoacán block

Colima Rift (Fig. 2; YANG *et al.* 2009; GARDINE *et al.* 2007). The low velocity anomaly separates two high-velocity zones beneath the Jalisco and Michoacán blocks, which are consistent with the location at depth of the slabs of the Rivera and Cocos plates (Fig. 2). The tomographic models indicate the width of the discontinuity is ∼100 km and suggest that there is a tear or gap between the slabs (YANG *et al.* 2009; GARDINE *et al.* 2007; SUHARDJA 2013).

According to further SKS splitting measurements by SOTO *et al.* (2009) and, more recently, by STUBAILO *et al.* (2012), seismic waves under the Michoacán block exhibit trench-normal fast polarizing directions whereas seismic waves passing under the mantle of the Jalisco block show fast split directions oblique to the trench (Fig. 3). Elsewhere it has been demonstrated that the crystallographic preferred orientation of olivines in peridotites align parallel to the flow of Earth's upper mantle during diffusion creep causing

shear-wave splitting in a similar manner (HESS 1964; TOMMASI *et al.* 1999). It thus appears arc corner flow under the Michoacán block is parallel to the convergence direction between the rapidly subducting Cocos plate and North America, whereas corner flow under the Jalisco block is highly oblique to the trench, which SOTO *et al.* (2009) interpreted as an evidence of toroidal flow around the northwestern edge of the Rivera slab and the Rivera–Cocos gap (Fig. 3).

3. From Mantle Plumes to Enriched Mantle Sources: A Review on the Proposed Genesis for OIB-Like Lavas in the Western Trans Mexican Volcanic Belt

The apparent paradox of the coexistence of magma with different petrologic affinity in continental volcanic arcs has been a constant challenge for geoscientist. The western part of the TMBV

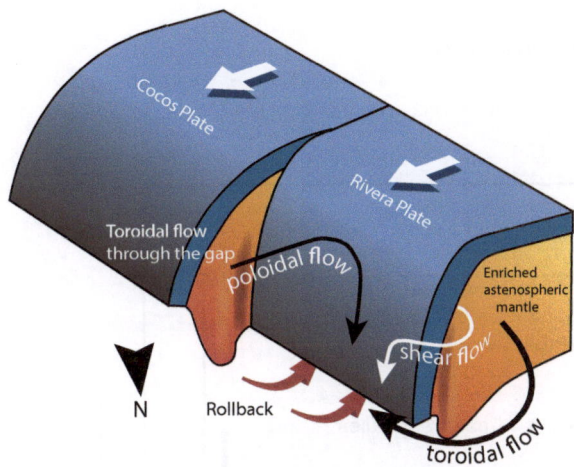

Figure 5

Conceptual model of the mantle flow patterns around the Rivera slab, western Middle America subduction zone. In the model, slab rollback induces the formation of two horizontal toroidal flow cells that advect enriched asthenosphere mantle around the north lateral edge of the slab and through the gap between the slabs in the south. This conceptual model can potentially explain the occurrence of oceanic island basalts in the western Trans-Mexican Volcanic Belt

represents a classic case. The arc is thought to be the expression of the subduction Rivera microplate, yet lavas with the chemical signature of intraplate volcanism have been emplaced in close spatial and temporal association with the more abundant calc-alkaline rocks (Ferrari *et al.* 2001 and all the references therein). Intraplate lavas are usually Na-rich and have been referred to as oceanic-island basalts (OIBs) (Moore *et al.* 1994) or intraplate alkali basalt (Righter 2000). Several authors showed that the trace element and isotopic compositions of these basalts cannot be the result of melting hydrated mantle wedge, motivating the development of a number of theories to explain this apparently contrasting volcanism and OIB peculiar genesis:

1. Based on the rare earth element (REE) patterns displayed by OIB-type basalts in the belt, Moore *et al.* (1994) and Márquez *et al.* (1999) favored a plume-related source arguing a close similarity with the geochemical signatures of well-known mantle plumes (Galápagos, Azores and Hawaii). According to this model a propagating rift opening developed from west to east in response to plume activity. It is important to note that in recent years the "Wilson–Morgan" hypothesis (Wilson 1973)

of hot spots or convection plumes rising from the lower mantle has been used to explain the origin of intraplate-type volcanism in a variety of tectonic settings. The Holocene San Quintín volcanic province in northern Baja California Mexico, constitutes a perfect example of this association since the lavas of this small volcanic field apparently unique within the Baja region, are spinel-lherzolite bearing alkalic basalts and show incompatible trace element abundances similar to OIB (Basu 1975; Storey *et al.* 1989). These authors suggested that alkali basalts from San Quintín share many of the trace element characteristics and EMU and HIMU signatures observed in OIBs consistent with asthenospheric sources and are compositionally indistinguishable from some ocean island, plume-associated basalts such as Hawaii and the Azores. Furthermore, Basu (1975) suggested that San Quintin xenoliths were accidental mantle fragments brought up by their host basalts, a conclusion that clearly favored the diapiric uprise in the mantle as the origin of xenoliths in San Quintín alkali basalts. However, subsequent studies dismissed the idea of real hot spot basalts and instead called the same rocks as high-Nb basalts and Nb-enriched basalts produced by adakite melt metasomatism (Calmus *et al.* 2003, 2011; Aguillón-Robles *et al.* 2001). Based on their geochemical tracers, Castillo (2008) further noted a chemical and isotopic similarity between high-Nb and mildly alkalic intraplate OIB lavas that commonly cap off-axis seamounts and fossil spreading centers in the eastern Pacific. Castillo suggested that these lavas are produced from enriched heterogeneities in the mantle or plums in the pudding (Morris and Hart 1983).

2. Hochstaedter *et al.* (1996) suggested OIB at the TMBV were the result of decompression melting of a heterogeneous upper mantle, resulting from asthenosphere advection into the mantle wedge by subduction poloidal flow (Luhr 1997).

3. Using the same upper mantle source, a comprehensive model for the origin of the Late Miocene mafic episode was proposed by Ferrari (2004) and detailed by Orozco-Esquivel *et al.* (2007) in which an eastward propagating slab detachment at ca. 7 Ma and 70–90 km depth facilitated the influx

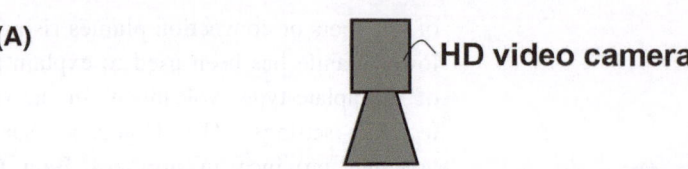

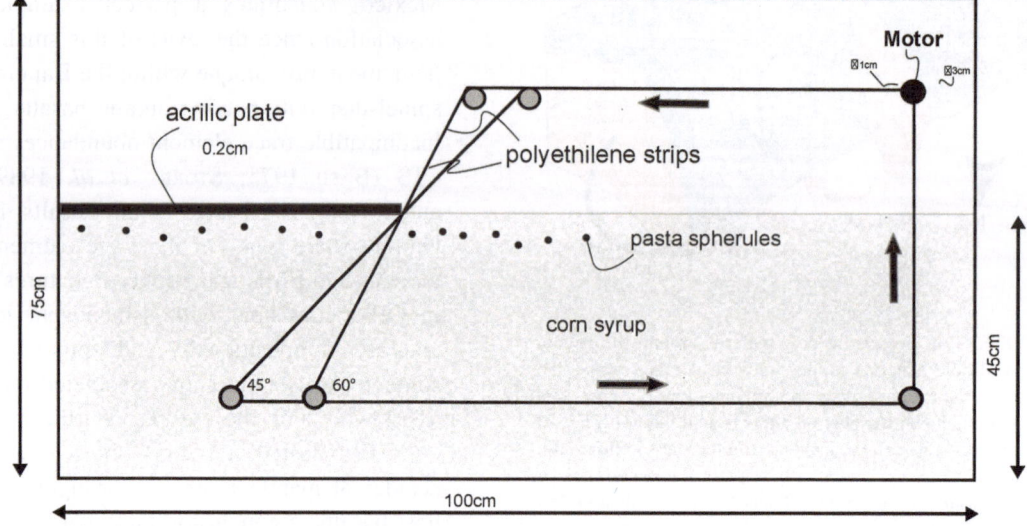

Figure 6
a Sketch of the experimental apparatus in which we perform scaled mantle flow simulations of the mantle wedge of the Middle America subduction zone. See details of the experimental setup in the main text. **b** Photograph of the actual subduction device

of asthenosphere into the mantle wedge from a mantle vortex in the Gulf of California sometime between the Late Miocene and Early Pliocene (FERRARI et al. 2012). More recently, a new conjecture following a similar line of reasoning

has gained significant consensus (Fig. 4, FERRARI 2004; FERRARI et al. 2012; YANG et al. 2009). The basic model imagines the retreat of the Rivera slab during the late Miocene to be the apparent source of vorticity, which forced asthenosphere mantle through the gash between the slabs. This, in turn, resulted in the mixing of compositionally enriched mantle with depleted, metasomatized shallow mantle beneath the arc. In this regard, CASTILLO (2008) pointed out that the source of these rocks as well as the Baja California post-subduction OIB and high-Nb basalts is the Pacific asthenospheric mantle. Moreover, in Castillo's theory, these rocks are fundamentally the same, sharing a common source in HSFE-enriched, metasomatized peridotite (CALMUS et al. 2011). TIAN et al. (2011) demonstrated that post-spreading lavas form abandoned spreading centers offshore Baja California Sur along with those from other fossil spreading axes define a compositional continuum ranging from normal mid-ocean ridge basalts to OIB. These authors propose that the compositional spectrum of these intraplate volcanic lavas is controlled by different degrees of partial melting of and heterogeneous mantle: vigorous mantle upwelling produces normal mid-ocean melts, whereas weak mantle upwelling produces melts of OIB affinity.

4. Experimental Setup

Our experimental method centers on a simplified simulation of mantle flow patterns which incorporates the subduction of both the Rivera and Cocos plates and

their first-order features. In our experiments we have excluded the effects of slab retreat. We decided to leave this effect out because there is evidence that the retreat was too slow to have mobilized a significant volume of mantle around the edges of the slab: during the last 8 million years the Rivera slab retreated ~50 km towards the trench at an average rate of ~0.5 cm/year (FERRARI et al. 2001). In a sense, what we are trying to do with our experiments is falsify this theory with a simpler mechanism of poloidal flow consistent with the observed boundary conditions.

A diagram of our apparatus is shown in Fig. 6 and details of the modeling materials, their rheological parameters, and scaling properties are provided in Table 1. Our scaled subduction experiments were housed in a transparent acrylic tank 1 m long, 25 cm wide, and 75 cm deep with a model/prototype scale ratio of 1:4,000,000, i.e., 1 cm in the model represents 40 km in nature (Table 1). For viscous mantle we use corn syrup, a fluid that at constant temperature behaves in a Newtonian fashion in which stress in the fluid depends linearly on strain rate. Given corn syrup has very little inertia compared to viscous forces (see below), and other advantages like being transparent, this substance is well suited for usage in analog and fluid dynamic experiments to represent linear-viscous strain independent and shear rate independent rheologies to model geological processes (HEURET et al. 2007; SCHELLART et al. 2011). The material properties of the employed corn syrup are as follows: a density of about 1.4 g/cm^3, and a viscosity of 6.5 Pa s (Table 1).

Table 1

List of parameters and values used in the model

Parameter	Units	Nature	Scaled model
Mantle thickness (H)	m	700,000	0.17
Length scale factor (L_{model}/L_{nature})	None	2.4×10^{-7}	–
Density			
ρ_c, Continental lithosphere	kg m^{-3}	3140	1180
ρ_m, Upper mantle		3220	1353
Upper mantle viscosity (η)	Pa s	5×10^{19}	6.5
Characteristic time (τ)	s	2.5×10^{14}	60
Characteristic velocity ($\hat{v}$)	cm/year	10	18
	cm/min		
Rivera plate velocity	cm/year	2	3
	cm/min		
Cocos plate velocity	cm/year	5	9
	cm/min		
Reynold's number (Re)	none	$\sim 10^{-20}$	$\sim 10^{-7}$

In the experimental apparatus, the corn syrup is set in forced motion by two polyethylene strips that simulated the subduction of the Rivera and Cocos slabs under the western edge of the North America plate. The analog experiment also considers the effects of the latter plate, which is represented by means of an acrylic plate 2 mm thick overriding the simulated subduction zone. The strip standing for the Rivera plate dips at an angle of 60°, whereas the strip acting as the Cocos plate dips 45° (Fig. 6). The strips form a conveyor belt inside the tank which is driven by a roller connected to a step motor. Sandpaper was glued to the roller to avoid frictional slipping of the polyethylene strips. To replicate in our scaled subduction experiments the slow motion of the Rivera plate and the rapid convergence rate of the Cocos plate, the roller was fabricated from two sections of different diameters with a 1:3 ratio, similar to the ratio of the convergence velocity of those plates (1:2.75).

5. Similarity and Dimensional Analysis

Besides being geometrically akin to the MASZ, our scaled experiment must undergo rates of motion and time intervals similar to those of the Rivera and Cocos plates. To establish a convection regime in the corn syrup satisfying this more stringent form of kinematic similarity, we need to consider a characteristic flow timescale and a characteristic motion rate, which are intrinsic to all fluid flows. A representative inertial time of the flow, τ, can be taken to be the ratio of inertial forces to viscous forces

$$\tau = g\rho L/\eta, \qquad (1)$$

where g is the gravitational acceleration, ρ is the density of the fluid, L is a characteristic size and η is the viscosity of the fluid. Thus, the Earth's upper mantle has a characteristic time $\tau_M \sim 10^{-10}$ s, whereas our scaled experiment has a characteristic time $\tau_m \sim 10^3$ s (Table 1). On the other hand, the simplest choice of a representative fluid motion rate, $\hat{v}$, is the characteristic length L divided by the characteristic inertial time τ

$$\hat{v} = L/\tau. \qquad (2)$$

Once again, the characteristic velocity for the Earth's upper mantle is $\hat{v}_M \sim 10$ cm/year, while the

experimental apparatus has a characteristic velocity $\hat{v}_m \sim 18$ cm/min (Table 1). Now, kinematic similarity exists between the scaled model and the subduction of the Cocos and Rivera plate if the ratio of flow velocities equals the ratio of the characteristic velocities

$$v_m/v_p = \hat{v}_m/\hat{v}_M. \qquad (3)$$

Substituting the values for $\hat{v}_m$ and $\hat{v}_M$ discussed above, and the rate of convergence, v_p, for the Rivera and the Cocos plates in relation (3), we can solve for the flow velocity v_m in the model. According to expression (3), the condition at which the scaled experiment becomes kinematically similar with the real subduction process is when the strip representing the Cocos plate moves at a rate $v_m \sim 9$ cm/min, and the strip representing the Rivera plate moves at a rate of $v_m \sim 3$ cm/min. By setting the motor driving the motion of the strips to an angular velocity of 0.04 s^{-1} (2.5 r.p.m) we were able to replicate those motion rates.

Perhaps what is more important is that at this angular velocity the Reynolds number of our experiments is small enough to guarantee kinematic and dynamic similarity, which is a further condition the experimental method must satisfy. The Reynolds number, Re, is a dimensionless group of physical parameters, which is given by the ratio of inertial forces to viscous forces

$$\text{Re} = \rho L v/\eta, \qquad (4)$$

where v the velocity of the fluid, and ρ is the fluid's density. For two flows to be dynamically similar they must have comparable Reynolds numbers (LANDAU and LIFSHITZ 1987). For example, subduction in the Earth's mantle occurs at Re $\sim 10^{-20}$; by contrast, it occurs at Re $\sim 10^{-7}$ in our scaled experiments (Table 1). While a cursory look of these figures clearly shows our experimental setup does not satisfy dynamic similarity, a closer inspection will reveal it is impractical. The fact is that full dynamic similarity requires the use of a fluid with the viscosity of pitch (10^7 Pa s). It can be demonstrated, however, that a Re small enough is all that is needed to ensure dynamic similitude. To illustrate this, consider the Navier–Stokes equations in their non-dimensional form, which describe the motion of a fluid substance

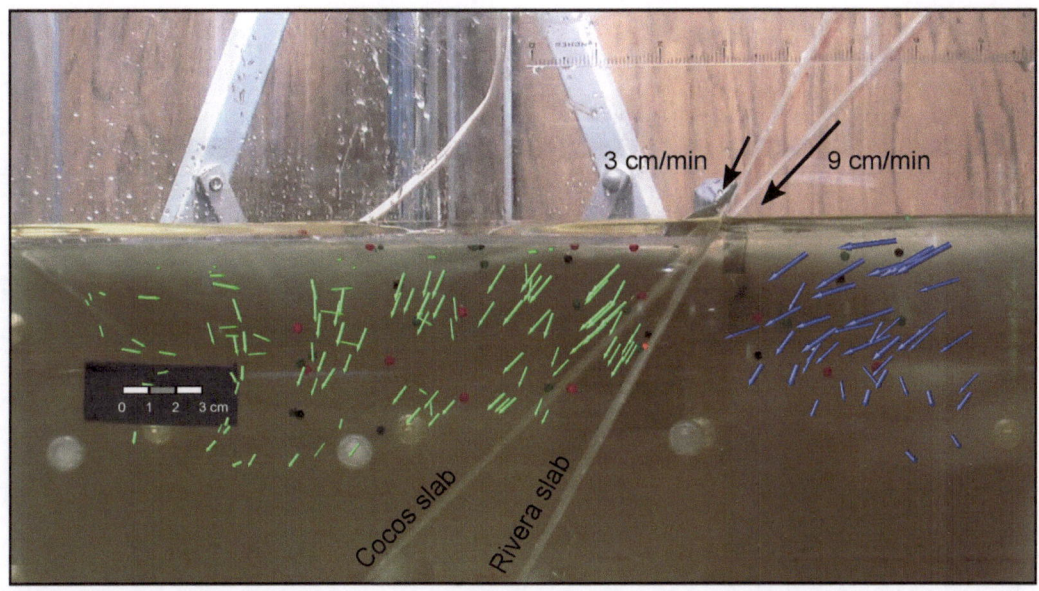

Figure 7
To bring into view the flow patterns in our subduction experiments we used the two-dimensional particle tracking velocimetry technique. We manually embedded neutrally buoyant pasta beads in the corn syrup simulating the fluid mantle whose motion faithfully reveal the experiments flow dynamics. In this side-view picture we can see the past beads (3 mm diameter), which previous to the experiment were *inked red*, *blue* and *black*, and flow vectors described by the particles. Vectors were calculated from two consecutive video frames 20 s apart. Notice that we have superimposed the flow vectors of several experiment runs to increase the resolution of the flow field

(see LANDAU and LIFSHITZ 1987, for more details on how to obtain these equations)

$$\mathrm{Re}\frac{\partial v^*}{\partial t^*} + \mathrm{Re}\ v^* \cdot \nabla v^* = \mathrm{Re}\ \mathrm{Eu}\ \nabla p^* - \nabla^2 v^*. \quad (5)$$

The left-hand side of the previous equation depicts changes in inertial forces, whereas the right-hand side represents pressure and dissipative forces acting in the fluid. In Eq. (5) v^*, p^*, and t^* are non-dimensional versions of the fluid velocity, dynamical pressure, and time, respectively. Additionally, Eu is the Euler number, which is the ratio of dynamic pressure to inertial forces. What Eq. (5) tells us is that for increasingly smaller Re, viscous forces dominate fluid motion more and more, until inertial forces are negligible at which point the flow becomes stationary and no longer depends on time (Stokes flow) other than through time-dependent boundary conditions. Thus, in the analog modeling of Stokes flow it is not necessary to achieve dynamic similarity since streamline patterns are geometrically similar for both the model and the original system.

6. Lagrangian Markers Motion Capture, and Computer Rendering of Flow Pathlines

Many kinds of flow visualization have been employed to make the physics of fluid flows visible (SMITS and LIM 2012). In our case, given the low Re of our experiments, we decided to use the two-dimensional particle tracking velocimetry technique in which an ensemble of neutral density particles travels with the fluid medium (Lagrangian markers or LM). From the motion of the particles it is possible to quantify accurately the flow velocity field and by numerical integration, flow pathlines (Fig. 7). In our simulations we manually seeded in the corn syrup LM consisting of pasta beads ~ 3 mm in diameter which were previously inked for better contrast during the recording of results. Motion of the LM was recorded with the help of two digital, HD video cameras: one mounted above the tank whose field of view was the horizontal, xy-plane, while the other camera was placed on the side of the tank with a field of view covering the vertical, xz-plane. Problems with

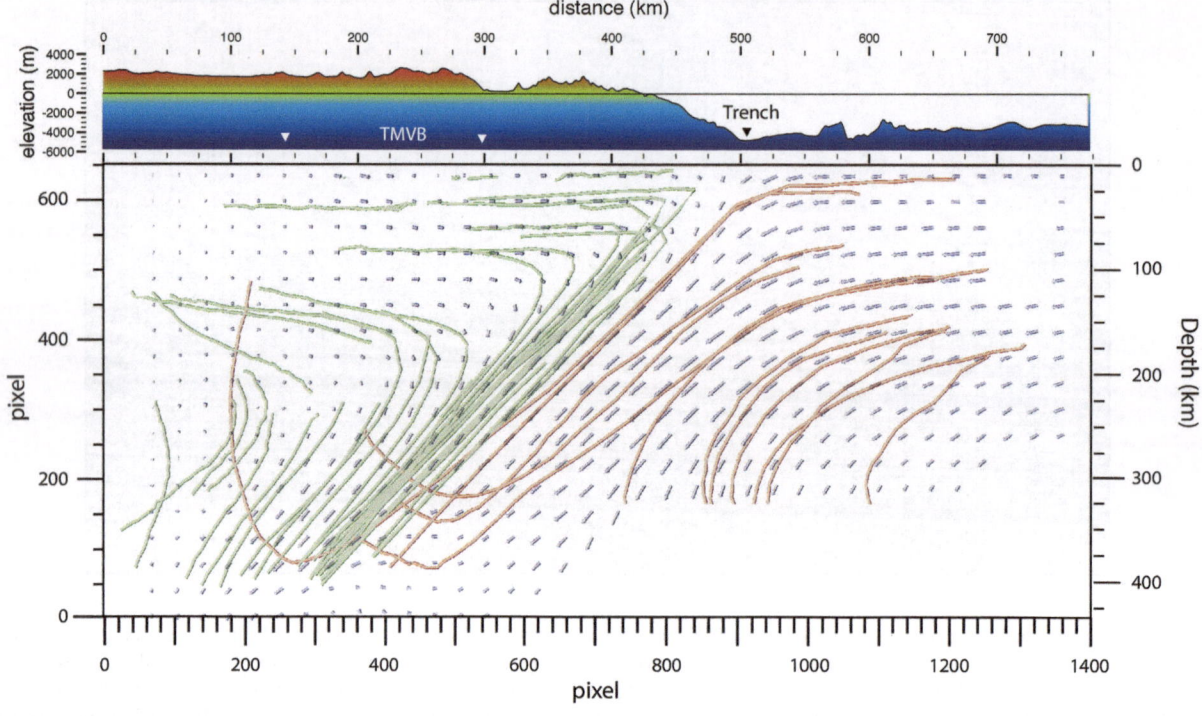

Figure 8

Side-view visualization of the flow pattern induced by the simulated Cocos slab revealed by our analog experiments. The topographic profile on top illustrate changes in elevation and bathymetry across Middle America subduction zone and we have added the main topographic landmarks for reference. *Green lines* correspond to the poloidal flow on the arc side of the subduction zone whereas *blue lines* highlight the poloidal flow on the ocean side of the subduction zone. *Red vectors* is the interpolated flow field derived from the Lagrangian markers

particle motion-capture did not allow us to reconstruct the pathlines in three dimensions (3D) directly.

On the problem of whether pasta beads with their relative large size faithfully reveal the flow dynamics, it turns out that the behavior of particles suspended in a fluid flow is controlled by the Stokes number, Stk, given by

$$Stk = \tau_r v / d_p, \qquad (6)$$

where τ_r is the relaxation time of the LM, and d_p the diameter of the particles. For acceptable tracing accuracy, the Stokes number of the particles should be <0.1, otherwise, they will deviate from the true pathlines, particularly where the flow velocity, v, changes abruptly. For our analog experiments, the relaxation time of the pasta beads is $\tau_r = 7 \times 10^{-5}$ s, their diameter is $d_p = 3$ mm, and the average velocity of the flow is $v = (3 + 9)/2$ cm/min (Table 1), which yields Stk $= 2 \times 10^{-5}$. This means that our choice of LM closely follow the flow.

On average, scaled experiments lasted 10 min each tracking the motion of ~ 30 particles. From the experiment's videos we extracted individual frames every 20 s for their subsequent particle motion-capture using two different methods (Fig. 7). For shallow particles and those located near the tank's walls we used an automatized motion capture software developed by SBALZARINI and KOUMOUTSAKOS (2005) which is available at http://mosaic.mpi-cbg.de/. The software calculates the position of the particles in the camera's field of view with the help of signal processing and autocorrelation/cross-correlation algorithms. The software, however, was not able to identify LM deep in the tank or away from the walls, in which case we manually captured the particles motion frame by frame.

Computer rendering of the flow was performed using the public-domain visualization software Open Data Explorer (http://www.opendx.org), which permit the mapping of flow attributes such as velocity,

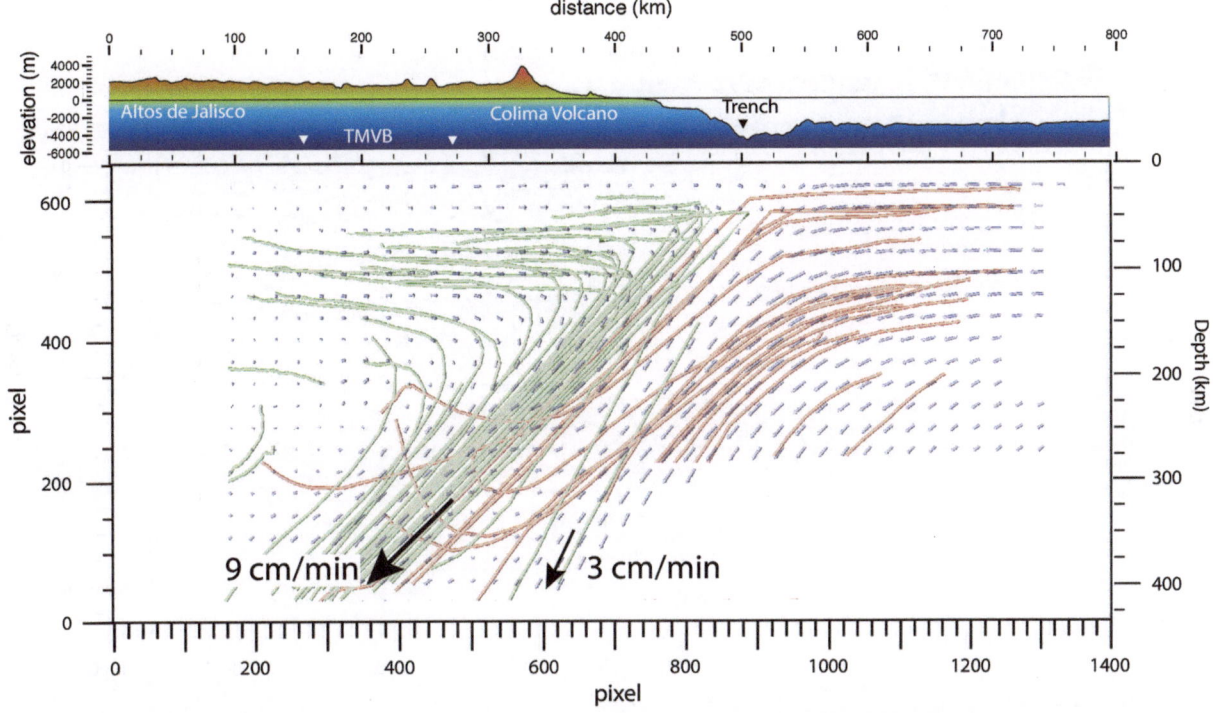

Figure 9

Side-view visualization of the flow pattern in the gap region of the subduction model. Pathlines colors are as in Fig. 8. *Black arrows* indicating the velocity of the slabs are not drawn at scale. Notice how the arc flow (*green pathlines*) is split in two, trailing both the Rivera and Cocos subducted slabs. On the other hand, the red pathlines (poloidal cell on the ocean side of the subduction zone), suggest a mantle upwelling is formed in the gap between the slabs. We hypothesize this is due to the development of a hydraulic jump in response to the abrupt change in velocity the flow experiences as it enters the mantle wedge. More details are provided in the main text

streamlets, or temperature to color and geometrical elements like arrows and tubes (Fig. 7). Once all the positions and times of the particles were captured, a discrete Lagrangian velocity field $\mathbf{v}^i$ was calculated by means of a simple first-order finite differences numerical scheme

$$\mathbf{v}^i(\mathbf{p}^i) = \left(\mathbf{p}^i_{j+\Delta t} - \mathbf{p}^i_j\right)/\Delta t, \qquad (7)$$

where $\mathbf{p}^i$ is the family of pathlines described by the LM. Then the stationary velocity field was interpolated by means of a Delaunay triangulation to a regular grid. All experiment's results are plotted in pixels and pixel/s and the relation to the natural scale is as follows: for images in the horizontal xy plane, 120 px represent 100 km, whereas for images in the vertical xz plane, 150 px represent 100 km.

7. Results

Six experiments were performed during the development of the present work (Figs. 8, 9, 10, 11). In each experiment we carefully seeded LM at specific locations in the apparatus to bring into view the flow dynamics of different regions of the simulated subduction zone. A further reason was purely practical: as we describe below, it proved too cumbersome to track down a large ensemble of particles often crisscrossing each other's trajectories (see Fig. 11). Two of the experiments were aimed to bring forth the two poloidal cells on both sides of the Cocos slab (arc and oceanic corner flow; Fig. 8); two more were meant to reveal flow perturbations in the oceanic mantle and mantle wedge caused by the gap between the Cocos and Rivera slabs (Fig. 9); and

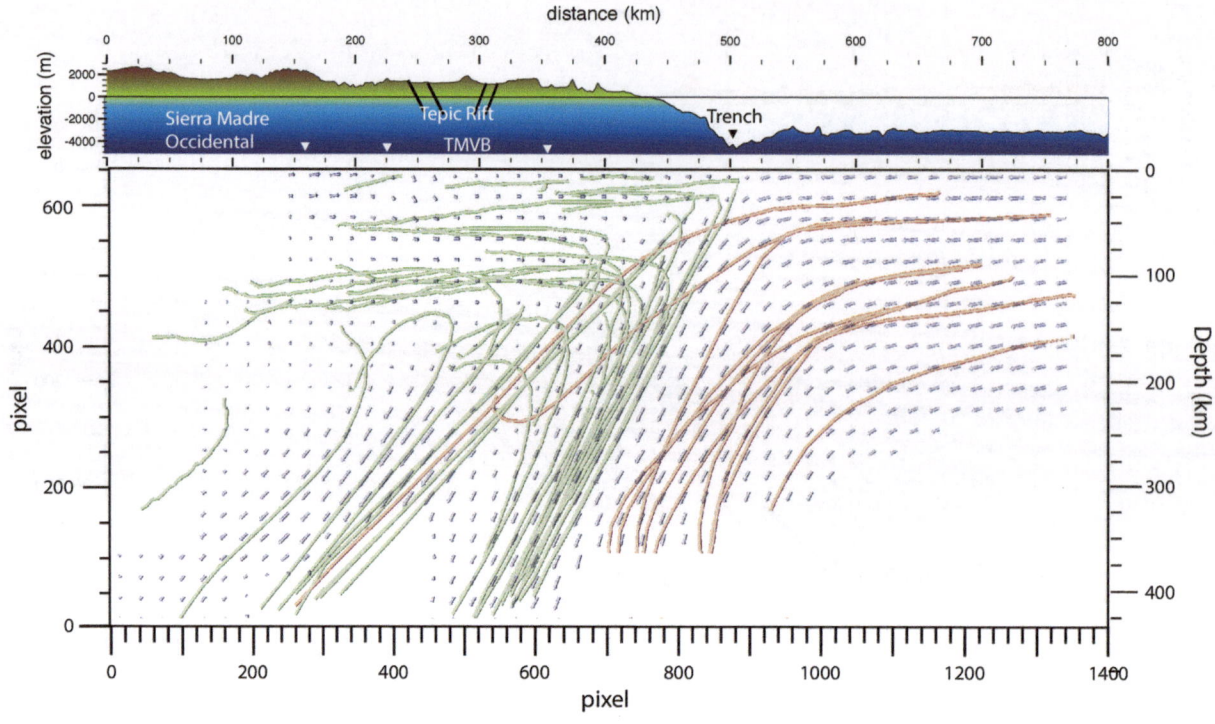

Figure 10
Side-view visualization of the flow pattern in around the Rivera slab of the subduction model. Pathlines colors are as in Fig. 8. Observe, once again, the arc poloidal flow (*green pathlines*) is split in two, trailing both the shallow-dipping Cocos slab and the steep-dipping Rivera slab. Also note the upward trajectories of the material markers located underneath the Tepic-Zacoalco rift, at a depth of 200 km, which suggests the development of a convective cell in the mantle wedge above the Rivera slab

with the last two experiments meant to image the circulation pattern around the Rivera slab (Fig. 10). Bear in mind that the vector and streamline fields in the two-dimensional (2D) plots described next do not necessarily imply a plane strain state of fluid motion, i.e., the out of plane component of motion is not null. On the contrary, the flow pattern is highly 3D; the graphs in Figs. 8, 9, 10 and 11 are only side-views projections onto the camera's field of view.

7.1. Flow Pattern Around the Cocos Slab

Figure 8 shows a side-view of the interpolated flow field (blue vectors) from the first two experiments as well as the pathlines described by the LM. Green pathlines in this figure are for particles located in the mantle wedge, whereas red pathlines are for LM initially located in the oceanic mantle beneath the subducting plate. Additionally, we have included in the figure and subsequent ones a topographic profile

across the subduction zone in which we have marked the location of the TMVB for further reference. In the first two experiments material markers have strong, normal-to-trench motion and display the characteristic arc and oceanic corner flow pattern of analytical 2D models of subduction zones (e.g., TURCOTTE and SCHUBERT 2002). Initially, the particles on both sides of the trench move parallel to the free surface of the syrup due to the transfer of linear momentum from the polyethylene strip to surrounding fluid. Then, they are dragged down near the trench to the bottom of the apparatus by the descending poloidal cells that develop on both sides of the simulated subduction zone.

The flow visualization in Fig. 8 we also shows a couple of stray pathlines originating in the oceanic mantle that cross the mantle wedge streamlines, ascending directly under the arc region of the simulated subduction zone. A toroidal cell which developed around the gap shot the particles into the

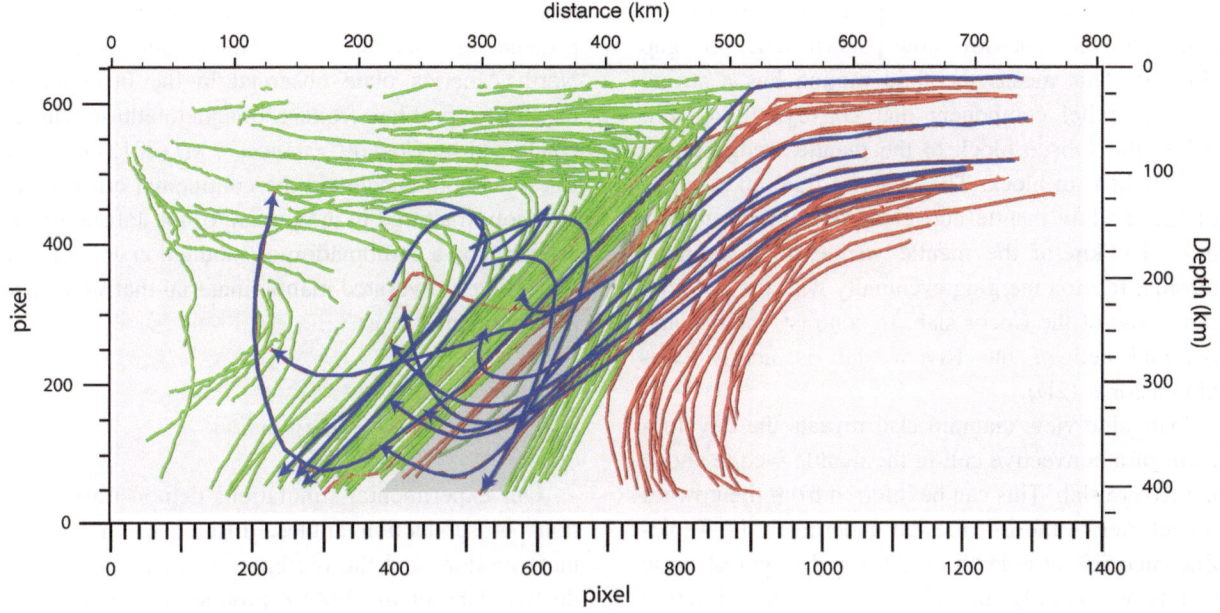

Figure 11

In this side-view diagram we have grouped the pathlines from Figs. 8, 9 and 10, giving a more detailed picture of the flow dynamics in the simulated subduction zone. Notice the oceanic asthenosphere passing through the tear between the slabs (*blue pathlines*) is conveyed to a depth of 250–350 km, then rises slowly trough the mantle wedge to a depth of ∼100 km, and sinks again into the lower mantle, forming a closed loop. A three-dimensional artistic rendering of the flow motion is provided in Fig. 13

mantle wedge through the opening between the slabs (see also results below).

7.2. Flow Pattern Around the Tear Between the Slabs

We now review the results of two experiments that highlight perturbations in fluid motion brought about by the tear directly under the Colima graben (Fig. 9). The color of the velocity and stream fields in the side-view diagram presented in this figure, are similar to those used in Fig. 8. As expected, the arc corner flow in these experiments splits between the Rivera and Cocos slabs, resulting in two groups of pathlines: one group trails the steep-dipping strip simulating the Rivera slab, whereas another set of particles plunge at a shallower angle traveling with the strip representing the Cocos slab. In contrast, our analog experiments illustrate important changes in the oceanic poloidal cell. Most evident is that numerous pathlines find their way to the mantle wedge by means of the tear in the subducting slabs. What is more puzzling is the way in which the

particles cease their downward motion about halfway down the acrylic tank to ascend directly below the arc (Fig. 9).

One possible explanation to this behavior appears to be the abrupt way in which the fluid loses momentum. As it enters the wedge the fluid sheds velocity from 9 to 3 cm/min, which compresses the fluid, forcing it to rise (Fig. 9). Thus, our experimental results provide support for the contention that fertile peridotite flowed into the mantle wedge beneath the Jalisco block following the breakup of the Farallon plate into smaller plates, but more importantly, our experiments also illustrate mechanisms of mantle upwelling by which OIB melts can be produced.

7.3. Flow Pattern Around the Rivera Plate

We now describe our two last corner flow experiments simulating the circulation regime of the Rivera slab (Fig. 10). As can be appreciated in the side-view visualization in this figure, the flow is rather intricate. This stems from the fact the mantle

197

flow under the Jalisco block splits into two streams similar to the side-view flow pattern under the gap (Fig. 9). This means the fluid motion has a strong trench-parallel component that conveys fluid from below the Jalisco block to the mantle wedge under the Michoacán block. The material flow takes place in the shallow mantle above the tear, in the uppermost 100 km of the mantle wedge, in a counter-toroidal fashion merging eventually with the poloidal (arc) flow of the Cocos slab. In contrast, the oceanic poloidal cell of the Rivera slab is nearly two-dimensional (2D).

The side-view diagram also reveals the development of a convective cell in the mantle wedge above the Rivera slab. This can be inferred from the upward trajectories of the LM located underneath the Tepic-Zacoalco rift, at a depth of 200 km. In light of these results, we initially entertain the idea of a correlation between these two processes, insinuating western Mexico is in a state of incipient rifting produced by the development of a convective cell. However, since our model results indicate the upwelling is too deep and slow (the ascent rate is 1 cm/min or, equivalently, 1 cm/Myear in the scaled real mantle), it is unlikely such a convective system can cause enough shear stress at the base of the crust to explain the brittle failure of the arc (d'Acremont et al. 2003). Besides, as we saw in our cursory review of the geodynamics of the MASZ, what drives the intra-arc extension of the TMVB is abrupt changes in convergence rate (Contreras 2013).

On the other hand, the vertical asthenosphere flow does clarify the presence of hot, low-density asthenosphere beneath the western section of the TMVB (Fig. 2; Yang et al. 2009), and the magmatic diversity of the volcanic arc (e.g., Ferrari et al. 2001; Gómez-Tuena et al. 2011; Petrone et al. 2003). The flow could have perturbed the thermal boundary layer under the arc enhancing magmatic production. Also note that shallow LM display a significant upward component of motion; particles underneath the arc, at a depth of ~50 km, become shallower as they approach the subducting slab reaching a near zero depth at the trench. This model predicts the observed large-scale deformation pattern in the Jalisco block as well. These upward directed components of displacement could explain a thinning

towards the trench of the continental crust and the presence of a mantle diapir in the leading edge of the North America plate observed in the inversion of resistivity profiles from a magnetotelluric survey conducted by Corbo-Camargo et al. (2013) (Fig. 4). These authors concluded the continental crust seems intensely fractured in the forearc of the Jalisco block, probably by a combination of subduction erosion and upwelling of hydrated mantle material that weakened the continental crust.

8. Discussion

Our experimental simulations demonstrate variations in subduction angles, differences in rates of plate motion, and the presence of a tear in the subducted slabs of the MASZ produce a self-sustained flow of oceanic asthenosphere into the mantle wedge underneath the Jalisco block. The differential motion of the slabs (~3.5 cm/year) produces energetic mantle stirring without the need of rollback and is quite vigorous when contrasted with the rate of retreat of the Rivera slab documented by geological observations (0–0.5 cm/year; Ferrari 2004). Moreover, our model results also illustrate the development of a convective cell in the mantle wedge of the simulated subduction zone, giving rise to a mantle upwelling under the arc (Figs. 9, 10). The ascent rate is 1 cm/min, which scaled to the real mantle is ~1 cm/Myear. These phenomena can explain a wealth of observations in the TMVB, which we discuss further below, but for the time being, let us consider first the details and intricacies of the flow structure.

8.1. Three-Dimensional Flow Structure

To bring into view the complexity of the flow, we combined all the pathlines recorded in our experiments in a single side-view diagram (Fig. 11). In the figure we have highlighted those trajectories from the oceanic poloidal cell, passing through the gap, which amalgamate with those in the mantle wedge with upward motion trajectories. It can be observed that pathlines passing through the gash have a significant downward motion component, which is in agreement with previous experimental findings. KINCAID and GRIFFITH (2003) scaled model demonstrated the combination of poloidal flow driven by downdip sinking and toroidal flow around the plate due to retreat produce a shear flow within the wedge (Fig. 5); as a result fluid in the vicinity of the plate edge moves horizontally around the border and then downward with the plate. This is an important result

because current conceptual models (e.g., FERRARI 2004; SOTO *et al.* 2009; YANG *et al.* 2009) do not factor in the strong shear flow created by the slab borders on both sides of the gap.

To get a clearer picture of the circulation in our experimental device, we made a top-view diagram (Fig. 12). In the diagram blue pathlines correspond to markers originally placed in the ocean side of the

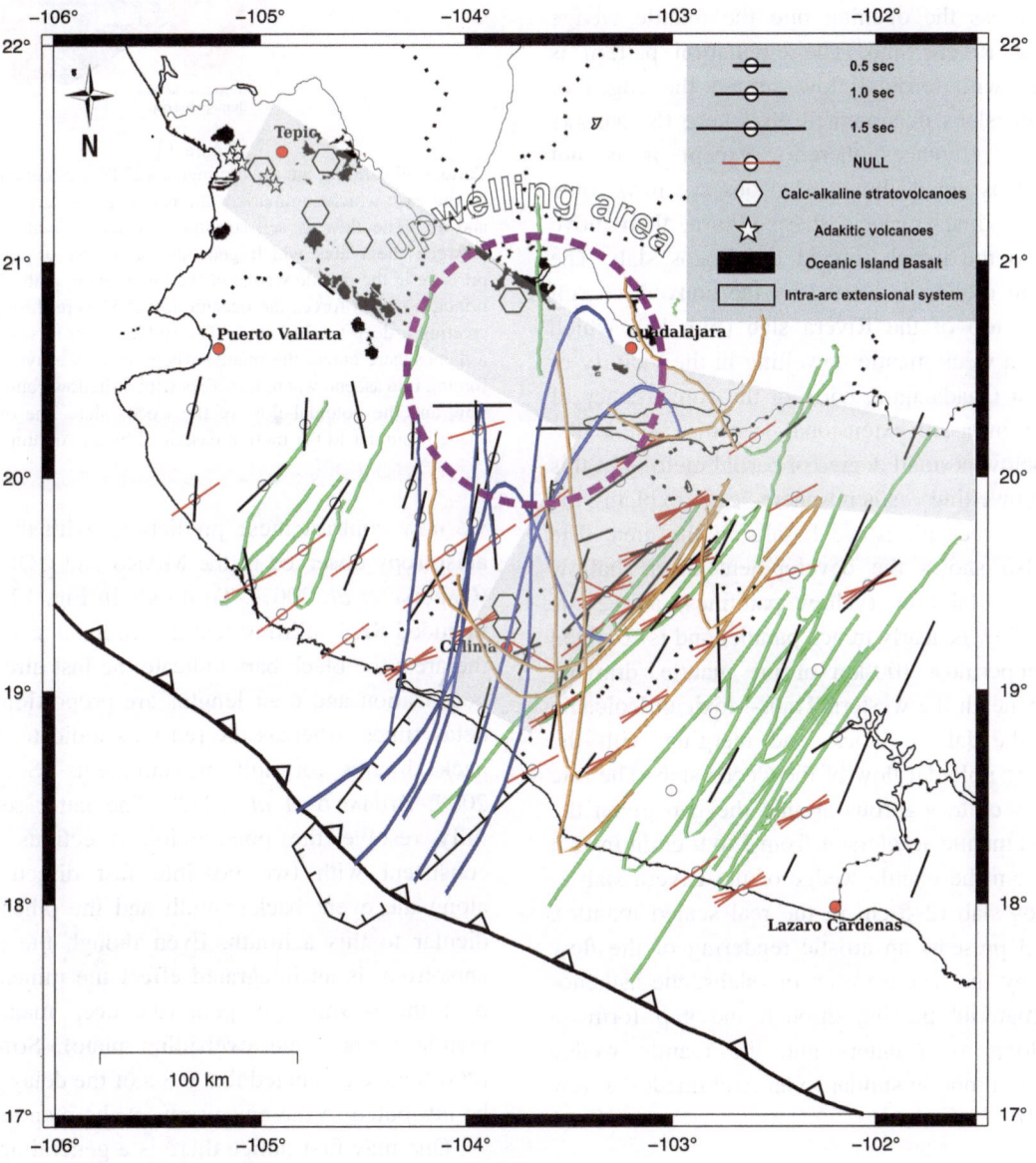

Figure 12

Top-view diagram of the pathlines described by the material markers in our subduction experiments. The pathlines were upscaled using the model/prototype scale ratio of 1:4,000,000. *Black bars* indicate the anisotropy directions in the shallow mantle (see main text for further details). *Balck areas* indicate the outcrop of oceanic island basalts (FERRARI 2004). The *oval* indicates the overall region where the mantle upwelling occurs in the experimental device

subducting plate which pass through the gap; yellow pathlines are for material markers initially placed in the mantle wedge above the Rivera slab which are drawn by the poloidal flow of the Cocos slab; and, green trajectories correspond to markers placed in the mantle wedge that follow the motion of the downgoing slabs. Notice that we have scaled our top-view diagram to the real subduction zone. Figure 12 reveals there is a deep influx of Pacific asthenosphere mantle across the opening into the mantle wedge above the Rivera slab. The circulation pattern is consistent with toroidal flow around the edges of subduction slabs documented elsewhere (SCHELLART 2004 and references therein), except it is not horizontal as noted above. Also observe how some of the flow lines make a sharp U-turn, then move parallel to the trench toward the Cocos slab. The reversal in motion is caused by the convective cell sitting on top of the Rivera slab (Fig. 11), which results in a weak mantle upwelling in the vicinity of the city of Guadalajara, north of the convergence of the three intra-arc extensional systems. As is discussed below, a small degree of partial melting of this mixed upwelling asthenosphere can explain the presence of oceanic-island basalts in the area. The figure also shows the development of a shallow counter-toroidal flow (yellow pathlines) above the gap. The flow is nearly trench parallel and is confined to the uppermost 100 km of the mantle, drawing mantle beneath the western Trans-Mexican Volcanic Belt to the Jalisco block, and merging with the descending poloidal flow of the Cocos slab. The rate of flow is quite vigorous around the gap given the simulated mantle accelerates from 3 to 9 cm/min as it moves from the mantle wedge of the Rivera slab to the Cocos slab (2–5 cm in the real scaled mantle). Figure 13 presents an artistic rendering of the flow induced by the tear between the slabs: the asthenosphere material passing through the gap forms a closed loop as it enters into the mantle wedge describing a motion similar to an Archimedes' screw (helicoid).

8.2. Comparison with Anisotropy Observations

Having, in the above section, made a detailed analysis of the flow pattern emerging the simulations,

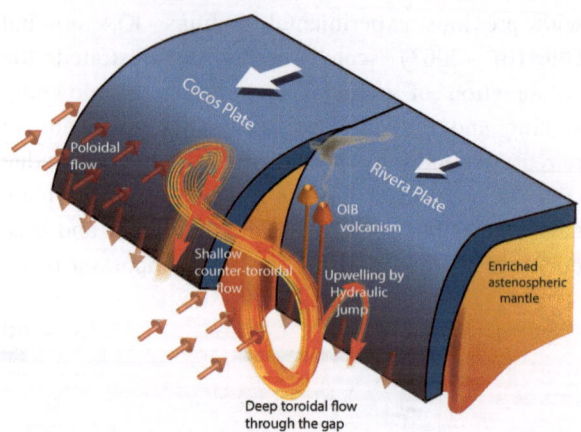

Figure 13

Cartoon illustrating the three-dimensional flow circulation induced by the gash which summarizes the results presented in Figs. 9, 10 and 11. The flow is self-sustained by the velocity differential between the plates which generates a region of low dynamic pressure in the mantle wedge of the fast moving plate (Bernoulli's principle). This forces the oceanic asthenosphere through the gap creating a deep toroidal flow. The sudden drop in velocity as the asthenosphere enters the mantle wedge creates a hydraulic jump, forcing it to ascend where it merges with a shallow counter-toroidal flow and the poloidal flow of the Cocos slab. The overall flow shape is similar to the motion described by an Archimedes' screw

we now contrast these predictions with the seismic anisotropy observed in the MARS and CODEX data (GARDINE et al. 2007). To do so, in Fig. 12 we have included SKS splitting fast directions documented in the area: the black bars indicate the fast directions of polarization and their lengths are proportional to the delay times, whereas the red bars indicate the event backazimuths for null measurements (SOTO et al. 2009; STUBAILO et al. 2012). The latter set cannot fully resolve the polarization directions and are consistent with two possible fast directions: one along the event backazimuth and the other perpendicular to this azimuth. Even though the observed anisotropy is an integrated effect the mineral fabric over the seismic ray path (the deep mantle, slab, mantle wedge, and overriding plate), SOTO et al. (2009) have estimated that 0.8 s of the delay time can be attributed to the anisotropy of the mantle wedge.

One may first notice there is a general agreement with the model predictions as they appear to explain to first order the variability in seismic anisotropy (Fig. 12). Firstly, the split orientations are parallel with the flow velocity away from the tear and near the trench; these regions are dominated by poloidal flow from the downgoing slabs (green pathlines in Fig. 12), hence, the expected orientation of the anisotropic fast axis. Secondly, the circular arrangement SKS fast split orientations just north of the tear has a strong resemblance to the toroidal mantle flow through the slab window. Further notice that the flow abruptly diverges from the strong trench normal poloidal flow beneath the Colima rift, where the gap is located. Finally, the model predictions for the arc region deserve special attention. The SKS splitting fast directions observed the City of Guadalajara change from being predominantly trench-normal to trench-parallel (Fig. 12). The basic theory, however, predicts these regions to have SKS split orientations aligned with the absolute plate motion of North America, which has been advancing against the Rivera and Cocos plates during the last ~ 10 Ma (DeMets and Traylen 2000). The scaled experiments, in contrast, may explain some of the variability observed in the arc region; according to our model, one possible explanation, to these trench-parallel measurements appears to be the shallow counter-toroidal flow toward the forearc of the Michoacán block.

8.3. On the Forces Driving the Flow

Since our simulations are isothermal, and inertial effects negligible, there are only two kind of forces acting on the fluid: viscous and pressure stresses. Under Stokes flow conditions these two forces must cancel out at every point of the fluid, i.e., $\nabla p - \eta \nabla^2 v = 0$, where p is the dynamic pressure and $\eta \nabla v$ the viscous stress. This means changes in dynamic pressure go hand in hand with changes in velocity gradients (Bernoulli's principle). Therefore, the abrupt change in tangential velocity across the gap, from 3 cm/min to 9 cm/min, causes a large pressure gradient. This is what drives the deep toroidal flow from the ocean side of the subducting plate down the discontinuity. At the same time, the subduction of the Rivera and Cocos slabs produce two rotating fluid rolls around a common line vortex (i.e., the trench); these two rotating structures create, in turn, low dynamic pressure systems in their cores whose magnitudes scale as the subduction velocity

(Turcotte and Schubert 2002). Thus, once again, the different subduction velocities produce a pressure gradient beneath the Michoacán block that propels the shallow counter-toroidal flow.

The previous results suggest that the circulation pattern in the mantle wedge prompted the Rivera slab to retreat, not the other way around. The experiments demonstrate the deep toroidal flow make the poloidal flow near the southern edge of the Rivera slab sluggish, allowing the shallow counter-toroidal to be practically trench-parallel. The weakened poloidal flow, in turn, would have resulted in a loss of dynamic support for the Rivera slab, facilitating its rollback (e.g., Zandt and Humphreys 2008). This theory, however, requires the Farallon slab to tear in two initially, assumption that invites the obvious question: what caused the tear in the slab in the first place? Here, we speculate that as the Pacific ridge approached the trench, subduction became unstable and a vertical gap opened between the slab propagating vertically downward (Burkett and Billen 2010). The separation caused the Rivera microplate to subduct more independently and started to retreat by the aforementioned mechanism.

Next, we focus on the forces driving the upwelling of mantle under the arc. Because the experiments are isothermal and chemically homogenous, the upwelling develops necessarily from the solid/fluid coupling along the subduction tank's walls and moving surfaces (slabs). Specifically, the mechanical coupling between the fluid and the fast moving surface representing the Cocos slab produces a high velocity flow around its neighborhood. Where the slab discharges fluid into the slow moving mantle wedge of the Rivera slab, the rapidly flowing liquid is abruptly slowed causing it to compress. This results in a liquid buildup which forces it to rise (Fig. 14); the effect is known as hydraulic jump and often is observed in the form standing waves or tidal bores in open channel flows. From conservation arguments it is possible to estimate the height h_2 of the jump by means of the relation described in the caption of Fig. 14: $h_2 = h_1 \times (v_1/v_2)$, where h_1 is the initial height of the flow, and v_1 and v_2 are the velocities of the flow before and after the jump. If we take $h_1 = 100$ km as the thickness of the flow passing through the gap (Fig. 11), $v_1 = 5$ cm/year,

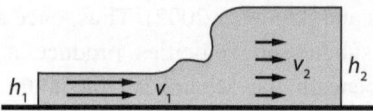

Figure 14

A hydraulic jump develops in a rapidly moving liquid where it enters an area of low discharge. This causes the fluid to compress and to increase its height. The height of the jump can be established from conservation arguments as follows: the mass flux upstream ($\rho v_1 h_1$) must be equal to mass flux downstream ($\rho v_2 h_2$), where ρ is the density of the fluid, which leads to $v_1 h_1 = v_2 h_2$. From this expression we can solve for the height h_2: $h_2 = h_1 (v_1/v_2)$

$v_2 = 2$ cm/year, then we obtain $h_2 = 250$ km, which is roughly the height of the jump observed in the experimental device once it has been scaled to the real mantle.

8.4. Implications for the Genesis of OIB Volcanism

As discussed in Sect. 3, several authors have reported OIB in the northwestern Pacific margin of Mexico along abandoned spreading centers, and the interior of the Baja California peninsula, at the San Quintín and Santa Clara volcanic fields (CASTILLO 2008; CALMUS et al. 2011; NEGRETE-ARANDA and CANON-TAPIA 2008). Attempts to explain the origin of these intriguing observations fall under two broad categories. On one hand, OIB-like (HNB) are claimed to be melts of mantle peridotites which have been metasomatized through stabilization of amphibole in HSFE by slab-derived melts (e.g., DEFANT et al. 1992; SAJONA et al. 1996; DEFANT and KEPEZHINKAS 2001; AGUILLÓN-ROBLES et al. 2001; BENOIT et al. 2001). On the other hand, recent studies have proposed lowering the degree of partial melting plays a major role in creating the general evolution of fossil spreading axes from normal mid-ocean ridge basalts (NMORB) to alkalic OIB-like basalts (TIAN et al. 2011). These authors have proposed a heterogeneous sub-oceanic mantle source along with a decreasing degree of partial melting control the evolutionary trend observed in the fossil spreading centers offshore Baja California.

We consider the general mechanism proposed by TIAN et al. (2011) appropriate to explain the nature of the OIB source beneath the western TMVB because of two important similarities between the two

tectonic settings: both appear to have a source in the Pacific asthenosphere mantle and are places of weak upwelling. We have seen in preceding paragraphs that the differential motion between the slabs and the presence of a gap at depth produce energetic mantle mixing, which resulted in a heterogeneous mantle beneath the TMVB. Thus, we consider the mantle wedge beneath the TMVB replaced in certain places by the asthenosphere (Figs. 11, 12, 13) containing less-refractory enriched patches with lower solidus lodged in a depleted, MORB-source less prone to melting. Such mantle patches have compositional structure similar to a "plum pudding" (MORRIS and HART 1983). Following the ideas of TIAN et al. (2011) we further consider OIB-like lavas originated from small degree of partial melting that preferentially fuses the enriched components in the heterogeneous source, producing a small amount of melt. The small degree of melt is further controlled by the slow rate of upwelling (1 cm/year) produced by the hydraulic jump in the mantle wedge of the Rivera slab.

At the same time the NMORB-like mafic pulses that coexist with OIB, most likely result from large degree of partial melting that fuse both the enriched and more dominant NMORB-matrix melts (MORRIS and HART 1983; TIAN et al. 2011). The OIB signature in the resultant melt is diluted during mixing and aggregation processes with the NMORB matrix melts. Thus, by adopting TIAN et al. (2011) ideas to explain small volume OIB-like magmatism during the last 5 Ma at the western TMVB it is possible to conceptually explain, the compositional diversity in the arc lava record. Further implications of this model are related to the localization of volcanism along the western TMVB; spatial distribution is then governed by the thermal regime and possible release of volatiles from the subducted slab, leading to localized diapirism of the "plum pudding" in the mantle wedge. Fresh "pudding" presumably can be convected/transported into the wedge to permit long-term production of enriched magmas.

In Fig. 12 we put to the test the soundness of our simple conceptual model. The basic idea of the test is very simple. Take the pathlines calculated from our laboratory scaled models and contrast the upwelling zone predicted by the model, where conditions are best suited for the production of OIB, with the actual location of the OIB outcrops in the western TMVB. Notice that in the plot we only included a limited set of pathlines to avoid cluttering. Essentially what we have is a model that can explain the eruption of OIB lavas around the city of Guadalajara, north of the convergence of the three rift systems, and OIB outcrops in the south shoulder in the central part of the Tepic-Zacoalco rift system. Notice the former lie outside the upwelling area but the development of preferential permeability paths along the shoulder can easily explain why these lava fields circumscribe the upwelling area. On the other hand, the model is not able to explain the occurrence of OIB in the westernmost section of the TMVB (Fig. 12). Perhaps, as suggested by Petrone *et al.* (2003), the presence of these rocks is linked to a lateral flow of the deep asthenosphere into the mantle wedge from the nearby Gulf of California region.

9. Conclusions

Mantle flow beneath the Jalisco and Michoacán block has been studied using a simple experimental model that takes into account differences in subducting angles and velocities observed in the western part of the MASZ. Experiments reveal the presence of a complex flow in the mantle wedge below those tectonic blocks. Toroidal flow occurs at two different depths: one at shallow depth in which flow is directed from the arc region of the Jalisco block towards the forearc of Michoacán block; a deeper flow transports fertile mantle from the oceanic asthenosphere mantle to the mantle wedge through the gap between the slabs. The latter flow may explain the heterogeneous compositions of lavas erupted in the TMVB. Model results are also in agreement with seismic anisotropy observed in the western TMVB. Finally, the model develops a hydraulic jump causing a mantle upwelling under the arc/backarc region, which provides with a consistent explanation of the occurrence of OIB around the general area where the intra-arc extensional systems converge, and the south shoulder of the central Tepic-Zacoalco rift system in the western TMVB.

Acknowledgments

We acknowledge the financial support of the Mexican Council of Science (CONACYT), which provided F. Neumann with the scholarship 242956, and J. Contreras with the Grant 60647. Further financial support was provided by CICESE's internal project 644143 and 644149 to J. Contreras and R. Negrete, respectively. Technical support was provided by Teodoro Hernandez and Sergio Arregui-Ojeda. The manuscript benefited greatly from two anonymous reviewers and additional comments by Patricia Persaud; we are grateful to them. The Generic Mapping Tools (http://www.soest.hawaii.edu/gmt) was used to create the majority of the maps. We also thank Sandy Suhardja for providing us with the high-resolution images used in Fig. 2.

References

Aguillon-Robles, A., Calmus, T., Benoit, M., Bellon, H., Maury, R.C., Cotten, J., Bourgois, J., and Michaud, F. (2001), *Late Miocene adakites and Nb-enriched basalts from Vizcaino Peninsula, Mexico: indicators of East Pacific rise subduction below southern Baja California?*, Geology. 29, 531–534.

Allan, J.F. (1986), *Geology of the Northern Colima and Zacoalco Grabens, southwest Mexico: Late Cenozoic rifting in the Mexican Volcanic Belt*, GSA Bulletin. 97, 473–485.

Basu, A.R. (1975), Geochemestry of the ultramafic xenoliths from San Quintin volcanic field, Baja California: in Boyd F.R., and Meyer H.O.A, eds., The mantle sample: Inclusions in kimberlites and other volcanics: American Geophysical Union, International Kimberlite Conference, 2d, Santa Fe, N.M., 1977, Proceedings, v. 2, 391–399.

Benoit, M., Aguillon-Robles, A., Calmus, T., Maury, R.C., Bellon, H., Cotten, J., Bourgois, J., and Michaud, F. (2001), *Geochemical diversity of Late Miocene volcanism in southern Baja California, Mexico: implication of mantle and crustal sources during the opening of an asthenospheric window*, The Journal of Geology. *110*, 627–648.

Burkett, E. R., and Billen, M. I. (2010). *Three-dimensionality of slab detachment due to ridge-trench collision: Laterally simultaneous boudinage versus tear propagation.* Geochemistry, Geophysics, Geosystems, *11*(11).

Calmus, T., A. Aguillón Robles, R. C. Maury, H. Bellon, M. Benoit, J. Cotten, J. Bourgois and F. Michaud (2003). "*Spatial and temporal evolution of basalts and magnesian andesites ("bajaites") from Baja California, Mexico: the role of slab melts.*" Lithos 66: 77–105.

Calmus, T., Pallares, C., Maury, R.C., Aguillon-Robles, A., Bellon, H., Benoit, M., and Michaud, F. (2011), *Volcanic markers of the post-subduction evolution of Baja California and Sonora, Mexico: Slab tearing versus lithospheric rupture of the gulf of California*, Pure and Applied Geophysics. *168*, 1303–1330.

CASTILLO, P.R. (2008), *Origin of the adakite-high-Nb basalt association and its implications for postsubduction magmatism in Baja California, Mexico*, GSA Bulletin. *120*, 451–462.

CORBO-CAMARGO, F., ARZATE-FLORES, J.A., ALVAREZ-BEJAR, R., ARANDA-GOMEZ, J.J., and YUTSIS, V. (2013), *Subduction of the Rivera plate beneath the Jalisco block as imaged by magnetotelluric data*, Revista Mexicana de Ciencias Geológicas. *30*, 268–281.

CONTRERAS, J. (2013), *A model for the state of brittle failure of the western Trans-Mexican Volcanic Belt*, International Geology Review. 1–12.

D'ACREMONT, E., LEROY, S., and BUROV, E.B. (2003), *Numerical modelling of a mantle plume: The plume head-lithosphere interaction in the formation of an oceanic large igneous province*, Earth and Planetary Science Letters. *206*, 379–396.

DEFANT, M.J., KEPEZHINSKAS P. (2001), *Evidence suggest slab melting in arc magmas*, EOS Trans. Am. Geophys. Union. *82*, 65–69.

DEFANT, M.J., JACKSON, T.E., DRUMMOND, M.S., DE BOER, J.Z., BELLON, H., FEIGENSON, M.D., MAURY, R.C., and STEWART, R.H. (1992), *The geochemistry of young volcanism throughout western Panama and southeastern Costa Rica: an overview*, Geological Society of London. *149*, 569–579.

DEMETS, C., and WILSON, D.S. (1997), *Relative motions of the Pacific, Rivera, North American, and Cocos plates since 0.78 Ma*, Journal of Geophysical Research. *102*, 2789–2806.

DEMETS, C., and TRAYLEN, S. (2000), *Motion of the Rivera plate since 10 Ma relative to the Pacific and North America plates and the mantle*, Tectonophysics. *318*, 119–159.

FERRARI, L. and ROSAS-ELGUERA, J. (1999), Late Miocene to Quaternary extension at the northern boundary of the Jalisco block, western Mexico: *in* Delgado-Granados, H., Aguirre-Díaz, G., and Stock, J.M., eds., The Tepic-Zoacalco rift revisited in Cenozoic Tectonics and Volcanism of Mexico, Geological Society of America, Special Paper. 334, 1-23.

FERRARI L., PETRONE, C.M., and FRANCALANCI, L. (2001), *Generation of oceanic-island basalt-type volcanism in the western Trans-Mexican volcanic belt by slab rollback, asthenosphere infiltration, and variable flux melting*, Geology. *29*, 507–510.

FERRARI, L. (2004), *Slab detachment control on mafic volcanic pulse and mantle heterogeneity in central Mexico*, Geology. *32*, 77–80.

FERRARI, L., OROZCO-ESQUIVEL, T., MANEA, V., and MANEA, M. (2012), *The dynamic history of the Trans-Mexican Volcanic Belt and the Mexico subduction zone*, Tectonophysics. *522–523*, 122–149.

GARDINE, M.D., DOMINGUEZ, T., WEST, M.E., GRAND, S.P., and SUHARDJA, S.K. (2007). *The deep seismic structure of Volcan de Colima, Mexico*, AGU Spring Meeting Abstracts. *1*, 2.

GÓMEZ-TUENA, A., MORI, L., GOLDSTEIN, S.L., and PÉREZ-ARVIZU, O. (2011). *Magmatic diversity of western Mexico as a function of metamorphic transformations in the subducted oceanic plate*. Geochimica et Cosmochimica Acta, *75*(1), 213–241.

HESS, H.H., (1964), *Seismic anisotropy of the uppermost mantle under oceans*: Nature. *204*, 629–631.

HEURET, A., FUNICIELLO, F., FACCENNA, C., and LALLEMANDA, S. (2007), *Plate kinematics, slab shape and back-arc stress: A comparison between laboratory models and current subduction zones*, Earth and Planetary Science Letters. *256*, 473–483.

HOCHSTAEDTER, A.G., RYAN, J.G., LUHR, J.F., and HASENAKA, T. (1996). *On B/Be ratios in the Mexican volcanic belt*, Geochimica et Cosmochimica Acta. *60*, 613–628.

KINCAID, C., and GRIFFITHS, R. W. (2003). *Laboratory models of the thermal evolution of the mantle during rollback subduction*. Nature, *425*(6953), 58–62.

LANDAU, L.D. and LIFSHITZ, E.M., Fluid Mechanics (Second Edition) (Pergamon 1987).

LONSDALE, P. (2005), *Creation of the Cocos and Nazca plates by fission of the Farallon plate*, Tectonophysics. *404*, 237–264.

LUHR, J.F., NELSON, S.A., ALLAN, J.F., and CARMICHAEL, I.S.E. (1985), *Active rifting in southwestern Mexico: Manifestations of an incipient eastward spreading-ridge jump*, Geology. *13*, 54–57.

LUHR, J.F. (1997), *Extensional tectonics and the diverse primitive volcanic rocks in the western Mexican Volcanic Belt*, The Canadian Mineralogist. *35*, 473–500.

MANN, P. (2007), *Global catalogue, classification and tectonic origins of restraining- and releasing bends on active and ancient strike-slip fault systems*, Geological Society, London Special Publications. *290*, 13–142.

MARQUEZ, A., OYARZUN, R., DOBLAS, M. and VERMA, S.P. (1999), *Alkalic (ocean-island basalt type) and calc-alkalic volcanism in the Mexican volcanic belt: A case for plume-related magmatism and propagating rifting at an active margin?*, Geology. *27*, 51–54.

MOORE, G., MARONE, C., CARMICHAEL, I.S.E. and RENNE, P. (1994), *Basaltic volcanism and extension near the intersection of the Sierra Madre volcanic province and the Mexican Volcanic Belt*, GSA Bulletin. *106*, 383–394.

MORRIS, J.D., and HART, S.R. (1983), *Geochemical and isotopic variability in lavas from the eastern Trans-Mexican Volcanic Belt: slab detachment in a subduction zone with varying dip*, Geochimica et Cosmochimica Acta. *47*, 2015–2030.

NEGRETE-ARANDA, R. and CANON-TAPIA, E. (2008), *Post-subduction volcanism in the Baja California Peninsula, Mexico: the effects of tectonic reconfiguration in volcanic systems*, Lithos. *102*, 392–414.

OROZCO-ESQUIVEL, T., PETRONE, C.M., FERRARI, L., TAGAMI, T. and MANETTI, P. (2007), *Isotopic and incompatible element constraints on the genesis of island arc volcanics from Cold Bay and Amak Island, Aleutians, and implications for mantle structure*, Lithos. *93*, 149–174.

PARDO, M., and SUAREZ, G. (1995). *Shape of the subducted Rivera and Cocos plates in southern Mexico: Seismic and tectonic implications*, Journal of Geophysical Research. *100*, 12,357–12,373.

PETRONE, C. M., FRANCALANCI, L., CARLSON, R. W., FERRARI, L., and CONTICELLI, S. (2003). *Unusual coexistence of subduction-related and intraplate-type magmatism: Sr, Nd and Pb isotope and trace element data from the magmatism of the San Pedro–Ceboruco graben (Nayarit, Mexico)*. Chemical Geology, *193*(1), 1–24.

RIGHTER K. (2000). *A comparison of basaltic volcanism in the Cascades and western Mexico: compositional diversity in continental arcs*, Tectonophysics. *318*, 99–117.

SAJONA, F.G., MAURY, R.C., BELLON, H., COTTEN, J., and DEFANT, M.J. (1996), High Field Strength Element Enrichment of Pliocene—Pleistocene Island Arc Basalts, Zamboanga Peninsula, Western Mindanao (Philippines), Oxford Univ Press. *37*, 693–726.

SBALZARINI, I.F., and KOUMOUTSAKOS, P. (2005), *Feature point tracking and trajectory analysis for video imaging in cell biology*, Journal of Structural Biology *151*, 182–95.

SCHELLART, W.P. (2004), *Kinematics of subduction and subduction-induced flow in the upper mantle*, Journal of Geophysical Research. *109*, 1–19.

SCHELLART, W.P., STEGMAN, D.R., FARRINGTON, R.J., and MORESI, L. (2011), *Influence of lateral slab edge distance on plate velocity, trench velocity, and subduction partitioning*, Journal of Geophysical Research: Solid Earth. *116*.

SCHOLZ, C.H. and CAMPOS, J. (1995), *On the mechanism of seismic decoupling and back arc spreading at subduction zones*, Journal of Geophysical Research. *100*, 22,103–22,115.

SCHOLZ, C.H. and CAMPOS, J. (2012), *The seismic coupling of subduction zones revisited*, Journal of Geophysical Research. *117*, B05310.

SMITS, A.J. and LIM, T.T., Flow Visualization: Techniques and Examples (Second Edition) (Imperial College Press 2012).

SOTO, L.G., NI, J.F., GRAND, S.P., SANDVOL, E., VALENZUELA, R.W., GUZMAN-SPEZIALE, M., GOMEZ-GONZALEZ, J.M., and DOMiNGUEZ-REYES, T. (2009), *Mantle flow in the Rivera–Cocos subduction zone*, Geophysical Journal International. *179*, 1004–1012.

STOCK, J.M, and LEE, J. (1994), *Do microplates in subduction zones leave a geological record?*, Tectonics. *13*, 1472–1487.

STOREY, M., ROGERS, G., SAUNDERS, A.D. and TERRELL, D.J. (1989), *San Quintin volcanic field, Baja California, Mexico:'within-plate'magmatism following ridge subduction,* Terra Nova. 1, 195–202.

STUBAILO, I., BEGHEIN, C., and DAVIS, P. M. (2012), *Structure and anisotropy of the Mexico subduction zone based on Rayleigh-wave analysis and implications for the geometry of the Trans-Mexican Volcanic Belt*, Journal of Geophysical Research. *117*, 1–16.

SUHARDJA, S. K. (2013), Mapping the Rivera and Cocos Subduction Zone, (Ph.D. Thesis) University of Texas, Austin.

SUTER, M., LEGORRETA-QUINTERO, O., LOPEZ-MARTINEZ, M., AGUIRRE-DIAZ, G., and FARRAR, E. (1995), *The Acambay graben: Active intraarc extension in the trans-Mexican volcanic belt, Mexico*, Tectonics. 14, 1245–1262.

SUTER, M., LOPEZ-MARTINEZ, M., LEGORRETA-QUINTERO, O., and MARTINEZ-CARILLO, M. (2001), *Quaternary intra-arc extension in the central Trans-Mexican volcanic belt*, GSA Bulletin. *113*, 693–703.

TIAN, L., CASTILLO, P.R., LONSDALE, P.F., HAHM, D. and HILTON, D.R. (2011), *Petrology and Sr-Nd-Pb-He isotope geochemistry of postspreading lavas on fossil spreading axes off Baja California Sur, Mexico*, Geochemistry, Geophysics, Geosystems. *12*.

TOMMASI, A. TIKOFF, B. and VAUCHEZ, A. (1999), *Upper mantle tectonics: three-dimensional deformation, olivine crystallographic fabrics and seismic properties*, Earth and Planetary Science Letters. *168*, 173–186.

TURCOTTE, D.L., and SCHUBERT, G., Geodynamics (Cambridge University Press 2002).

TURNER, S., and HAWKESWORTH, C. (1998). *Using geochemistry to map mantle flow beneath the Lau Basin*. Geology, *26*(11), 1019-1022.

VAN HUNEN, J., VAN DEN BERG, A.P. and VLAAR, N.J. (2002), *On the role of subducting oceanic plateaus in the development of shallow flat subduction*, Tectonophysics. *352*, 317–333.

VAN HUNEN, J., VAN DEN BERG, A.P. and VLAAR, N.J. (2004), *Various mechanisms to induce present-day shallow flat subduction and implications for the younger Earth: a numerical parameter study*, Physics of the Earth and Planetary Interiors. *146*, 179–194.

VERMA, S.P. (2002), *Absence of Cocos plate subduction-related basic volcanism in southern Mexico: a unique case on Earth?*, Geology. *30*, 1095–1098.

VERMA, S. P. (2009), *Continental Rift Setting for the Central Part of the Mexican Volcanic Belt: A Statistical Approach.* The Open Geology Journal. *3*, 8–29.

WILSON, J.T. (1973), *Mantle plumes and plate motions,* Tectonophysics. 19,149–164.

YANG, T., GRAND, S.P., WILSON, D., GUZMAN-SPEZIALE, M., GOMEZ-GONZALEZ, J.M., DOMINGUEZ-REYES, T., and NI, J. (2009), *Seismic structure beneath the Rivera subduction zone from finite-frequency seismic tomography*, Journal of Geophysical Research. *114*, 1–12.

ZANDT, G., and HUMPHREYS, E. (2008). *Toroidal mantle flow through the western US slab window*. Geology, *36*(4), 295–298.

(Received July 2, 2015, revised November 17, 2015, accepted November 25, 2015, Published online December 22, 2015)

Pure Appl. Geophys. 173 (2016), 3419–3443
© 2016 Springer International Publishing
DOI 10.1007/s00024-015-1213-8

❙ Pure and Applied Geophysics

Active Crustal Faults in the Forearc Region, Guerrero Sector of the Mexican Subduction Zone

KRZYSZTOF GAIDZIK,[1] MARIA TERESA RAMÍREZ-HERRERA,[1,2] and VLADIMIR KOSTOGLODOV[3]

Abstract—This work explores the characteristics and the seismogenic potential of crustal faults on the overriding plate in an area of high seismic hazard associated with the occurrence of subduction earthquakes and shallow earthquakes of the overriding plate. We present the results of geomorphic, structural, and fault kinematic analyses conducted on the convergent margin between the Cocos plate and the forearc region of the overriding North American plate, within the Guerrero sector of the Mexican subduction zone. We aim to determine the active tectonic processes in the forearc region of the subduction zone, using the river network pattern, topography, and structural data. We suggest that in the studied forearc region, both strike-slip and normal crustal faults sub-parallel to the subduction zone show evidence of activity. The left-lateral offsets of the main stream courses of the largest river basins, GPS measurements, and obliquity of plate convergence along the Cocos subduction zone in the Guerrero sector suggest the activity of sub-latitudinal left-lateral strike-slip faults. Notably, the regional left-lateral strike-slip fault that offsets the Papagayo River near the town of La Venta named "La Venta Fault" shows evidence of recent activity, corroborated also by GPS measurements (4–5 mm/year of sinistral motion). Assuming that during a probable earthquake the whole mapped length of this fault would rupture, it would produce an event of maximum moment magnitude Mw = 7.7. Even though only a few focal mechanism solutions indicate a stress regime relevant for reactivation of these strike-slip structures, we hypothesize that these faults are active and suggest two probable explanations: (1) these faults are characterized by long recurrence period, i.e., beyond the instrumental record, or (2) they experience slow slip events and/or associated fault creep. The analysis of focal mechanism solutions of small magnitude earthquakes in the upper plate, for the period between 1995 and 2008, revealed that frequent normal faults, sub-parallel to the trench, could be reactivated in the current stress field related to the Cocos subduction. Moreover, these features could also be reactivated by subduction megathrust earthquakes.

Key words: Forearc deformation, active tectonics, Guerrero sector, Middle America Subduction Zone, river network pattern, upper plate faults.

1. Introduction

Forearc regions being a part of one of the most active plate-boundary environments are characterized by destructive earthquakes and concurrent uplift or subsidence (e.g., PLAFKER 1969; LI 1993; MELTZNER et al. 2006; NATAWIDJAJA et al. 2006; SUBARYA et al. 2006; REHAK et al. 2008; MELNICK et al. 2009; ITO et al. 2011; SIMONS et al. 2011; VIGNY et al. 2011; TAJIMA et al. 2013), as well as lateral movement mainly in oblique subduction zones (e.g., FITCH 1972; JARRARD 1986; KIMURA 1986; DeMETS 1992; McCAFFREY 1992, 2009; DELOUIS et al. 1998; McCAFFREY et al. 2000; CHOY and KIRBY 2004; ONUR and SEEMANN 2004; CORTI et al. 2005; CÁSERES et al. 2005; LANGE et al. 2007, 2008; MORENO et al. 2008; FERNANDEZ 2009; MELNICK et al. 2009; BERGLAR et al. 2010; MÉTOIS et al. 2012; VARGAS et al. 2013). Indeed, some of the largest active strike-slip faults causing catastrophic earthquakes are related to oblique subduction zones (GUTSCHER and LALLEMAND 1999; MORENO et al. 2008; McCAFFREY 2009; MELNICK et al. 2009; MOLNAR and DAYEM 2010; VARGAS et al. 2013).

In forearc mountains, the influence of ongoing deformation on landscape development dominates and usually overlaps, and masks the effect of passive structural control (BIERMAN and MONTGOMERY 2014). However, even in such environments active deformation and passive tectonic control can be equally important in the drainage network development (STOKES et al. 2008; ŻABA et al. 2012). Active forearc regions inherit upper plate structures that may influence the long-term landscape evolution (ECHTLER et al. 2003). Generally, tectonic features and

[1] Laboratorio Universitario de Geofísica Ambiental and Instituto de Geografía, Universidad Nacional Autónoma de México, Ciudad Universitaria, Coyoacán, 04510 Mexico, DF, Mexico. E-mail: gaidzik@igg.unam.mx; tramirez@igg.unam.mx; ramirez@seismo.berkeley.edu

[2] Berkeley Seismological Laboratory, Department of Earth and Planetary Science, University of California Berkeley, Berkeley, CA, USA.

[3] Instituto de Geofísica, Universidad Nacional Autónoma de México, Ciudad Universitaria, Coyoacán, 04510 Mexico, DF, Mexico.

differences in lithology can exert an important control on fluvial pattern (STOKES *et al.* 2008), whereas active tectonic deformation such as faulting and regional and/or local uplift deflects river's courses and affects a river's morphology, controls the incision and erosion rates, and increases local relief (e.g., OUCHI 1985; BURBANK 1992; JACKSON *et al.* 1996; DEMOULIN 1998; RAMÍREZ-HERRERA 1998; BURBANK and ANDERSON 2001; MORELL *et al.* 2008; STOKES *et al.* 2008; ŻABA *et al.* 2012; GASPARINI and WHIPPLE 2014).

Many features of the active Mexican subduction forearc, within and outside the Guerrero sector, also known as the Guerrero seismic gap, have not yet been fully understood. Questions on deformation events and active tectonics in the forearc have not been fully addressed. Studies of active faults and their influence on the landscape in this area are lacking, with the exception of the coastal zone, where the relation between morphology and active tectonic deformation—mainly based on studies on marine terraces and paleoseismology—has been reported (RAMÍREZ-HER-RERA and URRUTIA-FUCUGAUCHI 1999; RAMÍREZ-HERRERA *et al.* 2007, 2009, 2010).

We aim to understand and determine active tectonic processes, using the river network pattern, topography and structural data, in the forearc of the active Mexican subduction zone, within the Guerrero sector of south-west Mexico. We focused on: (1) distinguishing evidence of active tectonic deformation using the river network pattern and topography; (2) evaluating the relationship between the geometry and orientation of the drainage network and fault systems in each of the studied drainage basins; and (3) recognizing tectonic structures that have the potential of being reactivated within the current stress field related to the Mexican subduction zone. With this purpose, we used geomorphic and structural methods correlating them with available data on the current stress regime from focal mechanism solutions and GPS measurements.

2. Study Area

2.1. Tectonic Setting

The study area is located in the forearc of the Cocos-North America subduction zone, between Zihuatanejo and Acapulco, in the sector known as

Figure 1 ▶

A Tectonic, seismic (according to RAMÍREZ-HERRERA *et al.* 2009) and tectonostratigraphic (according to CAMPA and CONEY 1983) setting of the study area: *MAT* Middle American Trench, *F.Z.* Fracture zones, *X* Xolapa terrane. *Symbols*: *shaded circles* indicate rupture areas and numbers the years of most important subduction seismic events of last century; *small dashed lines* indicate fracture zones; *thick dash* with *arrowheads* shows subduction zone; *thick dash* shows location of East Pacific Rise. *Arrows* indicate direction of convergence and numbers indicate convergence rate in cm/year (DEMETS *et al.* 2010). B Sketch map of the study area. Convergence vectors, according to MORVEL 2010 model (DEMETS *et al.* 2010), presented with the *3 sigma error ellipses*, indicate sinistral obliquity of the Cocos/North America subduction

the NW Guerrero seismic gap, and partially in the sector known as the SE Guerrero seismic gap (Fig. 1). The NW Guerrero seismic gap has experienced no significant thrust earthquakes since 1911 (ANDERSON *et al.* 1989; KOSTOGLODOV and PONCE 1994). However, large slow earthquakes have occurred on the subduction interface in 1998, 2001–2002, 2006, 2009–2010, and 2014 (KOSTOGLODOV *et al.* 2003, 2015). The southwest–northeast Cocos-North America plate subduction stress regime initiated ca. 22 Ma as a result of Farallón plate breaking up into the Cocos and Nazca plates (WORTEL and CLOETINGH 1981; BARCKHAUSEN *et al.* 2001). The oceanic lithosphere of the Cocos plate subducts beneath the North America continental plate with convergence rates ranging from 6.4 to 6.7 cm/year along the Guerrero coast (DEMETS *et al.* 2010). Relative motion is slightly oblique to the trench normal, yielding a trench-parallel sinistral component of ~8 mm/year (LOWRY *et al.* 2006). The angle of obliquity, even though small, is shown by every applied plate-motion model: 9–10° (MORVEL; DEMETS *et al.* 2010), ~12° (MORVEL 56; DEMETS *et al.* 2010), 13–14° (GSRM v. 1.2; KREEMER *et al.* 2003), ~12° (HS3-NUVEL 1A; GRIPP and GORDON 2002).

2.2. Regional Geology

The study area is located almost entirely within the tectonostratigraphic Xolapa terrane (Fig. 1A). Only the northernmost and north-eastern parts are within the Guerrero arc terrane. The Xolapa terrane (CAMPA and CONEY 1983) also known as the Chatino terrane (SEDLOCK *et al.* 1993) is a crustal block,

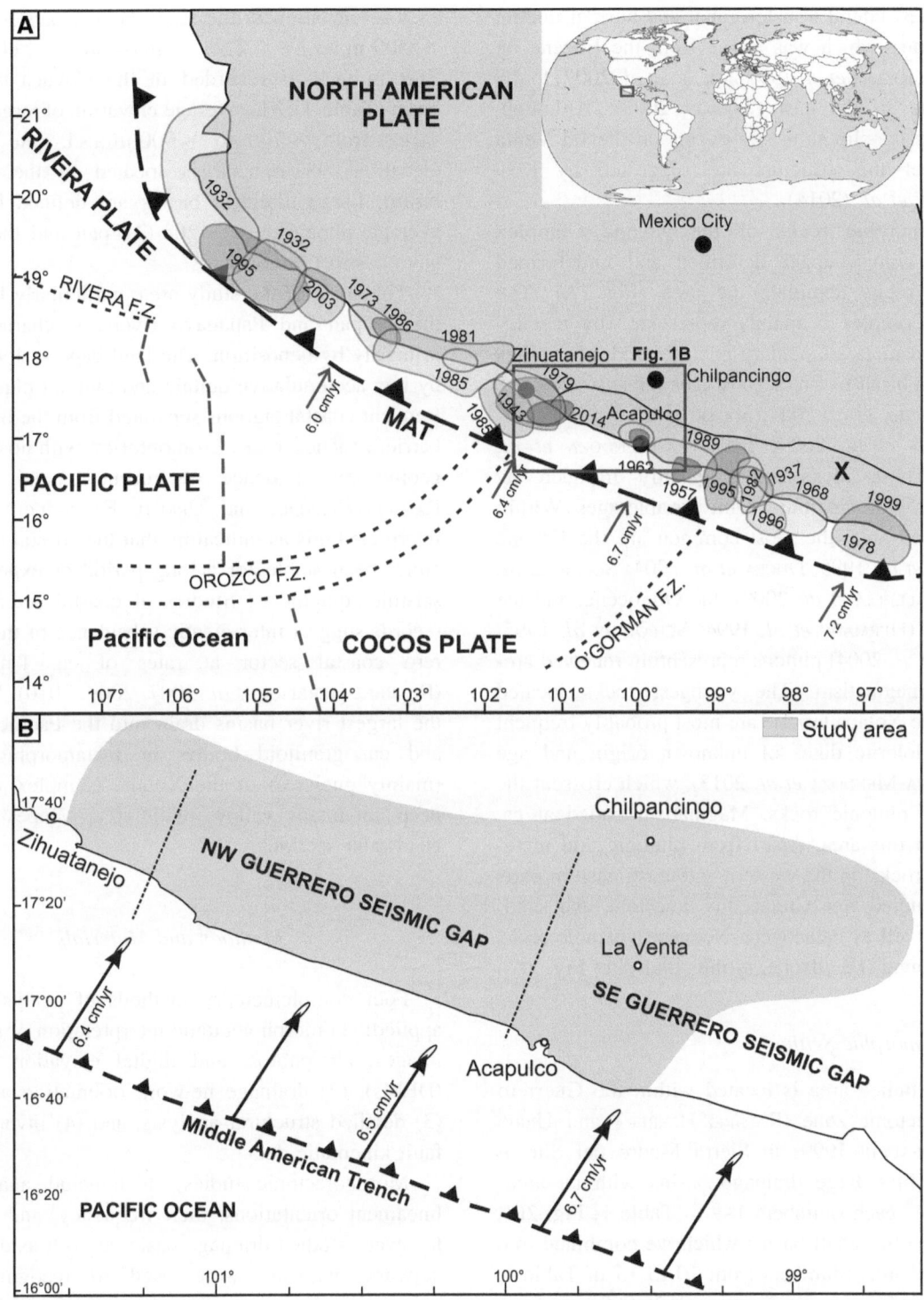

extending along the Pacific coast of southwestern Mexico (Fig. 1A). The contact of the Xolapa terrane with the neighboring terranes is complex. It includes both brittle and ductile faults, and parts of the boundary are covered by Cenozoic plutons. The most prominent boundary structure is the Chacalapa-La

Venta—left-lateral transtensional and normal ductile shear zone—which was active from the Eocene on the west (RILLER *et al.* 1992; SOLARI *et al.* 2007) to the Oligocene in the east (TOLSON 2005). Although, current strike-slip slow slip events on the La Venta section of this structure are suggested by KOS-TOGLODOV *et al.* (2014).

Metamorphic rocks of the Xolapa Complex together with younger deformed and undeformed plutonic rocks dominate the area (Fig. 2A). The Xolapa Complex is mainly represented by metaigneous and metasedimentary gneisses and migmatites of Precambrian to Cretaceous ages (HERRMANN *et al.* 1994; DUCEA *et al.* 2004; SOLARI *et al.* 2007; PÉREZ-GUTIÉRREZ *et al.* 2009; TALAVERA-MENDOZA *et al.* 2013). These rocks are frequently intruded by numerous plutonic bodies with variable ages. Within the studied area, the most common are the Eocene (SCHAAF *et al.* 1995; DUCEA *et al.* 2004; SOLARI *et al.* 2007; VALENCIA *et al.* 2009), the Oligocene, and the Miocene (HERMANN *et al.* 1994; SCHAAF *et al.* 1995; DUCEA *et al.* 2004) plutons representing renewed arc-related magmatism. The youngest rocks located within the Xolapa terrane are most probably frequent narrow dolerite dikes of unknown origin and age (TALAVERA-MENDOZA *et al.* 2013), which crosscut the youngest plutonic rocks. Magmatism ended at ca. 25 Ma in this area. Apart from plutonic and metamorphic rocks, in the western and north-eastern parts of the studied area Cretaceous limestone and sandstone as well as Palaeogene-Neogene volcanic rocks can be found (i.e.: dacite, ryolite, andesite; Fig. 2A).

2.3. Geomorphic Setting

The studied area is located within the Guerrero morphotectonic zone (RAMÍREZ-HERRERA and URRU-TIA-FUCUGAUCHI 1999) in Sierra Madre del Sur. It includes nine large drainage basins with an area >450 km^2 each (numbers 1–9 in Table 1; Fig. 2B), and numerous small basins which we combined into six larger units (numbers from 10 to 15 in Table 1; Fig. 2B), within an area of almost 25,000 km^2 from Zihuatanejo to Acapulco, in which the main rivers flow into the Pacific Ocean. For chosen areal units, we used names that correspond with the name of the main river in each unit (Table 1; Fig. 2B).

The elevation in the study area ranges from 0 to >3500 m a.s.l. The maximum elevation, 3526 m a.s.l., is recorded in the Coyuca drainage basin (Table 1). The average elevation of large basins varies from ~700 to ~900 m a.s.l. (the highest elevation—932 m a.s.l. is located in the Tecpan basin). Large drainage basins are defined by high average slope values ~20° (Tecpan and the Papagayo basin) (Table 1).

The coast of the study area, particularly between the Tecpan and Papagayo rivers, is characterized primarily by deposition. The landscape is dominated by low accumulative deltaic and alluvial plains with frequent coastal lagoons separated from the ocean by barrier beaches, rocky promontories, with no obvious geomorphic evidence of recent tectonic uplift. RAMÍREZ-HERRERA and URRUTIA-FUCUGAUCHI (1999) interpreted this as indicating that the coastal area has either been stable for a long period or experienced seismic 'quietness'. Studies of coastal sedimentary records suggest interseismic subsidence of the Guerrero coastal sector at rates of ca. 1 mm/year (RAMÍREZ-HERRERA *et al.* 2007, 2009, 2010). Most of the largest river basins drain into the Pacific Ocean and cut granitoid bodies or metamorphic rocks (mainly gneisses) of the Xolapa Complex forming deep mountain valleys with steep slopes in the headwater sections.

3. Methods and Materials

Four complementary methods of analysis were applied: (1) morphotectonic interpretation of satellite images, air photos, and digital elevation models (DEMs), (2) drainage network orientation analysis, (3) detailed structural analysis, and (4) inversion of fault kinematic data.

Morphotectonic studies, which include analysis of lineament orientations, their frequency, and lengths for every studied drainage basin, as well as drainage network analysis, were used to recognize the response of the river network pattern to active tectonic processes and to select sites for detailed structural studies during fieldwork. These analyses were conducted using digital maps and satellite image data. We used topographic maps (1:50,000,

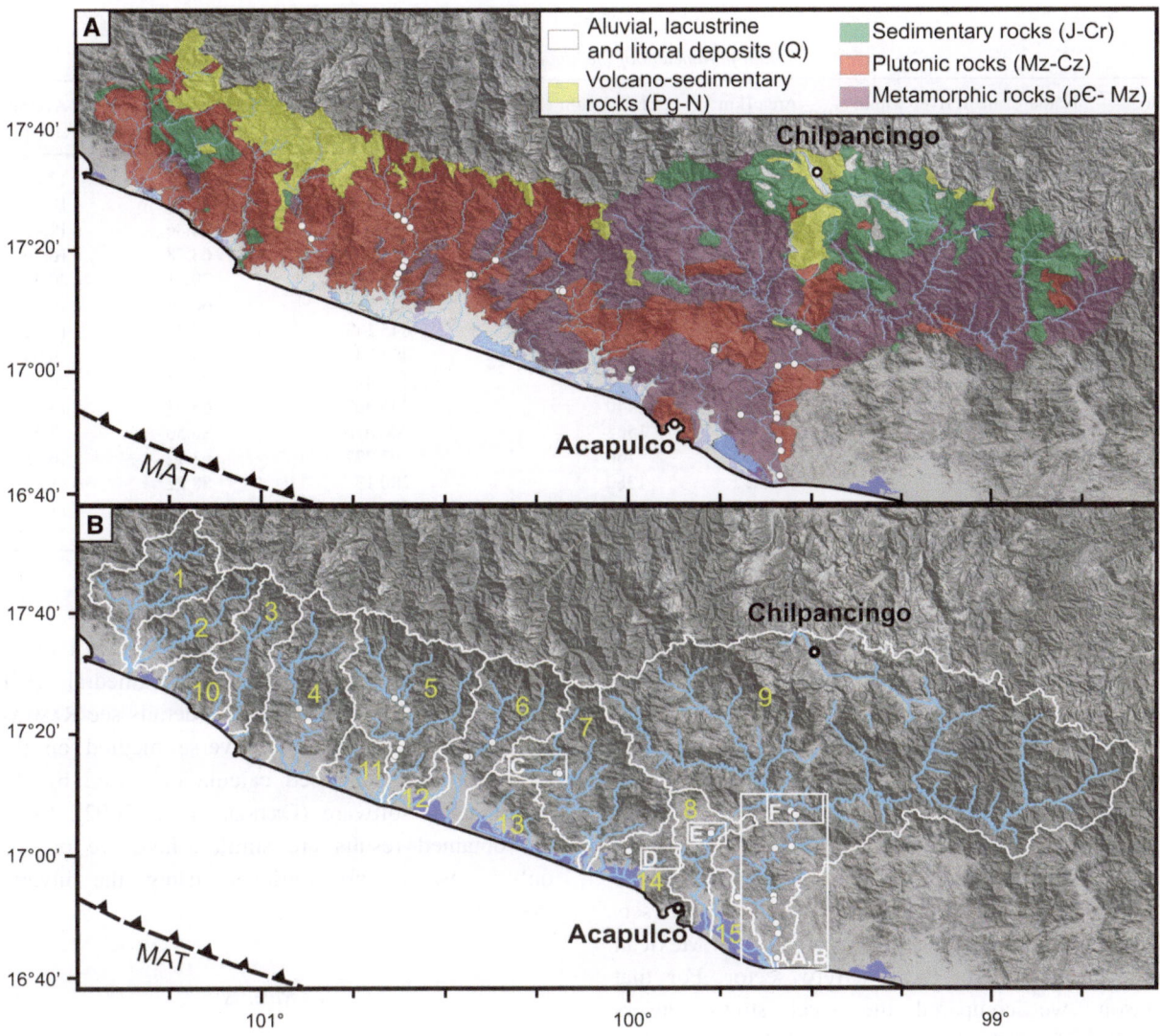

Figure 2

A Simplified geological map of the study area with the location of studied bedrock outcrops shown with dots (based on: CAMPA URANGA *et al.* 1998; CARRANZA *et al.* 1999; CRUZ LÓPEZ *et al.* 2000). **B** *Shaded relief map* of the forearc region of the Guerrero sector of the Mexican subduction zone based on 15 m resolution DEM of the Guerrero state (INEGI, http://www.inegi.org.mx). *White lines* mark water divides of studied drainage basins. *Numbers* from *1* to *15* are drainage basins ordered from west to east (for main parameters see Table 1): *1* San Jeronimito, *2* Coyuquilla, *3* Petatlan, *4* San Luis, *5* Tecpan, *6* Atoyac, *7* Coyuca, *8* La Sabana, *9* Papagayo, *10* Juluchuca, *11* Nuxco, *12* Del Tular, *13* Cacalutla, *14* Conchero, *15* Tres Palos. *Boxes* show the location of subsequent inserts in the Fig. 3

INEGI, 2014), geologic maps (1: 250,000, Mexican Geological Survey; CAMPA URANGA *et al.* 1998; CARRANZA *et al.* 1999; CRUZ LÓPEZ *et al.* 2000), satellite images, aerial photos, and 15-m resolution shaded relief DEM (obtained from INEGI).

Geometric, kinematic, dynamic and statistical structural analyses focused on brittle faults allowing us to define the regional fault pattern and to distinguish the youngest fault populations, based on relative age relations. These structural studies within selected outcrops along the main rivers of the chosen drainage basins (see Fig. 2 for studied outcrops location) were carried out based on the standard techniques described in the literature (e.g., RAMSAY 1967; RAMSAY and LISLE 2000; FOSSEN 2010 and literature therein). The fault data were collected during

Table 1

Main parameters of the studied drainage basins

No.	Name of main river within drainage basin	Area (km^2)	Maximum elevation (m a.s.l.)	Average elevation (m a.s.l.)	Maximum slope (°)	Average slope (°)
1	San Jeromito	812.16	2540	671.863	85.73	17.62
2	Coyuquilla	582.45	2560	840.477	62.30	18.86
3	Petatlan	548.64	2680	686.629	74.86	19.43
4	San Luis	1061.8	2760	754.826	63.78	16.36
5	Tecpan	1338.38	3040	932.299	79.11	20.37
6	Atoyac	902.7	3329	910.96	66.26	19.26
7	Coyuca	1300.77	3526	845.143	69.34	18.51
8	La Sabana	466.03	2260	493.749	68.93	9.94
9	Papagayo	7571.4	3323	1112.4	73.52	20.19
10	Juluchuca	254.79	1440	335.307	65.03	13.61
11	Nuxco	291.33	1568	206.676	86.56	7.58
12	Del Tular	141.72	760	97.777	63.12	6.55
13	Cacalutla	535.92	1380	280.13	88.22	11.44
14	Conchero	429.03	1600	290.912	62.8	11.77
15	Tres Palos	275.36	420	59.366	48.13	3.42

For location of drainage basins see Fig. 2B

two field campaigns (April and August, 2014) and include spatial orientation of fault surfaces and striations (only if measurable), turn of relative movement (based on the analysis of minor structures on fault surface, accompanying fractures and displacement of older structures) and information about lithology.

The stress analysis was applied to determine the possibility of reactivation of defined fault sets within the current stress field related to the Mexican subduction zone in the Guerrero sector. For that reason, we compared the local stress tensors resulting from the inversion of fault kinematic data recorded in the field with the current stress regime obtained from focal mechanism solutions of shallow, crustal earthquakes. For paleostress analysis, we used homogeneous field-based fault data sets derived from structural analysis. To separate homogeneous fault sets from a heterogeneous population, we utilized field observations of relative age relations and fluctuation histograms. To obtain the recent stress field, we used focal mechanism solutions of small magnitude, shallow earthquakes in the upper plate, which occurred within the studied area between 1995 and 2008 (data from PACHECO and SINGH 2010). Two independent analysis procedures were used to infer current and paleostress tensors: (1) geometric methods

supported by TectonicsFP software (ORTNER *et al.* 2002), which include the right dihedral (P/T method) and P-B-T methods (for details see RAMSAY and LISLE 2000), and (2) inverse method on the basis of computer-based calculations used by the TectonicsFP software (ORTNER *et al.* 2002). Since the obtained results are similar, here we present only stress tensors retrieved using the inverse method.

4. Results

4.1. Drainage Network and Aligned Morphological Features

The drainage network pattern in general can be characterized as dendritic (Fig. 2B). This might be related to the uniformly resistant crystalline rocks and, secondarily, to the frequent veins and dikes within metamorphic and plutonic rocks (HOWARD 1967; SCHUMM *et al.* 2000). Locally, especially within the small drainage basins, transitions to trellis (e.g., Cacalutla) or parallel (e.g., Del Tular, La Sabana) pattern can be observed (Fig. 2B).

The main river azimuths, calculated as the azimuth of a straight line from the headwater towards the river mouth, referring to the general orientation of a river (BARHOLOW 1989), are constant within the

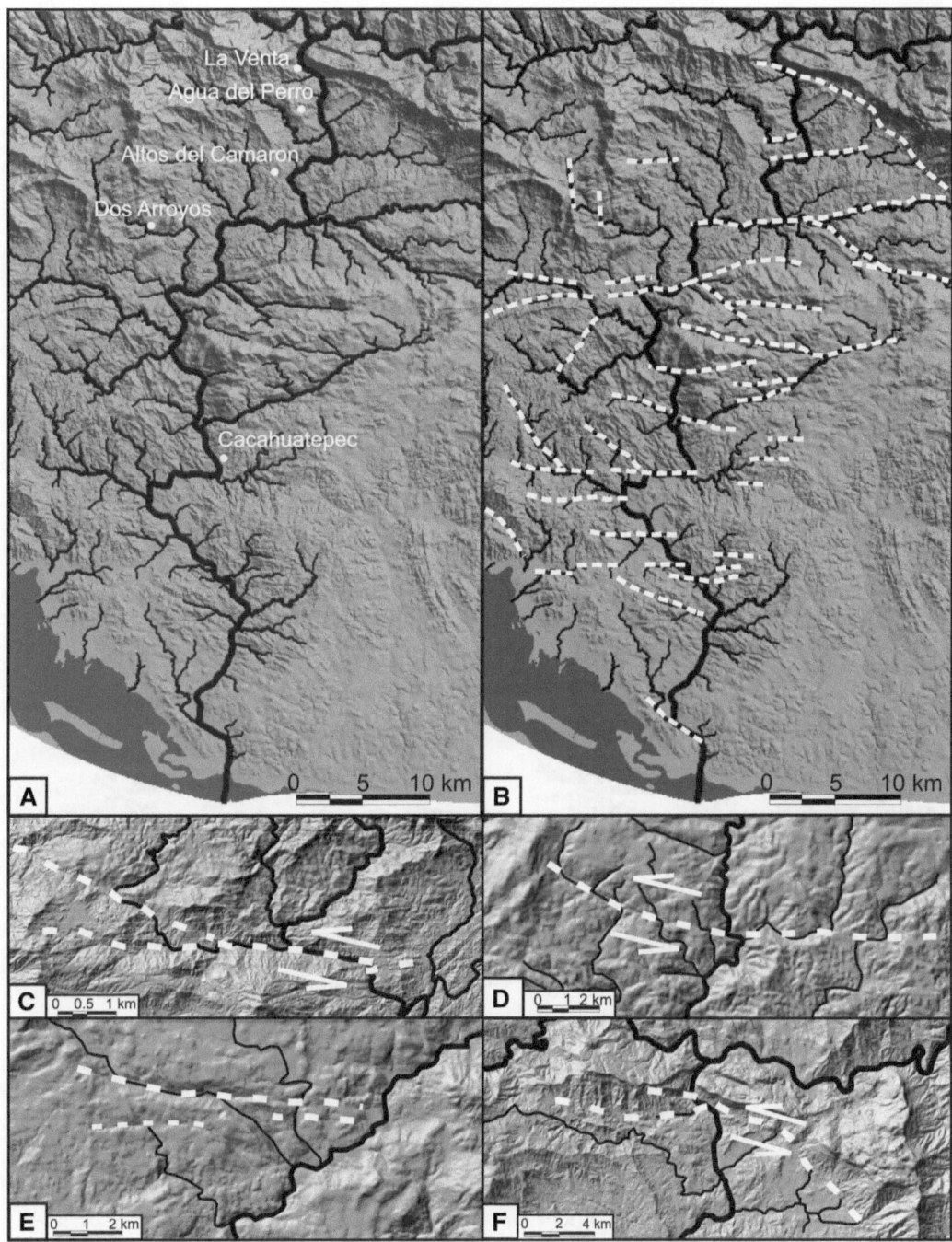

Figure 3
The relationship between the river network pattern and the main sub-latitudinal lineaments in the forearc region of the Mexican subduction in the Guerrero sector. **A** Drainage network pattern in the southern part of the Papagayo drainage basin. **B** Tectonic control on the river network development in the southern part of the Papagayo drainage basin: *white, dashed lines* lineaments influencing the river network. **C** W–E strike-slip fault deflecting the main direction of the Coyuca River. **D** WNW–ESE left-lateral fault offsetting the Conchero River and its tributaries. **E** W–E normal faults in the La Sabana basin forming scarps and waterfalls. **F** Sinistral deflection of the Papagayo River along the La Venta Fault in the Agua del Perro segment (named in this study)

213

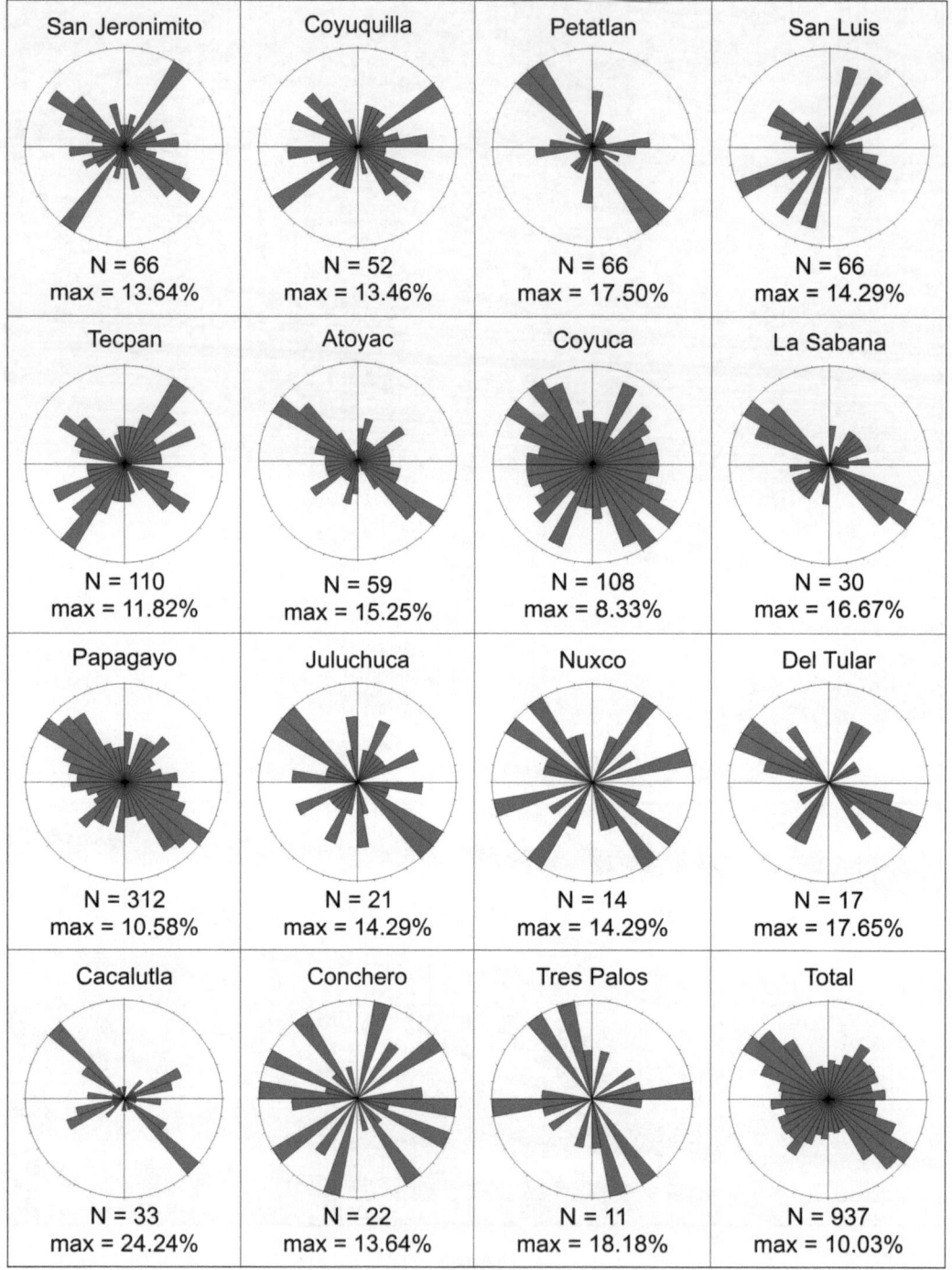

Figure 4
Spatial orientation of lineaments within each studied drainage basin (numbered as in Fig. 2B; Table 1) and the whole study area

entire study area. Rivers flow generally towards the SSW direction (azimuth of ~200°) in the western and central parts of the study area, and towards the S in the eastern part of the same area (Fig. 2B). However, local anomalies and deflections from this general direction of gravitational slope commonly occur, particularly in the central and eastern parts of the study area (from San Luis to Papagayo basin; Fig. 2B). Sharp bends and rapid changes in the direction of the river flow are especially clear in the basins of Papagayo, Conchero, La Sabana and Coyuca (Fig. 3). The analysis of satellite images and DEM revealed that they are mainly related to sub-latitudinal features parallel to the general orientation of the subduction zone.

The lineament pattern obtained for each drainage basin, in terms of both frequency and lengths, is generally consistent with the entire study area (Fig. 4). It is characterized by two main directions: (1) parallel (WNW–ESE to NW–SE) and (2) perpendicular (NE–SW) to the orientation of the Cocos subduction zone and the general trend of the mountain front (Fig. 4). Such a pattern generally agrees with the results of HERNÁNDEZ-SANTANA and ORTIZ-PÉREZ (2005) study conducted on the southern part of the Papagayo drainage basin.

Predominant sub-latitudinal to NW–SE lineaments form usually the longest structures. In the central and eastern parts of the study area, particularly in the Papagayo, La Sabana, Conchero and Coyuca drainage basins, these sub-latitudinal features offset the main river directions (Fig. 3). Their control on the river network is especially evident in the southern part of the Papagayo drainage basin (Fig. 3A, B). Here, a series of W–E features produce deflections and very sharp bends along the Papagayo River and its tributaries. Some of those features suggest left-lateral motion, whereas others—right-lateral kinematics. Sections, where the Papagayo River flows perpendicular to the direction of gravitational slope, delimited with sharp bends, can exceed 5 km (5.4 km close to Dos Arroyos town, 4 km close to Cacahuatepec town, 2.8 km close to Alto de Camarón; Fig. 3A, B). Close to La Venta, the Papagayo River is deflected by 1.5 km (Fig. 3F). Similarly, a large (4.3 km) left-lateral offset can be observed along the Coyuca River (Fig. 3C). In the Conchero drainage basin, where the main river and its tributaries are deflected, the left-lateral offset varies from 100 m to almost 500 m (Fig. 3D). Locally, as in the La Sabana and Coyuca basins, W–E features suggest also a substantial contribution of the vertical component in the total deformation. Latitudinal

lineaments in the northern part of the La Sabana basin are reflected in the landscape by clear scarps and waterfalls. The influence of these features gradually decreases to the west of the study area (Fig. 2B)—west of the San Luis basin, their control on the river network is unnoticeable with the exception of very local drainage anomalies.

Lineaments belonging to the second set (NE–SW) are frequent, particularly in the western part of the study area and in some central drainage basins. They usually form two sub-sets: WSW–ENE and NE–SW (Fig. 4). Very straight sections of many eastern tributaries of Tecpan, San Luis and Coyuquilla are formed along this direction. Contrary to the previously described main set, these features do not seem to deflect the main rivers.

4.2. Faults

Meso-faults were recorded in the field within chosen outcrops of crystalline rocks in drainage basins: Tecpan, Atoyac, Coyuca, Conchero, La Sabana and Papagayo (for location of studied outcrops see Figs. 2, 5). They are represented by both inherited (exhumed) structures and potentially active or those that could be reactivated. Sub-vertical and very steep faults dominate; almost half of recorded fault surfaces dip at angles >80° (Figs. 5, 6). Among them, vertical to very steep strike-slip faults and steep to moderate normal faults predominate. Within the studied outcrops, we recorded only two meso-faults for which kinematic indicators suggest a reverse sense of movement. Generally, the studied meso-faults can be divided into four sets according to their spatial orientation (Fig. 6): (1) vertical to very steep sub-latitudinal faults, represented mainly by strike-slip structures, locally also by normal faults, (2) NNE–SSW, NE–SW steeply dipping normal faults, (3) N–S vertical strike-slip faults, and (4) NW–SE structures represented by both strike-slip and normal faults.

Sub-latitudinal strike-slip faults are the most common fault structures in almost all the studied drainage basins (Figs. 5, 6). They usually form large surfaces with well-developed associated structures that unequivocally indicate the kinematics of faulting. These kinematic indicators suggest both senses of

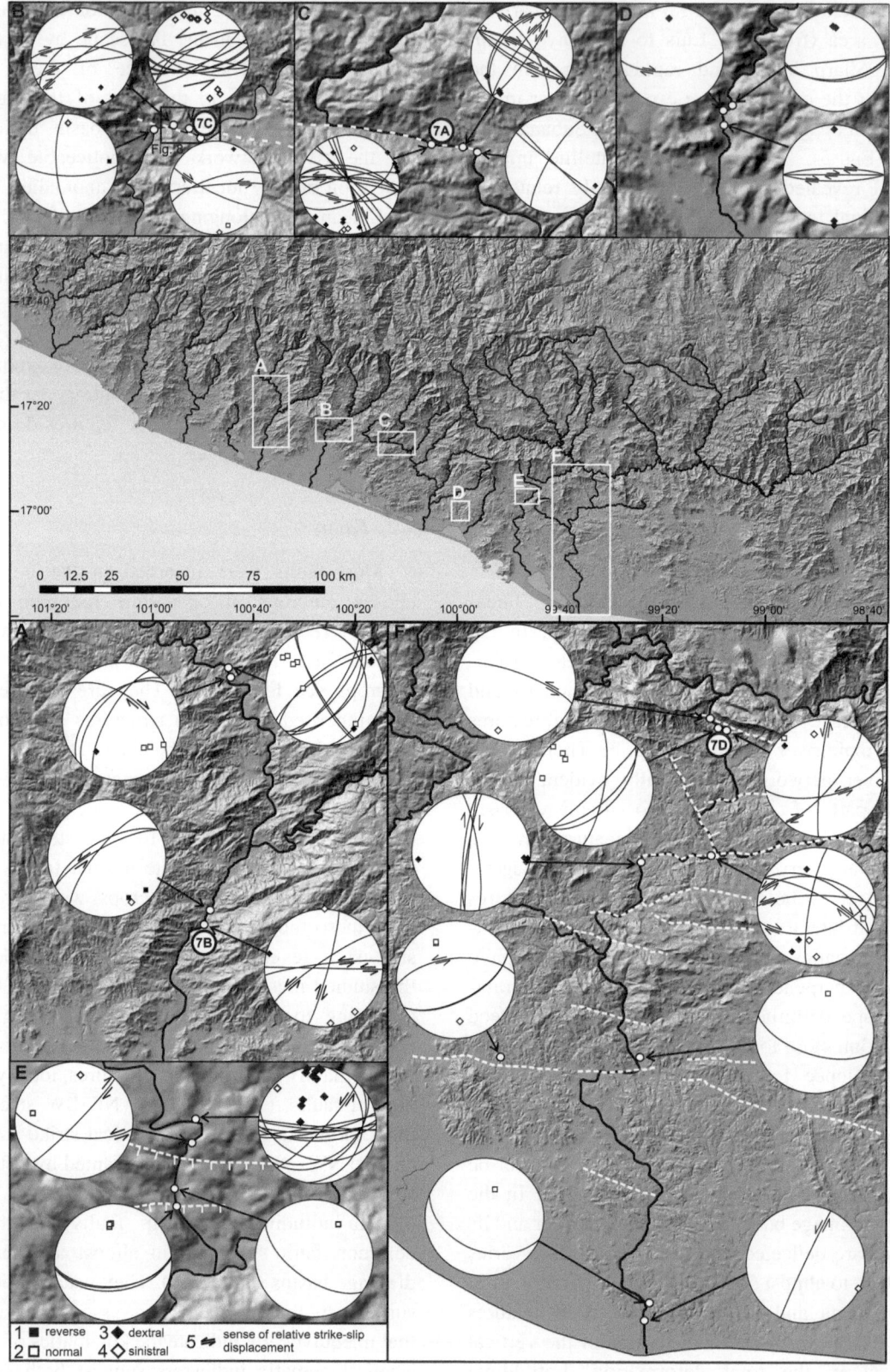

Figure 5
Spatial distribution and kinematics of meso-faults recorded in the field in crystalline rocks within studied drainage basins. **A** Meso-faults observed in granitoids, locally also in metamorphic rocks (second field-point from the north) within the Tecpan drainage basin. Note the predominance of NE–SW normal meso-faults scarcely reflected in the landscape. The control of sub-latitudinal left-lateral strike-slip faults, which are common in the southern outcrops, is also limited. **B** Secondary left-lateral strike-slip meso-faults observed in the left-lateral zone of higher rank that control the direction of the Atoyac River (granitoids). **C** Secondary strike-slip faults in the latitudinal fault zone of higher rank used by the Coyuca River (granitoids). **D** Sub-latitudinal right-lateral meso-faults with no expression in the landscape, Embarcadero River (gneiss). **E** W–E normal faults in the La Sabana basin forming scarps and waterfalls (gneiss). **F** Sub-latitudinal and sub-longitudinal strike-slip faults and W–E to NE–SW normal faults in the southern part of the Papagayo drainage basin and their relation to river anomalies (gneiss and granitoids). *Numbers* in *circles* from 7A to 7D indicate locations of structures presented on photographs in Fig. 7. *Box* in *insert B* indicates the location of structures shown in Fig. 8

relative displacement: (a) right-lateral determined mainly by slickenlines, irregularites on fault surface, subordinate fractures, mainly low-angle Riedel shears (R) and displacements of small, old veins (Fig. 7A); and (b) left-lateral indicated by minor structures occurring on faults planes, slickenlines, subordinate fractures, particularly low-angle Riedel shears (Fig. 7B), but also R′ and P shears, and extensional T fractures, as well as displacement of veins (Fig. 7C). Dextral faults can be found mainly within gneisses and migmatites of the Xolapa Complex, for example along the rivers Embarcadero and La Sabana (Fig. 5D, E), and locally in granitoids (Coyuca basin, see Fig. 5A). Their influence on the direction of the river flow seems to be limited. For example along the Embarcadero River these features, being sub-perpendicular to the direction of gravitational slope and, therefore, to the main direction of the river, do not change its direction. Locally they form small and short bends, but without rapid changes in the river direction. Left-lateral strike-slip faults, in contrast, are observed in both main rock types in the study area: metamorphic rocks of the Xolapa Complex (Papagayo drainage basin; Fig. 5F) and Cenozoic granitoids (Atoyac, Tecpan and Coyuca basins; Figs. 5A–C, 7B, C). They also crosscut and displace the youngest veins (Fig. 7C). Moreover, in two locations, we found these structures crosscutting

and displacing previously described dextral sub-latitudinal faults. These sinistral faults correlate very well with W–E lineaments, which have the most prominent impact on river pattern. We found them in the field in outcrops located along these lineaments, e.g., in basins of Atoyac, Coyuca, Papagayo (Fig. 5). They produce deflections of the main channels directions, particularly in the central and eastern part of the study area—i.e., the following rivers: Papagayo, Coyuca, Atoyac, Tecpan and San Luis (Figs. 3, 5). In those outcrops, the sub-latitudinal left-lateral meso-faults are usually parallel to the main W–E lineaments producing river offset (Fig. 5B, C, F) and/or they form a subordinate set of parallel faults at an angle of 10–20° to the main structure (Figs. 5B, 8).

Based on the crosscutting relationships observed in the field between the above-mentioned strike-slip faults and their occurrence in rocks of different ages, we interpreted the right-lateral structures as older, probably formed before the formation of the youngest granitoids, whereas the left-lateral structures, observed in all geological units from Precambrian to Cenozoic, as younger, developed after the formation of the youngest veins, which they crosscut and displace. This suggestion is also supported by the influence of these faults on the drainage network pattern. The effect on river flow of the first group is limited, whereas sinistral faults of the second group are usually related to sharp bends and long W–E sections of the main rivers (e.g., Atoyac, Coyuca and Papagayo; Fig. 5B, C, F). Therefore, dextral faults probably represent inherited, old structures, whereas sinistral faults of similar orientation are young structures that could be still active.

Moderately to steeply dipping normal faults, usually oriented from W–E to NE–SW and dipping towards the S to SE, are frequent within Tecpan (mainly in the northern outcrops located in the metamorphic rocks), Atoyac (granitoids), Coyuca (granitoids), La Sabana (gneisses) and Papagayo drainage basins (particularly in the northern outcrops, located close to La Venta, within both, gneisses and granitoids; Fig. 5). They also displace younger veins as do the sinistral sub-latitudinal faults. In the Atoyac and La Sabana drainage basins, these features present a clear influence on landscape. They produce scarps, waterfalls (La Sabana; Fig. 5E), and meso-grabens (Fig. 8 presents the example of such features

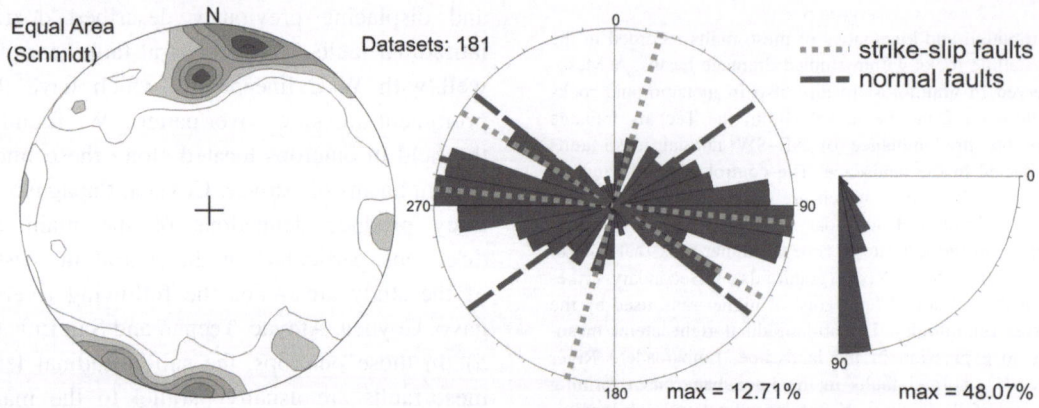

Figure 6
Spatial orientation of meso-faults recorded in the field within the selected drainage basins (Tecpan, Atoyac, Coyuca, Embarcadero, La Sabana and Papagayo) in the forearc region of the Guerrero sector

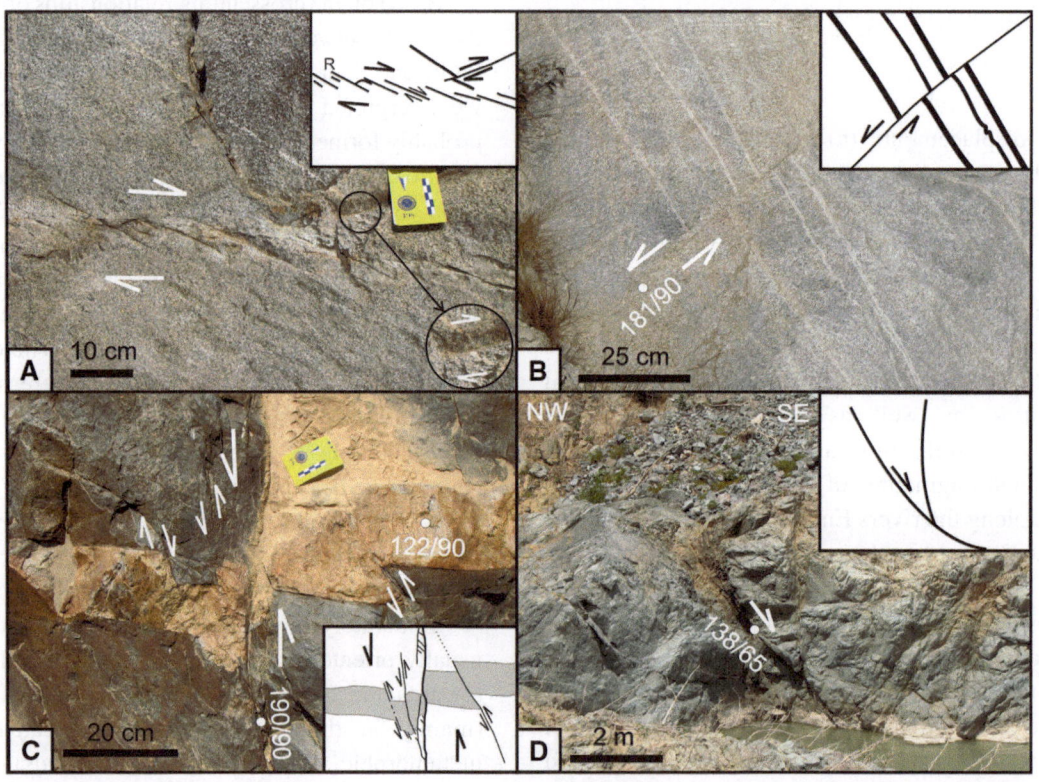

Figure 7
Studied meso-faults, examples from the field (for the location of presented structures see Fig. 5): **A** sub-latitudinal dextral strike-slip zone with low-angle Riedel shears and displaced quartz vein in granitoids, the Coyuca drainage basin; **B** latitudinal sinistral strike-slip fault shown by left-lateral displacement (~15 cm) of longitudinal quartz veins (granitoids, Tecpan drainage basin); **C** latitudinal, sinistral strike-slip fault with accompanying low-angle and high-angle Riedel shears displacing vein in the Atoyac drainage basin; **D** NE–SW normal fault in crystalline rocks near the La Venta dam, Papagayo drainage basin

observed in Atoyac basins). In the field, they are characterized by large, smooth surfaces, with poorly developed striation. Associated fractures are usually scarce and not clear. At least, some of the NE–SW normal faults are secondary features related to left-lateral fault zones (Fig. 8).

The last two fault sets recorded in the field—N–S strike-slip faults and NW–SE normal and strike-slip faults—are much scarcer. They occur only locally and represent old structures, often displaced by sub-latitudinal strike-slip faults.

Based on structural field data (meso-faults with kinematic indicators recorded in the field: slicken-lines, subordinate R R′, P and T fractures, and

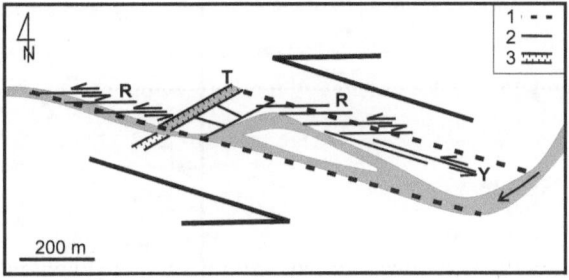

Figure 8
Interpretational scheme of the sub-latitudinal left-lateral strike-slip fault zone in the Atoyac drainage basin. *1* main fault zone (traced according to field structural data and morphotectonic interpretation of satellite images and DEM), *2* secondary meso-faults recorded in the field (*Y* synthetic, parallel to the main fault, *R* synthetic low-angle Riedel shears), *3* mesograbens produced by normal meso-faults located in the position of extensional *T* structures. For location of structures presented here see Fig. 5B

displacement of veins) and morphotectonic interpretation of DEM and satellite images (offset of the main river directions), we traced the most prominent left-lateral strike-slip fault structure in the study area, which we named La Venta Fault (LVF; Fig. 9). The nearly 200-km long fault is a complex structure including smaller segments reflected differently in the landscape, step-over, small subsidiary fault strands and small variations in strike within the defined segments. The La Venta Fault is segmented and divided by a ca. 8 km wide extensional step-over (named Coyuca), and a ca. 4 km wide compressional step-over (named Papagayo) (Fig. 9). These segments are from west to east: (1) Tecpan segment is 97.4 km long, and extends from the coast to the Coyuca step-over; (2) Coyuca segment is 65.1 km long from the Coyuca step-over to the Papagayo step-over; and (3) Agua del Perro segment—a 33.5-km long segment from the Papagayo step-over to the eastern termination of the study area (Fig. 9). Generally, the Agua del Perro and Coyuca segments are well constrained in landscape features, whereas the Tecpan segment, with the exception of clear lineament in Coyuca drainage basin and meso-faults recorded in Atoyac

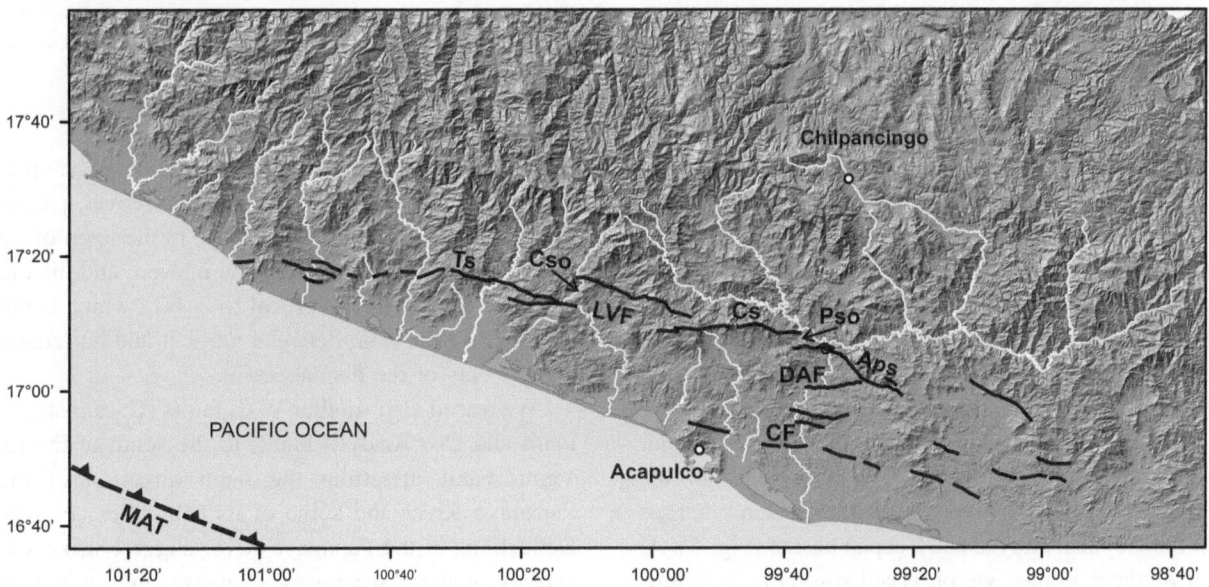

Figure 9
Major crustal strike-slip faults recognized in the Guerrero sector of the Mexican subduction zone. *LVF* La Venta Fault, *CF* Cacahuatepec Fault, *DAF* Dos Arroyos Fault; defined segments of the La Venta Fault: *Ts* Tecpan segment, *Cs* Coyuca segment, *APs* Agua del Perro segment; main step-overs: *Cso* Coyuca step-over, *Pso* Papagayo step-over; *MAT* Middle American Trench

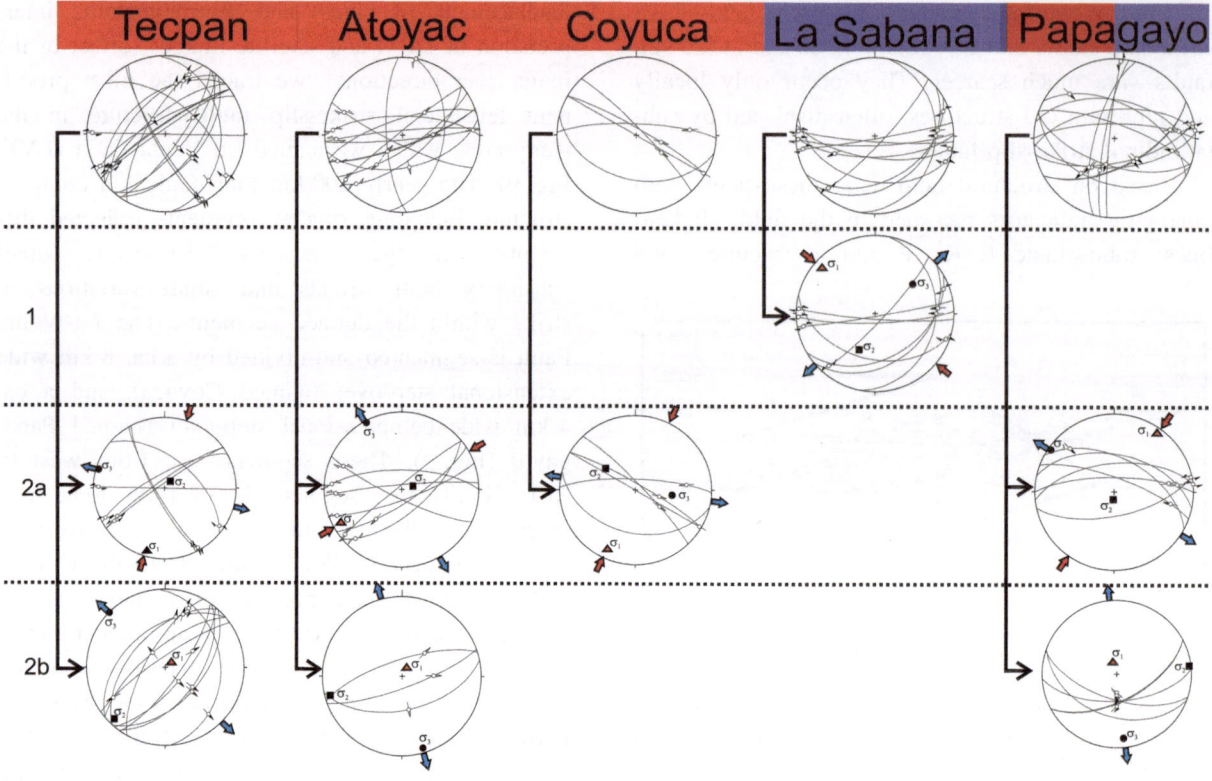

Figure 10
Angelier diagrams and stress tensor reconstruction for meso-faults recorded in the field within selected drainage basins, ordered from west to east. The predominant lithology in each of the basins is marked with *color*: *red* granitoids (mainly Eocene to Miocene), *purple* metamorphic rocks of the Xolapa Complex (Precambrian—Creataceous). Maximum, intermediate and minimum principal stress axes are marked with σ_1 (*triangle*), σ_2 (*square*) and σ_3 (*circle*), respectively. The *arrows* show the direction of shortening (*red arrows* pointing inward) and widening (*blue arrows* pointing outward). *Numbers* on the *left* define the relative age of calculated stress fields: *1* oldest strike-slip regime, *2a and 2b* younger strike-slip and normal stress regimes

basin, is poorly expressed in the landscape. Left-lateral strike-slip faults recorded in granitoids in Tecpan, even though they correlate well with the trace of La Venta Fault and its kinematics (as secondary Y and R faults synthetical with the main feature), do not seem to be well reflected in the morphology of the area, and they do not deflect the main river, nor the tributaries (Fig. 5A).

In the field, we studied the La Venta Fault in four locations: close to the town of La Venta in the Papagayo drainage basin and in the chosen outcrops in the Coyuca, Atoyac and Tecpan basins (Fig. 5). At all of these points, we observed sub-latitudinal left-lateral strike-slip meso-faults with the ratio of horizontal slip to vertical slip ranging from ∼6:1 to 1.8:1. In other words, the vertical component of these

strike-slip structures, even though small, can be up to ∼35 % of the total displacement. Moreover, particularly in the outcrops located close to the town of La Venta, the surface of this fault is uneven, and the dip can vary greatly from vertical to ∼60°, which could also modify the ratio between vertical and horizontal components of the displacement.

We traced also smaller W–E faults (Cacahuatepec Fault and Dos Arroyos Fault) to the south of the La Venta Fault offsetting the main direction of the Papagayo River and some of its tributaries (Fig. 9). Deflections of the Papagayo River suggest strike-slip activity of these structures. However, they might be connected with W–E structures in the La Sabana basin defined as normal faults, producing clear scarps and waterfalls along the La Sabana River.

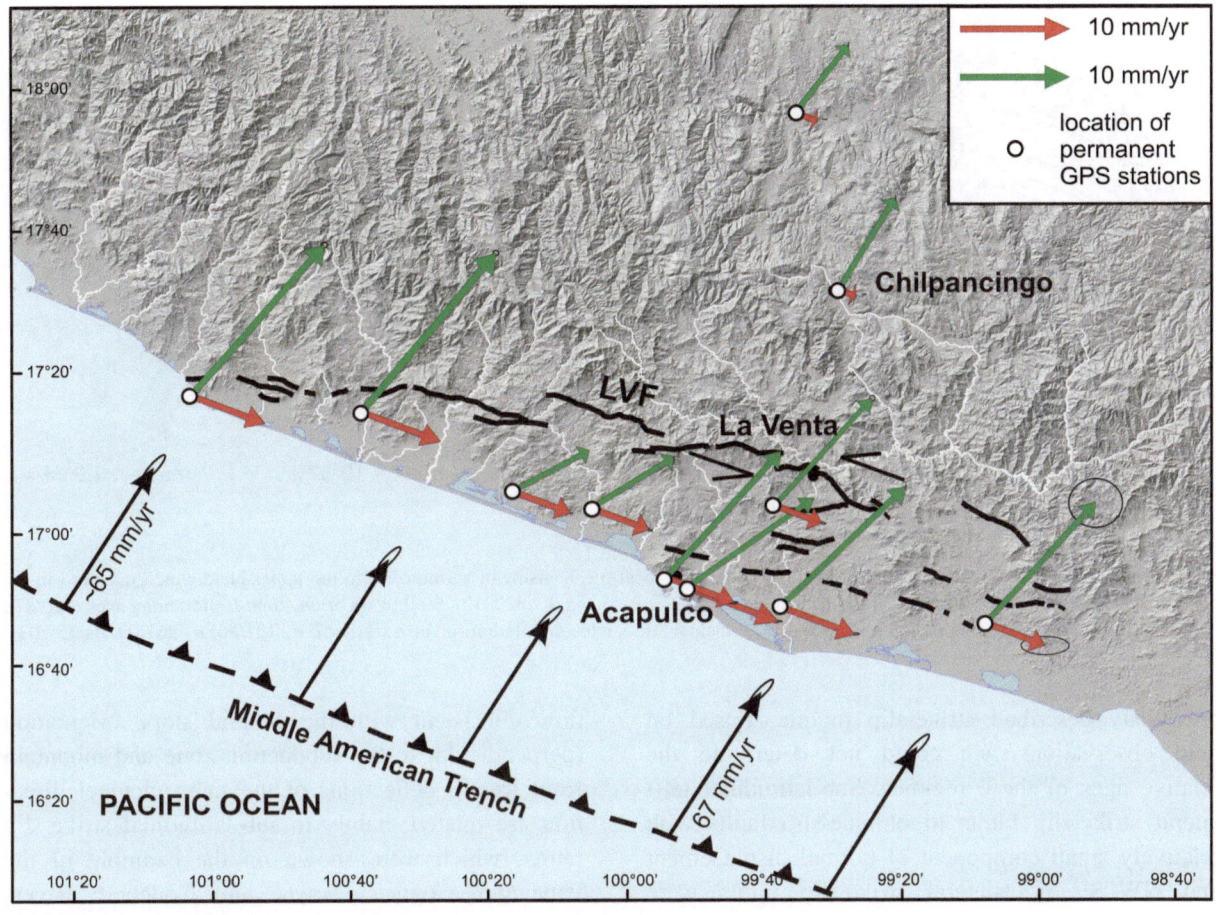

Figure 11

GPS secular vectors (*green arrows*) and sinistral, parallel to the Middle American Trench, components (*red arrows*) in the Guerrero sector of the Mexican subduction zone. *Black arrows* indicate convergence vectors (not scaled with the GPS velocity vectors). All vectors are plotted with the 3 sigma error ellipses. Note the abrupt change (4–5 mm/year) in GPS velocities across the Agua del Perro segment of the La Venta Fault (LVF)

4.3. Paleostress Reconstruction

Three main stress fields were determined based on paleostress analysis conducted for brittle meso-faults recorded in the field within five drainage basins: Tecpan, Atoyac, Coyuca, La Sabana and Papagayo (Fig. 10): (1) Strike-slip stress regime with the main compression axis (σ_1-axis) oriented NW–SE, estimated only for faults recorded in the La Sabana drainage basin; (2a) Strike-slip stress regime with the main compression axis (σ_1-axis) oriented NE–SW. Faults formed in this regime were encountered in all drainage basins with the exception of La Sabana; (2b) Normal stress regime with sub-vertical σ_1-axes and sub-horizontal NW–SE to N–S σ_3-axes, recorded in Tecpan, Atoyac and Papagayo drainage basins.

The first strike-slip regime (marked as 1 in Fig. 10) most likely can be considered as the oldest, because it was calculated based on measurements taken within rocks of the Xolapa Complex (p€-Mz; Fig. 2A). Right-lateral sub-latitudinal strike-slip faults, previously described as one of the oldest structures, were formed in this regime. Even though we encountered them also in other drainage basins, we estimated the stress field only for faults in the La Sabana basin due to insufficient fault data within other basins.

Two other estimated stress fields (2a and 2b in Fig. 10), according to field observations, crosscutting relationships and ages of rocks in which these features were observed, are younger than the

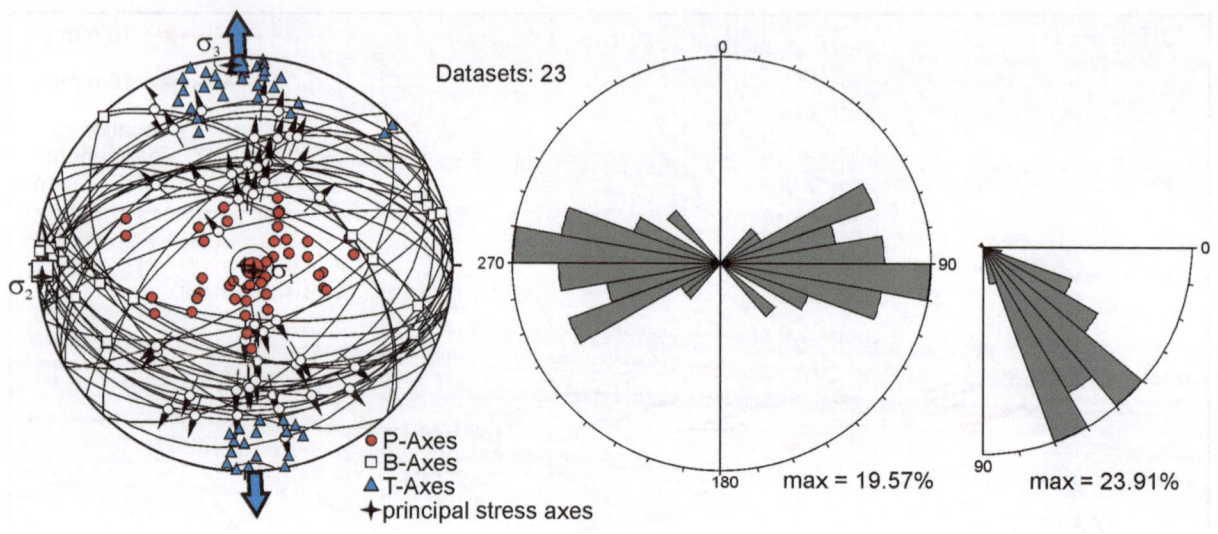

Figure 12

Stress tensor reconstruction and orientation of nodal planes for focal mechanisms of earthquakes in the *upper* plate which occurred in the study area between 1995 and 2008 (focal mechanism data from PACHECO and SINGH 2010). Strain axes orientation: *P* shortening axes (120/87), *B* neutral axes (268/03), *T* extension axes (358/02). Principal stress axes orientation (inverse method): σ_1 121/86, σ_2 267/03, σ_3 357/02

previously described strike-slip regime. Based on field observations, we could not determine the relative ages of these regimes. Sub-latitudinal left-lateral strike-slip faults to oblique-slip faults with relatively small component of normal displacement and NW–SE right-lateral strike-slip faults were formed under conditions of strike-slip regime marked as 2a in Fig. 10. Normal stress regime (marked as 2b in Fig. 10) is responsible for the formation of sub-latitudinal to NE–SW normal to oblique-slip faults.

5. Discussion

5.1. River Network Pattern and Active Tectonics

Active tectonic control is the most important factor influencing the river network development in the studied Guerrero sector of the Mexican subduction zone. The strongest influence is related to sub-latitudinal, left-lateral strike-slip faults, which might be active since they are offsetting the main-stream courses of the largest river basins: Papagayo, La Sabana, Coyuca, Atoyac, Tecpan and San Luis, as well as smaller rivers, such as Conchero (Fig. 4). Even though the main stream direction is dictated by

flow consistent with the general slope orientation (perpendicular to the subduction zone and mountain front trend), deflections of the main channel direction are related mainly to sub-latitudinal strike-slip faults, which were shown on the example of the Papagayo, Coyuca, Atoyac and Conchero Rivers (Fig. 3A, B). The amount of lateral offset related to those features varies greatly from dozens of meters (because of the resolution of applied DEM smaller offsets were not observable) up to a few kilometers, reflecting multiple faulting events. The activity of W–E to NE–SW normal faults is reflected in the landscape by scarps, waterfalls and several meso-graben. The influence of these structures systematically decreases to the west of the studied Guerrero sector, and becomes imperceptible to the west of the San Luis drainage basin.

Based on deflections and offset of the main river directions, as well as field structural data, we uncovered a new regional sub-latitudinal strike-slip fault that we named La Venta Fault, and that extends over a distance of almost 200 km in the study area (Fig. 9). Its left-lateral strike-slip activity is constrained by field evidence such as its geomorphic expression and kinematic indicators (reported here), and by slow slip events (KOSTOGLODOV *et al.* 2014) on the Agua del Perro segment (Fig. 9). GPS secular velocities parallel to the Middle American Trench suggest ~4–5 mm/year present day strike-slip, left-lateral motion along this segment (Fig. 11; KOS-TOGLODOV et al. 2014). Here, the La Venta Fault coincides with an Eocene–Oligocene inactive, ductile shear zone known as the Chacalapa-La Venta Fault Zone (RILLER et al. 1992; TOLSON 2005; SOLARI et al. 2007). Based on DEM's and satellite images, we also interpreted the trace of the La Venta Fault further west from the Agua del Perro section. This trace differs from that of the non-active Chacalapa-La Venta Fault Zone that has no expression on the landscape.

5.2. Stress Regimes

Three main stages of deformation were recognized based on fault measurements taken within the metamorphic rocks of the Xolapa Complex, as well as Eocene and Oligocene–Miocene granitoid rocks (Fig. 10). The oldest stress regime (marked as 1 in Fig. 10) coincides with the paleostress tensor obtained by MESCHEDE et al. (1996) for structures older than 120 Ma, which might be related to the Jurassic or Early Cretaceous deformation. This is in agreement with our field observations, since faults belonging to this stage were recorded mainly within rocks of the Xolapa Complex (Fig. 10), which according to SOLARI et al. (2007) and PÉREZ-GUTIÉRREZ et al. (2009) are older than ~130 Ma. These exhumed old structures are represented mainly by sub-latitudinal right-lateral strike-slip faults. Their influence on drainage network is limited, since in the majority of locations their occurrence did not influence the direction of the river valley (see Fig. 5D). Locally, we found also similar dextral structures in younger units (i.e., Cenozoic plutonic rocks in Coyuca and Atoyac basins; see Fig. 5). We interpreted these as secondary antithetic structures within left-lateral strike-slip faults of higher rank or as the effect of local variations in stress field.

Younger deformation stages were determined based on meso-faults which crosscut and displace the youngest veins (TALAVERA-MENDOZA et al. 2013) and granitoids related to Oligocene–Miocene magmatism (~25–34 Ma; HERRMANN et al. 1994; SCHAAF

et al. 1995; DUCEA et al. 2004) and older units. Thus, the obtained paleostresses are younger than ~25–34 Ma and, consequently, are strictly related to the Cocos plate subduction, which initiated ca. 22 Ma (WORTEL and CLOETINGH 1981; BARCKHAUSEN et al. 2001). They also agree with the two youngest paleostress tensors (younger than 25 Ma) proposed by MESCHEDE et al. (1996). Both sub-latitudinal normal and left-lateral strike-slip faults were formed under conditions of these two regimes. The NE- to NNE-directed σ_1-axes of the strike-slip paleostress tensors (marked as 2a in Fig. 10) correlate with the relative convergence direction of the Cocos/North America subduction in the last 20 Ma (MESCHEDE et al. 1996). They are also coherent with the NE-trending plate convergence from 120 to 70 Ma and from 40 to 20 Ma (MESCHEDE et al. 1996). This implies that for sub-latitudinal left-lateral strike-slip faults recorded in the Eocene-Miocene plutonic rocks, the age of their formation is <20 Ma. For meso-faults observed in older units, we cannot exclude the possibility that they were formed even in Mesozoic, especially those that suggest brittle–ductile conditions during their formation. Extensional deformation (stress regime marked as 2b in Fig. 10) could be related to the Miocene uplift of southern Mexico (MESCHEDE et al. 1996) and/or active normal crustal faulting in the upper plate above the subducting slab of the Cocos plate.

5.3. Current Stress Field and Possibility of Reactivation

The current stress tensor (Fig. 12) was determined based on focal mechanisms of earthquakes (magnitude ≥3.0) in the upper plate, which occurred within the Guerrero sector between 1995 and 2008 (focal mechanism data according to PACHECO and SINGH 2010). These were mainly very shallow (with maximum depth of 15 km), small magnitude earthquakes (with the exception of Mw = 5.8 normal fault Coyuca earthquake (PACHECO et al. 2002) all the other events were of magnitude ≤5), which occurred along sub-latitudinal steeply and moderately dipping normal faults—locally oblique, with a small strike-slip component (generally not exceeding 15 % of total displacement). The resulting normal stress regime is characterized by sub-vertical σ_1-axis and horizontal sub-longitudinal σ_3-axis (Fig. 12). It is coherent with the normal stress regime obtained for the meso-faults recorded in the field for Atoyac and Papagayo and roughly coherent with results obtained for Tecpan (marked as 2b in Fig. 10), as well as with the stress field reported by MESCHEDE et al. (1996) for faults younger than 25 Ma. Therefore, sub-latitudinal (orientation of strike from W–E to NW–SE and NE–SW) normal faults recorded in the field might be reactivated within the current extensional stress field in the upper plate related to the Cocos subduction. The occurrence of destructive, Mw > 7, normal-faulting earthquakes related to the Cocos subduction is not unique. Within the last 100 years, two such events were recorded in the Oaxaca sector on January 15, 1931 (Mw = 7.8) and September 30, 1999 (Mw = 7.5; SINGH et al. 2000). In the Guerrero sector, a shallow, crustal earthquake of smaller magnitude (Mw = 5.8) related to normal faulting occurred on the October 8, 2001 (MENDOZA 2004). The reactivation of pre-existing trench-parallel upper plate normal faults related to the subduction earthquake cycle was studied in a seismic gap in northern Chile by CORTÉS-ARANDA et al. (2015). These authors proposed that subduction megathrust earthquakes promote the most effective stress perturbations required to reactivate the pre-existing upper plate normal faults, particularly when they are located right above the megathrust. Such triggered reactivation of normal pre-existing crustal faults occurred during the last major megathrust earthquakes: the 2010 Mw 8.8 Maule Earthquake (FARÍAS et al. 2011; RYDER et al. 2012; ARON et al. 2013) and the 2011 Mw 9 Tohoku Oki Earthquake (KATO et al. 2011; IMANISHI et al. 2012; TODA and TSUTSUMI 2013). The Guerrero sector is favored for such a mechanism of normal fault reactivation, since we observed normal, shallow, small magnitude earthquakes here, proving, at least locally, a tensional stress regime in the upper plate (PACHECO and SINGH 2010). A similar situation was observed before the Tohoku Oki Earthquake (IMANISHI et al. 2012). Moreover, the latest calculations show that this sector of the Mexican subduction zone has the potential for a large subduction earthquake of Mw = 7.9–8.0 (LOWRY et al. 2006) to Mw ~8.15 (BEKAERT et al. 2015) and Mw = 8.1–8.4 (SINGH and

MORTERA 1991), which might reactivate trench-parallel pre-existing upper plate, young (<25 Ma) normal faults. Moreover, shallow crustal seismicity with normal focal mechanisms may occur during the repeating subduction thrust slow slip events, when a temporal extension of the forearc is observed (KOSTOGLODOV et al. 2003).

Sinistral strike-slip activity along trench-sub-parallel faults has been confirmed by our field study and by focal mechanism solutions, but only for deep, intraslab earthquakes (PACHECO and SINGH 2010). However, as mentioned above, the instrumental data available cover only a short period of time (1995–2008; PACHECO and SINGH 2010). The existence of active strike-slip faults, either in the form of major structures or a series of smaller features, in the forearc region of subduction zones, is common. In many areas of oblique subduction, such activity occurs as a result of plate motion partitioning into trench-parallel strike-slip deformation and convergent trench-normal contraction (FITCH 1972; AVÉ LALLEMONT and OLDOW 2000). Such a situation has been observed in many subduction zones with different amounts of obliquity: Chile (DELOUIS et al. 1998; LANGE et al. 2007, 2008; MORENO et al. 2008; MELNICK et al. 2009; MÉTOIS et al. 2012; VARGAS et al. 2013), Central America (CHOY and KIRBY 2004; CORTI et al. 2005; CÁSERES et al. 2005; FERNANDEZ 2009), Japan (DEMETS 1992), Sumatra (MCCAFFREY 1992, 2009; MCCAFFREY et al. 2000; BERGLAR et al. 2010), and Cascadia (ONUR and SEEMANN 2004; MAZZOTTI et al. 2014). In some cases (especially in Sumatra, Japan, and northern Chile), strike-slip faults are major structures, which can produce high seismic hazards. In others (i.e.: Central America, Cascadia, or other parts of the Chilean subduction zone), such structures are rather small, forming en-echelon structures or sets of sub-parallel faults. According to JARRARD (1986), the strike-slip rate increases with greater convergence obliquity. However, even though the Cocos subduction zone in the Guerrero sector is only slightly oblique (9–14°; GRIPP and GORDON 2002; KREEMER et al. 2003; DEMETS et al. 2010), it results in the upper plate strike-slip sinistral component of ~8 mm/year (LOWRY et al. 2006). This sinistral component could be accommodated by the La Venta Fault, which we interpret as a sinistral strike-slip structure (Fig. 13). This hypothesis is supported also by ∼4–5 mm/year present day strike-slip, left-lateral motion along the Agua del Perro segment of the LVF, estimated by KOSTOGLODOV et al. (2014) with GPS data (Figs. 11, 13). Moreover, our results suggest that at least some of that long-term motion can be accommodated also on smaller W–E features (e.g., Cacahuatepec Fault and Dos Arroyos Fault), located between the San Luis and Papagayo basins (mainly to the south of the La Venta Fault), probably forming en-echelon structures. Small change in GPS sinistral velocities can be also observed across those structures (Fig. 11). Nevertheless, this is not clear as it is based on the data collected at sparsely distributed GPS stations. Therefore, a denser grid of permanent GPS stations is needed to verify this hypothesis. Although strike-slip focal mechanism solutions are absent, we cannot exclude the possibility of activity of W–E sinistral strikes-slip faults, because the time period for which focal mechanism data are available is very short. Moreover, such activity could be related to slow slip transient events (SSE) or associated fault creep, as suggested by LOWRY et al. (2006) and, more recently, confirmed by KOSTOGLODOV et al. (2014). These works are based on continuous GPS measurements conducted during the last 10–15 years within the Guerrero sector of the Mexican subduction zone and corroborate the sinistral strike-slip movements in this zone, related to aseismic SSE. There is also

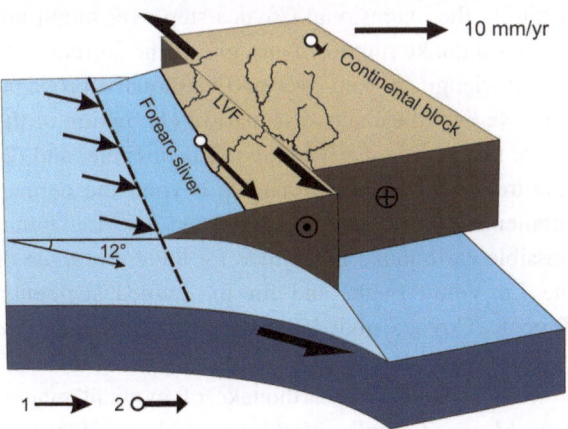

10 mm/yr

Continental block

LVF

Forearc sliver

12°

1 → 2 ○→

Figure 13

Diagram presenting a plate-boundary sub-parallel left-lateral strike-slip La Venta Fault (LVF) and a forearc sliver produced by oblique Mexican subduction (the angle of obliquity = 12°) in the Guerrero sector. *Symbols*: *1* direction of convergence, *2* GPS sinistral secular velocities

evidence that crustal strike-slip, like pre-existing normal faults, can be reactivated as a response to megathrust earthquakes (SHIRZAEI et al. 2012). Nevertheless, even if they are not reactivated, they still can play an important role in controlling the co-seismic deformation and local ground ruptures during major inter-plate earthquakes (ARRIAGADA et al. 2011). The strike-slip to reverse reactivation can also occur in the forearc of oblique convergence subduction margins due to changes in the strain partitioning into parallel strike-slip faulting and orthogonal thrusting (SOTO et al. 2007), which is the case in the Guerrero sector.

5.4. Maximum Expected Earthquake Magnitudes Along the Major Crustal Faults

As discussed earlier, studied strike-slip to oblique-slip faults are probably characterized by long (hundreds to thousands of years) recurrence periods or their activity is related to slow slip events and/or creeping (see Sect. 5.3). However, even if the activity of traced major crustal structures is characterized by creeping, they still could produce moderate to large-size earthquakes, depending on the depth of creeping and/or ratio of creeping rate to geologically determined slip rate (MCCALPIN 2009). Those rates for the study area are not available. Therefore, we cannot exclude a possibility that those faults may generate large-magnitude earthquakes. Assuming that for the maximum possible earthquake magnitude produced by those faults, the surface rupture length would be equal to the traced fault length, we calculated main parameters for such events (maximum moment magnitude Mw, maximum downdip rupture width, maximum rupture area, maximum displacement and maximum average displacement), using formulas proposed by WELLS and COPPERSMITH (1994). We analyzed three major potentially active structures: La Venta Fault and its defined segments (Tecpan, Coyuca and Agua del Perro), Cacahuatepec Fault and Dos Arroyos Fault (Table 2). We estimated each parameter based on two formulas: 1—for a fault without any assumption about its kinematics, 2—for strike-slip faults. Because we do not know the continuation of the traced faults towards the east, the numbers in Table 2 are the best estimates to our

Table 2

Maximum values of earthquake parameters calculated for the major crustal faults (for the location of faults and their segments see Fig. 9) according to formulas proposed by WELLS and COPPERSMITH (1994)

Parameter Fault	Fault length (km) = Maximum surface rupture length SRL (km)	Maximum moment magnitude Mw[a]	Maximum downdip rupture width RW[b] (km)	Maximum rupture area RA[c] (km²)	Maximum displacement MD[d] (m)	Maximum average displacement AD[e] (m)
1. La Venta Fault	185.5	7.7	28.7	3366.8	7.30	3.32
		7.7*	20.9*	3239.2*	7.97*	4.08*
1a. Tecpan segment	97.4	7.4	22.6	1705.2	3.95	1.98
		7.4*	17.2*	1691.7*	3.79*	2.13*
1b. Coyuca segment	65.1	7.2	19.4	1114.9	2.70	1.43
		7.2*	15.2*	1127.4*	2.38*	1.42*
1c. Agua del Perro segment	33.5	6.9	15.2	553.1	1.43	0.84
		6.9*	12.4*	577.3*	1.11*	0.73*
2. Dos Arroyos Fault	16.6	6.5	11.7	262.6	0.73	0.48
		6.5*	10.0*	283.5*	0.49*	0.36*
3. Cacahuatepec Fault	14.7	6.4	11.2	231.5	0.65	0.44
		6.5*	9.7*	251.3*	0.43*	0.32*

[a] $Mw = 5.08 + 1.16 \times \log(SRL)$; *$Mw = 5.16 + 1.12 \times \log(SRL)$

[b] $\log(RW) = 0.32 \times Mw - 1.01$; *$\log(RW) = 0.27 \times Mw - 0.76$

[c] $\log(RA) = 0.91 \times Mw - 3.49$; *$\log(RA) = 0.9 \times Mw - 3.42$

[d] $\log(MD) = 0.82 \times Mw - 5.46$; *$\log(MD) = 1.03 \times Mw - 7.03$

[e] $\log(AD) = 0.69 \times Mw - 4.8$; *$\log(AD) = 0.9 \times Mw - 6.32$

*Parameters calculated assuming that the faults are purely strike-slip

knowledge. Further studies to the east of the study area are needed to assess more accurate maximum values.

There are two different positions on how step-over influence earthquake rupture propagation on a fault. First, historic observations and numerical rupture models (HARRIS *et al.* 1991; HARRIS and DAY 1993; WESNOUSKY 2006; ELLIOTT *et al.* 2009) suggest that step-over >4 km wide, regardless of its nature, would stop earthquake rupture. Second, other studies suggest that rupture could jump a step-over up to 8 km wide, if it is a dilatational one (KNUEPFER 1989). In other words, earthquake rupture can jump a wider extensional step-over than a compressional one (HARRIS *et al.* 1991; HARRIS and DAY 1993). According to the first position, the ∼8 km Coyuca step-over would prevent rupture propagating from one segment towards another. The ∼4 km wide compressional Papagayo step-over, being at the eastern end of the fault, would either stop the rupture or not, depending on other factors, i.e., increase or decrease in normal

stress, dynamic stress drop, and slip gradient near step-over (HARRIS *et al.* 1991; HARRIS and DAY 1993; HAEUSSLER *et al.* 2004; WESNOUSKY 2006; ELLIOTT *et al.* 2009; LOZOS *et al.* 2015). Assuming the second position, the extensional Coyuca step-over might not stop earthquake rupture. Thus, giving the current state of knowledge on step-over and earthquake propagation, we hypothesize two scenarios: (1) rupture of the entire trace of the La Venta Fault is possible, and (2) rupture that does not propagate beyond the defined segments. Therefore, we estimated the maximum possible earthquake magnitude for the entire trace of the La Venta Fault, and for its defined segments: Tecpan, Coyuca and Agua del Perro. Thus, if the entire mapped length of the La Venta Fault ruptured, it would produce an earthquake of moment magnitude Mw = 7.7 and a rupture area above 3000 km² (Table 2), which is >20 % of the study area. While, the segments of the La Venta Fault are capable of producing earthquakes of maximum moment magnitude ranging from 6.9 for the Agua del Perro segment, 7.2 for the Coyuca segment, and 7.4 for the Tecpan segment (Table 2). The maximum displacement for large segments would be ~4 m (Tecpan segment) and ~3 m (Coyuca segment), and from 1 to 1.5 m for the Agua del Perro segment (Table 2). Smaller features located to the south of the La Venta Fault represented in calculations by Cacahautepec and Dos Arroyos Faults could produce events of maximum magnitude Mw ~ 6.4–6.5 (Table 2). The maximum displacements would not exceed 1 m.

6. Conclusions

Our studies suggest that river network pattern, topography and structural data in the studied forearc drainage basins prove the activity of tectonic faulting. We conclude that both normal and left-lateral strike-slip crustal faults sub-parallel to the orientation of the subduction zone might be active. Based on focal mechanism solutions of small magnitude earthquakes in the upper plate, sub-latitudinal to NW–SE and NE–SW normal faults might be reactivated within the current stress field in the upper plate related to the Cocos subduction. Their reactivation might also be triggered by megathrust earthquakes and large slow slip events. Left-lateral sub-latitudinal strike-slip faults are considered as potentially active because they are offsetting the main stream courses of the largest river basins. Kinematic indicators identified in the field also confirm their left-lateral strike-slip motion. We uncovered the most remarkable left-lateral sub-latitudinal structure in the study area—a regional fault that we named La Venta Fault. Its left-lateral activity was confirmed by geomorphic, structural, and GPS data in the Agua del Perro section. Strike-slip activity is a common phenomenon in oblique subduction zones, such as the active Cocos subduction in the Guerrero sector. The sinistral sliver motion in this area has also been observed with GPS measurements. We thus hypothesize that these faults are active and probably (1) are characterized by long, non-instrumental recurrence period or, 2) they experience slow events or associated fault creep. The La Venta Fault may be capable of producing earthquakes of maximum moment magnitude Mw = 7.7, whereas smaller Cacahautepec and Dos Arroyos Faults

located to the south could produce events of maximum magnitude Mw ~ 6.4–6.5. Although strike-slip activity in oblique subduction zones was already demonstrated for other active zones, further studies are needed to solve the problem of short-span instrumental data of strike-slip displacements.

Acknowledgments

Ramirez-Herrera acknowledges funding provided by a CONACYT-SEP Ciencia Básica Grant No. 129456, PAPIIT IN110514 grant, PASPA-UNAM 2015–2016. K. Gaidzik acknowledges postdoctoral fellowship by DGAPA-UNAM program. Thanks to the local community of La Remontita and Cacahuatepec, and several communities in the Guerrero State for kindly giving access to us and for help in the field, R. Basili for help in one of the field-work trips during a visit granted by the 2014 Programa de Intercambio Académico, Dirección General de Cooperación e Internacionalización (DGECI), Universidad Nacional Autonoma de México, México. We thank the editor W. Bandy and four anononimous reviewers for their comments that help to improve our manuscript.

REFERENCES

ANDERSON, J.G., SINGH, S.K., ESPINDOLA, J.M., and YAMAMOTO J. (1989). *Seismic strain release in the Mexican subduction thrust.* Phys. Earth Planet. Inter. *58*, 307–322.

ARON, F., ALLMENDINGER, R.W., CEMBRANO, J., GONZÁLEZ, G., and YÁÑEZ, G. (2013). *Permanent fore-arc extension and seismic segmentation: Insights from the 2010 Maule earthquake, Chile.* J. Geophys. Res. B: Solid Earth. *118*(2), 724–739.

ARRIAGADA, C., ARANCIBIA, G., CEMBRANO, J., MARTÍNEZ, F., CARRIZO, D., VAN SINT JAN, M., SÁEZ, E., GONZÁLEZ, G., REBOLLEDO, S., SEPÚLVEDA, S.A., CONTRERAS-REYES, E., JENSEN, E., and YÁÑEZ, G. (2011). *Nature and tectonic significance of co-seismic structures associated with the Mw 8.8 Maule earthquake, central-southern Chile forearc.* J. Struct. Geol. *33* (5), 891–897.

AVÉ LALLEMONT, H.G., and OLDOW, J.S. (2000). *Active displacement partitioning and arc-parallel extension of the Aleutian volcanic arc based on Global Positioning System geodesy and kinematic analysis.* Geology 28(8), 739–742.

BARCKHAUSEN, U., RANERO, C.R., VON HUENE, R., CANDE, S.C., and ROESER, H.A. (2001). *Revised tectonic boundaries in the Cocos Plate off Costa Rica: Implications for the segmentation of the convergent margin and for plate tectonic models.* J. Geohys. Res. B. *106*, 19207–19220, doi:10.1029/2001JB000238.

Reprinted from the journal

BARTHOLOW, J.M., 1989. *Stream temperature investigations: field and analytic methods. Instream Flow Information Paper No. 13.*, U.S. Fish Wildl. Serv. Biol. Rep. *89* (17), 139 pp.

BEKAERT, D.P.S., HOOPER, A., and WRIGHT, T.J. (2015). *Reassessing the 2006 Guerrero slow slip event, Mexico: implications for large earthquakes in the Guerrero Gap.* Journal of Geophysical Research: Solid Earth, In Press.

BERGLAR, K., GAEDICKE, CH., FRANKE, D., LADAGE, S., KLINGELHOE-FER, F., and DJAJADIHARDJA, Y.S. (2010). *Structural evolution and strike-slip tectonics off north-western Sumatra.* Tectonophysics *480*(1–4), 119–132.

BIERMAN, P.R., and MONTGOMERY, D.R., Key Concepts in Geomorphology (W. H. Freeman Publisher 2014).

BURBANK, D.W. (1992). *Causes of recent Himalayan uplift deduced from deposited patterns in the Ganges basin.* Nature *357*, 680–682.

BURBANK, D.W., and ANDERSON, R.S., Tectonic Geomorphology (Blackwell Scientific, Oxford 2001).

CAMPA, M.F., and CONEY, P.J. (1983). *Tectono-stratigraphic terranes and mineral distributions in Mexico.* Can. J. Earth Sci. *20*, 1040–1051.

CAMPA URANGA, M.F., GARCÍA DÍAZ, J.L., GARCÍA, J.B., TORRE-BLANCA CASTRO, T. DE J., AGUILERA MARTÍNEZ, M.A., and MARTÍNEZ, A.V. (1998). Carta Geológico-Minera Chilpancingo E14-8, Guerrero, Oaxaca y Puebla. Servicio Geológico Mexicano and Universidad Autónoma de Guerrero, carta E14–8, scale 1:250,000.

CARRANZA, E.R., AGUILERA MARTÍNEZ, M.A., and MARTÍNEZ, A.V. (1999). Carta Geológico-Minera Zihuatanejo E14-7-10, Guerrero. Servicio Geológico Mexicano, carta E14–7-10, scale 1:250,000.

CÁSERES, D., MONTERROSO, D., and TAVAKOLI, B., 2005. *Crustal deformation in northern Central America.* Tectonophysics *404*, 119–131.

CHOY, G.L., and KIRBY, S.H. (2004). *Apparent stress, fault maturity and seismic hazard for normal-fault earthquakes at subduction zones.* Geophys. J. Int. *159*, 991–1012.

CORTÉS-ARANDA, J., GONZÁLEZ, L.G., RÉMY, D., and MARTINOD, J. (2015). *Normal upper plate fault reactivation in northern Chile and the subduction earthquake cycle: From geological observations and static Coulomb Failure Stress Change (CFS).* Tectonophysic. *639*, 118–131, doi:10.1016/j.tecto.2014.11.019.

CORTI, G., CARMINATI, E., MAZZARINI, F., and OZIEL GARCIA, M. (2005). *Active strike-slip faulting in El Salvador, Central America.* Geology *33* (12), 989–992.

CRUZ LÓPEZ, D.E., SÁNCHEZ ANDRACA, H.R. and BUSTOS, O.L. (2000). Carta Geológico-Minera Acapulco E14-11, Guerrero y Oaxaca. Servicio Geológico Mexicano, carta E14–11, scale 1:250,000.

DELOUIS, B., PHILIP, H., DORBATH, L., and CISTERNAS, A. (1998). *Recent crustal deformation in the Antofagasta region (northern Chile) and the subduction process.* Geophys. J. Int. *132*, 302–338.

DEMETS, C., 1992. *Oblique convergence and defroamtion along the Kuril and Japan Trenches.* J. Geophys Res. *97* (B12), 17,615–17,625.

DEMETS, C., GORDON, R. G., and ARGUS, D. F. (2010). *Geologically current plate motions*, Geophys. J. Int. *181* (1), 1–80, doi:10.1111/j.1365-246X.2009.04491.x.

DEMOULIN, A. (1998). *Testing the tectonic significance of some parameters of longitudinal river profiles: the case of the Ardenne (Belgium, NW Europe).* Geomorphology *24*, 189–208.

DUCEA, M.N., GEHRELS, G.E., SHOEMAKER, S., RUIZ, J., and VALENCIA, V.A. (2004). *Geologic evolution of the Xolapa Complex, southern Mexico: evidence from U–Pb zircon geochronology.* Geol. Soc. Am. Bull. *116*, 1016–1025.

ECHTLER, H.P., GLODNY, J., GRÄFE, K., ROSENAU, M., MELNICK, D., SEIFERT, W., and VIETOR, T. (2003). Active tectonics controlled by inherited structures in the long-term stationary and non-plateau south–central Andes, EGU/AGU Joint Assembly, Nice, EAE03-A-10902.

ELLIOTT, A.J., DOLAN, J.F., and OGLESBY, D.D. (2009). *Evidence from coseismic slip gradients for dynamic control on rupture propagation and arrest through stepovers.* J. Geophys. Res.: Solid Earth, *114* (B2).

FARÍAS, M., COMTE, D., ROECKER, S., CARRIZO, D., and PARDO, M. (2011). *Crustal extensional faulting triggered by the 2010 Chilean earthquake: The Pichilemu Seismic Sequence.* Tectonics. *30*(6), TC6010, doi:10.1029/2011TC002888.

FERNANDEZ, M. (2009). *Seismicity of the Pejibaye-Matina, Costa Rica, region: a strike-slip tectonic boundary?* Geofis. Int. *48* (4), 351–364.

FITCH, T.J. (1972). *Plate convergence, transcurrent faults, and internal deformation adjacent to southeast Asia and the western Pacific*, J. Geophys. Res. *77*, 4432–4460.

FOSSEN, H., Structural Geology (Cambridge University Press, Cambridge, 2010).

GASPARINI, N.M., and WHIPPLE, K.X. (2014). *Diagnosing climatic and tectonic controls on topography: Eastern flank of the northern Bolivian Andes.* Litosphere, doi:10.1130/L322.1.

GRIPP, A.E., and GORDON R.G. (2002). *Young tracks of hotspots and current plate velocities.* Geophys. J. Int. *150*, 321–361.

GUTSCHER, M.-A., and LALLEMAND, S. (1999). *Birth of a major strike-slip fault in SW Japan.* Terra Nova *11*, 203–209.

HAEUSSLER, P. J., SCHWARTZ, D. P., DAWSON, T. E., STENNER, H. D., LIENKAEMPER, J. J., SHERROD, B., CINTI F.R., MONTONE P., CRAW P.A., CRONE A.J., and PERSONIUS, S. F. (2004). SURFACE RUPTURE AND SLIP DISTRIBUTION OF THE DENALI AND TOTSCHUNDA FAULTS IN THE 3 NOVEMBER 2002 M 7.9 EARTHQUAKE, ALASKA. B. Seismol. Soc. Am. *94* (6B), S23–S52.

HARRIS, R.A., and DAY, S.M. (1993). *Dynamics of fault interaction: Parallel strike-slip faults.* J. Geophys. Res. *98*(B3), 4461–4472.

HARRIS, R.A., ARCHULETA, R.J., and DAY, S.M. (1991). *Fault steps and the dynamic rupture process: 2-D numerical simulations of a spontaneously propagating shear fracture.* Geophys. Res. Let. *18*(5), 893–896.

HERRMANN, U., NELSON, B.K., and RATSCHBACHER, L. (1994). *The origin of a terrane: U/Pb zircon geochronology and tectonic evolution of the Xolapa complex (southern Mexico)*: Tectonics *13*, 455–474, doi:10.1029/93TC02465.

HERNÁNDEZ-SANTANA, J.R., and ORTIZ-PÉREZ, M.A. (2005). *Análisis morfoestructural de las cuencas hidrográficas de los ríos Sabana y Papagayo (Tercio Medio-Inferior), Estado de Guerrero.* Investigaciones Geográficas, Bol del Inst. de Geogr., UNAM *56*, 7–25.

HOWARD, A.D. (1967). *Drainage analysis in geologic interpretation; a summation.* AAPG Bull. *51*, 2246–2259.

IMANISHI, K., ANDO, R., and KUWAHARA, Y. (2012). *Unusual shallow normal-faulting earthquake sequence in compressional northeast*

Japan activated after the 2011 off the Pacific coast of Tohoku earthquake. Geophysical Research Letters, *39*(9).

ITO, Y., TSUJI, T., OSADA, Y., KIDO, M., INAZU, D., HAYASHI, Y., TSUSHIMA, H., HINO, R., and FUJIMOTO, H. (2011). *Frontal wedge deformation near the source region of the 2011 Tohoku-Oki earthquake.* Geophys. Res. Lett. *38*, L00G05, doi:10.1029/2011GL048355.

JACKSON, J., NORRIS, R., and YOUNGSON, J. (1996). *The structural evolution of active fault and fold systems in central Otago, New Zealand: evidence revealed by drainage patterns.* J. Struct. Geol. *18*, 217–234.

JARRARD, R. D. (1986). *Terrane motion by strike-slip faulting of forearc slivers.* Geology *14*, 780–783.

KATO, A., SAKAI, S. I., and OBARA, K. (2011). *A normal-faulting seismic sequence triggered by the 2011 off the Pacific coast of Tohoku Earthquake: Wholesale stress regime changes in the upper plate.* Earth, planets and space, *63*(7), 745–748.

KIMURA, G. (1986). *Oblique subduction and collision: Forearc tectonics of the Kuril arc.* Geology *14*, 404–407, doi:10.1130/0091-7613(1986)14<404:OSACFT>2.0.CO;2.

KNUEPFER, P.L.K. (1989). Implications of the characteristics of endpoints of historical surface fault ruptures for the nature of fault segmentation. Fault Segmentation and Controls of Rupture Initiation and Termination, 89–315.

KOSTOGLODOV, V., and PONCE, L. (1994). *Relationship between subduction and seismicity in the Mexican part of the Middle America trench.* J. Geophys. Res. *99*, 729–742, 1994.

KOSTOGLODOV, V., COTTE, N., WALPERSDORF, A., HUSKER, A., and SANTIAGO, J.A. (2014). Mysterious SSE of the Guerrero land. in: Proceedings of the Annual Meeting of the Mexican Geophysical Union, 2–7 Novemeber, 2014, Puerto Vallarta, Mexico.

KOSTOGLODOV V., HUSKER A., SANTIAGO J.A., CRUZ-ATIENZA V.M., COTTE N., and WALPERSDORF A. (2015). Three types of Slow Slip Events in Guerrero, Mexico. in: Tectonic Tremor and Silent Seismicity, International Workshop, 25–27 February, 2015, Mexico, Abstract Book, 14p.

KOSTOGLODOV, V., SINGH S. K., SANTIAGO J. A., FRANCO S. I., LARSON K. M., LOWRY A. R., and BILHAM R. (2003). *A large silent earthquake in the Guerrero seismic gap, Mexico.* Geophys. Res. Lett., *30* (15), 1807, doi:10.1029/2003GL017219.

KREEMER, C., HOLT W.E., and HAINES A.J. (2003). *An integrated global model of present-day plate motions and plate boundary deformation.* Geophys. J. Int., *154*, 8-34.

LANGE, D., RIETBROCK, A., HABERLAND, C., BATAILLE, K., DAHM, T., TILMANN, F., and FLÜH, E. (2007). *Seismicity and geometry of the south Chilean subduction zone (41.5°S–43.5°S): implications for controlling parameters.* Geophys. Res. Lett. *34*, L06311. doi:10.1029/2006GL029190.

LANGE, D., CEMBRANO, J., RIETBROCK, A., HABERLAND, C., DAHM, T., and BATAILLE, K. (2008). *First seismic record for intra-arc strike-slip tectonics along the Liquiñe-Ofqui fault zone at the obliquely convergent plate margin of the southern Andes.* Tectonophysics *455*, 14–24.

LI, C. (1993). *Forearc Structures and Tectonics in the Southern Peru-Northern Chile Continental Margin.* Mar. Geophys. Res. *17*, 97–113.

LOWRY, A.R., LARSON, K.M., KOSTOGLODOV, V., and SANCHEZ, O. (2006). *The fault slip budget in Guerrero, southern Mexico.* Geophys. J. Int., *200*, unpublished: http://aconcagua.geol.usu.edu/~arlowry/Papers/Budget.pdf.

LOZOS, J.C., OGLESBY, D.D., BRUNE, J.N., and OLSEN, K.B. (2015). *Rupture Propagation and Ground Motion of Strike-Slip Stepovers with Intermediate Fault Segments.* B. Seismol. Soc. Am. *105*(1), 387–399.

MAZZOTTI, S., DRAGERT, H., HYNDMAN, R.D., MILLER, M.M., and HENTON, J.A. (2014). *GPS deformation in a region of high crustal seismicity: N. Cascadia forearc.* Earth Planet. Sci. Lett. *198* (1–2), 41–48.

MCCAFFREY, R. (1992). *Oblique plate convergence, slip vectors, and forearc deformation.* J. Geophys. Res. *97*, 8905–8915.

MCCAFFREY, R. (2009). *The Tectonic Framework of the Sumatran Subduction Zone.* Annu. Rev. Earth Planet. Sci. *37*, 345–366.

MCCAFFREY, R, ZWICK, P, BOCK, Y, PRAWIRODIRDJO, L, GENRICH, J, STEVENS, C.W., PUNTODEWO, S.S.O., and SUBARYA, C. (2000). *Strain partitioning during oblique plate convergence in northern Sumatra: geodetic and seismologic constraints and numerical modeling.* J. Geophys. Res. *105*, 28363–76.

MCCALPIN, J.P. (Ed.), Paleoseismology (Academic press, Vol. 95, 2009).

MELNICK, D., BOOKHAGEN, B., STRECKER, M.R., and ECHTLER, H.P. (2009). *Segmentation of megathrust rupture zones from fore-arc deformation patterns over hundreds to millions of years, Arauco peninsula, Chile.* J. Geophys. Res. *114*, B01407, doi:10.1029/2008JB005788.

MELTZNER, A.J., SIEH, K., ABRAMS, M., AGNEW, D.C., HUDNUT, K.W., AVOUAC, J.-P, and NATAWIDJAJA, D.H. (2006). *Uplift and subsidence associated with the great Aceh-Andaman earthquake of 2004.* J. Geophys. Res. *111*, B02407, doi:10.1029/2005JB003891.

MENDOZA A.I. (2004). Algunos eventos recientes asociados a la brecha sísmica de Guerrero: Implicaciones para la sismotectónica y el peligro sísmico de la región. PhD thesis, UNAM, Mexico City.

MESCHEDE, M., FRISCH, W., HERRMANN, U., and RATSCHBACHER, L. (1996*). Stress transmission across an active plate boundary: An example from southern Mexico.* Tectonophysics *266*, 81–100, doi:10.1016/S0040-1951(96)00184-9.

MÉTOIS, M., SOCQUET, A., and VIGNY C. (2012). *Interseismic coupling, segmentation and mechanical behavior of the central Chile subduction zone.* J. Geophys. Res. *117*, B03406, doi:10.1029/2011JB008736.

MOLNAR, P., and DAYEM, K.E. (2010). *Major intracontinental strike-slip faults and contrasts in lithospheric strength.* Geosphere *6* (4), 444–467.

MORELL, K.D., FISHER, D.M., and GARDNER, T.W. (2008). *Inner forearc response to subduction of the Panama Fracture Zone, southern Central America.* Earth Planet. Sci. Lett. *265*, 82–95.

MORENO, M.S., KLOTZ, J., MELNICK, D., ECHTLER, H., and BATAILLE, K. (2008). *Active faulting and heterogeneous deformation across a megathrust segment boundary from GPS data, south central Chile (36–39 S),* Geochem. Geophys. Geosyst. *9*, Q12024, doi:10.1029/2008GC002198.

NATAWIDJAJA, D.H., SIEH, K., CHLIEH, M., GALETZKA, J., SUWARGADI, B.W., CHENG, H., EDWARDS, R.L., AVOUAC, J.-P., and WARD, S.N. (2006). *Source parameters of the great Sumatran megathrust earthquakes of 1797 and 1833 inferred from coral microatolls,* J. Geophys. Res. *111*, B06403, doi:10.1029/2005JB004025.

ONUR, T., and SEEMANN, M.R. (2004). Probabilities of significant earthquake shaking in communities across British Columbia: implications for emergency management. 13th World

Conference on Earthquake Engineering, Vancouver, B.C., Canada, August 1–6, 2004, no. 1065.

ORTNER, H., REITER, F., and ACS, P. (2002). *Easy handling of tectonic data: the programs TectonicsVB for Mac and TectonicsFP for Windows*. Comput. Geosci. 28, 1193–1200.

OUCHI, S. (1985). *Response of alluvial rivers to slow active tectonics movement*. Geol. Soc. Am. Bull. 96, 504–515.

PACHECO, J.F., and SINGH, S.K. (2010). *Seismicity and state of stress in Guerrero segment of the Mexican subduction zone*. J. Geophys. Res. 115, doi:10.1029/2009JB006453.

PACHECO, J. F., IGLESIAS, A., and SINGH, S.K. (2002). *The 8 October Coyuca, Guerrero, Mexico earthquake (Mw 5.9): A normal fault in the expected compressional environment*. Seism. Res. Lett. 73(2), 263.

PÉREZ-GUTIÉRREZ, R., SOLARI, L.A., GÓMEZ, T.A., and MARTENS, U. (2009). *Mesozoic geologic evolution of the Xolapa migmatitic Complex north of Acapulco, southern Mexico, and its tectonic significance*. Revista Mexicana de Ciencias Geológicas, 26, 201–221.

PLAFKER, G. (1969). Tectonics of the March 27, 1964 Alaska earthquake: U.S. Geological Survey Professional Paper 543–I, 74 p., 2 sheets, scales 1:2,000,000 and 1:500,000, http://pubs.usgs.gov/pp/0543i/.

RAMÍREZ-HERRERA, M.T. (1998). *Geomorphic assessment of active tectonics in the Acambay Graben, Mexican Volcanic Belt*. Earth Surf. Process. Landf. 23, 317–332.

RAMÍREZ-HERRERA, M.T. and URRUTIA-FUCUGAUCHI, J. (1999). *Morphotectonic zones along the coast of the Pacific continental margin, southern Mexico*. Geomorphology 28, 237–250.

RAMÍREZ-HERRERA, M.T., CUNDY, A., KOSTOGLODOV, V., CARRANZA-EDWARDS A., MORALES E., and METCALFE, S. (2007). *Sedimentary record of late Holocene relative sea-level change and tectonic deformation from the Guerrero Seismic Gap, Mexican Pacific Coast*. Holocene 17/8, 1211–1220.

RAMÍREZ-HERRERA, M.T., CUNDY, A., KOSTOGLODOV, V., and ORTIZ, M. (2009). *Late Holocene tectonic land-level changes and tsunamis at Mitla lagoon, Guerrero, México*. Geofis. Int. 48, 195–209.

RAMÍREZ-HERRERA, M.T., KOSTOGLODOV, V., and URRUTIA-FUCUGAUCHI, J. (2010). *Overview of Recent Tectonic Deformation in the Mexican Subduction Zone*. Pure Appl. Geophys. PAAG-320, doi:10.1007/s00024-010-0205-y.

RAMSAY, J.G. (1967). Folding and Fracturing of Rocks. McGraw-Hill Book Co. Inc., New York.

RAMSAY, J.G. and LISLE, R.J., The Techniques of Modern Structural Geology, 3: Applications of Continuum Mechanics in Structural Geology (Academic Press, London 2000, 702–1061).

REHAK, K, STRECKER, M., and ECHTLER, H. (2008). *Morphotectonic segmentation of an active forearc, 37°–41°S, Chile*. Geomorphology 94, 98–116.

RILLER, U., RATSCHBACHER, L., and FRISCH, W. (1992). *Left-lateral transtension along the Tierra Colorada deformation zone, northern margin of the Xolapa magmatic arc of southern Mexico*. J. S. Am. Earth Sci. 5, 237–249, doi:10.1016/0895-9811(92)90023-R.

RYDER, I., RIETBROCK, A., KELSON, K., BÜRGMANN, R., FLOYD, M., SOCQUET, A., VIGNY, C., and CARRIZO, D. (2012). *Large extensional aftershocks in the continental forearc triggered by the 2010 Maule earthquake, Chile*. Geophys. J. Int. 188(3), 879–890.

SCHAAF, P., MORÁN-ZENTENO, D.J., HERNÁNDEZ-BERNAL, M.S., SOLÍS-PICHARDO, G., TOLSON, G., and KOHLER, H. (1995). *Paleogene continental margin truncation in southwestern Mexico:*

Geochronological evidence. Tectonics 14, 1339–1350, doi:10.1029/95TC01928.

SCHUMM, S.A., DUMONT, J.F., HOLBROOK, J.M., Active Tectonics and Alluvial Rivers (Cambridge University Press, Cambridge, 2000).

SEDLOCK, R.L., ORTEGA-GUTIÉRREZ, F., and SPEED, R.C. (1993). Tectonostratigraphic terranes and tectonic evolution of Mexico. Geological Society of America, Special Paper 278.

SHIRZAEI, M, BÜRGMANN, R., ONCKEN, O., WALTER, T.R., VICTOR, P., and EWIAK, O. (2012). *Response of crustal faults to megathrust earthquakes cycle: InSAR evidence from Mejillones Peninsula, northern Chile*. Earth Planet. Sci. Lett. 333–334, 157–164, doi:10.1016/j.epsl.2012.04.001.

SIMONS, M., MINSON, S.E., SLADEN, A., ORTEGA, F., OWEN, S.E. MENG, L., AMPUERO, J-P., WEI, S., CHU, R., HELMBERGER, D.V., KANAMORI, H., HETLAND, E., MOORE, A.W., and WEBB, F.H. (2011). *The 2011 Magnitude 9.0 Tohoku-Oki Earthquake: Mosaicking the Megathrust from Seconds to Centuries*. Science 332, 1421–1425. doi:10.1126/science.1206731.

SINGH, S.K., and MORTERA, F. (1991). *Source-time functions of large Mexican subduction earthquakes, morphology of the Benioff zone and the extent of the Guerrero gap*. J. Geophys. Res. 96, 21,487–21,502.

SINGH, S. K., ORDAZ, M., ALCÁNTARA, L., SHAPIRO, N., KOSTOGLODOV, V., PACHECO, J. F., ALCOCER, S., GUTIÉREZ, C., QUAAS, R., MIKUMO, T., and OVANDO, E. (2000). *The Oaxaca Earthquake of 30 September 1999 (Mw = 7.5): a normal-faulting event in the subducted Cocos plate*. Seismol. Res. Lett. 71(1), 67–78.

SOLARI, L.A., TORRES DE LEÓN, R., HERNÁNDEZ PINEDA, G., SOLÉ, J., SOLÍS-PICHARDO, G., and HERNÁNDEZ-TREVIÑO, T. (2007). *Tectonic significance of Cretaceous–Tertiary magmatic and structural evolution of the northern margin of the Xolapa Complex, Tierra Colorada area, southern Mexico*. GSA Bulletin 119, 9/10, 1265–1279; doi:10.1130B26023.1.

SOTO, M.D., MANN, P., ESCALONA, A., and WOOD, L.J. (2007). *Late Holocene strike-slip offset of a subsurface channel interpreted from three-dimensional seismic data, eastern offshore Trinidad*. Geology 35 (9), 859–862.

STOKES, M., MATHER, A.E., BELFOUL, A., and FARIK, F. (2008). *Active and passive tectonic controls for transverse drainage and river gorge development in a collisional mountain belt (Dades Gorges, High Atlas Mountains, Morocco)*. Geomorphology 102, 2–20.

SUBARYA, C., CHLIEH, M., PRAWIRODIRDJO, L., AVOUAC, J.P., BOCK, Y., SIEH, K., MELTZNER, A.J., NATAWIDJAJA, D.H., and McCAFFREY, R. (2006). *Plate-boundary deformation associated with the great Sumatra-Andaman earthquake*. Nature 440, 46–51.

TAJIMA, F., MORI, J., and KENNETT, B.L.N. (2013). *A review of the 2011 Tohoku-Oki earthquake (Mw 9.0): Large-scale rupture across heterogeneous plate coupling*. Tectonophysics 586, 15–34. doi:10.1016/j.tecto.2012.09.014.

TALAVERA-MENDOZA, O., RUIZ, J., CORONA-CHAVEZ, P., GEHRELS, G.E., SARMIENTO-VILLAGRANA, A., GARCÍA-DÍAZ, J.L., and SALGADO-SOUTO, S.A. (2013). *Origin and provenance of basement metasedimentary rocks from the Xolapa Complex: New constraints on the Chortis–southern Mexico connection*. Earth Planet. Sci. Lett. 369–370, 188–199.

TODA, S., and TSUTSUMI, H. (2013). *Simultaneous Reactivation of Two, Subparallel, Inland Normal Faults during the Mw 6.6 11 April 2011 Iwaki Earthquake Triggered by the Mw 9.0 Tohoku-oki, Japan, Earthquake*. Bull. Seismol. Soc. Am. 103(2B), 1584–1602.

TOLSON, G. (2005). *La falla Chacalapa en el sur de Oaxaca*. Bol. Soc. Geol. Mex. *57*, 111–122.

VALENCIA, V.A., DUCEA, M., TALAVERA-MENDOZA, O., GEHRELS, G., RUIZ, J., and SHOEMAKER, S. (2009). *U-Pb geochronology of granitoids in the north-western boundary of the Xolapa Terrane*. Revista Mexicana en Ciencias Geologicas *26* (1), 189–200.

VARGAS, G., REBOLLEDO, S., SEPÚLVEDA, S., LAHSEN, A., THIELE, R., TOWNLEY, B., PADILLA, C., RAULD, R., HERRERA, M., and LARA, M. (2013). *Submarine earthquake rupture, active faulting and volcanism along the major Liquiñe-Ofqui Fault Zone and implications for seismic hazard assessment in the Patagonian Andes*. Andean Geol. *40* (1), 141–171.

VIGNY, C., SOCQUET, A., PEYRAT, S., RUEGG, J-C., MÉTOIS, M., MADARIAGA, R., MORVAN, S., LACASSIN, R., CAMPOS, J., CARRIZO, D., BEJAR-PIZARRO, M., BARRIENTOS, S., ARMIJO, R., ARANDA, C., VALDERAS-BERMEJO, M-C., ORTEGA, I., BONDOUX, F., BAIZE, S., LYON-CAEN, H., PAVEX, A., VILOTTE, J.P., BEVIS, M., BROOKS, B., SMALLEY, R., PARRA, H., BAEZ, J-C., BLANCO, M., CIMBARO, S., and KENDRICK, E. (2011). *The 2010 M_w 8.8 Maule Megathrust Earthquake of Central Chile, Monitored by GPS*. Science *332*, 1417–1421, doi:10.1126/science.1204132.

WELLS, D.L., and COPPERSMITH, K.J. (1994). *New empirical relationships among magnitude, rupture length, rupture width, rupture area, and surface displacement*. B. Seismol. Soc. Am. *84*(4), 974–1002.

WESNOUSKY, S.G. (2006). *Predicting the endpoints of earthquake ruptures*. Nature *444*(7117), 358–360.

WORTEL, R., and CLOETINGH, S. (1981). *On the origin of the Cocos-Nazca spreading center*. Geology *9*, 425–430.

ŻABA, J., MAIOLEPSZY, Z., GAIDZIK, K., CIESIELCZUK, J. and PAULO, A. (2012). *Faults network in the Rio Colca valley between Maca and Pinchollo, Central Andes, Southern Peru*. ASGP *82* (3), 279–290.

WEB REFERENCES

National Institute of Statistics and Geography: http://www.inegi.org.mx (December, 2014).

Mexican Geological Survey: http://www.sgm.gob.mx (June, 2014).

(Received April 20, 2015, revised November 19, 2015, accepted November 21, 2015, Published online January 7, 2016)

Pure Appl. Geophys. 173 (2016), 3445–3465
© 2015 Springer Basel
DOI 10.1007/s00024-015-1211-x

Pure and Applied Geophysics

Slow Slip History for the MEXICO Subduction Zone: 2005 Through 2011

Shannon Graham,[1] Charles DeMets,[2] Enrique Cabral-Cano,[3] Vladimir Kostoglodov,[3] Baptiste Rousset,[4] Andrea Walpersdorf,[4] Nathalie Cotte,[4] Cécile Lasserre,[4] Robert McCaffrey,[5] and Luis Salazar-Tlaczani[3]

Abstract—To further our understanding of the seismically hazardous Mexico subduction zone, we estimate the first time-dependent slip distributions and Coulomb failure stress changes for the six major slow slip events (SSEs) that occurred below Mexico between late 2005 and mid-2011. Slip distributions are the first to be estimated from all continuous GPS data in central and southern Mexico, which better resolves slow slip in space and time than was previously possible in this region. Below Oaxaca, slip during previously un-modeled SSEs in 2008/9 and 2010/11 extended farther to the west than previous SSEs. This constitutes the first evidence that slow slip accounts for deep slip within a previously noted gap between the Oaxaca and Guerrero SSE source regions. The slip that we estimate for the two SSEs that originated below Guerrero between 2005 and 2011 agrees with slip estimated in previous, mostly static-offset SSE modeling studies; however, we show that both SSEs migrated eastward toward the Oaxaca SSE source region. In accord with previous work, we find that slow slip below Guerrero intrudes up-dip into the potentially seismogenic region, presumably accounting for some of the missing slip within the well-described Guerrero seismic gap. In contrast, slow slip below Oaxaca between 2005 and 2011 occurred mostly down-dip from the seismogenic regions defined by the rupture zones of large thrust earthquakes in 1968 and 1978 and released all of the slip deficit that accumulated in the down-dip region during this period.

Key words: Slow slip events, earthquake cycle, Mexico subduction zone, global positioning system.

Electronic supplementary material The online version of this article (doi:10.1007/s00024-015-1211-x) contains supplementary material, which is available to authorized users.

[1] Department of Earth and Planetary Sciences, Harvard University, Cambridge, MA, USA. E-mail: shannonegraham@fas.harvard.edu
[2] Department of Geoscience, University of Wisconsin-Madison, Madison, WI 53706, USA.
[3] Instituto de Geofísica, Universidad Nacional Autónoma de México, Mexico City, Mexico.
[4] ISTerre, CNRS, Univ. Grenoble Alpes, 38041 Grenoble, France.
[5] Department of Geology, Portland State University, Portland, OR, USA.

1. Introduction

With the increase in continuous GPS (cGPS) stations deployed at subduction zones worldwide, our understanding of how plate motion is accommodated at these convergent margins is evolving rapidly. Slow slip observed in cGPS position time series, in conjunction with seismically detected tectonic tremor, have helped to define a complex transition zone between stick–slip and creep behavior along the subduction zone interface (Dragert *et al.* 2001; Lowry *et al.* 2001; Ohta *et al.* 2004, 2006). The location of many slow slip events (SSEs) immediately down-dip from seismogenic zones suggests they could trigger large thrust earthquakes. For example, the 11 March 2011 $M_w = 9.0$ Tohoku-Oki earthquake in Japan (Ito *et al.* 2012), 20 March 2012 $M_w = 7.4$ Ometepec earthquake on the Mexico subduction zone (Graham *et al.* 2014a), and 2014 $M_w = 8.1$ Iquique earthquake in Chile (Ruiz *et al.* 2014) were all preceded by SSEs close to or overlapping the eventual earthquake rupture zones. Numerical simulations of slow slip suggest that repeated SSEs in the transition zone between stick–slip and creep concentrate stress at the down-dip limit of the seismogenic zone, which in turn increases the probability that a future SSE will evolve into a dynamic rupture (Segall and Bradley 2012). SSEs also relieve interseismic strain accumulated along the plate interface, thereby limiting the size of coseismic rupture patches and consequently reducing subsequent earthquake magnitudes (Dixon *et al.* 2014). Measurements and modeling of SSEs are thus important for seismic hazard analysis and for an improved understanding of how subduction zones accommodate convergence.

SSE characteristics vary globally, exhibiting a wide range of durations, locations on the plate interface, recurrence intervals, magnitudes, and slip amplitudes (e.g., PENG and GOMBERG 2010; BEROZA and IDE 2011). Slow slip has also been shown to migrate along the plate interface, in some places up to 300 km along-strike (e.g., SCHMIDT and GAO 2010; DRAGERT and WANG 2011). In addition to slip migration along a fault, simulations of SSEs using rate-and-state-dependent friction indicate that slow slip can also nucleate simultaneously on distant regions of a fault (COLELLA et al. 2011, 2012). Moreover, WALLACE et al. (2012) postulate that SSEs can interact with one another and infer from Coulomb failure stress calculations that a deep slow slip event along the Hikurangi subduction zone in New Zealand may have triggered a series of subsequent, shallower SSEs.

Between 1993, when the first continuous GPS receiver was installed along the Mexico subduction zone (MSZ), and late 2012, at least 15 SSEs were recorded along the subduction interface below the Mexican states of Guerrero and Oaxaca (Fig. 1) (e.g., KOSTOGLODOV et al. 2003; BRUDZINSKI et al. 2007; CORREA-MORA et al. 2008, 2009; RADIGUET et al. 2012). In addition, smaller SSEs, may have occurred, presently below the threshold of GPS detection (VERGNOLLE et al. 2010). Before ∼2005, continuous GPS stations were clustered primarily in Guerrero and in central Oaxaca, with only a handful of stations located in the ∼200-km gap between the station clusters. Continuous receivers operating in Guerrero have detected SSEs every 3–4 years within a region that extends ∼250 km along-strike and up-dip into the seismogenic zone of the Guerrero seismic gap (Fig. 1) (RADIGUET et al. 2012). The magnitude of slow slip on the interface has reached a maximum of ∼200 mm, accounting for the slip equivalent of an $M \sim 7.5$ earthquake (e.g., LOWRY et al. 2001; KOSTOGLODOV et al. 2003; IGLESIAS et al. 2004; YOSHIOKA et al. 2004; LARSON et al. 2007; RADIGUET et al. 2012; CAVALIÉ et al. 2013). Along the Oaxaca segment, slow slip has occurred every 1–2 years with smaller maximum slip amplitudes of ∼100 mm and smaller

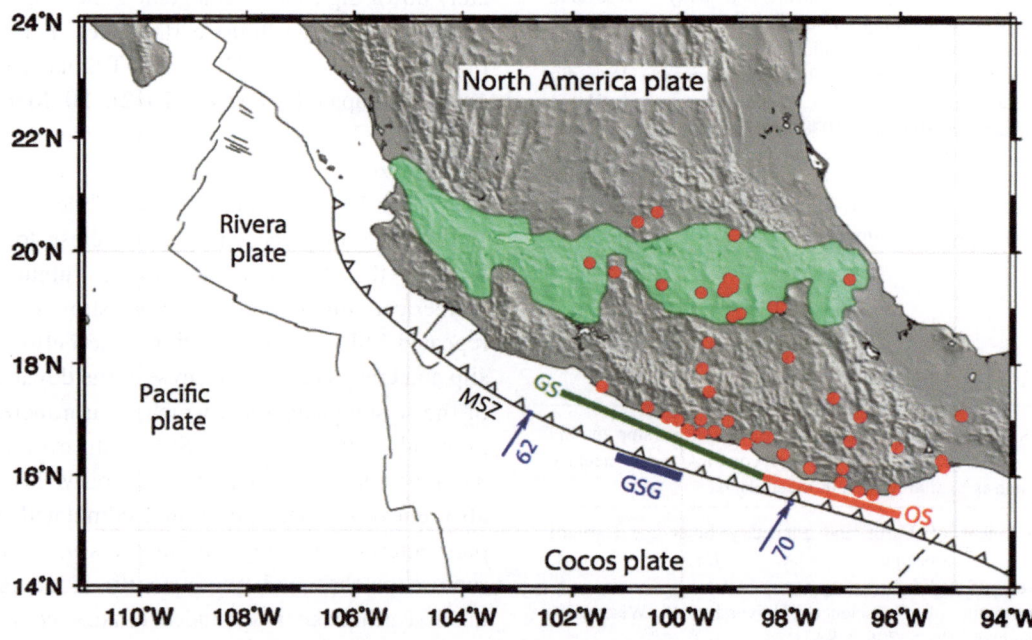

Figure 1

Tectonics of the study area. Convergence velocities for the Cocos plate relative to the North America plate are shown by blue vectors. Convergence rates are given in mm/year (DeMETS et al. 2010). The *green* and *red lines* demarcate the along-strike limits of the Guerrero segment (GS) and Oaxaca segment (OS), respectively. The *thick blue line* illustrates the northwestern Guerrero seismic gap (GSG) after RADIGUET et al. (2012). *Red circles* denote locations of GPS sites used to model the SSEs. *Green area* shows the Mexican Volcanic Belt. *MSZ* Mexico Subduction Zone

moment releases of M 6.6–6.9 than for Guerrero (CORREA-Mora *et al.* 2008, 2009; GRAHAM *et al.* 2014a).

The principal goal of this study is to undertake the first time-dependent modeling of a complete sequence of slow slip events along the Mexico subduction zone, including previously un-modeled SSEs in 2008/9 and 2010/11 beneath Oaxaca and four additional SSEs below Guerrero and Oaxaca. RADIGUET *et al.* (2011) completed the only other SSE time-dependent modeling in Mexico for the 2006 Guerrero SSE. Our focus is on SSEs that have occurred since 2005, when the GPS network geometry in southern Mexico became sufficiently dense to resolve the location and migration of SSEs along the subduction interface. In particular, we seek to determine whether slow slip migrates across the several-hundred-km-wide gap between the Guerrero and Oaxaca regions, as proposed by FRANCO *et al.* (2005), or whether slow slip in the two regions is spatially and temporally independent, as suggested by CORREA-MORA *et al.* (2009). The distinction between localized slow slip regions and wide spread slow slip has important seismic hazard implications for the Mexico subduction zone given that slow slip can load the seismogenic zone or evolve into dynamic rupture. Our analysis includes calculations of Coulomb failure stress changes in response to each SSE to evaluate possible cause-and-effect relationships between the six SSEs that are modeled herein and the 2012 Ometepec earthquake. Our SSE modeling results are the first to be determined from all available continuous GPS stations in southern and central Mexico, which is important for maximizing the spatiotemporal coverage of SSEs in this region.

2. Data and Methods

2.1. GPS Data and Analysis

2.1.1 Data, Processing, and Post-Processing Methods

For this analysis, we use data from 56 continuous GPS stations in central and southern Mexico (Fig. 2) spanning the period January 2005 through October 2011. GPS data were processed with Release 6.1 of the GIPSY software suite from the Jet Propulsion Laboratory (JPL). No-fiducial daily GPS station coordinates were estimated using a precise point-positioning strategy (ZUMBERGE *et al.* 1997), including constraints on a priori tropospheric hydrostatic and wet delays from Vienna Mapping Function (VMF1) parameters (http://ggosatm.hg.tuwien.ac.at), elevation- and azimuthally dependent GPS and satellite antenna phase center corrections from IGS08 ANTEX files (available via ftp from http://sideshow. jpl.nasa.gov), and corrections for ocean tidal loading (http://holt.oso.chalmers.se). Phase ambiguities were resolved for all the data using GIPSY's single-station ambiguity resolution feature. The no-fiducial station location estimates were transformed to IGS08 using daily seven-parameter Helmert transformations from JPL, thereby yielding daily point-positioned station coordinates that conform to the International Terrestrial Reference Frame 2008 (ITRF08) (ALTAMIMI *et al.* 2011). We applied methods described by MÁRQUEZ-AZÚA and DeMETS (2003) to estimate the common-mode noise for stations in southern Mexico and remove it from the position time series. Further details of this procedure are provided in GRAHAM *et al.* (2014a).

2.1.2 Example GPS Position Time Series

Figure 3 shows the position time series for two long-operating cGPS sites, one along the Oaxaca segment of the Mexico subduction zone (OAXA/2) and the other along the Guerrero segment (CAYA). The time series highlight obvious differences between the recurrence interval, timing, and relative amplitudes of SSEs along the two segments. At both sites, steady interseismic motion towards the north, representing elastic shortening of the upper North America plate in response to Cocos plate subduction, is interrupted by several-month-long periods of southward motion interpreted as slow slip on the plate interface. Of the six SSEs that occurred between 2005 and 2011, four were recorded at GPS sites in Oaxaca (2005/6, 2007, 2008/9, and 2010/11) and two in Guerrero (2006 and 2009/10) (Fig. 3b). The Guerrero SSEs exhibit longer durations and greater displacements than events in Oaxaca (Fig. 3).

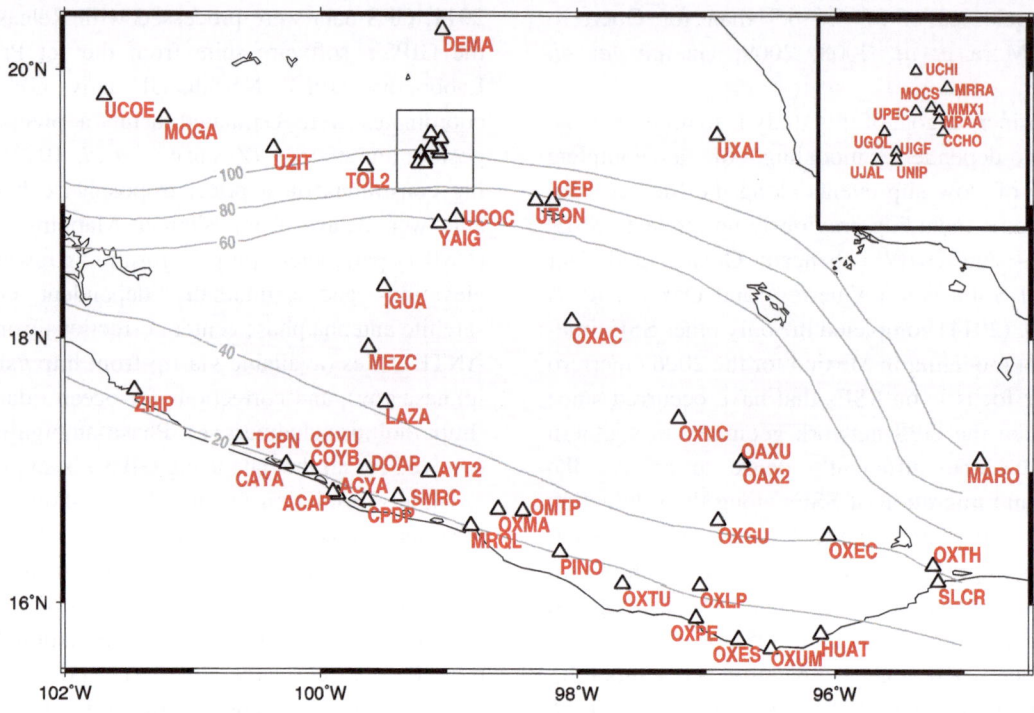

Figure 2
Continuous GPS station locations denoted by *black triangles*. Site names are denoted by their *four-letter* code. *Black rectangle* shows area for inset

2.2. Inversion Method: TDEFNODE

We use TDEFNODE (McCaffrey 2009) to invert the cGPS time series to solve simultaneously for the spatial and temporal evolution of slip associated with the six SSEs. In order to isolate the signal of the SSE from the secular interseismic rate at each site, we simultaneously solve for the inter-SSE slope and parameters describing slow slip. Slip is presumed to occur on the subduction zone interface, represented in an elastic half-space using an interface geometry based on depth contours from Radiguet et al. (2012). The model space extends along-strike from 94°W to 105°W and from the surface to 80 km depth. Using the Okada (1992) elastic half-space dislocation algorithm, Green's functions are calculated for fault nodes spaced 5 km along-strike and 3 km down-dip. Within the inversion, the spatial slip distribution is estimated at fault nodes with a spread smoothing constraint, where slip is penalized for distance from the slip centroid, and the time evolution of slip-rate per node is modeled with a Gaussian function. Smoothing parameters were adjusted systematically to identify the best tradeoff between the

model and data variance. We experimented with other spatial and temporal parameterizations for slip within TDEFNODE, but found that the above combination works best for modeling our data. During the inversion, we estimate the slip-rate amplitude at each node, the nominal event onset time, the rate and azimuth of slip migration, the time constant for the Gaussian function representing slip rate, and the inter-SSE slope. The rake of the slow slip is constrained to be opposite the N32°E direction of Cocos–North America plate convergence (DeMets et al. 2010). Further details on our inversion approach are given in Graham et al. (2014a).

3. Slow Slip History for 2005 Through 2011

Inversions of data from the 56 continuous GPS sites that were operating in central and southern Mexico from 2005 through 2011 result in spatio-temporal slip distributions for each SSE that are broadly consistent in location, magnitude, and duration with those estimated by previous authors, who used the cumulative SSE offsets to solve for slip (e.g.,

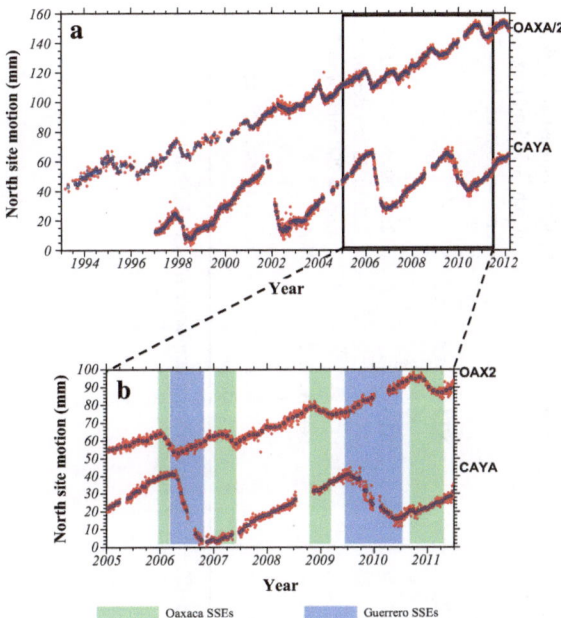

Figure 3

a Complete time series for the north components of cGPS stations in Oaxaca (OAXA/2) and Guerrero (CAYA) reduced by the motion of the North America plate at each site. Site locations are shown in Fig. 2. *Red circles* indicate daily positions and *open blue circles* show 30-day averages. **b** Same as **a**, but for the time period used in this study. *Green* and *blue rectangles* highlight SSEs below Oaxaca and Guerrero, respectively

CORREA-MORA *et al.* 2008; RADIGUET *et al.* 2011, 2012). For the six events and time series we modeled, the overall weighted root-mean-square (WRMS) misfit to the daily 3-D GPS site positions is 2.4 mm, close to the scatter in the observed daily station positions. The estimated slip on the subduction interface thus captures the north, east, and vertical displacement histories recorded by the cGPS stations.

Table 1 summarizes the main characteristics of the SSEs modeled herein as well as those modeled by other authors covering all SSEs recorded in southern Mexico between 2000 and 2012. The results and time series fits of our inversion for each of the six SSEs are presented below. In addition, we provide in the supplementary material an animation that depicts the slow slip history from 2005 to mid-2011.

3.1. 2005/6 Oaxaca SSE

Slip during the 2005/6 SSE below Oaxaca was focused primarily from 98°W to 97°W and at depths

of 20–40 km, with a maximum slip amplitude of 120 mm. Smaller amounts of slip extended ∼200 km along-strike (Fig. 4a) and up-dip into seismogenic portions of the subduction zone (Fig. 4a). This slip distribution represents the first evidence that slow slip may extend to shallower depths below Oaxaca than found by CORREA-MORA *et al.* (2008). The large displacement of the near-coastal site PINO (located in Fig. 2 and shown in Fig. 4a), data unavailable to CORREA-MORA *et al.* (2008), is the key evidence that this SSE extended up-dip to depths as shallow as 15 km. Our modeling suggests that slip migrated slowly up-dip and eastward (N120°E) at a rate of 1.5 km per day (Fig. 5 and supplementary animation). The cumulative geodetic moment for the SSE was 6.0×10^{19} N m ($M_w = 7.1$), the same as estimated by CORREA-MORA *et al.* (2008).

The evolution of the surface deformation predicted by our inversion for this SSE matches the GPS position time series at the sites that recorded it (Fig. 6). For example, the position time series for sites PINO, OAX2, OAXU, OXUM, and OXPE are all fit within their observational scatter (Fig. 6). Observations and predicted time series for the entire suite of sites that recorded this SSE are shown in Supplemental Figure A1.

3.2. 2006 Guerrero SSE

Between mid-March and late May of 2006, during the final stages of the 2005/6 SSE below Oaxaca, slow slip began beneath Guerrero ∼200 km WNW of the Oaxaca SSE (Fig. 5c). Over a period of 8 months, a region of high slip (∼150 mm) migrated southeast towards Oaxaca (Fig. 5c–f), giving rise to cumulative horizontal and vertical surface displacements as high as 50 mm (Figs. 6, A1). The cumulative slip distribution (Fig. 4b) is broadly consistent with both the location and amplitude estimated by RADIGUET *et al.* (2012) and the spatiotemporal evolution of slip (local slip duration, migration direction, and speed,) is in very good agreement with that found by RADIGUET *et al.* (2011). The majority of slip occurred between 20 and 45 km depth along the plate interface (Figs. 4, 5), with the shallowest slip located in the Guerrero seismic gap (Fig. 4). The peak SSE slip amplitude is 270 mm and the geodetic moment is 10.2×10^{19} N m, equivalent

Table 1

Characteristics of slow slip events (SSEs) along the Mexico subduction zone for the period 2001–2012

SSE	Location along-strike	Location down-dip (km)	Duration in months	Max slip amplitude (mm)	Equivalent magnitude (M_w)	Migration direction	Previous results
2001/02 Guerrero*	~99° to ~101.5°W	20 to 50	9	200	7.65	N/A	RADIGUET et al. (2012)
2004 Oaxaca*	~97.5° to ~101.5°W	~30	~4	115	7.3	N/A	CORREA-MORA et al. (2008)
2005/6 Oaxaca	~96.5° to ~98.5°W	–18 to 45	~6	120	7.1	N120E, at 1.5 km/day	CM08: 100 mm max amp, –96° to ~97.75°W, ~22 to 35 km depth; CM09: same location, max amp = 60 mm, $M = 7.1$
2006 Guerrero	~98.5° to ~102°W	20 to 45	~8	270	7.3	N108E	R12: ~99° to ~101°W, –20 to 40 km depth, max amp = 200 mm, $M = 7.5$ Rll: $M = 7.5$, WtoEslip migration, location = R12
2007 Oaxaca	~96.5° to ~97.5°W (centered at 97°W)	Centered at 30	~5	80	6.5	None	CM09: centered at 97°W, depth ~30 km, max amp = 30 mm, $M = 7.0$
2008/9 Oaxaca	~97° to 98.5°W	–30 to 45	~6	120	7.2	N140E, at 4 km/day	N/A
2009/10 Guerrero	–99° to ~102°W	20 to 45	~12	280	7.4	N108E	R12: ~99° to ~101.75°W, –20 to 45 km depth, max amp = 200 mm, $M = 7.53$
2010/11 Oaxaca	–96.75° to ~99°W	Centered at 40	~6	120	7.2	N40W, at 2 km/day	N/A
2011/12 Oaxaca*	~95.5° to ~98°W	20 to 40	5, interrupted byEQ	105	6.9	N36W, at 2.6 km/day	GRAHAM et al. (2014a)

CM08 Correa-Mora et al. 2008, *CM09* Correa-Mora et al. (2009), *Rll* Radiguet et al. (2011), *R12* Radiguet et al. (2012), *max amp* maximum amplitude of slip, *EQ* earthquake, *N/A* not applicable

* Results from studies listed in the rightmost column

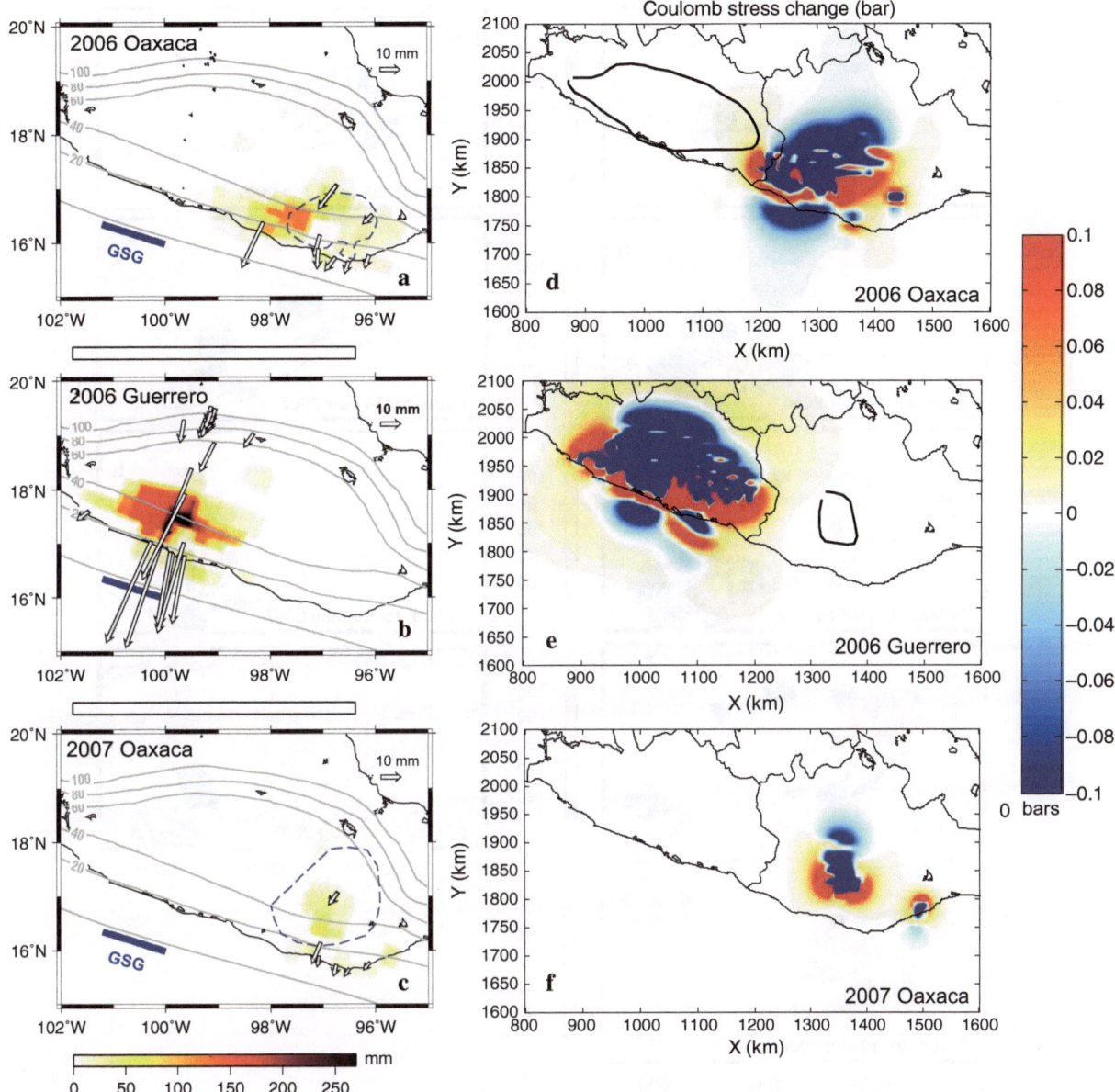

Figure 4

Cumulative slip during the 2005/6 (**a**), 2006 (**b**), and 2007 (**c**) SSEs from time-dependent inversions of the continuous GPS data. *White arrows* show total surface displacements predicted by the modeled slip. *Blue line marks* the along-strike location of the Guerrero seismic gap (GSG). *Blue dash contour* shows slip source for the 2006 (**a**) and 2007 (**c**) Oaxaca SSEs as modeled by CORREA-MORA *et al.* (2008, 2009), respectively. **d–e** Coulomb failure stresses calculated from the slow slip for each event. *Black contours* show the location of the subsequent SSE where applicable

to an $M_w = 7.3$ earthquake (Table 1). Slip continued until the onset of the 2007 SSE beneath Oaxaca.

The predicted displacements closely match the observed GPS displacements for the well-recorded 2006 SSE (Fig. 6). Measurements at sites PINO and OXPE, which are located progressively farther eastward along the coast from Guerrero (Fig. 2), clearly show a lack motion during the 2006 Guerrero SSE (Fig. 6). These observations indicate a limit to the eastern extent of the slow slip event.

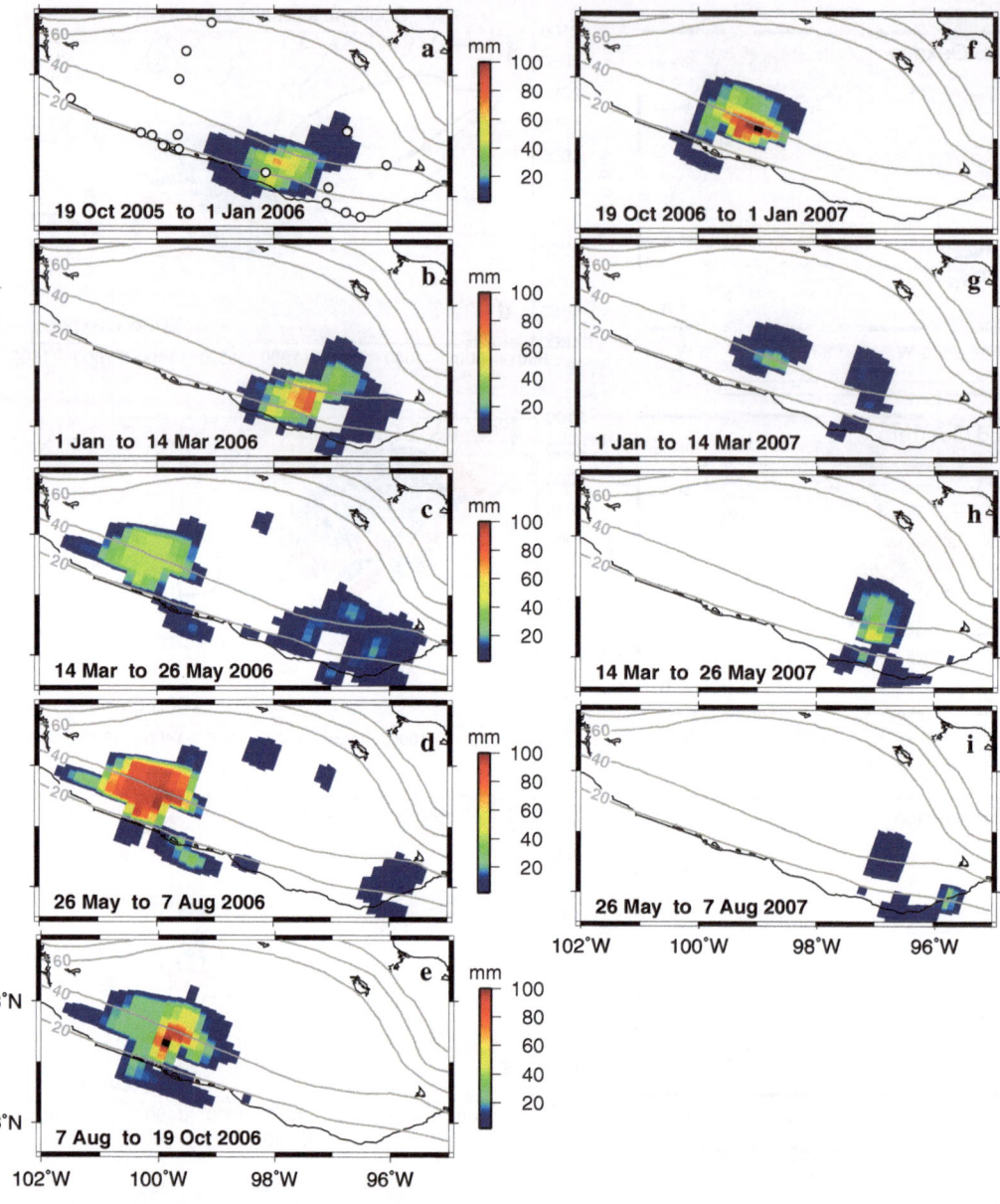

Figure 5
Evolution of slip from 2005 to 2007 at ∼2-month-long intervals. The supplementary information includes an animation of the same slip sequence at 2-week-long intervals. The 2006 SSE in Guerrero (**c–g**) began during the final stage of slow slip in Oaxaca (**a–c**). The eastward-migrating Guerrero SSE is followed in 2007 by a small SSE below Oaxaca (**g–i**). The *open circles* in **a** indicate the locations of GPS stations that were used to determine the slip

3.3. 2007 Oaxaca SSE

The smallest SSE detected between 2005 and 2011 occurred below Oaxaca in 2007, ∼100 km east of the slip region for the 2006 Guerrero SSE. Relative to the other five SSEs, the 2007 event was smaller in magnitude, spatial extent, and duration (Table 1; Fig. 4). Slip was centered at 97°W and 30 km depth, consistent with previous results (Table 1). The maximum slip amplitude reached 80 mm and the geodetic moment was 5.7×10^{18} N m, equivalent to $M_w = 6.5$. Due to hardware failures and other operational problems, our record of the 2007 SSE is less complete than for the other events. As a result, we cannot resolve with confidence whether there was any geographic or temporal overlap between the 2007 SSE and the waning stages of the 2006 SSE below Guerrero.

3.4. 2008/9 Oaxaca SSE

Between late 2006 and mid-2008, a significant number of new GPS stations were installed between Oaxaca and Guerrero to improve SSE detection. The upgraded cGPS network in southern Mexico detected the onset of slow slip below Oaxaca in October of 2008 following 15 months of uninterrupted inter-SSE elastic shortening in Guerrero and Oaxaca (Fig. 3b). Most slip during the 2008/9 SSE was focused from 30 to 45 km depth (Fig. 7), deeper than the 2005/6 SSE, while slip migrated along-strike (Fig. 8; Table 1). The 2008/9 SSE had nearly the same along-strike location, peak slip amplitude, and geodetic moment as the 2004 and 2005/6 SSEs below Oaxaca (Table 1; Fig. 7a). We note that the difference in the distribution of slip with earlier events could simply be a result of additional stations, particularly those further from the trench (compare GPS displacements in Figs. 4a, 7a). The predicted displacements match the observed cGPS position time series at all the sites (Fig. 9, Supplemental Fig. A2).

3.5. 2009/10 Guerrero SSE

The 2009/10 SSE below Guerrero was unique with respect to all other SSEs that were previously recorded in southern Mexico in that the GPS time series show two overlapping pulses of deformation, as noted by WALPERSDORF et al. (2011) (see plots for sites ACAP, ACYA, CAYA, IGUA, and MEZC in Fig. 9). We thus modeled the 2009/10 Guerrero SSE with two sub-events, the second of which began approximately 5 months after the first. Results indicate that the two sub-events are distinct in space and time (Figs. 7, 8d–j). Slip for the first initiated in late May of 2009 (Fig. 8d), increased rapidly in amplitude between late-May and mid-October (Fig. 8e), and diminished through mid-March of

2010 (Fig. 8f, g). Shortly thereafter, the second sub-event began at the eastern end of the earlier sub-event and accommodated as much as 160 mm of cumulative slip before its conclusion in October of 2010 (Figs. 7, 8h–j). The 2009/10 SSE ruptured much of the same part of the plate interface as did the 2006 Guerrero SSE and had similar fault-slip magnitudes (Table 1; Fig. 7). The overall geodetic moment was 12.3×10^{19} N m, equivalent to an $M_w = 7.4$ earthquake.

Using two sub-events, the predicted GPS time series match the observations closely (Fig. 9). If we instead model the SSE as a single event, the two sub-events merge into a single eastward-migrating slow slip event with two high-slip regions. Although the SSE characteristics remain largely unchanged, the time series are fit more poorly. We speculate that the two sub-events are expressions of a single eastward-migrating slow slip event with two high-slip regions (Figs. 7, 8). Modeling by RADIGUET et al. (2012) of the cumulative, static offsets for the 2009/10 SSE also shows two regions of high slip. The good agreement between our results, which were determined using different modeling techniques and fundamentally different approaches to GPS data processing, suggests that the occurrence of two SSEs in 2009/10 in close proximity in space and time is a robust result.

3.6. 2010/11 Oaxaca SSE

The source region of the 2010/11 Oaxaca SSE was located immediately east of slip during the 2009/10 Guerrero SSE (Fig. 7; Table 1). Slip during this $M_w = 7.2$-equivalent event is remarkably similar to the SSE below Oaxaca in 2008/9 (compare Fig. 7a, d), with the principal difference being that slip in 2010/11 extended ~50 km farther west than in 2008/9 (Fig. 7). Much of the area that slipped in 2008/9 and 2010/11 also slipped during the 2011/12 SSE that preceded the 20 March 2012 $M_w = 7.4$ Ometepec earthquake (GRAHAM et al. 2014a), although the 2011/12 SSE appears to have occurred ~5 to 10 km farther up-dip on the subduction interface.

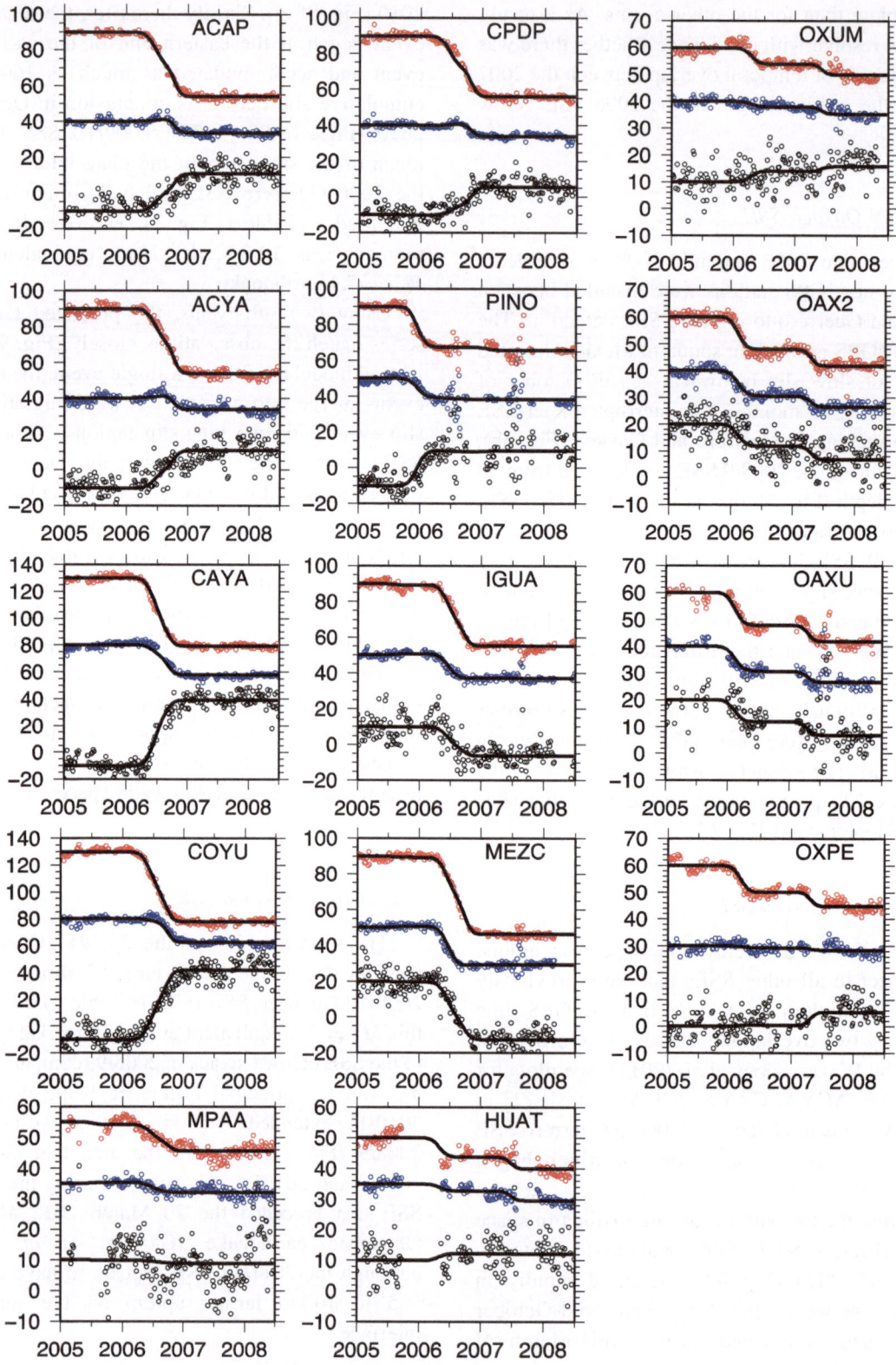

◄

Figure 6

Model fits for selected cGPS stations for the 2006 and 2007 SSEs. *Open circles* are north (*red*), east (*blue*), and vertical (*black*) daily station positions after removing their best-fitting slopes to emphasize the deformation associated with the SSEs. *Black lines* are model predictions

4. Coulomb failure Stress Changes

In order to test the hypothesis that stress changes caused by an SSE could trigger subsequent slow slip, we used Coulomb 3.3 (TODA et al. 2005; LIN and STEIN 2004) to determine the Coulomb failure stress (CFS) changes from the cumulative slip for each SSE at similar depths along the fault. We limited the calculation to fault nodes where the cumulative slow slip exceeded 30 mm, thereby omitting stress variations caused by the more poorly resolved low-slip areas. The effective coefficient of friction was set to 0.4, although changing it to 0.2, corresponding to higher pore-fluid pressures associated at SSE depths, had little impact on the results.

With one exception, our CFS change calculations offer no clear support for the hypothesis that SSEs in Guerroro and Oaxaca trigger SSE elsewhere along-strike (Figs. 4d–f, 7e–h). The predicted CFS changes are all 0.1 bars or smaller except for the first sub-event of the $M_w = 7.4$ 2009/10 Guerrero SSE, which caused up to 1.0 bar of positive CFS change in the source region of the second sub-event (Supplemental Fig. A3). Given that static stress changes as small as 0.1 bars have been correlated with triggered seismicity (KING et al. 1994), our CFS results are consistent with the possibility that the second sub-event was triggered by the first. Alternatively, ZIGONE et al. (2012) find that seismic waves from the 27 February 2010 Maule earthquake triggered tremor in Guerrero and propose that they also triggered the second sub-event of the 2009/10 Guerrero SSE. ZIGONE et al. (2012) speculate, and our results suggest, that the first sub-event increased the stresses on the region of the second sub-event. That area was subsequently destabilized by the passing seismic waves and evolved into slow slip (ZIGONE et al. 2012).

5. Mexico Subduction Zone Slip Budget

Prior to the 20 March 2012 Ometepec earthquake along the MSZ, the deformation measured at GPS sites in southern Mexico for the past decade consisted largely of a superposition of elastic strain that accumulated in the rocks surrounding the seismogenic areas of the plate interface and elastic strain that accumulated in the rocks surrounding deeper portions of the fault, partly relieved by slow slip events every 1–4 years (Fig. 3). Separating these two processes is essential in order to discriminate between strongly coupled areas of the subduction interface that are likely to slip during a future, large thrust earthquake and areas of the plate interface where the interseismic slip deficit and SSE slip may be in balance during the SSE cycle. To accomplish this goal, we combined an independent estimate of the spatial distribution of inter-SSE coupling along the plate interface [ROUSSET et al. (2015) submitted (this volume)] with the SSE slip sources from this study and CORREA-MORA et al. (2008), as follows.

Along the Guerrero segment, we determined the cumulative slip deficit between July 1, 2002 and October 1, 2010, spanning two complete SSE cycles (i.e. from the end of the 2002 Guerrero SSE to the end of the 2009/10 SSE as shown by Fig. 3), from the product of the 8.25-year length of the time period, the inter-SSE coupling per fault node (Fig. 10c) (ROUSSET et al. 2015) and the Cocos–North America plate convergence rate predicted at each node (DEMETS et al. 2010). Along the Oaxaca segment, we similarly determined the cumulative inter-SSE slip deficit between October 1, 2002 and May 26, 2011, spanning five complete SSE cycles, from the product of the 9.15-year period (Fig. 3), the inter-SSE coupling distribution from ROUSSET et al. (2015) and Cocos–North America convergence rate per fault node.

To find the slip that was relieved by SSEs during these same intervals, we summed the best-fitting slow slip distributions for the 2006 and 2009/10 SSEs below Guerrero and the 2004 (CORREA-MORA et al. 2008), 2005/6, 2007, 2008/9, and 2010/11 SSEs below Oaxaca (Fig. 10). Subtracting the accumulated slow slip from the unrelieved slip per fault node gives

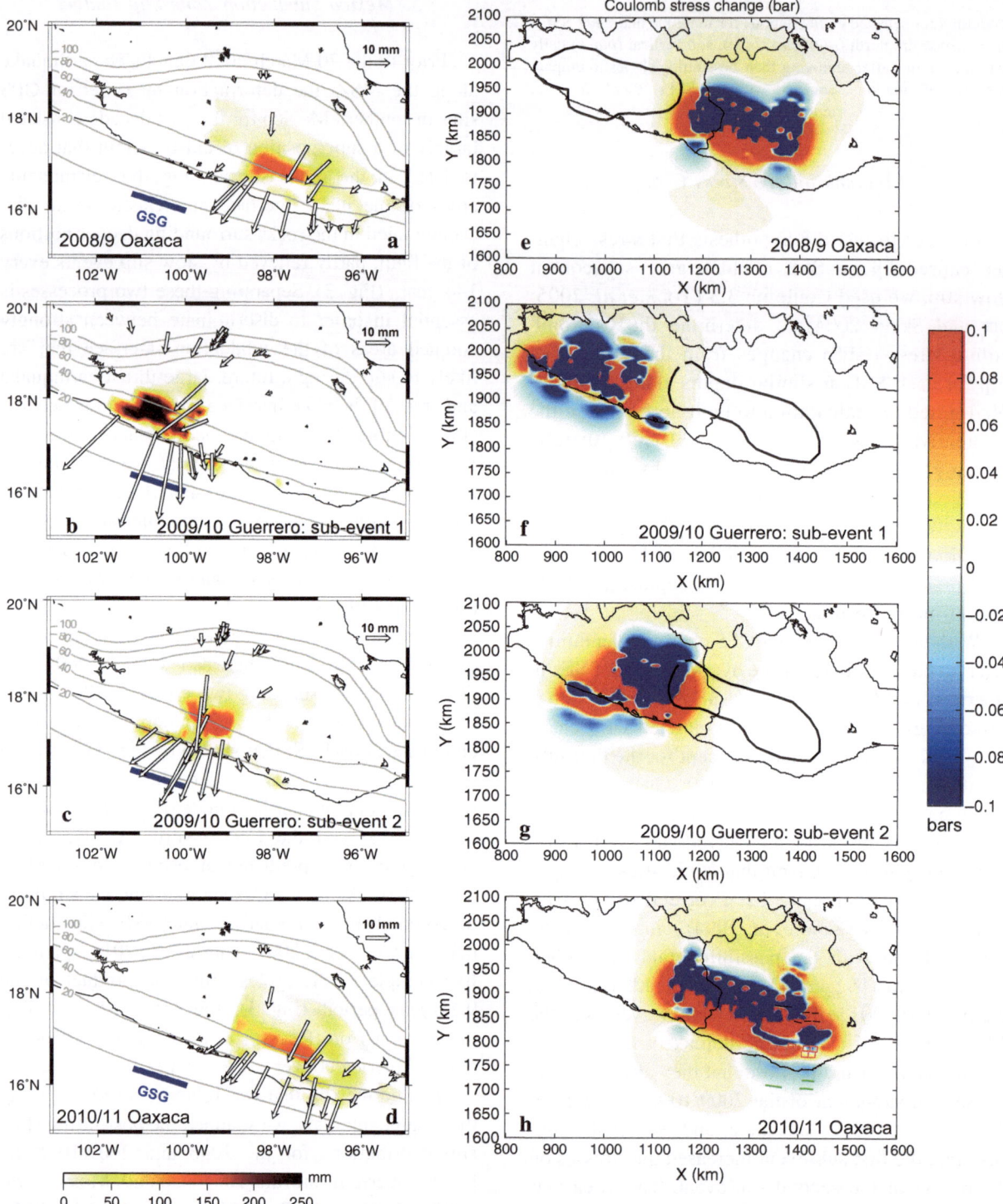

Figure 7
Cumulative slip during the **a** 2008/9, **b, c** 2009/10, and **d** 2010/11 SSEs. *White arrows* show surface displacements predicted by the modeled slip. *Blue line* marks the Guerrero seismic gap (GSG). **e–h** Coulomb failure stresses calculated from the slow slip for each event. *Black contours* show the location of the subsequent SSE where applicable

the distribution of unrelieved fault slip (Fig. 11). This represents our best estimate of where long-term interseismic slip deficit currently accumulates on the Mexico subduction interface. During the past decade, slow slip below Guerrero and Oaxaca not only relieved much of the slip deficit at depths below 20 km in both regions (Fig. 11), but also reduced the slip deficit in the Guerrero seismic gap by up to 75–100 % (Fig. 11). If deformation during the past decade is characteristic of the longer-term deformation, the small remaining slip deficit along the Guerrero seismic gap implies longer recurrence intervals for large earthquakes than for the surrounding areas, in accord with results reported by RADIGUET et al. (2012).

In Oaxaca, the cumulative slip deficit below 20 km depth was reduced from 200 to ∼0 mm after the cumulative slow slip is removed (Fig. 11). The deep slip deficit that accumulates during the 1–2 years between SSEs is thus approximately balanced by slow slip, in accord with results reported for Oaxaca by CORREA-MORA et al. (2008). Therefore, little or no slip deficit at depths greater than 20–25 km remains to be released during large earthquakes at the end of a SSE cycle. Slow slip from 2005 through 2011 did not, however, significantly reduce the large slip deficits at shallow depths, particularly near 97°W (Fig. 11), where a strongly coupled portion of the plate interface first identified by CORREA-MORA et al. (2008) and recently confirmed by ROUSSET et al. (2015) coincides with the 1978 Oaxaca earthquake rupture zone (Fig. 11). These results clearly identify this region as an important seismic hazard area.

6. Discussion

6.1. Comparison to Previous Results

The addition of new GPS stations between 99°W and 98°W since 2005 has improved the GPS network geometry of southern Mexico enough to show that SSEs on the Mexico subduction zone extend everywhere along-strike between Oaxaca and Guerrero. In particular, we find that slip during the previously unstudied 2008/9 and 2010/11 Oaxaca SSEs affected areas of the subduction interface farther west than was estimated for SSEs below Oaxaca in 2004 and 2005/6 (CORREA-MORA et al. 2009), possibly reaching the source regions of SSE below Guerrero (Fig. 12). In retrospect, the apparent absence of any slow slip between Oaxaca and Guerrero during the 2004 and 2005/6 SSEs (CORREA-MORA et al. 2009) may have been an artifact of the absence of observations at locations between Oaxaca and Guerrero.

Results for the 2009/10 SSE below Guerrero give slip distributions and slip amplitudes that are consistent with results presented by RADIGUET et al. (2012), but reveal details of the along-strike migration of the SSE that are missed with static offset modeling. We find that the 2009/10 SSE below Guerrero migrated eastward toward Oaxaca similar to the slip migration of the 2006 event. Observations from the Michoacan segment west of Guerrero, where security problems have precluded the installation of cGPS stations, are needed to determine whether SSEs originate in this region or migrate into it.

6.2. Slow Slip History and the 2012 Ometepec Earthquake

Our modeling adds to evidence described by GRAHAM et al. (2014a) that westward-migrating slow slip during the 2011/12 SSE, which originated below central Oaxaca, may have triggered the $M_w = 7.4$ Ometepec earthquake in March of 2012. In particular, Coulomb failure stress calculations (Sect. 4) indicate that the source region for the Ometepec earthquake (located at the intersection of the Oaxaca/Guerrero boarder and the Pacific coast) received positive CFS perturbations from the 2005/6, 2007, and 2008/9 SSEs below Oaxaca (Figs. 4d, f and 7e), the second sub-event of the Guerrero 2009/10 SSE (Fig. 7g), the 2010/11 Oaxaca SSE (Fig. 7h), and the 2011/12 SSE (GRAHAM et al. 2014a). Six SSEs in the 6 years preceding the earthquake thus caused static stress changes conducive to fault slip at the down-dip end of the eventual earthquake rupture zone. The combination of a fault segment potentially near failure, steady-state interseismic CFS changes conducive to fault slip, and the 2011/12 SSE that propagated toward the rupture zone in the months preceding the

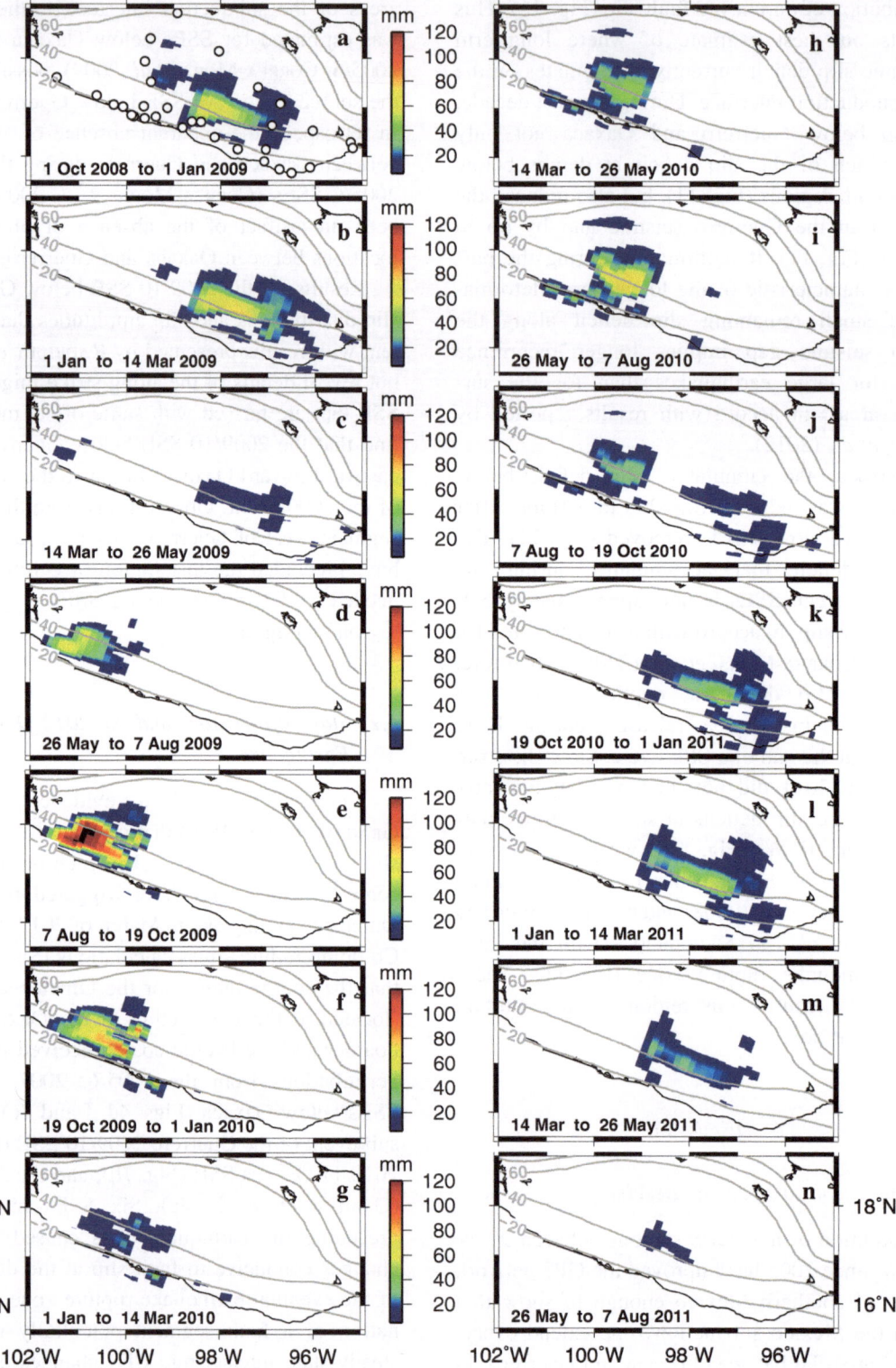

◀ Figure 8
Evolution of slip from 2008 through 2011 at ∼2 month-long intervals. The supplementary information includes an animation of the same slip sequence at 2-week-long intervals. The 2009/10 SSE in Guerrero (c–j) began during the final stage of slow slip in Oaxaca (a–c). CFS calculated for the eastward-migrating Guerrero SSE shows positive changes in the source region of the subsequent 2010/11 Oaxaca SSE (j–n). The *open circles* in **a** indicate locations of GPS stations that were used to determine the slip

earthquake bolster the case that SSE played a role in triggering the 2012 Ometepec earthquake (GRAHAM *et al.* 2014a). SIT (2013) presents evidence for the space–time evolution of microseismicity in the weeks preceding the earthquake that further supports this hypothesis.

6.3. Seismic Hazard Implications for Widespread Slow Slip

Our work has several implications for earthquake hazards in southern Mexico. First, the evidence that slow slip in the Guerrero seismic gap reduces the slip deficit at nominally seismogenic depths (Fig. 11), suggests that SSEs below Guerrero likely delay future earthquakes, in accord with results reported by RADIGUET *et al.* (2012). Second, the evidence for strong coupling across the 1978 rupture zone offshore from Oaxaca (Fig. 11) accompanied by slow slip events every 1–2 years immediately down-dip from the seismogenic zone argues for increased awareness of the seismic potential of this area during Oaxaca SSEs. Finally, our new evidence for trench-parallel migration of slow slip over distances of 100–200 km implies that Coulomb failure stress changes occur over a broader area than would be the case for the more localized slow slip source regions found by CORREA-MORA *et al.* (2009). There are thus more areas in which slow slip could evolve into dynamic rupture, depending on the state of existing stress along the coupled regions of the fault.

6.4. Comparison to Other Subduction Zones

Along the Cascadia subduction zone, slow slip and tremor occur at the same depth and location (e.g., ROGERS and DRAGERT 2003; SZELIGA *et al.* 2004, 2008; BRUDZINSKI and ALLEN 2007; BARTLOW *et al.* 2011). While in Oaxaca tremor has been noted only at

depths below SSEs (BRUDZINSKI *et al.* 2010), observations in Guerrero show two tremor regions (HUSKER *et al.* 2012). Persistent background tremor has been recorded in a 'sweet-spot' located ∼215 km from the trench in the flat slab region of the subduction interface down-dip from the Guerrero SSE region, while intermittent tremor occurs slightly up-dip of the sweet-spot and overlaps the down-dip edge of the SSE slip region (HUSKER *et al.* 2012; FRANK *et al.* 2015). These shallower, intermittent tremor episodes have been associated with short-term SSEs at the limit of detection with GPS (VERGNOLLE *et al.* 2010; FRANK *et al.* 2015). In this respect, SSE and tremor below Guerrero are more similar to those observed in the Bungo channel region of Japan, where tremor is also located both down-dip from long-term SSE slip patches (HIROSE and OBARA 2005; HIROSE *et al.* 2010) and coincides in space and time with short-duration, low magnitude SSEs detected with tiltmeters, but too small to be detected by GPS (HIROSE and OBARA 2005). It is possible that a similar relationship with shallower tremor bursts and small SSEs may also exist in Oaxaca; however, an investigation into this topic has yet to be conducted.

Our modeling results indicate that slow slip migrates steadily along-strike at rates of 1.5–4 km/day (Table 1) along the subduction zone interface, similar to the 2–15 km/day steady rates observed in Cascadia (DRAGERT and WANG 2011). In contrast, slip migration of shallow, short-term SSEs in New Zealand can be irregular and patchy, but may include periods during which SSE migrates at rates of 5–9 km/day (WALLACE *et al.* 2012). WALLACE *et al.* (2012) attribute the irregular migration pattern for New Zealand to the heterogeneity of the shallow portion of the subduction zone interface, where subducted seamounts and surrounding fluid rich sediments may give rise to a complex arrangement of velocity strengthening and velocity weakening fault patches.

Along the Mexico subduction zone, afterslip and SSEs occur at similar depths and afterslip may extend even farther down-dip (Fig. 12) (GRAHAM *et al.* 2014b). Although our modeling results indicate that slow slip has migrated across the region between the Oaxaca and Guerrero segments of the subduction zone, no modeled SSE has yet nucleated in the region

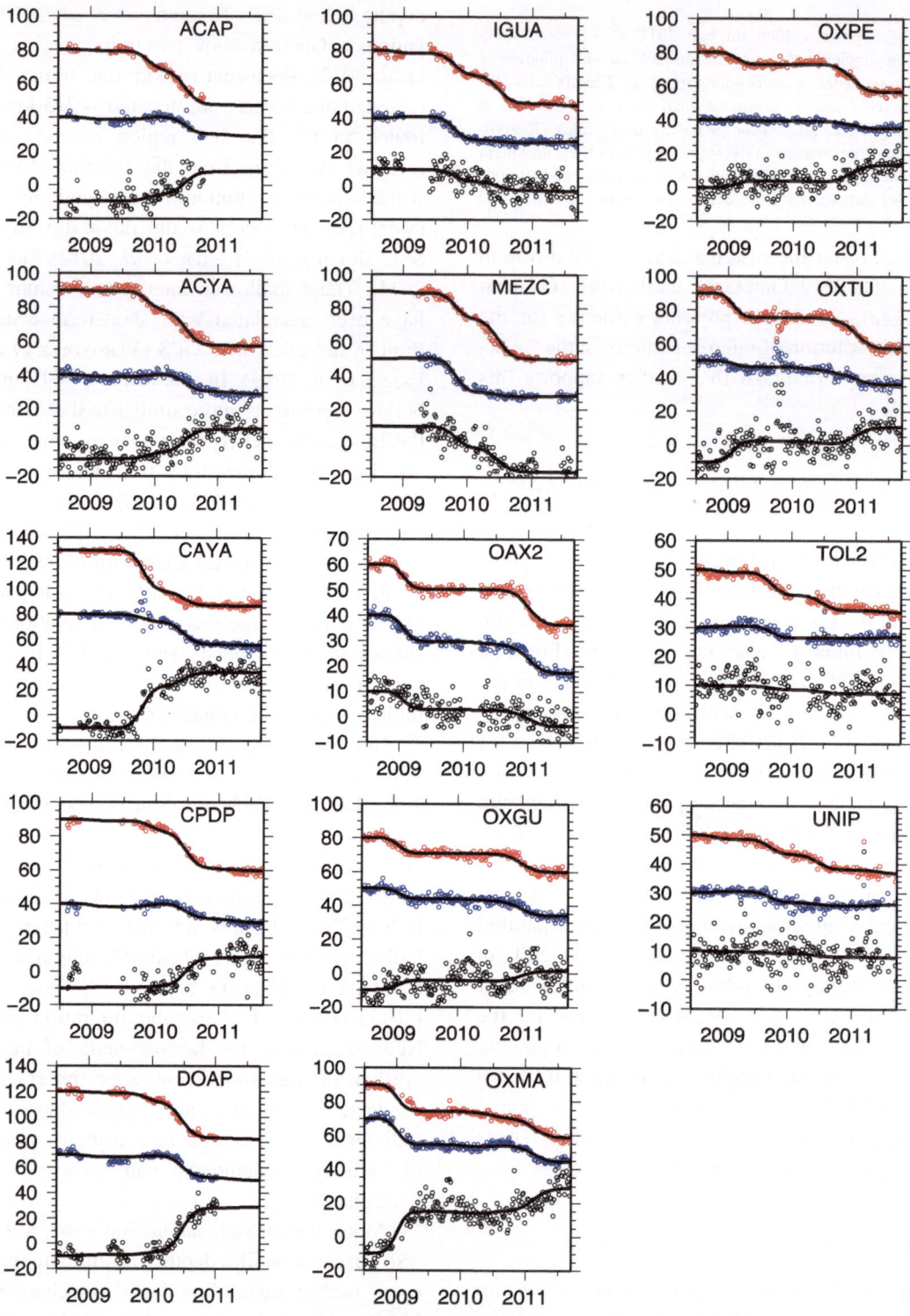

Figure 9

Model fits for selected cGPS stations for the 2008/9, 2009/10, and 2010/11 SSEs. *Open circles* show the north (*red*), east (*blue*), and vertical (*black*) daily station positions after removing their best-fitting slopes to emphasize the deformation associated with the SSEs. *Black lines* are model predictions

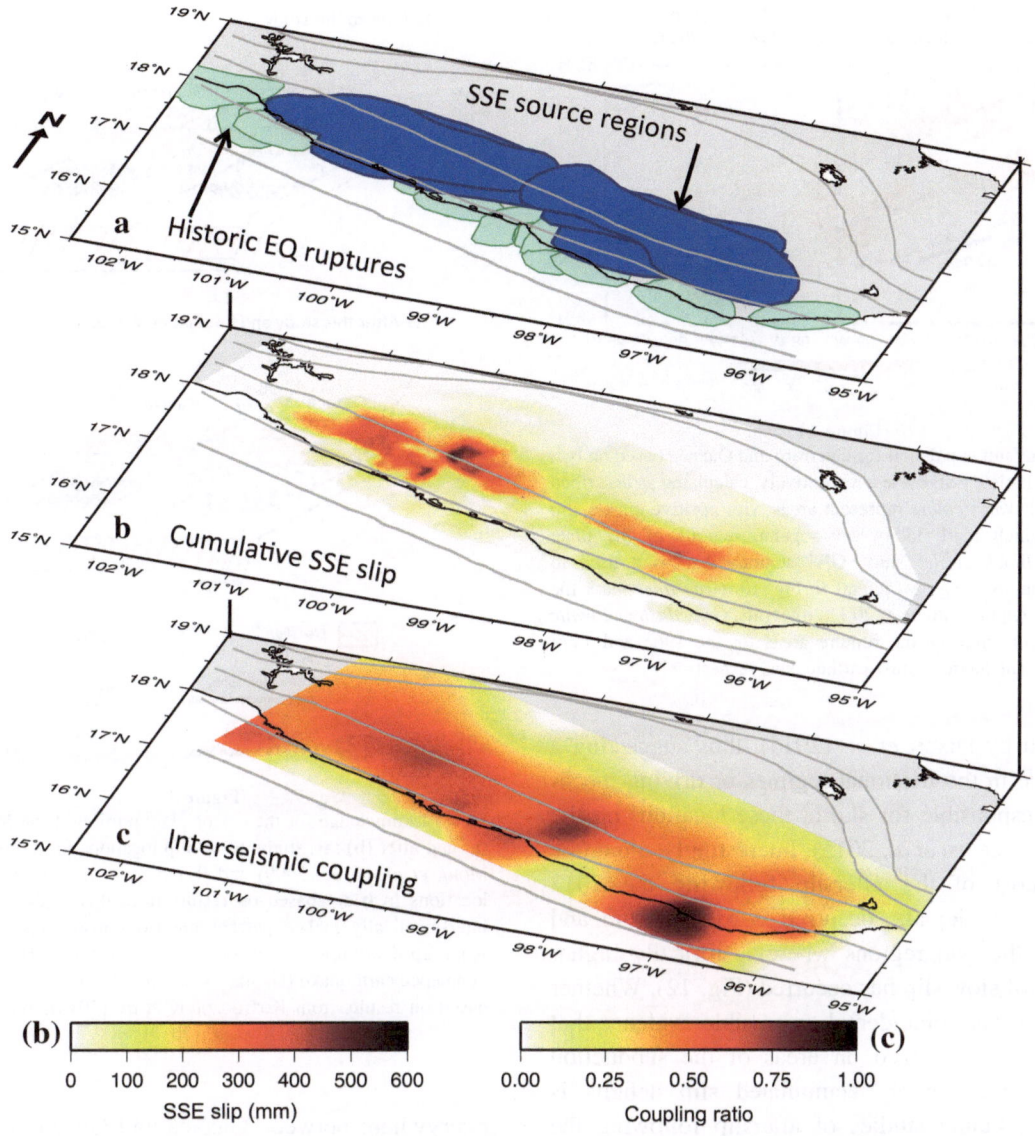

(b)

```
0   100  200  300  400  500  600
```
SSE slip (mm)

(c)

```
0.00   0.25   0.50   0.75   1.00
```
Coupling ratio

Figure 10

a Historical ruptures (*green*) compared with SSE source regions determined in this study (*blue*). **b** Cumulative slow slip from two complete SSE cycles beneath Guerrero (July 1, 2002–October 1, 2010), which encompasses the 2006 and 2009/10 SSEs, and five complete SSE cycles beneath Oaxaca (October 1, 2002–May 26, 2011) which includes the 2004 (Correa-Mora *et al.* 2008), 2005/6, 2007, 2008/9, 2010/11 SSEs. **c** Inter-SSE coupling determined by Rousset *et al.* (2015). *Gray lines* indicate subduction depth contours from 0 to 80 km depth

between 99°W and 98°W, where large amounts of afterslip followed the 2012 Ometepec earthquake (GRAHAM *et al.* 2014b). Further observations and modeling of SSEs below southern Mexico in 2013 and 2014 are needed to better understand whether the interface in this region differs from the SSE-prone, neighboring regions of the subduction interface or whether the absence of SSE nucleation in this region

since the mid-1990s is a merely a time-sampling artifact.

A recent study by MALSERVISI *et al.* (2015), which modeled the postseismic behavior following the 2012 Nicoya earthquake in Costa Rica, also compares the locations of earthquake afterslip and SSEs. Estimates for afterslip in the 2 years following the earthquake show little to no spatial overlap with the SSE patches

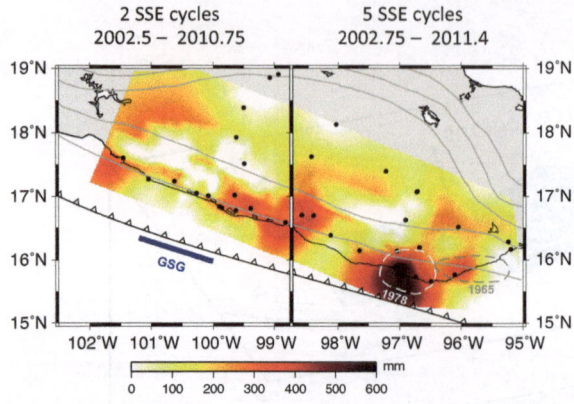

Figure 11

Cumulative fault slip deficit for Guerrero and Oaxaca based on two and five complete SSE cycles, respectively, calculated as described in the text. *Warm colors* represent areas with positive elastic slip deficits, which mark likely seismogenic regions of the plate interface. *Black circles* show GPS stations that were used to estimate the coupling coefficients in Fig. 10. *Gray lines* mark the subduction depth contours at 20 km intervals. *Gray dash* and *white dash contours* indicate the rupture areas for the 1965 and 1978 subduction-thrust earthquakes, respectively

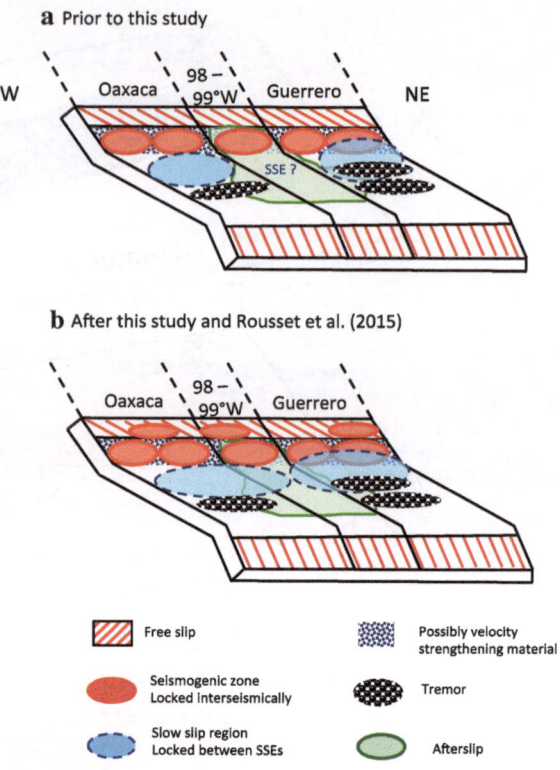

Figure 12

State of knowledge for the Cocos plate portion of the MSZ before (**a**) and after (**b**) this study. Slow slip locations for **a** after CORREA-MORA et al. (2008, 2009) and RADIGUET et al. (2012). Slow slip locations in **b** are based on results from this study and shallow interseismically locked patches are from ROUSSET et al. (2015). Afterslip distribution based on afterslip form the 20 March 2012 Ometepec earthquake (GRAHAM et al. 2014b). Tremor locations are based on results from KOSTOGLODOV et al. (2010) and BRUDZINSKI et al. (2010)

estimated by DIXON *et al.* (2014), thus suggesting a difference in the frictional regimes or driving mechanisms responsible for slip at these locations on the fault (MALSERVISI *et al.* 2015). Interestingly, along the MSZ, most of the afterslip following the 2012 Ometepec earthquake occurred down-dip from and between the two regions where historically higher amounts of slow slip has occurred (Fig. 12). Whether or not this was coincidental or possible evidence that afterslip is maximized on areas of the subduction interface with larger accumulated slip deficits is unknown. Future studies of afterslip following the recent 18 April 2014 $M = 7.2$ earthquake near the main Guerrero SSE patch might provide useful new information on this topic.

7. Conclusion

This study presents the first time-dependent modeling of a complete sequence of SSEs along the Mexico subduction zone using an improved GPS network geometry, as well as the first slip models for the 2008/9 and 2010/11 Oaxaca SSEs. We show that slow slip has affected the subduction interface everywhere between Oaxaca and Guerrero, indicating that stress changes from slow slip affect a larger region of the plate interface than was previously known. CFS calculations further suggest that the complex, long-duration slow slip below Guerrero in 2009/10 could have triggered the subsequent SSE below Oaxaca in 2010/11, although independent evidence for causality is absent. Slip estimates indicate that slow slip beneath Guerrero significantly reduces the slip deficit that accumulates across the Guerrero seismic gap, whereas SSEs beneath Oaxaca relieve little or no slip in the strongly coupled seismogenic zone along much of the Oaxaca coastline.

Acknowledgments

Funding was provided in part by National Science Foundation Grants EAR-1114174 (DeMets). Portions of the GPS network were supported by the National Science Foundation under award EAR-1338091; by UNAM-PAPIIT projects IN104213-2 and IN109315-3 (Cabral-Cano); by CONACYT 178058 and PAPIIT IN110514 (Kostoglodov); and by the Agence Nationale de la Recherche (France) under contract RA0000CO69 (G-GAP). Part of the GPS data was obtained by the Servicio Sismológico Nacional (SSN, México); we acknowledge Sara Franco Sanchez and the rest of SSN's personnel for its acquisition and data distribution support. Graphics were prepared with Generic Mapping Tools software (WESSEL and SMITH 1991). We thank the two anonymous reviewers for their helpful comments and suggestions.

REFERENCES

ALTAMIMI, Z., COLLILIEUX, X. and MÉTIVIER, L., 2011. *ITRF2008: an improved solution of the international terrestrial reference frame*, Journal of Geodesy, 85, 457–473. doi:10.1007/s00190-011-0444-4.

BARTLOW, N. M., S. MIYAZAKI, A. M. BRADLEY, and P. SEGALL, 2011. *Space-time correlation of slip and tremor during the 2009 Cascadia slow slip event*, Geophys. Res. Lett., 38, L18309, doi:10.1029/2011GL048714.

BEROZA, G. C., and IDE, S., 2011. *Slow earthquakes and nonvolcanic tremor*. Annual review of Earth and planetary sciences, 39, 271–296.

BOYARKO, D. C., and BRUDZINSKI, M. R., 2010. *Spatial and temporal patterns of nonvolcanic tremor along the southern Cascadia subduction zone*, J. Geophys. Res., 115, B00A22, doi:10.1029/2008JB006064.

BRUDZINSKI, M. R., and R. M. ALLEN, 2007. *Segmentation in episodic tremor and slip all along Cascadia*, Geology, 35, 907–910, doi:10.1130/G23740A.1.

BRUDZINSKI, M., CABRAL-CANO, E., CORREA-MORA, F., DEMETS, C. and MÁRQUEZ-AZÚA, B., 2007. *Slow slip transients along the Oaxaca subduction segment from 1993 to 2007*, Geophysical Journal International, 171, 523–538, doi: 10.1111/j.1365-246X.2007.03542.x.

BRUDZINSKI, M. R., H. R. HINOJOSA-PRIETO, K. M. SCHLANSER, E. CABRAL-CANO, A. ARCINIEGA-CEBALLOS, O. DIAZ-MOLINA, and C. DEMETS, 2010. *Nonvolcanic tremor along the Oaxaca segment of the Middle America subduction zone*, J. Geophys. Res., 115, B00A23, doi:10.1029/2008JB006061.

CAVALIÉ, O., PATHIER, E., RADIGUET, M., VERGNOLLE, M., COTTE, N., WALPERSDORF, A., KOSTOGLODOV, V. and COTTON, F., 2013. *Slow slip event in the Mexican subduction zone: Evidence of shallower slip in the Guerrero seismic gap for the 2006 event revealed by the joint inversion of InSAR and GPS data*, Earth and Planetary Science Letters, 367, 52–60, doi: http://dx.doi.org/10.1016/j.epsl.2013.02.020.

COLELLA, H.V., DIETERICH, J.H., RICHARDS-DINGER, K. and RUBIN, A.M., 2012. *Complex characteristics of slow slip events in subduction zones reproduced in multi-cycle simulations*, Geophysical Research Letters, 39, L20312, doi: 10.1029/2012gl053276.

COLELLA, H.V., DIETERICH, J.H. and RICHARDS-DINGER, K.B., 2011. *Multi-event simulations of slow slip events for a Cascadia-like subduction zone*, Geophysical Research Letters, 38, L16312, doi: 10.1029/2011gl048817.

CORREA-MORA, F., DEMETS, C., CABRAL-CANO, E., DIAZ-MOLINA, O. and MARQUEZ-AZUA, B., 2009. *Transient deformation in southern Mexico in 2006 and 2007: Evidence for distinct deep-slip patches beneath Guerrero and Oaxaca*, Geochemistry, Geophysics, Geosystems, 10, Q02S12, doi: 10.1029/2008gc002211.

CORREA-MORA, F., DEMETS, C., CABRAL-CANO, E., MARQUEZ-AZUA, B. and DIAZ-MOLINA, O., 2008. *Interplate coupling and transient slip along the subduction interface beneath Oaxaca, Mexico*, Geophysical Journal International, 175, 269–290, doi: 10.1111/j.1365-246X.2008.03910.x.

DEMETS, C., GORDON, R.G. and ARGUS, D.F., 2010. *Geologically current plate motions*, Geophysical Journal International, 181, 1–80, doi: 10.1111/j.1365-246X.2009.04491.x.

DIXON, T. H., JIANG, Y., MALSERVISI, R., McCAFFREY, R., VOSS, N., PROTTI, M., and GONZALEZ, V., 2014. *Earthquake and tsunami forecasts: Relation of slow slip events to subsequent earthquake rupture*. Proceedings of the National Academy of Sciences, 111(48), 17039–17044.

DRAGERT, H. and WANG, K., 2011. *Temporal evolution of an episodic tremor and slip event along the northern Cascadia margin*, Journal of Geophysical Research: Solid Earth, 116, B12406, doi: 10.1029/2011jb008609.

DRAGERT, H., WANG, K. and JAMES, T.S., 2001. *A Silent Slip Event on the Deeper Cascadia Subduction Interface*, Science, 292, 1525–1528, doi: 10.1126/science.1060152.

FRANCO, S.I., KOSTOGLODOV, V., LARSON, K.M., MANEA, V.C., MANEA, M. and SANTIAGO, J.A., 2005. *Propagation of the 2001–2002 silent earthquake and interplate coupling in the Oaxaca subduction zone, Mexico*, Earth Planets Space, 57, 973–985.

FRANK, W.B., RADIGUET, M., ROUSSET, B., SHAPIRO, N.M., HUSKER, A.L., KOSTOGLODOV, V., COTTE, N., and CAMPILLO, M., 2015, *Uncovering the geodetic signature of silent slip through repeating earthquakes*. Geophysical Research Letters, 42, 2774–2779, doi: 10.1002/2015GL063685.

FRY, B., CHAO, K., BANNISTER, S., PENG, Z. and WALLACE, L., 2011. *Deep tremor in New Zealand triggered by the 2010 Mw8.8 Chile earthquake*, Geophysical Research Letters, 38, L15306, doi: 10.1029/2011gl048319.

GRAHAM, S. E., DEMETS, C., CABRAL-CANO, E., KOSTOGLODOV, V., WALPERSDORF, A., COTTE, N., BRUDZINSKI, M., McCAFFREY, R., and SALAZAR-TLACZANI, L., 2014a. *GPS constraints on the 2011–2012 Oaxaca slow slip event that preceded the 2012 March 20 Ometepec earthquake, southern Mexico*. Geophysical Journal International, doi: 10.1093/gji/ggu019.

GRAHAM, S. E., DEMETS, C., CABRAL-CANO, E., KOSTOGLODOV, V., WALPERSDORF, A., COTTE, N., BRUDZINSKI, M., McCAFFREY, R., and SALAZAR-TLACZANI, L., 2014b. *GPS constraints on the Mw = 7.5 Ometepec earthquake sequence, southern Mexico: coseismic and post-seismic deformation.* Geophysical Journal International, *199*(1), 200–218, doi: 10.1093/gji/ggu167.

HIROSE, H., and OBARA, K., 2005. *Repeating short-and long-term slow slip events with deep tremor activity around the Bungo channel region, southwest Japan.* Earth, Planets, and Space, *57*(10), 961–972.

HIROSE, H., Y. ASANO, K. OBARA, T. KIMURA, T. MATSUZAWA, S. TANAKA, and T. MAEDA, 2010. *Slow earthquakes linked along dip in the Nankai sub- duction zone,* Science, *330*, 1502, doi:10. 1126/science.1197102.

HUSKER, A. L., KOSTOGLODOV, V., CRUZ-ATIENZA, V. M., LEGRAND, D., SHAPIRO, N. M., PAYERO, J. S., CAMPILLO, M., and HUESCA-PÉREZ, E. (2012). *Temporal variations of non-volcanic tremor (nvt) locations in the mexican subduction zone: Finding the nvt sweet spot.* Geochemistry, Geophysics, Geosystems, *13*(3).

IGLESIAS, A., SINGH, S., LOWRY, A., SANTOYO, M., KOSTOGLODOV, V., LARSON, K., FRANCO-SANCHEZ, S. and MIKUMO, T., 2004. *The silent earthquake of 2002 in the Guerrero seismic gap, Mexico (Mw = 7.6): Inversion of slip on the plate interface and some implications,* GEOFISICA INTERNACIONAL-MEXICO-, *43*, 309.

ITO, Y., HINO, R., KIDO, M., FUJIMOTO, H., OSADA, Y., INAZU, D., OHTA, Y., IINUMA, T., OHZONO, M., MIURA, S., MISHINA, M., SUZUKI, K., TSUJI, T. and ASHI, J. *Episodic slow slip events in the Japan subduction zone before the 2011 Tohoku-Oki earthquake,* Tectonophysics, doi: 10.1016/j.tecto.2012.08.022.

KING, G. C., STEIN, R. S., and LIN, J. (1994). *Static stress changes and the triggering of earthquakes.* Bulletin of the Seismological Society of America, *84*(3), 935–953.

KOSTOGLODOV, V., SINGH, S.K., SANTIAGO, J.A., FRANCO, S.I., LARSON, K.M., LOWRY, A.R. and BILHAM, R., 2003. *A large silent earthquake in the Guerrero seismic gap, Mexico,* Geophysical Research Letters, *30*, 1807, doi: 10.1029/2003GL017219.

KOSTOGLODOV, V., HUSKER, A., SHAPIRO, N. M., PAYERO, J. S., CAMPILLO, M., COTTE, N., and CLAYTON, R., 2010. *The 2006 slow slip event and nonvolcanic tremor in the Mexican subduction zone.* Geophysical Research Letters, *37*(24).

LARSON, K.M., KOSTOGLODOV, V., MIYAZAKI, S.i. and SANTIAGO, J.A.S., 2007. *The 2006 aseismic slow slip event in Guerrero, Mexico: New results from GPS,* Geophysical Research Letters, *34*, L13309, doi: 10.1029/2007GL029912.

LIN, J. and STEIN, R.S., 2004. *Stress triggering in thrust and subduction earthquakes and stress interaction between the southern San Andreas and nearby thrust and strike-slip faults,* J. Geophys. Res., *109*, B02303, doi:10.1029/2003JB002607.

LOWRY, A.R., LARSON, K.M., KOSTOGLODOV, V. and BILHAM, R., 2001. *Transient fault slip in Guerrero, southern Mexico,* Geophysical Research Letters, *28*, 3753–3756, doi: 10.1029/2001GL013238.

MALSERVISI, R., SCHWARTZ, S. Y., VOSS, N., PROTTI, M., GONZALEZ, V., DIXON, T. H., JIANG, Y., NEWMAN, A.V., RICHARDSON, J., WALTER, J. I., and VOYENKO, D., 2015, *Multiscale postseismic behavior on a megathrust: The 2012 Nicoya earthquake, Costa Rica,* Geochem. Geophys. Geosyst., *16*, 1848–1864, doi:10.1002/2015GC005794.

MÁRQUEZ-AZÚA, B. and DEMETS, C., 2003. *Crustal velocity field of Mexico from continuous GPS measurements, 1993 to June 2001: Implications for the neotectonics of Mexico,* Journal of

Geophysical Research: Solid Earth, *108*, 2450, doi: 10.1029/2002JB002241.

McCAFFREY, R., 2009. *Time-dependent inversion of three-component continuous GPS for steady and transient sources in northern Cascadia,* Geophysical Research Letters, *36*, L07304, doi: 10.1029/2008GL036784.

OBARA, K., 2010. *Phenomenology of deep slow earthquake family in south- west Japan: Spatiotemporal characteristics and segmentation,* J. Geophys. Res., *115*, B00A25, doi:10.1029/2008JB006048.

OHTA, Y., FREYMUELLER, J.T., HREINSDÓTTIR, S. and SUITO, H., 2006. *A large slow slip event and the depth of the seismogenic zone in the south central Alaska subduction zone,* Earth and Planetary Science Letters, *247*, 108–116, doi: http://dx.doi.org/10.1016/j.epsl.2006.05.013.

OHTA, Y., KIMATA, F. and SAGIYA, T., 2004. *Reexamination of the interplate coupling in the Tokai region, central Japan, based on the GPS data in 1997–2002,* Geophysical Research Letters, *31*, L24604, doi: 10.1029/2004gl021404.

OKADA, Y., 1992. *Internal deformation due to shear and tensile faults in a half-space,* Bull. Seism. Soc. Am, *82*, 1018–1040.

PENG, Z. and GOMBERG, J., 2010. *An integrated perspective of the continuum between earthquakes and slow-slip phenomena,* Nature Geosci, *3*, 599–607, doi: doi:10.1038/ngeo940.

PENG, Z., VIDALE, J.E., WECH, A.G., NADEAU, R.M. and CREAGER, K.C., 2009. *Remote triggering of tremor along the San Andreas Fault in central California,* Journal of Geophysical Research: Solid Earth, *114*, B00A06, doi: 10.1029/2008jb006049.

RADIGUET, M., F. COTTON, M. VERGNOLLE, M. CAMPILLO, B. VALETTE, V. KOSTOGLODOV, and N. COTTE, 2011. *Spatial and temporal evolution of a long term slow slip event: The 2006 Guerrero Slow Slip Event,* Geophys. J. Int., *184*, 816–828, doi:10.1111/j.1365-246X.2010.04866.x.

Radiguet, M., COTTON, F., VERGNOLLE, M., Campillo, M., WALPERSDORF, A., COTTE, N. and KOSTOGLODOV, V., 2012. *Slow slip events and strain accumulation in the Guerrero gap, Mexico,* Journal of Geophysical Research: Solid Earth, *117*, B04305, doi: 10.1029/2011jb008801.

ROGERS, G., and H. DRAGERT, 2003. *Episodic tremor and slip on the Cascadia subduction zone: The chatter of slient slip,* Science, *300*, 1942–1943, doi:10.1126/science.1084783.

ROUSSET, B., LASSERRE, C., CUBAS, N., GRAHAM, S., RADIGUET, M., DEMETS, C., SOCQUET, A., CAMPILLO, M., KOSTOGLODOV, V., CABRAL-CANO, E., COTTE, N., WALPERSDORF, A., 2015. *Lateral Variations of Interseismic Coupling along the Mexican Subduction Interface: Relationships with Long Term Morphology and Fault Zone Mechanical Properties.* Submitted to this issue Pure and Applied Geophysics.

RUBINSTEIN, J.L., GOMBERG, J., VIDALE, J.E., WECH, A.G., KAO, H., CREAGER, K.C. and ROGERS, G., 2009. *Seismic wave triggering of nonvolcanic tremor, episodic tremor and slip, and earthquakes on Vancouver Island,* Journal of Geophysical Research: Solid Earth, *114*, B00A01, doi: 10.1029/2008jb005875.

RUIZ, S., METOIS, M., FUENZALIDA, A., RUIZ, J., LEYTON, F., GRANDIN, R., VIGNY, C., MADARIAGA, R., and CAMPOS, J. (2014). *Intense foreshocks and a slow slip event preceded the 2014 Iquique Mw 8.1 earthquake.* Science, *345*(6201), 1165–1169.

SCHMIDT, D.A. and GAO, H., 2010. *Source parameters and time-dependent slip distributions of slow slip events on the Cascadia subduction zone from 1998 to 2008,* Journal of Geophysical

Research: Solid Earth, *115*, B00A18, doi: 10.1029/2008jb006045.

SEGALL, P. and BRADLEY, A.M., 2012. *Slow-slip evolves into megathrust earthquakes in 2D numerical simulations*, Geophysical Research Letters, *39*, L18308, doi: 10.1029/2012gl052811.

SIT S.M. Miami University; 2013. *New methods in geophysics and science education to analyze slow fault slip and promote active e-learning*. Doctoral dissertation (Geology).

SONG, T.-R.A., HELMBERGER, D.V., BRUDZINSKI, M.R., CLAYTON, R.W., DAVIS, P., PÉREZ-CAMPOS, X. and SINGH, S.K., 2009. *Subducting Slab Ultra-Slow Velocity Layer Coincident with Silent Earthquakes in Southern Mexico*, Science, *324*, 502–506, doi: 10.1126/science.1167595.

SZELIGA, W., T. I. MELBOURNE, M. M. MILLER, and V. M. SANTILLAN, 2004. *Southern Cascadia episodic slow earthquakes*, Geophys. Res. Lett., *31*, L16602, doi:10.1029/2004GL020824.

SZELIGA, W., T. MELBOURNE, M. SANTILLAN, and M. MILLER, 2008. *GPS constraints on 34 slow slip events within the Cascadia subduction zone, 1997–2005*, J. Geophys. Res., *113*, B04404, doi:10.1029/2007JB004948.

TODA, S., STEIN, R.S., RICHARDS-DINGER, K. and BOZKURT, S.B., 2005. *Forecasting the evolution of seismicity in southern California: Animations built on earthquake stress transfer*, J. Geophys. Res., *110*, B05S16, doi: 10.1029/2004jb003415.

VERGNOLLE, M., WALPERSDORF, A., KOSTOGLODOV, V., TREGONING, P., SANTIAGO, J., COTTE, N., and FRANCO, S., 2010. *Slow slip events in mexico revised from the processing of 11 year gps observations*. Journal of Geophysical Research: Solid Earth (1978–2012), *115*(B8).

WALLACE, L.M., BEAVAN, J., BANNISTER, S. and WILLIAMS, C., 2012. *Simultaneous long-term and short-term slow slip events at the Hikurangi subduction margin, New Zealand: Implications for processes that control slow slip event occurrence, duration, and migration*, Journal of Geophysical Research: Solid Earth, *117*, B11402. doi: 10.1029/2012jb009489.

WALPERSDORF, A., COTTE, N., KOSTOGLODOV, V., VERGNOLLE, M., RADIGUET, M., SANTIAGO, J. A., and CAMPILLO, M., 2011. *Two successive slow slip events evidenced in 2009–2010 by a dense GPS network in Guerrero, Mexico*. Geophysical Research Letters, *38*(15), doi:10.1029/2011GL048124.

WESSEL, P. and SMITH, W.H.F., 1991. *Free software helps map and display data*, Eos, Transactions American Geophysical Union, *72*, 441–446, doi: 10.1029/90eo00319.

YOSHIOKA, S., MIKUMO, T., KOSTOGLODOV, V., LARSON, K.M., LOWRY, A.R. and SINGH, S.K., 2004. *Interplate coupling and a recent aseismic slow slip event in the Guerrero seismic gap of the Mexican subduction zone, as deduced from GPS data inversion using a Bayesian information criterion*, Physics of the Earth and Planetary Interiors, *146*, 513–530, http://dx.doi.org/10.1016/j.pepi.2004.05.006.

ZIGONE, D., RIVET, D., RADIGUET, M., CAMPILLO, M., VOISIN, C., COTTE, N., WALPERSDORF, A., SHAPIRO, N.M., COUGOULAT, G., ROUX, P., KOSTOGLODOV, V., HUSKER, A., and PAYERO, J. S., 2012. *Triggering of tremors and slow slip event in Guerrero, Mexico, by the 2010 Mw 8.8 Maule, Chile, earthquake*. Journal of Geophysical Research: Solid Earth (1978–2012), *117*(B9).

ZUMBERGE, J.F., HEFLIN, M.B., JEFFERSON, D.C., WATKINS, M.M. and WEBB, F.H., 1997. *Precise point positioning for the efficient and robust analysis of GPS data from large networks*, J. Geophys. Res., *102*, 5005–5017, doi: 10.1029/96jb03860.

(Received July 5, 2015, revised November 10, 2015, accepted November 12, 2015, Published online December 1, 2015)

Reprinted from the journal

Pure Appl. Geophys. 173 (2016), 3467–3486
© 2015 Springer International Publishing
DOI 10.1007/s00024-015-1215-6

Pure and Applied Geophysics

Lateral Variations of Interplate Coupling along the Mexican Subduction Interface: Relationships with Long-Term Morphology and Fault Zone Mechanical Properties

BAPTISTE ROUSSET,[1] CÉCILE LASSERRE,[1] NADAYA CUBAS,[2] SHANNON GRAHAM,[3] MATHILDE RADIGUET,[1] CHARLES DEMETS,[4] ANNE SOCQUET,[1] MICHEL CAMPILLO,[1] VLADIMIR KOSTOGLODOV,[5] ENRIQUE CABRAL-CANO,[5] NATHALIE COTTE,[1] and ANDREA WALPERSDORF[1]

Abstract—Although patterns of interseismic strain accumulation above subduction zones are now routinely characterised using geodetic measurements, their physical origin, persistency through time, and relationships to seismic hazard and long-term deformation are still debated. Here, we use GPS and morphological observations from southern Mexico to explore potential mechanical links between variations in inter-SSE (in between slow slip events) coupling along the Mexico subduction zone and the long-term topography of the coastal regions from Guerrero to Oaxaca. Inter-SSE coupling solutions for two different geometries of the subduction interface are derived from an inversion of continuous GPS time series corrected from slow slip events. They reveal strong along-strike variations in the shallow coupling (i.e. at depths down to 25 km), with high-coupling zones (coupling >0.7) alternating with low-coupling zones (coupling <0.3). Coupling below the continent is typically strong (>0.7) and transitions to uncoupled, steady slip at a relatively uniform ~175-km inland from the trench. Along-strike variations in the coast-to-trench distances are strongly correlated with the GPS-derived forearc coupling variations. To explore a mechanical explanation for this correlation, we apply Coulomb wedge theory, constrained by local topographic, bathymetric, and subducting-slab slopes. Critical state areas, i.e. areas where the inner subduction wedge deforms, are spatially correlated with transitions at shallow depth between uncoupled and coupled areas of the subduction interface. Two end-member models are considered to explain the correlation between coast-to-trench distances and along-strike variations in the inter-SSE coupling. The first postulates that the inter-SSE elastic strain is partitioned between slip along the subduction interface and

homogeneous plastic permanent deformation of the upper plate. In the second, permanent plastic deformation is postulated to depend on frictional transitions along the subduction plate interface. Based on the location and friction values of the critical state areas identified by our Coulomb wedge analysis, we parameterise frictional transitions in plastic-static models of deformation over several seismic cycles. This predicts strong shear dissipation above frictional transitions on the subduction interface. The comparison of modelled surface displacements over a critical zone at a frictional transition and over a stable area with no internal wedge deformation shows differences of long-term uplift consistent with the observed along-strike variations in the coast-to-trench distances. Our work favours a model in which frictional asperities partly control short-term inter-SSE coupling as measured by geodesy and in which those asperities persist through time.

Key words: Middle America Trench, global positioning system, inter-SSE coupling, critical taper theory, plastic deformation, coastal morphology.

Electronic supplementary material The online version of this article (doi:10.1007/s00024-015-1215-6) contains supplementary material, which is available to authorized users.

[1] ISTerre, CNRS, Univ. Grenoble Alpes, 38041 Grenoble, France. E-mail: baptiste.rousset@ujf-grenoble.fr

[2] Institut des Sciences de la Terre de Paris, Pierre et Marie Curie University, Paris, France.

[3] Department of Earth and Planetary Sciences, Harvard University, Cambridge, MA, USA.

[4] Department of Geoscience, University of Wisconsin-Madison, Madison, WI, USA.

[5] Instituto de Geofísica, Universidad Nacional Autónoma de México, CU, Coyoacan, Mexico.

1. Introduction

Interseismic coupling maps derived from geodetic observations (e.g. CHLIEH *et al.* 2014) are now widely used to represent 2D variations in strain along seismogenic zones and are providing increasingly detailed views of interseismic strain accumulation along major subduction zones (e.g. MAZZOTTI *et al.* 2000; BÜRGMANN *et al.* 2005; FRANCO *et al.* 2012; NOCQUET *et al.* 2014). Although geodetic observations are now precise enough to distinguish fully coupled, locked sections of a fault from uncoupled or partially coupled sections, the physical origin, persistency through time, and hazard implications of geodetically derived coupling variations are still incompletely understood. Coupled areas appear to be roughly spatially correlated with rupture areas of major earthquakes, as shown for Mw8+ subduction earthquakes like in Sumatra (CHLIEH *et al.* 2008), Japan (LOVELESS and MEADE 2011) or the Andes (CHLIEH *et al.* 2011; METOIS *et al.* 2012). Along some subduction interfaces, areas that produce earthquakes, also called asperities, seem to be separated by areas of low coupling that may slow or arrest propagating rupture fronts. Large ruptures may break through such barriers (KONCA *et al.* 2008) due to dynamic effects (KANEKO *et al.* 2010; CUBAS *et al.* 2015). Most authors interpret variations in interseismic coupling, which may persist for a few decades or longer, as variations of frictional properties that may result from or influence other characteristics of a subduction zone such as its geology, morphology, and gravity, which may persist for millions of years. Features on the down-going plate such as seamounts or oceanic ridges may give rise to local variations in coupling along some subduction zones (e.g. SINGH *et al.* 2011), whereas morphological features of the overriding plate may control coupling variations elsewhere (BÉJAR-PIZARRO *et al.* 2013). A correlation between trench-parallel negative gravity anomalies and rupture areas of great earthquakes, as well as between positive gravity anomalies and aseismic creeping areas, is also observed (SONG and SIMONS 2003). All of these observations suggest that the seismogenic behaviour of subduction zones may be stationary over long time scales. Consistent with this idea, critical taper theory (DAHLEN 1984) postulates that the structure and morphology of a forearc prism depend on its mechanical properties, including its basal and internal friction angles and pore fluid pressure. Applied to subduction wedges in Chile and Japan (CUBAS *et al.* 2013a, b), this theory shows that parts of the wedge that are in a stable state are located above past earthquake rupture zones and strongly coupled areas of subduction interface, whereas areas that are at critical state (i.e. parts of the wedge that are affected by internal deformation) are associated with lower interseismic coupling and contour the stable areas.

In this study, we further explore the mechanical link between interseismic coupling, the characteristics of the slab interface and accretionary prism, and the long-term deformation of the overriding plate, possibly accumulated through non-elastic processes.

We focus on the Mexican subduction zone where the Cocos plate subducts below the North American plate. This area has been extensively studied from geodetic and seismological observations since the discovery of slow slip events (hereafter abbreviated "SSE") and tremors both in Guerrero and Oaxaca regions (e.g. LOWRY *et al.* 2001; BRUDZINSKI *et al.* 2007; CORREA-MORA *et al.* 2008, 2009; KOSTOGLODOV *et al.* 2010; WALPERSDORF *et al.* 2011; RADIGUET *et al.* 2012; HUSKER *et al.* 2012; GRAHAM *et al.* 2014a). To date, most studies of the spatio-temporal behaviour of slip along the Mexican subduction zone have focused separately on Guerrero and Oaxaca. With the increase of the GPS network density during the past years, especially in the Oaxaca region (GRAHAM *et al.* 2014b), a more regional analysis has become possible, as presented in this study.

Below, we use an interseismic GPS velocity field for southern Mexico measured in between SSE (inter-SSE velocity field) to retrieve a regional map of inter-SSE coupling from the Guerrero to the Oaxaca segments of the Mexico subduction zone. We first compute forward models to extract first order characteristics (downdip limit of coupling and data sensitivity to along-strike variations of coupling at shallow depth). We then invert for the 2D distribution of inter-SSE coupling on the subduction interface. We quantify along-strike variations in the shallow (depths of 0 to 25 km) coupling and compare them to along-strike variations in the coast-to-trench distances. Through a mechanical analysis of the long-term morphology of the accretionary wedge, following CUBAS *et al.* (2013a), we locate the critical state areas on the slab interface. This allows us to discuss, for the first time in this region, the persistency through time of inter-SSE coupling patterns and the mechanisms involved in the coastal long-term deformation of subduction zones.

2. *Seismic and Aseismic Behaviour*

Compared to other subduction zones, the Mexican subduction zone in the Guerrero-Oaxaca (Fig. 1) area has a narrow seismogenic zone, extending downdip by ~60 km (SUÁREZ and SÁNCHEZ 1996) to approximate depths of 25 km. Two areas devoid of

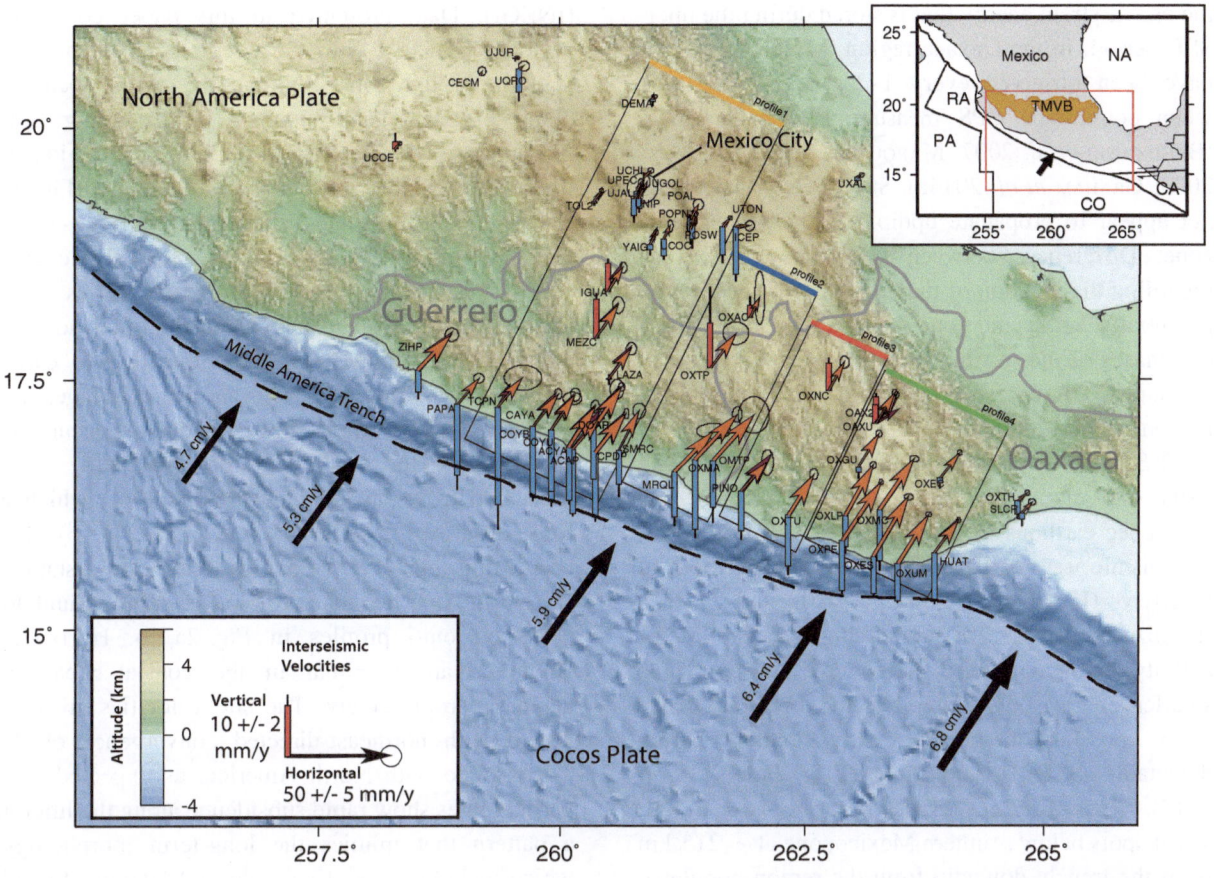

Figure 1

Inter-SSE GPS site velocities corrected for silent slip events as determined from GPS position time series. Velocities are relative to a stationary North America plate. *Orange* and *purple* vectors show horizontal velocities that are used in this study (54 stations) and by RADIGUET *et al.* (2012) (18 stations), respectively. Confidence ellipses ($1 - \sigma$) are shown for this study only. *Vertical thick bars* show the vertical site rates used in this study (*blue* for subsidence and *red* for uplift). The *thin, vertical, black lines* indicate the $1 - \sigma$ uncertainty. The *black* vectors offshore show convergence velocities predicted by the PVEL Cocos–North America angular velocity (DEMETS *et al.* 2010). The four *black rectangles* with a coloured side delineate the groups of GPS stations that are used for our forward modelling. The topography and bathymetry are combined from SRTM (FARR *et al.* 2007) and ETOPO1 (AMANTE and EAKINS 2009) measurements. *Inset* at *top right corner* shows geodynamic setting, with convergence of Cocos (CO) towards North America (NA) plates. Adjacent plates are the Pacific (PA), the Riviera (RA) and the Caribbean (CA) plates. The Trans-Mexican Volcanic Belt (TMVB) is shown in *orange*

seismicity were defined as seismic gaps: the Michoacan and the Guerrero gaps (ASTIZ and KANA-MORI 1984). The first was filled by the destructive M_w 8.1 Michoacan earthquake in 1985. The 200 km-long Guerrero Gap ruptured in 1957 and 1962 in its southeastern half (ORTIZ *et al.* 2000), while its remaining 100 km-long northwestern part has not broken since at least 1911 (KOSTOGLODOV *et al.* 1996).

Five SSEs have been detected in the Guerrero area since the installation of continuous GPS stations: in 1998, 2002, 2006, 2009 (LOWRY *et al.* 2001; KOS-TOGLODOV *et al.* 2010; VERGNOLLE 2010; WALPERSDORF

et al. 2011; RADIGUET *et al.* 2011), and the most recent event in 2014. They recur approximately every 4 years (COTTE *et al.* 2009) and are among the largest SSEs in the world, with equivalent magnitudes of 7.5 (RADIGUET *et al.* 2012). Models of the 2006 event from GPS and InSAR data (RADIGUET *et al.* 2011; CAVALIÉ *et al.* 2013; BEKAERT *et al.* 2015) show that the SSE is located at the transition between the stick-slip and steady slip areas, with the SSE partly intruding updip into the seismogenic zone. From an analysis of 12 years of interseismic strain, RADIGUET *et al.* (2012) show that SSEs release about 75 % of

257

the elastic strain energy that is stored during the inter-SSE period. In the Oaxaca region, M_w 6.6–6.9 SSEs have been observed every 1–2 years since 1993, when continuous GPS measurements began there (BRUDZINSKI et al. 2007; MARQUEZ-AZUA and DEMETS 2009; GRAHAM et al. 2014a). SSEs below Oaxaca do not appear to propagate updip into the seismogenic zone. Differences between Guerrero and Oaxaca regarding the location of the SSEs with respect to the potential seismogenic zone may be related to different phases of the seismic cycle for these two regions as suggested by seismic cycle modelling (e.g. LAPUSTA et al. 2000; LIU and RICE 2007). The 2011/2012 SSE below Oaxaca migrated ~200 km along strike and may have triggered the 2012, M_w 7.2, Ometepec earthquake (GRAHAM et al. 2014b).

Seismic tremors have also been detected both in Guerrero (KOSTOGLODOV et al. 2010) and Oaxaca (BRUDZINSKI et al. 2010). Although peak tremor activity below Guerrero is recorded during SSEs, smaller energy bursts are detected between major SSEs, possibly associated with smaller, barely detectable SSEs (VERGNOLLE 2010; FRANK et al. 2015). HUSKER et al. (2012) describe two tremor sweet spots below southern Mexico, one at ~215 km from the trench, downdip from the region populated by SSEs, where tremors are relatively continuous, and the other at 180–200 km from the trench, overlapping the SSE sweet spot and where tremor is more intermittent. Complementary studies of low frequency earthquakes (FRANK et al. 2013) localise the low frequency events within the two tremor areas on the subduction interface, with a temporal behaviour similar to that of tremors and thrust focal mechanisms compatible with the subduction motion.

3. Analysis of Inter-SSE Coupling in Southern Mexico

3.1. GPS Data

Several continuous GPS networks are running in Mexico, operated by the Geophysical Institute (IG) and the National Seismological Service (SSN) of the National Autonomous University of Mexico (UNAM), and one GPS network by the Mexican National Institute of Statistics and Geography (INEGI). Data presented in this paper cover the period 2005–2011, with a progressive densification of stations through time. Stations are mainly localised along the coast and along a profile going from Acapulco to Mexico City in the Guerrero region. In the Oaxaca region, stations are more sparse (Fig. 1). Extraction of the inter-SSE velocities, representing the strain accumulation between slow slip events, consists of fitting a site position time series via a linear regression that has been modified to also estimate and remove step functions due to earthquakes, slow slip events, and hardware changes and to mask periods when slow slip events occur. All continuous data employed for this study were processed using an identical methodology, which is fully described by GRAHAM et al. (2014a, b). The best-fitting inter-SSE velocities relative to a stationary North America plate are plotted in Fig. 1 and are projected onto profiles in Fig. 2a, b. Horizontal velocities are maximal in the coastal area and decrease progressively. The site velocities are consistent with northeast-directed convergence of the Cocos plate with North America, as expected. The vertical rates show rapid subsidence along the littoral, a pattern that mimics the long-term morphology, which includes several large coastal lagoons. Inland, a band of present-day uplift at ~125 km from the trench coincides with a high mountain range, the Sierra Madre del Sur. The inter-SSE vertical velocities taper to zero at ~250 km from the trench. Compared to RADIGUET et al. (2012), our study presents an increased number of GPS observations, and incorporates data from the Oaxaca region. At stations that are common to our studies, the GPS site velocities are consistent (compare orange and purple arrows on Fig. 1). We interpret this as evidence that the velocity estimates are robust given that different approaches were used in the two studies to extract the inter-SSE signal.

3.2. Forward Models

We begin with simple forward backslip modelling to approximate the downdip location of the transition between the coupled (or partially coupled) area and the deep, steady creeping area. We also test the sensitivity of the GPS velocities to along-strike

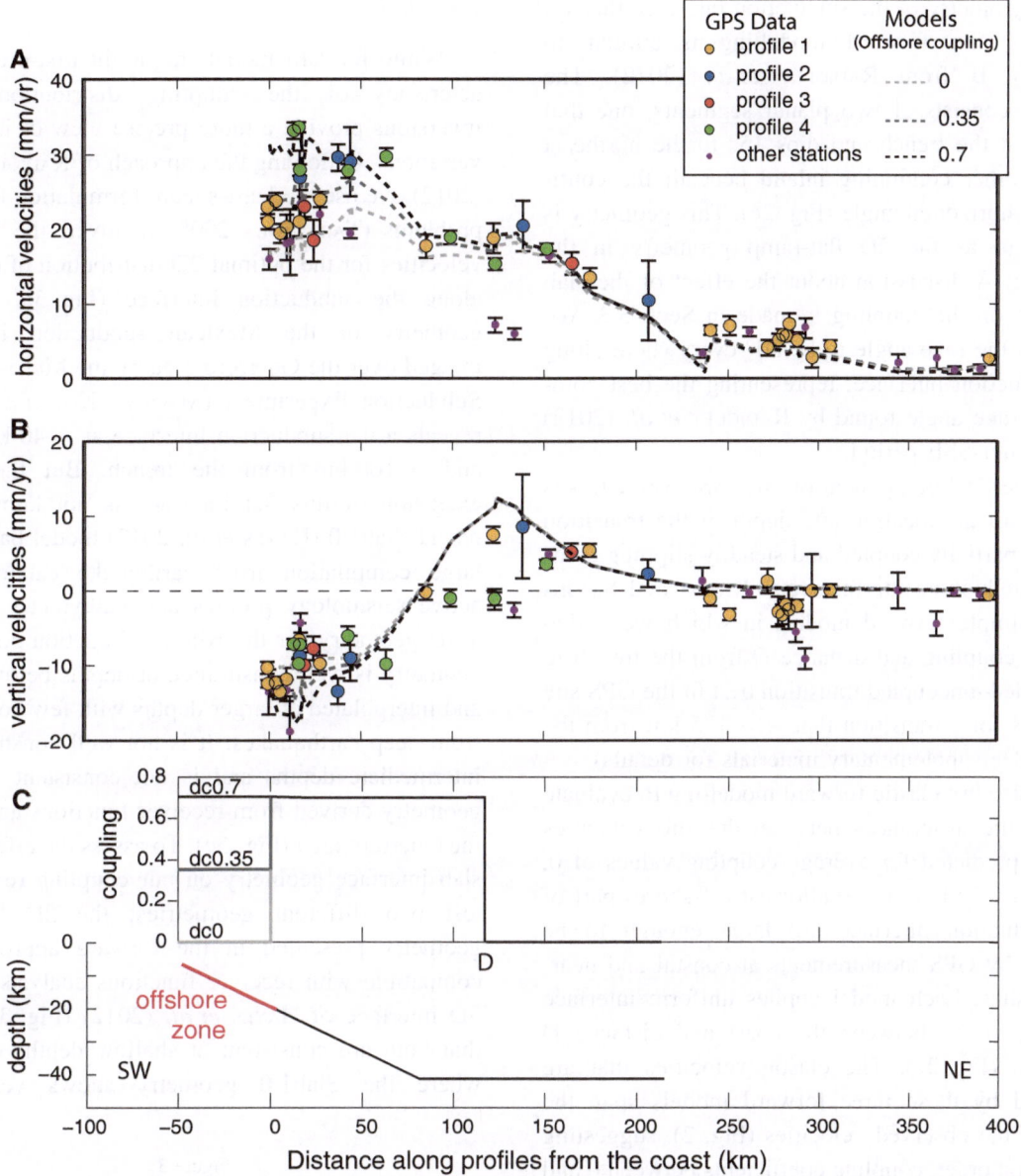

Figure 2

Trench-perpendicular profiles of the *horizontal* **a** and *vertical* **b** components of the 54 GPS site velocities used in this study and illustrated in Fig. 1. The *coloured dots* correspond to the groups of stations defined in Fig. 1. *Grey dashed lines* show three forward models with coupling values between the coast and trench of 0 (*light grey*), 0.35 (*mid grey*), and 0.7 (*black*). Velocities shown for those models are computed on GPS sites. **c** Forward modelling parameters along Profiles 1 to 4. The *top panel* shows the tested coupling values and the *bottom panel* shows a profile of the chosen fault plane geometry. Note that the lateral extension of these two planes is the same as the one used for static inversions in Fig. 3a. Offshore coupling values are set to 0, 0.35, or 0.7, as colour coded in **a**. The coupling from the coast inland to distance *D* is set to 0.7 and drops to 0 beyond point *D*

variations in the coupling offshore, a topic that we focus on in later sections (Fig. 2). To relate the slip on a given patch of the subduction interface to the surface displacements, we build Green's functions following the discrete wave number method (BOU-CHON 1981, 2003) in an elastic stratified medium, assuming the HERNANDEZ *et al.* (2001) velocity model and using AXITRA software (COUTANT 1989).

The geometry of the subduction interface that we adopt for the forward modelling is similar to Geometry B from RADIGUET *et al.* (2012). The interface consists of two planar segments, one that initiates at the trench and dips 15° to the northeast and the other continuing inland beneath the continent at a horizontal angle (Fig. 2c). This geometry is referred to as the 2D flat-ramp geometry in the following. A discussion about the effect of the slab geometry on the coupling is made in Sect. 3.3. We constrain the rake angle to be 80° everywhere along the subduction interface, representing the best compromise rake angle found by RADIGUET *et al.* (2012) for the inter-SSE period.

The vertical component of inter-SSE velocities is sensitive to the location and depth of the transition between partially coupled and steadily slipping areas of the subduction interface (e.g. KANDA and SIMONS 2012). Simple forward models in which we varied both the coupling and distance D from the trench to the coupled–uncoupled transition best fit the GPS site velocities for a transition that is ~ 175 km from the trench (See supplementary materials for details).

We also use elastic forward modelling to evaluate whether the differences between the site velocities that are predicted for average coupling values of 0, 0.35, and 0.7 along the shallowest (offshore) part of the subduction interface are large enough to be resolved by GPS measurements at coastal and near-coastal sites. Each model applies uniform interface coupling of 0.7 between the coast and distance D (175 km) (Fig. 2c). The elastic velocities that are predicted by these three forward models span the range of the observed velocities (Fig. 2), suggesting that to first order, coupling coefficients between 0 and 0.7 are necessary to fit the horizontal velocities and, to a lesser degree, the observed vertical rates. Stations along Profiles 2 and 4 (in orange and purple in Fig. 2) are best matched for shallow coupling values of 0.35 to 0.7, whereas stations along Profiles 1 and 3 (in yellow and red in Fig. 2) are best matched by coupling values smaller than 0.35. We conclude that, despite the parameter cross-correlation not investigated in those forward models, significant variations in coupling across the shallowest part of the subduction interface are detectable with the GPS velocities that are available for this study.

3.3. Static Inversions

While forward models highlight first-order characteristics of the coupling distribution, static inversions provide a more precise view of its spatial variations. Following the approach of RADIGUET *et al.* (2012), we use the least square formulation for linear problems (TARANTOLA 2005) to invert the 3D GPS velocities for the optimal 2D distribution of coupling along the subduction interface (Fig. 3a, b). The geometry of the Mexican subduction interface, imaged over the Guerrero area by the Meso-America Subduction Experiment (MASE) (KIM *et al.* 2012), reveals a flat subduction interface at ~ 40 km depth and ~ 100 km from the trench. But the lateral extension of this flat interface is not known. The recent Slab1.0 (HAYES *et al.* 2012) model based on a large compilation from earthquake catalogues to active seismology profiles and bathymetry presents a 3D geometry for the whole subduction zone. This geometry is well constrained at depths below 20 km and interpolated at larger depths with few constraints from deep earthquakes. It is not well constrained at intermediate depths and is not consistent with the geometry derived from receiver functions analysis in the Guerrero area (Fig. 3c). To assess the effect of the slab interface geometry on our coupling results, we test two different geometries: the 2D flat-ramp geometry presented in the forward approach and compatible with receiver functions analysis and the 3D interface of HAYES *et al.* (2012) (Fig. 3c). Note that both are consistent at shallow depths offshore, where the Slab1.0 geometry shows very little

Figure 3   ►
Best-fit inter-SSE coupling solutions for 2 slab geometries from our GPS site velocities inversion. **a** Best coupling solution for a uniform ramp-flat slab geometry. **b** Best coupling solution for the 3D Slab1.0 geometry from HAYES *et al.* (2012). The four *bold coloured lines* along trench coincide with the four GPS station groups that are defined in Fig. 1. Contours of historical earthquakes, GPS-recorded slow slip events and tremor activity are in *dark blue* (SONG *et al.* 2009), *light blue* (RADIGUET *et al.* 2012; BRUDZINSKI *et al.* 2007; GRAHAM *et al.* 2014a) and *purple* (BRUDZINSKI *et al.* 2010), respectively. Notable offshore bathymetric features are shown (*F.Z.* Fracture zone, *MAT* Middle America Trench). **c** Slab geometry profiles used in models **a** and **b**. The *black* profile shows the flat-ramp 2D geometry used in model **a**, while the four colour profiles show the 3D geometry used in model **b** (profiles correspond to the central section of *boxes* in Fig. 1, with same colour code)

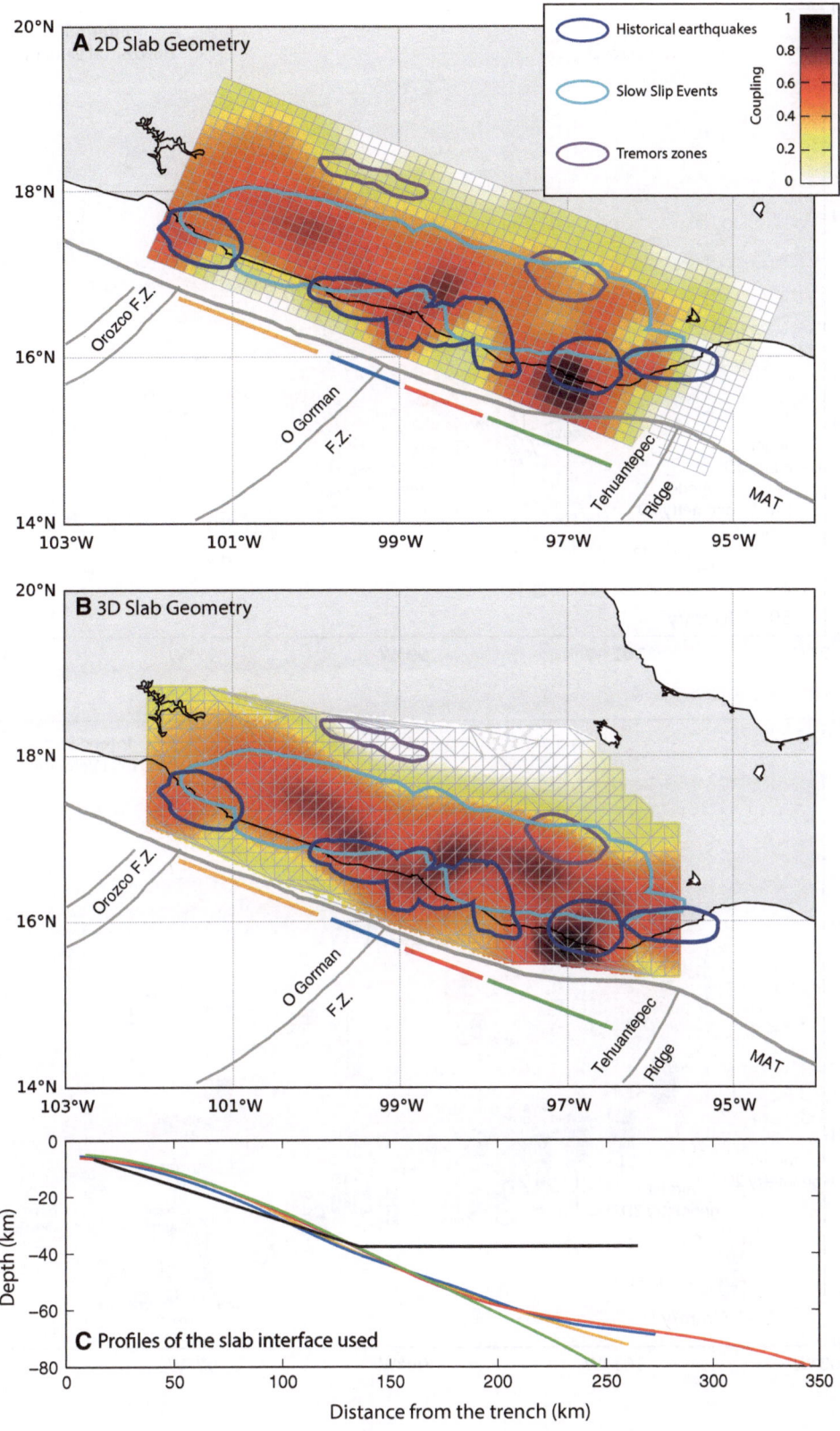

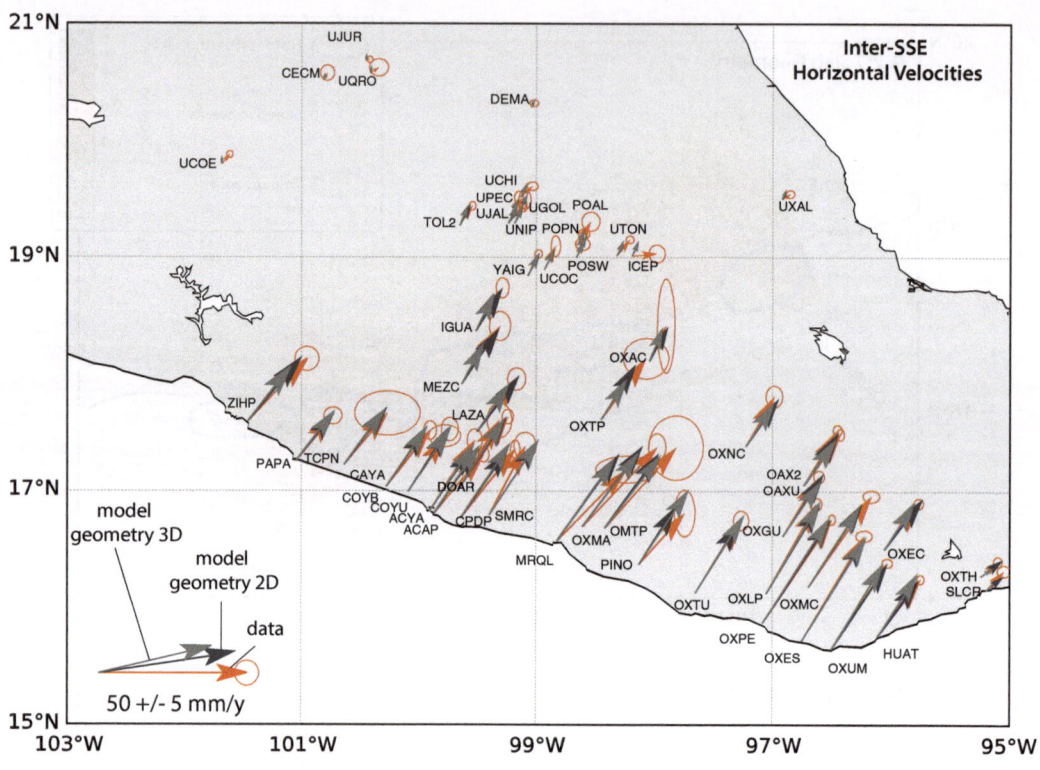

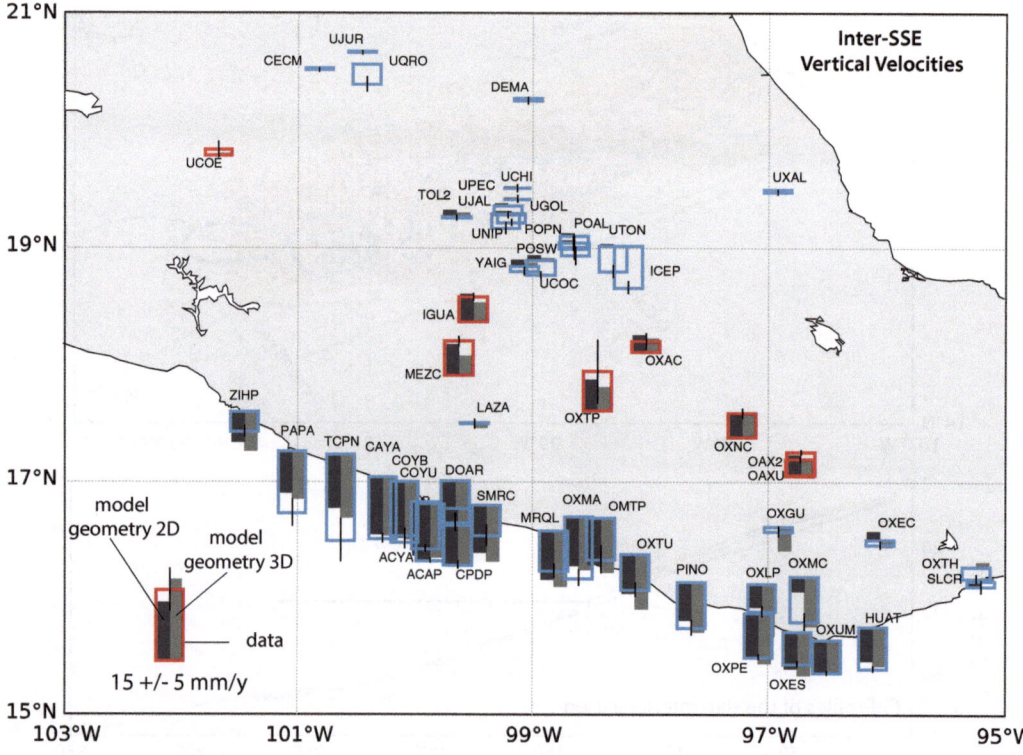

◄

Figure 4
Comparison of observed horizontal (*top*) and vertical (*bottom*) inter-SSE GPS velocities to those predicted by best-fitting coupling solutions shown in the previous figure. *Top Orange* and *grey* vectors show the observed and predicted horizontal velocities, respectively. *Light grey* corresponds to the 3D slab geometry solution while *dark grey* corresponds to the 2D one. *Bottom Red* and *blue boxes* show the observed uplift and subsidence rates, respectively. Errors on vertical rates are shown by a fine *black line* on *top of each box*. As for horizontal rates, *light grey boxes* correspond to the 3D slab geometry solution while *dark grey boxes* correspond to the 2D one

variations in dip angle along strike. The precise configuration and resolution of the inversion are detailed in supplementary materials.

Preferred coupling solutions are shown in Fig. 3a, b together with the location of SSEs, tremors and historical earthquakes. The horizontal and vertical components of the observed GPS velocities are well fit by both models (Fig. 4) except for stations in Mexico City that are subsiding due to non-tectonic ground water extraction and compaction of lake sediments, as documented by InSAR (LÓPEZ-QUIROZ *et al.* 2009). The estimated coupling values are relatively homogeneous at intermediate depths of the slab, but vary along strike at shallow depths. The location of the transition between deeper, uncoupled areas of the subduction interface and partly-to-fully coupled shallow areas is consistent with that found by forward models, at ~175 km from the trench. The shallowest part of the subduction interface consists of three strongly coupled and two weakly coupled areas (Fig. 3), consistent with our forward modelling results (Fig. 2). The westernmost strongly coupled area, however, is constrained by only one GPS station velocity and is thus less reliably determined (also see the resolution analysis presented in the supplementary materials). The two geometries tested show only slight differences. The coupling is more important at intermediate depths in Oaxaca with the 3D geometry and the downdip transition between highly coupled and fully uncoupled areas is sharpest with the 3D geometry due to its higher dip angle at depth.

Slow slip events occur below ~10 km depth in partially coupled areas (<0.6), slightly overlapping at places with historical earthquakes areas (Fig. 3). As SSEs are located only below the continent, the interseismic coupling distribution, sum of the inter-SSE coupling and SSEs contributions, still presents offshore coupling variations [GRAHAM *et al.* 2015, (Figure 11)], on which we focus on in the following. The most strongly coupled area in southern Mexico is off the coast of Oaxaca, where finite element modelling of inter-SSE velocities from a dense, mixed-mode GPS network (CORREA-MORA *et al.* 2008) previously demonstrated that the most strongly coupled part of the subduction interface coincides with the rupture area of the 1978 M_w 7.6 earthquake (STEWART *et al.* 1981). The area is therefore of high seismic hazard. Tremor activity located on the inland/deepest part of the flat slab coincides with low coupling (<0.3), presumably because the interface is silently sliding during the inter-SSE period.

4. Relation Between Short-Term Coupling and Long-Term Morphology

Inter-SSE surface deformation depends on variations of slip at depth that we model as lateral variations of coupling on the subduction interface. Earthquakes and SSEs induce surface displacements in the opposite direction as does inter-SSE, elastic shortening and release accumulated elastic strain. In the case of an homogeneous elastic medium, earthquakes and other phenomena that accommodate slip along the subduction interface release all of the accumulated elastic strain (SAVAGE 1983). No long-term surface deformation such as topography is thus created. Conversely, observations of surface morphology along subduction zones (coastal shape, crustal faults) show that lateral variations of these features coincide with geodetically derived coupling patterns at depth in some cases (e.g. BÉJAR-PIZARRO *et al.* 2013), indicating that some interseismic or inter-SSE deformation may be non-reversible.

Along the Mexican Subduction Zone, we compare the shape of the coastline with the distribution of inter-SSE coupling as modelled in the previous section. We compute distances between the trench and the coast in the direction of convergence, using the ETOPO1 Digital Elevation Model (DEM) at 1′ resolution (AMANTE and EAKINS 2009) for the offshore part, combined with the SRTM DEM at 90 m resolution (FARR *et al.* 2007) for the inland part. Figure 5a compares the trench-to-coast distances corrected from a linear trend with lateral

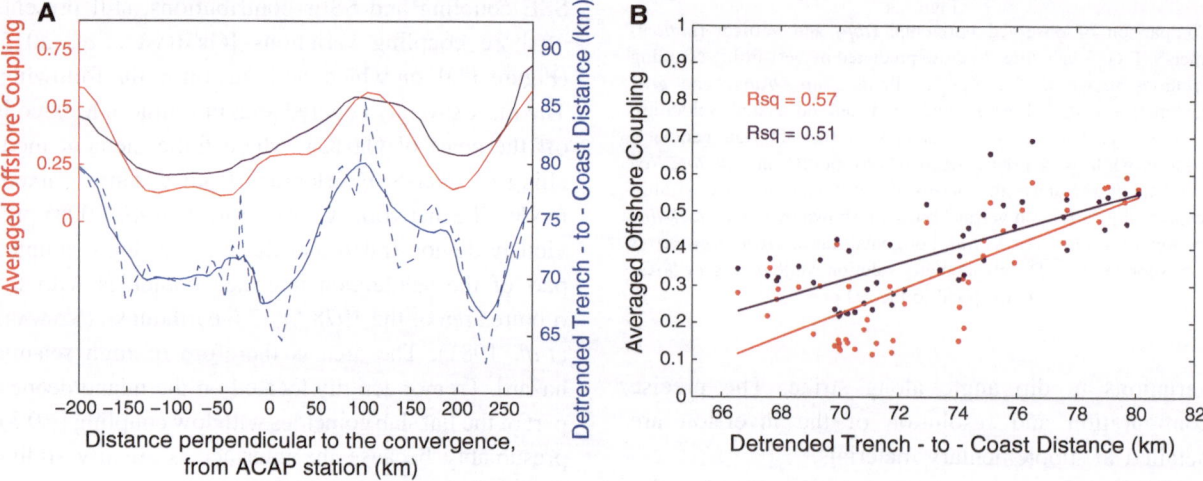

Figure 5

Comparison of the GPS-derived offshore coupling (*red* 2D slab geometry solution, burgundy: 3D slab geometry solution) with trench-to-coast distances estimated along profiles parallel to the convergence direction (*blue*). **a** The along-strike, offshore coupling is computed by averaging coupling values between the trench and the coast in the convergence direction. The trench-to-coast distance is detrended and shown at full resolution (*dashed line*) and after smoothing with a 40 km sliding window (*plain line*). **b** Correlation plots between averaged offshore coupling and trench-to-coast distances (*red* 2D slab geometry solution, burgundy: 3D slab geometry solution). Linear regression coefficients are indicated for both linear regressions

variations of coupling below the accretionary prism between the trench and the coast. The noteworthy correlation for long wavelengths shows that weakly coupled areas coincide with lesser trench-to-coast distances while strongly coupled areas correspond to greater trench-to-coast distances. The linear regression coefficients between trench-coast distances and offshore coupling values are on the order of 0.5 for both geometries used in the coupling modelling (Fig. 5b). The correlation between the coastal shape and lateral coupling variations may be evidence for a mechanical relation between slip on the subduction interface and long-term surface deformation. To test this hypothesis and explore potential mechanisms involved, we use an independent analysis based on the mechanics of the accretionary wedge to determine the relationships between the long-term deformation of the forearc, its frictional properties and the coupling pattern.

5. Contribution of the Critical Taper Theory

Coulomb wedge theory (DAVIS *et al.* 1983) relates the mechanics of a 2D wedge submitted to a constant force applied on the backstop to the morphology of the

wedge and its internal mechanical parameters. To preserve the static stress equilibrium, the wedge evolves into a critical geometry for which every point of the wedge is on the verge of failure (i.e. the Mohr circle is tangent to the Coulomb failure envelope on a normal stress versus shear stress diagram) which means that the wedge undergoes internal deformation while sliding. For the critical state, the stress in the wedge can be explained by ψ_b, the angle between the maximum principal stress axis and the bottom of the wedge (Fig. 6a), and ψ_t, the angle between the maximum principal stress axis and the top of the wedge. ψ_b is a function of the basal friction angle ϕ_b and the internal Hubbert–Rubey pore pressure ratio λ, as defined in CUBAS *et al.* (2013b), and ψ_t depends on the internal friction angle ϕ_{int} and λ. We consider that the pore pressure is the same in the bulk of the wedge and on the bottom interface. Frictional angles ϕ are related to the friction μ by $\mu = \tan\phi$ and to the effective friction by $\mu^{\text{eff}} = \tan\phi^{\text{eff}} = (1 - \lambda)\tan\phi$. DAHLEN (1984) and LEHNER (1986) derived the constitutive relation that holds for the stress equilibrium of the wedge, the Coulomb yielding and the frictional sliding along the basal interface. This relation links ψ_b, ψ_t, the topographic slope α and the basement slope β (Fig. 6a):

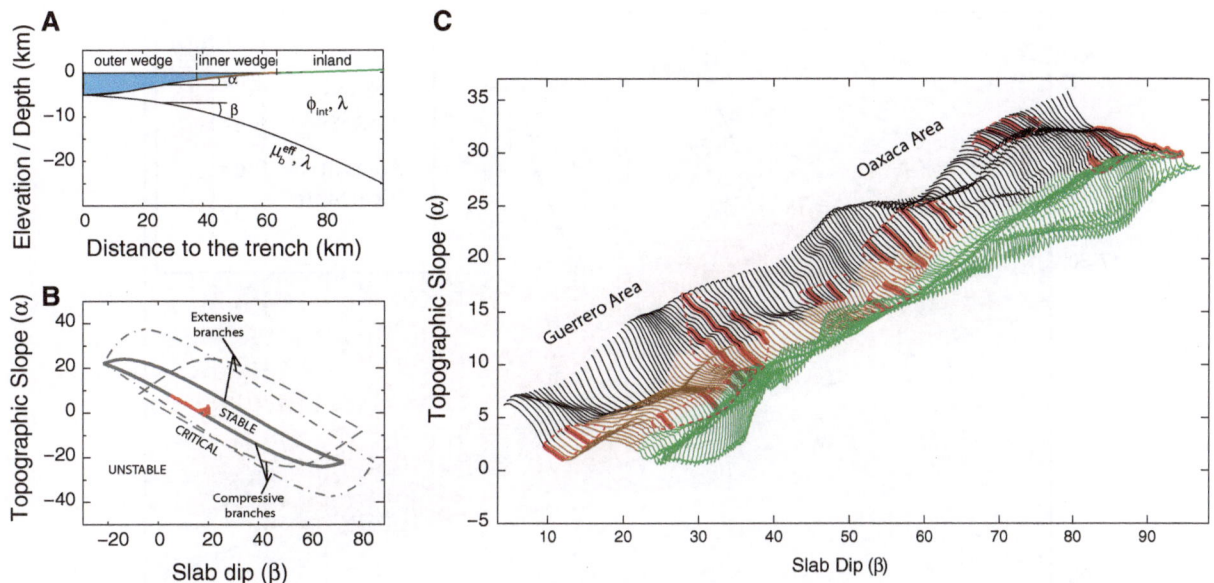

Figure 6

a Vertical cross section of the accretionary prism/upper plate that shows the model parameters used in critical taper theory (DAHLEN 1984). α and β are topographic and slab slopes, respectively. μ_b^{eff}, λ and ϕ_{int} are the effective friction on the subduction interface, the Hubbert–Rubey pore pressure and the internal friction angle, mechanical parameters considered in this study. *Black, orange and green lines* show the extents of the outer wedge, inner wedge and inland areas that are plotted in **c**. **b** Topographic slope as a function of slab dip derived from Dahlen's model for three various sets of mechanical properties (*continuous line* $\mu_b^{eff} = 0.25$, $\lambda = 0.65$; *dashed line* $\mu_b^{eff} = 0.09$, $\lambda = 0.65$; *pointed-dashed line* $\mu_b^{eff} = 0.25$, $\lambda = 0.40$). Stable, unstable and critical domains are shown. *Red line* is for profile of **a**. **c** Topographic slope as a function of slab dip for the 146, two-hundred-km-long trench-perpendicular profiles used for our study (see text for details). Each profile is shifted along *X* and *Y* axes from the previous one, by 0.2° and 0.5°, respectively (the bottom left profile is not shifted). *Red bold lines* identify the sections of the profiles that satisfy the conditions necessary for critical state areas. *Dashed red lines* contour critical state areas, where the mechanical parameters should be relatively uniform. Colours refer to location along profile as defined in **a**

$$\alpha + \beta = \psi_b - \psi_t \qquad (1)$$

Thus, given measurements of α and β, we can estimate the mechanical parameters of the wedge for morphologies consistent with critical state wedge conditions. For (α, β) couples located inside the critical enveloppe, the wedge is defined as being at stable state (Fig. 6b), which means that sliding can occur along the subduction interface without internal deformation.

Critical taper theory was first used to estimate average mechanical parameters for sub-aerial wedges such as Taiwan and submarine wedges (DAVIS *et al.* 1983; DAHLEN 1984; LALLEMAND *et al.* 1994). With improvements in resolution of the topography and slab geometry, CUBAS *et al.* (2013a, b) used the co-variation of α and β to retrieve mechanical properties along subduction zones and their spatial variations.

Here, we apply the method developed by CUBAS *et al.* (2013b) to the Mexican subduction forearc. We use the 1-min resolution ETOPO1 DEM (AMANTE and EAKINS 2009) to compute the topographic slope α. We estimate the basement megathrust slope β from the slab geometry model Slab1.0 (HAYES *et al.* 2012), also used for the coupling static inversion. Figure 6a displays α versus β for 146 two-hundred-km-long serial profiles perpendicular to the trench, where the slab geometry is well constrained. At distances greater than 200-km inland from the trench, the lack of micro-seismicity on the flat part of the slab, which is otherwise well described by receiver functions analysis (PÉREZ-CAMPOS *et al.* 2008; KIM *et al.* 2012), prevents an accurate determination of lateral variations of the slab geometry. We manually selected critical state segments for all 146 α versus β profiles. In the Mexico subduction zone, where β ranges from 0° to 20°, compressive branches of Dahlen's model correspond to straight lines with negative slopes. Thus, finding a critical state segment in our data consists in finding linear trends that match

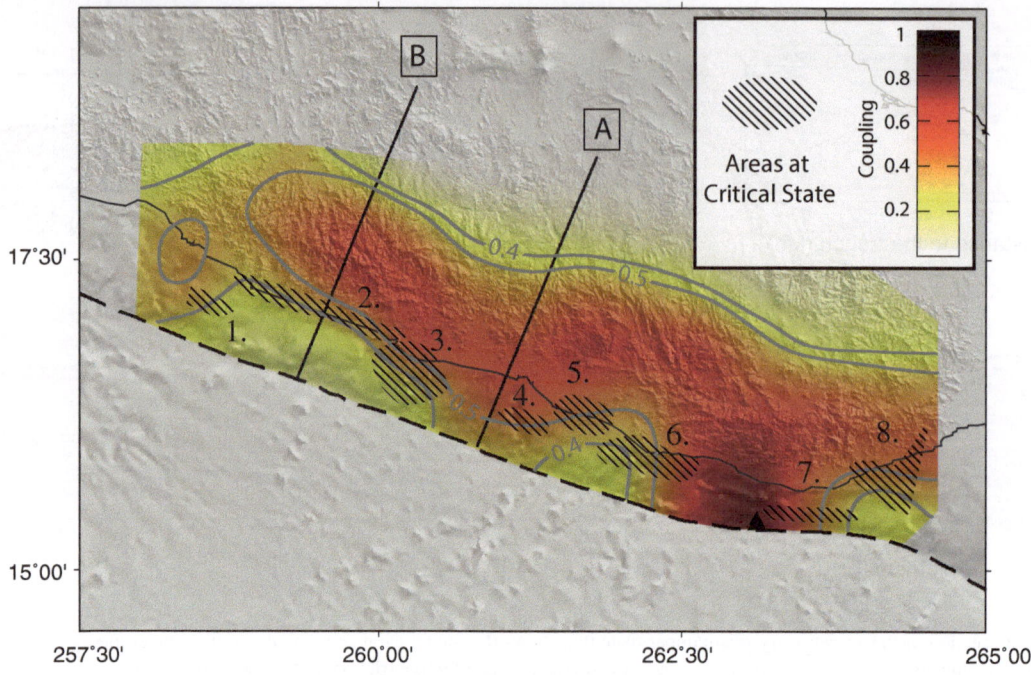

Figure 7
Overlay of the 8 critical state areas from Fig. 6 (*black stripes*) on the best-fitting coupling solution of Fig. 3b (3D slab geometry). Contours corresponding to transitional coupling values of 0.4 and 0.5 are shown by the *grey lines*. The two *black lines* show the location of Profiles **a** and **b** used for plastic deformation modelling (see Fig. 9)

Table 1

Best estimates of mechanical parameters for the eight critical state areas described in the text and identified in Fig. 7

Critical state areas	1.	2.	3.	4.	5.	6.	7.	8.
μ_b^{eff}	0.13	0.46	0.20	0.29	0.09	0.32	0.13	0.13
λ	0.67	0.47	0.78	0.45	0.73	0.58	0.68	0.73
μ_{int}	0.60	0.85	0.98	0.63	0.48	0.87	0.45	0.48

straight lines of Dahlen's model (Fig. 6b). After finding critical state segments in all profiles, we define 3D critical state areas by contouring compatible neighbours critical state segments (Figs. 6c, 7). The eight critical state areas that we identified are either narrow and elongated along the trench or are wider and extend from the trench to the coast, at morphological transitions. Critical state zones are all located at the transition between coupled and uncoupled areas, following a coupling contour of ~ 0.4–0.5.

To get a validation for the criticality of these areas as well as an estimate of their mechanical parameters μ_b^{eff}, λ and μ_{int}, we compute a misfit function for all critical state segments that we minimise following a L2 norm. Results are shown in Table 1. To evaluate

the robustness of the estimated parameters, we compute the probability density distribution of the model parameters and integrate it over one parameter to get 2D marginal probabilities. We compute them for each critical state segment on all profiles and average them to get the probability for a given critical state area (Fig. 8). The marginal probabilities are consistent between all eight critical state areas. Within 1σ, the ϕ_{int} probabilities range from 20° to 45°, the λ probabilities indicate that the parameter is poorly constrained, from 0.4 to 0.8, and ϕ_b^{eff} is better constrained and ranges from 3° to 25° depending on the critical area that is evaluated. The best fits for ϕ_{int} all lie within the 1σ contour, between 25° and 45° ($\mu_{int} = 0.47$ to 1). These agree with laboratory

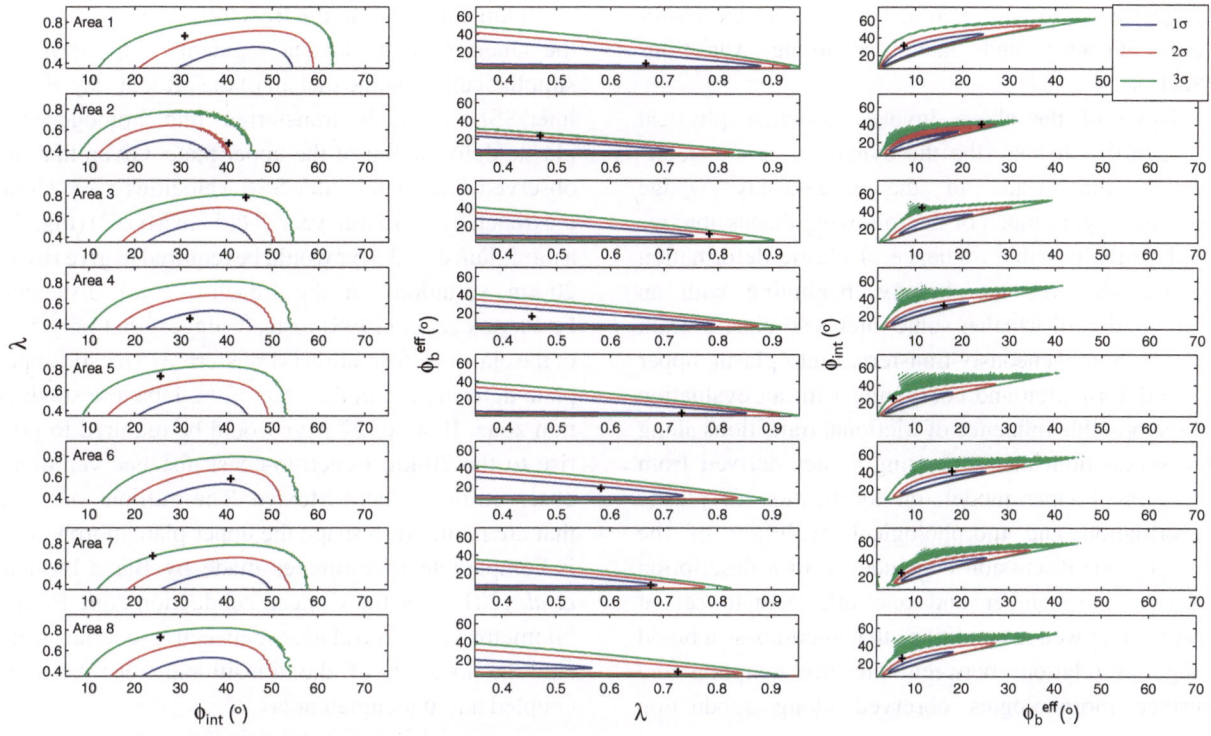

Figure 8

Marginal 2D probability distributions for all eight critical state areas. The three parameters estimated are λ: the Hubbert–Rubey pore fluid ratio, ϕ_b^{eff}: the effective basal friction angle and ϕ_{int}: the internal friction angle. The 1σ, 2σ and 3σ probability contours are shown in *blue, red and green*. The best fit is localised by a *black cross*

measurements of quartzo-feldspathic materials, which find internal frictions from 0.6 to 0.85 (BYER-LEE 1978). Though poorly constrained, the best fits for λ include values that are close to the hydrostatic pore fluid pressure ($\lambda = 0.4$). Best fits for ϕ_b^{eff} show significant lateral variations, with most values for ϕ_b^{eff} between 6° and 17° ($\mu_b^{\text{eff}} = 0.1$ to 0.3). ϕ_b^{eff} peaks at 25° ($\mu_b^{\text{eff}} = 0.46$) in critical area 2, which corresponds to the Guerrero seismic gap.

Applying critical taper theory to the Mexican subduction wedge thus reveals lateral variations of mechanical properties, with some parts of the wedge in the critical state, i.e. with internal deformation. The basal frictions are mainly low excepted in the Guerrero critical state area (area 2) that shows higher basal friction, in agreement with the presence of SSEs in this area. The long-term morphology of the Mexican margin might thus be driven by spatial variations of friction properties on the subduction interface, with critical state areas marking frictional transitions.

6. Discussion

In this study, we describe the Mexican subduction wedge with two models that describe deformation at different, but complementary timescales. Inter-SSE coupling (Sect. 3), which is a proxy for strain accumulating in the upper plate, operates over decadal time scales and varies significantly along strike. The morphological approach based on the critical taper theory (Sect. 5) describes average mechanical properties of the wedge on timescales that span several hundred seismic cycles (millennia or longer). The latter model enables to identify stable-state parts of the wedge, with no internal deformation, and critical state areas, where internal wedge deformation is likely to occur. All eight areas that we identified as critical state coincide with transitions between weakly coupled and strongly coupled areas of the subduction interface. This may indicate a persistency of mechanical properties through time, as is also suggested by the correlation between the trench-to-coast distance and lateral coupling variations (Sect. 4).

None of the above invoke particular physical mechanisms to describe the dynamics of the deformation that leads to the present-day wedge morphology. In much of the following discussion, we explore the possible influence of plastic deformation on the observed morphology, beginning with an examination of whether some inter-SSE elastic strain may be homogeneously transferred into plastic upper plate deformation and continuing with an evaluation of the possible influence of frictional transitions along the subduction interface, using values derived from the critical taper model, on localisation of plastic deformation and morphological evolution of the wedge. Our discussion concludes with a description of two end-member and one intermediate set of mechanical wedge conditions that encompass a broad range of relations between interface coupling and surface morphologies observed along subduction zones.

6.1. Distributed Transfer of Elastic Inter-SSE Strain into Plastic Deformation of the Accretionary Prism

One hypothesis to explain the correspondence between the present-day pattern of inter-SSE coupling in southern Mexico and the long-term coastline morphology is that inter-SSE strain that accumulates due to partial-to-full locking of the subduction interface is not completely released by slip along the interface, including coseismic rupture, postseismic fault afterslip, and SSE. A fraction of the strain may instead be transferred into non-reversible, plastic deformation of the upper plate (LE PICHON et al. 1998). Horizontal inter-SSE velocities at locations near and along the coast are ~ 3 times faster than vertical rates, as illustrated by forward models (Fig. 2). By implication, if long-term plastic deformation mimics the pattern of what inferred elastic inter-SSE deformation models, the trench-to-coast distance is more sensitive to the horizontal than vertical components of the elastic deformation (and its variations along the trench). A simple calculation can be made to estimate the minimum time required to give rise to the observed

~ 20 km variations in the trench-to-coast distance for the Guerrero and Oaxaca segments (Fig. 5) if we simplistically assume that all of the present-day elastic inter-SSE strain is transferred into homogeneous plastic deformation of the upper plate. Given that the observed horizontal inter-SSE velocities vary along the trench by ~ 15 mm year^{-1} (15 km Myr^{-1}) (Fig. 2), a minimum of 1.3 Myr would be required to give rise to 20-km variations in the trench-to-coast distances. LE PICHON et al. (1998) instead suggest that only 5 % of the elastic deformation is transferred to plastic upper plate deformation in the case of the Japanese subduction zone. If so, to 27 Myr would be required to give rise to the 20-km trench-to-coast distance variations observed in southern Mexico. The millions of years that are required to shape the upper plate morphology is comparable to estimates made by BÉJAR-PIZARRO et al. (2013) for the Chilean subduction zone from a kilometre-scale correlation between a coastal scarp and the location of the transitional zone between coupled and uncoupled areas.

6.2. Localised Strain Transfer Associated with Frictional Transitions

An alternative scenario for permanently deforming the upper plate is if plastic, upper plate deformation occurs during sliding episodes (coseismic or aseismic slip) in response to variations of mechanical properties on the slab interface or within the wedge and overriding plate. Such variations could be of geological or geometrical origin as discussed before. With no direct information on sub-surface geological and geometrical structures in our study area, our considerations are indirect, based on the results from the critical taper analysis. To quantify how frictional transitions might produce brittle deformation during slip episodes, as already explored by other studies (e.g. HU and WANG 2008), we ran a mechanical model based on the limit analysis method (CHANDRASEKHARAIAH and DEBNATH 1994; SALENÇON 2002). The limit analysis method is a static approach that predicts the localisation of brittle deformation from the frictional properties distribution. It is based on the weak formulation of the force equilibrium (the principle of virtual powers) and the theorem of maximum rock strength (MAILLOT and LEROY 2006).

Table 2

Frictional parameters used in plastic deformation models HM and TM

Model	ϕ_{int} (°)	μ_{int}	λ	$\phi_{b,hdf}^{eff}$ (°)	$\mu_{b,hdf}^{eff}$	$\phi_{b,ldf}^{eff}$ (°)	$\mu_{b,ldf}^{eff}$
HM	35	0.7	0.4	NA	NA	5.7	0.1
TM	35	0.7	0.4	20	0.4	5.7	0.1

HM (Homogeneous Model) and TM (Transitional Model) are representative of profiles A and B in Fig. 7, respectively. Indices ldf and hdf refer to low dynamic friction and high dynamic friction, respectively

For our simulations, the Coulomb criterion is used to define the maximum rock strength. The simulations were run using the Optum-Geo software (OptumGeo 2013). Two different setups were investigated. The first one assumes that the subduction interface is homogeneously, strongly coupled, representative of Profile A in Fig. 7. Below, we refer to this as Model HM (Homogeneous Model). For the second, representative of Profile B in Fig. 7, we assume that the shallowest area of the subduction interface is uncoupled and then transitions downdip to a region of strong coupling coinciding with the critical state area we identified (Area 2 in Fig. 7). We refer below to this as Model TM (Transitional Model). We use the 2D slab geometry for these models.

The mechanical properties assumed for each model (Table 2) are consistent with those estimated for the eight critical state areas identified earlier in the analysis. We distinguish effective frictions for aseismic sliding and seismic sliding, corresponding to weakly coupled areas and strongly coupled areas, respectively. For aseismic areas, the effective friction of the slab interface is constrained by critical taper results: we use the friction obtained within a critical area near the transition zone, and extrapolate it to the whole aseismic area at shallower depths. It corresponds to a quasi-static friction since the slip is very slow. For seismic areas, since deformation is assumed to be acquired during sliding episodes, we use dynamic friction values consistent with values found for other subduction earthquakes (e.g. KIMURA *et al.* 2012; CONIN *et al.* 2012; FULTON *et al.* 2013; UJIIE *et al.* 2013; CUBAS *et al.* 2013a, b). Exact values for quasi-static or dynamic friction are not well constrained, but the important point is that dynamic friction is always way lower than quasi-static friction (e.g. RICE 2006; TORO *et al.* 2011).

We first plot the shear dissipation (Fig. 9), showing areas where the wedge is on the verge of brittle failure. In the HM model, a hinge associated with the transfer of material from the horizontal to the steeper portion of the slab is developed. Models that assume progressively more gradual slope transitions predict progressively more diffuse shear zones. In the TM model, a back-thrust develops at the frictional transition in addition to the slab kink hinge. Since the basal effective friction is larger along the uncoupled, shallow area of the subduction interface than along the more strongly coupled interface farther downdip, the frontal part of the wedge is uplifted along the predicted back-thrust. Increasing the frictional value that is imposed along the strongly coupled area of the interface increases the dip angle of the back-thrust.

The limit analysis approach provides virtual velocities on all nodes of the model. To retrieve surface displacements, we assume that the ratio between virtual velocities of two segments projected along their fault is equal to the ratio of actual displacements between those segments. All surface displacements are normalised to the surface horizontal displacement at the backstop. In that case, considering 1 m of displacement at the backstop, i.e. for 1 m of convergence, the forearc is uplifted by 63 cm in the TM model and 27 cm in the HM model (top panel Fig. 9). Those uplifts should obviously be balanced by erosion which is not quantified here. However, considering an homogeneous erosion rate along the coast, the larger uplift built in the TM model in comparison with the HM model would contribute to reduce the trench-to-coast distance, which is consistent with the observations made in Fig. 5. Contrary to the first scenario presented in Sect. 6.1, the coastal shape is there mostly built from differential vertical displacements (differences in

269

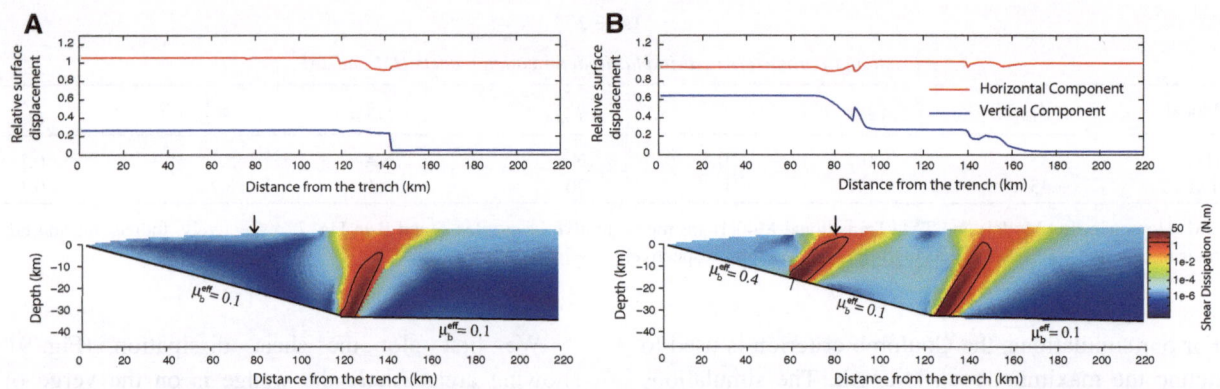

Figure 9
Deformation predicted by plastic modelling for two mechanical settings. Virtual displacement and virtual shear dissipation are obtained from limit analysis (OptumGeo 2013). **a** Is for a uniform low effective dynamic friction of 0.1 on the megathrust interface, while **b** is for an effective friction of 0.4 near the trench and a lower effective dynamic friction of 0.1 at depth. The density is set to 2600 kg/m^3. Slab geometry is identical as that used for the coupling modelling. Topography in **a** and **b** corresponds to that along Profiles **a** and **b** of Fig. 7. *Top panels* show the cumulative surface displacement due to brittle deformation (*red horizontal*, *blue vertical*) relative to the horizontal displacement at the backstop. *Bottom panels* represent the final shear dissipation in a logarithmic colourscale. The *black line* corresponds to the 5 N.m contour in which more than 99 % of the shear dissipation occurs. *Black arrows* indicate the location of the coastline

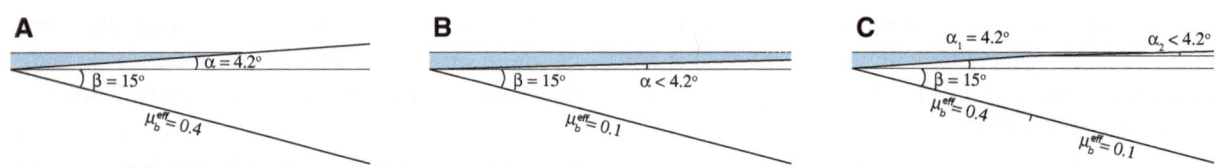

Figure 10
Sketches showing the influence of the effective basal friction μ_b^{eff} on the average topographic slope α. Other parameters are set constant ($\beta = 15°$, $\phi_{int} = 35°$, $\lambda = 0.4$). Cases **b** and **c** are similar to models **a** and **b** in Fig. 9

horizontal displacements between models TM and HM are very small, top panel Fig. 9).

Although homogeneous, upper plate plastic deformation that is proportional to the elastic strain accumulation (Sect. 6.1) and localised upper plate deformation due to frictional transitions on the subduction interface (Sect. 6.2) may both contribute to shaping coastlines, too few observational constraints are available to estimate their relative proportion, such as seismic profiles to identify splay faults which would attest of internal deformation in the prism.

6.3. Coastal Morphology as a Function of Frictional Variations Along Fault Dip

The evidence described above and shown in Fig. 5a for an along-strike correlation between areas of weak coupling on the shallow subduction interface

and diminished trench-to-coast distances agrees with a similar correlation described for other subduction zones, particularly in the eastern Pacific. Weakly coupled trench segments often coincide with prominent peninsulas (and hence very short trench-to-coast distances) such as the Arauco Peninsula in Chile (MELNICK *et al.* 2009) and the Piura peninsula in northern Peru (NOCQUET *et al.* 2014). Based on this correlation, we propose to distinguish three types of behaviours along subduction zones depending on how frictional properties vary downdip along the subduction interface (Fig. 10).

The first corresponds to subduction interfaces where coupling is weak between the trench and the downdip limit of the seismogenic zone (Fig. 10a). In this case, critical wedge theory predicts that the average topographic slope angle is high (4.2° for $\beta = 15°$ and $\mu_b^{eff} = 0.4$) and continual thickening of the upper plate (wedge), possibly including forethrust sequences. These processes favour the genesis of peninsulas and decrease trench-to-coast distances.

The second corresponds to a strongly coupled slab interface from the trench to the downdip limit of the seismogenic zone (Fig. 10b and Profile A of Fig. 7), as assumed for the HM model described in Sect. 6.2. In this case, the entire subduction interface may rupture at very low dynamic friction. This results in low average topographic slope angles ($<4.2°$ for $\beta = 15°$ and $\mu_b^{\text{eff}} = 0.1$), including negative slope angles that may be associated with extension of the forearc and the formation of sedimentary basins. Sedimentary basins in forearc settings have been detected by their free-air gravity anomalies and are spatially correlated with large seismic asperities that broke during the 20th century (WELLS *et al.* 2003; SONG and SIMONS 2003). Larger trench-to-coast distances are expected in this scenario.

The intermediate case consists of a subduction interface with frictional transitions within the seismogenic zone (Fig. 10c and Profile B of Fig. 7), as described for the TM model of Sect. 6.2. These frictional transitions may produce localised upper plate deformation, that may result in different topographic slope angles for the outer and inner wedges. The resultant trench-to-coast distance is intermediate between that expected for the two end-member scenarios described above.

7. Conclusion

Our analysis of the inter-SSE velocities of continuous GPS sites in southern Mexico reveals that significant variations in inter-SSE coupling occur along the Oaxaca and Guerrero segments of the Mexico subduction zone. The along-strike variations, alternating between areas where coupling is less than 0.3 and areas where it is more than 0.7, are mostly offshore. Coupling at depths below 25 km is more homogeneous and transitions to steady creep at distances of ~ 175 km from the trench. Variations in shallow coupling along the trench are strongly correlated with trench-to-coast distances, the latter of which may be a proxy for the mechanical state of the

subduction wedge. Using critical taper theory to relate the morphology of the subduction wedge to its average long-term mechanical properties, we identify eight critical state areas where internal deformation of the wedge is expected. All eight are located at transitions between weakly and strongly coupled areas of the subduction interface. An inversion of mechanical parameters for these critical state areas gives values that are consistent for 7 of the 8 critical state areas, with the lone exception coinciding with the Guerrero seismic gap, where anomalously high basal friction estimated from our inversion overlaps the region where silent slip events occur. Two mechanisms are proposed to explain how trench-to-coast distances have evolved to their present geometry in southern Mexico: (1) homogeneous plastic deformation of the upper plate may occur in response to inter-SSE elastic strain associated with the locked subduction interface (over time scales of Myrs), (2) frictional transitions on the subduction interface may produce localised shear zones in the upper plate and may cause differential uplift and/or subsidence along-strike in the upper plate. Our results suggest that the geometry/morphology of the subduction wedge can be related to its long-term mechanical parameters. To address in further detail how permanent deformation accumulates over one seismic cycle, dynamic modelling that incorporates rupture mechanics processes and of fault deformation could be appropriate.

Acknowledgments

This study is supported by the French National Research Agency (Agence Nationale de la Recherche, ANR G-GAP RA0000C069) and the USA National Science Foundation (Grant EAR-1114174). The GPS network maintenance and data acquisition were supported by Mexicos PAPIIT IN102105, IN103808 CONACYT 84544 and PAPIIT IN110514 Grants and by the French spatial agency CNES (project TOSCA SSEMEX). Some graphics were made with the Global Mapping Tool (GMT) software. We thank François Renard and Jean-Philippe Avouac for stimulating discussions about this study and the two anonymous reviewers for their constructive criticism which we found very helpful to improve the manuscript.

REFERENCES

AMANTE, C. and EAKINS, B. W. (2009). *ETOPO1 1 arc-minute global relief model: procedures, data sources and analysis*. US Department of Commerce, National Oceanic and Atmospheric Administration, National Environmental Satellite, Data, and Information Service, National Geophysical Data Center, Marine Geology and Geophysics Division.

ASTIZ, L. and KANAMORI, H. (1984). *An earthquake doublet in ometepec, guerrero, mexico*. Physics of the earth and planetary interiors, *34*(1):24–45.

BÉJAR-PIZARRO, M., SOCQUET, A., ARMIJO, R., CARRIZO, D., GENRICH, J., and SIMONS, M. (2013). *Andean structural control on interseismic coupling in the north chile subduction zone*. Nature Geoscience.

BEKAERT, D., HOOPER, A., and WRIGHT, T. (2015). *A spatially-variable power-law tropospheric correction technique for insar data*. Journal of Geophysical Research: Solid Earth.

BOUCHON, M. (1981). *A simple method to calculate green's functions for elastic layered media*. Bulletin of the Seismological Society of America, *71*(4):959–971.

BOUCHON, M. (2003). *A review of the discrete wavenumber method*. Pure and applied Geophysics, *160*(3-4):445–465.

BRUDZINSKI, M., CABRAL-CANO, E., CORREA-MORA, F., DEMETS, C., and MÁRQUEZ-AZÚA, B. (2007). *Slow slip transients along the oaxaca subduction segment from 1993 to 2007*. Geophysical Journal International, *171*(2):523–538.

BRUDZINSKI, M. R., HINOJOSA-PRIETO, H. R., SCHLANSER, K. M., CABRAL-CANO, E., ARCINIEGA-CEBALLOS, A., DIAZ-MOLINA, O., and DEMETS, C. (2010). *Nonvolcanic tremor along the oaxaca segment of the middle america subduction zone*. Journal of Geophysical Research, *115*(null):B00A23.

BÜRGMANN, R., KOGAN, M. G., STEBLOV, G. M., HILLEY, G., LEVIN, V. E., and APEL, E. (2005). *Interseismic coupling and asperity distribution along the kamchatka subduction zone*. Journal of Geophysical Research: Solid Earth (1978-2012), *110*(B7).

BYERLEE, J. (1978). *Friction of rocks*. Pure and applied Geophysics, *116*(4-5):615–626.

CAVALIÉ, O., PATHIER, E., RADIGUET, M., VERGNOLLE, M., COTTE, N., WALPERSDORF, A., KOSTOGLODOV, V., and COTTON, F. (2013). *Slow slip event in the mexican subduction zone: Evidence of shallower slip in the guerrero seismic gap for the 2006 event revealed by the joint inversion of insar and gps data*. Earth and Planetary Science Letters, *367*:52–60.

CHANDRASEKHARAIAH, D. and DEBNATH, L. (1994). *Continuum mechanics*. Academic press New York.

CHLIEH, M., AVOUAC, J.-P., SIEH, K., NATAWIDJAJA, D. H., and GALETZKA, J. (2008). *Heterogeneous coupling of the sumatran megathrust constrained by geodetic and paleogeodetic measurements*. Journal of Geophysical Research: Solid Earth (1978–2012), *113*(B5).

CHLIEH, M., MOTHES, P., NOCQUET, J.-M., JARRIN, P., CHARVIS, P., CISNEROS, D., FONT, Y., COLLOT, J.-Y., VILLEGAS-LANZA, J.-C., ROLANDONE, F., et al. (2014). *Distribution of discrete seismic asperities and aseismic slip along the ecuadorian megathrust*. Earth and Planetary Science Letters, *400*:292–301.

CHLIEH, M., PERFETTINI, H., TAVERA, H., AVOUAC, J.-P., REMY, D., NOCQUET, J.-M., ROLANDONE, F., BONDOUX, F., GABALDA, G., and BONVALOT, S. (2011). *Interseismic coupling and seismic potential along the central andes subduction zone*. Journal of Geophysical Research: Solid Earth (1978–2012), *116*(B12).

CONIN, M., HENRY, P., GODARD, V., and BOURLANGE, S. (2012). *Splay fault slip in a subduction margin, a new model of evolution*. Earth and Planetary Science Letters, *341*:170–175.

CORREA-MORA, F., DEMETS, C., CABRAL-CANO, E., DIAZ-MOLINA, O., and MARQUEZ-AZUA, B. (2009). *Transient deformation in southern mexico in 2006 and 2007: Evidence for distinct deep-slip patches beneath guerrero and oaxaca*. Geochemistry, Geophysics, Geosystems, *10*(2).

CORREA-MORA, F., DEMETS, C., CABRAL-CANO, E., MARQUEZ-AZUA, B., and DIAZ-MOLINA, O. (2008). *Interplate coupling and transient slip along the subduction interface beneath oaxaca, mexico*. Geophysical Journal International, *175*(1):269–290.

COTTE, N., WALPERSDORF, A., KOSTOGLODOV, V., VERGNOLLE, M., SANTIAGO, J.-A., and CAMPILLO, M. (2009). *Anticipating the next large silent earthquake in mexico*. Eos, Transactions American Geophysical Union, *90*(21):181–182.

COUTANT, O. (1989). Programme de simulation numerique axitra. *Rapport LGIT.*

CUBAS, N., AVOUAC, J., LEROY, Y., and PONS, A. (2013a). *Low friction along the high slip patch of the 2011 mw 9.0 tohoku-oki earthquake required from the wedge structure and extensional splay faults*. Geophysical Research Letters, *40*(16):4231–4237.

CUBAS, N., AVOUAC, J.-P., SOULOUMIAC, P., and LEROY, Y. (2013b). *megathrust friction determined from mechanical analysis of the forearc in the maule earthquake area*. Earth and Planetary Science Letters.

CUBAS, N., LAPUSTA, N., AVOUAC, J.-P., and PERFETTINI, H. (2015). *Numerical modeling of long-term earthquake sequences on the ne japan megathrust: comparison with observations and implications for fault friction*. Earth and Planetary Science Letters, *419*:187–198.

DAHLEN, F. (1984). *Noncohesive critical coulomb wedges: An exact solution*. Journal of Geophysical Research: Solid Earth (1978–2012), *89*(B12):10125–10133.

DAVIS, D., SUPPE, J., and DAHLEN, F. (1983). *Mechanics of fold-and-thrust belts and accretionary wedges*. Journal of Geophysical Research: Solid Earth (1978–2012), *88*(B2):1153–1172.

DEMETS, C., GORDON, R. G., and ARGUS, D. F. (2010). *Geologically current plate motions*. Geophysical Journal International, *181*(1):1–80.

DI TORO, G., HAN, R., HIROSE, T., DE PAOLA, N., NIELSEN, S., MIZOGUCHI, K., FERRI, F., COCCO, M., and SHIMAMOTO, T. (2011). *Fault lubrication during earthquakes*. Nature, *471*(7339):494–498.

FARR, T. G., ROSEN, P. A., CARO, E., CRIPPEN, R., DUREN, R., HENSLEY, S., KOBRICK, M., PALLER, M., RODRIGUEZ, E., ROTH, L., et al. (2007). *The shuttle radar topography mission*. Reviews of geophysics, *45*(2).

FRANCO, A., LASSERRE, C., LYON-CAEN, H., KOSTOGLODOV, V., MOLINA, E., GUZMAN-SPEZIALE, M., MONTEROSSO, D., ROBLES, V., FIGUEROA, C., AMAYA, W., et al. (2012). *Fault kinematics in northern central america and coupling along the subduction interface of the cocos plate, from gps data in chiapas (mexico), guatemala and el salvador*. Geophysical Journal International, *189*(3):1223–1236.

FRANK, W. B., RADIGUET, M., ROUSSET, B., SHAPIRO, N. M., HUSKER, A. L., KOSTOGLODOV, V., COTTE, N., and CAMPILLO, M. (2015).

Uncovering the geodetic signature of silent slip through repeating earthquakes. Geophysical Research Letters, *42*(8):2774–2779.

FRANK, W. B., SHAPIRO, N. M., KOSTOGLODOV, V., HUSKER, A. L., PAYERO, J. S., CAMPILLO, M., and PRIETO, G. A. (2013). *Low-frequency earthquakes in the mexican sweet 1 spot.* Geophysical Research Letters.

FULTON, P., BRODSKY, E., KANO, Y., MORI, J., CHESTER, F., ISHIKAWA, T., HARRIS, R., LIN, W., EGUCHI, N., TOCZKO, S., et al. (2013). *Low coseismic friction on the tohoku-oki fault determined from temperature measurements.* Science, *342*(6163):1214–1217.

GRAHAM, S. E., DEMETS, C., CABRAL-CANO, E., KOSTOGLODOV, V., ROUSSET, B., WALPERSDORF, A., COTTE, N., LASSERRE, C., MCCAFFREY, R., and SALAZAR-TLACZANI, L. (2015). *Slow slip history for the mexico subduction zone: 2005 through 2011.* Pure and Applied Geophysics.

GRAHAM, S. E., DEMETS, C., CABRAL-CANO, E., KOSTOGLODOV, V., WALPERSDORF, A., COTTE, N., BRUDZINSKI, M., MCCAFFREY, R., and SALAZAR-TLACZANI, L. (2014a). *Gps constraints on the 2011-2012 oaxaca slow slip event that preceded the 2012 march 20 ometepec earthquake, southern mexico.* Geophysical Journal International, page ggu019.

GRAHAM, S. E., DEMETS, C., CABRAL-CANO, E., KOSTOGLODOV, V., WALPERSDORF, A., COTTE, N., BRUDZINSKI, M., MCCAFFREY, R., and SALAZAR-TLACZANI, L. (2014b). *Gps constraints on the mw= 7.5 ometepec earthquake sequence, southern mexico: coseismic and post-seismic deformation.* Geophysical Journal International, *199*(1):200–218.

HAYES, G. P., WALD, D. J., and JOHNSON, R. L. (2012). *Slab1. 0: A three-dimensional model of global subduction zone geometries.* Journal of Geophysical Research: Solid Earth (1978–2012), *117*(B1).

HERNANDEZ, B., SHAPIRO, N., SINGH, S., PACHECO, J., COTTON, F., CAMPILLO, M., IGLESIAS, A., CRUZ, V., GÓMEZ, J., and ALCÁNTARA, L. (2001). *Rupture history of september 30, 1999 intraplate earthquake of oaxaca, mexico (mw= 7.5) from inversion of strong-motion data.* Geophysical research letters, *28*(2):363–366.

HU, Y. and WANG, K. (2008). *Coseismic strengthening of the shallow portion of the subduction fault and its effects on wedge taper.* Journal of Geophysical Research: Solid Earth (1978-2012), *113*(B12).

HUSKER, A. L., KOSTOGLODOV, V., CRUZ-ATIENZA, V. M., LEGRAND, D., SHAPIRO, N. M., PAYERO, J. S., CAMPILLO, M., and HUESCA-PÉREZ, E. (2012). *Temporal variations of non-volcanic tremor (nvt) locations in the mexican subduction zone: Finding the nvt sweet spot.* Geochemistry, Geophysics, Geosystems, *13*(3).

KANDA, R. V. and SIMONS, M. (2012). *Practical implications of the geometrical sensitivity of elastic dislocation models for field geologic surveys.* Tectonophysics.

KANEKO, Y., AVOUAC, J.-P., and LAPUSTA, N. (2010). *Towards inferring earthquake patterns from geodetic observations of interseismic coupling.* Nature Geoscience, *3*(5):363–369.

KIM, Y., MILLER, M. S., PEARCE, F., and CLAYTON, R. W. (2012). *Seismic imaging of the cocos plate subduction zone system in central mexico.* Geochemistry, Geophysics, Geosystems, *13*(7).

KIMURA, G., HINA, S., HAMADA, Y., KAMEDA, J., TSUJI, T., KINOSHITA, M., and YAMAGUCHI, A. (2012). *Runaway slip to the trench due to rupture of highly pressurized megathrust beneath the middle trench slope: the tsunamigenesis of the 2011 tohoku earthquake off the east coast of northern japan.* Earth and Planetary Science Letters, *339*:32–45.

KONCA, A. O., AVOUAC, J.-P., SLADEN, A., MELTZNER, A. J., SIEH, K., FANG, P., LI, Z., GALETZKA, J., GENRICH, J., CHLIEH, M., et al. (2008). *Partial rupture of a locked patch of the sumatra megathrust during the 2007 earthquake sequence.* Nature, *456*(7222):631–635.

KOSTOGLODOV, V., BANDY, W., DOMINGUEZ, J., and MENA, M. (1996). *Gravity and seismicity over the guerrero seismic gap, mexico.* Geophysical Research Letters, *23*(23):3385–3388.

KOSTOGLODOV, V., HUSKER, A., SHAPIRO, N. M., PAYERO, J. S., CAMPILLO, M., COTTE, N., and CLAYTON, R. (2010). *The 2006 slow slip event and nonvolcanic tremor in the mexican subduction zone.* Geophysical Research Letters, *37*(24).

LALLEMAND, S. E., SCHNÜRLE, P., and MALAVIEILLE, J. (1994). *Coulomb theory applied to accretionary and nonaccretionary wedges: Possible causes for tectonic erosion and/or frontal accretion.* Journal of Geophysical Research: Solid Earth (1978–2012), *99*(B6):12033–12055.

LAPUSTA, N., RICE, J. R., BEN-ZION, Y., and ZHENG, G. (2000). *Elastodynamic analysis for slow tectonic loading with spontaneous rupture episodes on faults with rate-and state-dependent friction.* Journal of Geophysical Research: Solid Earth (1978–2012), *105*(B10):23765–23789.

LE PICHON, X., MAZZOTTI, S., HENRY, P., and HASHIMOTO, M. (1998). *Deformation of the japanese islands and seismic coupling: an interpretation based on gsi permanent gps observations.* Geophysical Journal International, *134*(2):501–514.

LEHNER, F. (1986). *Comments on "noncohesive critical coulomb wedges: An exact solution" by fa dahlen.* Journal of Geophysical Research: Solid Earth (1978–2012), *91*(B1):793–796.

LIU, Y. and RICE, J. R. (2007). *Spontaneous and triggered aseismic deformation transients in a subduction fault model.* Journal of Geophysical Research: Solid Earth (1978–2012), *112*(B9).

LÓPEZ-QUIROZ, P., DOIN, M.-P., TUPIN, F., BRIOLE, P., and NICOLAS, J.-M. (2009). *Time series analysis of mexico city subsidence constrained by radar interferometry.* Journal of Applied Geophysics, *69*(1):1–15.

LOVELESS, J. P. and MEADE, B. J. (2011). *Spatial correlation of interseismic coupling and coseismic rupture extent of the 2011 mw= 9.0 tohoku-oki earthquake.* Geophysical Research Letters, *38*(17):L17306.

LOWRY, A. R., LARSON, K. M., KOSTOGLODOV, V., and BILHAM, R. (2001). *Transient fault slip in guerrero, southern mexico.* Geophysical Research Letters, *28*(19):3753–3756.

MAILLOT, B. and LEROY, Y. M. (2006). *Kink-fold onset and development based on the maximum strength theorem.* Journal of the Mechanics and Physics of Solids, *54*(10):2030–2059.

MARQUEZ-AZUA, B. and DEMETS, C. (2009). *Deformation of mexico from continuous gps from 1993 to 2008.* Geochemistry, Geophysics, Geosystems, *10*(2).

MAZZOTTI, S., LE PICHON, X., HENRY, P., and MIYAZAKI, S.-I. (2000). *Full interseismic locking of the nankai and japan-west kurile subduction zones: An analysis of uniform elastic strain accumulation in japan constrained by permanent gps.* Journal of Geophysical Research: Solid Earth (1978–2012), *105*(B6):13159–13177.

MELNICK, D., BOOKHAGEN, B., STRECKER, M. R., and ECHTLER, H. P. (2009). *Segmentation of megathrust rupture zones from fore-arc deformation patterns over hundreds to millions of years, arauco peninsula, chile.* Journal of Geophysical Research: Solid Earth (1978–2012), *114*(B1).

METOIS, M., SOCQUET, A., and VIGNY, C. (2012). *Interseismic coupling, segmentation and mechanical behavior of the central chile subduction zone*. Journal of Geophysical Research: Solid Earth (1978–2012), *117*(B3).

NOCQUET, J., VILLEGAS-LANZA, J., CHLIEH, M., MOTHES, P., ROLANDONE, F., JARRIN, P., CISNEROS, D., ALVARADO, A., AUDIN, L., BONDOUX, F., et al. (2014). *Motion of continental slivers and creeping subduction in the northern andes*. Nature Geoscience, *7*(4):287–291.

OPTUMGEO (2013). Optum computational engineering. *http://www.optumce.com*.

ORTIZ, M., SINGH, S., KOSTOGLODOV, V., and PACHECO, J. (2000). *Source areas of the acapulco-san marcos, mexico earthquakes of 1962 (m 7.1; 7.0) and 1957 (m 7.7), as constrained by tsunami and uplift records*. GEOFISICA INTERNACIONAL-MEXICO-, *39*(4):337–348.

PÉREZ-CAMPOS, X., KIM, Y., HUSKER, A., DAVIS, P. M., CLAYTON, R. W., IGLESIAS, A., PACHECO, J. F., SINGH, S. K., MANEA, V. C., and GURNIS, M. (2008). *Horizontal subduction and truncation of the cocos plate beneath central mexico*. Geophysical Research Letters, *35*(18).

RADIGUET, M., COTTON, F., VERGNOLLE, M., CAMPILLO, M., VALETTE, B., KOSTOGLODOV, V., and COTTE, N. (2011). *Spatial and temporal evolution of a long term slow slip event: the 2006 guerrero slow slip event*. Geophysical Journal International, *184*(2):816–828.

RADIGUET, M., COTTON, F., VERGNOLLE, M., CAMPILLO, M., WALPERSDORF, A., COTTE, N., and KOSTOGLODOV, V. (2012). *Slow slip events and strain accumulation in the guerrero gap, mexico*. Journal of Geophysical Research: Solid Earth (1978–2012), *117*(B4).

RICE, J. R. (2006). *Heating and weakening of faults during earthquake slip*. Journal of Geophysical Research: Solid Earth (1978–2012), *111*(B5).

SALENÇON, J. (2002). *De l'élasto-Plasticité au Calcul à la Rupture*. Editions Ecole Polytechnique.

SAVAGE, J. (1983). *A dislocation model of strain accumulation and release at a subduction zone*. Journal of Geophysical Research: Solid Earth (1978–2012), *88*(B6):4984–4996.

SINGH, S. C., HANANTO, N., MUKTI, M., ROBINSON, D. P., DAS, S., CHAUHAN, A., CARTON, H., GRATACOS, B., MIDNET, S.,

DJAJADIHARDJA, Y., et al. (2011). *Aseismic zone and earthquake segmentation associated with a deep subducted seamount in sumatra*. Nature Geoscience, *4*(5):308–311.

SONG, T.-R. A., HELMBERGER, D. V., BRUDZINSKI, M. R., CLAYTON, R. W., DAVIS, P., PÉREZ-CAMPOS, X., and SINGH, S. K. (2009). *Subducting slab ultra-slow velocity layer coincident with silent earthquakes in southern mexico*. Science, *324*(5926):502–506.

SONG, T.-R. A. and SIMONS, M. (2003). *Large trench-parallel gravity variations predict seismogenic behavior in subduction zones*. Science, *301*(5633):630–633.

STEWART, G. S., CHAEL, E. P., and MCNALLY, K. C. (1981). *The november 29, 1978, oaxaca, mexico, earthquake: A large simple event*. Journal of Geophysical Research: Solid Earth (1978–2012), *86*(B6):5053–5060.

SUÁREZ, G. and SÁNCHEZ, O. (1996). *Shallow depth of seismogenic coupling in southern mexico: Implications for the maximum size of earthquakes in the subduction zone*. Physics of the earth and planetary interiors, *93*(1):53–61.

TARANTOLA, A. (2005). *Inverse problem theory and methods for model parameter estimation*. Society for Industrial & Applied.

UJIIE, K., TANAKA, H., SAITO, T., TSUTSUMI, A., MORI, J. J., KAMEDA, J., BRODSKY, E. E., CHESTER, F. M., EGUCHI, N., TOCZKO, S., et al. (2013). *Low coseismic shear stress on the tohoku-oki megathrust determined from laboratory experiments*. Science, *342*(6163):1211–1214.

VERGNOLLE, M., WALPERSDORF, A., KOSTOGLODOV, V., TREGONING, P., SANTIAGO, J., COTTE, N., and FRANCO, S. (2010). *Slow slip events in mexico revised from the processing of 11 year gps observations*. Journal of Geophysical Research: Solid Earth (1978–2012), *115*(B8).

WALPERSDORF, A., COTTE, N., KOSTOGLODOV, V., VERGNOLLE, M., RADIGUET, M., SANTIAGO, J. A., and CAMPILLO, M. (2011). *Two successive slow slip events evidenced in 2009–2010 by a dense gps network in guerrero, mexico*. Geophysical Research Letters, *38*(15).

WELLS, R. E., BLAKELY, R. J., SUGIYAMA, Y., SCHOLL, D. W., and DINTERMAN, P. A. (2003). *Basin-centered asperities in great subduction zone earthquakes: A link between slip, subsidence, and subduction erosion?* Journal of Geophysical Research, *108*(B10):2507.

(Received June 30, 2015, revised November 18, 2015, accepted November 22, 2015, Published online December 29, 2015)

Pure Appl. Geophys. 173 (2016), 3487–3511
© 2015 Springer Basel
DOI 10.1007/s00024-015-1183-x

Pure and Applied Geophysics

P-Wave Velocity Tomography from Local Earthquakes in Western Mexico

Juan A. Ochoa-Chávez,[1] Christian R. Escudero,[1] Francisco J. Núñez-Cornú,[1] and William L. Bandy[2]

Abstract—In western Mexico, the subduction of the Rivera and Cocos plates beneath the North America plate has deformed and fragmented the overriding plate, forming several structural rifts and crustal blocks. To obtain a reliable subsurface image of the continental crust and uppermost mantle in this complex area, we used P-wave arrivals of local earthquakes along with the Fast Marching Method tomography technique. We followed an inversion scheme consisting of (1) the use of a high-quality earthquake catalog and corrected phase picks, (2) the selection of earthquakes using a maximum location error threshold, (3) the estimation of an improved 1-D reference velocity model, and (4) the use of checkerboard testing to determine the optimum configuration of the velocity nodes and inversion parameters. Surprisingly, the tomography results show a very simple δVp distribution that can be described as being controlled by geologic structures formed during two stages of the separation of the Rivera and Cocos plates. The earlier period represents the initial stages of the separation of the Rivera and Cocos plates beneath western Mexico; the later period represents the more advanced stage of rifting where the Rivera and Cocos plates had separated sufficiently to allow melt to accumulate below the Colima Volcanic complex. During the earlier period (14 or 10–1.6 Ma), NE–SW-oriented structures/lineaments (such as the Southern Colima Rift) were formed as the two plates separated. During the second period (1.6 Ma to the present), the deformation is attributed to magma, generated within and above the tear zone between the Rivera and Cocos plates, rising beneath the region of the Colima Volcanic Complex. The rising magma fractured the overlying crust, forming a classic triple-rift junction geometry. This triple-rift system is confined to the mid- to lower crust perhaps indicating that this rifting process is still in an early stage. This fracturing, along with fluid circulation and associated heat advection within the fractures, can easily explain the observed distribution of δVp, as well as many of the results of previous seismological studies. Also surprisingly, we find no evidence at deep crustal depths to support either a trenchward migration of the volcanic arc or toroidal asthenospheric flow through the slab tears bounding the Jalisco Block to the NW and SE.

Key words: Local tomography, subduction.

¹ Centro de Sismología y Volcanología de Occidente (SisVOc), Universidad de Guadalajara-Centro Universitario de la Costa, Avenida Universidad 203, Puerto Vallarta, Jalisco 48280, Mexico. E-mail: escudero.sisvoc@gmail.com

² Dept. de Geomagnetismo y Exploración, Instituto de Geofísica, Universidad Nacional Autónoma de México, Mexico, D.F. CP 04510, Mexico.

1. Introduction

Since Aki and Lee (1976) introduced the seismic tomography technique to determine three-dimensional velocity anomalies, it has become an important tool for imaging subsurface seismic structure using different kinds of datasets. Such analyses include the use of local earthquakes (Foulger *et al.* 2003; Graeber and Ash 1999; Haslinger *et al.* 1999; Kissling *et al.* 1994; Kissling 1998; Koulakov *et al.* 2006, 2007, 2009, 2010; Paul *et al.* 2001; Sallarès *et al.* 2000; Steck *et al.* 1998; Tryggvason *et al.* 2002; Zhao *et al.* 1992), teleseismic datasets (Alinaghi *et al.* 2007; Koulakov and Sobolev 2006; Koulakov *et al.* 2011; Rawlinson and Fishwick 2011; Rawlinson *et al.* 2011; Rawlinson and Kennett 2008; Rawlinson and Urvoy 2006; Thurber 2003; Yang *et al.* 2009), refraction and reflection data (Rawlinson *et al.* 2001a; Zelt *et al.* 2006; Zelt 1999), and simultaneous inversion of active and passive sources (Rawlinson and Urvoy 2006). These tomography techniques are applied at different scales (local, regional, and global) and in different tectonic environments (volcanic zones, subduction zones, etc.). In this study, we invert P-waves travel times of local earthquakes recorded by a temporary seismic array for imaging velocity anomalies related to the subsurface seismic structure. The crossing path coverage of the earthquakes P-wave within our dataset is mostly crustal, with some penetration of the uppermost mantle due to the earthquakes produced within the subducting slabs.

Previous studies in western Mexico are abundant, they include geological (Allan 1986; Bandy *et al.* 1998, 2001; Ferrari and Rosas-Elguera 1999; Frey *et al.* 2007; Luhr *et al.* 1985; Rosas-Elguera *et al.* 1996), magnetic (Bourgois and Michaud 1991), gravity (Bandy *et al.* 1999, 1995; Serrato-Díaz *et al.*

2004), oceanographic sea-floor (BANDY *et al.* 2008, 2011), palaeomagnetism (BANDY *et al.* 2000; DEMETS and TRAYLEN 2000; DEMETS and STEIN 1990; DEMETS and WILSON 1997), geodetic (MARQUEZ-AZUA and DEMETS 2009; SELVANS *et al.* 2011), among many others. However, seismic studies to determine tectonic structures have been limited to specific areas with few short period stations (ANDREWS *et al.* 2011; NAVA *et al.* 1999; NUÑEZ-CORNU *et al.* 2002; RUTZ-LÓPEZ and NUÑEZ-CORNU 2004), seismic reflection experiments (BANDY *et al.* 2005; BARTOLOMÉ *et al.* 2011; MICHAUD *et al.* 2000), and studies involving regional networks with poorly constrained locations of small earthquakes (PARDO and SUÁREZ 1995; SANCHEZ and NUNEZ-CORNU 2009).

Detailed seismic studies were made possible with the deployment of the Mapping the Rivera Subduction zone (MARS) array in January 2006. These studies include finite differences algorithm to study the seismic structure (DOUGHERTY *et al.* 2012), anisotropy analysis to study mantle flow (LEÓN SOTO *et al.* 2009), receiver function analysis to study crust and subduction structure (SUHARDJA *et al.* 2015), finite-frequency teleseismic tomography to study the seismic structure beneath the Rivera subduction zone (YANG *et al.* 2009), and seismic analysis to study the subduction geometry (GUTIERREZ *et al.* 2015). However, there are no previous seismic tomography studies in this area at the scale of our study.

Interpretation of other analysis common in local earthquake tomography, i.e., S-wave tomography, *Vp/Vs*, and source location, will be subject of future papers, as we only wish to analyze the p-wave velocity anomalies. However, the p-wave velocity distribution recovered helped unravel the tectonic configuration in the JB and MB region.

2. *Tectonic Setting*

The fragmented Rivera microplate and the Cocos plate are subducting beneath the North America plate producing substantial deformation and fragmentation of the overriding continental plate, forming segmented structural systems and crustal blocks (BOURGOIS and MICHAUD 1991; STOCK and LEE 1994). Within the study area, the Tepic-Zacoalco (TZR), the

Colima Rift (CR), and the Chapala-Tula (CTR) rifts form a rift–rift–rift triple junction that delimits the tectonic units known as the Jalisco and the Michoacán Block (Fig. 1), where the angles between the rifts are, going clockwise from the Colima rift, 100°, 115°, and 145° (LUHR *et al.* 1985). The Jalisco Block and the Michoacán Block have a complex interaction within the triple junction system (ALLAN 1986; BOURGOIS and MICHAUD 1991; FERRARI and ROSAS-ELGUERA 1999; FREY *et al.* 2007; LUHR *et al.* 1985; ROSAS-ELGUERA *et al.* 1996, 1997, 2003).

The Mesoamerican Trench (MAT) is the southwest border of the Jalisco and Michoacan Blocks (BANDY *et al.* 1995; BOURGOIS *et al.* 1988). The shallow structure of the subducting Rivera and Cocos plates was recently constrained by Gutierrez *et al.* (submitted to

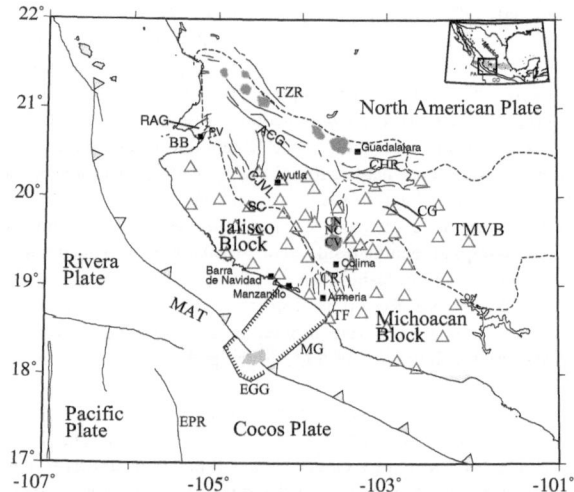

Figure 1

Main tectonic units within the studied area: The Pacific plate (*PA*), North American plate (*NA*), Rivera plate (*RI*), Cocos plate (*CO*), the Jalisco Block (*JB*), the Michoacán Block (*MB*), the Middle America trench (*MAT*), the Tepic-Zacoalco Rift (*TZR*), the Chapala Rift (*CHR*), the Colima Rift (*CR*), East Pacific Rift (*EPR*), El Gordo Graben (*EGG*), and the Manzanillo Graben (*MG*). The Trans-Mexican Volcanic Belt (*TMVB*) comprises the area within the dashed line. Major volcanoes include the Colima Volcano (*CV*), Nevado de Colima (*NC*), and the Cantaro Volcano (*CN*), denoted as areas of dark gray. Other features are Bahía de Banderas (*BB*), Chapala Lake (*CHL*), Tamazula Fault (*TF*), Rio Ameca Graben (*RAG*), Amatlan de Cañas (*ACG*), Sierra Cacoma (*SC*), Central Jalisco Volcanic Lineament (*CJVL*), and the Cotija Graben (*CG*). Major cities are Puerto Vallarta (*PV*), Colima (*COL*), Guadalajara (*GDL*), Manzanillo (*MAN*), and Barra de Navidad (*BN*), denoted in black squares. Stations deployed during of the Mapping the Rivera Subduction Zone (*MARS*) experiment, triangles. Inset shows regional location of the main map. [Background map is modified from DEMETS and TRAYLEN (2000) and FERRARI et al. (2000)]

BSSA). The important results of their study include (1) the Rivera and Cocos plates are tilted toward the CR, (2) the Rivera plate has approximately 4° of subduction angle near the trench then dives into the mantle with an angle of approximately 37°, and (3) the Cocos plate has oblique geometry with subduction angle of 18° at the southern end to 30° at the northern end. Below our area of resolution, structures deeper than 100 km were imaged by teleseismic tomography using data from the MARS array (YANG *et al.* 2009). Beneath the CR, they observe a gap between the Rivera and Cocos slabs starting at about 150-km depth that increases in size with increasing depth. The Rivera slab reaches ~100-km depth beneath the Colima volcanoes and is subducting more steeply than the adjacent Cocos Plate (YANG *et al.* 2009). Beneath the TZR and CTR, YANG *et al.* (2009) determined that the seismic velocity anomalies associated with both subducting plates are deeper than 250 km.

3. Data and Method

The initial dataset considered in this study is composed by first arrivals of P-waves from 2100 local earthquakes that occurred in western Mexico between 2006 and 2007 (Fig. 2). Stations deployed during the MARS experiment recorded these events. The temporary seismic network consisted of 50 stations equipped with Streckeisen STS-2 and Quanterra Q330 100 samples per second instruments (Fig. 1). Gutierrez *et al.* (2015) developed the catalog in two stages, i.e., automatic location and manual corrections. To identify and locate earthquakes within the seismic records, they implemented the Short-Time-Average/Long-Time-Average trigger (STA/LTA) algorithm, a spatial grid search and event locator using the iasp91 velocity model. However, due to important differences between the automatic detections and phase picks with the actual data. They performed manual corrections of phase picks, eliminated non-earthquake detections, and detected phases previously not detected by the automatic algorithms. The final earthquake locations were estimated using HYPO71 and a local velocity model. Dañobeitia *et al.* (1997) using data from the Crustal Offshore Research Transect by Extensive Seismic Profiling (CORTES-P96) experiment developed this a priori local velocity model. The equipment deployed during the CORTES-P96 experiment was multichannel

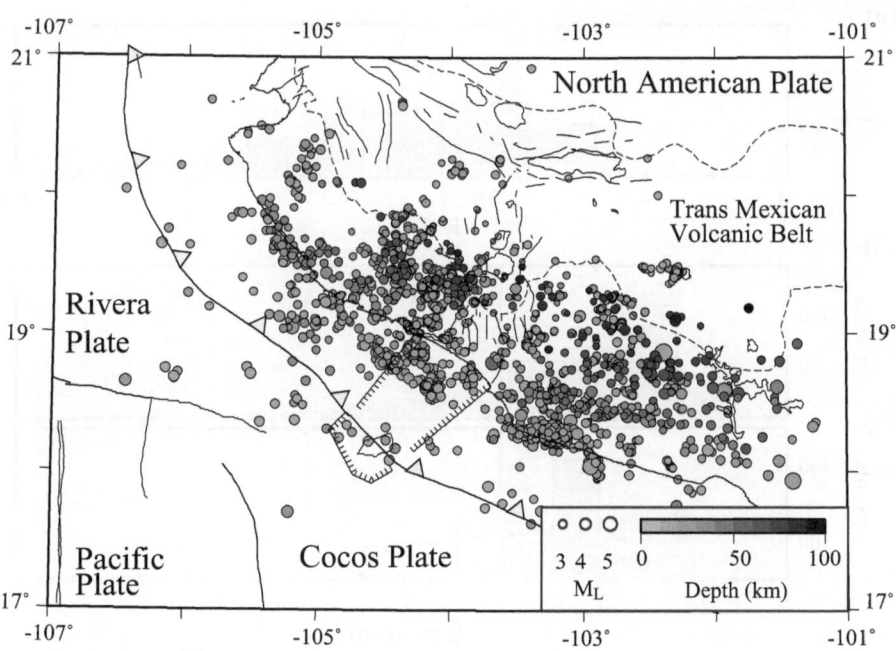

Figure 2
Local seismicity from GUTIERREZ *et al.* (in press). Map shows hypocenters of 2100 earthquakes recorded from 2006 to 2007 by MARS temporary network, depths shown in *gray tones* and magnitudes in *circle* of relative sizes

seismic profiling, deployment of 35 ocean bottom seismometers, active sources, 50 temporary stations and more than 3000 km of gravity and magnetic potential field measurements. DAÑOBEITIA *et al.* (1997) estimated the velocity model using wide-angle seismic reflection and the analysis of direct and refracted P-phases.

The local magnitude range of the located earthquakes in this catalog is within the 2.0 to 5.3 M_L, the average location error is 0.8 km horizontally and 1.0 km vertically, while the origin time has a 0.2 s mean error (GUTIERREZ *et al.* 2015).

The selection of an adequate dataset for the tomographic inversion guarantees the convergence and stability in the inversion of hypocentral coordinates (SALLARÈS *et al.* 2000). Therefore, we impose several restrictions to the dataset used in the tomographic inversion; specifically, we required a minimum of 10 readings of P-phases for each event; a maximum azimuthal gap of 180°; origin time root mean square (rms) error lower than 2.5 s; a horizontal localization error threshold of 1.5 km, both horizontally and vertically (Fig. 3); and the event must have

been recorded in at least four stations. Additionally, we only use phase picks where the distance between the station and the earthquake location is less than the crossover distance to avoid selecting first arrivals of other P-wave phases. The final tomographic inversion was performed using approximately 1000 events and 9630 P-phase readings.

3.1. 1-D reference Model

KISSLING *et al.* (1994) shows the importance of a satisfactory reference velocity model prior to the 3-D tomographic inversion. Such a reference model is necessary to deal with the non-uniqueness of the solution of the linearized inverse problem during the 3-D tomographic inversion and to avoid inaccurate and biased results (KISSLING *et al.* 1994; SALLARÈS *et al.* 2000; THURBER *et al.* 2003). The reference model was determined by analyzing the parameters of several 1-D models with a varying number of layers, thicknesses, and initial velocities. SALLARÈS *et al.* (2000) propose that the best reference model must minimize the averaged residual time, must

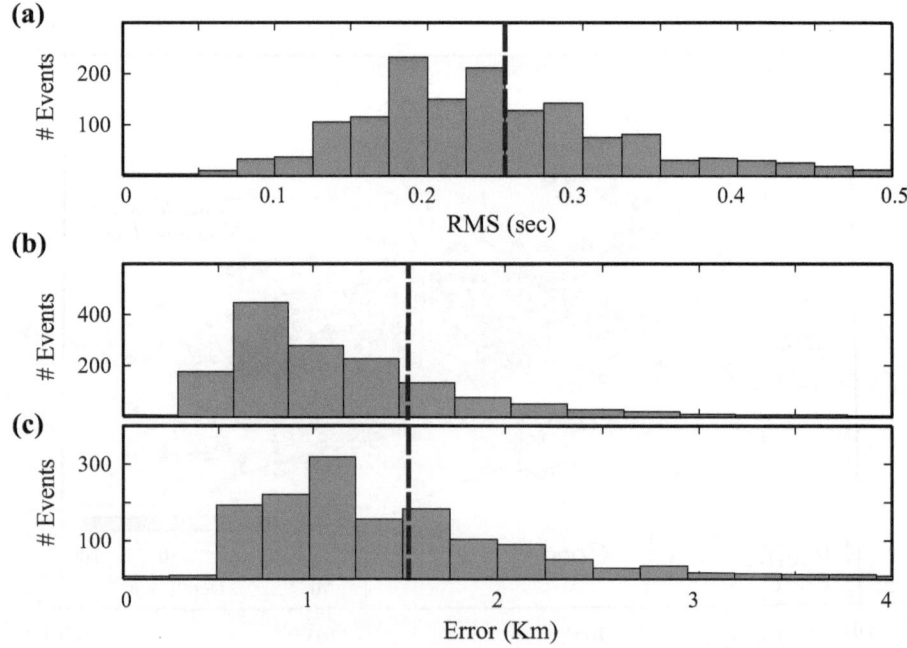

Figure 3
Earthquake hypocenters error. **a** Origin time error as root mean square, **b** horizontal location error, and **c** depth error. *Dashed lines* indicated maximun error threshold for earthquakes used in the tomographic inversion. Figure modified from GUTIERREZ *et al.* (in press)

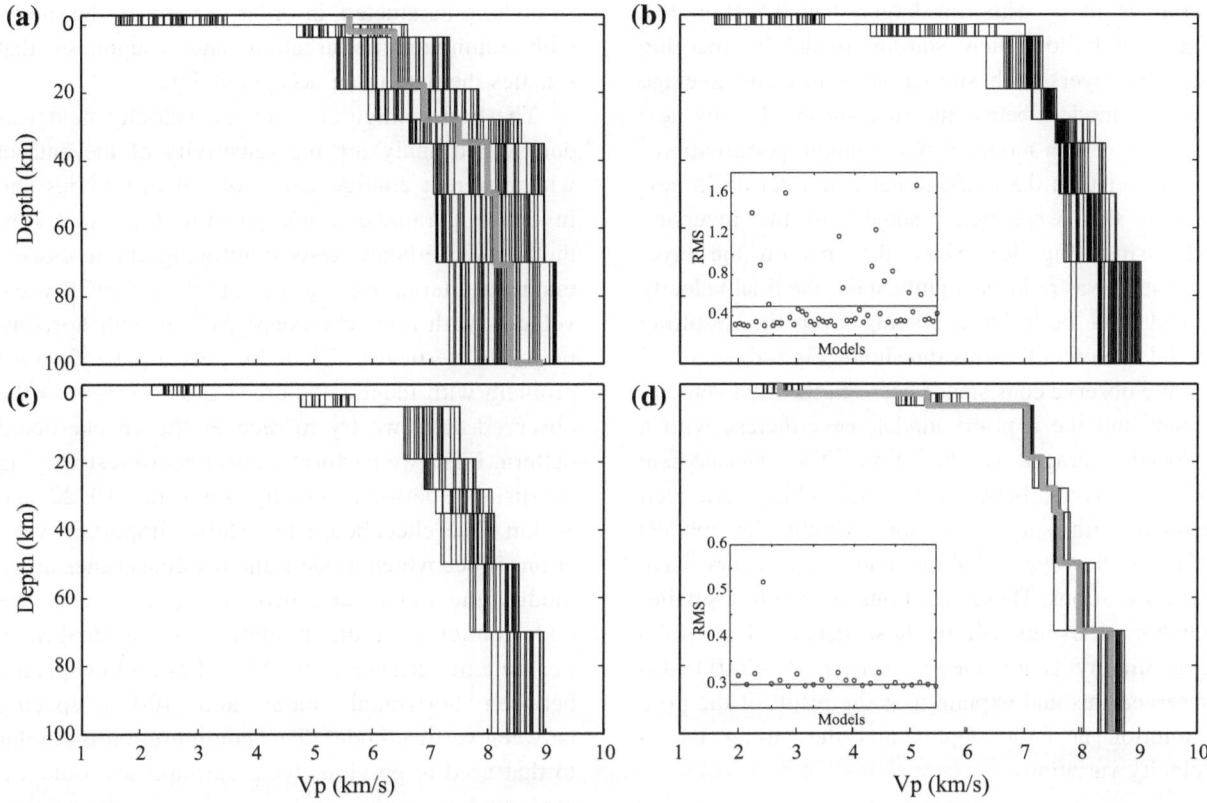

Figure 4
Estimation of the 1-D velocity model. **a** Updated a priori model and 50 velocity models with randomly introducing perturbations, >1 km/s. **b** 1-D velocity models after first successive inversions and 0.5-RMS threshold (inset). **c** New 1-D model and 25 new velocity models with randomly introducing perturbations, >0.5 km/s. **d** 1-D velocity models after second successive inversions and 0.3-RMS threshold (inset). Best reference 1-D velocity model show in *think gray line*

weakly depend on the initial model, and must be consistent with a priori information.

The starting velocity model used in the 3D tomographic inversion was computed using a modified procedure of KISSLING *et al.* (1994) proposed by PAUL *et al.* (2001). This procedure guarantees the independence of the final 1-D model with respect to the updated a priori model by probing a large part of the solution space (PAUL *et al.* 2001). We implemented 1-D inversions by using the VELEST algorithm (KISSLING *et al.* 1994), which assumes constant velocities within layers. First, we consider previous structural information (RUTZ-LOPEZ and NUÑEZ-CORNU 2004) to be used as an a priori velocity model. Such a model was determined from seismic refractions, using seismic records from seven temporary stations deployed from 1996 to 1998. Second, we

update the a priori model by conducting a series of successive 1-D inversions and earthquake relocations that produce a homogeneous dataset that eliminates errors and inconsistencies between event clusters of dissimilar origin. Using a trial and error procedure, we randomly include events until we minimize the error and maximize the number of events used in the inversions. The updated a priori model consisted of new velocities layers, station corrections, event hypocenters, and smaller variance than the initial a priori model. Then, we generated 50 velocity models by introducing random perturbations, no greater than 1 km/s, to the velocity layers of the updated a priori model (Fig. 4a). Along with station corrections and hypocenter parameters, the new perturbed velocity models were used as the initial model in successive inversions. From the 50 velocity models, we eliminate those with rms larger than 0.5 (Fig. 4b). Then we build a new starting model by merging adjacent layers with similar velocities and average over all models below the rms threshold. This new model was then modified with random perturbations, no greater than 0.5 km/s, to get a new set of 25 new models that were again subject to the inversion processing (Fig. 4c). Since the rms on the layer velocities was reduced significantly, the final velocity model was built by averaging over the resulting models below a 0.3-rms threshold (Fig. 4d).

We observe consistence between the final velocity model and the a priori model, nevertheless with a reduced averaged residual time. The intermediate velocity layers, between 3.6 and 50 km, are well defined with small variations within the models (Fig. 4); however, shallow and deep layers have large variations. These variations are due to a smaller number of events within those depths. The Moho discontinuity is not clear; Paul et al. (2001) also observed this and explain it as the result of the poor resolution at large depths and the strong lateral velocity variations.

3.2. 3-D Inversion and Resolution Analysis

We apply the Fast Marching Method (FMM), described in detail in Rawlinson and Sambridge (2004a, 2005), and Rawlinson et al. (2006, 2010), which is a tomographic inversion scheme designed for body wave datasets. The FMM uses a grid-based eikonal solver for the forward problem of travel-time prediction that allows inversion for velocity, interface depth, and source location parameters due to its iterative non-linear inversion scheme. The FMM is combined with a rapid subspace inversion method, where in order to account for the non-linearity of the inverse problem, the forward and inverse steps are applied iteratively. With our 3D parameterization scheme, we required six iterations to stabilize the model variance. In this study, we focus on the velocity inversion. The method requires independent layers defined by a 3-D regular grid of velocity nodes and boundary layers defined by 2-D grids. The algorithm implements cubic B-splines to describe a smoothly varying velocity field within the nodes. The method allows the user to modify the damping and

smoothing parameters in order to identify the model with minimum perturbation and roughness that satisfies the data to an acceptable level.

The best configuration of the velocity field was determined analyzing the sensitivity of the dataset with different configurations of velocity nodes and inversion parameters. We perform the usual synthetic checkerboard tests employing input models that have alternating regions of high and low P-wave velocity with a checkerboard pattern both horizontally and vertically. Then by solving the forward problem with identical sources and receivers of the observed data, we try to recover the checkerboard pattern. First, we perform a checkboard test varying the distance between velocity nodes, i.e., 10, 20, and 30 km. The checkboard test shows important variations in recovered models and residuals times as we modify the model and inversion parameters. The node configuration that minimizes the residual time, i.e., data fit variance of 0.003 s², has 20-km spacing between horizontal nodes and 10-km spacing between vertical nodes. This configuration is similar to that used in previous local earthquakes tomography studies with similar station coverage and datasets (DeShon et al. 2006; Paul et al. 2001). Additionally, we performed a series of inversions with different values of the damping and smoothing parameters in order to identify the model with minimum perturbation and roughness that satisfies the dataset (Fig. 5). The values that minimize the data variance with respect to model variance and roughness are damping, $\eta = 1$, and smoothing, $\varepsilon = 15$. Figure 6 shows the histogram with the distribution of travel-time residuals for the final model (after the inversion). Over 95 % of the travel times have residuals smaller than 0.51 s and 70 % are smaller than 0.26 s. Most of the large residuals are for hypocentral distances greater than 60 km and they are not related to a specific station or event. There is a variance reduction of about 86 % between the final and initial model.

The well-resolved zones in the checkerboard tests are used as reference to qualitatively compare with the final tomographic results. This procedure allows discerning between well and poorly resolved nodes. Using the well-resolved checkboard test (Fig. 7), we conclude that (1) the intermediate layers, between 10

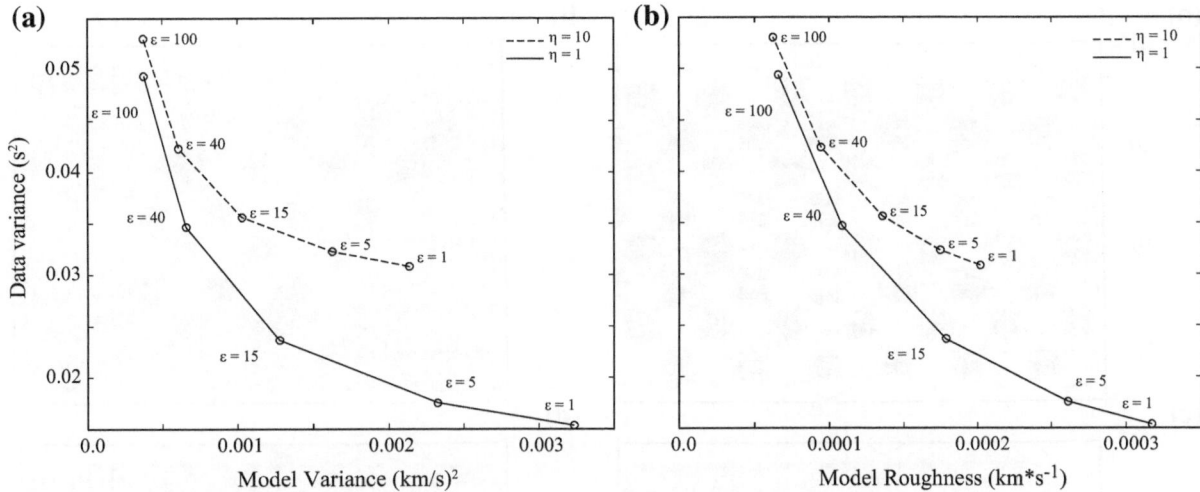

Figure 5
Regularization curves for trade-off analysis. Several tomographic inversions were made using different values of the damping, η, and smoothing, ε, parameters. The values that minimize the data variance with respect to model variance (**a**) and roughness (**b**) are damping, $\eta = 1$, and smoothing, $\varepsilon = 15$

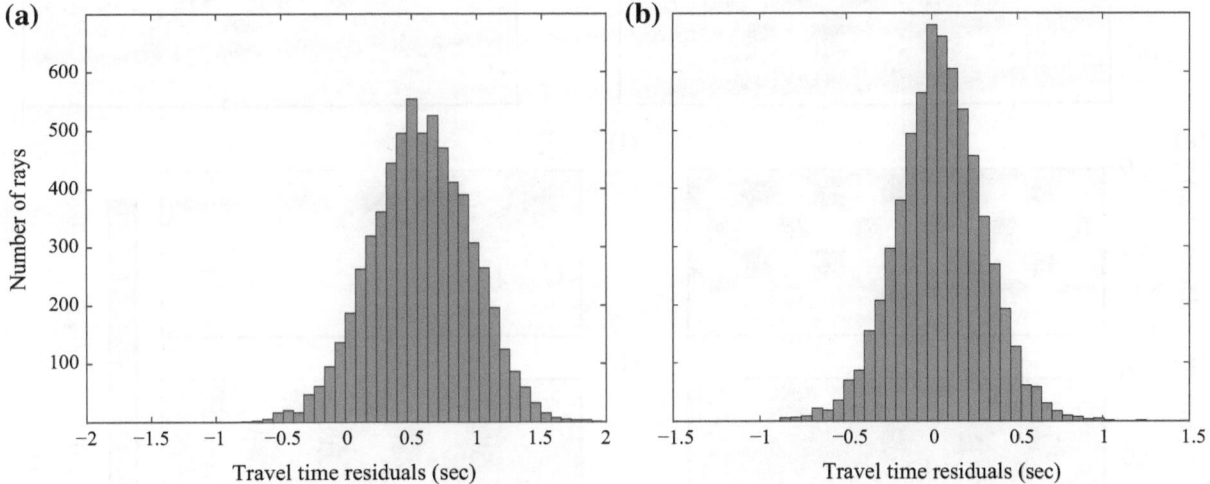

Figure 6
Histogram of travel-time residuals for the solution model; **a** initial distribution and **b** distribution after tomographic inversion

and 40 km, have the best approximation where the velocity anomalies are properly recovered, (2) the velocity anomalies are only recovered within the area covered by seismic stations, (3) smearing is observed at the edges of the area covered by the seismic network, and (4) the well-recovered area gets smaller with increasing depth due to localization of the earthquakes since most have less than 35-km depth.

4. Results

This section includes a summary of the main P-wave velocity anomalies from the map view and vertical cross sections oriented perpendicular and parallel to the trench. Discussion on the geological and tectonic interpretation of the anomalies will be presented in Sect. 5. Figure 8 shows the tomography

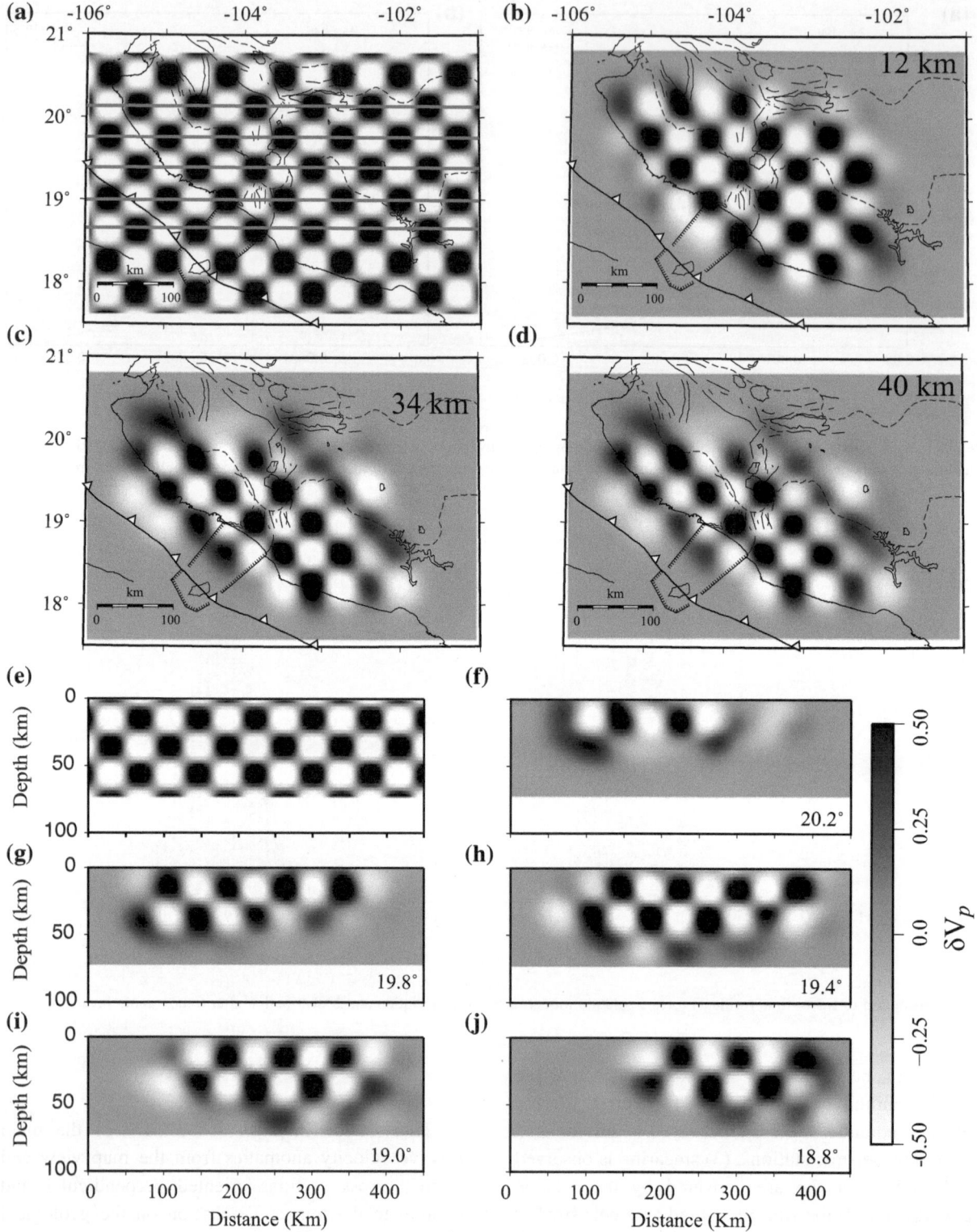

Figure 7
Results of checkerboard test with the better inversion parameters. Initial horizontal pattern (**a**) and initial cross-section pattern (**e**). Recover patterns after tomographic inversion at depth of 12, 34, and 40 km, (**b**, **c**), and (**d**), respectively. Recover patterns after tomographic inversion for cross sections as indicated in map are show in (**f**, **g**, **h**, **i**, and **j**). Velocity deviations are indicated in δVp (*scale bar*) with respect to the best reference model (Fig. 4d)

results in horizontal cross sections at depth of 5, 10, 15, 20, 25, 30, 35, and 40 km. The plots only show anomalies located within the well-resolved zones (Fig. 7) as indicated by the checkerboard tests. We also present the results in cross sections, five perpendicular to the trench and three approximately parallel to the trench (Fig. 9). In the perpendicular cross section, we projected earthquakes located within a volume of 80-km depth and 50 km wide centered at a line of 250-km length (Fig. 10). In the approximately parallel cross sections, we projected earthquakes located within a volume of 80-km depth and 30 km wide centered at a line of 500-km length (Fig. 11). The azimuth and length of the parallel to the trench cross sections are arbitrarily chosen to span most of the well-resolved area of the tomography. Additionally, to show a better view of the Colima Rift, we present cross sections perpendicular to the preferred rift strike (Fig. 12).

4.1. Horizontal Variations of P-Wave Velocity at Different Depths

The pattern of anomalies imaged at the upper layers contrast from deeper layers (Fig. 8). Low-velocity anomalies cover most of the studied area at shallow layers, i.e., 5 and 10 km, intermediate layers, i.e., 20–30 km, have a unique conspicuous low-velocity anomaly, and deeper layers, i.e., 35–40 km, have high-velocity anomalies.

Horizontal cross sections at 5- (Fig. 8a) and 10-km (Fig. 8b) depths show similar velocity anomalies pattern; large regions of low-velocity anomalies characterize the velocity field within these layers. However, the 5-km layer has more isolated and easy to identify anomalies, where the most northern anomaly at 5-km layer correlates to the south end of the Central Jalisco Volcanic Lineament. This anomaly is also imaged at the 10-km layer; however, at this depth it extends southeast along the southern

border of the Trans-Mexican Volcanic Belt. Other low-velocity anomalies are imaged at 5-km depth, i.e., a low-velocity anomaly with circular shape is located south of Lake Chapala at the Cotija Graben, and a strip of discontinuous low-velocity anomalies along the coast. However, these anomalies merge at 10-km depth to form a continuous strip along the coast. A low-velocity anomaly with circular shape is visible at 5 km beneath the south of the Colima Volcano Complex (CVC). This anomaly extends to 10-km depth to form a broad area of discontinuous anomalies difficult to correlate with a specific tectonic feature.

At 15- (Fig. 8c) and 20-km (Fig. 8d) depths, the discontinuous low-velocity anomalies that cover most of the studied area attenuates, while a conspicuous large-scale low-velocity anomaly becomes evident. Two low-velocity anomalies remain along the coast as deep as 30 km, one toward the northwest and the other at the most southeast area. A southeast-northwest elongated low-velocity anomaly beneath the Michoacán Block is found as deep as 25 km. However, the most important low-velocity anomaly that become visible within these intermediated layers, from the 15-km depth horizontal cross section and extends at depth down to the deepest layer at 40 km. This major low-velocity anomaly is centered at the Colima Volcano Complex (CVC) and has a clear Y-pattern. While at 15-km depth low-velocity areas surround this anomaly, at 20 km low-velocity areas become more isolated and easy to spot. The center of the anomaly is offset to the west of the CVC, with one limb trending south along the Colima Rift, while the other two limbs trend approximately southeast-northwest along the limit of the TMVB. Although the northwest trending limb follows the direction of the Central Jalisco Volcanic Lineament, it is located to the SE beneath the Sierra Cacoma. The southeast limb trends toward the Cotija Graben. The anomaly is located in a well-resolved area (Fig. 7); therefore, we discard that the shape is the result of smearing or other inversion effects. This anomaly is very interesting and its origin is not clear at first. The Y-shape certainly looks like a triple-rift system commonly associated with continental rifts. However, it is located south of the commonly accepted position of the triple-rift junction, and there is no surface

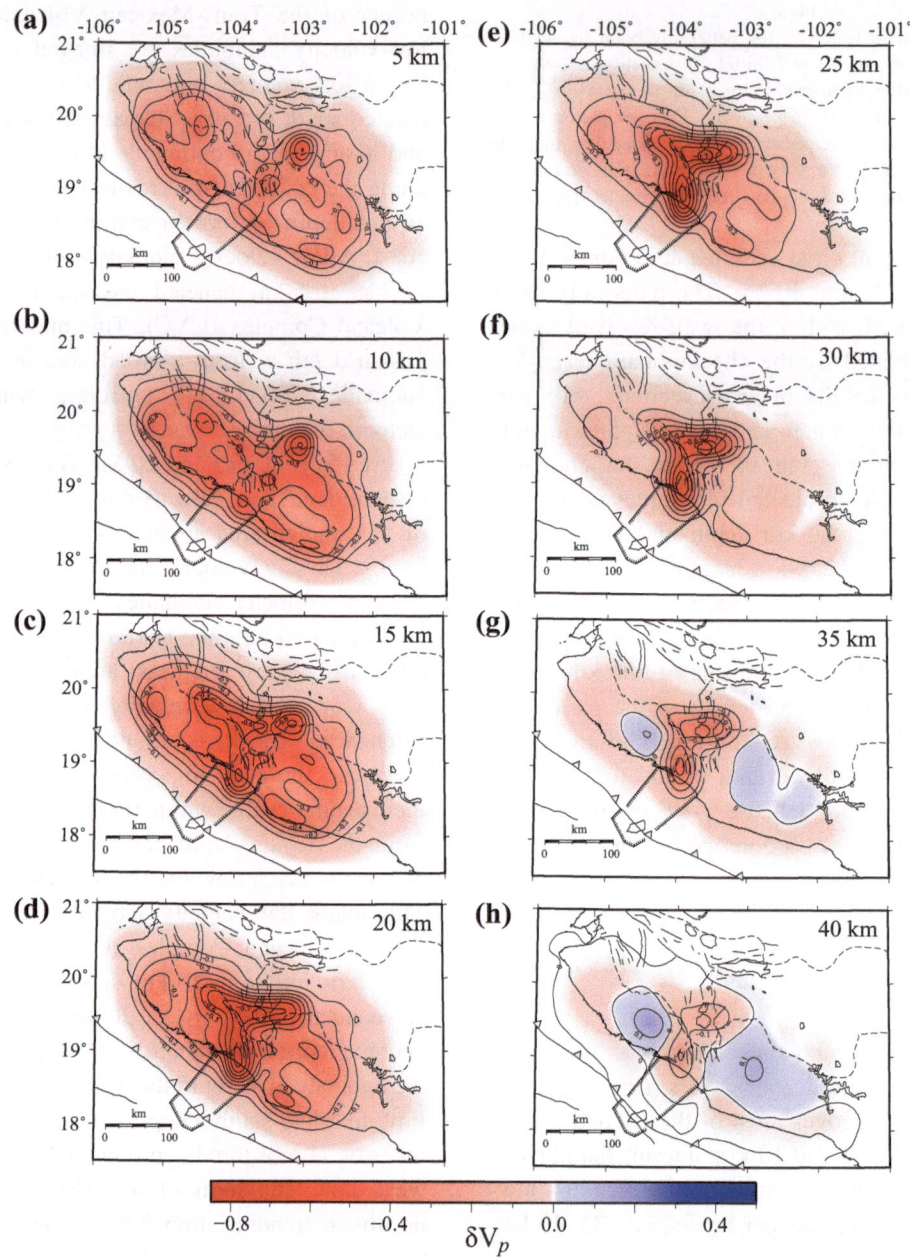

Figure 8
Final tomographic results for layers at 5-, 10-, 15-, 20-, 25-, 30-, 35- and 40-km depths. Velocity deviations are indicated in δVp (*scale bar*) with respect to the best reference model (Fig. 4d). Isodepth contours of the Rivera and Cocos plate subduction, and depth seismicity (**h**) (GUTIERREZ *et al.* in Press)

evidence that indicates that such tear is present in the crust.

At layers 25 (Fig. 8e) and 30 km (Fig. 8f), the dominant feature is the Y-pattern low-velocity anomaly, while the other small and isolated low-velocity anomalies almost disappear. At 30-km depth, the northwest limb of the anomaly becomes smaller. At this depth, the north section of the anomaly correlates with the CVC and the south section with the southern Colima Rift.

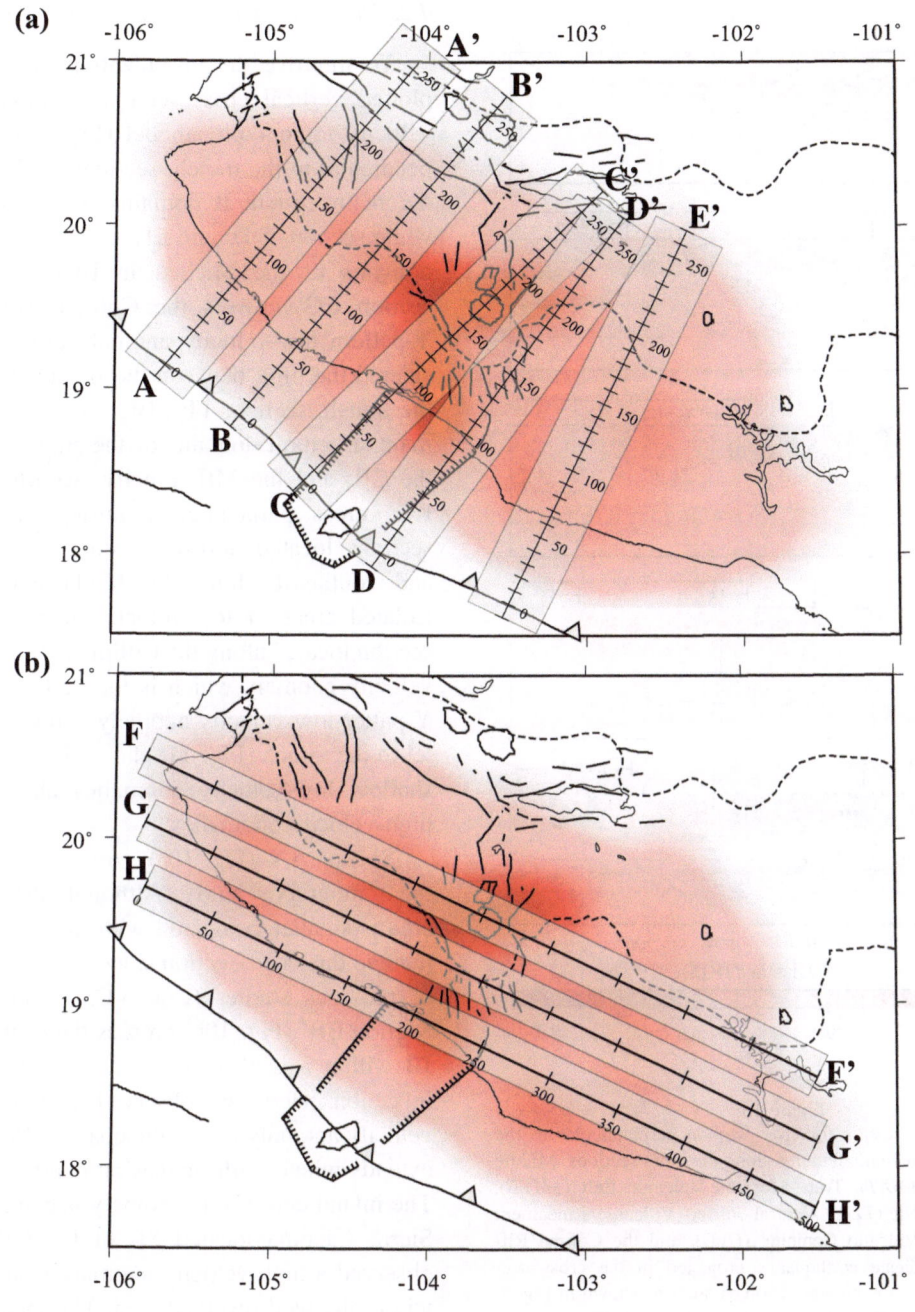

Figure 9

Map view of perpendicular to the trench cross sections (**a**) and approximately parallel to the trench (**b**) cross sections. *Tick marks* along the *lines* indicated distance in km, *rectangles* indicate the volume of seismicity projected in the cross sections showed in Figs. 10 and 11; the cross section are labeled with capital case. Figures also show the 25-km tomographic layer for reference

The main features at 35- (Fig. 8g) and 40-km (Fig. 8h) depths are two areas of high-velocity anomalies, one north at the JB and the other southeast at the MB. These two high-velocity anomalies are divided by the Y-pattern low-velocity anomaly. These high-velocity anomalies are located in a well-constrained region of the model, and therefore are not the result of artifacts of the inversion process. Within these depths, the northern limb of the Y-pattern low-velocity anomaly reduces its size to finally correlate

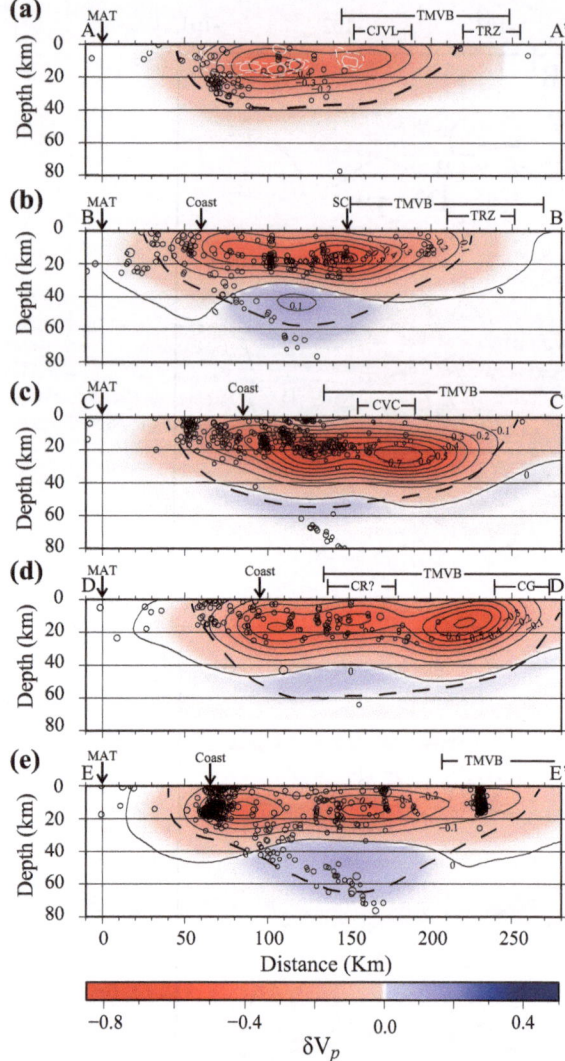

4.2. Vertical Variations of P-Wave Velocity

To improve the visualization of the results, we plotted vertical cross sections through the P-wave velocity tomography model (Figs. 10, 11, 12). Perpendicular to the trench, sections AA' and BB' cross the main structural domains of the Jalisco Block, while sections DD' and EE' cross the main structural domains of the Michoacán Block. Note that the section CC' crosses the Colima Rift through the Y-pattern low-velocity anomaly along an area considered the limit between the JB and MB. Parallel to the trench, sections FF', GG', and HH' cross all the main structural domains of the studies area, i.e., JB, the CR, and the MB. Finally, sections II', JJ', and KK' cross perpendicular through the CF. Cross sections located northwest along the Jalisco Block and southeast along the Michoacán Block have isolated areas of low-velocity anomalies, the cross section located along the Colima Rift has a large low-velocity anomaly which is the vertical image of the Y-patter low-velocity anomaly. Cross sections parallel to the trench (Fig. 11) show the interface between shallow low-velocity anomalies above and deeper high-velocity anomalies.

Section AA' (Fig. 10a) crosses the northern part of the tomography study. Although the well-resolved area is smaller compared with the other profiles, we include this cross section since it is located in an area of previous studies (CORBO-CAMARGO et al. 2013). Section BB' (Fig. 10b) crosses the central part of the JB within a well-resolved area of the tomographic inversion. Here we observed a conspicuous low-velocity anomaly below the coast at 15-km depth and extends inland with increasing depth, size, and δVp. The inland end of this anomaly is centered under the Sierra Cacoma located SE of the CJVL. We also observed a high-velocity anomaly from 38 to 60 km within the well-resolved area. The interface between the low velocity above and the high velocity below has a concave shape. Sections CC' and DD' (Fig. 10c, d) cross through the Colima Rift, where a prominent low-velocity anomaly in a well-resolved region at 40-km depth is observed. Section EE' (Fig. 10c) crosses the central part of the MB within a well-resolved area of the tomographic inversion. We observed two conspicuous low-velocity anomalies,

Figure 10

Final tomographic results for cross section perpendicular to the trench. Important tectonic features along the cross sections: Middle America Trench (*MAT*), Trans-Mexican Volcanic Belt (*TMVB*), Tepic-Zacoalco Rift (*TZR*), Central Jalisco Volcanic Lineament (*CJVL*), Colima Volcano Complex (*CVC*), and the Colima Rift (*CR*). *Circles* indicate earthquakes projected in the cross section. *Dashes lines* indicate well-resolved areas as shown in Fig. 8. *White dashed lines* indicate high-resistivity magnetotelluric anomalies (CORBO-CAMARGO et al. 2013). Velocity deviations are indicated in δVp (*scale bar*) with respect to the best reference model (Fig. 4d). Cross sections are labeled with capital case as shown in Fig. 9

exactly with the CVC at 35-km depth. On the other hand, the south trending limb of the low-velocity anomaly corresponds with the southern Colima Rift.

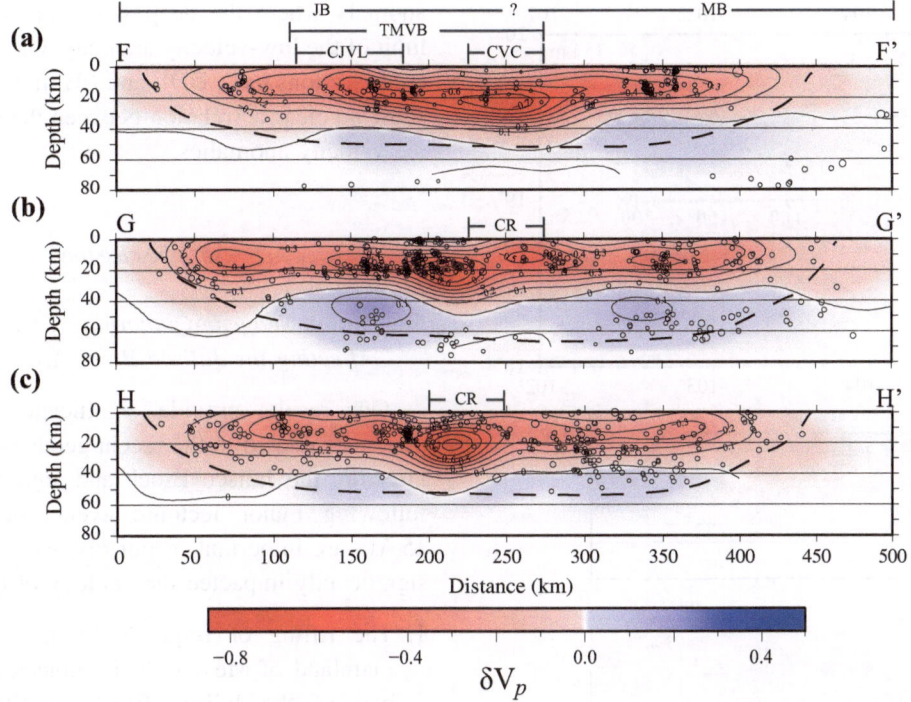

Figure 11

Final tomographic results for cross section approximately parallel to the trench. Important tectonic features along the cross sections: Jalisco Block (*JB*), Michoacan Block (*MB*), Trans-Mexican Volcanic Belt (*TMVB*), Central Jalisco Volcanic Lineament (*CJVL*), and the Colima Rift (*CR*). *Circles* indicate earthquakes projected in the cross section. *Dashes lines* indicate well-resolved areas as shown in Fig. 8. Velocity deviations are indicated in δVp (*scale bar*) with respect to the best reference model (Fig. 4d). Cross sections are labeled with capital case as shown in Fig. 9

one beneath the coast and the other 150-km inland. We also observed a high-velocity anomaly from 40 to 60 km within the well-resolved area; this interface between low and high anomalies has horizontal shape.

Sections FF' (Fig. 11a) crosses through the southern limit of the TMVB and the CVC. We observe an elongate low-velocity anomaly below the TMVB that thickens below the CVC. However, we do not observe a clear image of the magma chamber, at least not for the resolution of our tomographic study. We observed no correlation between the low-velocity anomalies and the seismicity. The cross section GG' (Fig. 11b) shows the lateral variations of the interface between the low-velocity (above) and the high-velocity (below) regions. Although we observed the interface along cross sections FF' and HH' (Fig. 11c), the GG' cross section has a wider and deeper well-constrained region of the tomographic results. We observe an average interface depth of 35 km beneath the Jalisco and Michoacán Blocks.

However, in the area beneath the Colima Rift, we observed that the interface has 45-km depth. This feature is observed northeast and southwest in the FF' and HH' cross sections, respectively; moreover, cross sections perpendicular to the trench, CC' and DD' suggest that this anomaly extends into the offshore region of the southern Colima Rift. Cross sections II', JJ', and KK' (Fig. 12) show a more detailed perspective of velocity anomalies beneath the CVC and the CR. These cross sections are horizontally oriented such that they cross perpendicular to the direction of the CR. Cross section II' crosses through the CVC along the northeast limb of the Y-pattern low-velocity anomaly. In cross-section view, this anomaly has a curved shape, it is deeper and thicker below the CVC, and its shallow part is at the inland end below the CG. Cross sections JJ' and KK' cross perpendicular through the Colima Rift. We observed that in this region the low-velocity anomaly has a depth of 40 km and is approximately 50 km wide.

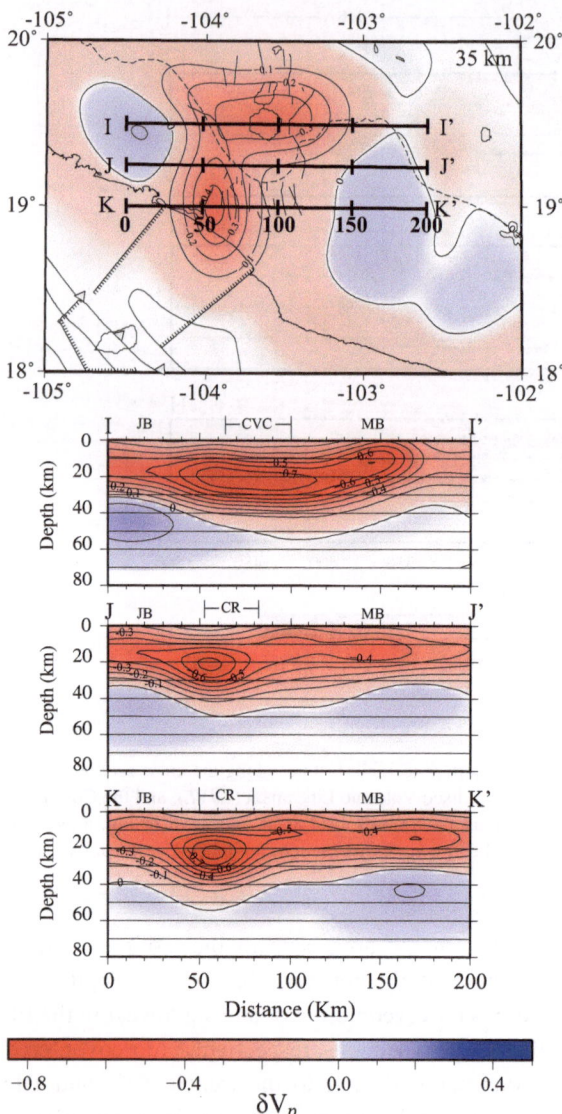

Figure 12

Final tomographic results for cross section perpendicular to the dominant strike of the Colima Rift. Important tectonic features along the cross sections: Jalisco Block (*JB*), Michoacán Block (*MB*), Trans-Mexican Volcanic Belt (*TMVB*), Central Jalisco Volcanic Lineament (*CJVL*), and the Colima Rift (*CR*). *Dashes lines* indicate well-resolved areas as shown in Fig. 8. Velocity deviations are indicated in δVp (*scale bar*) with respect to the best reference model (Fig. 4d)

No clear correlation between velocity anomalies and the seismicity within the studied area is observed. Nevertheless, we can indicate some important features. Cross section BB' at the central part of the JB has shallow seismicity, above 20–30 km. Here the crustal earthquakes are confined in the low-velocity

anomaly where the deepest earthquakes follow the limit of the low-velocity anomaly. On the other hand, cross sections CC', DD', and EE' in the CR and MB show no clear correlation between the seismicity and the velocity anomalies.

5. Discussion

5.1. Major Neogene-Quaternary Tectonic Events Affecting the Jalisco Block Area

Our results provide significant constraints on proposals related to the recent geodynamic processes affecting the Jalisco Block and adjacent areas. The following major tectonic events occurring since 15 Ma are important to our discussion as they have significantly impacted the geology of this area:

1. The rifting of Baja California away from the mainland of Mexico. This subjected the northern part of the Jalisco Block to NW–SE-directed extension which produced several grabens (e.g., Bahía de Banderas, Río Ameca Graben, and the Amatlan de Cañas Graben) as well as transtensional faulting and associated topographic depressions within the Tepic-Zacoalco Graben (e.g., JOHNSON and HARRISON 1989; BARRIER *et al.* 1990; ALLAN *et al.* 1991; GARDUÑO and TIBALDI 1991; MICHAUD *et al.* 1991, 1993; FERRARI 1995, 2004; FERRARI *et al.* 1994, 2000, 2001; FERRARI and ROSAS-ELGUERA 1999; GÓMEZ TRUÑA *et al.* 2007).

2. The tearing of the subducting oceanic plate beneath the Gulf of California (e.g., CALMUS *et al.* 2003, 2011) which was concurrent with the opening of the Gulf of California. This tearing appears to have progressed eastward beneath the Trans-Mexican Volcanic Belt where it produced an eastward progressing pulse of mafic magmatism between 11.5 and 6.5 Ma. The Los Altos de Jalisco lavas (Fig. 13a) may be related to this plate tearing (FERRARI 2004). Several studies (e.g., FERRARI *et al.* 2001; LEÓN SOTO *et al.* 2009) propose that mantle material may be flowing through this tear and into the adjacent mantle wedge beneath the Jalisco Block, producing the recent alkaline volcanism within the Jalisco Block.

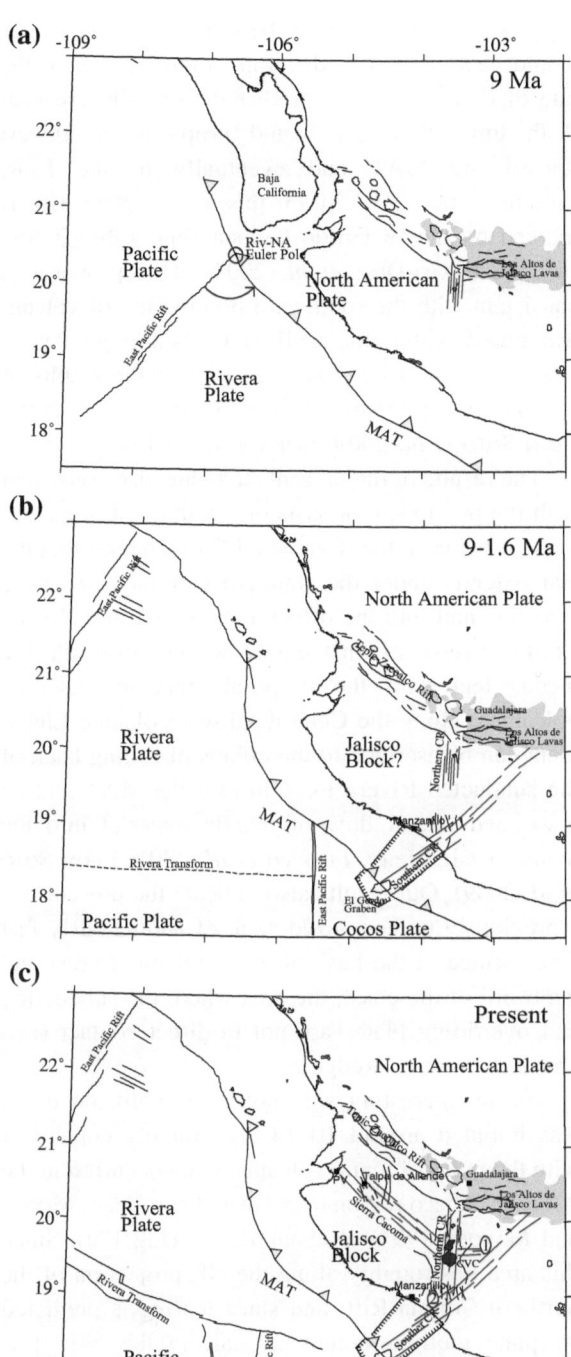

(a)

(b)

(c)

◄

Figure 13

Proposed chronology of the development of structural trends associated with the southward propagating rifting along the subducted Rivera-Cocos plate boundary and related ascent of magma generated by this rifting. **a** Structural trends and emplaced mafic magmas present at 9 Ma. [Background map is modified from DeMets and Traylen (2000) and Ferrari et al. (2000)]. Also shown is the approximate location of the collision of the East Pacific Rise (EPR) with the Middle America Trench (MAT) off the southern tip of what is now Baja California that occurred between 14 and 9 Ma. Note: we place the RIV-NA Euler pole at the ridge-trench intersection only for the sake of illustration. However, this may be close to reality as the present day Euler pole determinations, which are better constrained, indicate that the Rivera-NA pole is presently close to this intersection (e.g., Suárez et al. 2013). The curved arrow indicates the pivoting motion of the Rivera plate relative to the North American plate about this Euler pole. **b** Structural trends developing sometime between 9 and 1.6 Ma. These structures include (1) the system of NE–SW-oriented lineations located just east of Manzanillo (the present day southern Colima Rift) and east of Guadalajara, (2) the El Gordo Graben located within the oceanic lithosphere just west of the MAT, (3) a new segment of the EPR which collided with the MAT off Manzanillo at about 1.8 Ma, and (4) probably the newly developed Rivera Transform which connects the new segment of the EPR and the older segment which bounds the Rivera plate to the west. Note that we propose that the southern Colima Rift and the Northern Colima Rift were not connected during this time as they were formed by two separate geodynamic processes, both related to the separation of the Rivera and Cocos plates. **c** Present day structural trends. Magma emplacement in the Colima Volcanic Complex was initiated at 1.6 Ma and continues at present. Fracturing associated with the rising magma fractures the lower and mid-crust into a triple-rift/fractured system centered just west of the Volcán Colima. See text for an explanation of the *circled number 1* and 2

3. The separation of the Jalisco Block from the North American Plate. Although controversial (e.g., Bandy and Pardo 1994; Gómez-Tuena et al. 2007; Frey et al. 2007; Selvans et al. 2011), the complex pattern of transtensional deformation

within the Tepic-Zacoalco Rift and the extension within the Colima Rift has been proposed to be related to this plate fragmentation event (e.g., Luhr et al. 1985). This separation event also appears to have been initiated concurrent with the opening of the Gulf of California (e.g., Ferrari and Rosas-Elguera 1999).

4. The rollback of the Rivera Plate beneath the Jalisco Block. This may have produced the recent extension within the Tepic-Zacoalco Rift (Ferrari et al. 1994, 2001; Yang et al. 2009; León Soto et al. 2009) and an associated trenchward migration of the volcanic front (Ferrari et al. 1994).

5. The collision of the EPR with the Middle America Trench (MAT) off the southern tip of Baja California between 14 and 9 Ma, the effects of which were felt throughout the entire western margin of the Mexican Pacific (e.g., Handchu-macher 1976; Mammerickx and Klitgord 1982).

Of importance to the present study is that this collision fragmented the Guadalupe Plate into the Rivera and Cocos plates at around 10 Ma (KLITGORD and MAMMERICKX 1982; MAMMERICKX and KLITGORD 1982; DEMETS and TRAYLEN 2000), after which the Rivera plate began to pivot, relative to the North American plate, about an Euler pole located near the point of collision (Fig. 13a) via the mechanism first proposed by MENARD (1978). In the area of the Jalisco Block, this event had at least two major effects. First, this pivoting motion resulted in oblique convergence along the Jalisco Subduction Zone, at which the Rivera Plate subducts beneath North America (e.g., BANDY 1992; KOSTOGLODOV and BANDY 1995). Second, the pivoting motion produced divergence between the Rivera and Cocos plates which resulted in the formation of a southwest propagating tear between the two subducted plates beneath North America (BANDY 1992).

5.2. Implications of the Results on the Processes Producing the Deformation in the Central and Southern Parts of the Jalisco Block

The idea that the Colima Rift and Volcán Colima are somehow related to deformation of the subducting oceanic lithosphere has been around for quite some time (e.g., MOOSER and MALDONALDO 1961; MOOSER 1972; STOIBER and CARR 1973). More recently, NIXON (1982) proposed that the Colima Rift formed in response to vertical motions along the subducted Rivera Fracture zone due to age/buoyancy differences. BARRIER et al. (1990) noted that "the rifting [in the Colima Rift] is propagating southward..." and like Nixon they stated that this "suggests an opening related to the deformation of the subducted Cocos plate." From these early ideas in conjunction with plate motion models, BANDY (1992) proposed that the Colima Rift and Volcán Colima formed in response to divergence between the subducted portions of the Rivera and Cocos plates, the divergence being a result of the pivoting of the Rivera Plate away from the Cocos Plate due to the collision of the East Pacific Rise and the MAT off the southern tip of Baja California (Fig. 13a). The plate motion models

suggest that this tearing between the two plates would have started to develop to the NE where the rate of divergence was predicted to be the greatest. With time, this tear would propagate southwest toward the MAT and eventually to the EPR. Presently, the SW tip of this propagating rift is marked by the El Gordo Graben (Fig. 13b) (BANDY 1992; SERRATO-DÍAZ et al. 2004). This proposal is consistent with the southward progression of volcanism noted within the Colima Graben (e.g., ALLAN 1986; BARRIER et al. 1990), as well as the results of several recent seismic studies (YANG et al. 2009; LEON-SOTO et al. 2009; SPIKA et al. 2014).

The results of the present study are also consistent with the previous proposals of a southwest propagating tear between the Rivera and Cocos plates, that the tear extends under the southern Colima Rift to the coastline and into the offshore area, and that Volcán Colima is related to this tear. However, our results are inconsistent with the proposals that the alkaline volcanism along the Central Jalisco Volcanic Lineament can be ascribed to the effects of rolling back of the subducted Rivera Plate toward the MAT and an associated lateral flow of mantle material into the mantle wedge (e.g., FERRARI et al. 2001; LEON SOTO et al. 2009). Our results also indicate the presence of a previously unrecognized zone of anomalously fast δV_p located at the base of the continental crust and landward of the coast, the area where the subducting and overriding plates are not in direct contact (i.e., above the mantle wedge).

As an interesting side note, this collision event was initiated around 10–14 Ma, roughly coincident with the pulse of mafic volcanism that occurred in the Altos de Jalisco region near Guadalajara (e.g., MOORE and MARONE 1994) at about 10 Ma (Fig 13a). Since this area lies roughly along the NE projection of the northern Colima Rift, and since tearing is predicted by plate motion models to start in the NE, it is conceivable to us that this mafic volcanism was generated during the initial rifting of the subducted part of the Guadalupe Plate due to this ridge-trench collision (Fig. 13a). Thus, the Altos de Jalisco lavas, and perhaps the triple-rift system near Guadalajara (which includes the northern Colima Graben), might also be related to the pivoting of the Rivera Plate away from the Cocos Plate due to this ridge-trench collision event. If so, the formation of the Colima Volcanic Complex represents the second major episode of melt generation and melt rise associated with this tearing (Fig. 13c). Similar to our proposal, FERRARI (2004) also proposed that these lavas were produced by tearing of the subducted slab beneath western Mexico; however, in their proposal the tearing started in the Gulf of California and propagated eastward along the Trans-Mexican Volcanic Belt. We feel that the two proposals are complimentary as the eastward propagating tear in the subducted slab proposed by FERRARI (2004) would have made it easier for the Rivera Plate to pivot away from the Cocos plate.

Getting back to the implications of our results, the distributions of δVp found in the present study can be described very simply as being controlled by geologic structures formed during two stages of the separation of the Rivera and Cocos plates. The earlier period represents the initial stages of the separation of the Rivera and Cocos plates beneath western Mexico; the later period represents the more advanced stage of rifting where the Rivera and Cocos plates had separated sufficiently to allow melt to accumulate below the Colima Volcanic complex. During the earlier period (9–1.6 Ma), NE–SW-oriented structures/lineaments formed in response to the separation of the Rivera and Cocos plates beneath western Mexico (Fig. 13b). It is unclear why these structures took on a NE–SW orientation, perhaps it was due to the combined effects of the direction of Rivera-Cocos relative motion present at that time and the trends of pre-existing structures within the subducting lithosphere (BANDY 1992). The structures formed within the continental lithosphere during this period are still recognizable; they include the NE–SW-oriented southern Colima Rift and the NE–SW lineaments present in the area located along the NE projection of the southern margin of the southern Colima Rift (i.e., in the region east of the northern Colima Rift and south of Lake Chapala). The deformation produced during the second period (1.6 Ma to the present; Fig. 13c) can be described as the result of magma, generated within and perhaps above the tear zone between the Rivera and Cocos plates, rising beneath the region of the Colima Volcanic Complex. The oldest of these volcanoes, Volcan Cantaro, was active

between 1 and 1.6 Ma (ALLAN 1986). The rising magma appears to have fractured the overlying crust (Fig. 13c), forming a classic triple-rift junction geometry (e.g., BURKE and DEWEY 1973).

The distribution of δVp at 40-km depth is quite important in our understanding of the mantle processes affecting the Jalisco Block. Given the complicated proposals concerning the evolution of the Jalisco Block, this distribution is surprisingly simple: Landward of the coast, it is simply a low-velocity zone located beneath the central and southern parts of the Colima Rift which is sandwiched between two zones of fast velocities. However, only the low-velocity zone extends seaward of the coast. According to recent estimates of the depth to the top of the subducting lithosphere (Gutierrez *et al.* in Press) in conjunction with the crustal thickness estimates of SUHARDJA *et al.* (2015), the fast zones are only located at the base of the continental crust where the subducting and overriding plates are not in direct contact, and they do not correspond to the subducting oceanic plates. (i.e., the depth to the top of the subducted slab is greater than 40 km). Fast velocity zones at the base of the crust above a mantle wedge have been observed in other subduction zones such as the Kuril Arc (e.g., THYBO and ARTEMIEVA 2013) where it was proposed that the basal crustal layers were due to magmatic underplating.

Although the low δVp zone observed on the 40-km depth slice does not lie within the subducting slab north of the southern Colima Rift, it does lie within the subducting slab seaward of the coast where the depth to the top of the subducting plate is less than 40 km. Consequently, like the results of SPICA *et al.* (2014), our results support the proposal of a tear (coinciding with the low-velocity zone) between the Rivera and Cocos plates located beneath the southern Colima rift near the coast; however, our results indicate that the tear extends further seaward than previously indicated, roughly midway between the coast and the MAT.

The low-velocity zone extending southward from Volcán Colima comprises two distinct lows, one centered under Volcán Colima and the other centered in the southern Colima Rift, about 7 km north of the town of Armería, near the intersection of the NE–SW and N–S lineaments. Could this indicate a recent southward relocation of the magmatic center due to a southward propagating tear in the subducting plates?

Surprisingly, there are no other major anomalies present in the 40-km, nor in the 35-km, depth slice. This lack of anomalies at the base of the crust/uppermost mantle argues against the proposal of (1) a trenchward migration of the volcanic front, (2) the proposal of slab dehydration beneath the Sierra Cacoma (CORBO-CAMARGO et al. 2013), and (3) a lateral flow of asthenospheric mantle into the mantle wedge beneath the Jalisco Block (except for one area adjacent to Volcán Colima where a small zone of low δVp extends for a short distance into the Jalisco Block). It is inconceivable that these processes, if present, would not produce a zone of anomalous δVp at the base of the crust given that Vp depends on water content, composition, and temperature.

The distribution of δVp becomes more complicated within the middle continental crust (15 to 25 km). The anomaly pattern for values of δVp < −0.4 at these depths consists of three branches, all of which extend away from a junction located about 30 km west of Volcán Colima. This triple low-δVp junction system, henceforth called the TLVJ, is very distinct at crustal depths greater than 20 km, but becomes diffuse at 15-km depth and is not recognizable at depths <10 km. Such a geometry immediately brings to mind the classical triple-rift junction observed for many continental rifts throughout the world (e.g., BURKE and DEWEY 1973), the triple-rift system being the result of the ascent of magma beneath the junction of the three rifts. It is somewhat surprising that the junction is not centered directly beneath Volcán Colima. However, a recent crustal study (SPICA et al. 2014) using ambient seismic noise shows a low Rayleigh-wave group velocity zone centered under Volcán Colima. The reason for this discrepancy is not clear.

The diffuse nature of the anomalies with decreasing crustal depths suggests a progressive diffusion/advection of heat due to fracturing and fluid circulation as the magma rises through the crust. It may also indicate that our reference velocities were slightly too high in the shallow crust; however, consistent with our results, SUHARDJA et al. (2015) similarly proposed, based on high Vp/Vs ratios, that the crust in this area is being "broadly heated, with

possible partial melting, or has had the addition of fluids trenchward of the recent volcanism."

The southern branch of the TLVJ trends NS and terminates within the southern Colima Rift. Interestingly, its NS orientation coincides with the trend of the northern Colima Rift and not the southern Colima Rift, which has a NE–SW orientation. We take this to be evidence that the NE–SW trending structures and lineaments predate the formation of the present day low-velocity zone. Further, the location of the SW terminus of this branch is the same for all depths below 15 km, suggesting that its SW terminus is being controlled by the older NE–SW system of lineaments.

The eastern branch of the TLVJ is oriented E-W and extends about 50 km east of Volcán Colima. Like the southern branch, the eastern branch also terminates against the NE–SW system of lineaments, again suggesting the NE–SW system of lineaments is older than the TLVJ. The terminus coincides with both a prominent gravity low (marked 1 on Fig. 13c) (BANDY et al. 1995) and high Vp/Vs ratios (SUHARDJA et al. 2015). Given the granitic nature of the crust in the area, SUHARDJA et al. (2015) concluded that the high Vp/Vs ratios were most likely due to partial melt or high fluid content within the crust. This is consistent with the results of our study; specifically, the fracturing of the crust would most likely result in an increase of the crustal fluid content. Further, this provides an alternative explanation for the low gravity anomaly observed by BANDY et al. (1995) since it is likely that the fluid filled fractures would significantly lower the bulk density of the granites.

Of particular interest is that the low-δVp zone at the eastern terminus of the eastern branch extends to shallow depths (see the 5- and 10-km depth slices) where it appears as an isolated circular anomaly located along a NE–SW alignment of three isolated, circular low-δVp anomalies. Our results in conjunction with those of SUHARDJA et al. (2015) and BANDY et al. (1995) strongly suggest that fluids are circulating within the eastern low-δVp branch of the LVTJ, most likely through fractures related to the magma rising beneath the Colima Volcanic Complex, and that these fluids are being diverted upward when they reach the older NE–SW-oriented fracture system.

The western branch of the TLVJ extends from the triple junction, northwestward (azimuth of about 304°) beneath the Sierra Cacoma (Fig. 13c). From the distribution of values of $\delta Vp < -0.3$, four important characteristics of this branch of the TLVJ are noted, namely, (1) the NW extent of this branch of the TLVJ is not constant with depth; instead it extends progressively to the NW as crustal depths decrease, (2) the axis of this branch lies trenchward of the CJVL, although the low-δVp zone covers much of the CJVL, (3) the present day alkaline volcanism along the CJVL (i.e., Los Volcanes, Talpa de Allende, and Mascota volcanic fields) is located along the NW extension of this low-δVp zone at shallow crustal depths, and (4) the low-δVp zone becomes more diffuse with decreasing crustal depths, covering much of the Jalisco Block on the 15-km and 10-km depth slices.

The results of the studies of Corbo-Carmargo et al. (2013) and SUHARDJA et al. (2015) can be explained by an influx of intercrustal water in this area, which Corbo-Carmargo et al. (2013) propose is due to metamorphic dehydration of the subducted Rivera Plate. However, as mentioned previously, this proposal is hard to reconcile with the lack of a δVp anomaly at the base of the crust. It is also hard to reconcile with the result that the NW–SE low-δVp branch of the TLVJ is not observed to extend along the entire Jalisco Block at deep crustal depths as should be the case if it was due to slab dehydration. We propose, instead, that the broad heating, high crustal fluid content, and low-δVp values in this area are due to fluid circulation and associated heat advection in the fracture systems within this branch of the TLVJ, accompanied by the diffusion of heat into the adjacent areas of the Jalisco Block. In our proposal, the fractures are being formed by the ascent of magma through the crust near Volcán Colima, hydraulic fracturing may play a role in extending the fractures away from Volcán Colima along the three branches of the TLVJ.

At 10-km depth, the distribution of δVp values differs significantly from that observed in the lower and mid-crustal levels. The TLVJ which was so prominent at deeper crustal depths is no longer observed. Instead, the low-velocity anomalies are aligned along three trends. The first two trends

parallel the trench. One of these follows the coast along its entire extent. The other is aligned with the western branch of the triple-rift junction system in the Jalisco Block; however, it lies south of the eastern branch of the TLVJ in the area of the southern Mexico Block. Note also that the appearance of these two trends under the southern Mexico Block is gradual with depth.

Specifically, they are observed in the 25-, 20-, and 15-km depth slices, where their magnitudes increase with decreasing crustal depths. The third trend is oriented NE–SW and lies along the southeast boundary of the southern Colima Rift and along its NE prolongation and is most likely related to fluids flowing along the fractures of the older system of NE–SW-oriented lineaments (Fig. 13b).

At 5-km depths, the three trends are of lower magnitude but are still recognizable as alignments of isolated circular low-velocity anomalies.

The origins of the two zones of low-δVp values which parallel the trench are not clear. The low-δVp zone located along the coast may relate to the formation of the bounding fault of a forearc sliver (e.g., FITCH 1972), the formation of the fault being the result of the oblique subduction along both the Rivera-North America and Cocos-North America subduction zones (SERPA 1989 DEMETS and STEIN 1990; BANDY et al. 2005). However, the anomalous zone is observed at deep crustal depths (30 km) only to the southeast of the Colima Rift. The NW part of the second zone correlates with the western limb of the deeper TLVJ and most likely is just the upward continuation of this limb of the TLVJ. However, the SE part of this zone is less easily explained as it lies south of the eastern limb of the deeper TLVJ. This part may also correspond to faults produced by oblique subduction, indeed this anomalous zone extends to deep crustal depths (25 km). Such a fault, the Atoyac fault, has been observed south of Acapulco and Oaxaca, but this fault is not known to extend further NW into our study area (Lowery et al., unpublished manuscript).

To summarize, we propose that the magma rising beneath the Colima Volcanic Complex is being generated due to divergence between the subducted Rivera and Cocos plates. The mechanism for this melting is unclear but several possibilities exist. For example, melting may be due to some form of pressure release/decompressional melting as the subducting plate tears (BANDY et al. 1995), or alternatively, the magma may arise due to melting of the mantle wedge due to an increased temperature produced by the upwelling of hot asthenosphere material into the gap formed by the divergence between the subducting plates (FERRARI 2004; YANG et al. 2009), or a complex combination of several melt generation processes (VIGOUROUX et al. 2008). To account for the lack of volcanoes above the entire length of the tear, we propose that the melt generation is confined to an area slightly behind the tip of the SW propagating rift where the amount of extension just becomes sufficient to generate a substantial amount of magma. After the tear equilibrates (i.e., after the void space in the tear zone fills with material and the temperature becomes equilibrated with its surroundings), the generation of magma stops, until a new void is generated further to the southwest. Our results suggest that this zone of magma generation/accumulation, presently fully developed under Volcán Colima, might be starting to shift SW to the area just north of Armaría, Colima. The upwelling of magma in the area of Volcán Colima uplifts and fractures the overlying lower crust into a classic triple-rift junction geometry and that this fracturing is more prominent in the mid-crustal levels. Fluids are circulating within the fractures and that, associated with this fluid circulation, heat advection is broadly heating the TLVJ and the surrounding areas. The reason for why the fracturing is more intense at mid-crustal depths than at deeper depths is unclear; perhaps it is the result of reduced confining pressure or perhaps it is controlled by the characteristics of the crustal rheology in this area. The reason for why present day intraplate alkaline volcanism is confined to the NW tip of the low-velocity anomaly and why the intraplate alkaline volcanism has been progressing to the NW (BANDY et al. 2001) is also unclear; perhaps this is due to some kind of pressure release melting mechanism due to fracturing of the crust ahead of a NW progressing tear.

5.3. Implications of the Results on the Processes Producing the Deformation in the Northern Parts of the Jalisco Block

During the opening of the Gulf of California, a tearing or truncation of the subducting oceanic plate occurred beneath the Gulf of California (e.g., CALMUS et al. 2003, 2011). This tear within the subducted plate has been proposed to have extended eastward beneath the Trans-Mexican Volcanic belt where it produced an eastward progressing pulse of mafic magmatism from 11.5 to 6.5 Ma (FERRARI 2004). Unfortunately, the study area does not incorporate the area of the proposed tear; however, several studies (e.g., FERRARI et al. 2001; LEÓN SOTO et al. 2009) have suggested that mantle material may be flowing through the tear and into the adjacent mantle wedge beneath the Jalisco Block, producing the recent alkaline volcanism within the interior of the Jalisco Block. Several seismic stations (stations 42, 43, and 55) of LEÓN SOTO et al. (2009) which were used as evidence for such a flow lie within the region of good resolution of our study. All of these stations show a fast direction of polarization for the SKS and SKKS shear waves oriented roughly north–south, counterclockwise of those observed at stations within the interior of the Jalisco Block to the southeast. Stations 42 and 43 of that study are located within the low-δVp zone associated with the Mascota and Talpa de Allende grabens at shallow crustal depths. However, this anomaly does not extend to deep crustal depths (>25 km) which does not support the proposal of mantle flow coming in under the Jalisco Block from the north. Further, Station 55 of that study, is not associated with a low-δVp zone at any depth. Given this, we question the proposal that mantle material is flowing through the tear zone and beneath the northern end of the Jalisco Block. Instead, we propose that it is more likely that the fast directions of shear wave polarization determined in the study of LEÓN SOTO et al. (2009) indicate the direction that the slab is subducting into the mantle (i.e., the direction that mantle material is being dragged down by the subducting slab).

6. Conclusions

The main conclusions of this study include the following:

1. The distribution at deep crustal depths (35–40 km) shows a very simple geometry, namely, a NE-oriented low-δVp zone sandwiched between two high-δVp zones. The low-velocity region consists of two closed lows, one centered under Volcán Colima and the other centered in the southern Colima rift near the town of Armería. The high-δVp zones do not extend to mid- or upper crustal depths (<30 km). Further, these two zones lie above the subducting oceanic lithosphere and therefore may indicate the presence magmatic underplating of the continental crust in this region.

2. The distribution of δVp is dominated at the middle and lower crustal depth (15–35 km) by three elongated zones of low velocities within which δVp of less than −0.6 km/s are present. These three zones converge in the vicinity of the Colima Volcanic Complex forming a geometric pattern analogous to the triple-rift systems associated with continental breakup. The NW limb of this triple junction extends further to the NW as crustal depths decrease. We propose that this distribution of δVp values is the result of the upwelling of magma under the Colima Volcanic which uplifts and fractures the overlying crust into a classic triple-rift junction geometry and that this fracturing is more prominent in the mid-crustal levels. We further propose that fluids are circulating within the fractures and that, associated with this fluid circulation, heat advection is broadly heating the TLVJ and the surrounding areas.

3. The distribution of δVp at upper crustal depths (<10 km) is characterized by large regions of low-velocity anomalies. However, the 5-km layer has more isolated and easy to identify anomalies. These anomalies appear to show three alignments, two parallel to the coast and one perpendicular to the coast. The anomaly perpendicular to the coast correlates with the SE margin of the southern Colima rift and its NE prolongation, and is probably due to fluids/heat rising along the old NE-oriented structures formed during the initial stages of the tearing between the Rivera and Cocos plates as the Rivera plate pivoted counterclockwise away from the Cocos plate.

4. Our results do not support previous proposals of upward asthenospheric flow through tears in the subducting plates, the flow extending beneath the Jalisco Block, neither to the NW nor to the SE of the block as no anomalous velocities are observed in those areas at deep crustal depths.

5. Our results do not support the previous proposals that the alkaline basalts located within the Central Jalisco Volcanic Lineament are the result of trenchward rollback of the subducted Rivera plate and an associated trenchward migration of the volcanic arc as no anomalous velocities are observed in that area at deep crustal depths. We conclude that the origin of the intraplate alkaline volcanism (or Ocean-Island Basalts) along the Central Jalisco Volcanic lineament needs to be re-examined.

6. Our results do not support the previous proposal of slab dehydration beneath the Sierra Cacoma as no anomalous velocities are observed in that area at deep crustal depths.

Acknowledgments

This research was supported in part through CONACYT-Gobierno del Estado de Jalisco (CONACYT-FOMIXJAL), project: Red Sismológica Telemétrica de Jalisco (RESAJ), project ID: 2008-09-96538 and "Caracterización del Peligro Sísmico Tsunamigénico asociado a la estructura cortical del contacto Placa de Rivera-Bloque Jalisco" (TSUJAL), project ID: 2012-08-189963. Juan Alejandro Ochoa Chavez was also supported in part through grants from CONACYT-Gobierno del Estado de Jalisco (CONACYT-FOMIXJAL), project: Maestría en Ciencias en Geofísica, project ID: 2008-04-96567. We thank the Incorporated Research Institutions for Seismology (IRIS) Data Management Center for providing open access to the global digital data used in this study.

References

AKI K., and LEE W.H.K. (1976), *Determination of Three-Dimensional Velocity Anomalies Under a Seismic Array using First P Arrival Times from Local Earthquakes*, J. Geophys. Res. *81*, 4381–4399.

ALINAGHI A., KOULAKOV I., and THYBO H. (2007), *Seismic tomographic imaging of P- and S-waves velocity perturbations in the upper mantle beneath Iran*, Geophysical Journal International *169*:1089–1102, doi: 10.1111/j.1365-246X.2007.03317.x.

ALLAN J. F. (1986), *Geology of the Northern Colima and Zacoalco Grabens, southwest Mexico: Late Cenozoic rifting in the Mexican Volcanic Belt*, Geol. Soc. America Bull., *97*:473–485, doi: 10.1130/0016-7606(1986).

ALLAN, J.F., NELSON, S.A., LUHR, J.F., CARMICHAEL, I.S.E., WOPAT, M., and WALLACE, P.J., Pliocene-recent rifting in SW Mexico and associated volcanism: an Exotic terrain in the making, *in the* Gulf and Peninsular Province of the California's, AAPG Memoir 47, (ed. Dauphin J.P. and B.R.T. Simoneit) (AAPG, Tulsa, Ok. 1991) pp. 425-445.

ANDREWS V., STOCK J., RAMÍREZ VÁZQUEZ C.A., and REYES-DÁVILA G. (2011), *Double-difference Relocation of the Aftershocks of the Tecomán, Colima, Mexico Earthquake of 22 January 2003*, Pure Appl. Geophys. *168*:1331–1338, doi: 10.1007/s00024-010-0203-0.

BANDY, W.L. (1992), Geological and Geophysical Investigation of the Rivera-Cocos Plate Boundary: Implications for Plate Fragmentation, Ph.D. Dissertation, Texas A&M University, College Station, pp. 195.

BANDY W.L., HILDE T.W.C., and YAN C.Y. (2000), *The Rivera-Cocos plate boundary: Implications for Rivera-Cocos relative motion and plate fragmentation*, Geol. Soc. America Bull., *334*:1–28.

BANDY W.L., KOSTOGLODOV V., HURTADO-DÍAZ, and MENA M. (1999), *Structure of the southern Jalisco subduction zone, Mexico, as inferred from gravity and seismicity*, Geofisica Internacional *38*:127–136.

BANDY W.L., KOSTOGLODOV V., and MORTERA-GUTIÉRREZ C.A. (1998), *Southwest migration of the instantaneous Rivera-Pacific Euler pole since 0.78 Ma.*, Geofisica Internacional, *37*:153–169.

BANDY W.L., MICHAUD F., BOURGOIS J., et al. (2005), *Subsidence and strike-slip tectonism of the upper continental slope off Manzanillo, Mexico*, Tectonophysics, *398*:115–140, doi: 10.1016/j.tecto.2005.01.004.

BANDY W.L., MICHAUD F., DYMENT J., et al. (2008), *Multibeam bathymetry and sidescan imaging of the Rivera Transform–Moctezuma Spreading Segment junction, northern East Pacific Rise: New constraints on Rivera–Pacific relative plate motion*, Tectonophysics, *454*:70–85. doi: 10.1016/j.tecto.2008.04.013.

BANDY W.L., MICHAUD F., MORTERA-GUTIÉRREZ C.A., et al. (2011), *The Mid-Rivera-Transform Discordance: Morphology and Tectonic Development*, Pure Appl. Geophys., *168*:1391–1413, doi: 10.1007/s00024-010-0208-8.

BANDY W.L., MORTERA-GUTIÉRREZ C.A., JAIME U.F., and HILDE T.W.C. (1995), *The Subducted Rivera-Cocos Plate Boundary*, Geophys. Res. Lett., *22*:3075–3078.

BANDY, W.L., and PARDO, M. (1994), *Statistical examination of the existence and relative motion of the Jalisco and southern Mexico blocks*, Tectonics, *13*, 755–768.

BANDY W.L., URRUTIA-FUCUGAUCHI J., McDOWELL F.W., and MORTON-BERMEA O. (2001), *K-Ar ages of four mafic lavas from the Central Jalisco Volcanic Lineament: Supporting evidence for a NW migration of volcanism within the Jalisco block, western Mexico*, Geofisica Internacional, *40*:259–269.

BARTOLOMÉ R., DAÑOBEITIA J.J., MICHAUD F., CORDOBA D., and DELGADO-ARGOTE L.A. (2011), *Imaging the Seismic Crustal Structure of the Western Mexican Margin between 19°N and 21°N*, Pure Appl. Geophys., *168*:1373–1389, doi: 10.1007/s00024-010-0206-x.

BARRIER, E., BOURGOIS, J., and MICHAUD, F., (1990), *The active Jalisco triple junction rift system*, C.R. Acad. Sciences Paris, Serie II, *310*, 1513–1520.

BOURGOIS, J., RENARD V., AUBOUIN J., BANDY W., BARRIER E., CALMUS T., CARFANTAN J.C., GUERRERO J., MAMMERICKX J., MERCIER DE LEPINAY B., MICHAUD F., and SOSSON M. (1988), Active fragmentation of the North American plate: Offshore boundary of the Jalisco block off Manzanillo, Comptes Rendues, Acadadémie des Sciences Paris, 307(Serie II), p. 1121–1130.

BOURGOIS J., and MICHAUD F. (1991), *Active fragmentation of the North America plate at the Mexican triple junction area off Manzanillo*, Geo-Marine Letters, *11*:59–65.

BURKE K., and DEWEY J.F. (1973), *Plume-Generated Triple Junctions: Key Indicators in Applying Plate Tectonics to Old Rocks*, The Journal of Geology, *81*:406–433, doi: 10.2307/30070631.

CALMUS T., AGUILLÓN-ROBLES A., MAURY R.C., BELLON H., BENOIT M., COTTON J., BOURGOIS J., and MICAUD F. (2003), *Spatial and temporal evolution of basalts and magnesian andesites ("'bajaites'") from Baja California, Mexico: the role of slab melts*, Lithos, *66*:77–105.

CALMUS T., PALLARES C., MAURY R.C., AGUILLÓN-ROBLES A., BELLON H., BENOIT M., and MICAUD F. (2011), *Volcanic Markers of the Post-Subduction Evolution of Baja California and Sonora, Mexico: Slab Tearing Versus Lithospheric Rupture of the Gulf of California*, Pure Appl. Geophys., *168*:1303–1330, doi: 10.1007/s00024-010-0204-z.

CORBO-CAMARGO F., ARZATE-FLORES J.A., ALVAREZ-BEJAR R., ARANDA-GOMEZ J.J., and YUTSIS V., (2013), *Subduction of the Rivera plate beneath the Jalisco block as imaged by magnetotelluric data*, Revista Mexicana de Ciencias Geologicas, *30*:268–281.

DEMETS C., and STEIN S. (1990), *Present-day Kinematics of the Rivera Plate and Implications for Tectonics in Southwestern Mexico*, J. Geophys. Res., *95*:21931–21948.

DEMETS C., and TRAYLEN S. (2000), *Motion of the Rivera plate since 10 Ma relative to the Pacific and North American plates and the mantle*, Tectonophysics, *318*:119–159, doi: 10.1016/S0040-1951(99)00309-1.

DEMETS C., and WILSON D.S. (1997), *Relative motions of the Pacific, Rivera, North American, and Cocos plates since 0.78 Ma.*, J. Geophys. Res., *102*:2789–2806.

DESHON H.R., SCHWARTZ S.Y., NEWMAN A.V., et al (2006), *Seismogenic zone structure beneath the Nicoya Peninsula, Costa Rica, from three-dimensional local earthquake P- and S-wave tomography*, Geophysical Journal International, *164*:109–124, doi: 10.1111/j.1365-246X.2005.02809.x.

DOUGHERTY S.L., CLAYTON R.W., and HELMBERGER D.V. (2012), *Seismic structure in central Mexico: Implications for fragmentation of the subducted Cocos plate*, J. Geophys. Res., *117*, B09316, doi: 10.1029/2012JB009528.

FERRARI L. (2004), *Slab detachment control on mafic volcanic pulse and mantle heterogeneity in central Mexico*, Geol., *32*, 77–5, doi: 10.1130/G19887.1.

FERRARI L. (1995), *Miocene shearing along the northern boundary of the Jalisco block and the opening of the southern Gulf of California*, Geol, *23*, 751–5, doi: 10.1130/0091-7613(1995)023<0751:MSATNB>2.3.CO;2.

FERRARI L., PASQUARÈ G., VENEGAS SALGADO S., et al. (1994), *Regional tectonics of western Mexico and its implications for the northern boundary of the Jalisco block*, Geofisica Internacional, *33*, 139–151.

FERRARI L., PASQUARE, G., VENEGAS-SALGADO, S., and ROMERO-RIOS, F., Geology of the western Mexican Volcanic Belt and adjacent Sierra Madre Occidental and Jalisco block, Cenozoic Tectonics and Volcanism of Mexico: Geological Society of America Special Paper 334 (ed. Delgado-Granados, H., Aguirre-Díaz, G., and Stock, J.M., eds)(Boulder, Colorado 2000) pp. 65–83.

FERRARI L., PETRONE C.M., FRANCALANCI L. (2001), Generation of oceanic-island basalt–type volcanism in the western Trans-Mexican volcanic belt by slab rollback, asthenosphere infiltration, and variable flux melting, Geol., *29*, 507–510, doi: 10.1130/0091-7613(2001)029<0507:GOOIBT>2.0.CO;2.

FERRARI, L., and ROSAS-ELGUERA, J., Late Miocene to Quaternary extensión at the northern boundary of the Jalisco block, western Mexico: The Tepic-Zacoalco rift revised, Cenozoic Tectonics and Volcanism of Mexico: Geological Society of America Special Paper 334 (ed. Delgado-Granados, H., Aguirre-Díaz, G., and Stock, J.M.) (Boulder, Colorado 1999) pp. 41–63.

FITCH T.J. (1972) *Plate convergence, transcurrent faults, and internal deformation adjacent to southeast Asia and the western Pacific*, J. Geophys. Res., *77*, 4432–4460.

FOULGER G.R., JULIAN B.R., PITT A.M., and HILL D.P. (2003), *Three-dimensional crustal structure of Long Valley caldera, California, and evidence for the migration of CO2 under Mammoth Mountain*, J. Geophys. Res., *108*, 2147, doi: 10.1029/2000JB000041.

FREY H.M., LANGE R.A., HALL C.M., et al. (2007), *A Pliocene ignimbrite flare-up along the Tepic-Zacoalco rift: Evidence for the initial stages of rifting between the Jalisco block (Mexico) and North America*, Geol. Soc. America Bull., *119*, 49–64, doi: 10.1130/B25950.1.

GARDUÑO, V., and TIBALDI, A. (1991), *Kinematic evolution of the continental active triple junction of the western Mexican Volcanic Belt*, Comptes Rendus de lÁcadámie des Sciences, Paris Serie II, *312*, 135–142.

GÓMEZ-TEUNA, A., OROZCO-ESQUIVEL, M. T., and FERRARI, L., Igneous petrogenesis of the Trans-Mexican Volcanic Belt, Geology of Mexico: Celebrating the Centenial of the Geological Society of Mexico, Geological Society of America Special Paper 422 (ed. Alaniz-Álvarez, S.A., and Nieto-Samaniego, Á.F., eds.) (Boulder, Colorado 2007) pp. 129–181.

GRAEBER F.M., and ASH (1999), *Three-dimensional models of P wave velocity and P-to-S velocity ratio in the southern central Andes by simultaneous inversion of local earthquake data*, J. Geophys. Res., *104*, 20237–20256.

HASLINGER F., KISSLING E., and ANSORGE J., et al. (1999), *3D crustal structure from local earthquake tomography around the Gulf of Arta (Ionian region, NW Greece)*, Tectonophysics, *304*, 201–218, doi: 10.1016/S0040-1951(98)00298-4.

JOHNSON, C.A., and HARRISON, C.G.A. (1989), *Tectonics and volcanism in western Mexico: A landsat thematic mapper perspective*, Remote Sens. Environ., *28*, 273–286.

KISSLING E. (1998), *Geotomography with Local Earthquakes*, Review of Geophysics, *26*, 659–698.

KISSLING E., ELLSWORTH W.L., EBERHART-PHILLIPS D., and KRADOLFER U. (1994), *Initial reference models in local earthquake tomography*, J. Geophys. Res. *99*, 19635–19646.

KLITGORD, K.D., and MAMMERICKX, J. (1982), *Northern East Pacific Rise: magnetic anomaly and bathymetric framework*, J. Geophys. Res., *87*, 6725–6750.

KOSTOGLODOV V., and BANDY W.L. (1995), *Seismotectonic constraints on the convergence rate between the Rivera and North American plates*, J. Geophys. Res., *100*, 17977–17989.

KOULAKOV I., BINDI D., PAROLAI S., et al. (2010), *Distribution of Seismic Velocities and Attenuation in the Crust beneath the North Anatolian Fault (Turkey) from Local Earthquake Tomography*, Bulletin of the Seismological Society of America, *100*, 207–224, doi: 10.1785/0120090105.

KOULAKOV I., BOHM M., ASCH G., et al. (2007), *Pand Svelocity structure of the crust and the upper mantle beneath central Java from local tomography inversion*, J. Geophys. Res., *112*, B08310–19, doi: 10.1029/2006JB004712.

KOULAKOV I., KABAN M.K., TESAURO M., and CLOETINGH S. (2009), *P- and S-velocity anomalies in the upper mantle beneath Europe from tomographic inversion of ISC data*, Geophysical Journal International, *179*, 345–366, doi: 10.1111/j.1365-246X.2009.04279.x.

KOULAKOV I, SOBOLEV S.V. (2006), *A tomographic image of Indian lithosphere break-off beneath the Pamir-Hindukush region*, Geophysical Journal International, *164*, 425–440, doi: 10.1111/j.1365-246X.2005.02841.x.

KOULAKOV I., SOBOLEV S.V., ASCH G. (2006), *P- and S-velocity images of the lithosphere-asthenosphere system in the Central Andes from local-source tomographic inversion*, Geophysical Journal International, *167*, 106–126, doi: 10.1111/j.1365-246X.2006.02949.x.

KOULAKOV I.Y., DOBRETSOV N.L., BUSHENKOVA N., YAKOVLEV A.V. (2011), *Slab shape in subduction zones beneath the Kurile–Kamchatka and Aleutian arcs based on regional tomography results*, Russian Geology and Geophysics, *52*, 650–667.

LEÓN SOTO G., NI J.F., GRAND S.P., et al. (2009), *Mantle flow in the Rivera-Cocos subduction zone*, Geophysical Journal International, *179*, 1004–1012, doi: 10.1111/j.1365-246X.2009.04352.x.

LUHR J.F., NELSON S.A., ALLAN J.F., CARMICHAEL I.S.E. (1985), *Active rifting in southwestern Mexico: Manifestations of an incipient eastward spreading-ridge jump*, Geol., *13*, 54–57, doi: 10.1130/0091-7613(1985)13<54:ARISMM>2.0.CO;2.

MARQUEZ-AZUA B., and DeMETS C. (2009), *Deformation of Mexico from continuous GPS from 1993 to 2008*, Geochem Geophys Geosyst, doi: 10.1029/2008GC002278.

MAMMERICKX, J., and KLITGORD, K.D. (1982), *Northern East Pacific Rise: Evolution from 25 m.y.B.P. to the present*, Journal of Geophysical Research, *87*, 6751–6759.

MENARD, H. W. (1978), *Fragmentation of the Farallon plate by pivoting subduction*, J. Geol., *86*, 99–110.

MICHAUD F., DANOBEITIA J.J., CARBONELL R., et al. (2000), *New insights into the subducting oceanic crust in the Middle American trench off western Mexico (17–19°N)*, Tectonophysics, *318*, 187–200.

MICHAUD, F., QUINTERO, O., BARRIER, E., and BOURGOIS, J. (1991), *The northern boundary of the Jalisco block (western Mexico): location and evolution from 13 Ma to present*, Comptes Rendus de Íacadámie des Sciences, Paris Serie II, *312*, 1359–1365.

MICHAUD, F., QUINTERO, O., CALMUS, T., BOURGOIS, J., and BARRIER, E. (1993), *The Amatlan de cañas depression (western Mexico): Neogene distension in the northern part of the Jalisco block*, Comptes Rendus de Íacadámie des Sciences, Paris Serie II, *317*, 251–258.

MOORE G., and MARONE C. (1994), *Basaltic volcanism and extension near the intersection of the Sierra Madre volcanic province and the Mexican Volcanic Belt*, Geol. Soc. America Bull., *106*, 383–394.

MOOSER, F. (1972), *The Mexican Volcanic belt: Structure and Tectonics*, Geofisica Internacional, *12*, 55–70.

MOOSER, F., and MALDONADO, M. (1961), *Penecontemporaneous tectonics along the Mexican Pacific Coast*, Geofísica Internacional, *1*, 1–20.

NAVA A.F., GARCÍA-ARTHUR R., CASTRO R.R., et al. (1999) *S wave attenuation in the coastal region of Jalisco–Colima, Me*. Physics of the Earth and Planetary Interiors, *115*, 247–257.

NIXON G.T. (1982), *The relationship between Quaternary volcanism in central Mexico and the seismicity and structure of subducted ocean lithosphere*, Geol. Soc. America Bull., *93*, 514–11, doi: 10.1130/0016-7606(1982)93<514:TRBQVI>2.0. CO;2.

NUÑEZ-CORNU F.J., RUTZ LOPEZ M., NAVA P.F.A., et al (2002), *Characteristics of seismicity in the coast and north of Jalisco Block, Mexico*, Physics of the Earth and Planetary Interiors, *132*, 141–155.

PARDO M., and SUÁREZ G. (1995), *Shape of the subducted Rivera and Cocos plates in southern Mexico: Seismic and tectonic implications*, J. Geophys. Res., *100*, 12357–12373.

PAUL A., THOUVENOT F., SPALLAROSSA D., et al. (2001), *A three-dimensional crustal velocity model of the southwestern Alps from local earthquake tomography*, J. Geophys. Res., *106*, 19367–19389.

RAWLINSON N, DE KOOL M, and SAMBRIDGE M (2006), *Seismic wavefront tracking in 3D heterogeneous media: applications with multiple data classes*, Explor. Geophys., *37*, 322–330.

RAWLINSON N., and FISHWICK S. (2011), *Seismic structure of the southeast Australian lithosphere from surface and body wave tomography*, Tectonophysics, doi: 10.1016/j.tecto.2011.11.016.

RAWLINSON N., HOUSEMAN G.A., and Collins C.D.N. (2001), *Inversion of seismic refraction and wide-angle reflection traveltimes for three-dimensional layered crustal structure*, Geophysical Journal International, *145*, 381–400.

RAWLINSON N., HOUSEMAN G.A., COLLINS C.D.N., and DRUMMOND B.J. (2001), *New evidence of Tasmania's tectonic history from a novel seismic experiment*, Geophys. Res. Lett. *28*, 3337–3340.

RAWLINSON N., and KENNETT B.L.N. (2008), *Teleseismic tomography of the upper mantle beneath the southern Lachlan Orogen, Australia*, Physics of the Earth and Planetary Interiors, *167*, 84–97, doi: 10.1016/j.pepi.2008.02.007.

RAWLINSON N., and KENNETT B.L.N. (2004), *Rapid estimation of relative and absolute delay times across a network by adaptive stacking*, Geophysical Journal International, *157*, 332–340, doi: 10.1111/j.1365-246X.2004.02188.x.

RAWLINSON N., KENNETT B.L.N., and Heintz M. (2006), *Insights into the structure of the upper mantle beneath the Murray basin from 3D teleseismic tomography*, Australian Journal of Earth Sciences, *53*, 595–604. Doi, 10.1080/08120090600686751.

RAWLINSON N., KENNETT B.L.N., VANACORE E., et al. (2011), *The structure of the upper mantle beneath the Delamerian and Lachlan orogens from simultaneous inversion of multiple teleseismic datasets*, Gondwana Research, *19*, 788–799, doi: 10. 1016/j.gr.2010.11.001.

RAWLINSON N., POZGAY S., and FISHWICK S. (2010), *Seismic tomography: A window into deep Earth*, Physics of the Earth and Planetary Interiors, *178*, 101–135, doi: 10.1016/j.pepi.2009.10. 002.

RAWLINSON N., READING A.M., and KENNETT B.L.N. (2006), *Lithospheric structure of Tasmania from a novel form of teleseismic tomography*, J. Geophys. Res., doi: 10.1029/2005JB003803.

RAWLINSON N., and SAMBRIDGE M. (2004), *Multiple reflection and transmission phases in complex layered media using a multistage fast marching method*, Geophysics, *69*, 1338–1350, doi: 10.1190/ 1.1801950.

RAWLINSON N., and SAMBRIDGE M. (2004), *Wave front evolution in strongly heterogeneous layered media using the fast marching method*, Geophysical Journal International, *156*, 631–647, doi: 10.1111/j.1365-246X.2004.02153.x.

RAWLINSON N., and SAMBRIDGE M. (2003), *Seismic Traveltime tomography o fthe Crust and Lithosphere*, Advances in Geophysics, *46*, 81–197.

RAWLINSON N., and SAMBRIDGE M. (2003), *Irregular interface parametrization in 3-D wide-angle seismic traveltime tomography*, Geophysical Journal International, *155*, 79–92.

RAWLINSON N., and SAMBRIDGE M. (2005) *The fast marching method: an effective tool for tomographic imaging and tracking multiple phases in complex layered media*, Explor. Geophys., *36*, 341–350.

RAWLINSON N., and URVOY M. (2006), *Simultaneous inversion of active and passive source datasets for 3-D seismic structure with application to Tasmania*, Geophys. Res. Lett., *33*, L24313–5, doi: 10.1029/2006GL028105.

ROSAS-ELGUERA J., ALVA-VALDIVIA L.M., GOGUITCHAICHVILI A., et al. (2003), *Counterclockwise Rotation of the Michoacan Block: Implications for the Tectonics of Western Mexico*, International Geology Review, *45*, 814–826, doi: 10.2747/0020-6814.45.9. 814.

ROSAS-ELGUERA J., FERRARI L., GARDUÑO-MONROY V.H., and Urrutia-Fucugauchi J. (1996), *Continental boundaries of the Jalisco block and their influence in the Pliocene-Quaternary kinematics of western Mexico*, Geol., *24*, 921–924, doi: 10.1130/0091-7613(1996)024<0921:CBOTJB>2.3.CO;2.

ROSAS-ELGUERA J., FERRARI L., MARTINEZ M.L., and URRUTIA-FUCUGAUCHI J. (1997), *Stratigraphy and Tectonics of the Guadalajara Region and Triple-Junction Area, Western Mexico*, International Geology Review, *39*, 125–140, doi: 10.1080/ 00206819709465263.

RUTZ-LÓPEZ M., and NUÑEZ-CORNU F.J. (2004), *Sismotectónica del Norte y Oeste del Bloque de Jalisco Usando Datos Sísmicos Regionales*, Geofisica Internacional, *24*, 2–13.

SALLARÈS V., DAÑOBEITIA J.J., and FLUEH E.R. (2000), *Seismic tomography with local earthquakes in Costa Rica*, Tectonophysics, *329*, 61–78.

SANCHEZ J.J., and NUNEZ-CORNU F.J. (2009), *Seismicity and Stress in a Tectonically Complex Region: The Rivera Fracture Zone, the Rivera-Cocos Boundary, and the Southwestern Jalisco Block,*

Mexico, Bulletin of the Seismological Society of America, *99*, 2771–2783, doi: 10.1785/0120080350.

SELVANS M.M., STOCK J.M., DeMETS C., et al. (2011), *Constraints on Jalisco Block Motion and Tectonics of the Guadalajara Triple Junction from 1998–2001 Campaign GPS Data*, Pure Appl. Geophys., *168*, 1435–1447, doi: 10.1007/s00024-010-0201-2.

SERPA, L., KATZ, C., and SKIDMORE, C. (1989), *The southeastern boundary of the Jalisco block in west-central Mexico (abstract)*, EOS Trans, AGU, *43*, 1319.

SERRATO-DÍAZ G.S., BANDY W.L., and MORTERA GUTIÉRREZ C.A. (2004), *Active rifting and crustal thinning along the Rivera-Cocos plate boundary as inferred from Mantle Bouguer gravity anomalies*, Geofisica Internacional, *43*, 361–381.

SPICA, Z., CRUZ-ATIENZA, V.M., REYES-ALFARO, G., LEGRAND, D., and IGLESIAS, A. (2014), *Crustal imaging of western Michoacán and the Jalisco block, Mexico, from Ambient Seismic Noise*, Journal of Volcanology and Geothermal Research, *289*, 193–201.

STECK L.K., THURBER C.H., FEHLER M.C., et al. (1998), *Crust an upper mantle P wave velocity structure beneath Valles caldera, New Mexico: Results from the Jemez teleseismic tomography experiment*, J. Geophys. Res., *103*, 24301–24320.

STOCK J.M., and LEE J. (1994), *Do microplates in subduction zones leave a geological record?*, Tectonics, *13*, 1472–1487.

STOIBER, R.E., and CARR, M.J., 1973, *Quaternary volcanic and tectonic segmentation of central America*, Bulletin of Volcanology, *37*, 304–325.

SUHARDJA S.K., GRAND S.P., WILSON D., et al. (2015), *Crust and subduction zone structure of Southwestern Mexico*, J. Geophys. Res., doi: 10.1002/2014JB011573.

THURBER C.H. (2003), *Seismic Tomography of the Lithosphere with Body Waves*, Pure Appl. Geophys., *160*, 717–737.

THYBO, H., ARTEMIEVA, I.M. (2013), *Moho and magmatic underplating in continental lithosphere*, Tectonophysics, *609*, 605–619, Doi:10.1016/j.tecto.2013.05.032.

TRYGGVASON A., RÖGNVALDSSON S.T., FLÓVENZ Ó. (2002), *Three-dimensional imaging of the P- and S-wave velocity structure and earthquake locations beneath Southwest Iceland*, Geophysical Journal International, *151*, 848–866.

VIGOUROUX N., WALLACE P.J., and KENT A.J.R. (2008), *Volatiles in High-K Magmas from the western Trans-Mexican Volcanic Belt: Evidence for Fluid Fluxing and Extreme Enrichment of the Mantle Wedge by subduction processes*, Journal of Petrology, *49*, 1589–1618, doi:10.1093/petrology/egn039.

YANG T., GRAND S.P., WILSON D., et al. (2009), *Seismic structure beneath the Rivera subduction zone from finite-frequency seismic tomography*, J. Geophys. Res., doi: 10.1029/2008JB005830.

ZELT C.A. (1999), *Modelling strategies and model assessment for wide-angle seismic traveltime data*, Geophysical Journal International, *139*, 183–204.

ZELT C.A., ELLIS R.M., ZELT B.C. (2006), *Three-dimensional structure across the Tintina strike-slip fault, northern Canadian Cordillera, from seismic refraction and reflection tomography*, Geophysical Journal International, 167, 1292–1308, doi: 10.1111/j.1365-246X.2006.03090.x.

ZHAO D., HASEGAWA A., HORIUCHI S. (1992), *Tomographic Imaging of P and S Wave Velocity Structure Beneath Northeastern Japan*, J. Geophys. Res., *97*, 19909–19928.

(Received May 6, 2015, revised September 22, 2015, accepted September 25, 2015, Published online November 19, 2015)

Pure Appl. Geophys. 173 (2016), 3513–3524
© 2015 Springer Basel
DOI 10.1007/s00024-015-1167-x

Pure and Applied Geophysics

Seismic Characteristics of the Vulcanian Explosions from the 2003–2005 Eruption at Colima Volcano, Mexico

FRANCISCO JAVIER NÚÑEZ-CORNÚ,[1] JUAN MANUEL ESPÍNDOLA,[2] FIDENCIO ALEJANDRO NAVA PICHARDO,[3] and
CARLOS SUÁREZ-PLASCENCIA[1]

Abstract—Colima Volcano (19.512°N, 103.617°W, 4000 m.a.s.l.), located on the border between the states of Jalisco and Colima in western Mexico, is the most active volcano in the country. Its activity has taken place through diverse styles of eruption, from very explosive to effusive. In the last decades it has presented frequent vulcanian eruptions with episodes of dome construction–destruction. Four of these cycles occurred from 1990 to 2005, the last one from July 2003 to September 2005. We focus on this last period, for which we analyzed seismic phases and coupled pressure airwaves from high dynamic range seismograms, both in the time and frequency domains, to determine characteristic features, propagation velocities, and origin times for both deep seismic sources and the associated explosions. The results show that the sources of the P-waves associated with the explosions are not located at the summit, but instead at different shallow locations for the different explosions, suggesting the presence of various magmatic pathways within the volcano.

Key words: Colima Volcano, Vulcanian explosions, Magmatic pathways, Volcanic seismic signals.

1. Introduction

Colima Volcano (19.512°N, 103.617°W, 4000 m.a.s.l.; Fig. 1) is well known for its frequent and varied activity. In the last decades, it has presented several cycles of vulcanian eruptions characterized by dome construction and destruction

¹ Centro de Sismología y Volcanología de Occidente (Sis-VOc), Universidad de Guadalajara, Av. Universidad 203, 48280 Puerto Vallarta, Jal, Mexico. E-mail: pacornu77@gmail.com; carlos.csuarez@gmail.com

² Depto. Volcanología, Instituto de Geofísica, UNAM. Cd. Universitaria, Mexico, D.F., Mexico. E-mail: jmec@unam.mx

³ Depto. Sismología, Centro de Investigación Científica y Educación Superior de Ensenada, (CICESE), Cta. Ensenada-Tijuana 3918, 22860 Ensenada, BC, Mexico. E-mail: fnava@cicese.mx

(LUHR and CARMICHEL 1990; ROBIN *et al.* 1987). One of these cycles, which occurred in the 1903–1904 period (DÍAZ 1906), preceded the explosive 1913 eruption. More recently Colima has presented four cycles of this type of activity (1991–1994; 1999–2001; 2001–2003; 2004–2005). The characteristics of the first three cycles and, to a certain extent those of the last one, were presented in a preceding paper (NÚÑEZ-CORNÚ *et al.* 2010). In this work, we present the results of our analyses of the vulcanian explosions occurring in 2003 and 2005 corresponding to that cycle. This type of explosive activity is similar to that preceding the 1913 eruption, adding importance to the detailed analysis of the observations of this modern day recurrence.

The general characteristics of the seismicity accompanying this period can be summarized as follows: three explosions occurred during July and August 2003; no further explosions occurred until 2005. In that year, 13 significant explosions occurred from March to September. Contrarily to the behavior observed during previous periods (NÚÑEZ-CORNÚ *et al.* 2010), no relevant precursory activity was detected before the 16 explosions (Fig. 2), suggesting that the volcano is currently in the post brittle/ductile seal-breakage state in the model of FOURNIER (1999). In this work, we focus on the signals associated with these explosions, which have multiple sources such as those arising from the volcanic edifice, from the ground-coupled atmospheric propagation due to the explosion blast, and from the impact of the ejecta.

The complexity of the phenomena taking place during vulcanian eruptions leads to the generation of various types of seismic signals. The waveforms and spectral content of those seismic records are related

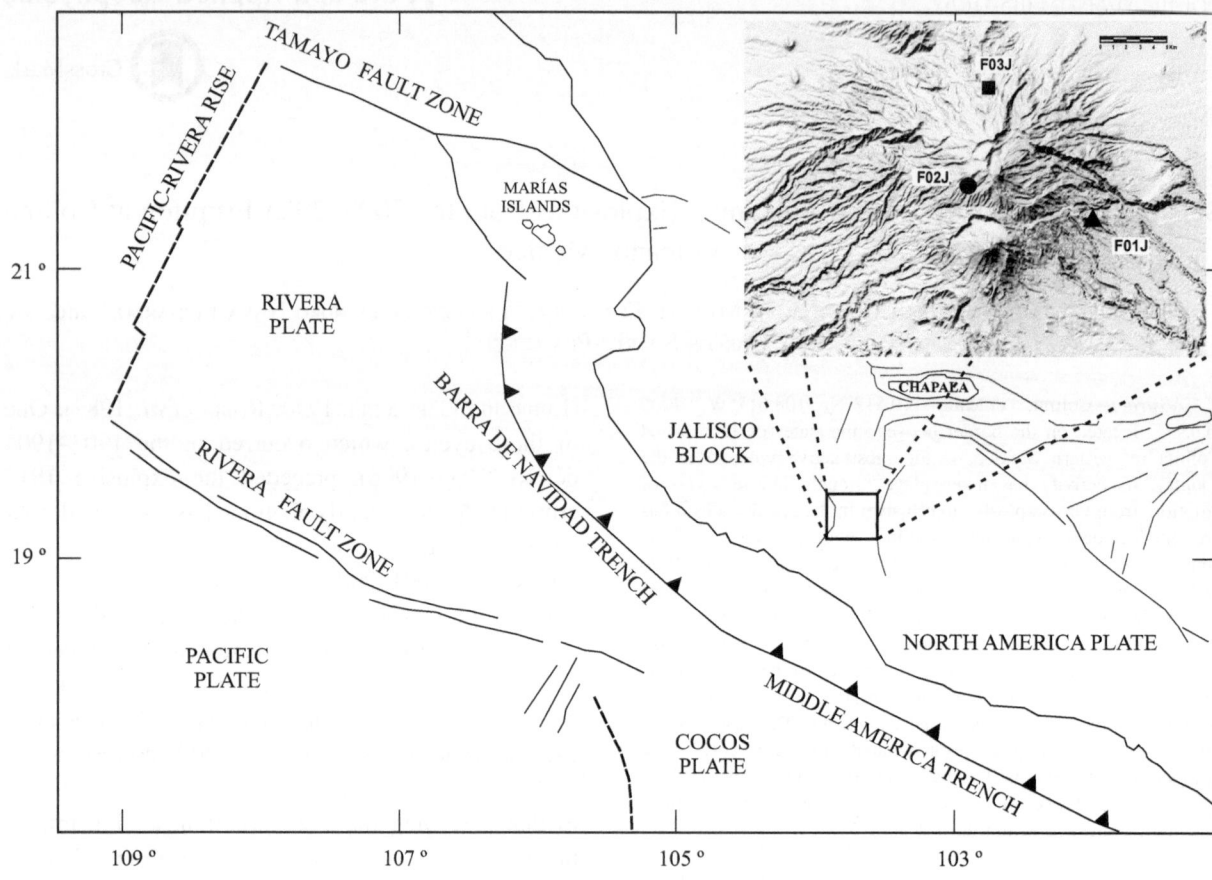

Figure 1
Tectonic setting of Jalisco block region and location of Colima Volcano

as well to the condition of the volcanic system; that is, in terms of its being closed, partially open, or open. Volcanic earthquakes have been recognized to arise from processes related to faulting in the rock matrix, to the dynamics of the volcanic fluids during magma pressurization or a combination of both mechanisms. The sudden gas decompression produced when the pressure exceeds the strength of the rock cap generates seismic waves that are followed by those that propagate through the atmosphere (NÚÑEZ-CORNÚ *et al.* 2010).

Due to our lacking sonic or pressure sensor data for our analysis we use only the signal registered at the seismometers. It is clear that the frequency observed for the sonic airwave coupling at the seismic stations is a function of the response of the seismometer and the ground at the site. NÚÑEZ-CORNÚ

et al. (2010) observed several frequency peaks for explosions that occurred between 1999 and 2003: in seismic stations located at distances between 6 and 10 km, frequency peaks between 11 and 14 Hz associated with a coupled air shock wave (audible, as reported by observers) called shock airwave (SAW), and frequency peaks between 31 and 33 Hz associated with the pressure (infrasonic) airwave coupled airwave, called pressure airwave (PAW).

The complete seismic signal that we observe, generated by a vulcanian explosion is thus characterized by the following process:

(a) The injection of magmatic material (fluids, gas, ash, debris, etc.), which later reaches the vent.

(b) The ejection of magmatic constituents into the atmosphere, which generates a PAW, and

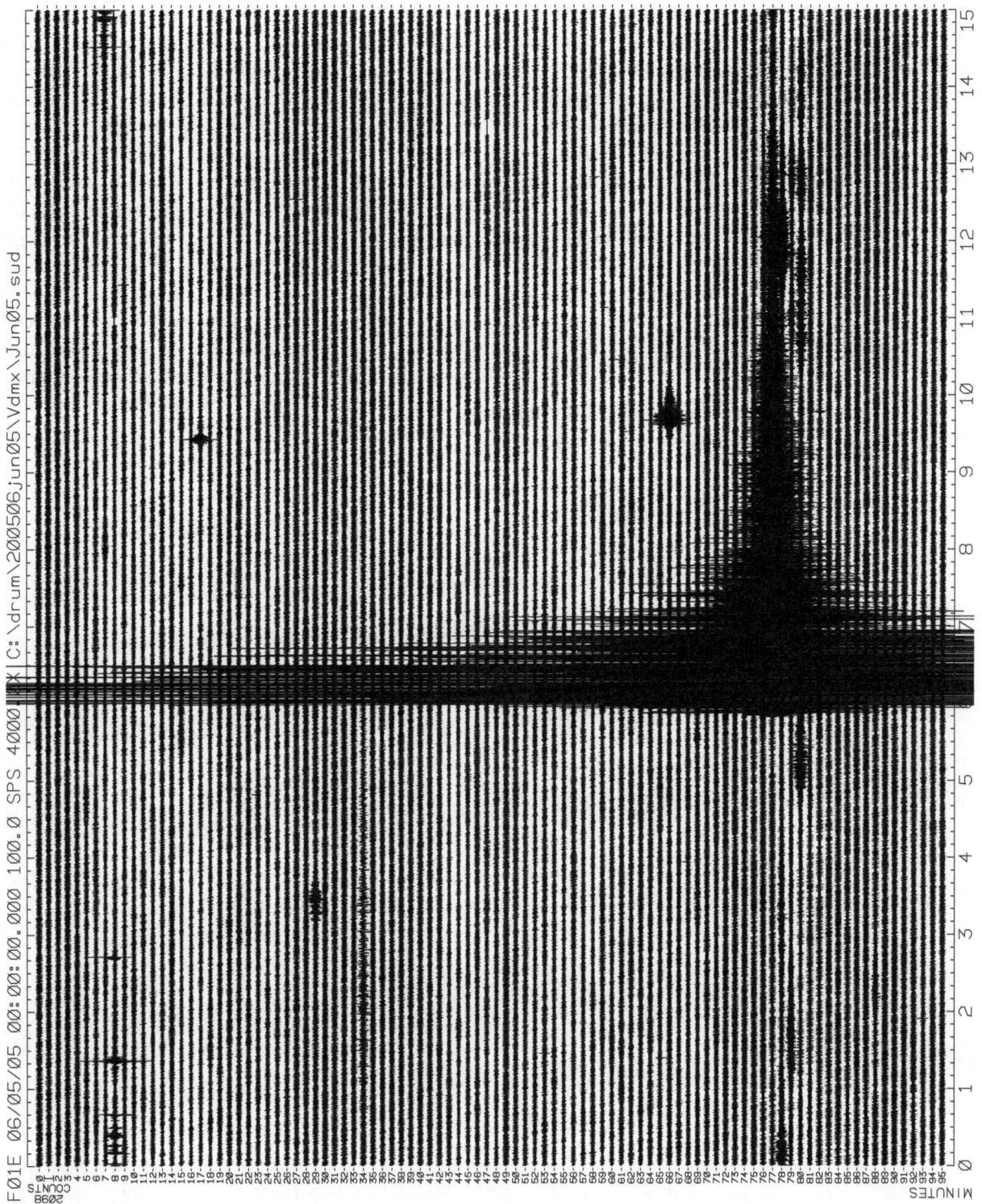

Figure 2
Sample helicorder seismogram at station F01J (component EW) showing the absence of precursory activity before the explosive event number 12 (May 6, 2005)

possibly a SAW. The ejection process involves coupling between the PAW and SAW (or only PAW) and the ground at the summit.

(c) The flux of magmatic material through the conduit and the fall of ejecta from the eruptive column.

(d) The travel path of the seismic signal, and finally

(e) The ground coupling of the PAW and SAW (or PAW) at the seismic station.

(f) We illustrate these below, processes through the analysis of the main eruptions of the 2003–2005 cycle.

2. Data and Methods of Analysis

The activity of 2003–2005 was recorded by the RESJAL (*Red Sísmica Telemétrica de Jalisco*) network, operated by *Protección Civil Jalisco* and *SisVOc, Universidad de Guadalajara*. The network consists of three telemetered stations on the volcano: F01J, F02J, F03J (Fig. 3); RESJAL operated four more stations in northern Jalisco. Each station is equipped with a 3-component 1 Hz Lennartz Le3D seismometer (NÚÑEZ-CORNÚ *et al.* 2010) and a 24-bit Everest-Kinemetrics digitizer recording at 100 samples per second. We studied the 16 larger explosions (Table 1) recorded by RESJAL using data mainly from station F03J located 12 km north of the summit crater. This station was selected because it operated continuously during the study period. Data from the other stations were used to construct record sections for individual explosions to obtain S–P times.

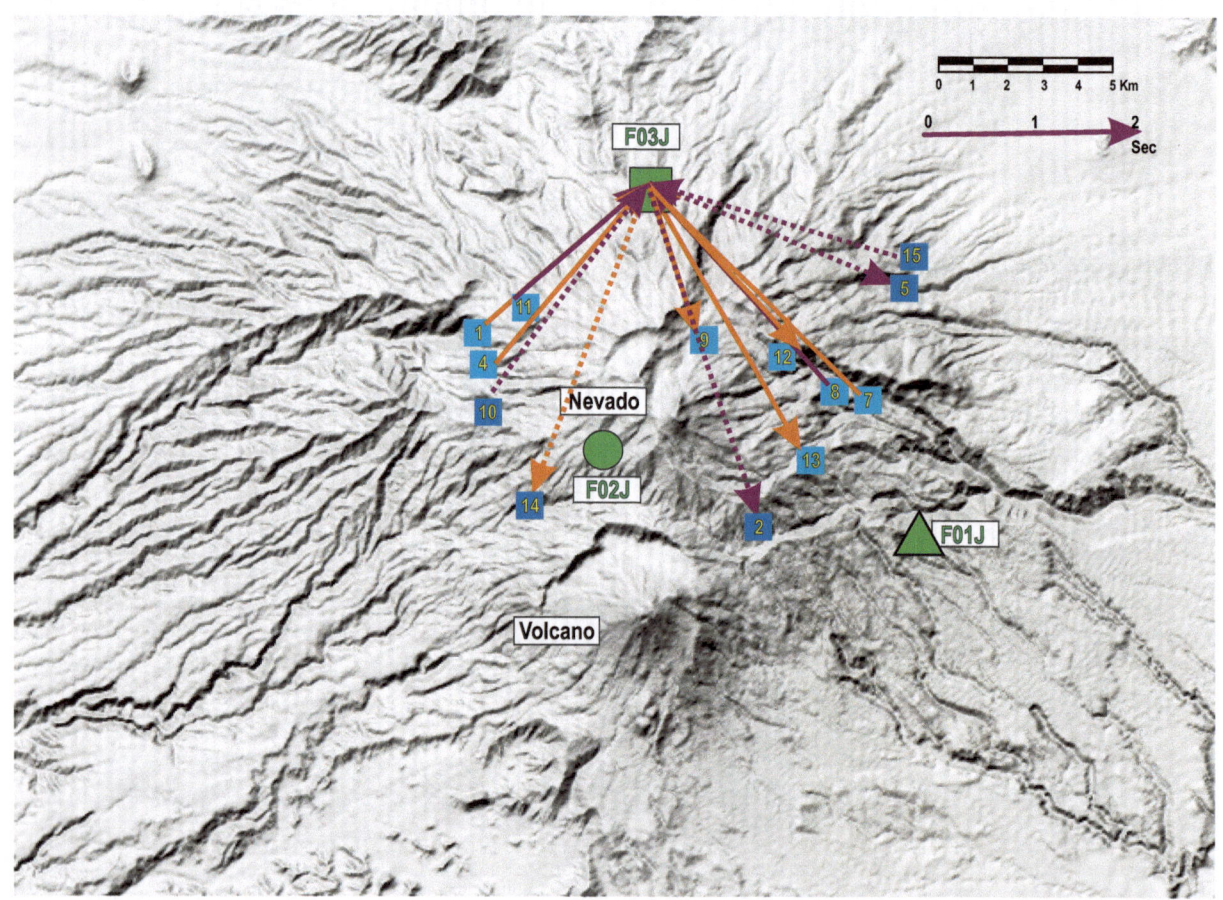

Figure 3

Digital elevation model of the Colima Volcano area displaying seismic station locations. *Solid green triangle* station F01J; *solid green circle* station F02J; *solid green square* station F03J. *Numbers* refer to explosions listed in Table 1. *Arrows* show azimuths calculated from first arrivals at F03J; *Orange* P-wave polarized *vertically*; *Purple* P-wave polarized *horizontally*; *Dotted arrows* dilatation in the *vertical* component

Table 1

Vulcanian explosions at Colima Volcano 2003–2005

No.	Year	Month	Day	Hour	SoundVel (m/s)	PAW T0 (s)	Vp (km/s)	P-wave T0 (s)	F03-Pol P-wave	Dif T0 (s)	F03 Azm	F03 S–P	F02 S–P	F01 S–P	F03NS Freq (Hz)	F03NS (PAW) Freq (Hz)
1	2003	July	17	10:27	325	30.03	8.00		V		228	2.04			1.08	31.93
2	2003	August	2	20:41	310	34.08	2.35	33.06	H	1.02	162*	3.32		1.31	1.21	32.13
3	2003	August	29	04:52											0.59	29.88
4	2005	March	13	21:27	300	55.00	4.00	56.58	V	−1.58	200	2.23		2.18	1.21	10.84
5	2005	March	26	03:40					H		112*	2.51			1.20	27.73
6	2005	April	20	01:56											0.56	10.84
7	2005	May	10	14:26	334	56.10	5.50	56.48	V	−0.38	135	2.82	2.26		1.21	28.42
8	2005	May	16	02:01	320	34.20	5.20	34.98	H	−0.78	138	2.55	1.65	1.60	0.89	10.94
9	2005	May	24	00:09	328	45.50	5.80	43.22	V	2.28	162	1.51	1.13		0.96	10.94
10	2005	May	30	08:37	338	21.00	8.00		H		216*	2.53	1.76		0.98	10.64
11	2005	June	2	04:59	336	60.15	6.50		H		228	1.66	0.90	3.58	1.20	28.91
12	2005	June	6	19:20	342	45.99	5.00		V		138	2.13	2.15	2.98	1.20	28.12
13	2005	June	10	02:52	338				V		150	2.91	3.00	2.96	0.54	30.76
14	2005	July	27	09:14	328	32.80	8.00		V		200*	3.13		3.66	1.20	31.30
15	2005	September	16	15:45	345	50.00	7.00		H		106*	2.43	1.16	1.51	0.90	11.04
16	2005	September	27	10:06								3.07	2.47	1.9	1.06	11.72

SoundVel Sound velocity as obtained from record sections, *Vp* P-wave velocity as obtained from record sections, *PAW* origin time of the pressure airwave as obtained from the record section, *P-wave T0* origin time of the P-wave obtained from the record section, *F03-Pol P-wave* P-wave polarity at station F03, *Dif T0* time difference at source between P-wave and PAW, *F03 Azm* arrival direction from particle motion at F03, *F03 S–P* S–P time at F03, *F02 S–P* S–P time at F02, *F01 S–P* S–P at F01, *F03NS* peak frequency of the complete signal in the NS component, *F03NS (PAW)* is the PAW peak frequency in the NS component

* dilatation on vertical component

The N–S and E–W components at F03J were not rotated, as it is deployed approximately along the radial and transverse directions with respect to the summit crater. We used 120 s samples beginning 10 s before the onset of the P-wave. For each record, we calculated the raw spectrum and spectrogram.

Separation of the different trains that compose the seismograms was achieved by band-passing the N–S component between 0.2–1 and 20–40 Hz with a Butterworth recurrent filter (HAVSKOV and OTTE-MÖLLER 1990). For events recorded at more than one station, we constructed record sections to estimate propagation velocities, as well as origin times from the various sources, seismic and explosive (acoustic).

The onset times of the different seismic phases were determined manually from screen-enhanced images of the filtered signals (picking precision ≤0.04 s).

3. Results

Figure 4 shows the raw spectrogram, the spectrum, and a section of the spectral frequency window

between 10 and 40 Hz, for the 16 explosions analyzed. The energy from the SAW and PAW components (NÚÑEZ-CORNÚ et al. 2010) can be clearly seen. It is also clear that each event exhibits different patterns. The seismic signals have dominant frequencies in two bands, one centered at about 1.20 Hz and the second below 1.00 Hz. The frequency of the coupled SAW ranges between 10 and 14 Hz, and that of the coupled PAW between 28 and 33 Hz. In six cases the frequency peaks are at 10–14 and 28–33 Hz (events 1, 2, 5, 7, 12, and 13), corresponding to the SAW and PAW phases which are clearly distinguished (NÚÑEZ-CORNÚ et al. 2010). In the events 4, 6, 8, 10 and 16 they are barely seen.

Figure 5 shows raw and band-passed seismograms from the N–S component of F03J. The seismograms were filtered between [0.2, 1.0] and [20, 40] Hz. Filtering allows identification of common elements in the events. No conspicuous similarities are observed in the raw waveforms, except for the SAW phase; which is clearly discernible for the 10 May 2005 and 10 June 2005 events. The low-pass

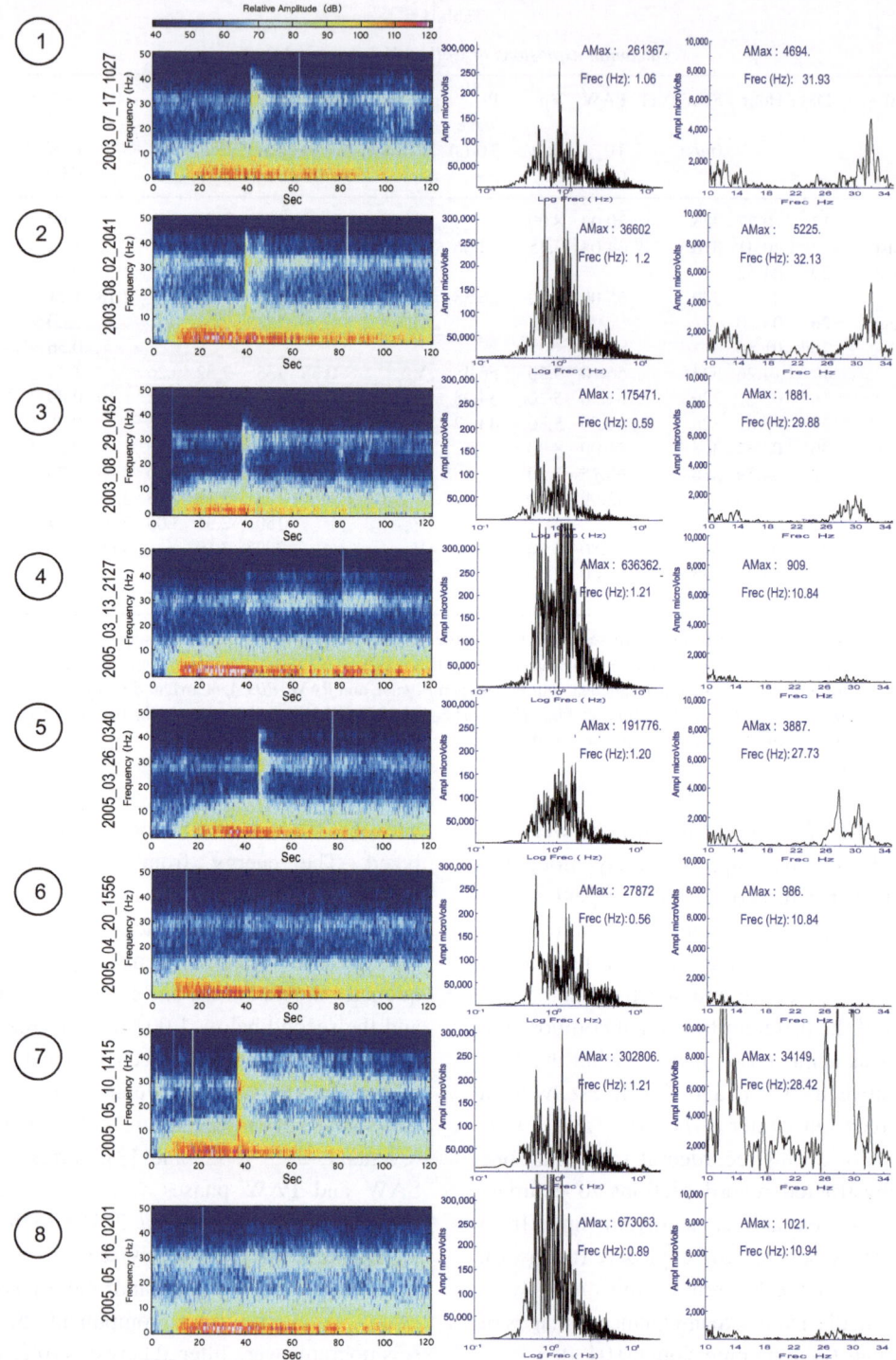

Figure 4

Raw spectrogram, spectrum and close-up of the spectrum for the 16 vulcanian explosions (window between 10 and 40 Hz). Numbers in the *left column* are dates in yyyy_mm_dd_hhmm format; numbers above spectra are maximum amplitude and peak frequency

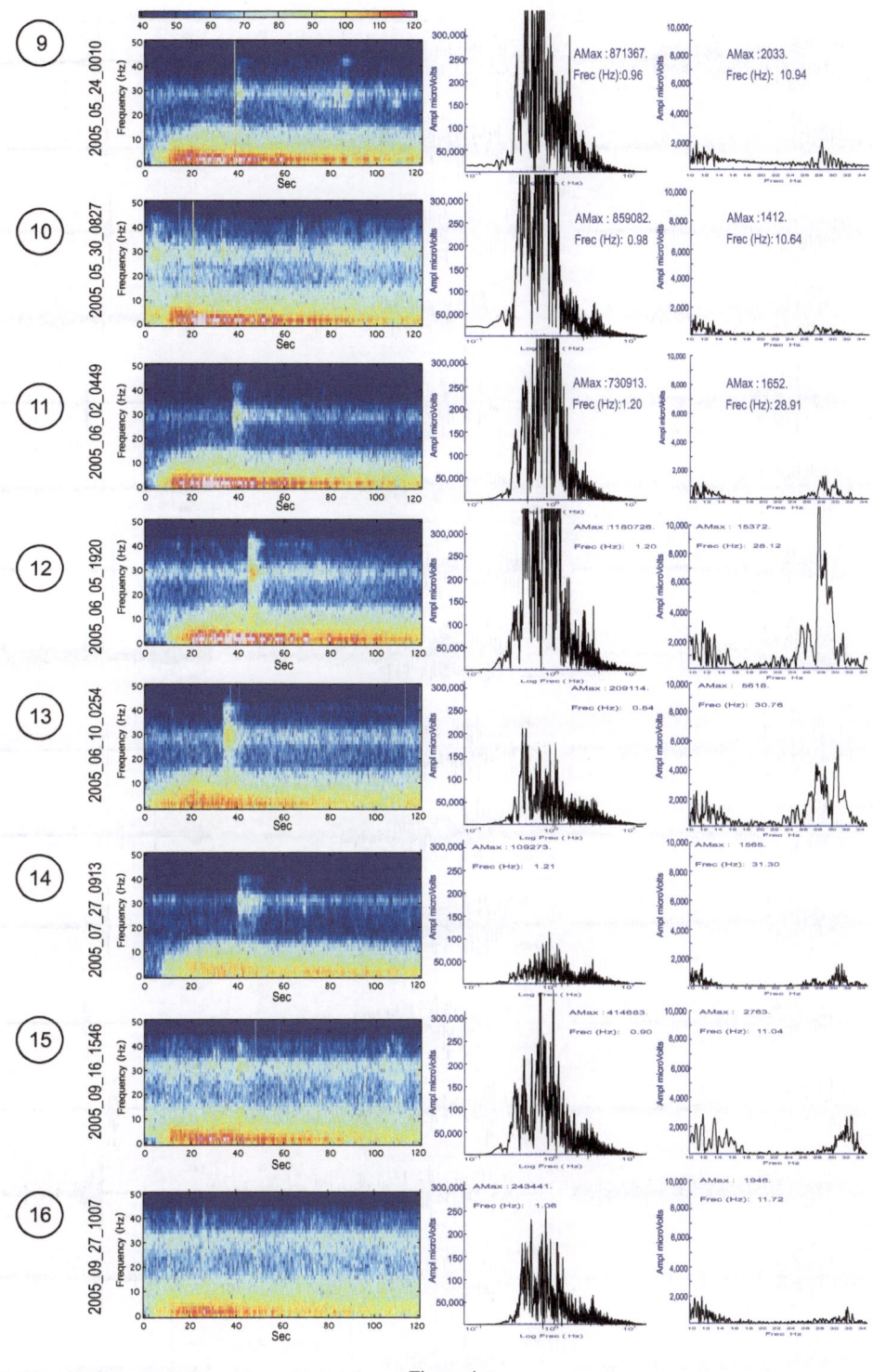

Figure 4
continued

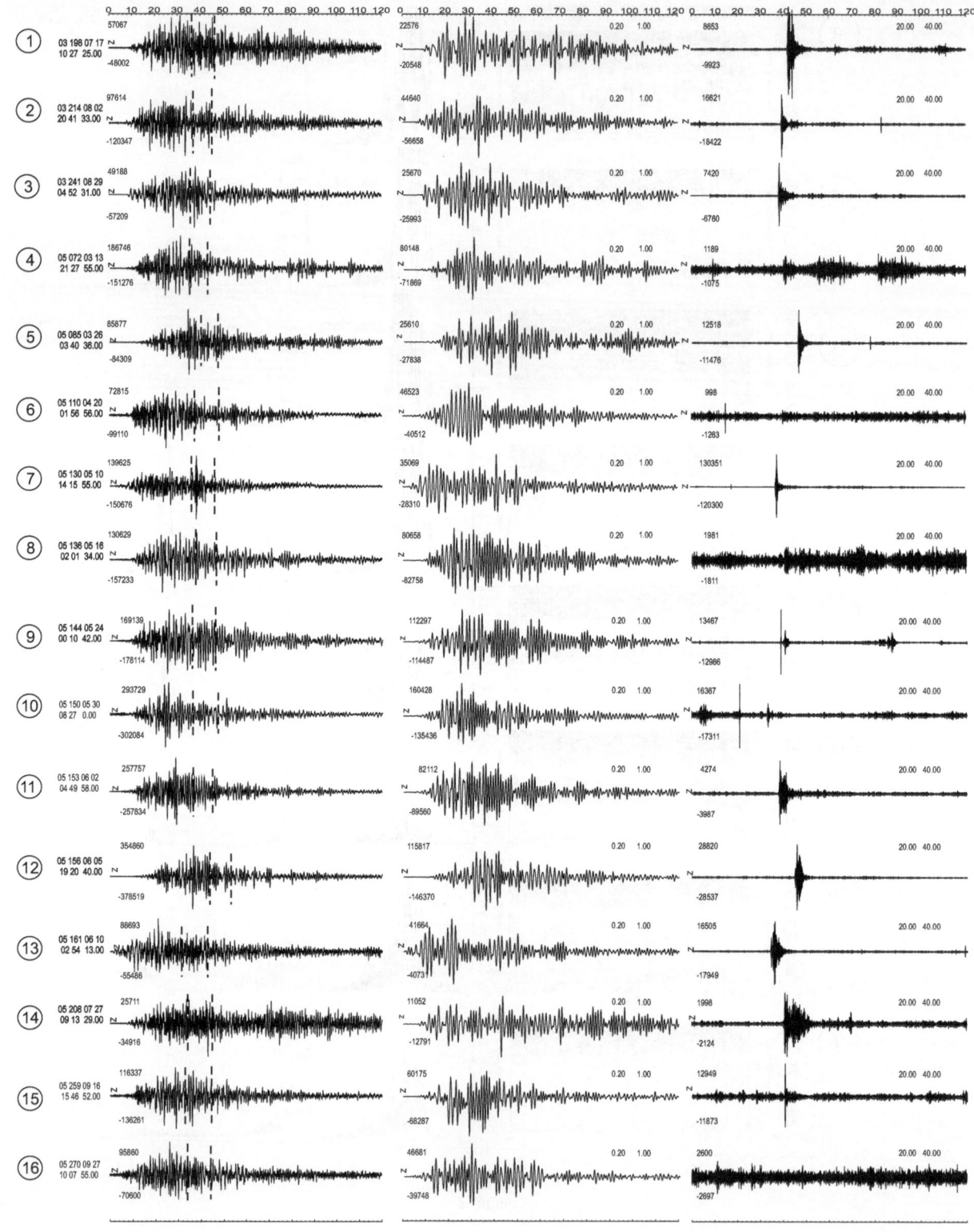

Figure 5

N–S component seismograms recorded at F03J for the 16 explosions, raw and band-pass filtered [0.2, 1.0] and [20.0, 40.0] Hz. Numbers in the *left column* are dates in yyyy_mm_dd_hhmm format; *dashed lines* indicate estimated PAW arrival interval on non filter seismogram; *above* and *below* each seismogram are the maximum and minimum amplitudes in microvolts, below the maximum amplitudes are shown the band-pass filter limits

filtered signals show multiple episodes of energy release. The high-passed waveforms clearly display the PAW in events 1, 2, 3, 5, 7, 11, 12, 13, and 14; while it is absent (or below noise level) for events 4, 6, 8, and 16. It is probably present in the remaining events, albeit with small amplitude.

We made record sections of the coupled airwaves (PAW and/or SAW) recorded at our seismic stations, assuming the crater summit as the location of the source (Fig. 6). With this procedure, we were able to estimate the velocity of sound, using the reduction velocity around the volcano at the time of the explosions, and the explosion origin time (McNUTT 1986; NÚÑEZ-CORNÚ *et al.* 2010). The estimated sonic velocities are displayed in Table 1. While the estimated sound velocities are reasonable considering the atmospheric conditions around the volcano (JOHNSON 2003), the values obtained from the P-wave arrivals are not, except for event 2 (2.35 km/s), which agrees with a reasonable velocity model for the volcano (NÚÑEZ-CORNÚ *et al.* 2010); a result which can be easily explained assuming different locations of the seismic sources as opposed to those of the explosion sources (PAW and SAW) if we assume the explosion take place at the summit crater.

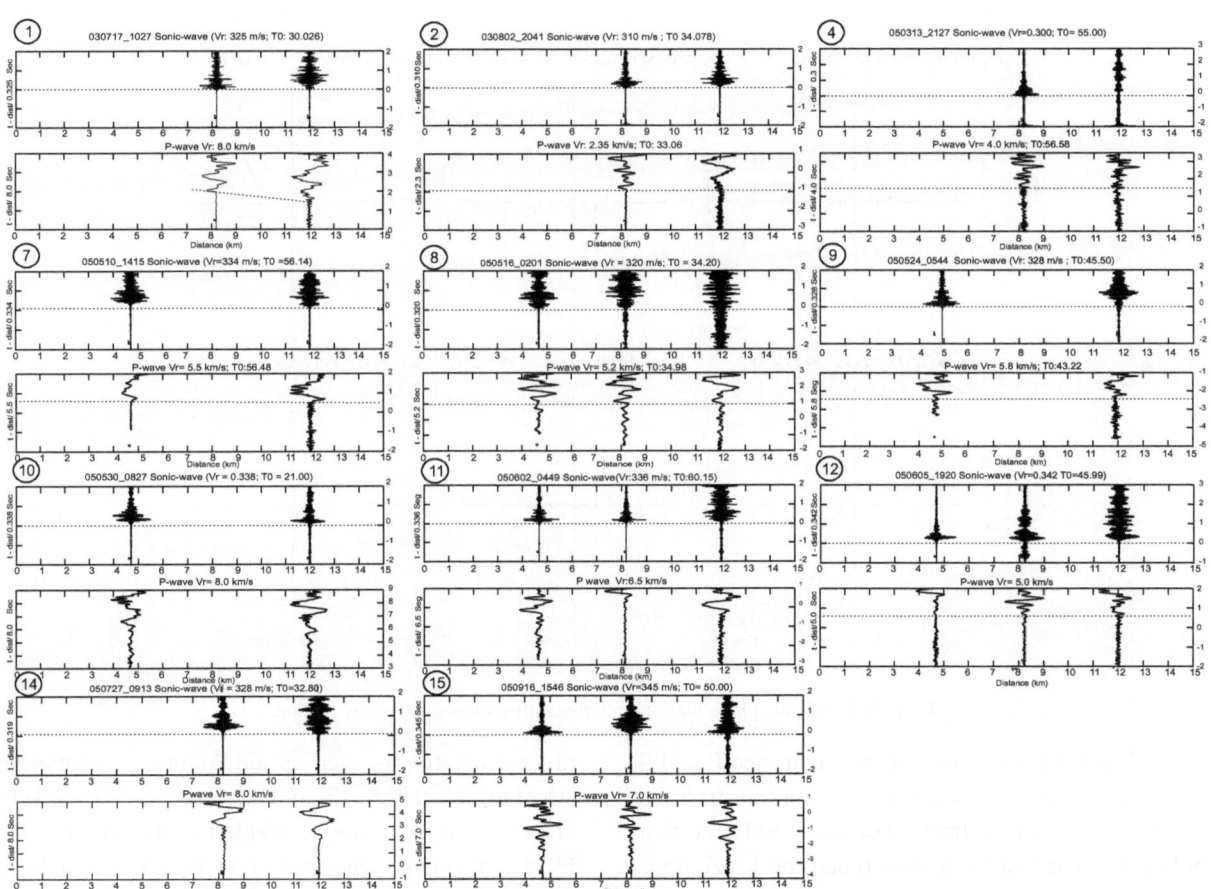

Figure 6

Record sections for P-waves and explosions (PAW) recorded at more than one station

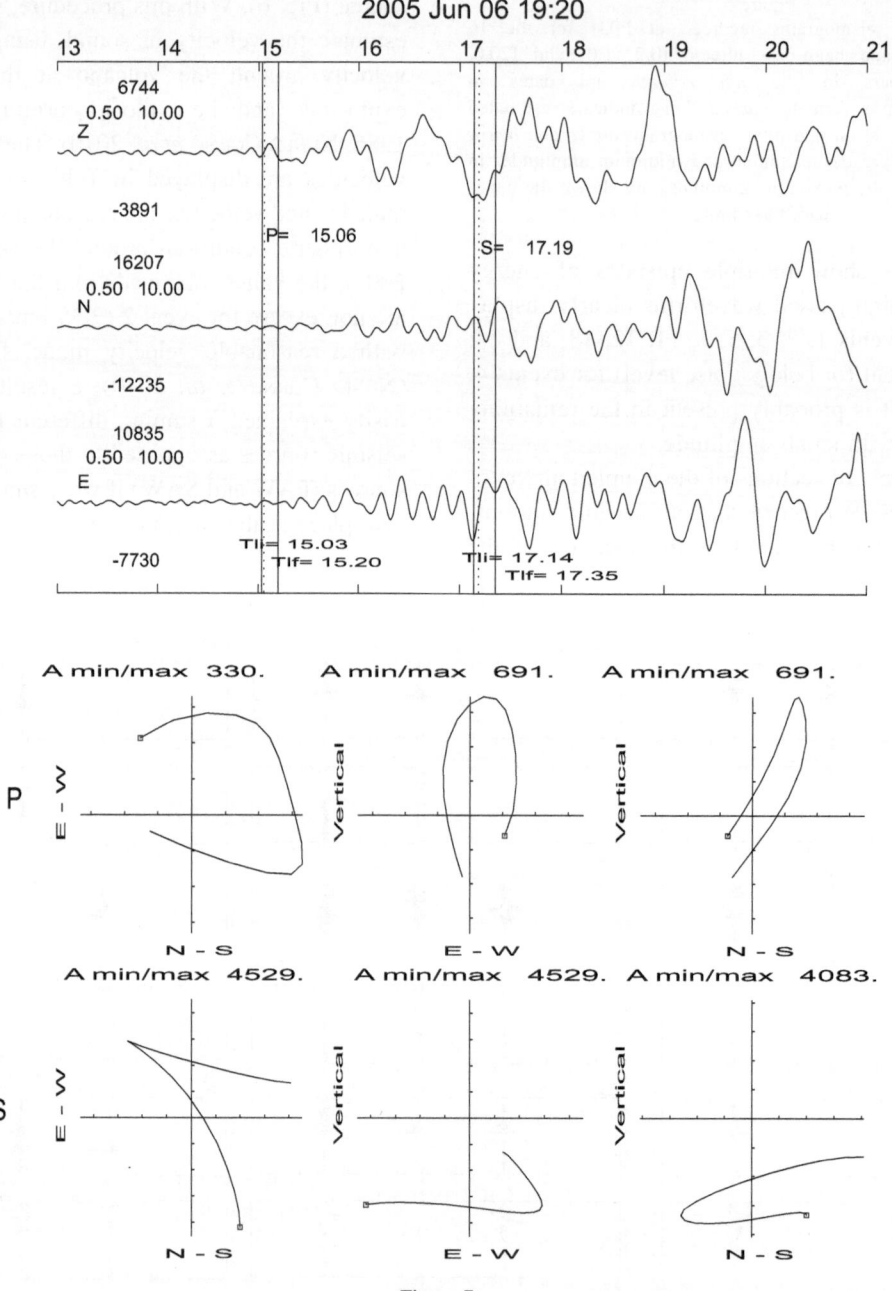

Figure 7
Example of seismic phase and polarity determination using particle motion

To support this assumption, we identified P and S wave arrivals (Fig. 7) using particle motion to identify polarity within an interval of 0.2 s, and used first motion P-wave amplitude to estimate back azimuths and the approximate direction of source locations (Table 1). This procedure is sometimes of limited reliability due to the influence on particle motion of diverse conditions such as topography, and internal structure of the medium (MÉTAXIAN *et al.* 2009); however, in our case it yielded consistent results. F03J S–P times range from 1.51 to 3.32 s (Table 1) implies a range distance to the source, epicentral and depth combined, of 6.83 to 15.01 km, assuming a $Vp = 3.3$ km/s (NÚÑEZ-CORNÚ *et al.* 2010). In the

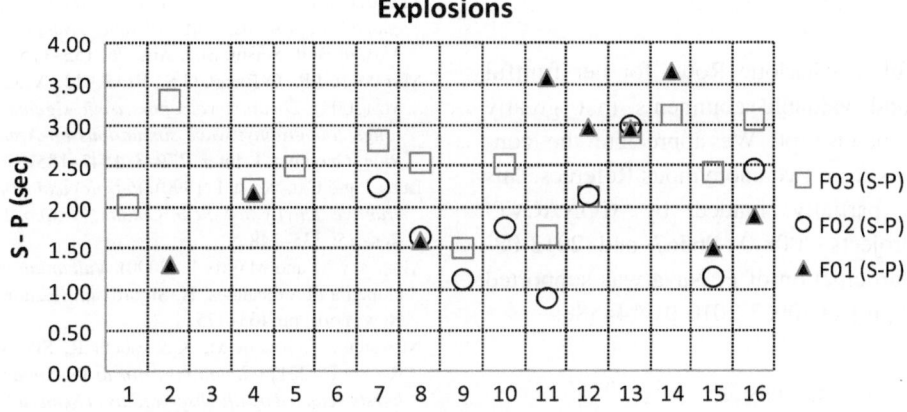

Figure 8
Distribution of S–P times recorded for each explosions at the stations used in this study

case of F02J the range distance, epicentral and depth combined, to the source is 4.07–13.56 km. We note that the different polarities of the P-wave arrivals (on the three components) and the polarization of the P-wave particle motions agrees with the existence of different fault orientations or sources proposed by HILL (1977). In Fig. 8, we plot the S–P times available for the three stations and each explosion, from which it is possible to conclude that the source of the P-waves is different for each explosion.

4. Discussion and Conclusions

At many volcanoes, it is assumed that P-waves and explosions have a common source. Our results indicate that each vulcanian explosion is preceded by a P-wave generated at a different location from that of the explosion center. These sources operate at different places for the various explosions, as follows from the differences in the P-wave to sonic-wave source time offsets. This observation suggests multiple different fluid pathways discharging to the summit crater area, where the over-pressurized fluids can cause explosions.

Our results support diverse scenarios corresponding to various conditions of fluid pressure: The presence of SAW phases documents conditions of over-pressurized fluid (MORRISEY and MASTIN 2000), while PAW generation accompanied by ash and gas emissions, or the simple emission of ash and gas with no coupled wave signal at the seismic station, indicate low pressure conditions. Such variation in pressure conditions and the ensuing degassing phenomena are observed in the activity of several volcanoes; for example, NISHIMURA et al. (2012) found different pressure conditions between gas bursts and vulcanian eruptions occurring in 2007 and 2010 at Semeru Volcano, Indonesia. At Colima, the distinct energy release pulses are revealed in the low-pass filtered seismograms, which indicate that magmatic parcel fluxes are episodic, rather than single events.

The azimuthal distribution of the foci obtained from P-wave first arrivals, S–P times, and the distance of F03J to the hypocentral area reported by NÚÑEZ-CORNÚ and SANCHEZ-MORA (1999) suggest that the P-wave sources have a shallow distribution beneath the volcano. This agrees with a post brittle/ductile seal breakage state described in the FOURNIER (1999) model, in which the hypocenter depth resides at the depth of the transition where the outer, brittle material provides a seal to the pressurized gases evolving from the magma below.

Vulcanian explosions are the ending phase of a series of processes that begin with the forcing of volcanic fluids through the rock matrix; this process involves the opening or propagation of fractures or faulting. Therefore, the magmatic parcels ejected from the crater originate from various shallow sources beneath the edifice, with pathways that merge at the summit.

Acknowledgments

We are grateful to Charlotte Rowe for her fruitful observations and valuable comments that greatly improved the manuscript. We appreciate the constructive reviews from two anonymous Referees.This research was partially funded by CONACyT-FOMIXJAL projects 2008-04-96567 and 2012-08-189963. The participation of F. Nava was supported by CONACyT project I0007-2010-01-144588.

REFERENCES

DÍAZ, S. (1906). Efemérides del Volcán de Colima, Observatorios de Zapotlan y Colima, de 1893 a 1905. Imprenta y Fototipia de la Secretaría de Fomento. MEXICO, 199 pp.

FOURNIER, R. (1999). *Hydrothermal processes related to movements of fluid from plastic into brittle rock in the magmatic-epithermal environment*. Econo. Geol. *94*, 1193–1211.

HAVSKOV, J. and OTTEMÖLLER, L. (1990). *SEISAN earthquake analysis software*. Seism. Res. Let. *70*, 532–534.

HILL, D. (1977). *A model for earthquake swarms*. J. Geophys Res *82*: 1347–1352.

JOHNSON, J.B. (2003). *Generation and propagation of infrasonic airwaves from volcanic explosions*. J. Volcanol. Geoth. Res, *121*, 1–14.

McNUTT, SR, (1986). *Observations and Analisys of B-Type earthquakes, Explosions*, and *Volcanic Tremor at Pavlof Volcano, Alaska*. Bull. Seism. Soc. Am. *76*, 153–175.

MÉTAXIAN, J.P., O'BRIEN, G.S., BEAN, C.J., VALETTE, B. and MORA, M. (2009). *Locating volcano-seismic signals in the presence of rough topography: wave simulations on Arenal volcano, Costa Rica*. Geophys. J. Inter. *179–3*, 1547–1557.

LUHR J and CARMICHEL I (1990). *Petrological monitoring of cyclical eruptive activity at Volcán Colima, México*. J. Volcanol. Geoth. Res. *35*: 335–348.

MORRISEY M and MASTIN L (2000). *Vulcanian eruptions*. In Encyclopedia of Volcanoes, H. Sigurdsson (Editor), Academic Press, New York, pp 463–475.

NISHIMURA T, IGUCHI M, KAWAGUCHI R, SURONO, HENDRASTO M, ROSADI U (2012). *Inflations prior to Vulcanian eruptions and gas bursts detected by tilt observations at Semeru Volcano, Indonesia*. Bull. Volcanol. *74*, 903–911, doi:10.1007/s00445-012-0579-z.

NÚÑEZ-CORNÚ, F.J., SANCHEZ-MORA, C. (1999). *Stress Field Estimations for Colima Volcano, Mexico, Base on Seismic Data*. Bull. Volcanol. *60*, 568–580.

NÚÑEZ-CORNÚ, F.J., SUÁREZ-PLASCENCIA, C., RUTZ-LÓPEZ, M., VARGAS-BRACAMONTES, D., SÁNCHEZ, J.J., (2010). *Comparison of Seismic Characteristics of Four Cycles of Dome Growth and Destruction at Colima Volcano, Mexico, from 1991 to 2004*. Bull. Seism. Soc. Am. *100*, 5A, 1904–1927. doi:10.1785/0120080356.

ROBIN, C., MOUSSAND, P., CAMUS, G., CANTAGREL, J., GOURGAND, A. and VICENT, P. (1987). *Eruptive History of Colima Volcanic Complex (Mexico)*. J. Volcanol. Geoth. Res. *50*, 99–113.

(Received May 7, 2015, accepted August 13, 2015, Published online August 31, 2015)

Pure Appl. Geophys. 173 (2016), 3525–3551
© 2016 Springer International Publishing
DOI 10.1007/s00024-016-1384-y

Pure and Applied Geophysics

CrossMark

Bahía de Banderas, Mexico: Morphology, Magnetic Anomalies and Shallow Structure

Carlos A. Mortera Gutiérrez,[1] William L. Bandy,[1] Francisco Ponce Núñez,[2] and Daniel A. Pérez Calderón[1]

Abstract—The Bahía de Banderas lies within a tectonically complex area at the northern end of the Middle America Trench. The structure, morphology, subsurface geology and tectonic history of the bay are essential for unraveling the complex tectonic processes occurring in this area. With this focus, marine geophysical data (multi-beam bathymetry, high resolution seismic reflection and total field magnetic data) were collected within the bay and adjacent areas during four campaigns aboard the B.O. EL PUMA conducted in 2006 and 2009. These data image the detailed morphology of, and sedimentation patterns within, the Banderas Canyon (a prominent submarine canyon situated on the south side of the bay) as well as the shallow subsurface structure of the northern part of the bay and the submarine Marietas Ridge, which bounds the bay to the west. We find that the Marietas Ridge is presently a transtensional feature; the course of the Banderas Canyon is controlled by extensive turbidite fan sedimentation in its eastern extremity and by structural lineaments to the west; the canyon floor is filled by sediments and exhibits almost no evidence for recent tectonic movements; the southern canyon wall is quite steep and a few sediments are deposited as submarine fans at the base of the southern wall; and extensive turbidite fans form the lower part of the northern canyon wall, producing a gently sloping lower northern wall. We find no evidence for a regional east–west striking lineament between the bay and the Middle America Trench, which casts doubts on the previous assertion that the Banderas Canyon is unequivocally related to the presence of a regional half-graben. Finally, a N71°E oriented normal fault offsets the seafloor reflector by 15 m within the central part of the bay, suggesting that the bay is currently being subjected to NNW–SSE extension.

Key words: Bahía de Banderas, Banderas Canyon, marine geophysics, Canyon morphology, subsurface structure, multi-beam bathymetry.

1. Introduction

The Bahía de Banderas is a broad, tectonically active, coastal embayment located on the Pacific margin of Mexico offshore of Puerto Vallarta, Jalisco (Fig. 1). Geologically, the bay is important because it is the offshore extension of the tectonically active Rio Ameca Rift (e.g., Johnson and Harrison 1989, 1990; Núñez-Cornú et al. 2000, 2002; Arzate et al. 2006), which has been proposed to be the northern boundary of the Jalisco Block (e.g., Johnson and Harrison 1989); a small crustal block which may be in the process of slowly rifting away from the rest of North America (Luhr et al. 1985; Bandy and Pardo 1994; Selvans et al. 2011).

Given its tectonic importance, surprisingly few detailed marine geological and geophysical studies have been carried out within the bay and in the offshore area between the bay and the Middle America Trench (MAT). Existing studies include (1) several cursory bathymetric surveys using conventional wide-beam echo-sounders and satellite altimetry data (Fisher 1961; Dauphin and Ness 1991; Alvarez 2007) and a bathymetry map (Núñez-Cornú et al. 2016) constructed from multibeam data collected during the CORTES 96 and TsuJal projects (Dañobeitia et al. 1997; Córdoba et al. 2014), (2) geological and geochemical studies related to observations of present day submarine hydrothermal activity within the bay (Núñez-Cornú et al. 2000; Taran et al. 2002), and one cursory total field magnetic survey (Alvarez et al. 2010). In addition to these studies, several earthquake studies have been conducted in the area of the Bahía de Banderas (e.g., Núñez-Cornú et al. 2002; Rutz López 2007; Núñez-Cornù 2011; Rutz López et al. 2013) and presently the bay is covered by a local seismic network (Red Sísmica y Acelerométrica Telemétrica de Jalisco, RESAJ) (Núñez-Cornù et al. 2011) operated by the Universidad de Guadalajara,

[1] Instituto de Geofísica, Universidad Nacional Autónoma de México, Ciudad Universitaria, Delegación Coyoacan, 04510 Mexico, DF, Mexico. E-mail: cmortera@geofisica.unam.mx
[2] Instituto de Ciencias del Mar y Limnología, Universidad Nacional Autónoma de México, Ciudad Universitaria, Delegación Coyoacan, 04510 Mexico, DF, Mexico.

Reprinted from the journal

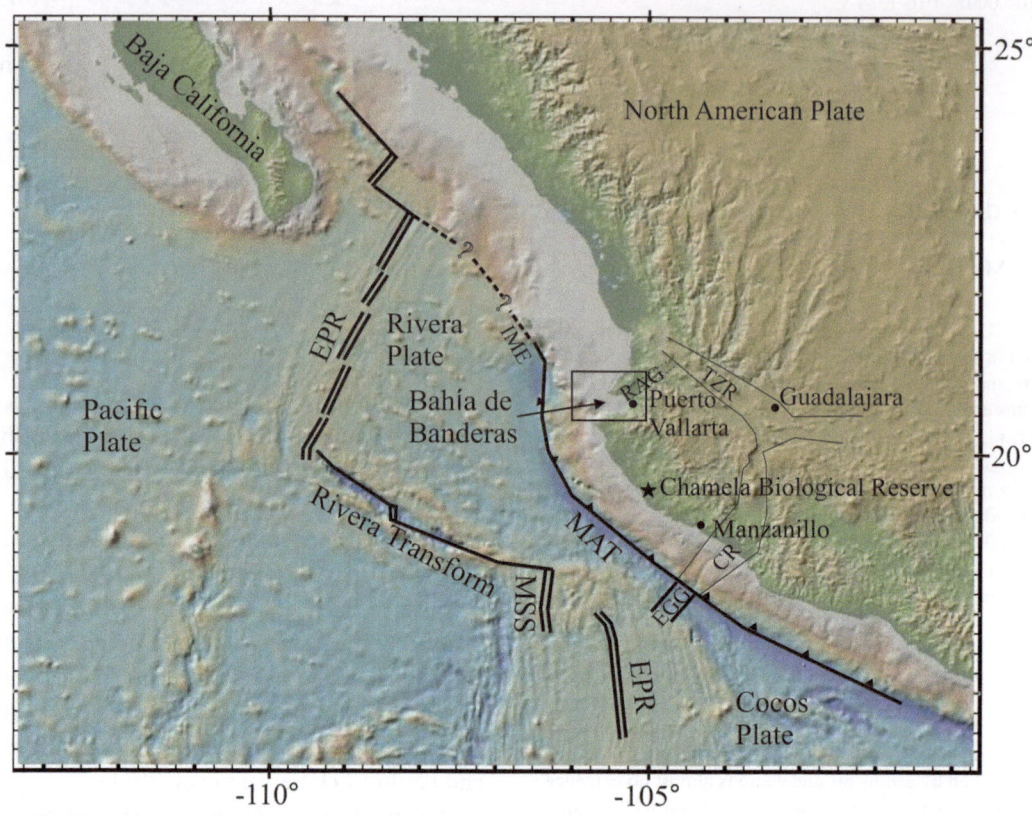

Figure 1

Location of study area (*rectangle*). *CR* Colima Rift, *EGG* El Gordo Graben, *EPR* East Pacific Rise, *IME* Islas Marias Escarpment, *MAT* Middle America Trench, *MSS* Moctezuma Spreading Segment, *RAG* Rio Ameca Graben, *TZR* Tepic Zacoalco Rift. Background map from http://www.geomapapp.org which was constructed with elevation data from http://ned.usgs.gov, http://asterweb.jpl.nasa.gov, and http://topex. ucsd.edu/WWW_html/mar_topo.html

Puerto Vallarta (SisVoc). These studies found many interesting features within the bay worthy of further investigation; for example, active hydrothermal vents along the northern margin of the bay, and a very prominent submarine canyon along the bay's southern margin, and seismic activity within the bay.

The initial formation of the bay has been proposed to be related to the opening of the Gulf of California (e.g., Ferrari 1995) in the middle to late Miocene. Since that time the bay has been subjected to several distinct tectonic environments. Thus, one might reasonably expect to observe a complex array of morphotectonic structures in this area and unraveling their development history will most likely require very detailed datasets.

With the aim of better defining the morphology, subsurface geology and tectonic history of the Bahía de Banderas and surrounding area, detailed total field

marine magnetic data, conventional and multibeam bathymetric data and sub-bottom seismic reflection data were collected from 2006 to 2009 during four marine geophysical campaigns of the B.O. EL PUMA which is owned and operated by the Universidad Nacional Autónoma de México (UNAM); these campaigns are the PMITA01, BABRIP06 and MORTTIC06 campaigns conducted in 2006 and the MORTIC08 campaign conducted in January 2009. Herein we present a detailed analysis of these previously unpublished data.

2. Tectonic and Geologic Setting

The Bahía de Banderas is situated in a tectonically complex region near the northern terminus of the Middle America Trench off Puerto Vallarta, Jalisco,

Mexico (Fig. 1). No clear evidence exists to determine the age of the initial formation of the bay. However, the orientation of the bay, roughly perpendicular to the transform faults of the southern Gulf of California, is consistent with the proposals that it initially formed in association with the opening of the southern Gulf of California in the middle Miocene, between 12 and 14 Ma (e.g., Lyle and Ness 1991; Ferrari 1995; Arzate et al. 2006). Since its initial formation, the area has been affected by a variety of stress regimes; these include: (1) stresses arising from the separation of the Jalisco Block from the rest of the North American Plate, most likely initiated during the early Pliocene (Luhr et al. 1985; Allan 1986; Allan et al. 1991), (2) stresses arising from plate motion changes associated with ridge-trench collisions and the resulting separation of the Rivera Plate from the Cocos Plate which appears to have been initiated in the mid to upper Miocene (Lonsdale 1991; DeMets and Traylen 2000), and (3) stresses arising from the highly oblique subduction between the Rivera Plate and the Jalisco Block along the northernmost part of the Middle America Trench (Kostoglodov and Bandy 1995). Further, Maillol et al. (1997) proposed that the Valle de Banderas graben has been subjected to stresses arising from a regional right-lateral shear couple affecting the NW part of the Jalisco Block. Additionally, there is growing evidence (e.g., Couch et al. 1991; Brown et al. 2009; Bartolomé et al. 2011) that subduction along the Middle America Trench is presently progressing to the northwest along the Islas Marias Escarpment (i.e. a new trench may be in the process of developing along the escarpment). If so, then the area of Bahía de Banderas presently may be subjected to stresses related to the bending of the Rivera Plate as it begins to subduct beneath the escarpment. This could explain the observed deepening of the trench as it approaches the escarpment (Bartolomé et al. 2011).

Structurally, the bay appears to be the offshore extension of the Rio Ameca Rift (Fig. 2) (the Rio Ameca Graben of Johnson and Harrison 1989) which is a regional, NE–SW zone of crustal extension located between the Tepic-Zacoalco Rift to the NE and the Pacific coast, NW of Puerto Vallarta, Jalisco, Mexico, where it is delineated by a broad alluvial plain. This alluvial filled valley has been called the Puerto Vallarta Graben (Ferrari and Rosas-Elguera 2000) or, alternatively, the Valle de Banderas Graben (Arzate et al. 2006). Herein, we will refer to the combination of the onshore, sediment filled, topographic depression and the Bahía de Banderas as the "Puerto Vallarta Graben". We will use the term "Rio Ameca Rift" to refer to the regional extensional zone identified by Johnson and Harrison (1989). Thus, the Puerto Vallarta Graben is the western part of the Rio Ameca rift. Several previous authors have made a distinction between the offshore and onshore parts of the Puerto Vallarta Graben (Arzate et al. 2006; Alvarez 2007). We will refer to the onshore part of the Puerto Vallarta Graben as the "Valle de Banderas", and the offshore part of the Puerto Vallarta Graben will be referred to as the "Bahía de Banderas". Note that, unless otherwise specified, we use the term "graben" in a generic sense, without regards to whether or not it is a half- or full-graben.

The Rio Ameca Rift has been proposed based on structural geology to mark the NW boundary of the Jalisco Block (e.g., Johnson and Harrison 1989; Alvarez 2002; Rutz-López and Núñez-Cornú 2004). However, other investigators, based on petrologic, lithologic and magnetic characteristics, place the limit somewhat further to the north within the Tepic-Zacoalco rift (e.g., Ferrari 1995; Rosas-Elguera et al. 1996; Ferrari and Rosas-Elguera 2000; Urrutia-Fucugauchi and González-Morán 2006). Young fault scarps, thermal springs and seismicity indicate that the western part of the rift (i.e. the Puerto Vallarta Graben) is tectonically active at present (Ferrari et al. 1994; Dañobeitia et al. 1997; Ferrari and Rosas-Elguera 2000; Núñez-Cornú et al. 2000; Alfonso et al. 2003; Taran et al. 2002, 2013; Canet and Prol-Ledesma 2007). However, it is noteworthy that, even with the installation of a local seismic red (RESJAL) in the area of the bay, the bounding faults of this proposed graben are not well defined by presently recorded seismicity (e.g., Núñez-Cornú et al. 2000, 2002; Rutz-López and Núñez-Cornú 2004; Rutz López et al. 2013).

Overall, the study area lies within the granitic Puerto Vallarta Batholith (e.g., Böhnel and Negendank 1988; Schaaf et al. 1993) (Fig. 2) and can be subdivided into six distinct physiographic provinces. Within the confines of the Bahía de Banderas, two

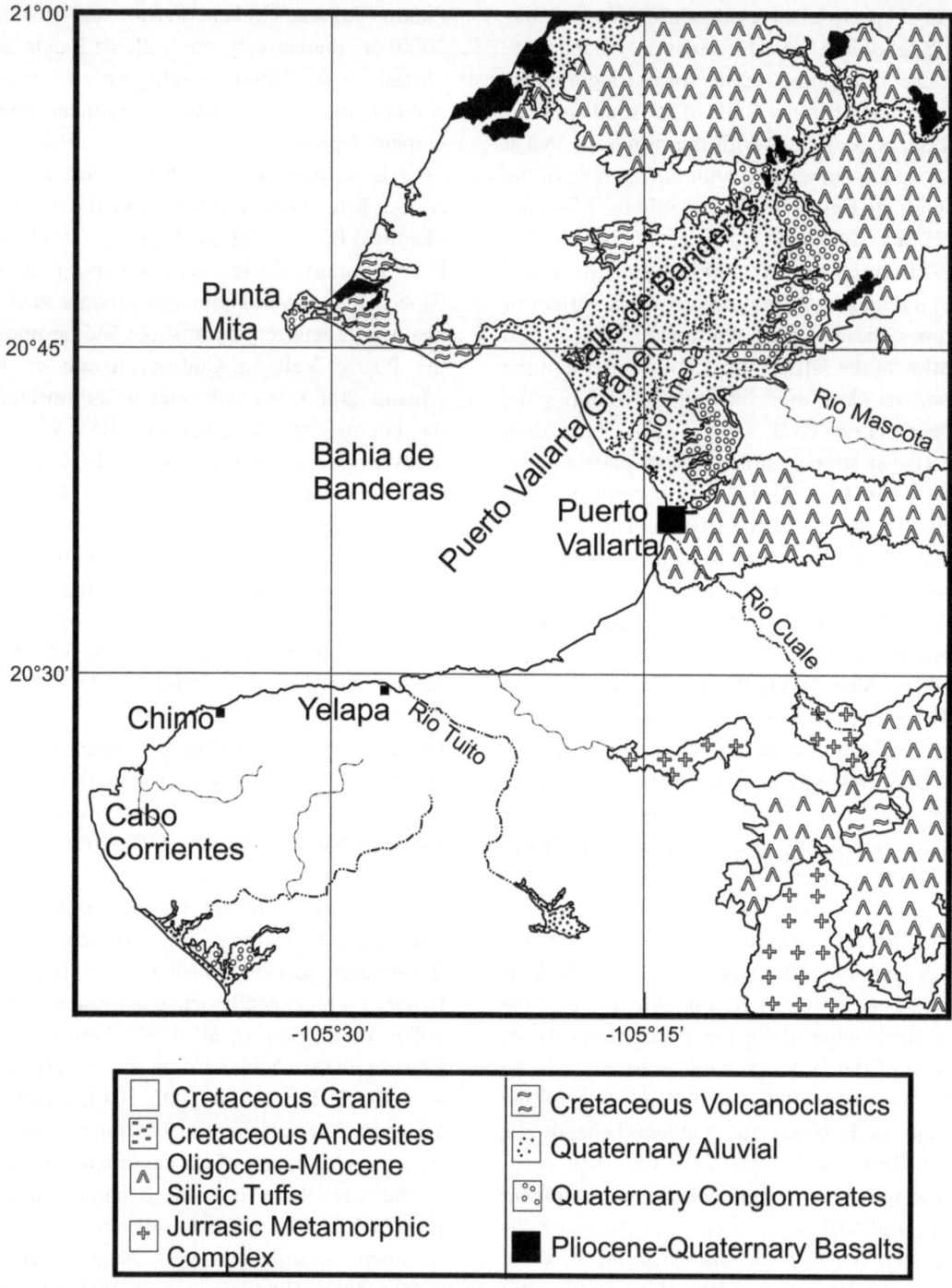

Figure 2
Geologic map of the study area Modified from INEGI (1988)

provinces are observed; herein referred to as the Northern and Southern Bay provinces. The Southern Bay Province is delineated by the presence of a deep

submarine canyon, which we refer to as the Banderas Canyon (Fig. 3) (e.g., Fisher 1961), within which depths reach in excess of 1.5 km at the mouth of the bay just north of Cabo Corrientes. The Banderas Canyon has been proposed, based on bathymetry and magnetic data (Alvarez 2007; Alvarez et al. 2010), to lie within a half-graben structure, the main fault located along the southern margin of the bay. However, no direct evidence (such as seismicity) for such faulting has been presented in the literature and if present, the main fault does not extend onshore where one could easily confirm its existence. In contrast to the Southern Bay Province, the Northern Bay Province is distinguished by a shallow platform (for the most part <100 m). The "Fisura de la Coronas" and associated hydrothermal springs are located within this province (Núñez-Cornú et al. 2000). Taran et al. (2013) found low $^3He/^4He$ isotope ratio values

(0.4 Ra) in this area. The Northern Bay Province is bounded to the west by the Marietas Ridge. Very little is known about the subsurface geology of these marine provinces including the Marietas Ridge.

The third province, the Granitic Highlands Province, bounds the bay to the south and consists almost exclusively of surface exposures of Cretaceous granitic rocks. However, an extensive, east–west trending band of Oligocene–Miocene volcanic tuffs is observed (Fig. 2) at 20°37′N; the tuffs being flanked on the north and south by the granites. No faults have been recognized at the contacts between the tuffs and granites, thus, these tuffs most likely infilled a topographic depression present at the time when the tuffs were deposited. These tuffs intersect the coastline

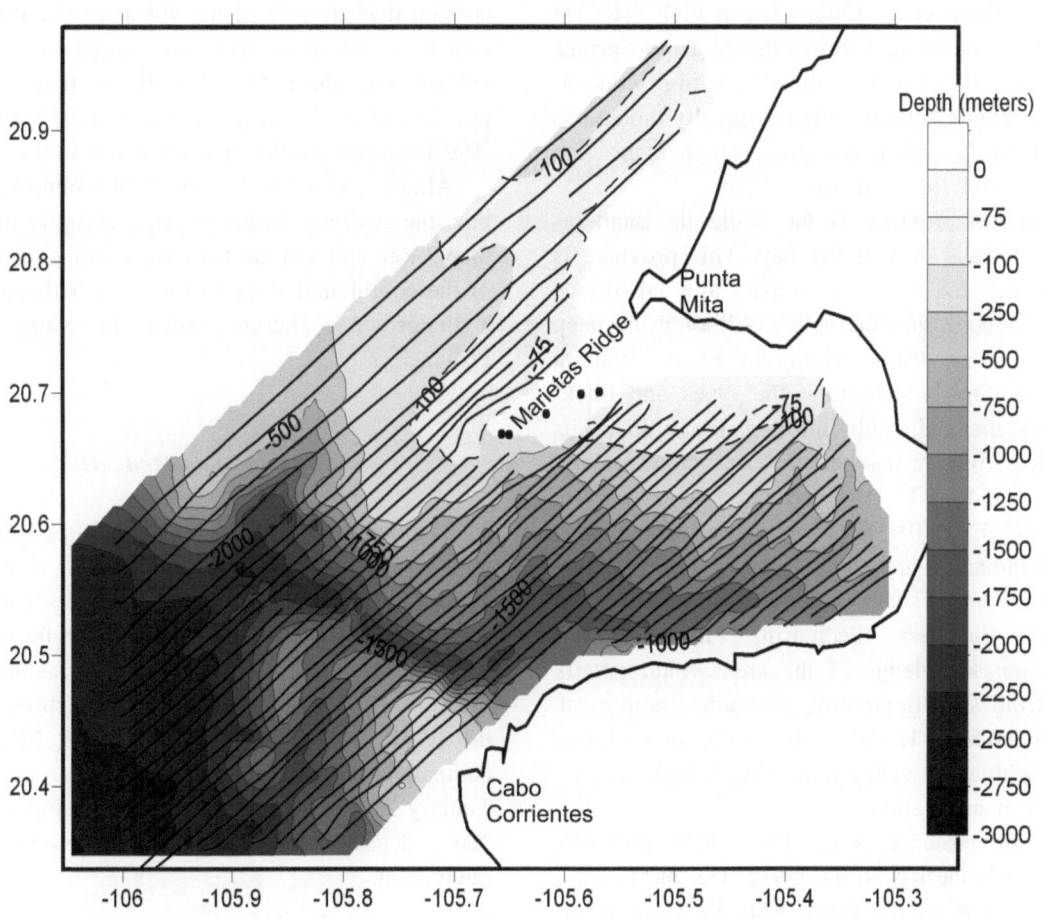

Figure 3

Ship tracks along which magnetic and single beam bathymetric data were collected during the 2006 PMITA campaign. The survey consisted of *41 lines* with an average spacing of 1.5 km. Background bathymetric map (contour interval = 250 m) was constructed using only the single beam echo sounder bathymetric data collected during the PMITA campaign

near the town of Puerto Vallarta, in the vicinity of the mouth of the Banderas Canyon, and, therefore, may be important to the development of the submarine canyon within the bay since the tuffs are more easily eroded than the surrounding granites. This province has undergone several episodes of uplift. Based on the lack of the Sierra Madre Occidental ignimbrites in the Puerto Vallarta Batholith, Rosas-Elguera et al. (1996) proposed that the batholith was uplifted and the ignimbrites eroded prior to the Neogene. Based on the geomorphology (wave-cut terraces, notches, etc.) along the Jalisco coast, Ramírez-Herrera et al. (2011) proposed that the Puerto Vallarta Batholith has been uplifted since at least the Pliestocene to the present, the rates of uplift being 0.7–0.9 m/ka during the Pliestocene and increasing to 3 m/ka during the Holocene. Taran et al. (2013) found high ^{3}He/^{4}He isotope ratio values (2.3 Ra) in the El Tuito Springs (20°22.2′N, 105°26.4′W) indicating a high concentration of mantle helium. This is typically thought to be due to deep crustal fractures, which makes the crust permeable to the mantle helium.

The fourth province is the Valle de Banderas Province located east of the bay. This province is readily delineated by the Quaternary alluvial surface deposits found within the valley. Although no deep wells have been drilled within the Puerto Vallarta Graben to directly determine the types and thicknesses of the sediments infilling the graben, its subsurface structure has been inferred from gravity, magnetic and MT data (Arzate et al. 2006; Alvarez et al. 2010). These data have been interpreted to indicate a graben/half-graben structure filled by up to 2.5 km of sediments near the coast, with the sediment thickness decreasing northeastward. Clearly, our knowledge of the area would greatly benefit from a drilling/coring program. Taran et al. (2013) found high ^{3}He/^{4}He isotope ratio values (up to 4.5 Ra) within the valley indicating a high concentration of mantle helium.

The fifth province is the Punta Mita Province, which bounds the bay to the north. The most distinguishing feature of the Punta Mita Province is the great variation in the types of rocks outcropping within the province. In addition to outcrops of the granitic rocks typical of the Granitic Highlands Province to the south, outcrops of Paleozoic metamorphics, marbles and silicic tuffs, and Miocene (approximately 10 Ma) basalts and basaltic dikes are observed (INEGI 1988; Fernández de la Vega-Márquez and Prol-Ledesma 2011). The basalts are reported to have "erupted in submarine conditions forming massive lava, pillow lava and pillow breccias intercalated with repetitiously and thinly bedded mudstone (turbidite deposits) and ash beds" (Jensky 1974; Sawlan 1991). K–Ar ages for these basalts range from 7.5 to 12.5 Ma (Sawlan 1991). The geology of the Marietas Ridge is poorly studied, however, there are reports that the Islas Marietas, which lie along the Marietas Ridge, are predominantly of volcanic origin (e.g., Cano Sánchez 2004). Given this and the location of the ridge near Punta Mita, we tentatively propose that the ridge is the offshore extension of the Punta Mita Province. If correct, then the age of the volcanics comprising the islands would most likely correspond to that of the volcanic episodes noted within the onshore part of the province (7.5–12.5 my). Taran et al. (2013) found low ^{3}He/^{4}He isotope ratio values (0.6 Ra) in this area.

Almost immediately outside the confines of the bay, the seafloor depths increase abruptly to greater than 3 km, and a broad flat terrace forms the majority of the continental slope in this area. This area is the sixth province, herein called the Slope Terrace Province.

3. Data and Methods

The data used in this study consists of previously unpublished total field magnetic data, single beam and multibeam bathymetric data, seafloor backscatter strength data and sub-bottom seismic reflection data. These data were collected during four campaigns of the B.O. EL PUMA conducted since 2006, namely, the PMITA01, BABRIP06 and the MORTIC06 campaigns in 2006 and the MORTIC08 campaign in January 2009. For all campaigns, the ship's location was determined using non-differential GPS navigation.

3.1. Magnetic Data

The total field magnetic data presented herein were collected along 41 profiles during a single cruise, PMITA01, conducted during 12–18 January 2006, using a GEOMETRICS G877 marine proton precession magnetometer. The data coverage is illustrated in (Fig. 3). The magnetic sensor was towed 250 m behind the ship to minimize the effects of the ship (a 50 m long, steel hulled vessel) on the measurements. Measurements were taken every 2 s. The location of the sensor behind the ship was calculated as the data were recorded using GEO-METRICS MAGLOG LITE software.

The recorded total field magnetic measurements were reduced to magnetic anomalies by first sub-tracting the reference value of the Earth's magnetic field, and then correcting for diurnal variations and the effects of the ship's heading. The IGRF11 model (IAGA, working Group V-MOD 2010) was used to calculate the reference value for each measurement. The calculated magnetic field values are definitive for dates prior to 2010 (i.e. for all our data).

To correct for diurnal variations, a permanent base station was installed onshore within the UNAM Biological Reserve located near Chamela, Jalisco (19°29′56.1″N, 105°02′32.1″W; ~120 km southeast of the bay). A site-selection magnetic survey was run to find a location where the local horizontal magnetic gradient was <0.1 nT/m. Measurements of the total field were made with a GEOMETRICS G856AX proton precession magnetometer at 1 min intervals. The diurnal variations (Fig. 4) for the survey dates were fairly regular with variations being for the most part less than ±20 nT. According to the Dst-Indice of the WDC for Geomagnetism, Kyoto, Japan, no magnetic storms occurred during the survey period.

After correcting for diurnal variations, the data were then corrected for ship's heading following the methods of Bullard and Mason (1961), Whitmarsh and Jones (1969) and Buchanan et al. (1996). Finally, the anomaly data were gridded (100 m × 100 m grid

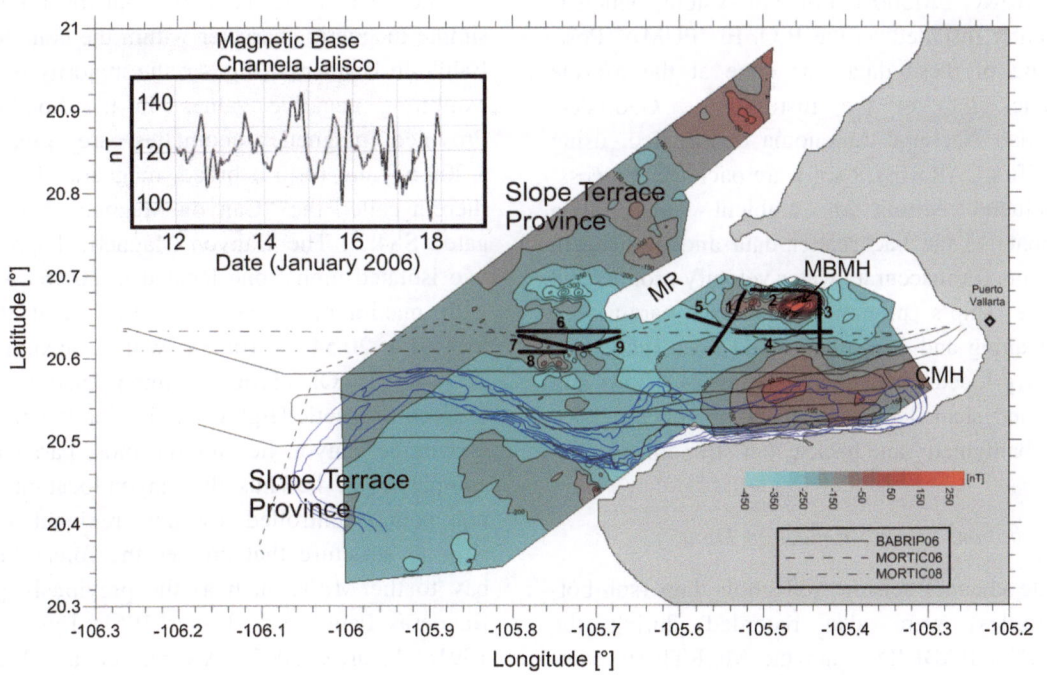

Figure 4

Map showing location of ship tracks along which multibeam bathymetric and subbottom seismic reflection data was collected during the BABRIP06, MORTIC06 and MORTIC08 campaigns of the B.O. EL PUMA superimposed on the magnetic anomaly contour map (contour interval = 50 nT) constructed from the data collected during the PMITA campaign. *Numbered bold lines* locate the seismic reflection profiles illustrated in the various *figures* of this article. The *solid line* contours (from the multibeam data) reveal the canyon floor. The *inset* is a plot of the diurnal variations recorded at the Chamela base for 12–18 January 2006. *CMH* canyon magnetic high, *MBMH* Mid-Bay magnetic high, *MR* Marietas Ridge

node spacing) and contour maps of the data were constructed.

3.2. Bathymetric and Seafloor Backscatter Strength Data

Depth measurements were also collected during the PMITA campaign using the Kongsberg ES60 (with a 38 kHz transducer) single beam echosounder. These data were also collected along the ship tracks shown in Fig. 3: however, only the data collected in the areas not covered by the multibeam data are used in the construction of the final bathymetry map of the bay.

Multibeam bathymetry and seafloor backscatter data were recorded during the MORTIC06 (12–13 October 2006), BABRIP06 (5–11 October 2006) and the MORTIC08 (3–23 March 2009) campaigns of the B.O. EL PUMA. The locations of the ship tracks along which these data were obtained are illustrated in Fig. 4. These data were obtained using the KONGSBERG EM300 multibeam system, which is permanently installed on the B.O. EL PUMA. Post-processing of these data was done at the Marine Geophysics Lab of the Instituto de Geofísica, Universidad Nacional Autonoma de Mexico, using IFREMER's CARAIBES software package. Processing included editing of ambient noise, gain adjustments to the backscatter data and, if needed, adjustments for inaccurate water velocity profiles and inaccurate ship's motion calibration parameters. After cleaning and adjustments had been made, the data was gridded (grid node spacing of 30 m) and contour and shaded relief maps were generated for both the bathymetry and backscatter strength (Fig. 5).

3.3. Sub-bottom Seismic Reflection Data

Single channel seismic reflection data (sub-bottom profiles) were also recorded during the MORTIC06, BABRIP06 and the MORTIC08 campaigns of the B.O. EL PUMA, concurrent with the collection of the multibeam bathymetric data (see Fig. 4 for profile locations). The sub-bottom seismic reflection data were collected using the Kongsberg TOPAS-PS18 Parametric Sub-bottom Profiler, which is also permanently installed aboard the B.O. EL

PUMA. The source pulse was a 1.5–5.5 kHz chirp waveform, 15 ms sweep. The sample rate for recording the returning signal was 33 μs. Although the p-wave velocity within the sediments is unknown, we estimate that the vertical resolution is less than 1 m (more details of the system specifications can be found at the Kongsberg web page). The data were processed during their collection (application of match filter, time varying gain and instantaneous amplitude processing) and analog (gif-files) displays of the resulting profiles were made and stored along with the raw field data. Post-cruise processing was limited to gain adjustment and redisplay using the TOPAS-REPLAY software. Depth sections were made using a constant velocity of 1450 m/s.

4. Results

4.1. Magnetic Signature of the Bahía de Banderas

The map of the magnetic data illustrates a very simple magnetic character within the confines of the Bahía de Banderas (Fig. 4); the majority of the area exhibiting negative values. In the Southern Bay Province magnetic anomalies are greater than −200 nT and form a broad, magnetically high area (herein called the "Canyon Magnetic High") elongated S84°E. The Canyon Magnetic High contains two isolated highs; one located at 20.56°, −105.48° with maximum value of −25 nT, and the other located at 20.54°, −105.38° with a maximum value of −47 nT. Of particular importance is that the Canyon Magnetic High is confined to the bay and the submarine canyon lies for the most part within the anomaly (Fig. 4). Thus, the canyon location is clearly not being controlled by any regional east–west striking structure that crosses the forearc from the bay to the MAT, such as the previously proposed Banderas Fault (e.g., Fisher 1961; Lyle and Ness 1991; Alvarez 2007; Alvarez et al. 2010). We interpret the Northern Bay Province to be an overall magnetically low area, with values lower than those of the Southern Bay Province, however, given the shallow depth in the northernmost part of the bay, we were unable to collect magnetic data in that area during the BABRIP06 campaign. This overall

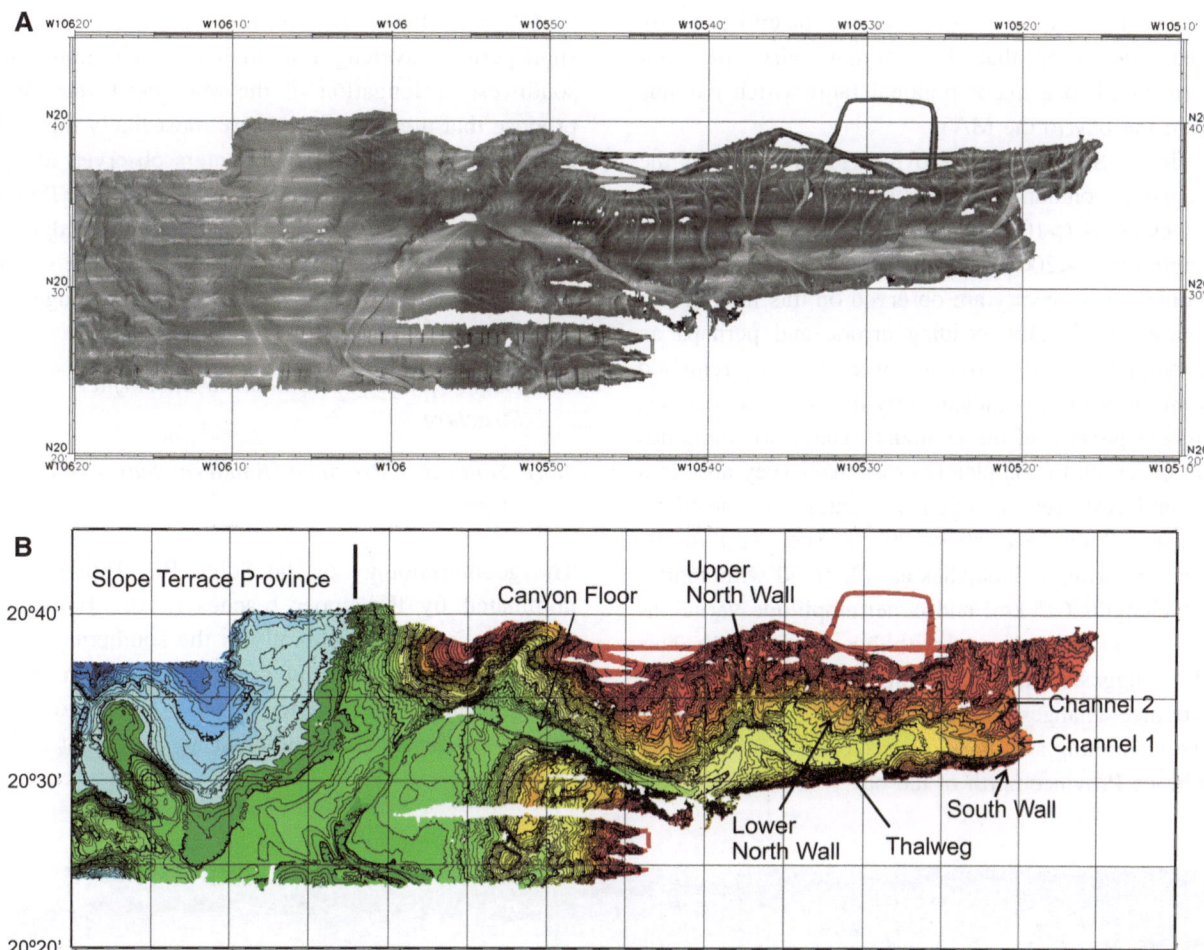

Figure 5
Seafloor backscatter strength image (*top*) and bathymetric contour map (*bottom*) constructed from the new multibeam data (contour interval for the bathymetry map is 50 m)

negative region is disrupted in its western half by a prominent east–west trending magnetic high (herein called the Mid-Bay Magnetic High) where amplitudes reach up to 413 nT. The Mid-Bay Magnetic High appears to extend northwestward across the Marietas Ridge and into the Slope Terrace Province where it may connect with a weak, NNW-SSE orientated, magnetic-high.

Between the Mid-Bay Magnetic High and the Canyon Magnetic High, one observes a magnetic low elongated east–west. Within the bay south of the Mid-Bay Magnetic High, this low magnetic anomaly extends along the boundary between the Southern and Northern Bay Province (i.e. along the upper part of the northern flank of the canyon). This magnetic low

appears to extend across the Marietas Ridge and into the Slope Terrace Province where a similar east–west oriented magnetic low is observed. However, the magnetic low in the Slope Terrace Province is shifted by about 2 km to the north relative to its counterpart within the bay. This is consistent with a small amount of northward translation of the continental slope region relative to onshore area noted to the south (Bandy et al. 2005; Urías Espinosa et al. 2016), as well as with a slight northward translation of a forearc block due to the highly oblique convergence of the Rivera plate with respect to the North American plate in this region (e.g., Kostoglodov and Bandy 1995). It is important to note that this prominent elongated magnetic low in the Slope

Terrace Province does not cut across the entire survey area, suggesting that this anomaly also does not correspond to a major regional fault which extends from the bay to the MAT.

In the Slope Terrace Province south of 20.5°N the magnetic contours are quite smooth exhibiting long wavelengths (>10 km) and fairly small variations in amplitudes (<200 nT) (note: some artifacts of the acquisition geometry are observed on this map in this region; specifically, heading errors, and perhaps an incomplete diurnal correction, were not fully removed from the data as is indicated by the small deflections, zig-zag pattern, of the contours). Here, no anomalies are observed to completely cross the survey area in a general east–west direction. In contrast, in the Slope Terrace Province north of 20.5°N (i.e. west of the Marietas Ridge), anomalies are observed with shorter wavelengths (<5 km) and larger amplitude variations (up to 400 nT) compared with those to the south. This also suggests that the geology in Slope Terrace Province changes seaward of the bay, most likely due to faulting and associated volcanism in the Slope Terrace Province north of the bay.

Of particular interest are the very small, circular, short period wavelength anomalies present along the southwest prolongation of the Marietas Ridge. We propose that these anomalies are most likely the SW continuation of the volcanic centers observed along the Marietas Ridge and in the onshore area near Punta Mita. Thus, we consider that the Marietas Ridge is part of the Punta Mita Province and that these small anomalies mark the southwest extent of the ridge.

4.2. Geomorphology and Shallow Subsurface Structure

4.2.1 South Bay Province (Banderas Submarine Canyon)

The geomorphology of the South Bay Province is dominated by the upper reaches of the Banderas Canyon (Fig. 6), which follows the southern shoreline of the Bahía de Banderas until Cabo Corrientes where it abruptly shifts to a WNW orientation, an orientation nearly perpendicular to the coast. The survey covers 100 % of the canyon between

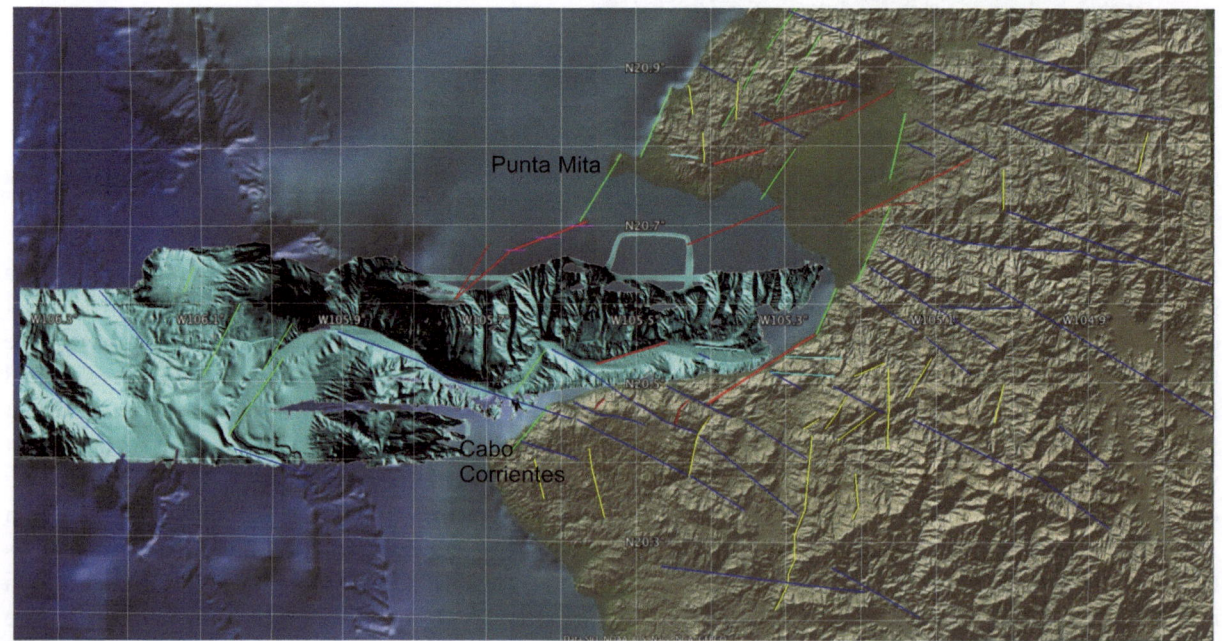

Figure 6
Shaded relief bathymetric map of the Bahia de Banderas Canyon constructed from the multibeam data of the EL PUMA campaigns. Background map, constructed from Google Earth and GeomapApp (http://www.geomapapp.org) illustrates the onshore structural lineaments of the area

105°20′W and 106°03′W (within the Slope Terrace province). The south side of the canyon has a single steep wall whose overall orientation is N85°E from 105°20′W to 105°50′W at which point the southern wall of the canyon abruptly changes to N110°E and continues along this azimuth until 106°05′W, the beginning of the flatter Slope Terrace Province. Relief of the south wall exceeds 1 km within the bay west of 105°20′W. In contrast, the northern wall of the canyon (Figs. 7, 8) has a steep upper wall that is cut by numerous dendritic channels, and a relatively gentle lower wall where the dendritic drainage pattern changes to a series of parallel linear channels, which empty into a flat canyon floor. Given the fan morphology of the gently dipping lower northern wall, it is most likely made up of unconsolidated turbidite fan deposits. This gross morphology is best illustrated in the 3D image of the seafloor backscatter strength image draped on the bathymetry (Fig. 8).

4.2.1.1 Deflections in Canyon Orientation The canyon exhibits sharp deviations in its course at several locations. To the east the deviations appear to be controlled by the presence of turbidite fans, whereas to the west they are structurally controlled (Figs. 6, 7, 9). These deviations can be divided into two groups based on the changing azimuths of the canyon segments: namely, those where the canyon is oriented N45E (canyon segments 3 and 5 have this

orientation) and those where the canyon is oriented N110E (canyon segments 4, 6 and the western part of segment 2 have this orientation). The N110E trending segments have been previously noted to be aligned parallel to major lineaments observed on the satellite images of the adjacent onshore area south of the canyon (Núñez-Cornú et al. 2016). The N110E lineaments dominate south of the bay, whereas the N45E lineaments are mainly found in the Punta Mita Province. Thus, it appears that these two groups of lineaments intersect in the western part of the canyon and that they control the canyons course. A more detailed analysis of these trends and their relation to the canyon can be found in Núñez-Cornú et al. (2016) who reported on the multibeam data collected in the Bahía de Banderas during the 2014 TsuJal project.

4.2.1.2 Canyon Floor The canyon floor is broad (up to 2 km wide) with a very gentle down-canyon dip. This, along with the low backscatter strength (Fig. 5a), indicates that the canyon floor is most likely formed by sediments ponded in the canyon axis. Between 105°20′W and 105°41′W, the canyon is made up of three arcuate segments, the eastern two segments being concave to the north and the western most being concave to the NW (Fig. 9). The floor of the canyon here gradually flattens from 1.8° in the east to 0.8° to the west. With the exception of the marked change in slope (from 3.7° to 1.8°) at the east end of the survey area, no reversals in slope nor

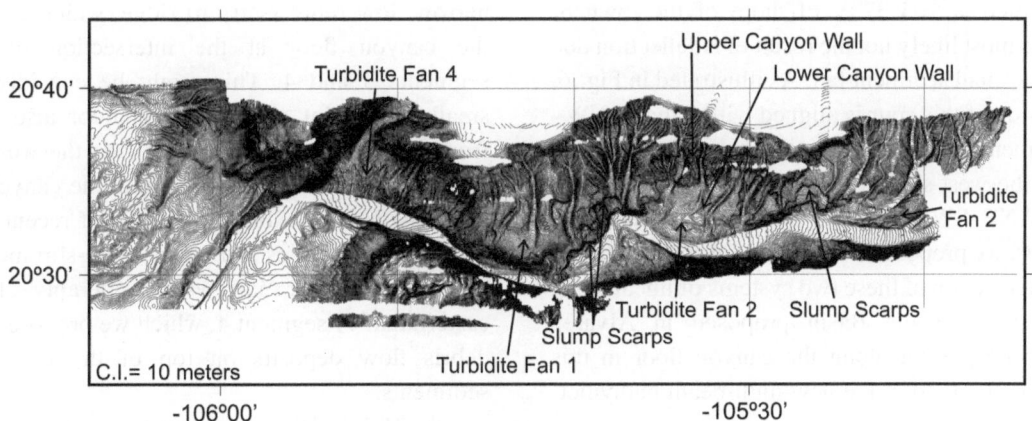

Figure 7
Detailed bathymetric map of the Banderas Canyon (contour interval 10 m) illustrating the extensive turbidite fans and slumps on the northern canyon wall

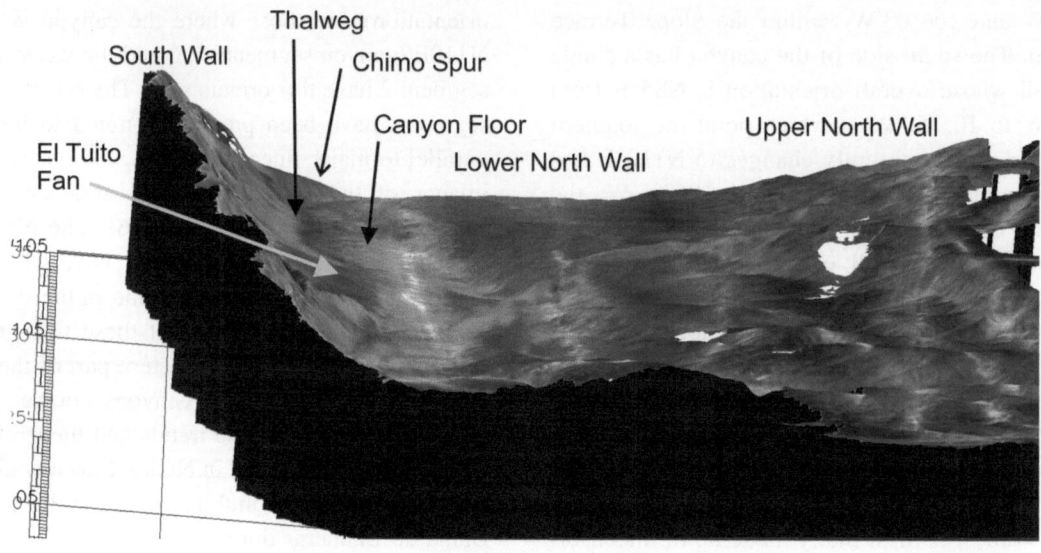

Figure 8

3-D image of the seafloor backscatter mosaic draped on bathymetry; view looking to the west from the east end of the survey area. Note that the present day active channel cuts into the sediments on the south side of the broad canyon floor. Note also that the northern canyon wall is made up of a steep highly eroded upper canyon wall and a gentle sloping, smoother, lower canyon wall

knickpoints [areas of anomalously steep slope (Mitchell 2006)] are noted, suggesting the lack of recent tectonic movements in this area of the canyon.

Two spurs in the canyon floor are located between the three arcuate segments. The easternmost of which (at 105°27′W) is due to a deflection of the channel by a submarine fan (herein called the El Tuito Fan) located within the canyon; the sediments of the fan originate from the El Tuito River, which intersects the coast near the town of Yelapa (Fig. 2). The second spur (located at 105°37′W offshore of the town of Chimo) is most likely not the result of a deflection due solely to a small sediment fan. As illustrated in Fig. 6, the east side of this spur is aligned with a major NW–SE lineament observed south of the canyon onshore, whereas the west side of the spur is aligned with the system of NE–SW lineaments found in the Punta Mita area. Thus, we propose that the western spur is formed by the intersection of these two systems of lineaments. Finally, the Majagua Basin proposed in Alvarez (2002) to be present along the canyon floor in this area is not observed in the new multibeam bathymetric map.

West of 105°41′W the canyon segments are more linear. The canyon floor exhibits a continuous gentle dip ranging from 1° along segment 4 (located between 105°41′W and 105°55′W) to 0.7° along canyon segment 5 (located between 105°55′W and 106°03′W). The Yalapa and Cabo Corrientes basins proposed in Alvarez (2002) to be present along the canyon floor in this area are not observed in the new multibeam bathymetric map. Segment 4 again has an orientation parallel to the southern system of lineaments and we propose that the course of the canyon along segment 4 is being controlled by this fracture system. Recent tectonic activity is evidenced by a narrow, low relief (<10 m) ridge, which cuts across the canyon floor at the intersection of canyon segments 3 and 4. This might be considered as a small knickpoint as the canyon floor also steepens from 0.8° to the east to about 1° to the west of this point. This is the only place along the canyon where the canyon floor exhibits any sign of recent tectonic activity, and implies some recent dip-slip movement. The canyon floor narrows considerably along the eastern half of segment 4, which we propose is due to debris flow deposits on top of the canyon floor sediments.

At 105°55′W, the course of the canyon is deflected sharply (almost 90°), taking on the NE–SW orientation that is parallel to the system of lineaments of the Punta Mita area. Thus, we propose

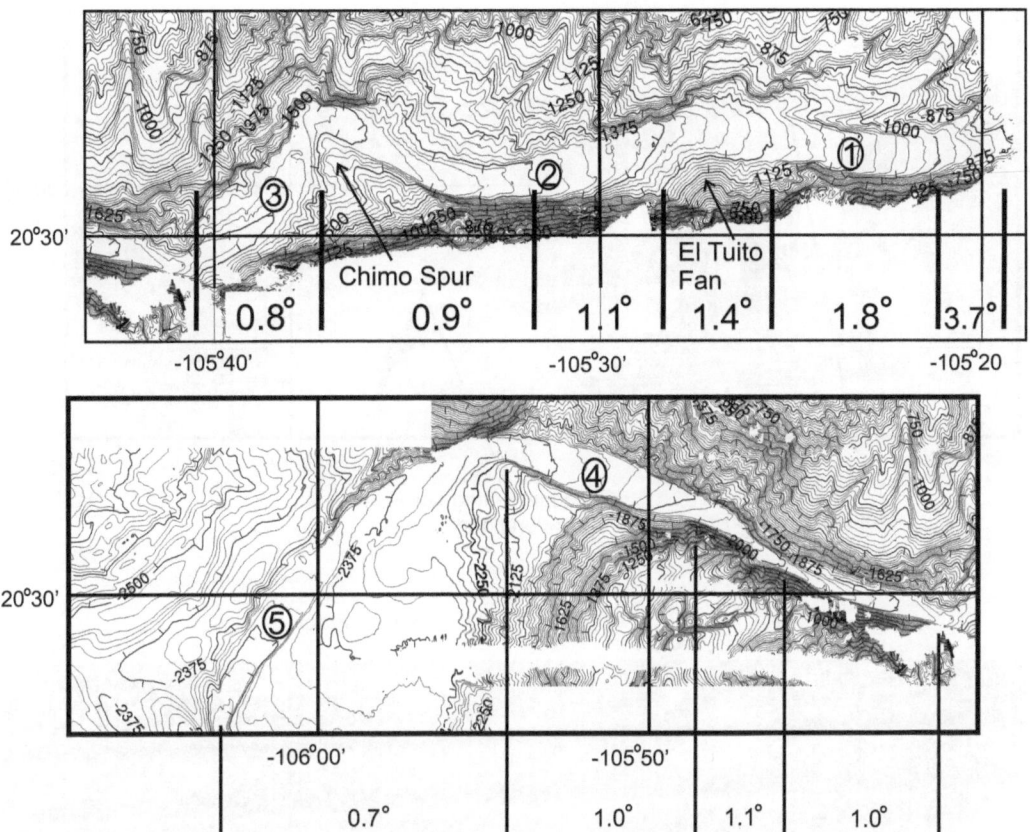

Figure 9
Contour map of the Banderas canyon (contour interval is 25 m). Numbers along the canyon are the slope (in degrees) of the canyon floor. Note that the canyon steepens east of 105°20′W where it begins bends northward towards its terminus (not imaged) near Puerto Vallarta. *Circled numbers* designate the canyon segment

that this abrupt deflection of the canyon is controlled by this fracture system. Indeed, segment 5 appears to run on the east flank of a NE–SW oriented bathymetric ridge. The slope of the canyon floor is 0.7° along segment 5, and the canyon runs out of the survey area at 106°04′W. Again, there are no slope reversals or knickpoints along segment 5 to indicate recent tectonic activity in this area.

4.2.1.3 Mass Wasting Features in the Canyon The northern wall of the canyon exhibits signs of abundant slope instabilities and mass wasting whereas, the southern wall appears to be quite stable (Fig. 7). Perhaps this is due to differences in the type of rock forming the two walls. Although no dredge sample have been collected in this area, it is likely that the southern canyon wall is made up of granites. Given the dendritic drainage pattern, the steep upper part of

the northern canyon wall may also be made up of granites or highly consolidated sediments. The geometry of the more gently sloping lower part of the northern wall suggests that this area is made up of several large turbidite fans formed by sediments flowing into the canyon from the north. Figure 10 illustrates that the lower canyon wall adjacent to the canyon floor on the north side is quite steep and appears to truncate the main channels of the turbidite fans, similar to that observed in the Capbreton Canyon off the coast of Spain and France (Mulder et al. 2004). The knickpoints within these channels formed by the truncation do not show any signs of northward retreat, as they should if these channels were active. This is good evidence that, recently, the main channel (i.e. the canyon thalweg) has had the most activity; activity sufficient to erode the frontal part of the fans (Neil Mitchell, personal communication). However,

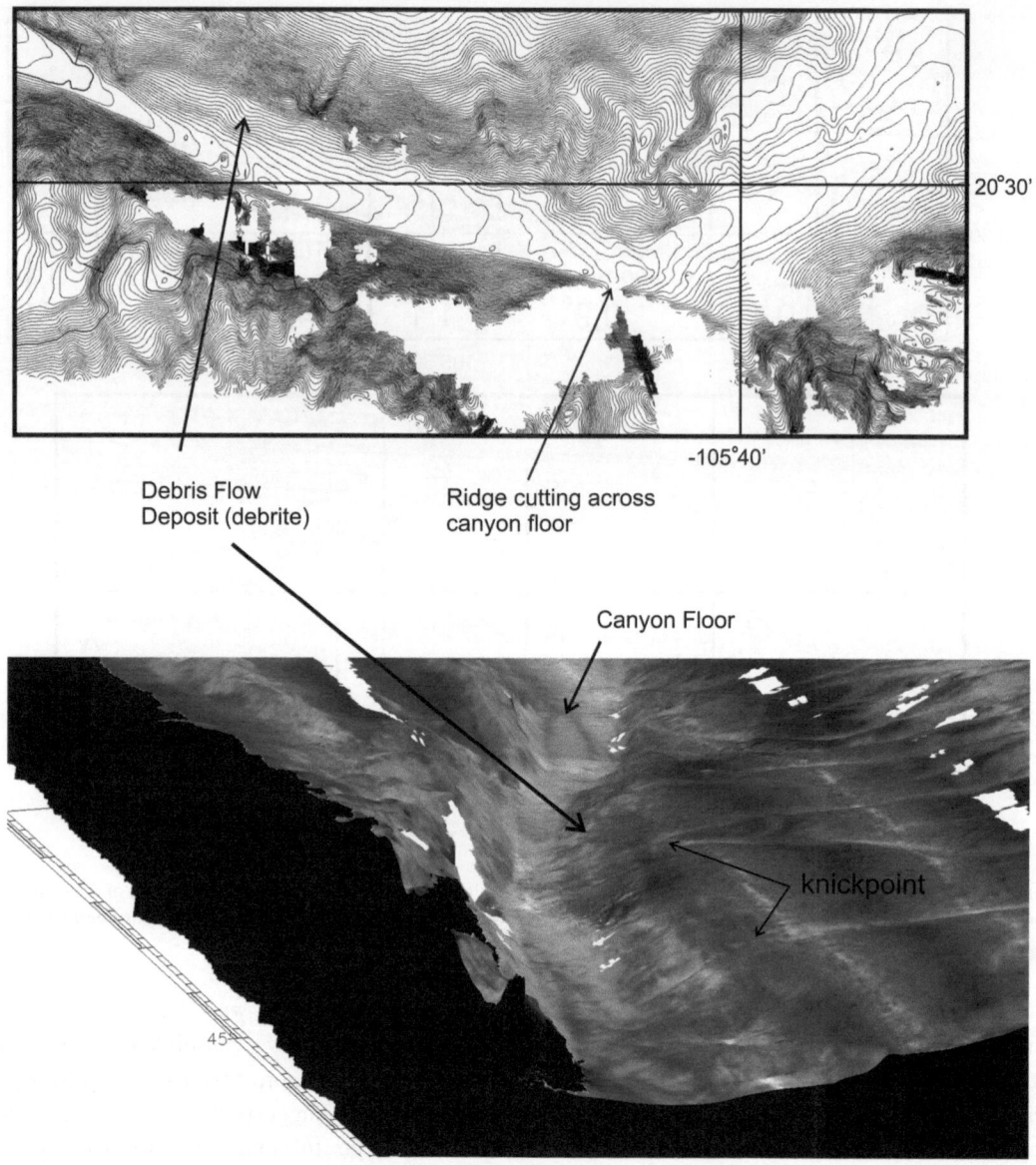

Figure 10

Top Bathymetric contour map (contour interval = 10 m), illustrating the low relief ridge cutting across the canyon floor between canyon segments 3 and 4. This is the only noticeable disruption of the canyon floor sediments in the survey area. *Bottom* 3-D image of seafloor reflectivity image draped on bathymetry illustrating (1) the sediments being deposited on top of the canyon floor, thus reducing the width of the canyon floor, and (2) the lack of a nothward regression of the knickpoints of the main drainage channels of the northern wall

sediments of the westernmost of the turbidite fans on the lower northern wall appear to have flowed over the canyon floor reducing its width at 105°45′W (Fig. 10). Thus, at least some of these channels have been recently active, perhaps activated by storm conditions as proposed for the Capbreton Canyon by Mulder et al. (2004) or by earthquakes. Numerous slump scarps are noted on the NW side of canyon segment 3 (Fig. 7) as well as adjacent to the El Tuito fan, further suggesting that the lower part of the northern wall is made up of unconsolidated, or loosely consolidated, sediments that are unstable.

4.2.2 Shallow Subsurface Geology of the Northern Bay Provence

The northernmost of the east–west oriented seismic reflection profiles (profile 2 located in water depths of about 55 m, Fig. 11) shows that the shallow subsurface (<50 m below the seafloor) geology within the western half of the Northern Bay Province consists of two distinct seismic (depositional) sequences separated by an erosional, angular unconformity.

In profile 2, the lower sequence exhibits high amplitude, continuous, parallel to subparallel, wavy, internal reflectors that form a series of anticlines and synclines. These structures are also observed on profile 1 (Fig. 11), and from these two profiles we determined that the strike of syncline SA is N70°E. The internal reflectors of the lower sequence are truncated at an angular unconformity at its upper boundary along the entire length of profile 2 (a distance of 10 km), suggesting that this sequence most likely underlays much of the Northern Bay Province. The lithology of the rocks comprising this sequence is unknown, however, the characteristics of the internal reflectors of this sequence suggests that these are deposits of neritic marine sediments (e.g., Sangree and Widmier 1977). The angular unconformity is not eroded uniformly, but instead, contains several pinnacles, some of which outcrop on the seafloor. This either indicates that the lower sequence consists of material which exhibit variable resistance to erosion or that these pinnacles may be small reefs. Dredging is planned for the future to determine the lithology of the outcrops of the lower sequence.

The thickness of the upper sequence is varied. In the middle part of profile 2, this sequence is less than 2 m thick, and in several places the rocks of the lower sequence appear to outcrop on the seafloor. In this central area the angular unconformity is for the most part horizontal. To the west, the angular unconformity dips ~0.5° to the west until it flattens at a point located just east of syncline SA where the maximum thickness of sediments (15 m) in the overlying sequence is observed. In this area, the upper sequence consists of two units: an overall higher reflectivity upper unit that is free of horizontal reflectors (indicating no significant variation in the type of sediment being deposited), and a lower unit with low amplitude continuous internal reflectors. These internal reflections show an eastward onlapping on the angular unconformity suggesting either tectonic uplift of the central area or subsidence and subsequent infilling of the western area. Since the eastern part of profile 2 crosses the north flank of the Mid-Bay Magnetic High, we favor a tectonic uplift due to magma emplacement in the central area of profile 2. Alternatively, instead of tectonic movements, the geometry of the unconformity (i.e. a step-like profile) could be explained by coastal erosion produced by an abrupt ~20 m rise in sea level during the last eustatic rise in sea level (Neil Mitchell, personal communication; Trenhaile 2002). More data is needed to distinguish between the two possibilities.

On the eastern end of profile 2, the thickness of the upper sequence increases compared to that over the high central area. However, only the upper unit of the upper sequence is present. The angular unconformity dips eastward and is offset about 4 m by a buried fault. No faults are observed to abruptly displace the seafloor along the entire extent of profile 2. The presence of the angular unconformity at water depths of 55 m indicates that either this unconformity formed during the last major drop in sea level or that the area has recently subsided by at least this amount.

Profile 4 (Fig. 12), also oriented east–west but located south of profile 2, nearer to the Banderas Canyon in water depths of around 140–200 m, shows thicker sediments relative to those observed on profile 2. The exact thickness of sediments is not determinable from the data since the angular unconformity noted on profile 2 is not observed on profile 4 (we assume that it is buried by these sediments). However, from profile 4 the minimum sediment thickness is 70 m. Like profile 2, no faults are observed to cut the seafloor on profile 4. Profile 3 (Fig. 13) illustrates that this increase in sediment thickness towards the canyon is not gradual, but instead occurs across a large fault (F4), downthrown to the south. This fault is the only fault noted to clearly offset the seafloor reflector in this area, the offset being 15 m. The multibeam bathymetric data (see inset of Fig. 13) that was collected concurrent with these seismic reflection profiles indicates that this fault strikes N71°E.

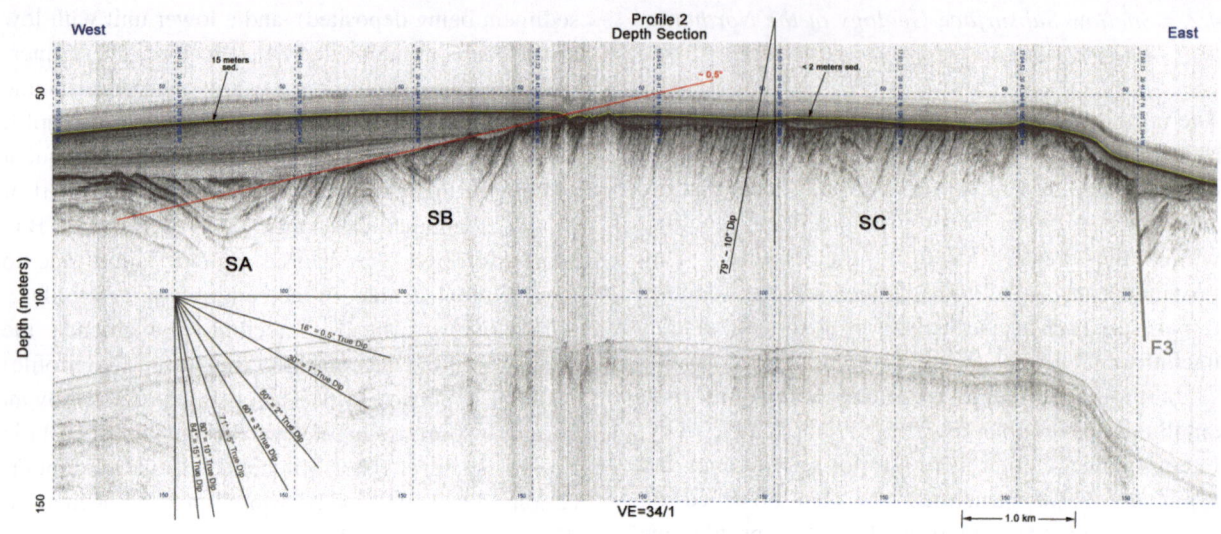

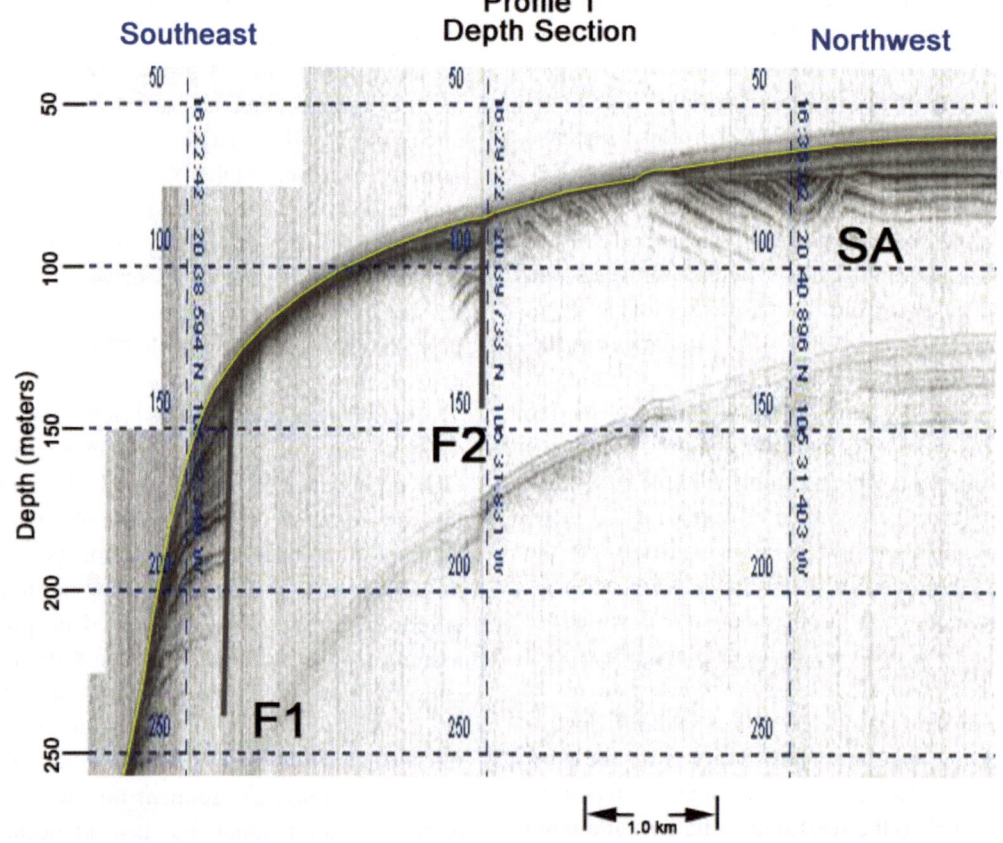

Figure 11

Top Seismic reflection profile 2 illustrating the shallow subsurface structure in the western part of the Northern Bay Province. *S* syncline, *F* fault. The vertical exaggeration of all profiles shown in this study is ~34:1. A *graph* is presented showing the relationship between observed and actual dips. *Bottom* Seismic reflection profile 1. See Fig. 4 for profile locations. *Vertical scale* is in meters calculated from the two-way travel time using a velocity of 1450 m/s. Note that the apparent sedimentary layer above the seafloor is the effect of the time varying gain and *bottom* detection *algorithm* and is not a real sedimentary layer (i.e. the time varying gain started too soon)

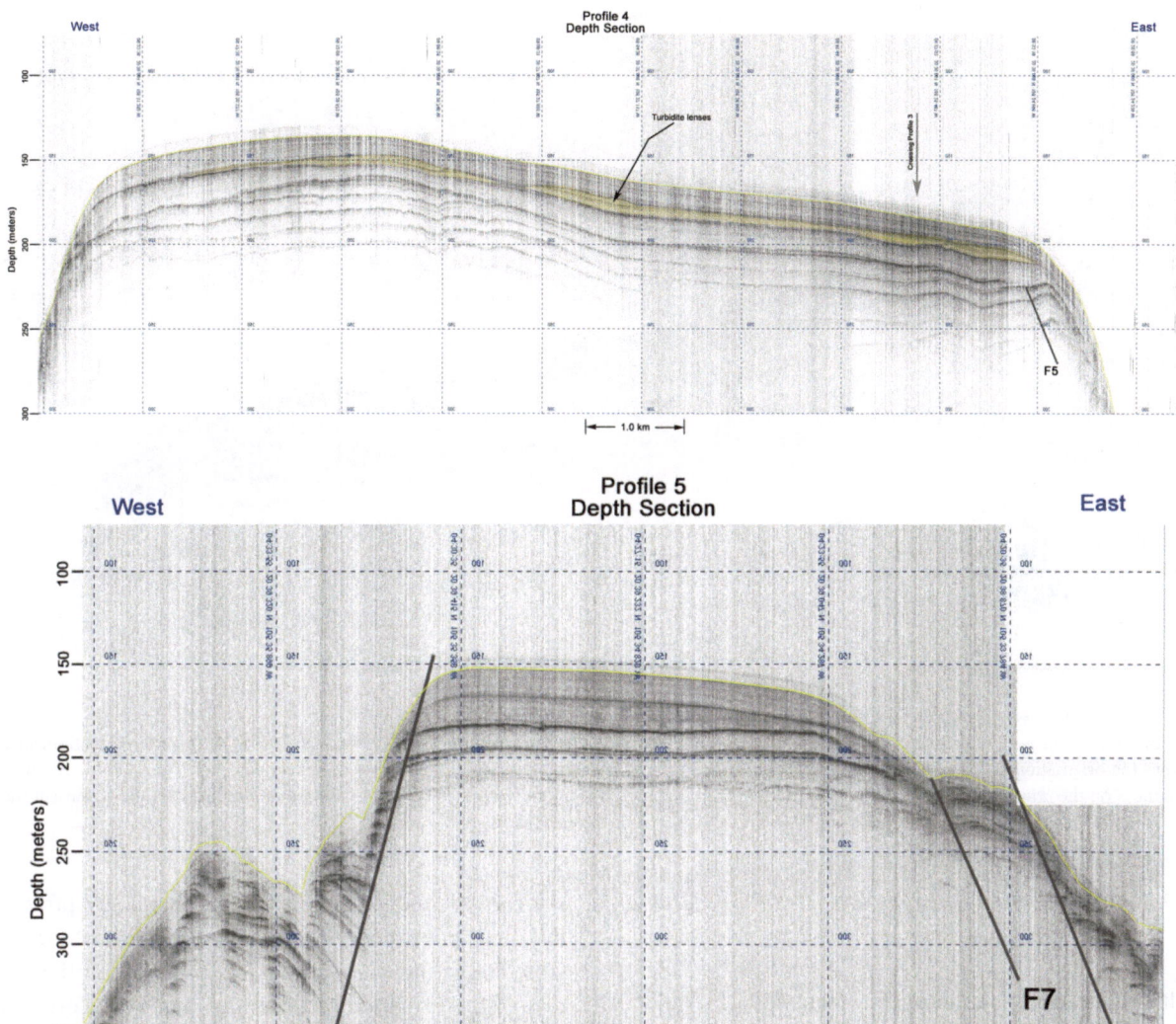

Figure 12
Seismic reflection profiles 4 (*top*) and 5 (*bottom*) further illustrating the shallow subsurface structure in the Northern Bay Province. See Fig. 4 for profile locations

Overall, the internal reflectors of the sediment sequence observed on profile 3 are continuous, parallel to subparallel, and wavy. Amplitudes are variable. These characteristics again indicate deposition with the neritic zone within which the energy alternates between high and low energy (Sangree and Widmier 1977). There is evidence for a higher energy depositional environment at about 10–15 m below the seafloor (see also profile 4) where one can observe several lenses of what appear to be massive turbidite deposits.

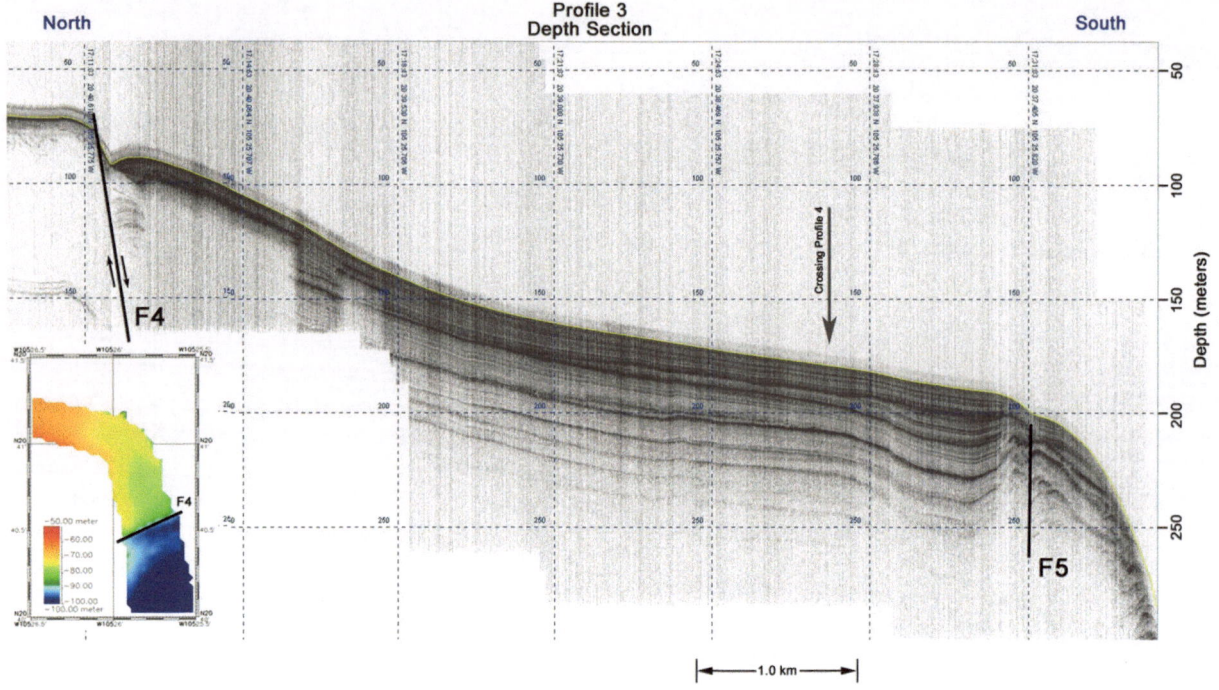

Figure 13
Seismic reflection profile 3 illustrating the shallow subsurface structure in the Northern Bay Province. See Fig. 4 for profile location. *Inset* shows the orientation of fault F4 as determined from multibeam bathymetric data. F4 offsets the seafloor reflector by 15 m and the sediment thickness on the *downthrown side* is about 60 m compared to about 2 m on the *upthrown side* indicating that the throw on the basement (not imaged) is at least 75 m

4.2.3 Shallow Structure of the Marietas Ridge (Punta Mita Province)

Although a continuation of the Punta Mita Province, the Marietas Ridge changes trend at the islands of Isla Larga and Isla Redonda located about 8 km SW of Punta Mita (Fig. 6). Specifically, between Punta Mita and these two islands, the Marietas Ridge strikes at an azimuth of ~212°. However, these two islands along with the EL Morro rock (located at 20°41′, −105°37′) and three more small rocks (herein called "Las Tres Tortugas", located at 20°40.1′, −105°39.3′) are aligned at an azimuth of ~247°. It is also of interest that the Isla Larga and Isla Redonda are aligned east–west, as are the Las Tres Tortugas. These alignments suggest that the Marietas Ridge has had a complex development history, which most likely includes the shallow intrusion of dykes along deep-seated faults. Magnetic modeling, planned for the future, could clarify the development history of the ridge.

The seismic reflection data of this study provide the first published images of the shallow crustal structure of the Marietas Ridge, in particular the SW part of the ridge located south of Las Tres Tortugas. There, the Marietas Ridge strikes at an azimuth of ~230° and is asymmetric, the ridge crest being located on the east side of the ridge (Figs. 14, 15). The recent sediments noted on the seismic profiles in the Northern Bay Province are absent over this ridge, with the exception some sediments infilling a few small seafloor depressions. Therefore, the internal reflectors essentially belong to one seismic sequence. This sequence is disrupted like the lower seismic sequence noted to the east, however, with the presently available data we cannot confirm that they are indeed the same sequence. The age and lithology of the rocks comprising this sequence are unknown, but since several of these rocks outcrop on the seafloor, the age and lithology could be determined in the future by dredging.

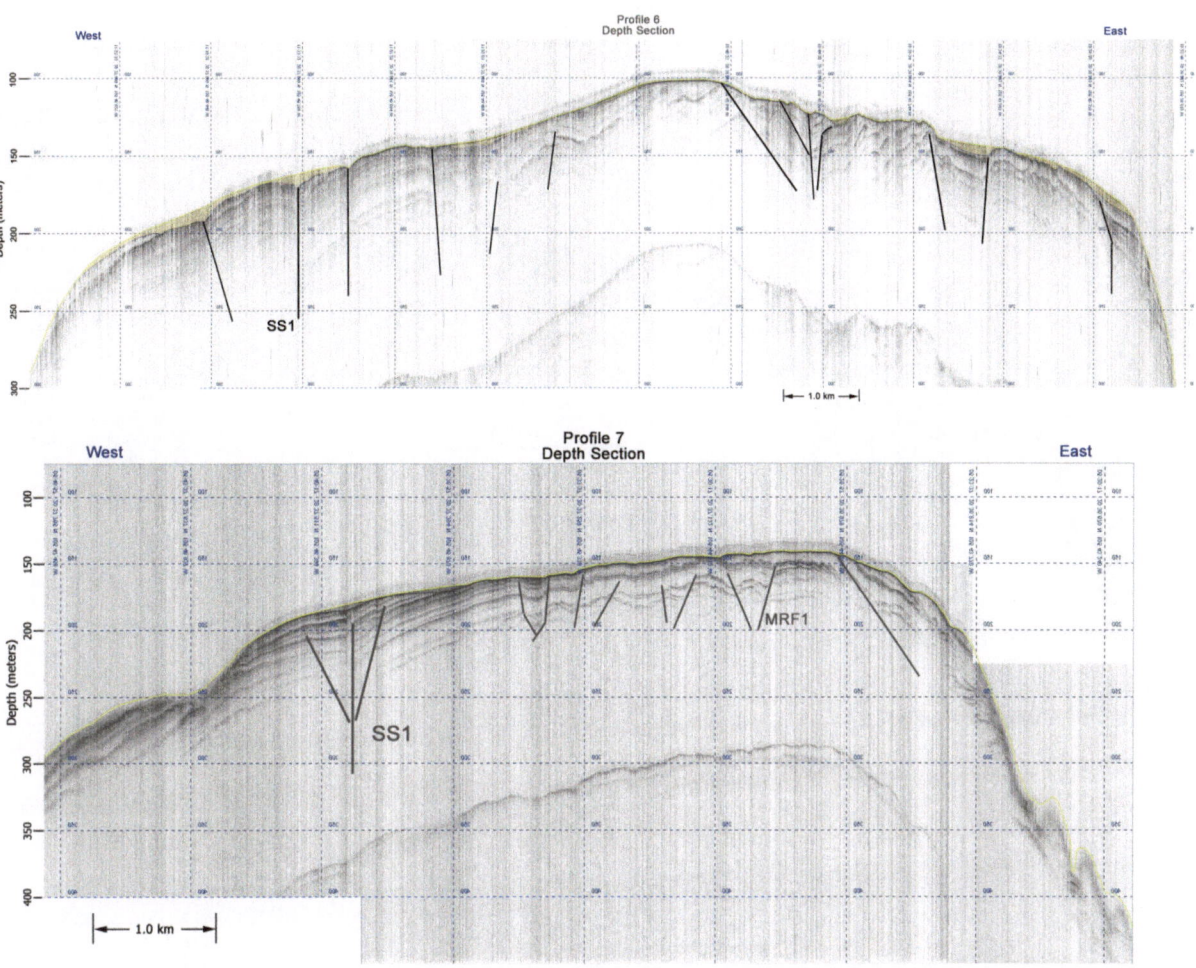

Figure 14

Seismic reflection profiles 6 (*top*) and 7 (*bottom*) illustrating the shallow subsurface structure of the SW end of the Marietas Ridge. *SS* for strike-slip fault, *MRF* Marietas Ridge Fault. Note the well-developed negative flower structure associated with fault SS1 west of the ridge crest. See Fig. 4 for profile location

The internal reflections of this sequence differ east and west of the ridge crest (see profile 6, Fig. 14). Specifically, under and to the east of the ridge crest the internal reflectors are parallel, wavy, and discontinuous with mixed high and low amplitudes. This character is typical of faulted neritic sediments, the faulting occurring after sediment deposition. In contrast, in the western part of the ridge, away from the ridge crest, although the internal reflectors are also parallel with mixed high and low amplitudes, they are more even and continuous than those found to the east. This indicates substantially less disruption of the sedimentary layers, however, the internal reflectors of the western area are

disrupted by a well-developed, negative flower structure (SS-1 on profiles 6, 7 and 8; Figs. 14, 15), which indicates that SS-1 is a transtensional fault (e.g., Harding et al. 1985). SS-1 strikes parallel to the ridge crest suggesting that its development is related to the development of the ridge. Another major fault, (normal fault MRF#1, profiles 7 and 8), which offsets the internal reflectors by about 20 m, strikes parallel to the ridge crest, leading us to conclude that, in general, the observed disruption of this sequence is concurrent with the formation of the ridge. Also in the western area, the upper reflectors of this sequence exhibit erosional truncation at the seafloor reflector indicating that the eastern area has been uplifted and

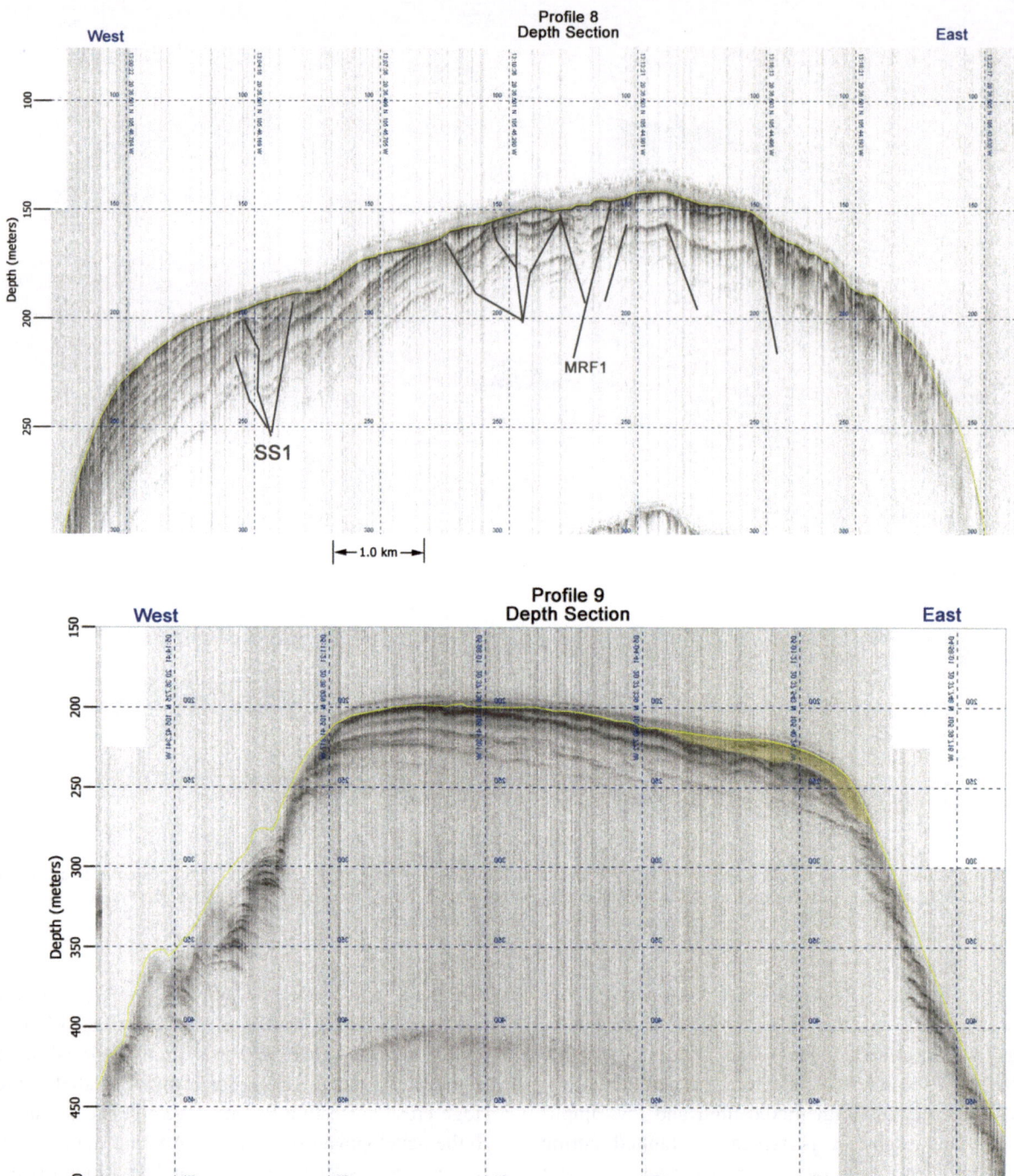

Figure 15
Seismic reflection profiles 8 (*top*) and 9 (*bottom*) illustrating the shallow subsurface structure of the SW end of the Marietas Ridge. *SS* strike-slip fault, *MRF* Marietas Ridge Fault. Again, note the well-developed negative flower structure associated with fault SS1 west of the ridge crest. See Fig. 4 for profile location

eroded. This is consistent with the lack of recent sediments over the ridge.

5. Discussion

5.1. Present Day Stress

As illustrated in Fig. 6 the area of the Bahía de Banderas contains at least five families of lineaments attesting to the complex tectonic history of this region. The lineaments have preferred orientations of north–south, N35°E–N45°E, N70°E, east–west, and N110°E.

Although the N110°E lineaments are observed throughout the region, they are the dominate lineament south of the bay. These lineaments control the course of the Banderas Canyon along the canyon segments 4 and 7 and the west half of canyon segment 5. Further, structural highs in the western part of Slope Terrace Province exhibit a similar strike (Fig. 6). Recent seismicity has been associated to these lineaments (e.g., Rutz López et al. 2013). The knickpoint at the east end of canyon segment 4 may also be the result of recent tectonic activity along at least one of these lineaments.

On the north side of the bay and within the Valle de Banderas, two families of lineaments dominate; one with a preferred orientation of N35°E to N45°E and another with a preferred orientation of N70°E. The N35°E–N45°E lineaments clearly form the NW and SE boundaries of the Valle de Banderas (e.g., Ferrari et al. 1994). Further, Arzate et al. (2006) also found a series of buried faults with the same orientation within the valley. The overall orientation of the Punta Mita Province (including the Marietas Ridge) also exhibits this orientation. The N70°E lineaments are observed on both sides of the Valle de Banderas (e.g., Ferrari et al. 1994) and within the Northern Bay Province (Núñez-Cornú et al. 2000) where they are associated with hydrothermal activity.

The remaining two families of lineaments are less dominant. A few short east–west lineaments are observed in the Punta Mita Province and along the southern margin of the bay. The north–south lineaments are sparse, being mainly observed south of the bay. No seismicity has been associated with these lineaments suggesting that they are older features.

Concerning the question of the present day stress field of the area, within the confines of the bay east of the Marietas Ridge, we observe only one major fault which we can unequivocally say is presently active; namely the large normal fault (F4) which trends N71°E and which is observed to displace the seafloor reflector by 15 m on seismic profile 3. If this fault has purely normal dip-slip, then its orientation indicates that the current stress field in the bay is extensional and that the tensional axis is oriented N19°W. Further, a similar trend was found for the main fracture of the active hydrothermal system "Fisura de la Coronas" located on the north side of the Bay (Núñez-Cornú et al. 2000). From these observations, along with the lack of unequivocal evidence for recent activity along the N35°W–N45°W family of lineations, we conclude that the N71°E family of lineations is indicative of the present day stress field at least within the bay, and probably also within the Valle de Banderas.

If our conclusion is correct and the entire Puerto Vallarta graben is presently being subjected to NNW–SSE directed tension, then the NE striking faults bounding, and located within, the Valle de Banderas represent a previous stress regime, and thus, the tensional axis has since rotated clockwise to a more northerly direction, one that is more parallel to the strike of the Middle America Trench west of the bay. This is consistent with the proposal of Kostoglodov and Bandy (1995) that the recent tectonic activity within this area is due to the highly oblique subduction of the Rivera plate with respect to the North American plate. The N19°W direction for the tensional axis is more parallel to the strike (roughly north–south) of the MAT in this area. Reorientation of the stress field in this area has been previously proposed by several investigators (e.g., Ferrari et al. 1994; Arzate et al. 2006).

The seismic reflection data also provide evidence for such a clockwise shift in the tensional axis. Specifically, these data image transtensional faulting within the SW part of the Punta Mita province (the Marietas Ridge), which is made up of structures belonging to the family of N45°E lineaments. If, as we propose, the N45°E trend is older, then a clockwise rotation of the tensional axis would produce transtension along pre-existing structures having this trend.

The gently sloping, undisrupted canyon floor also supports the proposal that the N110°E and N45°E lineations are representative of older stress fields and that the area of the canyon is not presently being subjected to intense tectonism. The only possible exception that we see in our data is the N110°E trending lineament that outcrops on the canyon floor, forming the 10 m high ridge. These observations lead us to conclude that the N110°E and N45°E lineaments are indicative of older stress regimes. However, it appears that some of these lineaments may have reactivated at present (Rutz López et al. 2013; Núñez-Cornú et al. 2016).

5.2. The Existence of the Banderas Fault

The idea of the existence of a major structural lineament passing through the Bahía de Banderas to the Middle America Trench dates back to the work of von Humboldt (as reported on by Mooser 1972) who proposed that the Trans-Mexican Volcanic belt marked a regional mega-shear, later proposed to be the continuation of the Clarion fracture zone (Menard 1955). More recently, Lyle and Ness (1991) presented such a lineament in their bathymetric map of the area and several subsequent investigators have presented, ad hoc, the lineament in their work (Alvarez 2007; Alvarez et al. 2010) calling it the Banderas Fault; although some have questioned its existence (e.g., Núñez-Cornú et al. 2000).

The data presented herein clearly indicate that there is no major morphotectonic or magnetic structure that can be traced extending from the bay completely across the study area. The Canyon Magnetic High (Fig. 4) is confined to the bay and the east–west magnetic low running through the center of the bay terminates within the survey area. The canyon itself bends southward prior to reaching the Middle America Trench, being deflected by NE–SW striking structures, so it is clearly not marking the presence of a major tectonic structure extending westward from the bay to the MAT. Further to the west, though still east of the trench, the structural highs and lows trend NW–SE (Fig. 5), not east–west.

The lack of evidence for a major block boundary between the bay and the MAT calls into question the proposals that the Rio Ameca Graben marks the northern boundary of the Jalisco Block. However, the difference in the magnetic signature of the Slope Terrace Province north and south of the latitude of the bay does indicate that a structural and/or compositional change of the crust occurs in the area of the bay.

5.3. Sediment Transport Characteristics

Although no direct observations of sediment transport were made during the study, several characteristics can be gleaned from the geophysical data. The lack of knickpoint retreat on the drainage channels at the base of the northern canyon wall (Fig. 10) indicates that these channels have experienced little activity since the time that the main channel was last entrenched, and that the majority of the recent activity has been within the canyon thalweg (Mitchell, person. comm.), similar to that described by Mulder et al. (2004) for the Capbreton Canyon. This, along with the broad, flat nature of the canyon floor, leads us to propose that most of the recent sediment transport within the canyon most likely occurred during major storms when the sediment content of the two main rivers (Rio Ameca and Rio Cuale) was high. Hyperpycnal flows from these rivers, produced during storms, transported the sediments into the canyon; most likely down the two large, flat floored, channels (channels 1 and 2, Fig. 5) that feed the canyon from the east. These flows most likely extended down the entire length of the canyon thalweg, producing the broad, flat, gently seaward dipping canyon floor morphology.

During dry periods, the discharge of the rivers is low, <1 m^{3}s^{-1} for the Rio Ameca (CNA-SEMARNAT 1999, as reported in Plata and Filonov 2007), thus during these times, fine grained sediments are probably distributed throughout the bay by hypopycnal flows emanating from these rivers.

The debris flow deposits within the canyon floor (Fig. 10) and the slump scarps on the canyoń's northern wall (Fig. 7) indicate that part of the sediment feeding the canyon originates from mass wasting of the canyon's northern wall. These mass-wasting events may be triggered by wave action or due to erosion at the base of the canyon wall by the hyperpycnal flows during storms. Alternatively, the mass-wasting events may be triggered by earthquakes.

The presence of the El Tuito Fan within the canyon floor indicates that, here, a significant amount of sediments are flowing down the steep southern wall, a source being the granitic highlands to the south. It is not known if the fan is a long-term feature (i.e. comprised of coarse grained sediments) or if it is a transient structure (comprised of finer grained sediments) that will be washed away during a future storm. However, given the steepness of the canyon's southern wall, coarse-grained sediments could be easily transported to the canyon floor, similar to the transport of coarse-grained sediments to the deep waters of fjord deltas via submarine chutes (Prior et al. 1981).

The distribution of recent sediments observed on the seismic reflection profiles appears to indicate that sediments are being reworked within the bay during storms, the depth of the wave base being greater at the mouth of the bay (i.e. over the Marietas Ridge). Specifically, on profiles 6 thru 9, crossing the Marietas Ridge, the wave base appears to be between 220 and 160 m. In contrast, the wave base within the bay appears to be only 60–70 m (profiles 2 and 3). This can be explained by a damping of the wave energy within the bay or by a divergence of the wave field due to the submarine canyon.

5.4. Origin of the Submarine Canyon

Several researchers (Arzate et al. 2006, 2010; Alvarez 2007) have recently proposed that the Banderas Canyon lies within a half graben, with the steep southern wall being the main fault and the more gentle western wall representing reverse drag of the basement layer into the main fault. Further, it was proposed that this main fault was of regional extent, extending completely across the continental slope to the MAT.

Several observations lead us to reconsider these claims about the possible origins of the Banderas Canyon, these are: (1) the results of this study contradict the previous claim that a major fault extends to the MAT along the westward prolongation of the Banderas Canyon (how can such a large fault, the main fault of the half-graben, be confined to only the bay?)

(2) the previous studies did not consider the alternative that the gentle northern canyon wall may be due to sediments being deposited at the base of a steep normal fault rather than reverse drag, (3) the quality of the previous magnetic data was poor (most likely the result of the non-conventional marine magnetic acquisition method employed), and the locations of the previous models were poorly selected (we feel that it would have been better had the modeled profiles been located within the confines of the bay, east of the Marietas Ridge), and (4) the disagreement between the proposed model and the results of Rutz López et al. (2013). Given our results, we propose two (Fig. 16) other possible origins of the Bandera Canyon in addition to the previously proposed half-graben model.

Figure 2 illustrates the presence of an east–west oriented silicic tuff unit, sandwiched between the cretaceous granites of the Puerto Vallarta batholith, located onshore where the canyon is projected to intersect the coast. The tuffs, being softer than the adjacent granites and volcanic in origin, would be more easily eroded than the granites. Thus, the tuff deposits may continue into the bay and the canyon may simply originate from preferential erosion of the tuffs, i.e. the overall course of the canyon is being controlled by the location of the tuffs (upper panel Fig. 16). In this scenario the steep southern wall of the canyon was formed by the uplift of the Puerto Vallarta Batholith to the south noted in previous studies and may also mark the contact between the granites and the tuff unit. The more gently sloping lower northern wall is the result of turbidite deposition in the northern part of the erosional canyon, the turbidites forming large coalescing fans.

Alternatively (lower panel in Fig. 16), the characteristics of the canyon may be the result of both uplift of the Puerto Vallarta Batholith south of the canyon in conjunction with a generally NW movement of the Punta Mita Province away from the rest of the Puerto Vallarta batholith. Like the previous scenario, in this scenario the steep southern canyon wall was formed by the uplift of the Puerto Vallarta batholith to the south. Tension within and north of the bay, perhaps related to the opening of the Gulf of California, was accommodated by at least one normal fault, the fault plane of which is the steeper, highly eroded upper part of the northern wall. Sediments

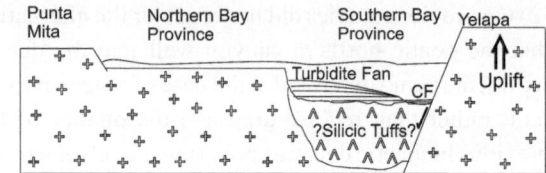

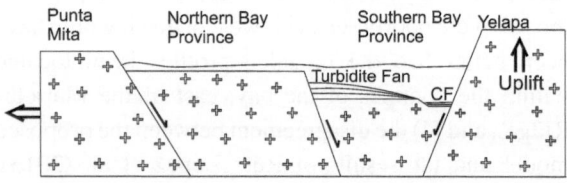

Figure 16

Cartoon illustrating two additional possible scenarios for the development of the Banderas Canyon, namely erosion of the tuffs (*top*), and uplift of the Puerto Vallarta batholith to the south in conjuction with a NW movement of the batholith to the north (*bottom*). *CF* channel floor

originating from the main rivers emptying into the bay to the north were transported over the fault and deposited on the downthrown block at the base of the fault as extensive sediment fans. In this scenario, the gentle lower part of the northern slope is due to the formation of these sediment fans and is not due to a southward bending of the basement as it approaches the southern canyon wall.

We propose that, given the available data, all three scenarios are possible. Plans are currently being made to acquire surface dredge samples and multi-channel seismic reflection data capable of penetrating the turbidite fan deposits within the northern part of the canyon to determine the thickness of these sediments as well as the attitude of the top of the basement block underlying the turbidite fans. From these data one should be able to distinguish which of the three scenarios is correct.

6. Conclusions

1. A N71°E striking fault offsets the seafloor reflector within the central part of the bay by about 15 m. The thickness of the upper sedimentary section increases from less than 2 m on the up thrown block to more than 65 m on the down thrown block indicating that it is a major, presently active, normal fault. The strike of this fault

is similar to the strike of the fractures associated with the Fisura de las Coronas located on the north side of the bay. This suggests that the bay is currently being subjected to NNW-SSE extension. This is almost parallel to the strike of the MAT in this area, consistent with the proposal that this area is being subjected to trench parallel stresses produced by highly oblique convergence between the Rivera and North American plates at the north end of the MAT.

2. The seismic reflection data reveals the presence of negative flower structures that disrupt the seafloor reflector on the west side of the Marietas Ridge indicating that the ridge is presently undergoing transtensional deformation. The numerous, small, high-frequency magnetic anomalies that lie along the southern prolongation of this ridge are consistent with the presence of volcanics, similar to those observed on the Islas Marietas and within Punta Mita. Thus, we propose that the Punta Mita province encompasses the Marietas Ridge and the area of these magnetic anomalies. As such it forms the western limit of the Northern Bay Province.

3. No evidence has been found in the magnetic or bathymetric data to confirm the existence of the previously proposed Banderas fault, a regional east–west striking fault proposed to extend from the bay to the Middle America Trench.

4. The course of the Banderas Canyon is controlled by extensive turbidite fan sedimentation in its eastern extremity and by N110°E and N45°E oriented structural lineaments to the west.

5. The canyon floor is filled by sediments and exhibits almost no evidence for recent tectonic movements.

6. The southern canyon wall is quite steep and very few sediment fan deposits are observed at the base of the southern wall. In contrast, extensive turbidite fans form the lower part of the northern canyon wall, producing a gently southward sloping northern wall.

7. The lack of evidence for the Banderas Fault along with the observation that turbidite fans are responsible for the gentle dip of the northern canyon wall indicates that the previous assertion that the Banderas Canyon is unequivocally related

to the presence of a regional half-graben needs to be re-evaluated. We propose two other alternatives for the development of the Banderas Canyon that are consistent with the available data. The first is that the Banderas Canyon is purely an erosional feature, cutting through a more easily eroded silicic tuff. The second is that it is a combination of uplift of the Puerto Vallarta Batholith south of the bay and a roughly northwestward movement of the area north of the bay.

Acknowledgements

We thank the captain and the crew of the B.O. EL PUMA for their help during the various campaigns. We also thank Dr. Neil Mitchell and an anonymous reviewer for their valuable and thorough review of the manuscript. We also thank Renata Dmowska for acting as editor on our contribution to this special volume. Ship time for the research cruises PMITA, MORTIC06, BABRIP06 and MORTIC08 carried out aboard the B/O EL PUMA was funded by the Universidad Nacional Autónoma de Mexico. The research received funding from Consejo Nacional de Ciencias y Tecnología (CONACyT) Grants 36681-T and 50235 and DGAPA Grants IN104707, IN114602, IX117504, IN104199, IN110897, IN108110 and IX111304.

REFERENCES

Alfonso, P., Prol-Ledesma, R. M., Canet, C., Melgarejo, J. C., & Fallick, A. E. (2003). Sulfer isotope geochemistry of the submarine hydrothermal coastal vents of Punta Mita, Mexico. *Journal of Geochemical Exploration, 78–79*, 301–304.

Allan, J. F. (1986). Geology of the northern Colima and Zacoalco grabens, southwest Mexico: Late Cenozoic rifting in the Mexican Volcanic Belt. *Geological Society of America Bulletin, 97*, 473–485.

Allan, J. F., Nelson, S. A., Luhr, J. F., Carmichael, I. S. E., Wopat, M., & Wallace, P. J. (1991). Pliocene-Holocene rifting and associated volcanism in southwest Mexico: An exotic terrane in the making. In J. P. Dauphin & B. R. T. Simoneit (Eds.), *The Gulf and Peninsular Province of the Californias, AAPG Memoir 47* (pp. 425–445). Tulsa: AAPG.

Alvarez, R. (2002). Banderas rift zone: A plausible NW limit of the Jalisco Block. *Geophysical Research Letters, 29*, 55-1–55-4. doi:10.1029/2002GL016089.

Alvarez, R. (2007). Submarine topography and faulting in Bahía de Banderas, Mexico. *Geofísica Internacional, 46*, 93–116.

Alvarez, R., López-Loera, H., & Arzate, J. (2010). Modeling the marine magnetic field of Bahía de Banderas, Mexico, confirms the half-graben structure of the bay. *Tectonophysics, 489*, 14–28.

Arzate, J. A., Álvarez, R., Yutsis, V., Pacheco, J., & López-Loera, H. (2006). Geophysical modeling of Valle de Banderas graben and its structural relation to Bahía de Banderas, Mexico. *Revista Mexicana de Ciencias Geológicas, 23*, 184–198.

Bandy, W. L., & Pardo, M. (1994). Statistical examination of the existence and relative motion of the Jalisco and southern Mexico Blocks. *Tectonics, 13*, 755–768.

Bandy, W. L., Michaud, F., Bourgois, J., Calmus, T., Dyment, J., Mortera-Gutiérrez, C. A., et al. (2005). Subsidence and strike-slip tectonism of the upper continental slope off Manzanillo, Mexico. *Tectonophysics, 398*, 115–140. doi:10.1016/j.tecto. 2005.01.004.

Bartolomé, R., Dañobeitia, J., Michaud, F., Córdoba, D., & Delgado-Argote, L. A. (2011). Imaging the seismic crustal structure of the western Mexican margin between 19°N and 21°N. In W. L. Bandy, Y. Taran, C. A. Mortera Gutierrez, & V. Kostoglodov (Eds.), *Geodynamics of the Mexican Pacific Margin* (Vol. Pageoph Topical Volumes, pp. 123–140). Basel: Birkhäuser.

Böhnel, H., & Negendank, J. F. W. (1988). Paleomagnetism of the Puerto Vallarta intrusive complex and the accretion of the Guerrero terrain, Mexico. *Physics of the Earth and Planetary Interiors, 52*, 330–338.

Brown, H., Holbrook, S., Paramo, P., Lizarralde, D., Axen, G. J., Fletcher, J., González-Fernández, A., Harding, A., Kent, G., & Umhoefer, P. (2009). *Seismic structure of the Rivera plate beneath the Jalisco block, western Mexico, from the PESCADOR experiment*, MARGINS program report for award 01-12152, 01-12149, 01-12058, 01-11983, 01-11738, 01-11738, p. 5, April, 2009.

Buchanan, S. K., Scrutton, R. A., Edwards, R. A., & Whitmartsh, R. B. (1996). Marine magnetic data processing in equatorial regions off Ghana. *Geophysical Journal International, 125*, 123–131.

Bullard, E. C., & Mason, R. G. (1961). The magnetic field astern of a ship. *Deep Sea Research, 8*, 20–27.

Canet, C., & Prol-Ledesma, R. M. (2007). Mineralizing processes at shallow submarine hydrothermal vents: Examples from Mexico. In S. A. Alinez-Álaniz & A. F. Nieto-Samaniego (Eds.), *Geology of Mexico: Celebrating the Centenary of the Geological Society of Mexico: Geological Society of America Special Paper 422* (pp. 359–376). Boulder: Geological Society of America.

Cano Sánchez, L. E. (2004). *Ficha Informativa de los Humedales de Ramsar (Islas Marietas)* (p. 14). San Blas: Comisión Nacional de Áreas Naturales Protegidas.

CNA-SEMARNAT (1999). Régimen de almacenamientos hasta 1999, Banco Nacional de Datos de Aguas Superficiales, in CD-ROM.

Córdoba, D., Núñez-Cornú, F. J., Dañobeitia, J., Bartolome, R., Bandy, W., Escudero, C., Cameselle, A. L., Espíndola Castro, J. M., Prada, M., Níñez, D., Zamora Camacho, A., Gomez, A., & Ortiz, M. (2014). TsuJal Project: New Geophysical Studies about Rivera Plate and Jalisco Block (Mexico), Agu Fall Meeting 2014, abstract T11c-4566.

Couch, R. W., Ness, G. E., Sanchez-Zamora, O., Calderón-Riveroll, G., Doguin, P., Plawman, T., et al. (1991). Chapter 3. Gravity anomalies and crustal structure of the gulf and Peninsular Province of the Californias. In J. P. Dauphin & B. R. T.

Simoneit (Eds.), *The Gulf and Peninsular Province of the Californias, AAPG Memoir 47* (pp. 25–45). Tulsa: AAPG.

Dañobeitia, J. J., Cordoba, D., Delgado-Argote, L. A., Michaud, F., Bartolomé, R., Farran, M., et al. (1997). Expedition gathers new data on crust beneath Mexican West Coast. *EOS Transactions of the American Geophysical Union, 78*, 565–572.

Dauphin, J. P., & Ness, G. E. (1991). Bathymetry of the Gulf and Peninsular province of the Californias. In J. P. Dauphin & B. R. T. Simoneit (Eds.), *The Gulf and Peninsular Province of the Californias, AAPG Memoir 47* (pp. 21–24). Tulsa: AAPG.

DeMets, C., & Traylen, S. (2000). Motion of the Rivera plate since 10 Ma relative to the Pacific and North American plates and the mantle. *Tectonophysics, 318*, 119–159.

Fernández de la Vega-Márquez, T., & Prol-Ledesma, R. M. (2011). Imágenes Landsat TM y modelo digital de elevación para la identificación de lineamientos y mapeo litológico en Punta Mita (México). *Boletín de la Sociedad Geológica Mexicana, 63*, 109–118.

Ferrari, L. (1995). Miocene shearing along the northern boundary of the Jalisco Block and the opening of the southern Gulf of California. *Geology, 23*, 751–754.

Ferrari, L., & Rosas-Elguera, J. (2000). Late Miocene to quaternary extensión at the northern boundary of the Jalisco block, western Mexico: The Tepic-Zacoalco rift revisited. In H. Delgado-Granados, G. Aguirre-Díaz, & J. M. Stock (Eds.), *Cenozoic tectonics and volcanism of Mexico: Geological Society of America Special Paper 334* (pp. 41–63). Bolder: Geological Society of America.

Ferrari, L., Pascuare, G., Venegas, S., Castillo, D., & Romero, F. (1994). Regional tectonics of western Mexico and its implications for the northern boundary of the Jalisco Block. *Geofísica Internacional, 33*, 139–141.

Fisher, R. L. (1961). Middle America Trench: Topography and structure. *Geological Society of America Bulletin, 72*, 703–720.

Harding, T. P., Vierbuchen, R. C., & Christie-Blick, N. (1985). Structural styles, plate-tectonic settings, and hydrocarbon traps of divergent (Transtensional) wrench faults. In K. T. Biddle & N. Christie-Blick (Eds.), *Strike-slip deformation, basin formation, and sedimentation, Society of Economic Paleontologists and Mineralogists Special Publication No. 37* (pp. 51–77). Tulsa: SEPM.

IAGA Working Group V-MOD. (2010). International geomagnetic reference field: The eleventh generation. *Geophysical Journal International, 183*, 1216–1230.

INEGI (1988). *Carta Geologica, Pto. Vallarta F13-11*, 1:250,000.

Jensky, W.A. (1974). *Reconnaisssance geology and geochronology of the Bahía de Banderas area, Nayarit and Jalisco, Mexico*. M.A. Thesis, University of California, Santa Barbara, California, p. 80.

Johnson, C. A., & Harrison, C. G. A. (1989). Tectonics and volcanism in central Mexico: A landsat thematic mapper perspective. *Remote Sensing of Environment, 28*, 273–286.

Johnson, C. A., & Harrison, C. G. A. (1990). Neotectonics in central Mexico. *Physics of the Earth and Planetary Interiors, 64*, 187–210.

Kostoglodov, V. V., & Bandy, W. L. (1995). Seismotectonic constraints on the convergence rate between the Rivera and North America plates. *Journal Geophysical Research, 100*, 17977–17989.

Lonsdale, P. (1991). Structural patterns of the Pacific Floor Offshore of Peninsular California. In J. P. Dauphin & B. R. T.

Simoneit (Eds.), *The Gulf and Peninsular Province of the Californias, AAPG Memoir 47* (pp. 87–125). Tulsa: AAPG.

Luhr, J. F., Nelson, J. F., Allan, J. F., & Carmichael, I. S. E. (1985). Active rifting in southwestern Mexico: Manifestations of an incipient eastward spreading-ridge jump. *Geology, 13*, 54–57.

Lyle, M., & Ness, G. E. (1991). The opening of the southern Gulf of California. In J. P. Dauphin & B. R. T. Simoneit (Eds.), *The Gulf and Peninsular Province of the Californias, AAPG Memoir 47* (pp. 403–423). Tulsa: AAPG.

Maillol, J. M., Bandy, W. L., & Ortega-Ramírez, J. (1997). Paleomagnetism of Plio-Quaternary basalts in the Jalisco block, western Mexico. *Geofísica Internacional, 36*, 21–35.

Menard, H. W. (1955). Deformation of the Northeastern Pacific Basin and the west coast of North America. *Bulletin of the Geological Society of America, 66*, 1149–1196.

Mitchell, N. C. (2006). Morphologies of knickpoints in submarine canyons. *Bulletin of the Geological Society of America, 118*, 589–605.

Mooser, F. (1972). The Mexican volcanic belt: Structure and tectonics. *Geofisica Internacional, 12*, 55–69.

Mulder, T., Cirac, P., Gaudin, M., Bourillet, J.-F., Tranier, J., Normand, A., et al. (2004). Understanding continent-ocean sediment transfer. *Eos Transactions American Geophysical Union, 85*, 257–264.

Núñez-Cornù, F. J. (2011). Peligro Sísmico en el Bloque de Jalisco, Mexico. *Física de la Tierra, 23*, 199–229.

Núñez-Cornú, F. J., Prol-Ledesma, R. M., Cupul-Magaña, A., & Suárez-Plascencia, C. (2000). Near shore submarine hydrothermal activity in Bahía Banderas, western Mexico. *Geofisica Internacional, 29*, 171–178.

Núñez-Cornú, F. J., Marta, R. L., Nava P., F. A., Reyes-Dávila, G. F., & Suárez-Plascencia, C. (2002). Characteristics of seismicity in the coast and north of Jalisco Block, Mexico. *Physics of the Earth and Planetary Interiors, 132*, 141–155.

Núñez-Cornù, F. J., Suárez Plascencia, Escudero, C. R., & Gomez, A. (2011). Jalisco regional Seismic Network (RESAJ). *Eos Transactions American Geophysical Union*, Abstract #S51A-2181.

Núñez-Cornú, F. J., Cordoba Barba, D., Dañobeitia Canales, J. J., Bandy, W. L., Ortiz Figueroa, M., Bartolome, R., et al. (2016). Geophysical studies across Rivera Plate and Jalisco Block, Mexico: TsuJal Project. *Seismological Research Letters, 87*(1), 59–72.

Plata, L., & Filonov, A. (2007). Internal tide in the northwestern part of Banderas Bay, Mexico. *Ciencias Marinas, 33*, 197–215.

Prior, D. B., Wiseman, W. J., Jr., & Bryant, W. R. (1981). Submarine chutes on the slopes of fjord deltas. *Nature, 290*, 326–328.

Ramírez-Herrera, M. T., Kostoglodov, V., & Urrutia-Fugugauchi, J. (2011). Overview of recent tectonic deformation in the Mexican subduction zone. In W. L. Bandy, Y. Taran, C. Mortera Gutierrez, & V. Kostoglodov (Eds.), *Geodynamics of the Mexican Pacific Margin* (pp. 165–183). Basel: Birkhäuser. ISBN 978-3-0348-0196-6.

Rosas-Elguera, J., Ferrari, L., Garduño-Monroy, V. H., & Urrutia-Fucugauchi, J. (1996). Continental boundaries of the Jalisco block and their influence in the Pliocene-Quaternary kinematics of western Mexico. *Geology, 24*, 921–924.

Rutz López, M. (2007). *Peligro Sísmico en Bahía de Banderas*. Thesis, Universidad de Guadalajara, May, 2007.

Rutz-López, M., & Núñez-Cornú, F. J. (2004). Sismotectonica del Norte y Oeste del bloque de Jalisco usando datos sísmicos regionales. *GEOS, 24*, 2–13.

Rutz López, M., Núñez Cornú, F. J., & Suárez Plascencia, C. (2013). Study of seismic clusters at Bahía de Banderas Region, Mexico. *Geofisica Internacional, 52*, 59–72.

Sangree, J. B., & Widmier, J. M. (1977). Seismic stratigraphy and global changes of sea level, Part 9: Seismic interpretation of clastic depositional facies. In C. E. Payton (Ed.), *Seismic stratigraphy—applications to hydrocarbon exploration, AAPG Memoir 26* (pp. 165–184). Tulsa: AAPG.

Sawlan, M. G. (1991). Magmatic evolution of the Gulf of California Rift. In J. P. Dauphin & B. R. T. Simoneit (Eds.), *The Gulf and Peninsular Province of the Californias, AAPG Memoir 47* (pp. 301–369). Tulsa: AAPG.

Schaaf, P., Köhler, H., Müller-Sohnius, D., & von Drach, V. (1993). The Puerto Vallarta Batholith—its anatomy displayed by isotopic fine structure. In F. Ortega-Gutierrez, E. Centeno-García, D. J. Morán-Centeno, & A. Gómez-Caballero (Eds.), *Proceedings of First Circum-Pacific and Ciurcum-Atlantic Terrane Conference* (pp. 921–924). México: Instituto de Geología, Universidad Nacional Autónoma de México.

Selvans, M. M., Stock, J. M., DeMets, C., Sanchez, O., & Marquez-Azua, B. (2011). Constraints on Jalisco Block motion and Tectonics of the Guadalajara triple junction from 1998 to 2001 Campaign GPS data. In W. L. Bandy, Y. Taran, C. Mortera Gutierrez, & V. Kostoglodov (Eds.), *Geodynamics of the Mexican Pacific Margin* (pp. 185–198). Basel : Birkhäuser. ISBN 978-3-0348-0196-6.

Taran, Y. A., Inguaggiato, S., Marin, M., & Yurova, L. M. (2002). Geochemistry of fluids from submarine hot springs at Punta de Mita, Nayarit, Mexico. *Journal of Volcanology and Geothermal Research, 115*, 329–338.

Taran, Y. A., Morán-Zenteno, D., Inguaggiato, S., Varley, N., & Luna-González, L. (2013). Geochemistry of thermal springs and geodynamics of the convergent Mexican Pacific margin. *Chemical Geology, 339*, 251–262.

Trenhaile, A. S. (2002). Modeling the development of marine terraces on tectonically mobile rock coasts. *Marine Geology, 185*, 341–361.

Urías Espinosa, J., Bandy, W. L., Mortera Gutiérrez, C. A., Núñez Cornú, F., & Mitchell, N. (2016). Multibeam bathymetric survey of the Ipala Submarine Canyon, Jalisco, Mexico (20°N): The southern boundary of the Banderas forearc block? *Tectonophysics, 671*, 249–263.

Urrutia-Fucugauchi, J., & González-Morán, T. (2006). Structural patterns at the northwestern sector of the Tepic-Zacoalco rift and tectonic implications for the Jalisco block, western Mexico. *Earth Planets Space, 58*, 1303–1308.

Whitmarsh, R. B., & Jones, M. T. (1969). Daily variation and secular variation of the Geomagnetic Field from shipboard observations in the Gulf of Aden. *Geophysical Journal International, 18*, 477–488.

(Received September 11, 2015, revised August 18, 2016, accepted August 20, 2016, Published online September 3, 2016)

Pure Appl. Geophys. 173 (2016), 3553–3573
© 2016 The Author(s)
This article is published with open access at Springerlink.com
DOI 10.1007/s00024-016-1388-7

Crustal Architecture at the Collision Zone Between Rivera and North American Plates at the Jalisco Block: Tsujal Project

JUANJO DAÑOBEITIA,[1] RAFAEL BARTOLOMÉ,[2] MANEL PRADA,[2,6] FRANCISCO NUÑEZ-CORNÚ,[3] DIEGO CÓRDOBA,[4]
WILLIAM L. BANDY,[5] F. ESTRADA,[2] ALEJANDRA L. CAMESELLE,[2] DIANA NUÑEZ,[3,4] ARTURO CASTELLÓN,[1]
JOSÉ LUIS ALONSO,[1] CARLOS MORTERA,[5] and MODESTO ORTIZ[7]

Abstract—Processing and analysis of new multichannel seismic records, coincident with wide-angle seismic profiles, acquired in the framework of the TsuJal project allow us to investigate in detail the complex structure of the oceanic domain in the collision zone between Rivera Plate and Block Jalisco at its northern termination. The subducting Rivera Plate, which is overridden by the North American Plate–Jalisco Block, is clearly identified up to 21.5°N (just south of Maria Magdalena Island) as a two clear reflections that we interpret as the interplate and Moho discontinuities. North of the Tres Marias Islands the seismic images display a different tectonic scenario with structures that are consistent with large faulting and rifted margin. A two-dimensional velocity approach for the crustal geometry is achieved using joint refraction/reflection travel time tomography, the uncertainty of the results is assessed by means of Monte Carlo analysis. Our results show an average oceanic crustal thickness of 6–7 km with a moderate increase towards the Jalisco Block, an anomalous thick layers (~3.0 km) displaying a relatively low velocity (~5.5 km/s) underneath Maria Magdalena Rise, and an estimated Moho depth deeper than 15 km in the collision zone between Rivera Plate and Jalisco Block. We have also determined an anomalous crust on the western flank of the Tres Marias Islands, which may be related to the initial phases of continental breakup of the Baja California Peninsula and Mexico mainland. High-resolution bathymetry provides remarkable images of intensive slope instabilities marked by relatively large slides scars of more than 40 km² extent, and mass-wasting deposits probably triggered by the intense seismicity in the area.

[1] Unidad de Tecnología Marina, CSIC, P. Marítimo de la Barceloneta 37-49, 08003 Barcelona, Spain. E-mail: jjdanobeitia@utm.csic.es

[2] Instituto de Ciencias del Mar, CSIC, P. Marítimo de la Barceloneta 37-49, 08003 Barcelona, Spain.

[3] Centro Universitario de la Costa (CUC), Av. Universidad 203, Del. Ixtapa, 48280 Puerto Vallarta, Mexico.

[4] Universidad Complutense de Madrid, Ciudad Universitaria, Plaza Ciencias, s/n, 28040 Madrid, Spain.

[5] Instituto de Geofísica, UNAM, Ciudad Universitaria, Delegación Coyoacán, C.P. 04510, Mexico, D.F., Mexico.

[6] *Present Address*: Dublin Institute for Advanced Studies, Dublin, Ireland.

[7] Centro de Investigacion Cientifica y de Educacion Superior de Ensenada, Mexico, Mexico.

1. Introduction

The crustal architecture of the western boundary of Mexico is strongly controlled by the Middle America Trench (MAT) which is the morphological expression of an active subduction zone involving the North American Plate overriding the relatively young Rivera and Cocos plates, and it is the location of significant tectonics, widespread seismicity and magmatism (Fig. 1). Both oceanic plates, Rivera and Cocos, are fragments of the Farallon Plate (e.g., Atwater and Stock 1998); the Rivera Plate acting independently since 10 Ma (DeMets and Traylen 2000). The Rivera Plate is a key structural element to understand the complex geodynamic interactions that take place at the west coast of Mexico (Yang et al. 2009); with accreting seafloor along its western boundary, the Pacific Rivera Rise (PRR), and subducting seafloor at the northern tip of the MAT along the Jalisco Block. Several authors have shown that the Rivera Plate is kinematically distinct from the North American and Cocos plates (e.g., Bandy and Yan 1989; DeMets and Stein 1990). Moreover, the eastern termination of the Rivera–Cocos boundary is still uncertain since no bathymetric expressions can be clearly associated with the plate boundary (Bourgois and Michaud 1991; Bandy et al. 1995; Michaud et al. 2001; Peláez Gaviria et al. 2013). The lithosphere of the Rivera Plate, dated as late Miocene near the MAT (Klitgord and Mammerickx 1982), is consumed at the trench at a convergence rate that varies from 5.0 cm/year near the Rivera Cocos Plate Boundary to 2.0–3.0 cm/year along the Tres Marias Escarpment (Kostoglodov and Bandy 1995).

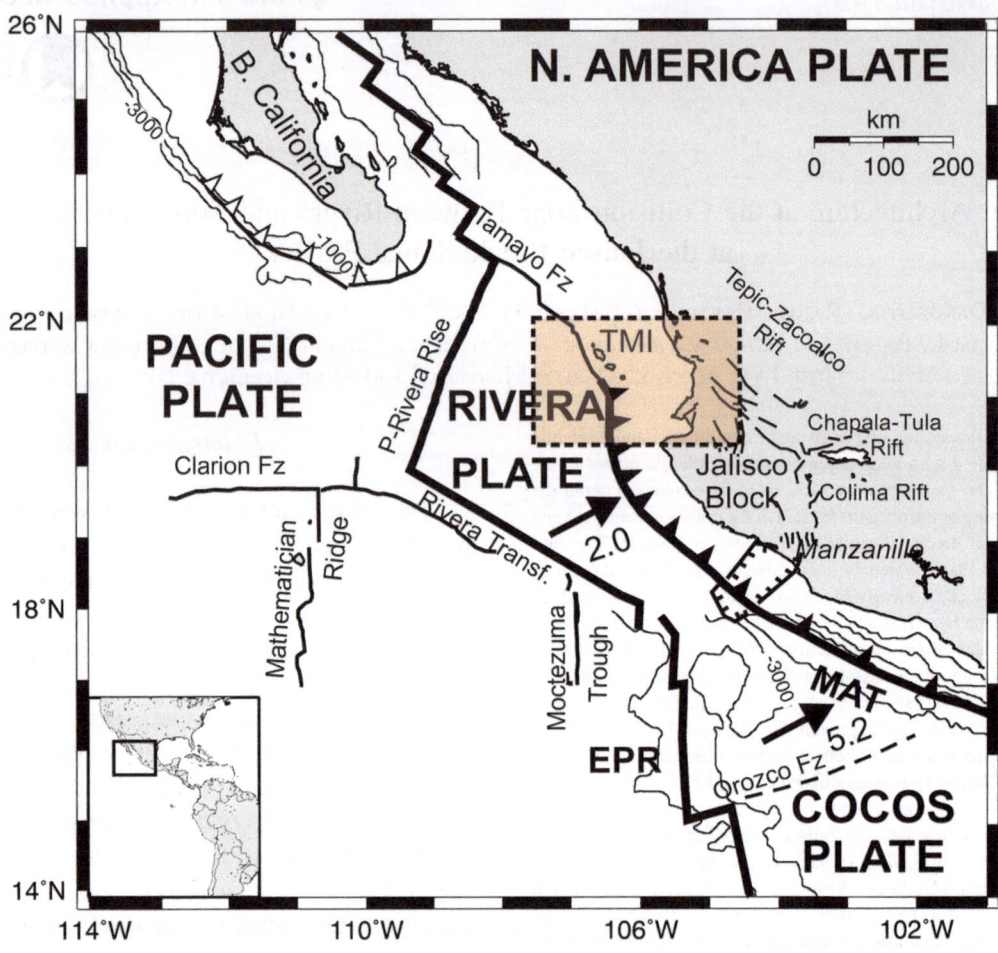

Figure 1

Tectonic framework map showing the main structural features in the interaction area of the Rivera, Pacific, North America and Cocos plates. *Pink frame* shows the study area around the Tres Marias Islands (TMI). *MAT* Middle America Trench, *EPR* East Pacific Rise

The studies of the internal structure of the subducted plate have relied primarily on potential field data (Bandy et al. 1993, 1999), and only few seismic profiles are reported in the literature (Bourgois et al. 1988; Bourgois and Michaud 1991; Khutorskoy et al. 1994; Michaud et al. 1996; Bandy et al. 2005). In 1996 the Spanish R/V Hespérides and the Mexican R/Vs Altair and Humboldt surveyed the northwestern Mexican margin between 16°N and 30°N in a geophysical experiment named Crustal Offshore Research Transect by Extensive Seismic Profiling, CORTES-P96 (Dañobeitia et al. 1997). During this research expedition swath bathymetry, backscattering and multichannel seismic reflection data and wide-angle profiles were acquired. These data (Bartolome et al. 2011) show a strong reflection, at 2 s (twtt)

underneath the sedimentary cover, interpreted as the Moho discontinuity and they computed a mean dip angle of $7° \pm 1°$ for the initial subduction of the Rivera Plate, the dip angle gently increases towards the south. The angle of subduction of the Rivera Plate has been obtained from studies of local seismicity and crustal structure, and displays a gentle subduction angle near the trench (Nuñez-Cornú et al. 2002, 2016), then dives more steeply into the mantle at an angle of 34° (Gutierrez et al. 2015). The Cocos Plate has a slightly curved subducted slab with oblique geometry dipping to the Colima Rift, ranging in dip from 18° in the south to 30° in the north (Gutierrez et al. 2015). This has led to the proposal of a step in the slab between the Cocos and Rivera Plates at the present time (Nixon 1982; Ferrari et al. 2001) formed during a period of very low convergence rate (19 mm/year on average) between 8.5 and 4.6 Ma (DeMets and Traylen 2000).

The seismicity is discontinuous and moderate in magnitude (Nuñez-Cornú et al. 2002; Núñez-Cornú et al. 2004) (Fig. 2), although some of the largest destructive earthquakes in western America are reported offshore of the Jalisco region (Singh et al. 1985; Courboulex et al. 1997). The largest instrumental record occurring in Mexico is the 1932 Jalisco $M_w = 8.2$ event (e.g., Eissler and McNally 1984; Singh et al. 1985; Okal and Borrero 2011). The main event of the 1995 earthquake was a subduction-related thrust earthquake that activated normal faults along the northwest margin of Manzanillo (19°N, 105°W, Fig. 2), clearly indicating present day plate convergence. Most of the shocks during the 1932 strike occurred at the Rivera–North America boundary, while the shocks/aftershocks of 1995 seem to happen at the rupture zone between Rivera–Cocos Plate interface (Pacheco et al.1997, Núñez-Cornú 2011).

During the spring of 2014, within the framework of TSUJAL project (Núñez-Cornú et al. 2016), Spanish and Mexican scientists investigated the western margin of Mexico at the collision zone between the Rivera, Cocos and North American plates. The main objective was to define the crustal architecture of this active margin and recognize potential structural sources that can trigger earthquakes and tsunamis at the Rivera–North America convergence zone. To achieve these goals a wide range of geophysical data were acquired aboard the RRS James Cook. Herein we present the northernmost post-stack time migrated multichannel seismic (MCS) sections together with coincident wide-angle (WA) seismic data (Fig. 3). These new data show seismic images from the Earth's surface down to the Moho that enables to seismically characterize the crustal structure of the overriding and subducting plates in the offshore area of the Jalisco Block. We also describe the spatial distribution of the incoming plates (sediment, crust and uppermost mantle lithosphere) and identify some of the geodynamic processes occurring in this area.

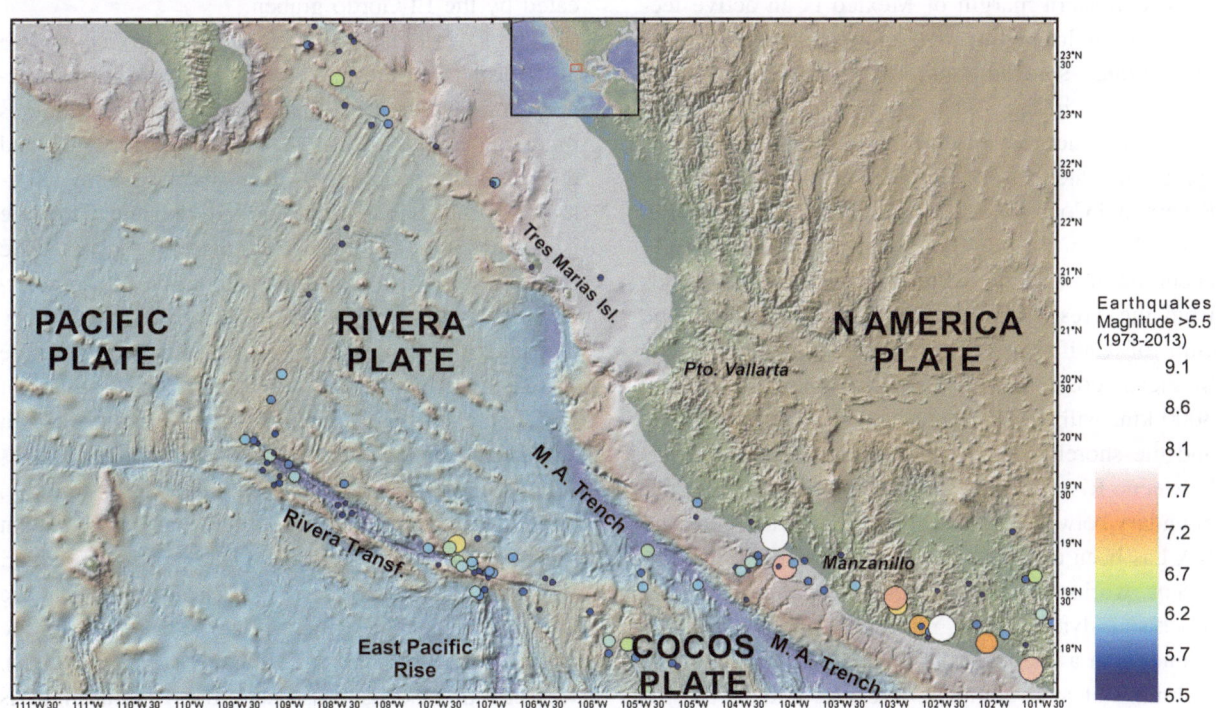

Figure 2

Seismicity from 1973 to 2013 for magnitude >5.5. Note the location of earthquakes with M_w larger than 7.0 along the Middle American Trench

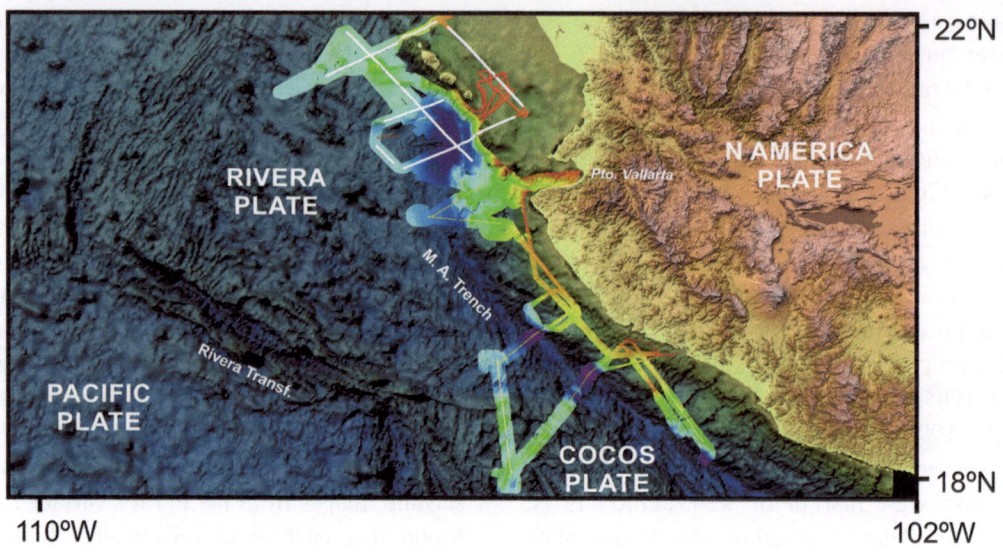

Figure 3
Track-chart and bathymetry acquired during the TsJual Survey along the seismic lines. *White lines* denote the seismic lines analyzed and interpreted in the northern Rivera Plate–Jalisco Block study of this paper

2. Geological Setting

The western margin of Mexico is an active tectonic region displaying ongoing, strong, tectonic plate interactions: spreading process in the East Pacific Rise (EPR) in the West boundary of the Rivera Plate (Mammerickx 1984) at an average half-spreading rate of 2.7 cm/year. Subduction of the Rivera and Cocos plates beneath the North American Plate (NA) in the Middle American Trench (MAT) occur on its eastern boundary. The trench extends from the Tres Marias Islands, located south of the Gulf of California at 21°N, southward along the Mexican coast for a distance of approximately 3000 km, with the distance between the trench axis and the shoreline of only 40–60 km. The boundary between Rivera and Cocos plates and the SE boundary between the three plates (Rivera, Cocos and NA has been classically defined as diffuse, although recent works based on Bouger anomaly gravity models (Alvarez and Yutsis 2015) located the boundary at a line from the MAT to the region of the Colima Volcanic complex. We can find in the SE boundary other processes, such as ridge–trench collisions between the East Pacific Rise (EPR) and the MAT, off Manzanillo, and slab windowing between the Rivera and Cocos plates (Bandy 1992), which is characterized by lithospheric extension and delineated by the El Gordo graben.

The Rivera Plate fragmented from the Cocos Plate at 10 Ma (DeMets and Traylen 2000). Its convergence rate with respect to the North American Plate is moderate, about 2–3 cm/year, although this value and the degree of obliquity (see discussion by Kostoglodov and Bandy 1995) are still debated. This rate is always slower than the adjacent Cocos Plate (5–8 cm/year). Rivera Plate subduction is characterized by a steeper angle and more northerly trajectory than the adjacent Cocos Plate, with a dip angle increasing at depths around 100 km. A summary of DSDP site 473 results and interpretations of a 287 m length core on the Rivera Plate, south of Tres Marias Islands, indicates that older sediments are 6–6.5 m.y. (Upper Miocene) and no older than 8 m.y. with igneous rocks at the end of the core. Moreover, Klitgord and Mammerickx (1982) published that the lithosphere of the Rivera Plate consumed at the trench dates as late as Miocene (approx. 9 Ma), whereas the Cocos crust near the Rivera Plate is 10 Ma and progressively becomes older to the SE, being 25 Ma old at 90°W (Pardo and Suárez 1995; Mammerickx and Kiltgord 1982). Terrigenous clay deposited in the early Pliocene to Quaternary and calcareous claystone deposited during early Pliocene are separated by a strong seismic reflector from the Upper Miocene sediments in the core. Sediments are unexpectedly terrigenous despite the distance from land and the MAT. Sediment accumulation rates are 40 m/m.y. for the last 3 million years and 20 m/m.y. from 3 to 6.5 million years. Sediment velocities are 1.5–1.6 km/s increasing to 1.98 km/s at the basal sediments. Igneous rocks below the sediments are mainly massive, altered diabase with a density of 2.7 g/cm^3 and a velocity of 5.2–5.3 km/s, but a fragment of pillow basalt was found at the bottom of the hole.

The margin along the Middle America trench from the Tres Marias Islands to the Manzanillo Area (18°N, 104°W), was uplifted and emergent before the late Miocene. The margin started to subside during the upper Miocene–lower Pliocene at least until the Pliocene–Quaternary limit (Mercier de Lépinay et al. 1997). From the Pliocene–Quaternary limit (approx. 2.5 Ma), the Manzanillo area continue subsiding whereas the Tres Marias Islands began to be uplifted (McCloy et al. 1988). Regional subsidence of an active margin is generally related to tectonic erosion and therefore, such long-term subduction-erosion regimen for the Manzanillo area implies trench retreat along the margin (Mercier de Lépinay et al. 1997). The recent overview of Ramirez-Herrera et al. (2011) confirm that, although spatial and temporal variability exists, coastal subsidence is occurring at the southern Colima Graben and the Guerrero seismic gap near Acapulco, whereas coastal uplift is occurring between Puerto Vallarta and Manzanillo and along the coast south of the Colima/El Gordo Graben until Lazaro Cardenas, South Acapulco.

The Jalisco Block is generally considered as part of the NA Plate may have some degree of independent motion. There are active faults along the Tepic-Chapala, Colima and Chapala rifts in southwestern Mexico that some authors considered related to Jalisco–North America tectonics (Allan 1986; Johnson and Harrison 1989; Allan et al. 1991; Fig. 1). Selvans et al. (2011) find also a slow motion of 2 mm/year to the southwest of the stations in the Jalisco Block relative to the NA Plate in a recent GPS campaign near the triple junction, which is compatible with such a tearing; however, motion could be related to the earthquake cycle behavior of subduction megathrusts and, therefore, the Jalisco Block might not have an independent motion.

Recent studies also evidence gas hydrates in the western (Minshull et al. 2005; Bartolome et al. 2011, 2016; Bandy and Mortera 2012) in the form of bottom simulating reflectors (BSR) on multichannel seismic reflection profile located on the continental slope area of the northern part of the Jalisco Subduction Zone, off Puerto Vallarta, between 20° and 20.5°N, and in the south off Manzanillo. Thus, there is evidence to suggest that the Pacific margin of Mexico may contain significant reserves of hydrocarbons in the form of gas hydrates.

3. TSUJAL Project

The global objective of the TSUJAL project included the characterization of the shallow and deep structure of the Rivera Plate and the Jalisco Block in order to investigate the recent tectonic deformation due to the subduction process affecting the area, and the extensional zones associated with transform faults (Núñez-Cornú et al. 2016). In addition, the data places constraints on the geophysical parameters related to the lithospheric structure in the western Mexican margin (Fig. 1). The integration of these parameters together with the available acquired geological information will allow us to obtain a complete image of the lithosphere that will be used for a seismic and tsunamigenic hazard assessment (Trejo-Gómez et al. 2015).

The specific objectives of TSUJAL survey include characterizing the subduction of the Rivera Plate along the western Mexican margin, clarifying the convergence direction and angle of subduction, in the oceanic crust, the trench, the continental slope and rise and up to the coast. In addition, to identify the active structures, mainly faults, with recent neotectonic deformations that are prone for generating earthquakes and tsunamis. There has been a substantial effort in the project to image the shallow geometry, pattern and style of deformation of the morphological structures using high-resolution methods, and the correlation with its seafloor topography. Finally, we plan to identify and typify mass transport deposits and submarine landslides associate to active faults, emphasizing the ones that can generate earthquakes and tsunamis. TSUJAL data were acquired (Bartolome et al. 2015) in the framework of a bi-lateral barter agreement between the National Environmental Research Council (NERC) and Consejo Superior de Investigaciones Cientificas (CSIC), under the European research alliance OFEG (Ocean Facility Exchange Group).

4. Methods

In this section, we outline the acquisition, processing and modeling of seismic and bathymetric data used in this work.

4.1. Multichannel Seismic Reflection Data

The TSUJAL survey focused on the acquisition of Multichannel Seismic Reflection (MCS) data using a 5.85-km-long digital streamer of the Unidad de Tecnología Marina (UTM—CSIC, Spain) and the airgun source system from the RRS James Cook (NERC, UK), which includes 12 Bolt® G-Guns 1500LL with 4 air compressors and a laboratory equipped to control the firing and synchronization of guns (Fig. 4). A big source array was designed to maximize the energy concentrated at the lowest frequency range. The shooting interval was defined at 50 m, as a compromise between maximum redundancy of data (CMP fold) and capacity of the air compressors (Table 1). During the entire marine seismic survey we applied the Precautionary Principle following the guide recommendations of Joint Nature Conservation Committee. Thus, the safety distance was determined to be 1000 m from the guns and only on a couple of occasions marine mammals and turtles were sighted causing the acquisition operations to be stopped.

A total of 14 MCS profiles were acquired between 108° and 104°W and 17.5° and 22°N. Herein we present the northern MCS network of profiles acquired around the Tres Marias Islands, at the NW of Bahia Banderas (Fig. 5a), which amounts to a total of seven MCS profiles of about 620 km length (Table 2).

4.2. Refraction and Wide-Angle Seismic Data

The TSUJAL experiment used 16 short period Ocean Bottom Seismometers OBS model LC2000SP (1 hydrophone and three components seismometer) designed and built by the Scripps Oceanographic Institution (La Jolla, USA) and provided by the Marine Technology Unit (UTM-CSIC). We used a source that consisted of 14 air guns, in a six array configuration, with a total volume of 6800 cubic inches, towed at 15 m depth, shooting every 120 s along each line. A network of 6 wide-angle (WA) profiles was recorded, in the region around the Maria Islands, by 16 OBSs at 32 locations, in Jalisco and Nayarit offshore regions, as well as on a terrestrial network (100 portable seismic stations) in 240 locations across 5 seismic profiles of 200–300 km in length combined with the Seismological Network of the State of Jalisco (RESAJ, Núñez-Cornú et al. 2016). Figure 5b shows the location of these lines and the location of the OBSs along each line. In this work, we present offshore WAS data of lines RTS-IM01 (95 km) and RTS-IM02 (225 km). WAS data along transect RTS-IM01 were acquired by 4 OBSs, whereas 6 OBSs were used to record data along line RTS-IM02. The average receiver spacing between OBS was ∼20 km. Processing of WAS data involved a band pass filter (3–7; 20–25 Hz), and a deconvolution and a wiener shaping filter (prediction lag of 60 ms) in order to retrieve far-offset signal. After the processing we were able to identify arrival times up to 100–120 km offset.

In Figs. 6 and 7, we display the records sections, together with the seismic phases identified and consequent ray tracing used along both lines RTS-IM01 and RTS-IM02 with a reduction velocity of 8 km/s. The phases used for the modeling correspond to P-waves refracted through the overriding plate (Pgc), oceanic crust (Pgo) (which is clearly identified as a first arrival until 40 km offset) and mantle (Pn) (first arrival are observed at far offsets until more than 100 km), and near offset reflections at the intraplate boundary (PiP), and the higher amplitude reflection at the Moho (PmP) some of which are observed at a relatively near offset range (∼30 km). Overall, a total of ∼1600 and 3800 travel times were manually selected from receiver gathers of lines

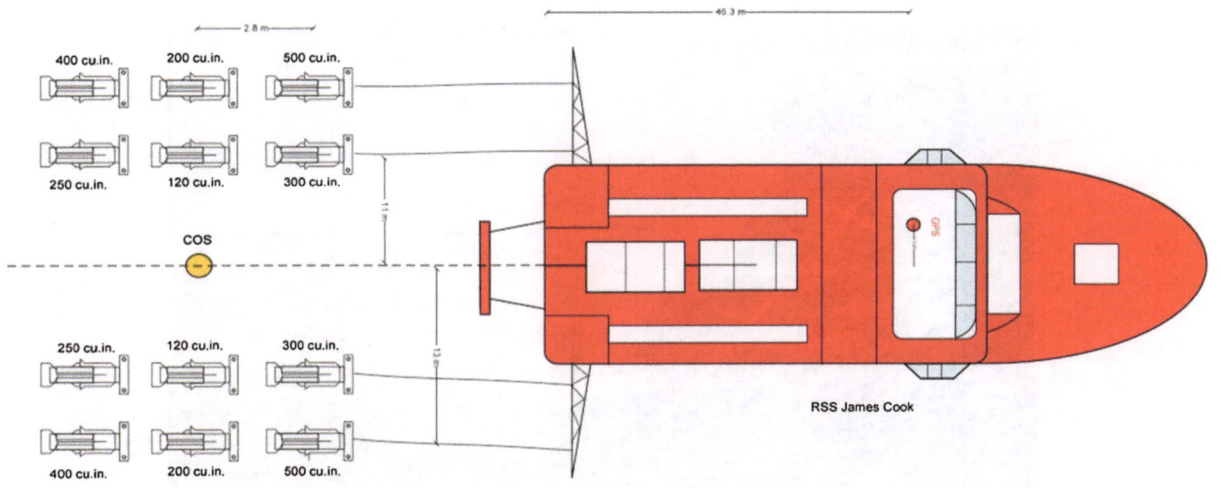

Figure 4
Total gun-array (3540 cu.in.) configuration used during acquisition of MCS profiles TS08 to TS14

Table 1

Source parameters used during TsuJal northern seismic survey, including multichannel (TS08-TS14) and wide-angle (WA) profiles used in this study

Source parameters	
Source controller	Big Shot®
Source type	Bolt® G.Guns 1500LL
Air pressure	2000 psi
Volume	3540 cu. in. (TS08-TS14)
	6800 cu. in. (WA)
Compressors	4 x Hamworthy® 4™ 565 W100
Number of arrays	4
Total number of guns	12 (3 guns for array)
Gun synchronization	± 0.1 ms
Deployment depth	8 m
Shot interval	50 m for mcs
	120 s for wa
Aiming point	50 ms

RTS-IM01 and RTS-IM02, respectively. For these picked travel times, errors are supposed to be half a period of the dominant frequency of the recorded signal, thus, that we use an uncertainty of 50 ms for the travel time of crustal refracted phases (Pg), and for longer offset; of 60–80 ms for mantle refractions (Pn), and of 90–100 ms for intraplate and Moho reflections (PiP and PmP). The uncertainty is chosen on the basis of the signal-to-noise ratio, which correlated with the offset (i.e., the signal-to-noise ratio decreases with offset).

The selected travel times were then inverted using the joint refraction-reflection travel-time tomography method published in Korenaga et al. (2000). This method uses a set of velocity and depth smoothing constraints defined by correlation lengths (CL) to stabilize the iterative inversion. After several tests, we used velocity CL that varies between 3 km at the top of the model to 15 km at the bottom for horizontal CL, and between 0.5 and 5 km for vertical CL. Depth CL was set at 5 km.

To obtain the final models (Fig. 8), we have followed a layer stripping strategy, which is commonly used in crustal tomography approaches as it allows to retrieve the geometry of the main geological interfaces in detail (e.g., intraplate and Moho; Sallarès et al. 2013a). This strategy consists of inverting seismic phases sequentially from near to far offset. The number of steps depends on the number reflections (i.e., layers) interpreted along each profile. This way, inversion of line RTS-IM02 has consisted of three steps: first we inverted both overriding crust first arrivals (Pgc) and reflections at the intraplate boundary (PiP), second we inverted Pgc together with oceanic crust first arrivals (Pgo) and Moho reflections (PmP), and finally, in the third and last step we inverted all the previous refractions together with Moho reflections and mantle refractions (i.e., Pgc, Pgo, Pn and PmP). The final root mean

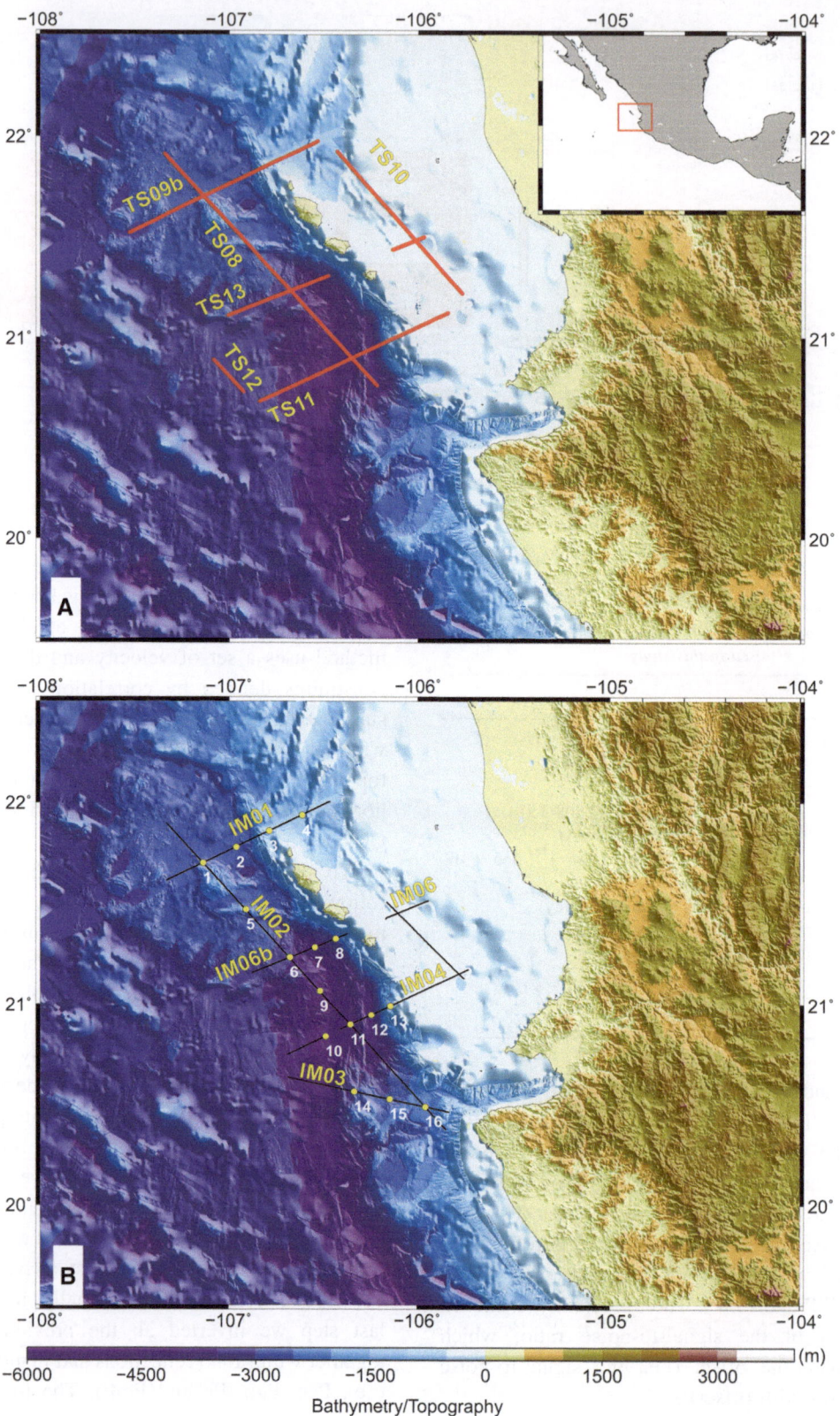

◄Figure 5
Location map of the Seismic Profiling around Tres Marias Islands acquired during TsuJal survey. **a** Displays the position and names of MCS profiles. **b** Displays the position of the wide-angle profiles, coincident with MCS location. *White* numbering indicates OBS locations in each profile

Table 2

MCS profile locations and length during TsuJal northern survey

Profile	Latitude (°) N Start of line	Longitude (°) W	Latitude (°) N End of line	Longitude (°) WE	Length
TS08	20,76	106,22	21,91	107,34	172,75
TS09b	21,52	107,52	21,98	106,53	115,25
TS010	21,92	106,43	21,22	105,77	103,9
TS11	21,12	105,84	20,68	106,84	115,1
TS12	20,73	106,93	20,89	107,08	24,05
TS13	21,11	107,01	21,31	106,47	59,55
TS14	21,44	106,14	21,50	105,97	18,55

square values for model RTS-IM01 and RTS-IM02 are ∼45 and ∼50 ms, respectively.

4.2.1 Model Parameter Uncertainty

The uncertainty of the model parameters (Vp, depth) in crust for both profiles (RTS-IM01 and RTS-IM02) has been evaluated by means of a Monte Carlo analysis following a similar approach as Korenaga et al. (2000). A total of 200 Monte Carlo different tests were used in this approach.

Initial Vp models were built following a two-layered structure. The upper layer velocity gradient was set by velocities of 1.5 km/s at the top (seafloor) to 5 km/s at its bottom, whereas the lower layer gradient was defined by velocities between 5.5 km/s at the top to 7 km/s at the bottom. The upper layer thickness was randomly chosen between values of 1 km to 3 km, while a range of 8 km and 12 km was set to choose the thickness of the lower layer. Upper and lower velocities of each layer were randomly varied within 10 %. The initial set of 200 flat Moho reflectors were randomly set at depths between 13 km and 17 km. Overall, the initial Vp uncertainty was set at 0.3 km/s, while the depth uncertainty was ∼2 km. The set of 200 perturbed data set were created by randomly modifying the selected travel time within their uncertainties, that is 50 ms for Pg and 90 ms for PmP. Additionally, velocity and depth

CL were also randomly chosen, similar to Korenaga and Sager (2012), varying between 3 km and 15 km, from top to bottom for CL horizontal, and between 0.5 km and 5 km for CL vertical (op. cit.)

According to Tarantola (1987), the velocity and depth standard deviation of the number of successfully inverted models can be taken as a measure of the model parameter uncertainty. The results of inverting 200 Monte Carlo realizations are shown in Fig. 9 for both models. The derivative weight sum (DWS) is also included in Fig. 9, which gives accurateness on the linear sensitivity of the inversion (Toomey and Foulger 1989).

The crustal Vp uncertainty for both lines RTS-IM01 and RTS-IM02 varies between 0.1 km/s and 0.2 km/s, being <0.1 km/s in areas with high ray density (Fig. 9). A high uncertainty exists (up to 0.5 km/s) in the first 2–3 km of the model. This is commonly observed in regions with high vertical Vp gradients like the crust–mantle or the sediments–basement boundary (Sallarès et al. 2013b). The depth uncertainty for the Moho reflector has uncertainties between 0.8 and 0.5 km along Line RTS-IM01, and 0.2–1.25 km along Line RTS-IM02. The overall crustal velocity structure and depth of the Moho along both models is significantly well resolved despite of the high receiver spacing (∼20 km).

4.3. Swath-Bathymetry Acquisition and Processing

During the TSUJAL (JC-098) cruise high-resolution swath-bathymetry and backscatter data were acquired with two Kongsberg Simrad's multibeam echosounder, hull-mounted in the RSS James Cook: EM120 and EM710. The EM120 was running during the whole cruise and the EM710 only in shallow waters (<500 m). The nominal sonar frequency for the EM120 is 12 kHz with an angular coverage sector of up to 150 degrees and 191 beams per ping as narrow as 1 degree. Achievable swath width on a flat bottom will normally be up to 5.5 times the water depth. The EM710 operates at sonar frequencies in the 70–100 kHz range with an angular coverage sector of up to 140° and 128 beams. Both echosounders were operated simultaneously with a hydrographic echosounders Simrad EA600. To avoid interferences all echosounders were triggered by a

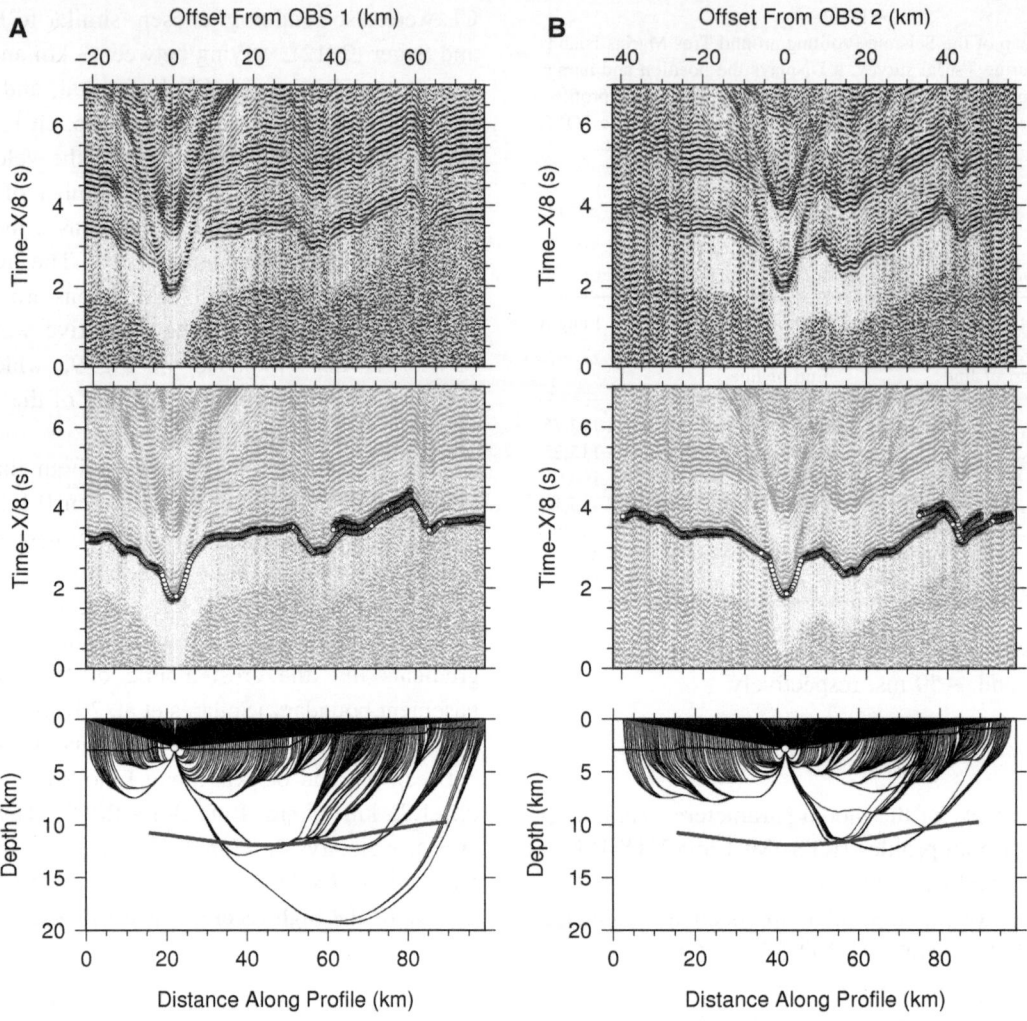

Figure 6

Record sections (*up*), identified phases (*middle*) and ray tracing (*bottom*) corresponding to the four OBSs deployed along the wide-angle seismic profile TS-IM01, north of TMI. *Blue line* indicates the location of the crust–mantle boundary (Moho reflector)

Kongsberg K-Sync V1.7.0 synchronization unit. We calibrated the sound velocity in the water column every day by performing XBT probes.

Due of the characteristics of this cruise, most of the lines were covered up to four times, including: during the deployment of OBS; while shooting refraction line; during recovery of OBS and finally while firing for the MCS seismic reflection data acquisition. This has conditioned the strategy of the multibeam acquisition data and data density, which can reach up to more than 2000 values/km^2 in lines,

sailed up to five times or when shallow bathymetric profiles were recorded with two echosounders simultaneously.

All bathymetric data recorded has been processed on board using CARIS (Hips & Sips v. 8.1). Processing steps include navigation editor for ping removal corresponding to the stop time during deployment and recovering of OBS sensors and filtering of sections acquired at speeds lower than 2 knots. Soundings were also filtered with the "Swath Editor" command to remove the most obvious bad or/

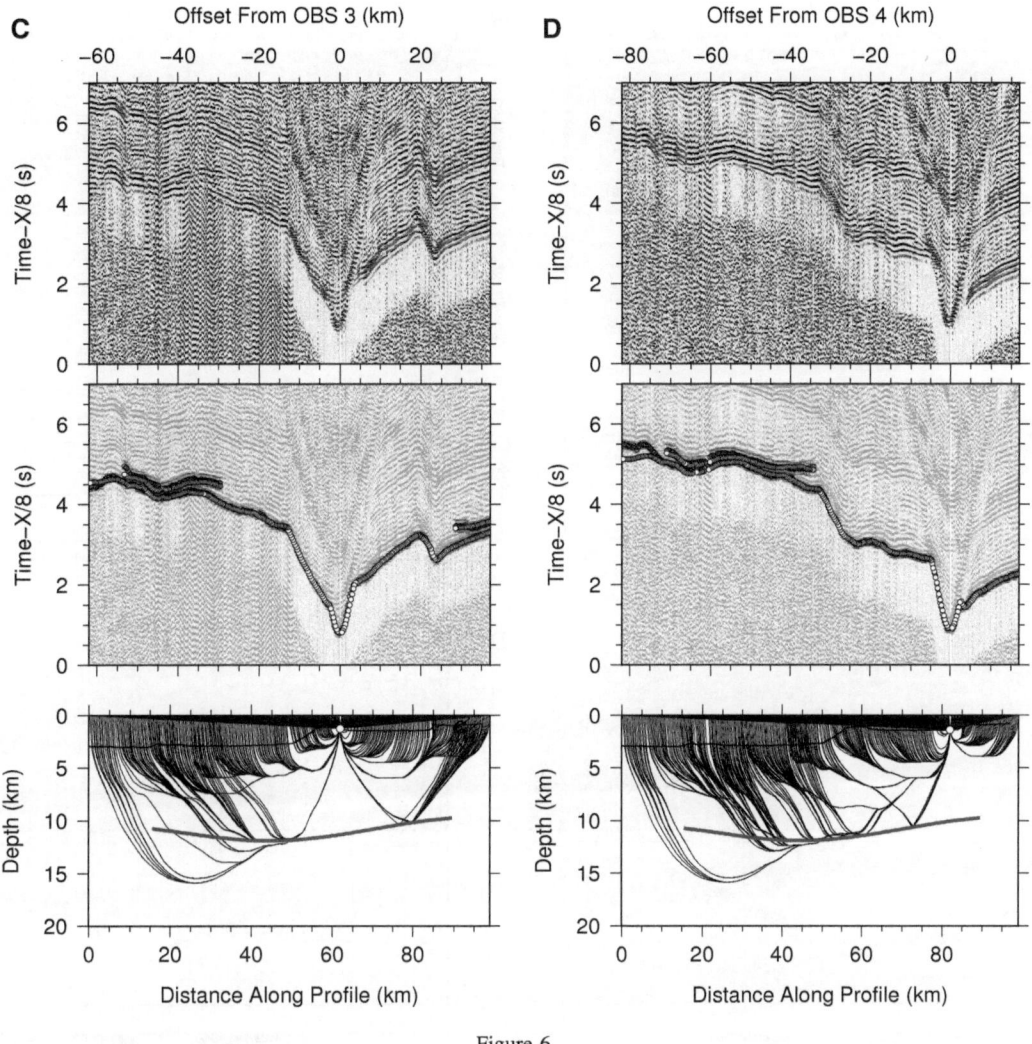

Figure 6
continued

and isolated sounds. Then, the whole base surface is systematically inspected and cleaned with the "2D Subset editor" command.

5. Oceanic Crustal Structure

5.1. Significance of the MCS, Refraction and Wide-Angle Data

The northernmost MCS profile, TS09b (Fig. 5a), strikes SW–NE for 120 km. It is located about 30 km north of the Tres Marias Islands (TMI) at the eastern flank of the Tres Marias escarpment (TME). In the migrated section of profile TS09b (Fig. 10) there is a

basin up (between marks 95–110 km) to 15 km wide, infilled by up to 1 s (twtt) of sediments, which runs parallel to the TME. From its location and shape, this basin appears to be formed by a half graben, the main fault located to the NE. This basin may have been developed coincident with the collision of the Rivera Plate as a pull-apart basin or by block rotation in an oblique subduction system, that is compatible with the structural transform regime during the last 5.5 my, and basically overlaps the older extensional tectonics (Moore and Curray 1982). Some compressional structures are observed near the eastern flank of the TME. In the western flank of TME we observe some faults that could be related to the faulting and

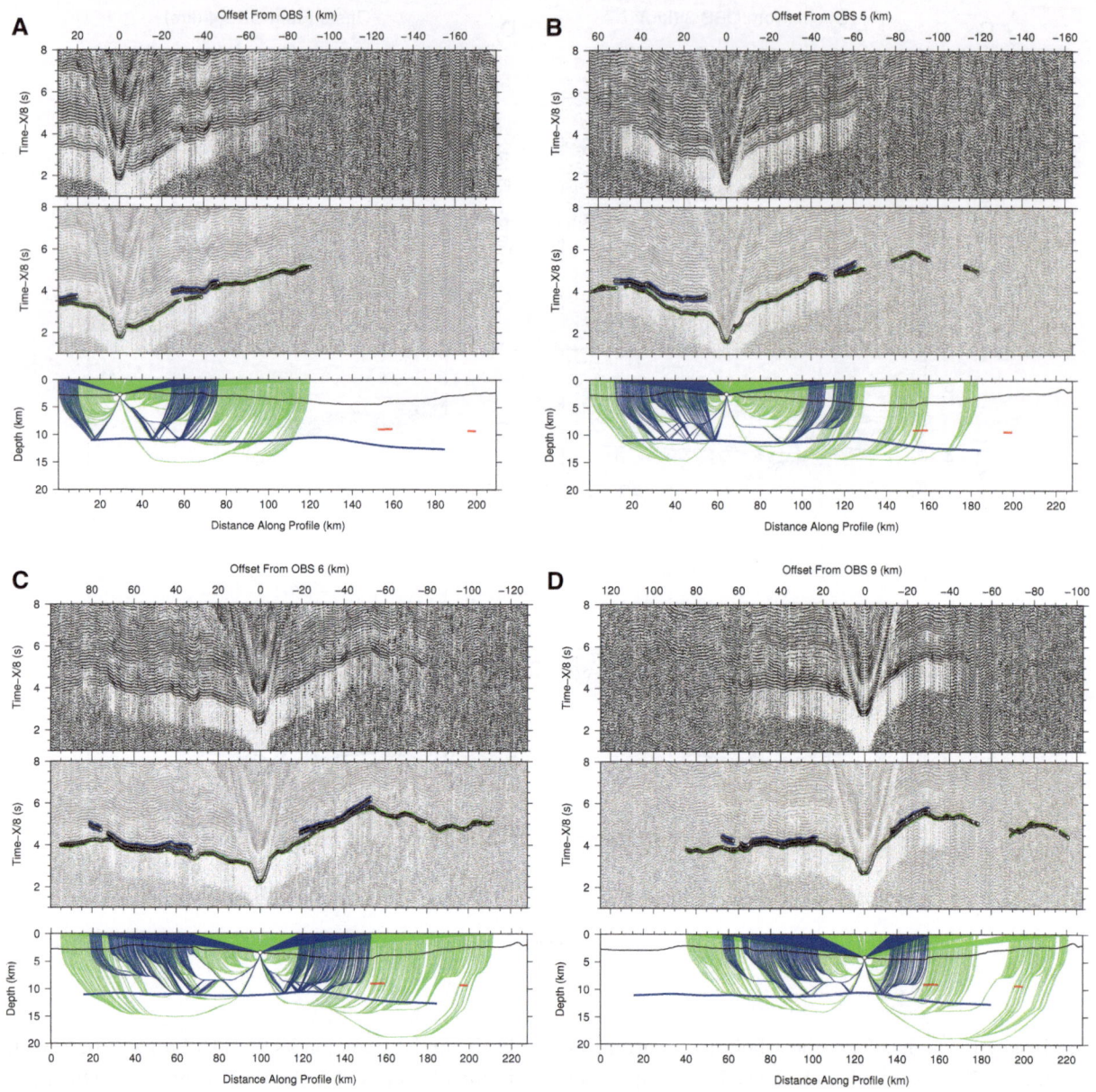

Figure 7
Record sections (*up*), identified phases (*middle*) and ray tracing (*bottom*) corresponding to the six OBSs deployed along the wide-angle profile RTS-IM02, west of TMI

rifting processes in the early stages of initial axis along the Maria Magdalena Rise (MMR), this is reinforced by NNE trending magnetic anomalies, which record the early opening between the southern tip of Baja California peninsula and western Mexico (Lonsdale 1991).

In this initial extensional period extensive rifting of the continental crust occurred at the tip of the

Peninsula and in the conjugate Tres Marias Block. Then, rifting aborted leaving remnants of continental fragments, dyke injection and rapid subsidence at the site of Marina Magdalena Rise. The velocity model obtained from the wide-angle profile, RTS-IM01, coincident with the MCS (Figs. 8, 9), shows a thin crust (∼6–7 km) in the TME with a slight progressive thickening underneath the TMI. The MCS profile

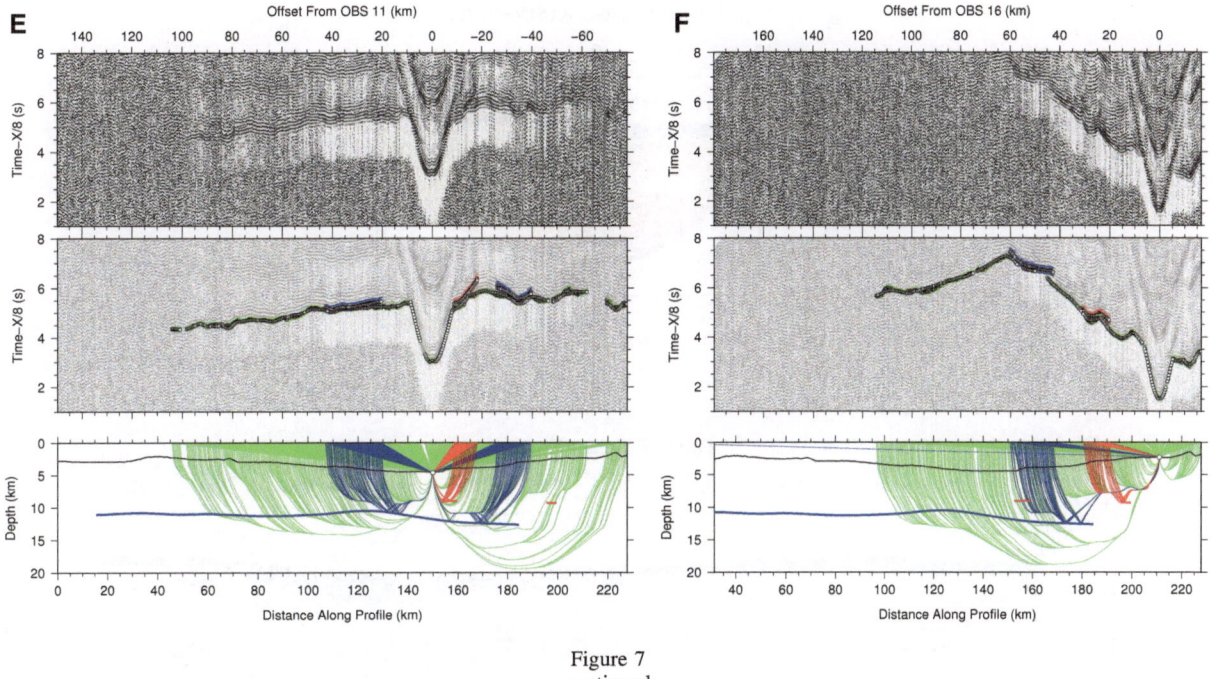

Figure 7
continued

TS13 (Fig. 11), which runs parallel to the TS09b 60 km southwards, is a short profile 62 km long ending at the mouth of the strait between Maria Magdalena and Maria Cleofas Islands. The noticeable feature of this profile is that we clearly observe an incipient subducted slab from Rivera Plate with a not yet completely developed accretionary prism, and a narrow trench of 6.0 km, providing a strong indication that the northern subduction of the Rivera Plate stops nearby.

The southern MCS stacked profile, TS11 (Fig. 12) of about 120 km length shows a complex deformation front, most probably due to the oblique subduction, profile TS11 crosscuts a lineated N–S tectonic structure (named Sierra Cleofas by Nuñez et al. 2016), that can be traced for almost 100 km in the bathymetry, from south of Marias Islands. This uplifted structure shows a segmented flower-structure consistent with strike slip faulting parallel to the margin. Nuñez et al. (2016) analyze most of the significant structural lineaments in the area, offshore and onshore, and conclude that the main observed trends are NW–SE and N–S. The interplate contact between the Rivera and NA Plates is observed as a strong dipping reflection at 6 s two-way travel time (TWTT).

Offshore Puerto Vallarta, the MCS data (TS06b) reveals high-amplitude reflections at around 7–8.5 s TWTT, roughly 2.5–3.5 s TWTT below the seafloor, that conspicuously define the bending morphology of the Rivera Plate at this latitude. These strong reflections have been interpreted as the Moho discontinuity.

The profile that crosses orthogonally all the three above-mentioned MCS profiles, TS08 (Fig. 13), runs parallel to the TMI for a length of 120 km and crosses the MMR. This seismic section clearly illustrates the bending of the Rivera Plate subducting underneath the Jalisco Block. The wide-angle profile RTS-IM02 is coincident with TS08 MCS section (Fig. 14) and shows a mean crustal thickness for the oceanic domain of less than 10 km, and noticeable low velocity layers immediately underneath the MMR (Figs. 8, 9), coinciding with high heat flow values (Prol-Ledesma et al. 1989). A similar relation between seismic velocities and heat flow values is observed in the mid-ocean ridges (e.g., Lau Basin; Dunn et al. 2013), where both low velocities and

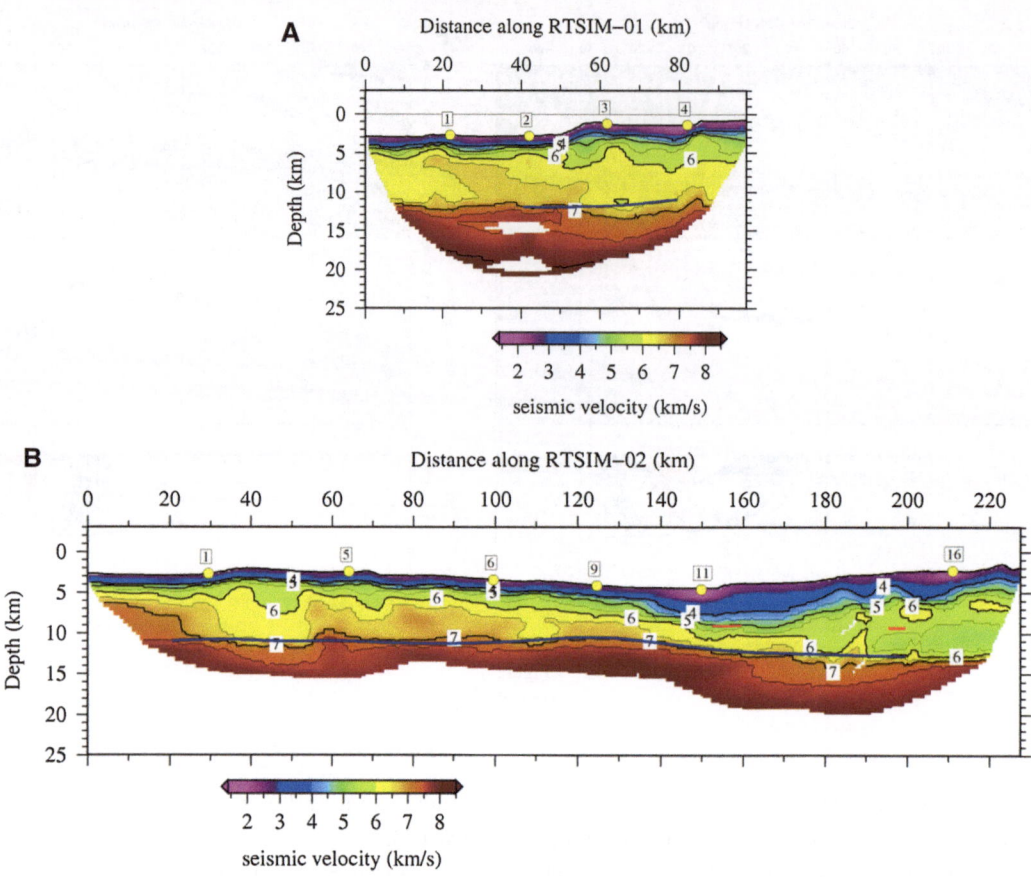

Figure 8
P-wave velocity models for WA profiles RTS-IM01 (*up*) and RTS-IM02 (*bottom*) obtained inverting the crustal phases identified in the record sections. *Numbers* at the surface indicate OBSs positions and *blue lines* show the location of the Moho reflector

thermal anomaly are related to the presence of melt. These anomalous layers, of a mean thickness of 3.0 km and low seismic velocities with respect to the neighboring crust could have been formed during the overlapping spreading center (OSC) between MMR and EPR ~ 3.5 Myr ago. The OSC could act as a trap where magma stagnates, which would lead to an increase of the thickness in the extrusive layer and the bulk of the porosity of the upper crust (Bazin et al. 2001) giving a plausible explanation to the anomalous layer geometry. An analogous pattern has been reported beneath overlap relict basins in the EPR (Canales et al. 2003).

5.2. Bathymetric Features

The surveyed continental slope westward of Tres Marias islands is characterized by steep slopes that range from about 500 m water depth to 4164 m and mean slope gradients of about 12.5° (Fig. 15). Locally, peaks of more than 25° are observed between the Maria Magdalena and Maria Cleofas islands presents several arcuate scarps developed in the upper slope and identified as blue lines in Fig. 15b. These structures indicate an area where failures are ubiquitous. The complexity of their headwalls and its relationship suggest a multi-event origin. The continental slope in front of the Maria Magdalena Island presents the biggest landslide scars reaching up to 7.2 km wide and covering an area of 41.5 km^2 (Fig. 15). Downslope, the seafloor are characterized by several lobular surfaces that represent mass-wasting deposits related with upslope scars (colored areas in Fig. 15b). The areal distribution and size of such deposits indicate that the slope off Maria Magdalena Island presents the larger instability

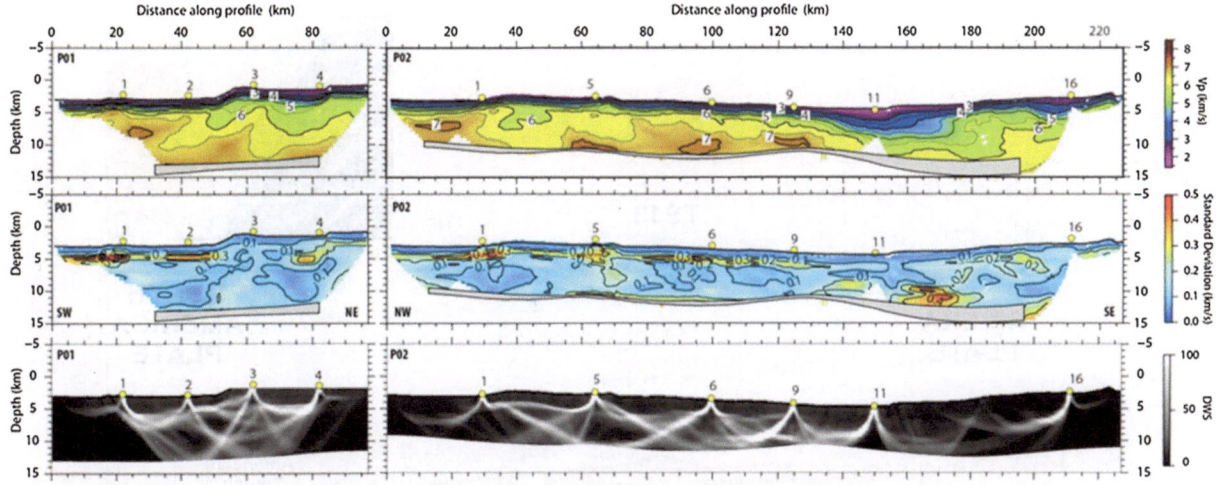

Figure 9

Upper panels show the average of 200 successful Monte Carlo realizations for lines RTS-IM01 (*left*) and RTS-IM02 (*right*). *Middle panel*'s show the Vp uncertainty value for each line, whereas the depth uncertainty for the Moho reflector is represented by the grey band in each panel. All the average RMS values for line RTS-IM01 and RTS-IM02 were 64 ms (χ^2 1.18) and 57 ms (χ^2 0.93), respectively. Lowermost panels show the derivative weight sum (DWS) average of the 200 Monte Carlo realizations that might be taken as a proxy of the ray density through the model

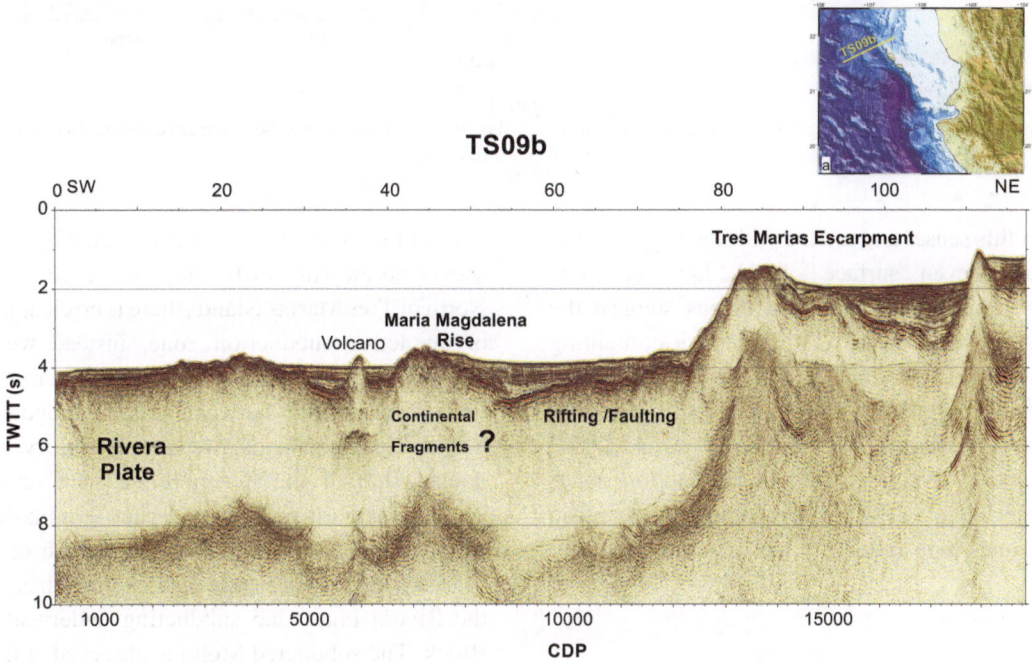

Figure 10

Migrated MCS profile TS09b, striking SW–NW with a length of 120 km. This northern profile shows the absence of subduction north of Tres Marias Islands and the morphology could be the consequence of the extension related to the opening of the Gulf of California

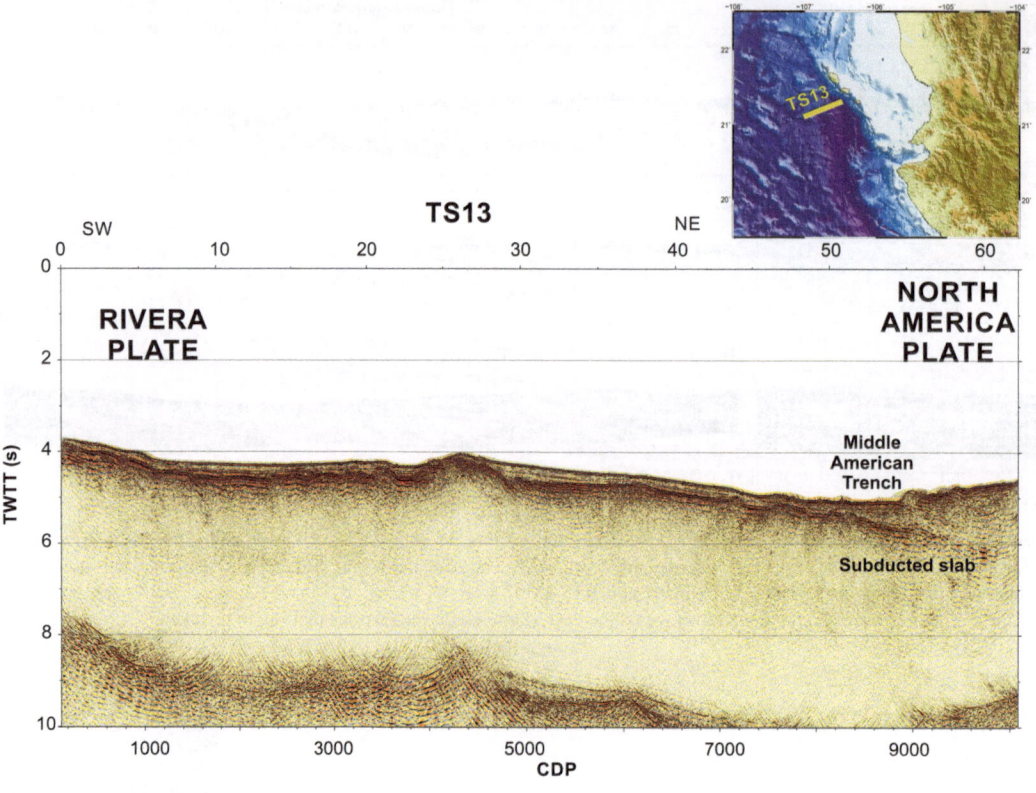

Figure 11

Migrated MCS profile TS13, striking SW–NW with a length of 63 km showing the presence of subduction between the two southern islands in the Tres Marias area

events. In this sense, and related with the biggest scar, a lobular uneven surface of 112 km^2 develops (Fig. 15). Its roughness and dimensions suggest the existence of a huge mass-wasting deposit containing several blocks (the biggest 1.7 × 0.9 km). This configuration resembles debris avalanche deposits identified in volcanic margins (Gee et al. 2001; Chiocci and de Alteriis 2006) or subduction zones (von Huene et al. 2004) where predominate steep gradients and slope instability are inherent.

6. Conclusions

Newly acquired seismic data illuminate the complex interactions between Rivera and North American plates, giving insights of the geodynamics at the northern edge of these plates.

We have identified in the MCS sections the collision zone between the Rivera and North American

plates in the region of the Tres Marias Islands, which shows noteworthy differences from north to south. North of Tres Marias Islands there is no clear indication of an active subduction zone, instead we observe faulting at the west flank of the Tres Marias Islands, while southwards between Maria Magdalena and Maria Cleofas Islands, we clearly observe the subducted slab of the Rivera Plate with an incipient accretionary prism indicating near termination of the subduction zone. At the southwest edge of the Tres Marias islands the seismic images show unequivocally the Rivera Plate slab subducting underneath Jalisco Block. The subducted Moho is observed at the Rivera Plate, in the southern profile TS11 perpendicular to the North American Plate as a clear bending interface from the outer trench at 8.0 s (TWTT) to 8.4 s TWTT underneath the MAT (Fig. 12). The crustal velocity structure and depth of the Moho along RTS-IM01 and RTS-IM02 profiles is significantly well resolved despite the high receiver spacing (~20 km), as shown

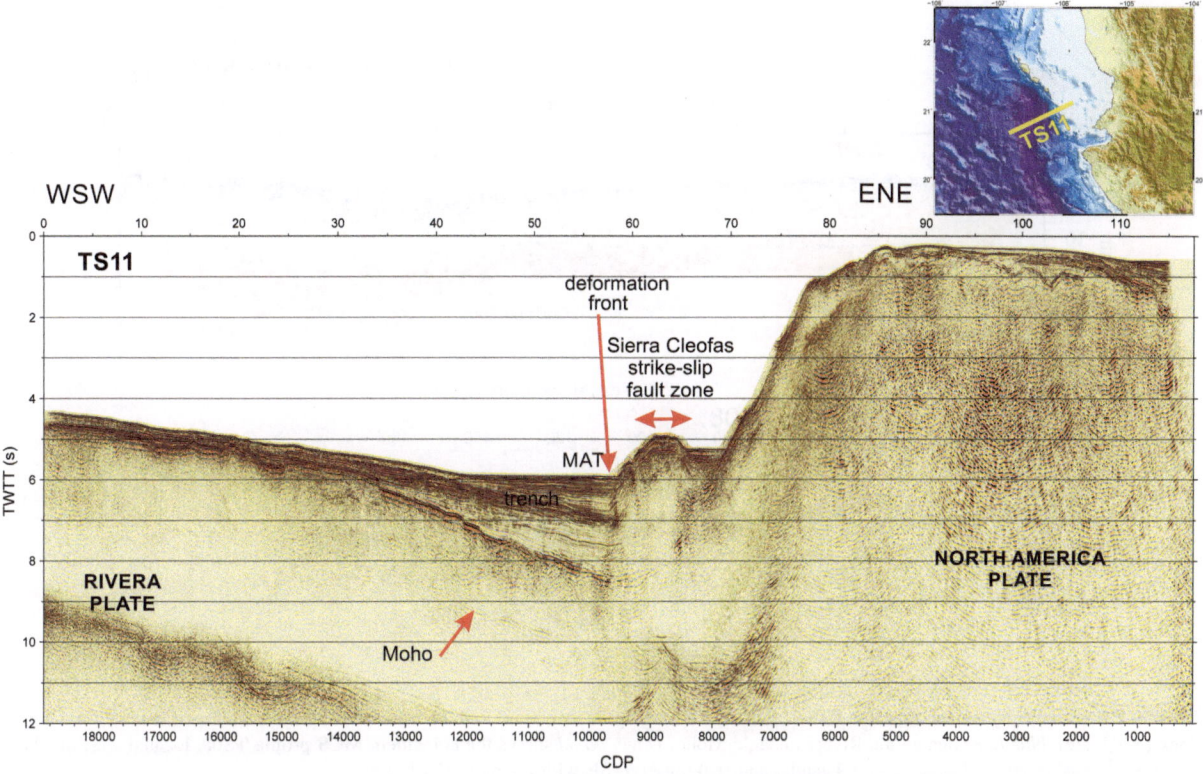

Figure 12
Migrated MCS profile TS11, striking SW–NW with a length of 120 km showing a clear subduction morphology of the Rivera Plate beneath North America Plate

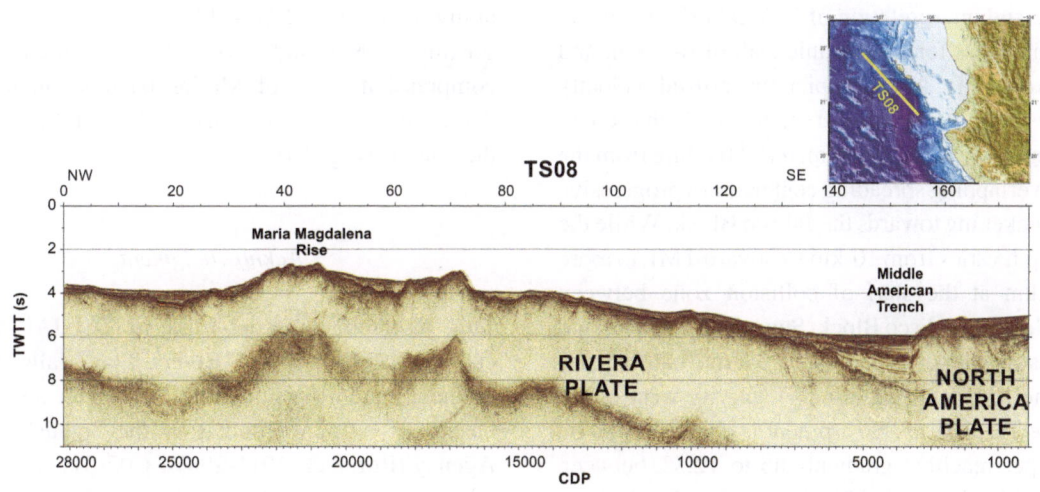

Figure 13
Migrated MCS profile TS08, striking NW–SE with a length of 120 km and running parallel to the margin

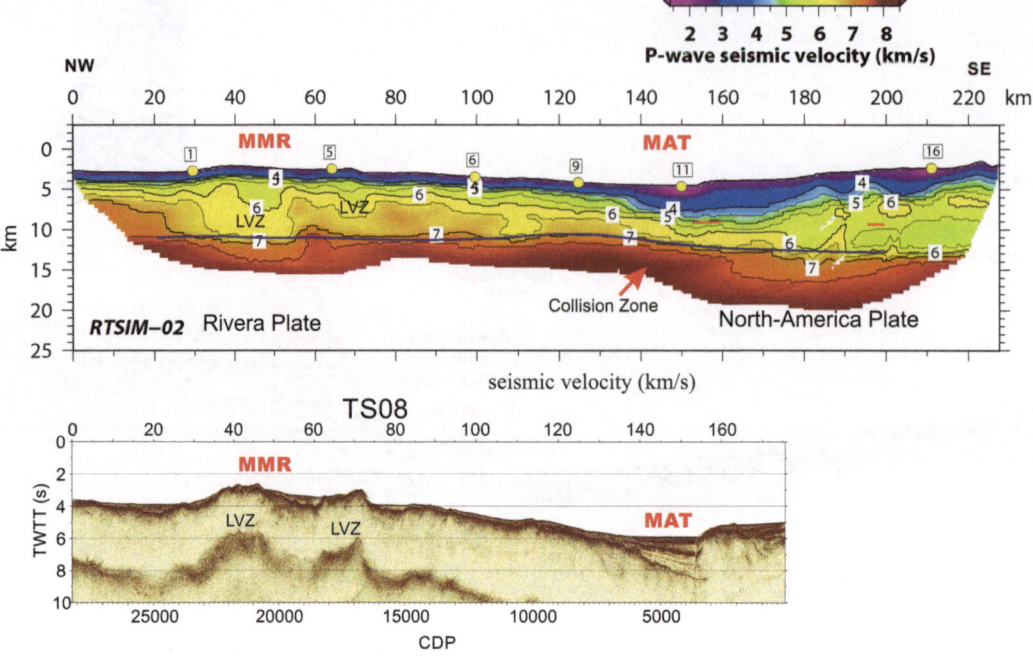

Figure 14

Up panel P-wave velocity model of profile RTS-IM02 from tomographic inversion. *Red line* shows interplay boundary of Rivera Plate–Jalisco Block (NA Plate), *blue lines* denote the Rivera oceanic Moho. *Down panel* shows the coincident MCS profile TS08, located west of Marias Islands, and striking NW–SE with a length of 120 km

in the Monte Carlo analysis. The RMS travel time misfit for the final models (i.e., RTS-IM01 and RTS-IM02) is 40 ms and 50 ms, respectively (Fig. 8). The modeling and interpretation of WA data show a mean crustal thickness for the oceanic slab of 6–7 km, and several relatively low conspicuous crustal velocity variations ($\lesssim 5.5$ km/s) underneath MMR, that could indicate an excess of melting material feeding from the extinct overlapping spreading center, and a progressive crustal thickening towards the Jalisco Block. While the Moho depth varies from 10 km westward TMI, to more than 15 km at the start of collision zone between Rivera Plate and Jalisco Block. Superficial faulting and mass-wasting deposits indicated by the bathymetric data, denotes recent (quaternary) activity west of the TMI. The Tres Marias Escarpment is characterized by steep slopes reaching gradients up to 12.5°, between the Maria Magdalena and Maria Cleofas islands arcuate scarps developed in the upper slope, favoring failures and significant landslides, as the identified landslide, such scars at the foot of Maria Magdalena Island that extends for almost 42 km², containing

blocks up to 2.5 km². At continental slope margins typically gradients of only 1°–5° can produce large scale failures, so slope instability in the area are certainly favored and possibly triggered by the local seismicity. MB and SBP data show indications of compression, west of Marias Islands, suggested by deformation of sedimentary wedges and elevation of the Islands (Fig. 15).

Acknowledgments

This research was supported by TSUJAL (Crustal characterization of the Rivera Plate–Jalisco Block boundary and its implications for seismic and tsunami hazard assessment, funded by the Spanish National Agency (Ref. CGL2011-29474-C02-01) and "Ramon y Cajal" program (R. Bartolome). The Government of Jalisco State has also funding for this project with various funds CONACYT–FOMIXJAL 2008–96567 (2009); CON- 848, CONACYT–FOMIXJAL 2008–96539 (2009); 849, CONACYT–FOMIXJAL

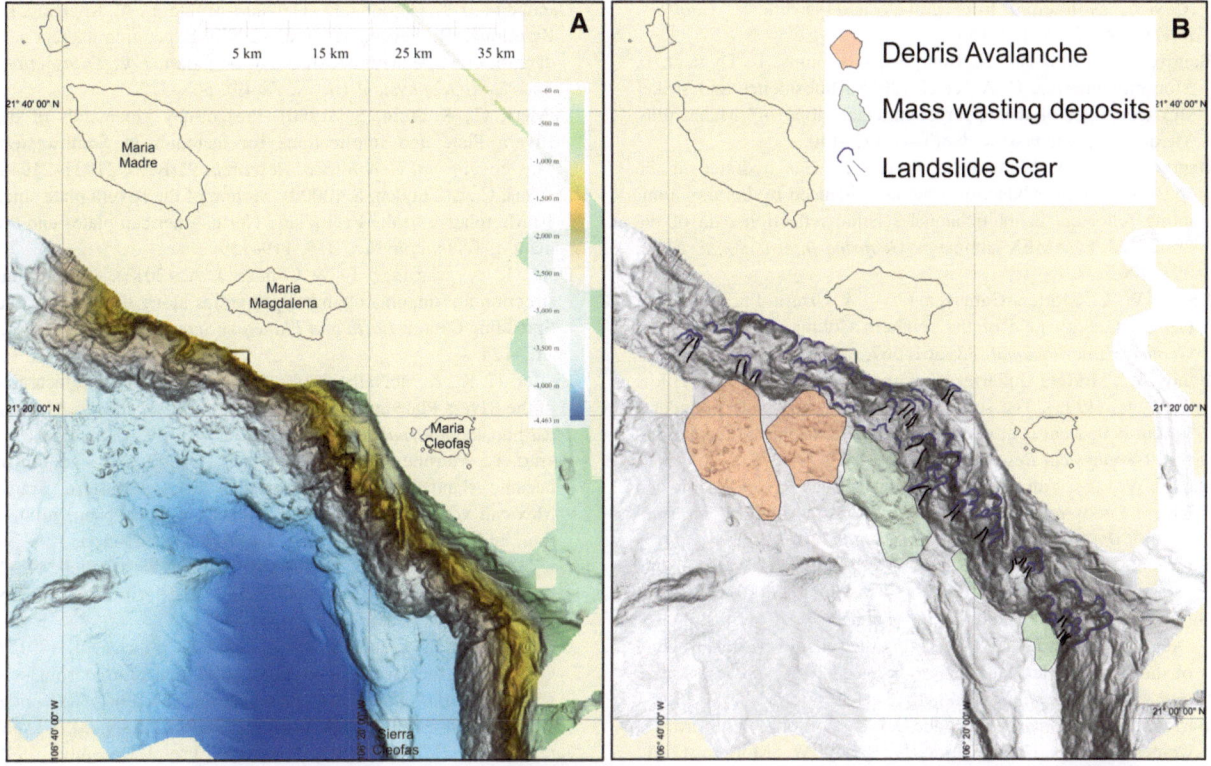

Figure 15

a Multibeam bathymetry of the continental margin westward off Tres Marias Islands. **b** Shaded relief with geomorphological interpretations showing margin instabilities characterized by landslide scars in the upper slope and mass-wasting deposits at the base of the slope

2010–149245 (2011); 850, CONACyT-FOMIXJal (2012-08-189963) and also 851, partial funding was provided by UNAM DGAPA 852 Grant #IN115513-3 and CABO Project JGAP2013,and UNAM DGAPA Grant #IN114602.. The Mexican Army provided the vessel ARM HOLZINGER for the environmental assessment and companion vessel for the multichannel seismic acquisition during the survey. We acknowledge the technical staff at UTM-CSIC, NERC, and crew of RSS JAMES COOK, Holzinger and the Mexican SEMAR and the support of the Mexican Instituto Oceanográfico del Pacífico in Manzanillo for their help during the entire cruise.

References

Allan, J. F. (1986). Geology of the Northern Colima and Zacoalco Grabens, southwest Mexico: Late Cenozoic rifting in the Mexican Volcanic Belt. *Geological Society of America Bulletin, 97,* 473–485.

Allan, J. F., Nelson, S. A., Luhr, J. F., Charmichael, I. S. E., Wopat, M., Wallace, P. J. (1991). Pliocene-Holocene rifting and associated volcanism in southwest Mexico: An exotic terrane in the making. In J. Dauphin & B. Simoneit (Eds.), *The Gulf and Peninsular Provinces of the Californias, AAPG Memoir 47* (pp. 425–445). American Association of Petroleum Geologists: Tulsa.

Alvarez, R., & Yutsis, V. (2015). The elusive Rivera-Cocos plate boundary: not diffuse. In T. J. Wright, A. Ayele, D. J. Ferguson, T. Kidane, & C. Vye-Brown (Eds.), *Magmatic Rifting and Active Volcanism* (vol 420). Geological Society: London.

Atwater, T. M., & Stock, J. (1998). Pacific–North America plate tectonics of the Neogene south western United States: An update. *International Geology Review, 40,* 375–402.

Bandy, W.L. (1992). *Geological and geophysical investigation of the Rivera Cocos plate boundary: Implications for plate fragmentation*, Ph.D. dissertation, Texas A&M University, College Station, p. 195.

Bandy, W. L., Kostoglodov, V., Hurtado-Díaz, A., & Mena, M. (1999). Structure of the southern Jalisco subduction zone,

Mexico, as inferred from gravity and seismicity. *Geofísica Internacional, 38*, 127–136.

Bandy, W. L., Michaud, F., Bourgois, J., Calmus, T., Dyment, J., Mortera-Gutiérrez, C. A., et al. (2005). Subsidence and strike-slip tectonism of the upper continental slope off Manzanillo, Mexico. *Tectonophysics, 398*(3–4), 115–140.

Bandy, W. L., & Mortera, C. A. (2012). Gas hydrates in the southern Jalisco subduction zone as evidenced by bottom simulating reflectors in multichannel seismic reflection data of the 2002 BART/FAMEX campaign. *Geofísica Internacional, 51*(4), 393–400.

Bandy, W. L., Mortera-Gutierrez, C. A., & Urrutia-Fucugauchi, J. (1993). Gravity field of the southern Colima graben, Mexico. *Geofísica Internacional, 32*, 561–567.

Bandy, W., Mortera-Gutierrez, C., Urrutia-Fucugauchi, J., & Hilde, T. W. C. (1995). The subducted Rivera-Cocos plate boundary: Where is it, what is it, and what is its relationship to the Colima rift? *Geophysical Research Letters, 22*, 3075–3078.

Bandy, W. L., & Yan, C. Y. (1989). Present-day Rivera-Pacific and Rivera-Cocos relative plate motions (abstract). *EOS. Transactions of the American Geophysical Union, 70*, 1342.

Bartolome, R., Dañobeitia, J. J., & Cordoba, D. (2015). Seismic research in Western Mexico. *Sea Technology, 56*(10), 25–29.

Bartolome, R., Dañobeitia, J. J., Michaud, F., Córdoba, D., & Delgado-Argote, L. (2011). Imaging the seismic crustal structure of the western Mexicana margin between 19°N and 21°N. *Pure and Applied Geophysics, 168*, 1373–1389.

Bartolome, R., Górriz, E., Dañobeitia, J.J, Cordoba, D., Martí, D., Cameselle, A.L., Núñez-Cornú, F., Bandy, W., Mortera, C. A., Castellón, A., Alonso, J.L. (2016). Multichannel seismic imaging of the Rivera Plate subduction at the seismogenic Jalisco Block area (Western Mexican margin). *Pure and Applied Geophysics* (this volume) (**in press**).

Bazin, S., Harding, A. J., Kent, G. M., Orcutt, J. A., Tong, C. H., Pye, J. W., et al. (2001). Three-dimensional shallow crustal emplacement at the at the 9°30′N overlapping spreading center on the East Pacific Rise: Correlations between and tomographic images. *Journal Geophysical Research, 106*, 16101–16117. doi:10.1029/2002GL015137.

Bourgois, J., & Michaud, F. (1991). Active fragmentation of the North American plate at the Mexicana triple junction area off Manzanillo (Mexico). *Geo-Marine Letters, 11*, 59–65.

Bourgois, J., Renard, D., Auboin, J., Bandy, W., Barrier, E., Calmus, T., et al. (1988). Fragmentation en cours du bord Ouest du Continent Nord Americain: Les frontieres sous-marines du Bloc Jalisco (Mexique). *Comptes rendus de l'Académie des sciences, Paris, 307*(II), 1121–1133.

Canales, J. P., Detrick, R. S., Toomey, D. R., & Wilcok, W. S. D. (2003). Segment-scale variations in the crustal structure of 150–300 kyr old fast spreading oceanic crust (East Pacific Rise, 8°15′N–10°5′N) from wide-angle seismic refraction profiles. *Geophysical Journal International, 152*, 766–794. doi:10.1046/j.1365-246X.2003.01885.x.

Chiocci, F. L., & de Alteriis, G. (2006). The Ischia debris avalanche: first clear submarine evidence in the Mediterranean of a volcanic island prehistorical collapse. *Terra Nova, 18*(3), 202–209.

Courboulex, F., Singh, S. K., Pacheco, J. F., & Ammon, C. J. (1997). The 1995 Colima-Jalisco, Mexico, Earthquake (M_w 8): A study of the rupture process. *Geophysical Reseach Letters, 24*, 1019–1022. doi:10.1029/97GL00945.

Dañobeitia, J. J., Córdoba, D., Delgado-Argote, L. A., Michaud, F., Bartolomé, R., Farrán, M., et al. (1997). Expedition gathers new data on crust beneath Mexicana West Coast. *Eos, Transactions American Geophysical Union, 78*(49), 565–572.

DeMets, C., & Stein, S. (1990). Present-day kinematics of the Rivera Plate and implications for tectonics in Southwestern Mexico. *Journal Geophysical Research, 95*(B13), 21931–21948.

DeMets, C., & Traylen, S. (2000). Motion of the Rivera plate since 10 Ma relative to the Pacific and North American plates and the mantle. *Tectonophysics, 318*, 119–159.

Dunn, R. A., Martinez, F., & Conder, J. A. (2013). Crustal construction and magma chamber properties along the Eastern Lau Spreading Center. *Earth and Planetary Science Letters, 371–372*, 112–124.

Eissler, H. K., & McNally, K. C. (1984). Seismicity and tectonics of the Rivera Plate and implications for the 1932 Jalisco, Mexico, earthquake. *Journal Geophysical Research, 89*, 4520–4530.

Ferrari, L., Petrone, C., & Francalanci, L. (2001). Generation of oceanic-island basalt-type volcanism in the western Trans-Mexican volcanic belt by slab rollback, asthenosphere infiltration, and variable flux melting. *Geology, 29*, 507–510.

Gee, M. J. R., Watts, A. B., Masson, D. G., & Mitchell, N. C. (2001). Landslides and the evolution of El Hierro in the Canary Islands. *Marine Geology, 177*, 271–293.

Gutierrez, Q. J., Escudero, C. R., Núñez-Cornú, F. J. (2015). Geometry of the Rivera-Cocos subduction zone inferred from local seismicity. *Bulletin of the Seismological Society of America, 105*(6), 3104–3113. doi: 10.1785/012010358. (ISSN: 0037-1106)

Jaramillo, S. H., & Suarez, G. (2011). The 4 December 1948 earthquake (M_w 6.4): Evidence of reverse faulting Beneath the Tres Marías escarpment and its implications for the Rivera-North American relative plate motion. *Geofísica Internacional, 50*(3), 313–317.

Johnson, C. A., Harrison, C. G. A. (1989). Tectonics and volcanism in central Mexico: A Landsat Thematic Mapper perspective. *Remote Sensing of Environment, 28*, 273–286.

Khutorskoy, M. D., Delgado-Argote, L. A., Fernandez, R., Kononov, V. I., & Polyak, B. G. (1994). Tectonics of the offshore Manzanillo and Tecpan Basins, Mexican Pacific, from heat flow, bathymetric and seismic data. *Geofísica Internacional, 33*(1), 161–185.

Klitgord, K. D., & Mammerickx, J. (1982). Northern east Pacific rise: Magnetic anomaly and bathymetric framework. *Journal Geophysical Research, 87*, 6725–6750. doi:10.1029/JB087iB08p06725.

Korenaga, J., Holbrook, W. S., Kent, G. M., Kelemen, P. B., Detrick, R. S., Larsen, H. C., et al. (2000). Crustal structure of the southeast Greenland margin from joint refraction and reflection seismic tomography. *Journal of Geophysical Research, 105*(B9), 21591–21614. doi:10.1029/2000JB900188.

Korenaga, J., & Sager, W. W. (2012). Seismic tomography of Shatsky rise by adaptive importance sampling. *Journal Geophysical Research, 117*, B0812. doi:10.1029/2012JB009248.

Kostoglodov, V., & Bandy, W. (1995). Seismotectonic constraints on the convergence rate between the Rivera and North American plates. *Journal Geophysical Research, 100*, 17977–17989. doi:10.1029/95JB01484.

Lonsdale, P. (1991). Structural patterns of the Pacific floor offshore of Peninsular California. In: J.P. Dauphin, B.R.T. Simoneit (Eds.), AAPG Memoir: The Gulf and Peninsular Province of the Californias. Am. Assoc. Petr. Geol., Tulsa. 47: 87–125.

Mammerickx, J. (1984). The Morphology of propagating spreading centers: New and old. *Journal Geophysical Research, 89*, 1817–1828.

Mammerickx, J., & Kiltgord, K. (1982). Northern East Pacific Rise: Evolution from 25 my B.P to the present. *Journal Geophysical Research, 87*, 6751–6759.

McCloy, C., Ingle, J. C., Barron, J. A. (1988). Neogene stratigraphy, foraminifera, diatoms, and depositional history of María Madre Island Mexico: Evidence of Early Neogene marine conditions in the southern Gulf of California. *Marine Micropaleontology, 13*, 193–212.

Mercier de Lépinay, B., Michaud, F., Calmus, T., Bourgois, J., Poupeau, G., Saint-Marc, P., The Nautimate Team (1997). Large Neogene subsidence event along the Middle America Trench off Mexico (18°-19°N): Evidence from submersible observations. *Geology, 25*, 387–390.

Michaud, M., Dañobeitia, J. J., Carbonell, R., Bartolomé, R., Córdoba, D., Delgado, L., et al. (2001). New insights into the subducting oceanic crust in the Middle American Trench off western Mexico (17–19°N). *Tectonophysics, 318*(1), 187–200.

Michaud, F., Royer, J.-Y., Bourgois, J., & Mercier de Lepinay, B. (1996). Comment on "Segmentation and disruption of the East Pacific rise in the Mouth of the Gulf of California" by Peter Lonsdale (Marine Geophysical Researches 17, pp. 323–359, 1995). *Marine Geophysical Researches, 18*, 597–599.

Minshull, T., Bartolomé, R., Byrne, S., & Dañobeitia, J. J. (2005). Low heat flow from young oceanic lithosphere at the Middle America Trench off Mexico. *Earth and Planetary Science Letters, 239*, 33–41. doi:10.1016/j.epsl.2005.05.045.

Moore, D. G., & Curray, J. R. (1982). Geologic and tectonic history of the Gulf of California. *Initial Reports of the Deep Sea Drilling Project, 64*, 1279–1294.

Nixon, G. T. (1982). The relationship between Quaternary volcanism in central Mexico and the seismicity and structure of subducted ocean lithosphere. *Geological Society of America Bulletin, 93*, 514–523.

Núñez-Cornú, F. J. (2011). Peligro Sísmico en el Bloque de Jalisco. *Física de la Tierra, 23*, 199–229. doi:10.5209/rev_FITE.2011. v23.36919.

Núñez-Cornú, F. J., Córdoba Barba, D., Dañobeitia, J. J., Bandy, W. L., Ortiz, M., Bartolome, R., et al. (2016). Geophysical studies across Rivera Plate and Jalisco Block, MEXICO: TsuJal project. *Seismological Research Letters, 87*(1), 59–72. doi:10.1785/0220150144.

Núñez-Cornú, F. J., Reyes-Dávila, G. A., Rutz López, M., Trejo-Gómez, R., Camarena-Garcia, M. A., & Ramírez-Vazquez, C. A. (2004). The 2003 Armería, México earthquake (M_w 7.4): Mainshock and early aftershocks. *Seismological Research Letters, 75*(6), 506–605.

Nuñez-Cornú, F., Rutz, M., Nava, F. A., Reyes-Davila, G., & Suarez-Plascencia, C. (2002). Characteristics of the seismicity in the coast and north of Jalisco Block, Mexico. *Physics of the Earth and Planetary Interiors, 132*, 141–155.

Okal, E. A., & Borrero, J. C. (2011). The 'tsunami earthquake' of 22 June 1932 in Manzanillo, Mexico: seismological study and tsunami simulations. *Geophysical Journal International, 187*, 1443–1459.

Pacheco, J., Singh, S. K., Domínguez, J., Hurtado, A., Quintanar, L., Jiménez, Z., et al. (1997). The October 9, 1995 Colima-Jalisco, Mexico earthquake (M_w 8): an aftrshock study and a comparison of this earthquake with those of 1932. *Geophysical Research Letters, 24*(17), 2223–2226.

Pardo, M., Suárez, G. (1995). Shape of the subducted Rivera and Cocos plates in southern Mexico: Seismic and tectonic implications. *Journal of Geophysical Research. 100*, 12357–12373.

Peláez Gaviria, J. R., Mortera Gutiérrez, C. A., Bandy, W. L., & Michaud, F. (2013). Morphology and magnetic survey of the Rivera-Cocos plate boundary off Colima, Mexico. *Geofísica Internacional, 52*, 73–85.

Prol-Ledesma, R.M., Sugrobov, V.M., Flores E.L. Juárez M.G., Smirnov, Y., Gorshkov, A.P., Bondarenko, V.G., Rashidov, V.A., Nedopekin, L.N., Gavrilov, V.A. (1989). *Component parts of the World Heat Flow Data Collection.* doi:10.1594/PANGAEA.805577.

Sallarès, V., Martinez-Loriente, S., Prada, M., Gràcia, E., Ranero, C., Gutscher, M. A., et al. (2013a). Seismic evidence of exhumed mantle rock basement at the Gorringe Bank and the adjacent Horseshoe and Tagus abyssal plains (SW Iberia). *Earth and Planetary Science Letters, 365*, 120–131. doi:10.1016/j.epsl.2013.01.021.

Sallarès, V., Meléndez, A., Prada, M., Ranero, C. R., McIntosh, K., & Grevemeyer, I. (2013b). Overriding plate of the Nicaragua convergent margin: Relationship to the seismogenic zone of the 1992 tsunami earthquake. *Geochemistry Geophysics Geosystems,*. doi:10.1002/ggge20214.

Selvans, M., Stock, J., DeMets, C., Sanchez, O., & Marquez-Azua, B. (2011). Constraints on Jalisco Block motion and tectonics of the Guadalajara triple junction from 1998–2001 Campaign GPS Data. *Pure and Applied Geophysics,*. doi:10.1007/s00024-010-0201-2.

Singh, S. K., Ponce, L., & Nishenko, S. (1985). The great jalisco, Mexico earthquake of 1932: subduction of the Rivera plate. *Bulletin of the Seismological Society of America, 75*, 1301–1313.

Tarantola, A. (1987). *Inverse problem theory: methods for data fitting and model parameter estimation.* New York: Elsevier Sci.

Toomey, D. R., & Foulger, G. R. (1989). Tomographic inversion of local earthquake data from the Hengill-Grensdalur central volcano complex, Iceland. *Journal of Geophysical Research, 94*, 17497–17510.

Trejo-Gómez, E., Ortíz, M., & Núñez-Cornú, F. J. (2015). Source Model of the October 9, 1995 Jalisco-Colima Tsunami as constrained by field re survey reports, and on the numerical simulation of the tsunami. *Geofísica Internacional, 54*(2), 149–159.

von Huene, R., Ranero, C. R., & Watts, P. (2004). Tsunamigenic slope failure along the Middle America Trench in two tectonic settings. *Marine Geology, 203*, 303–317.

Yang, T., Grand, S. P., Wilson, D., Guzmán-Speziale, M., Gómez-González, J. M., Domínguez-Reyes, T., et al. (2009). Seismic structure beneath the Rivera subduction zone from finite-frequency seismic tomography. *Journal Geophysical Research, 114*, B01302. doi:10.1029/2008JB005830.

(Received October 13, 2015, revised August 23, 2016, accepted August 25, 2016, Published online September 8, 2016)

Reprinted from the journal

Pure Appl. Geophys. 173 (2016), 3575–3594
© 2016 The Author(s)
This article is published with open access at Springerlink.com
DOI 10.1007/s00024-016-1331-y

Pure and Applied Geophysics

Multichannel Seismic Imaging of the Rivera Plate Subduction at the Seismogenic Jalisco Block Area (Western Mexican Margin)

Rafael Bartolome,[1] (iD) Estefanía Górriz,[1,7] Juanjo Dañobeitia,[2] Diego Cordoba,[3] David Martí,[4] Alejandra L. Cameselle,[1] Francisco Núñez-Cornú,[5] William L. Bandy,[6] Carlos A. Mortera-Gutiérrez,[6] Diana Nuñez,[5] Arturo Castellón,[2] and Jose Luis Alonso[2]

Abstract—During the TSUJAL marine geophysical survey, conducted in February and March 2014, Spanish, Mexican and British scientists and technicians explored the western margin of Mexico, considered one of the most active seismic zones in America. This work aims to characterize the internal structure of the subduction zone of the Rivera plate beneath the North American plate in the offshore part of the Jalisco Block, to link the geodynamic and the recent tectonic deformation occurring there with the possible generation of tsunamis and earthquakes. For this purpose, it has been carried out acquisition, processing and geological interpretation of a multichannel seismic reflection profile running perpendicular to the margin. Crustal images show an oceanic domain, dominated by subduction–accretion along the lower slope of the margin with a subparallel sediment thickness of up to 1.6 s two-way travel time (approx. 2 km) in the Middle American Trench. Further, from these data the region appears to be prone to giant earthquake production. The top of the oceanic crust (intraplate reflector) is very well imaged. It is almost continuous along the profile with a gentle dip (<10°); however, it is disrupted by normal faulting resulting from the bending of the plate during subduction. The continental crust presents a well-developed accretionary prism consisting of highly deformed sediments with prominent slumping towards the trench that may be the result of past tsunamis. Also, a bottom simulating reflector (BSR) is identified in the first half a second (twtt) of the section. High amplitude reflections at around 7–8 s twtt clearly image a discontinuous Moho, defining a very gentle dipping subduction plane.

Key words: Jalisco Block, Mexico, Rivera plate, crustal structure, seismic imaging, subduction, earthquake, tsunami, BSR, gas hydrate, trench infill.

1. Introduction

Subduction is the geodynamic process by which one tectonic plate converges with and slides beneath another tectonic plate, finally sinking into the mantle; its motion being driven by the higher density (colder) of the descending oceanic slab with respect to the surrounding mantle asthenosphere. At depths between 80 and 120 km, oceanic basalt is converted to eclogite, which further increases the slab density and the downward force. The interplate coupling is weaker for older plates (Kanamori 1977), and the friction along the main thrust interface increases with increasing velocity of the subducting lithosphere. As a consequence, high rates of earthquakes, defining the Wadati–Benioff zone, and tsunamis occur in subduction zones where the subducting slab is old. The cold environment of the slab depresses the local geothermal gradient and increases the portion of the Earth deforming in a brittle manner.

Recent research based mainly on the 2011 Tohoku event has suggested that the generation of huge tsunamis may require the release of gravitational energy as well as elastic energy (George et al. 2011). This gravitational energy comprises primarily crustal wedge uplift. When a megathrust earthquake occurs, gravitational energy is discharged by the formation of giant landslides or the rigid motion of a crustal block sliding though a normal fault or splay faults that may

[1] Instituto de Ciencias del Mar-CSIC, P. Marítimo de la Barceloneta 37-49, 08003 Barcelona, Spain. E-mail: rafael@icm.csic.es

[2] Unidad de Tecnología Marina-CSIC, P. Marítimo de la Barceloneta 37-49, 08003 Barcelona, Spain.

[3] Universidad Complutense de Madrid, Ciudad Universitaria, Plaza Ciencias, s/n, 28040 Madrid, Spain.

[4] Instituto de Ciencias de la Tierra Jaume Almera, C/Lluis Solé Sabaris s/n, 08028 Barcelona, Spain.

[5] Centro Universitario de la Costa (CUC), Av. Universidad 203, Del. Ixtapa, 48280 Puerto Vallarta, Mexico.

[6] Instituto de Geofísica, UNAM, Ciudad Universitaria, Delegación Coyoacán, C.P. 04510 México, DF, Mexico.

[7] *Present Address*: University of Barcelona, Marti i Franques s/n, 08028 Barcelona, Spain.

produce the subsequent tsunami generation (GRILLI *et al.* 2012). Therefore, monitoring bathymetry and gravity before and after such events, especially involving the crustal wedge, will help scientists to understand the phenomenology and estimate the associated risks. As a consequence, there is an obvious need to reevaluate earthquake and tsunami hazard assessments based on these data.

The Rivera plate is particularly a region where large earthquakes have occurred with very destructive consequences, including the generation of big tsunamis, e.g., the Mw >8.0 1932 and 1995, demonstrating that the Jalisco Block is a zone of high seismic potential as a consequence of the subduction dynamics. Research in subduction zones includes several geophysical techniques, such as multichannel reflection seismic and high-resolution bathymetry.

The main objective of this work is to characterize the internal structure of the Rivera Subduction Zone in the Jalisco Block area to understand the geodynamic and the recent tectonic deformation. A key point in the definition of sources (faults, landslides) is to conduct a study of the relationship between superficial, including the seafloor structure, and depth structures where the lithosphere subducts and major seismic activity occurs. Growing understanding of subduction dynamics suggests that crustal deformation near subduction zones will have to be monitored and investigated in detail to obtain the information needed for tsunami risk estimation. Understanding the lateral and vertical extent of the boundaries of the subducting slab will constrain plate tectonic configurations and the timing of events in the western Mexican margin.

To understand the processes involved in the subduction of the Rivera plate and to solve the lack of seismic imaging and bathymetry information, a multidisciplinary geophysical approach is needed to characterize the area from the surface, to the deep zones. This information will be of paramount importance for future seismic hazard assessment. We describe the acquisition, processing and the geological crustal interpretation of a multichannel seismic marine profile running perpendicular to the coast at the Jalisco Block area. These data, of unprecedented quality, were recently acquired in the framework of the TSUJAL (TSU-nami and JAL-isco) project (BARTOLOMÉ *et al.*

2015). The unprecedented quality of the data provides a brand new seismic image of the internal structure of the Rivera Subduction Zone beneath the North American plate from which the geodynamic context can be inferred. Furthermore, the characteristics of the interaction between the Jalisco Block and the sediments thickness of the trench clearly indicate the probability of the occurrence of large earthquake.

2. Geological Setting

The western margin of Mexico is an active region displaying ongoing tectonic plate interactions. Seafloor spreading is occurring along the northern most segment of the East Pacific Rise (EPR), commonly referred to as the Rivera Rise, which bounds the Rivera plate to the west. Spreading rates ranges from 5.3 cm/year at the northern end of the rise to 7.3 cm/year at it southern end (e.g., BANDY 1992). On its eastern boundary, the Rivera plate subducts beneath the North American plate (NA) in the Middle American Trench (MAT). South of the Colima Graben, the Cocos plate subducts beneath the NA at the MAT. The MAT extends from the Tres Marias Islands, located south of the Gulf of California at 21°N, southward along the Mexican coast for a distance of approximately of 3000 km. The continental shelf and slope along the MAT is quite narrow, being between 40 and 60 km. The southern boundary of the Rivera plate consists of two distinct segments. The western segment, west of 106.25°W, is the Rivera Transform zone, which is an active transform fault separating the Rivera and Pacific plates. The eastern segment, east of 106.25°W, is the poorly defined Rivera-Cocos plate boundary. Two closely spaced plate triple junction lie along this boundary, namely, the Pacific-Rivera-Cocos triple junction at its intersection with the EPR to the west and the Rivera-Cocos-North American triple junction at its intersection with the MAT. Several proposals exist as to the nature of this boundary. These include a hinge fault (i.e., pure vertical motion between the plates, NIXON 1982), a southwest propagating divergent boundary (BANDY 1992), and a diffuse left-lateral transform boundary (DEMETS and WILSON 1997). The location of this boundary beneath NA is also being debated. Nixon (1982) proposed that it extends beneath the southern and northern Colima rifts (Fig. 1). Bandy (1992) similarly proposed that it extends beneath the southern Colima rift but instead proposed that it was located west of the northern Colima rift. The gravity study of Bandy et al. (1995) supported the proposal of Bandy (1992), however, it raised the possibility that the tear zone between the Rivera and Cocos plates beneath NA could extend to the area beneath the northern Colima rift (i.e., to the NE the boundary was a wide tear zone). The proposal of a wide tear zone between the subducted Rivera and Cocos plates beneath the northern Colima Graben and adjacent areas

produced by divergence between the Rivera and Cocos plates is supported by the seismic studies of Yang et al. (2009), Leon Soto et al. (2009) and Ochoa-Chavez et al. (2015). More recently, Alvarez and Yutsis (2015) also place the boundary beneath the southern Colima rift. In contrast, DeMets and Wilson (1997) placed the boundary north of the El Gordo graben/southern Colima rift (Fig. 1).

The area has also been subjected to the effects of at least two Ridge–Trench collisions. The first collision occurred at around 10 Ma when the EPR collided with the trench off the southern tip of Baja California, the collision most likely was responsible for the separation

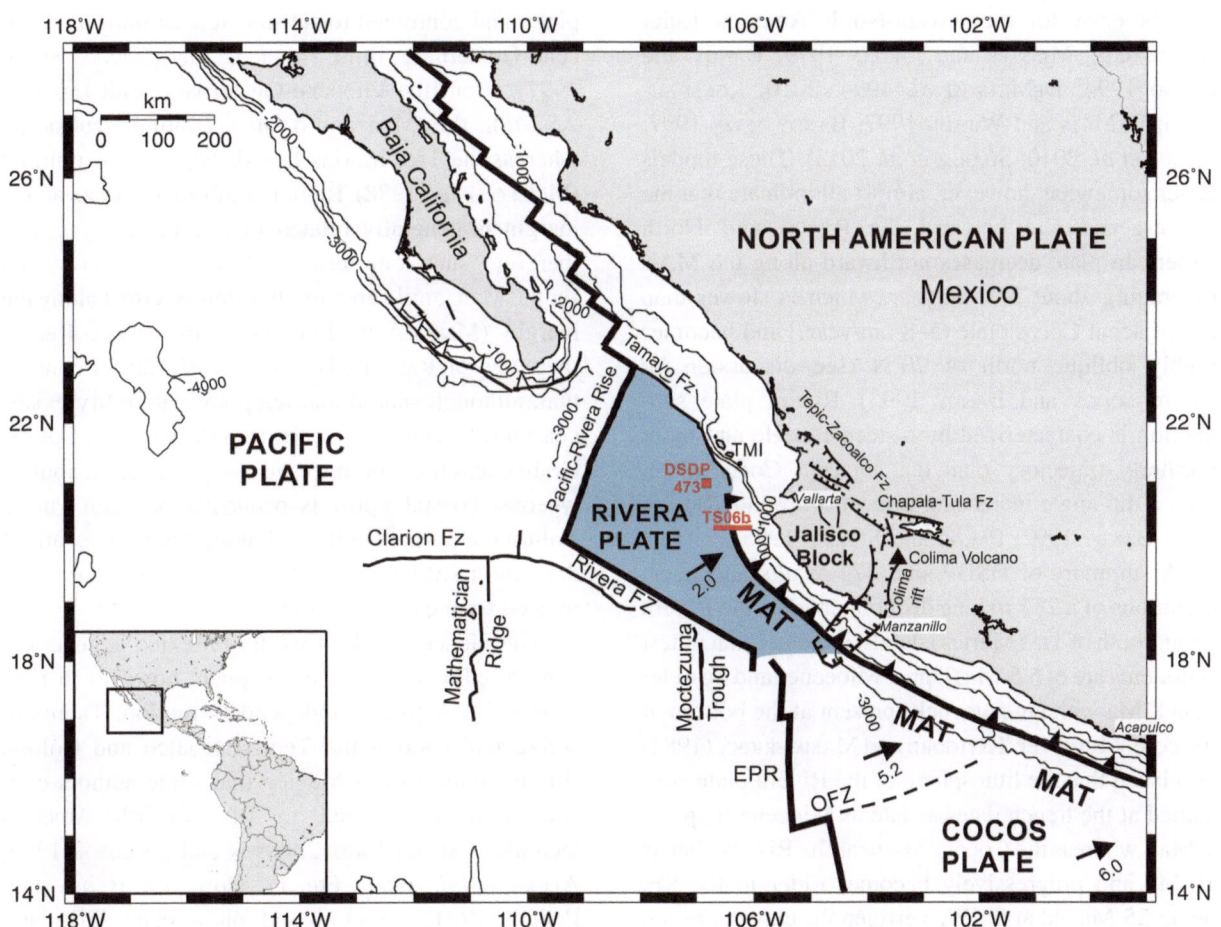

Figure 1

Location map of western Mexico. Four lithospheric plates act in the area: Pacific, Rivera, Cocos and North America. *Blue area* highlights the area of the Rivera plate. *Arrows* indicate relative convergence rates (cm/year) between the oceanic and continental plates (Pardo and Suárez 1995). *Red line* indicates the location of MCS section TS06b. *Red square* shows the DSDP drill 473 south Tres Marias Islands. *TMI* Tres Marias Islands, *MAT* Middle America Trench, *EPR* East Pacific Rise

of the Rivera plate from the rest of the plate to the south and the initiation of slab windowing between the Rivera and Cocos plates (e.g., KLITGORD and MAMMERICKX 1982; BANDY 1992). The second collision occurred at around 2 Ma when the northward propagating segment of the EPR collided with the MAT near Manzanillo/Chamela (BANDY 1992; LONSDALE 1995; MICHAUD et al. 2000, 2001; PELÁEZ GAVIRIA et al. 2013). The effects of the most recent collision in the Jalisco/Colima area have been largely unaddressed in the literature and deserve more attention.

The Rivera plate began to separate from the Cocos plate at about 10 Ma (DEMETS and TRAYLEN 2000) as evidenced by spreading velocities along the East Pacific Rise and Mathematician Ridge. Several models exist for the Rivera-North America Euler vector (e.g., MINSTER and JORDAN 1979; BANDY and PARDO 1994; DEMETS et al. 1994, 2010; LONSDALE 1995; DEMETS and WILSON 1997; BANDY et al. 1997; ARGUS et al. 2010; SUÁREZ et al. 2013). These models differ somewhat, however, almost all indicate that the convergence rate between the Rivera and North American plate decreases northward along the MAT [averaging about 2–3 cm/year, which is slower than the adjacent Cocos plate (5–8 cm/year)] and becomes highly oblique north of 20°N (see discussion by KOSTOGLODOV and BANDY 1995). Rivera plate subduction is characterized by a steeper angle and more northerly trajectory than the adjacent Cocos plate, with a dip angle increasing at depths around 100 km (e.g., BANDY 1992; PARDO and SUÁREZ 1995).

A summary of DSDP site, 473 results and interpretations of a 287 m long drill obtained on the Rivera plate, south of Tres Marias Islands, indicates that oldest sediments are 6–6.5 Ma (Upper Miocene) and no older than 8 Ma with igneous rocks present at the bottom of the core. Moreover, KLITGORD and MAMMERICKX (1982) published that the lithosphere of the Rivera plate consumed at the trench dates as late as Miocene (approx. 9 Ma), whereas the Cocos crust near the Rivera plate is 10 Ma and progressively becomes older to the SE, being 25 Ma old at 90°W. Terrigenous clay deposited in the early Pliocene to Quaternary and calcareous claystone deposited during early Pliocene are separated by a strong seismic reflector from the Upper Miocene sediments in the core. Sediments are unexpectedly terrigenous despite the distance from land and

the MAT. Sediment accumulation rates are 40 m/million years for the last 3 million years and 20 m/million years from 3 to 6.5 million years. Sediment velocities are 1.5–1.6 km/s increasing to 1.98 km/s at the basal sediments. Igneous rocks below the sediments are mainly massive, altered diabase with a density of 2.7 g/cm^3 and a velocity of 5.2–5.3 km/s, but a fragment of pillow basalt was found at the bottom of the drill.

The margin along the Middle America trench from the Tres Marias Islands to the Manzanillo Area (18°N, 104°W) was uplifted and emergent before the late Miocene. The margin started to subside during the upper Miocene-lower Pliocene (coincident with the initial formation of the Rivera plate as an independent plate) and continued to subside at least until the Pliocene-Quaternary limit (MERCIER de LEPINAY et al. 1997). From the Pliocene-Quaternary limit (approx. 2.5 Ma), the Manzanillo area continue subsiding whereas the Tres Marias Islands began to be uplifted (MCCLOY et al. 1988). Regional subsidence of an active margin is generally related to tectonic erosion, and therefore, such long-term subduction erosion regimen for the Manzanillo area implies trench retreat along the margin (MERCIER DE LEPINAY et al. 1997). Recent overview of RAMÍREZ-HERRERA et al. (2011) confirm that, although spatial and temporal variability exists, coastal subsidence is occurring at the southern Colima Graben and the Guerrero seismic gap near Acapulco, whereas coastal uplift is occurring between Puerto Vallarta and Manzanillo and along the coast south of the Colima/El Gordo Graben until Lazaro Cardenas, located southeast of Acapulco.

The Jalisco Block of western Mexico is generally considered as part of the NA plate; however, it may have some degree of independent motion. There are active faults along the Tepic-Zacoalco and Colima rifts in southwestern Mexico that some authors considered to be related to Rivera-North America tectonics (ALLAN 1986; JOHNSON and HARRISON 1989; ALLAN et al. 1991, Fig. 1). However, BANDY and PARDO (1994) found, based on a plate kinematic analysis that if the Jalisco Block is moving relative to NA, its motion must be less than 1 cm/year. Consistent with this result, SELVANS et al. (2011) find a slow motion of 2 mm/year to the southwest for the Jalisco Block relative to NA in a recent GPS campaign conducted near the triple junction near Guadalajara. This motion may be compatible with a tearing of the Jalisco Block away from NA, however, the authors point out that the motion could be related to the behavior of the earthquake cycle for this margin and if so, the Jalisco Block may not be an independent crustal block.

Recent studies also evidence gas hydrates (MINSHULL *et al.* 2005; BARTOLOME *et al.* 2011; BANDY and MORTERA GUTIÉRREZ 2012) in the form of bottom simulating reflectors (BSR) on multichannel seismic reflection profile located in the continental slope area of the northern part of the Jalisco Subduction Zone, off Puerto Vallarta, between 20° and 20.5°N, as well as off Manzanillo. Thus, there is evidence to suggest that the Pacific margin of Mexico may contain significant reserves of hydrocarbons in form of gas hydrates.

3. Earthquake History of the Jalisco Region

The macro seismic history of the Jalisco region dates back to the year 1544, and is considered one of the most active seismic zones in North America. In the last 120 years, ten major earthquakes were reported with Ms ≥7.5, including the destructive events occurring on the 3rd and 18th June 1932 with Ms = 8.2 and 7.8, respectively, and one with a magnitude of 8.0 occurring in 1995, all of them located offshore Jalisco. The 1932 event had a maximum tsunami run-up height of 3 m (OKAL and BORRERO 2011) and caused 400 casualties, whereas the 1995 earthquake triggered a tsunami with a run-up height of 5.1 m that affected a 200 km long stretch of coast with severe damage confined to areas with shallow shoreline topography (Fig. 2). Considering that the area of rupture of the 1995 event spans only the southern half of the area proposed for the 1932 events (NUÑEZ-CORNÚ *et al.* 2004) and the 77 years recurrence time estimated by SINGH *et al.* (1985) for earthquakes similar to the 1932 event on the coast of Jalisco, which includes the Bay of Banderas where Puerto Vallarta is located, the northern Jalisco coastal area is a zone of high seismic hazard potential and known as Vallarta Gap (YAGI *et al.* 2004).

Moreover, not only do earthquakes associated with subduction processes occur in the region, but also large intraplate earthquakes take place, such as the historical events of December 27, 1568 and

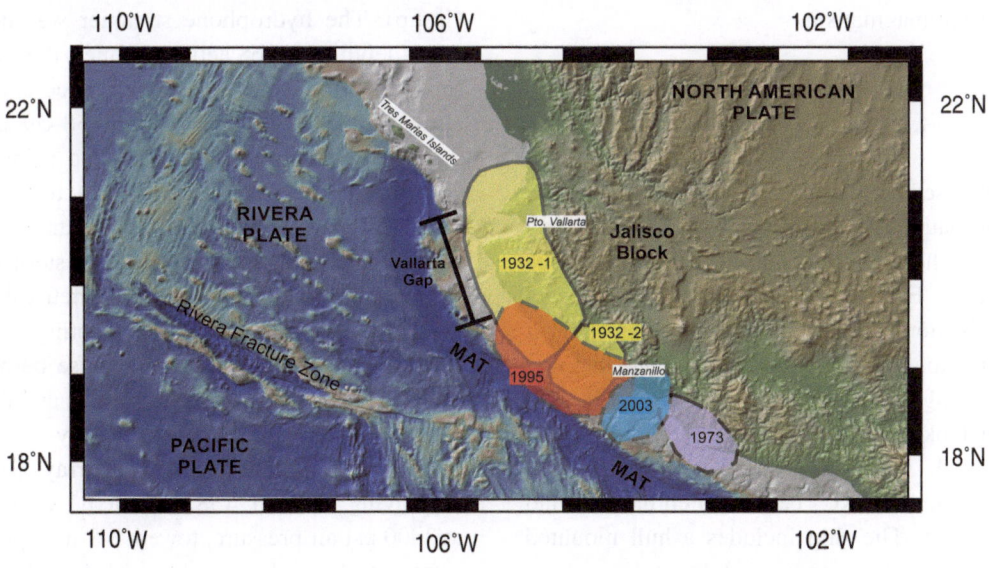

Figure 2

Approximate rupture areas for the two 1932 earthquakes (Mw 8.2 and Mw 7.8), 1973 (Mw 7.6), 1995 (Mw 8.0) from ANDREWS *et al.* (2011) and 2003 earthquake (Mw 7.4) from the SSN (Mexican Servicio Sismológico Nacional). The 1932 events are shown with a *solid line*, while the 1973, 1995 and 2003 events are shown with a *dashed line*. The Vallarta Gap shows the expected future rupture area offshore the Jalisco Block

February 11, 1872. However, there is a debate concerning the interplate (NUNEZ-CORNU et al. 2011) versus intraplate (ANDREWS et al. 2011; QUINTANAR ROBLES et al. 2011) origin of the Armería earthquake (Mw 7.4) of 2003, located near the diffuse triple junction between the Cocos, Rivera and NA plates.

Previous studies have shown some seismic images and models of the Rivera plate subducting under Mexico mainland. But in terms of seismic reflection data, a few single channel seismic profiles (BOURGOIS and MICHAUD 1991; KHUTORSKOY et al. 1994; MICHAUD et al. 1996), one threefold profile (BANDY et al. 2005), several threefold profiles acquired in the Rivera transform zone (BANDY et al. 2011), and five multichannel analog seismic data acquired in 1996 during the CORTES project (BARTOLOME et al. 2011; DAÑOBEITIA et al. 1997, Fig. 3) are available in the literature. These data show active shallow strike-slip tectonics and a subsidence in the upper continental slope area at 18.5°N, and the subducting Rivera plate crustal structure has been well identified near Puerto Vallarta at 21°N. Little is known about the depth and structure of the plate underneath the Jalisco margin at north of 20°N and the paths of the water transported within the subducting oceanic lithosphere into the Earth's interior, water that affects intraslab earthquakes and arc magmatism.

4. Data and Methods

The data used in this study consist of multibeam bathymetric data collected in the Rivera, Cocos and NA plate offshore of the Jalisco Block, and one 65.5 km long, ESE–WNW oriented, multichannel seismic reflection profile, TS06b, running almost perpendicular to the Rivera Subduction Zone that was collected during the TSUJAL project (Fig. 3). The experiment took place onboard the 90 m long RRS JAMES COOK from the National Environmental Research Council (NERC) between February 17 and March 19, 2014. The ship includes a hull mounted water mapping system composed by a Kongsberg EM120 and Kongsberg EM710 multibeam echosounders, and the raw data were processed using the CARIS software (v. 8.1) while onboard by ICM technicians to produce a 80 × 80 m grid of

Figure 3 ▶

Up 3D seafloor topography map of the western Mexican margin represented with a six times exaggeration vertical. TSUJAL high-resolution bathymetry at 80 m × 80 m grid space is superimposed to a background SANDWELL and SMITH (2009) bathymetry with a 20 % transparency. Red square indicates the area of Fig. 3b. Middle TSUJAL high-resolution bathymetry acquired offshore the Jalisco Block with the location of MCS profile TS06b and DSDP site 73 in red, and CORTES MCS sections in white (BARTOLOME et al. 2011) used in this work to help the interpretation. Vertical Exaggeration: 6. Down Synthetic bathymetric profile of TS06b in red, extended until the coast, showing the margin imaged by the multichannel seismic profile

bathymetric values. One Expendable Bathy Thermographs (XBT) per day was launched during the survey to measure water velocity profile and whose values have been integrated into the echosounder acquisition program.

Multichannel seismic data were acquired aboard the RRS James Cook using, for the first time, the 6 km long digital streamer belonging to the Spanish RV Sarmiento de Gamboa (BARTOLOMÉ et al. 2015; NÚÑEZ-CORNÚ et al. 2016; DAÑOBEITIA et al. 2015), in the framework of a bilateral barter agreement between the National Environmental Research Council (NERC) and Consejo Superior de Investigaciones Cientificas (CSIC), thanks to the European research alliance OFEG (Ocean Facility Exchange Group). The hydrophone streamer was deployed at 10 m depth with 468 active channels (5850 m length) separated 12.5 m apart. CMP distance is 6.25 m and allows a CMP nominal fold of 58–59 traces. The source airgun arrays were designed using Gundalf® commercial software to maximize the energy concentrated at the lowest frequency range. The arrays were towed from the center of the stern to the port and starboard sides getting a symmetrical configuration from the streamer. The shooting interval was chosen to be 50 m as a compromise between maximum redundancy of data (CMP fold) and the capacity of the air compressors. Source array was composed by 12 guns divided in 4 subarrays of 3 guns, employing 3540 in^3 (58 L) BOLT® GGuns 1500LL at 2000 psi air pressure, towed at 8 m depth. The data were sampled at 1 ms and recorded in SEG-D format. Shooting during the cruise was carefully carried out following established rules to avoid harming local marine mammals, using soft starts and monitoring the presence of marine mammals and turtles in the

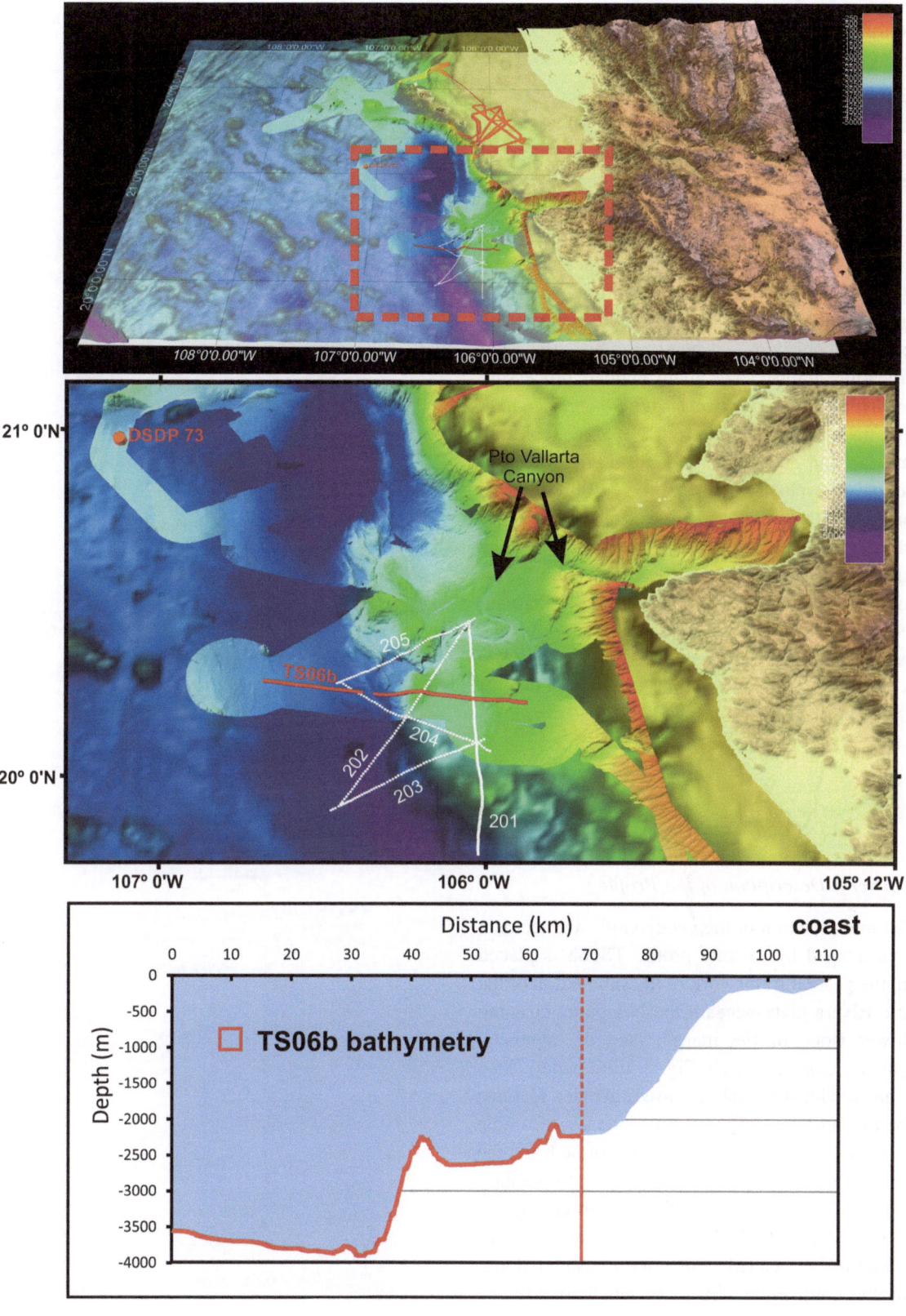

Figure 4
Processing flow applied to multichannel seismic profile TS06b

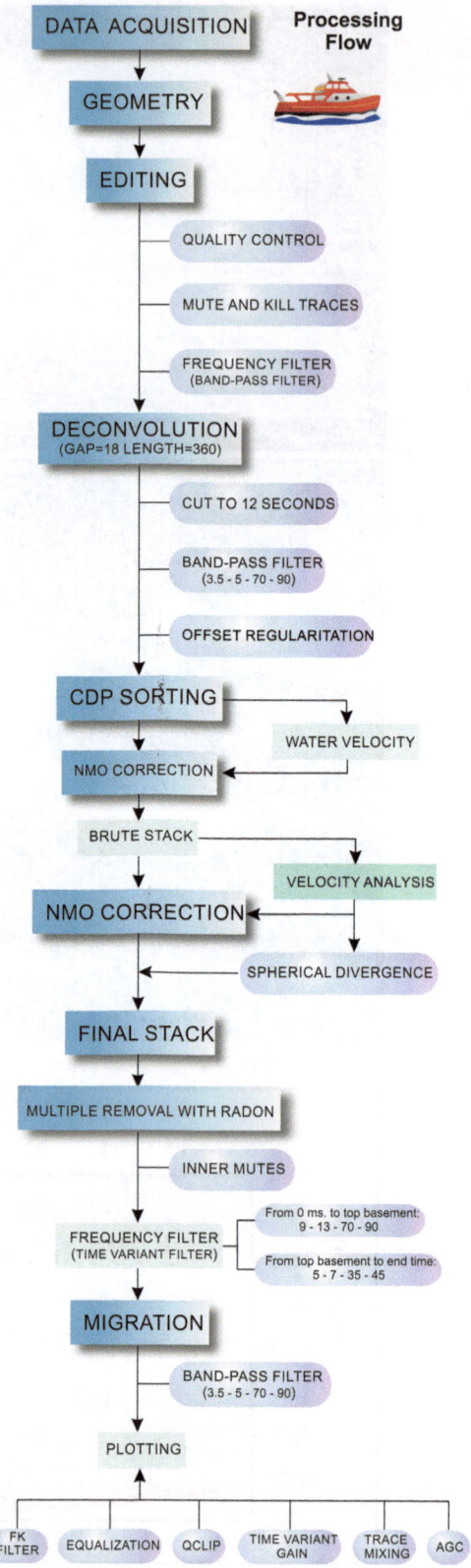

vicinity of the vessel, which caused us to stop shooting repeatedly.

To obtain the best seismic crustal images, several processing steps were applied to the seismic data to improve the seismic resolution and to enhance the signal to noise ratio. The real marine geometry has been calculated and added into the trace headers, using the real positioning information obtained from navigation files, which includes hydrophone and source GPS positions, channel number and shot identification number. The main applied processing steps include: (1) prestack signal enhancement (editing traces, mute and filtering), (2) deconvolution (18 ms gap and 360 ms length), to improve vertical resolution, (3) velocity analysis every 200 CMP, (4) Radon demultiple, to suppress multiple seafloor arrivals, (5) NMO correction and stack to increase the signal to noise ratio, and finally, (6) Kirchhoff time migration, which increases the horizontal resolution and collapse diffractions relocating in time the events. Time variant band pass filtering and different gains have been applied for final display. Details of parameters used during processing are shown in Fig. 4.

5. Results and Discussion

5.1. General Description of the Profile

The interpretation of the geodynamic and tectonic features imaged by seismic profile TS06b is placed within the general framework of the subduction of the oceanic Rivera plate beneath the NA plate, crossing the lower slope of the margin (see the synthetic bathymetric cross section in Fig. 3, third panel). Two large areas with different tecto-sedimentary features can be distinguished in the profile: the western area (from CDP 7000 to CDP 11000), formed by the subducting oceanic Rivera plate and sedimentary marine deposits above it; and the eastern area (from CDP 100 to´CDP 7000), formed by the accretionary prism and the associated sedimentary basins belonging to the continental NA plate which overly the

Figure 4
Processing flow applied to multichannel seismic profile TS06b

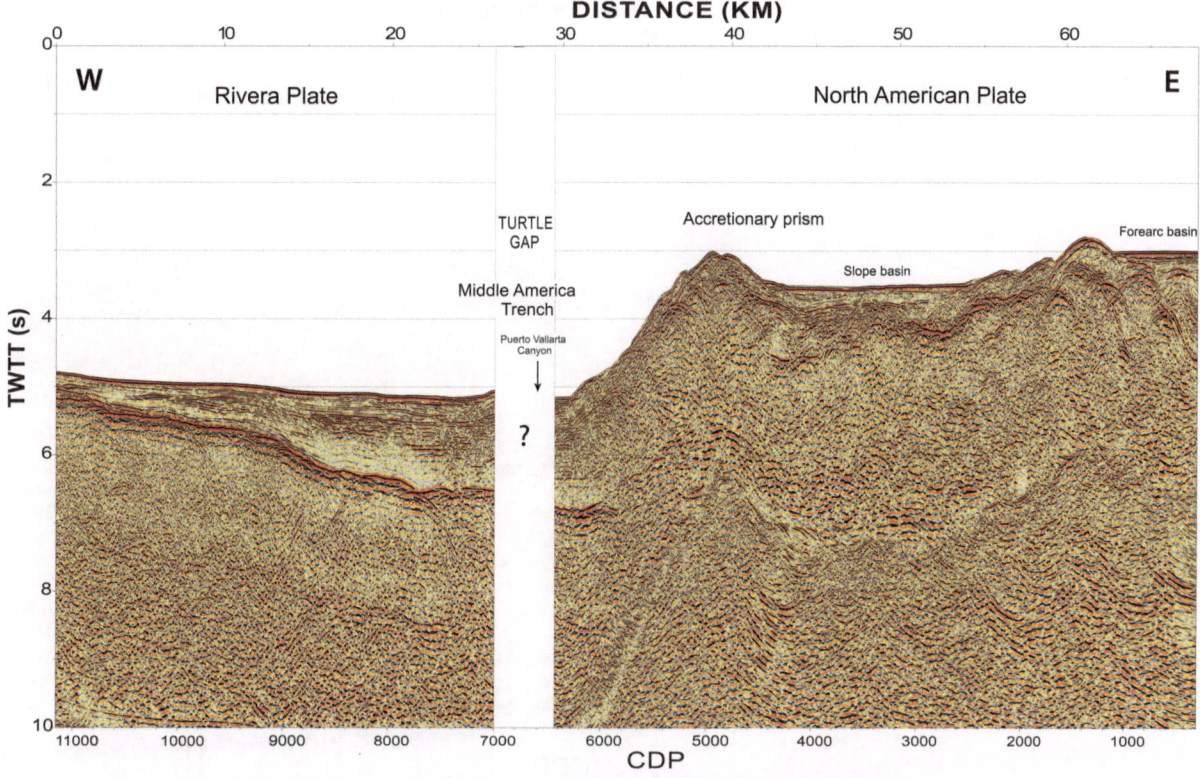

Figure 5

Time migration section of multichannel seismic profile TS06b after applying the processing flow of Fig. 4. See text for details

subducting oceanic plate. Both areas are delimited by a strong horizontal amplitude reflector, which marks the limit between the two tectonic plates, the interplate boundary (Fig. 5). This boundary can be followed almost continuously along the entire profile. Below the interplate reflector, at 7–8 s two-way travel time (twtt), the crust—mantle boundary (Moho) can be clearly distinguished along the profile as a discontinuous reflector that parallels, and is located 2 s twtt below, the interplate reflector (Fig. 6).

From west to east, the top of the Rivera plate imaged below the sedimentary cover indicates that the plate is entering at a gentle dip angle (<10°) in this part of the margin. We can distinguish normal faults, and the resulting horst and graben morphology, as a consequence of the extensional stresses produced as the plate bends down into the subduction zone (see from 0 to 25 km in Fig. 6).

Further to the east of the MAT, the top of oceanic basement may be traced continuously for about 20 km beneath the accretionary wedge where the Rivera plate is covered by about 3 s (twtt) of sediments, indicating subduction–accretion in this part of the margin, which is in agreement with previous studies (e.g., BARTOLOME *et al.* 2011). A pull up image effect of the Rivera plate under the accretionary prism (from CDP 4000 to CDP 6000, see Fig. 5) is observed. This pull up image effect is not geologically real and is caused by the relative seismic velocity differences of the acoustic waves between the lower velocity water column, left of the accretionary prism, respect to the higher sedimentary velocity cover in the prism, and their consequent representation in time (twtt) causing what appears to be a structural high.

5.2. Oceanic Crustal Structure

The MAT imaged in the section TS06b is about 3900 m deep, only 400 m deeper than the surrounding regional bathymetric oceanic lithosphere

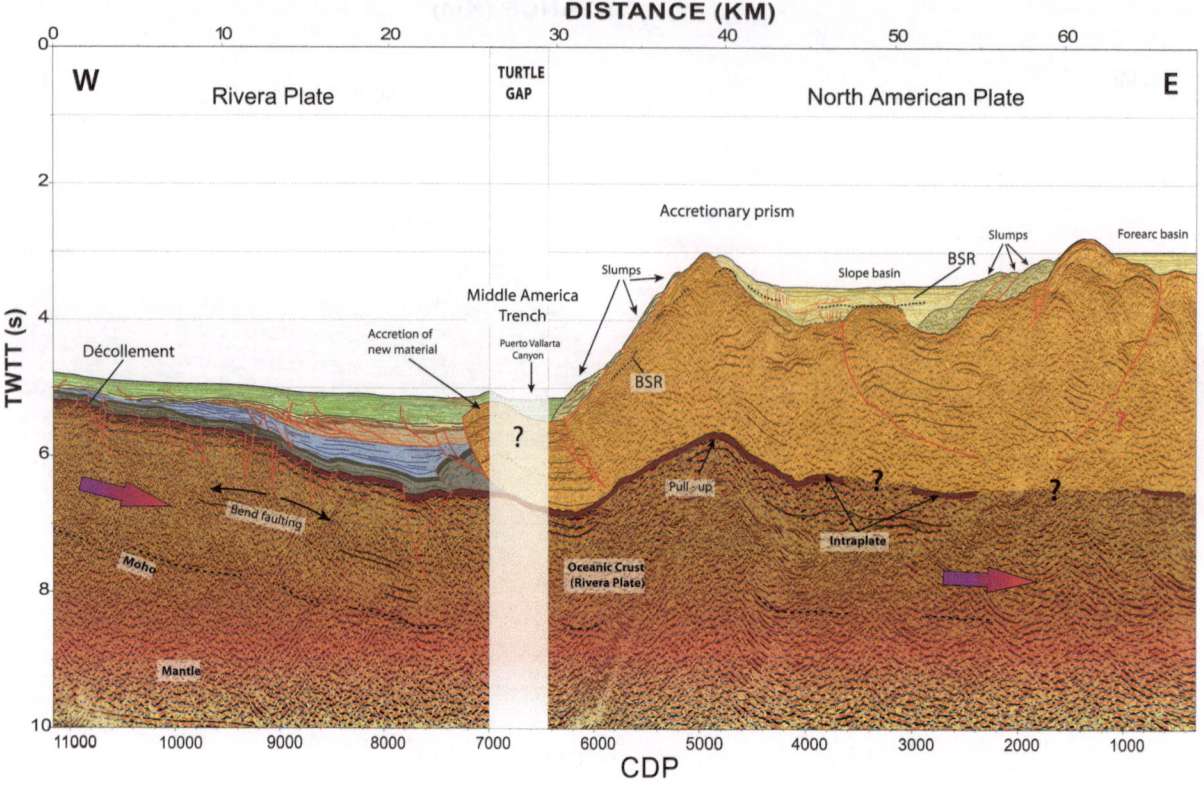

Figure 6
Interpretation of time migrated section TS06b, where the oceanic Rivera plate subducts beneath the NA plate. See text for details

(Fig. 3a). Sediment supply origin could be continental following the conclusions of Deep Sea Drilling Project (DSDP) 473. The trench bends sharply northward at 20°N, increasing the obliquity of subduction north of this latitude. There are several factors controlling the depth of the trenches, but possibly the most important is the supply of sediment, which fills the trench and reduces the bathymetric expression of the trench. The deepest trenches in the world, in general, deeper than 8000 m are all non-accretionary (no prism), such as the Mariana (11,000 m), Tonga (10,800 m), and Philippine (10,500 m) trenches. The age of the lithosphere at the time of subduction controls as well the depth of the trench: the older the seafloor (and cooler and thicker), the greater the subsidence of the trench. The slow convergence of the Rivera plate also causes the capacity of the convergent margin to dispose of (i.e., to subduct) sediment to be exceeded. Studies of the sediment fill in the northern part of the MAT are limited (e.g., Ross 1971; Renard *et al.* 1980; Mercier

de Lépinay *et al.* 1997). The Deep Sea Drilling Project (DSDP) and the Ocean Drilling Program (ODP) data in Mexico are restricted to two different areas. The first area is in front of Acapulco about 700 km further to the south of line TS06b. The second area is that of DSDP Site 473 drilled in 1978 (Fig. 3b) which is located about 70 km southwest of Islas Tres Marias and 90 km NW of seismic line TS06b.

Sediment transport is controlled in the Jalisco Block by submarine landslides, debris flows and submarine canyons such as the Puerto Vallarta Canyon (see Fig. 3). Sediments are transported down the Puerto Vallarta Canyon and into the trench, but sediments have yet to completely fill the trench. In addition to sedimentation and age of the lithosphere, three tectonic processes control the trench-fill at convergent plate boundaries: subduction accretion, sediment subduction and tectonic (subduction) erosion. Which of these processes is dominant depends on the state of stress at subduction zone; when the interplate coupling (normal stress across the subduction zone) is low then sediment subduction operates. Either accretion or erosion occurs when coupling is high (UYEDA and KANAMORI 1979). As the Rivera plate is a young plate, the descent is more difficult because the plate is light and hot and the interplate coupling becomes strong, and due to the existence of an accretionary prism, we can then conclude that subduction accretion is the dominant process at this latitude at the foot of the margin along the lower slope and that it is more important than off Manzanillo. This is opposite of what MANEA et al. (2003) conclude in the Jalisco Block area inferring a sediment fill in the MAT close to zero from gravity anomalies, suggesting that the dominant process is the sediment subduction. The assumption of null fresh sediment fill is due to the lack of seismic data at the time of the publication, but now the shallow sedimentary structure of the Rivera plate is well imaged with profile TS06b. The sedimentary cover has strong reflectivity, generally subparallel, and their continuity can be followed for about 25–30 km. Trench sediment thickness is about 1.6 s twtt, corresponding to a 2 km depth for an average sediment velocity of 2.5 km/s. The maximum thickness value of the trench-fill is compatible with previous works in the area based on seismic images (BARTOLOME et al. 2011, see Fig. 3 for location of the profiles) confirming a decreasing trend towards the south, with values of 2 s twtt in profile 205 and 1.5 twtt in profile 204, north and south of line TS06b, respectively. These deposits seem to be syn-tectonic to the normal faulting and subduction of the Rivera plate, mainly characterized by thrusts and reverse faulting. The sedimentary sequence can be divided into four subunits following their geometric relationships, presenting onlap, downlap and erosive truncations geometries (Fig. 7). This geometry would indicate different phases of sedimentation during the subduction of the Rivera plate, difficult to date due to a lack of information of wells, drills or cores in this unexplored area despite DSDP 473.

5.3. Continental Structure

East of the trench, the shallow morphology of the NA plate is represented by an accretionary prism formed by a sedimentary structure up to 4 s twtt of maximum thickness. This unit is characterized by a set of chaotic reflectors that lack lateral continuity (Fig. 8), indicating that the sediments of this unit have been highly deformed.

The top of the accretionary prism presents a sharp relief, forming two sediment filled basins (Fig. 8). The western basin, herein, referred to as the slope basin, is located in the center of the accretionary prism (between CDP 1600 and CDP 4800), and the sediments observed may be originated during the erosion of the accretionary prism. This took place during different episodes of uplift, generating slumps and normal faults related to gravitational movements that give rise to thrusts in the upper block (Fig. 8). The second basin, located on the eastern border of the profile (from CDP 1 to CDP 1200), is the forearc basin of the subduction system, whose sediments are onlaping the basement. Also, a system of slumps that reaches the base of the prism has been observed (from CDP 5000 to CDP 6400), covered by sediment units.

Three major zones of faulting can be identified in the accretionary wedge. The first one, located in the westernmost part (about CDP 7200), presents west vergence and constitutes the frontal thrust of the prism, being the boundary between the accretion prism and the sediments covering the Rivera plate (Fig. 8). The second fractured zone located in the center of the accretionary prism (around CDP 3600) has been interpreted as a thrust, with an East dipping orientation, which uplifts the top of the accretionary unit. The last major fracture is located in the eastern part of the profile at 65 km (about CDP 600) although other similar trending minor faults, 1.5 km to the west, could be also interpreted with a bathymetric escarpment of 160 m. This fault, difficult to detect in the seismic profile, has been identified through the 3-D high-resolution bathymetry acquired during the TSUJAL seafloor ocean mapping survey (Fig. 9). The western side of this fault appears to be uplifted relative to the eastern side (being part the forearc basin) suggesting a thrust fault (or reverse fault) associated with the accretionary prism, however, it dips down to the west instead of the expected east direction. This fault might be a candidate to generate a tsunami if a large part of it broke at once, or a

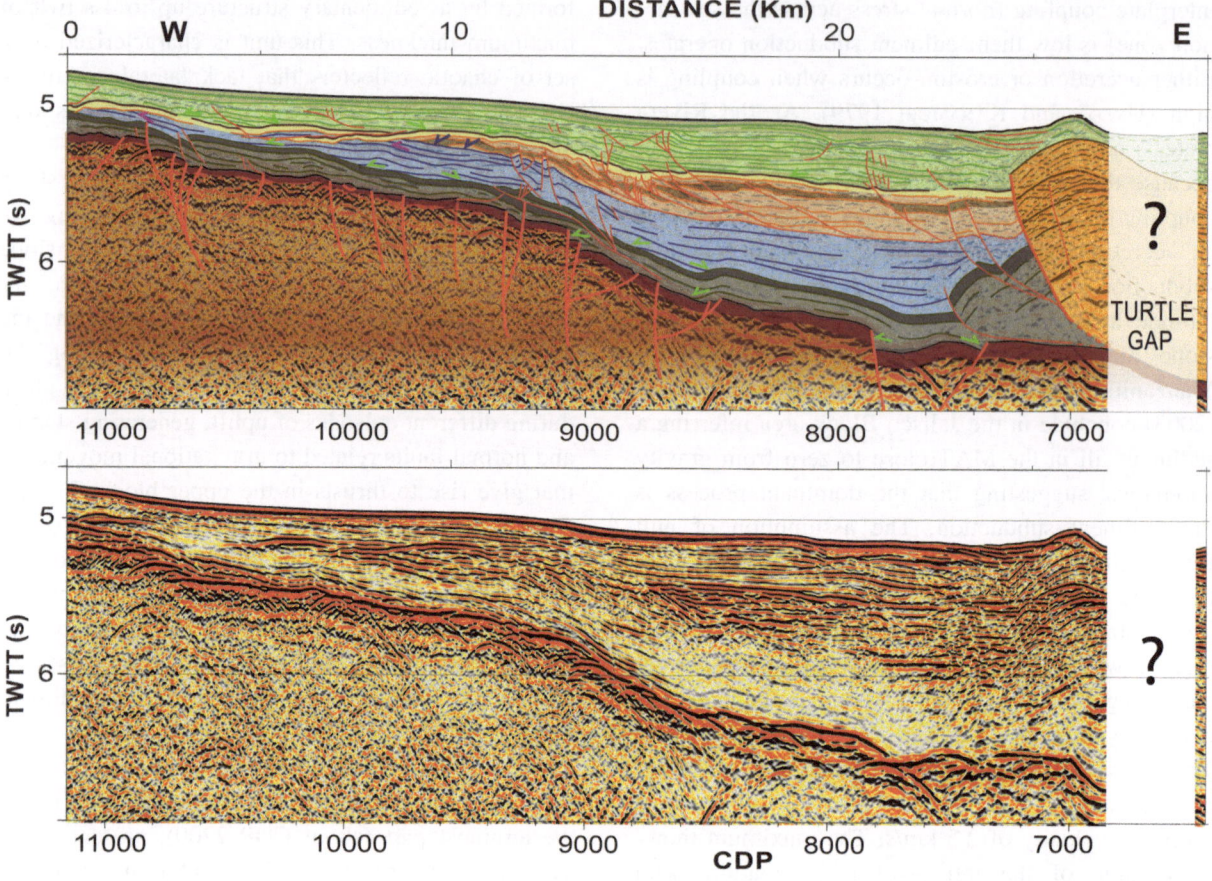

Figure 7
Up Interpretation of the oceanic Rivera plate section of profile TS06b. Onlap (*green arrows*), downlap (*blue arrows*) and erosional truncations (*pink arrows*) geometries are mixed along the sedimentary units. Note the normal faulting in the top of the oceanic crust below the sediments due to the bending of the plate during the subduction. Trench is filled by sediments about 1.6 s twtt of thickness, corresponding to a 2 km depth for an average sediment velocity of 2.5 km/s. *Down* multichannel seismic section of profile TS06b without interpretation

strong motion of the fault would produce sediment shaking which in turn could cause a landslide and a tsunami. Further bathymetric studies related with this regional fault should be conducted in the area for a tsunami risk assessment.

5.4. *Generation of Earthquakes*

Recent 2004 Sumatra–Andaman and 2011 Tohoku events violate the classical RUFF and KANAMORI (1980) direct relationship between convergence rate and age to the maximum earthquake magnitude (SHEARER and BÜRGMANN 2010). It is assumed that faster subducting lithosphere should increase friction at the interface, and that a younger

subducting plate (more buoyant and lighter) is strongly coupled with the continental plate. Thus, both faster and younger plates increase the magnitude of the expected event. But the Mw 9.2 Sumatra earthquake occurred in an area with a low convergence rate (3 cm/year) and where the age of the subducting plate is 55–90 Ma old. The Mw 9 Tohoku event occurred in a 130 Ma old subducting plate, which is older than most of the other ocean floors in the world, and where the convergence rate is 10 cm/year. Great earthquakes in the past like the Mw 9+ Chile and Alaska earthquakes occurred on the plate boundary with a convergence rate of 11 cm/year and plate age of 20 Ma, and 6 cm/year and 40 Ma, respectively (KANAMORI 2006).

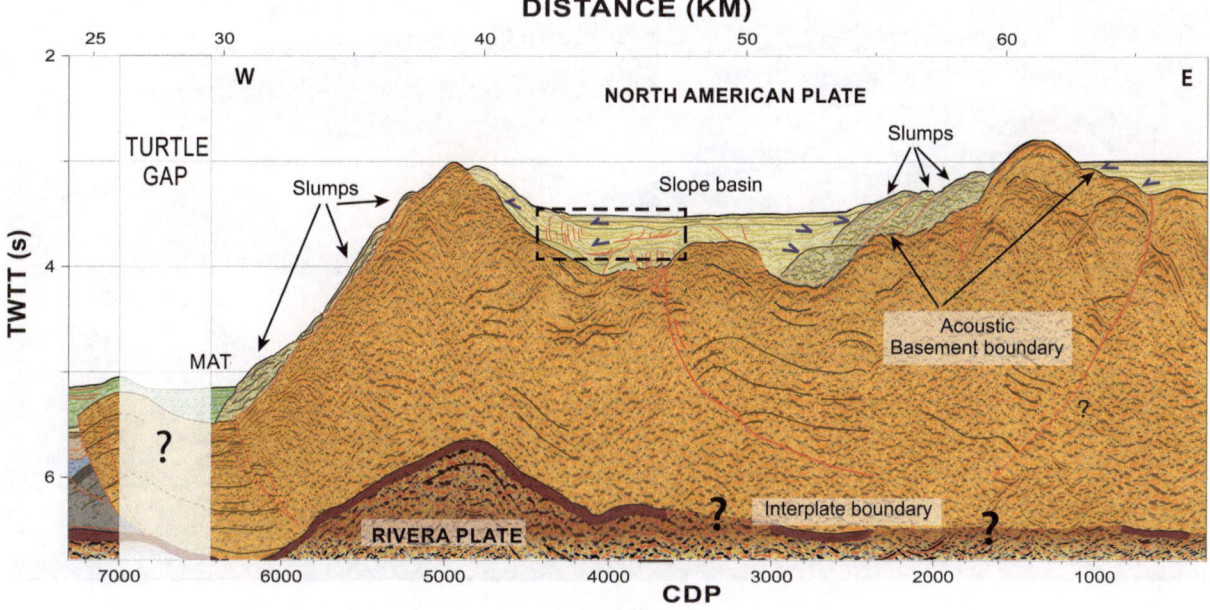

Figure 8
Accretionary prism, slope and forearc basins interpreted in multichannel profile TS06b. Basins show onlap (*blue arrows*) geometries. *Dashed box* may indicate a probably extensional gravitational collapse and the resulting thrust generation

Thus, new relationships are needed to relate the magnitude of an earthquake with geophysical parameters. For instance, both tectonic stresses and geometrical irregularities along the subduction interface have been empirically related to giant earthquakes (Mw ≥8.5). HEURET *et al.* (2012) found a statistical relationship between the occurrence of subduction megathrust earthquakes and the combination of thick trench infill (≥1 km) and neutral upper plate strain. In fact, the combination of these two factors is more highly correlated with the occurrence than either factor alone. When the propagation of rupture in the trench-parallel direction breaks a larger number of thrust fault segments then the magnitude of a seismic earthquake increases. The propagation seems to be controlled by the distribution of geometrical irregularities and thick trench infill would represent the geometrical irregularities along the subduction interface. In other words, trench infill acting as a proxy for smoothing of subducting plate relief by sediment input into the subduction channel and facilitating propagation. Although statistics are difficult due to the limited number of large

earthquakes, SCHOLL *et al.* (2011) demonstrated that trench sectors with axial deposits thicker than 1.0 km are associated with the occurrence of an unusually high number of giants earthquakes (67 % of Mw >8.5 in the Earth). Moreover, upper (in the Jalisco Block at the NA) plate strain would act as a proxy for the tectonic stresses applied to the subduction interface, and the tectonic stresses may be inferred from deformation in the back-arc. The propagation of an initial rupture to neighboring asperities is made difficult by compressive tectonics, easily associated with neutral back-arcs and never associated with extensional domains (HEURET *et al.* 2012). Recent GPS investigations found the Jalisco Block to be moving slowly (2 mm/year) to the southwest relative to NA plate (SELVANS *et al.* 2011) or perhaps not even moving as an independent block. Consequently, considering these two factors together: almost neutral strain present in the Jalisco Block (upper plate strain) and large (1.6 km) sediment thickness observed in our seismic profile, the Jalisco Block has the geophysical favorable conditions to generate giant earthquakes.

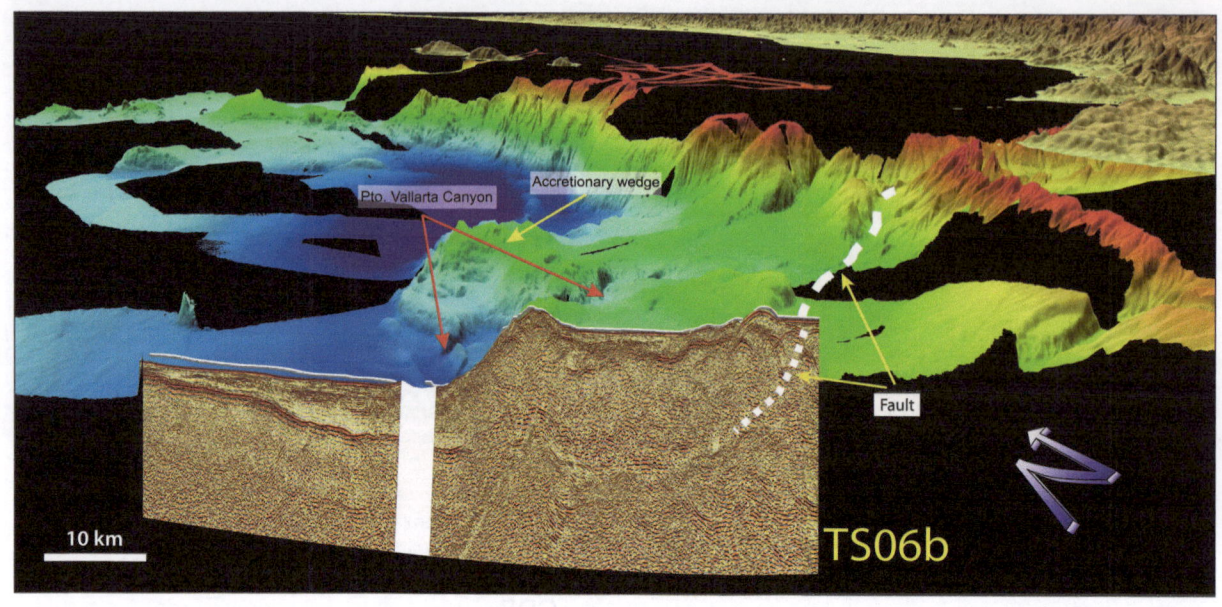

Figure 9
Location of the line TS06b in the 3-D high-resolution bathymetry map acquired during the TSUJAL survey. A major fault observed in the bathymetry grid, difficult to be interpreted in the seismic line, is shown in the eastern part of the profile

5.5. Evidence for Gas Hydrates

Gas hydrates have been recovered in sediment cores but their occurrence are usually inferred from a prominent identifiable high amplitude reflector in the seismic signal, commonly named bottom simulating reflector (BSR), although gas hydrates have also been encountered in regions without BSR (HAACKE et al. 2007). Multichannel seismic imaging is the best method to detect the presence of BSR. The BSR identified in seismic data in the offshore part of the Jalisco Block has been detected in the shallow continental domain and has a reversed polarity with respect to the seafloor. There are at least two types of origins for the occurrence of BSR. One is related to the presence of gas hydrates causing a negative acoustic impedance contrast between sediments containing gas hydrate extending from the seafloor to the BSR (which increases seismic velocity) and free gas underneath the gas hydrate stability zone (lower seismic velocity) in the pore space of accreted sediments (Fig. 10). Therefore, these BSR have reversed polarity. The other origin has been related with the strong positive acoustic impedance contrast between silicate rich sediments of the different diagenetic stages opal A, opal CT and quartz (KASTNER et al. 1977). Therefore, diagenesis-related BSRs have the same polarity as the seafloor reflection. The BSR identified in seismic data of the Jalisco Block have a reversed polarity with respect to the seafloor, and consequently, are related with the presence of natural gas hydrates.

The BSR imaged crosscuts sedimentary layers, and is easily recognized subparallel to the seaflor along the accretionary wedge and the forearc basin at 0.25–0.3 s (twtt) below the seaflor, extending for about 25 km along the profile (from CDP5800 to CDP2200). The BSR reflector is disrupted, and thus difficult to observe, by the presence of faulting, slumps and landslides, because gas hydrates may dissolve, and finally escape, causing the impedance contrast to disappear (Figs. 6, 10). This indicates that slope failure can cause the destruction of BSRs. But also rapid decompression of the underlying sediments in areas with occurrence of BSR (indicating methane hydrate) could lead to catastrophic release of methane and the resulting margin destabilization with the occurrence of mass transport deposits.

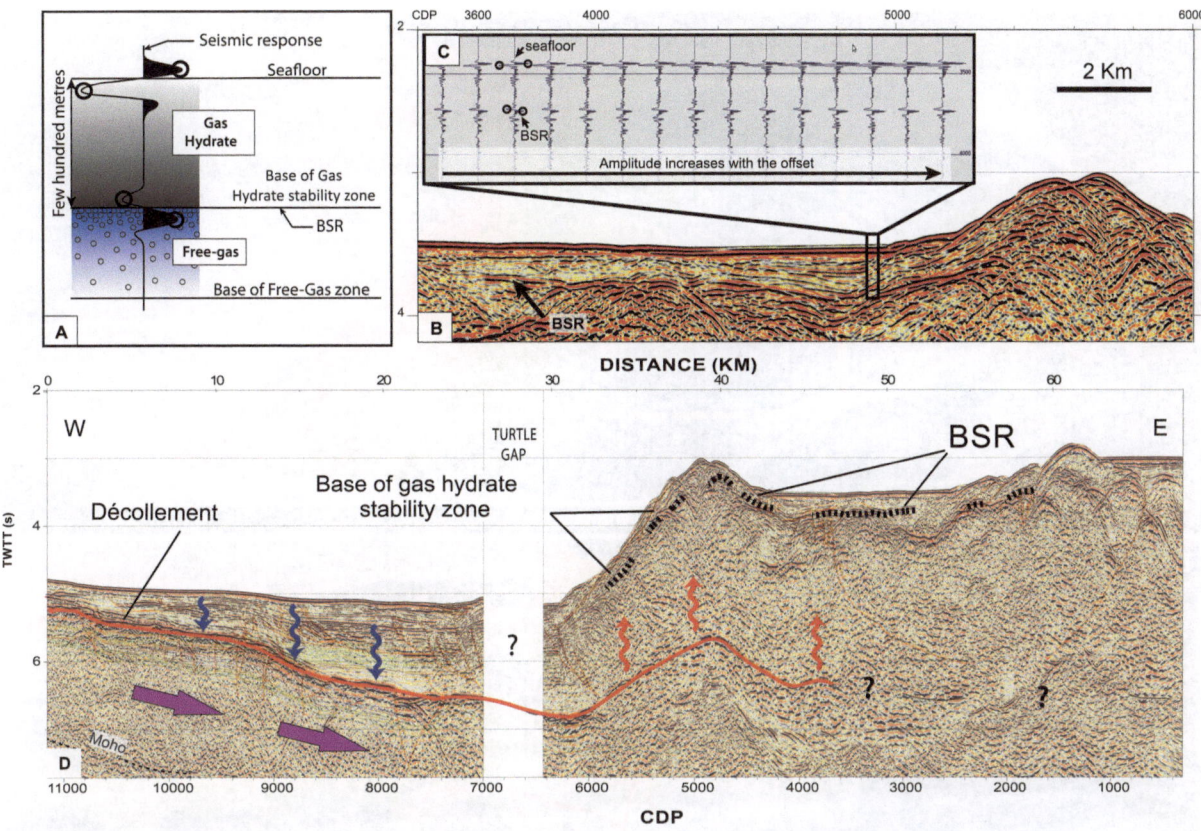

Figure 10

Gas hydrate occurrence in line TS06b. **a** Illustration of a submarine section containing the structure and location of gas hydrate (*above*) and free gas (*below*) of the bottom simulating reflector (BSR) (Image modified from HAACKE *et al.* 2007). **b** Detailed view of the accretionary wedge in profile TS06b showing a clear BSR at 200–300 twtt below the seafloor. **c** Traces (without AGC) of the CDP 4450 showing the high amplitude, reverse polarity and the amplitude increasing along offset of the BSR. **d** Structural interpretation of the line TS06b, from 2 to 8 s twtt, illustrating a theoretical transfer of fluids along the subduction plate and the formation of the gas hydrate

Significant evidence for the presence of gas hydrates, identified as BSR reflectors in multichannel seismic records, has been recognized in the same area in previous studies (BARTOLOME *et al.* 2011) and in the continental slope area of the southern Jalisco Subduction Zone off Manzanillo by BANDY and MORTERA GUTIÉRREZ (2012). Reflectors are found near Manzanillo at 0.4 s (two-way travel time) below the seafloor reflector. This result once again suggests that gas hydrates may exist in the continental slope region of the entire Jalisco Subduction Zone. However, more seismic reflection data needs to be collected to verify this assertion. These new findings are of value to evaluate the gas hydrate potential of the Jalisco

Subduction zone, an area that may contain a significant source of energy for the future (Fig. 11).

Deep faulting observed along the profile at the top of the oceanic Rivera plate can play an important role for circulation of fluids and the formation of gas hydrates in the accretionary prism, especially in two particular areas. First, in the décollement zone or basal thrust, which acts as a conduit for fluids producing wet sediments, and second in the accretionary wedge, where rapid rates of upward fluid flow and seafloor uplift occur (FOSSEN 2010, see Fig. 10). Therefore, it is more likely to occur serpentinization processes in the deeper part of the interplate contact due to the alteration process of olivine in the presence

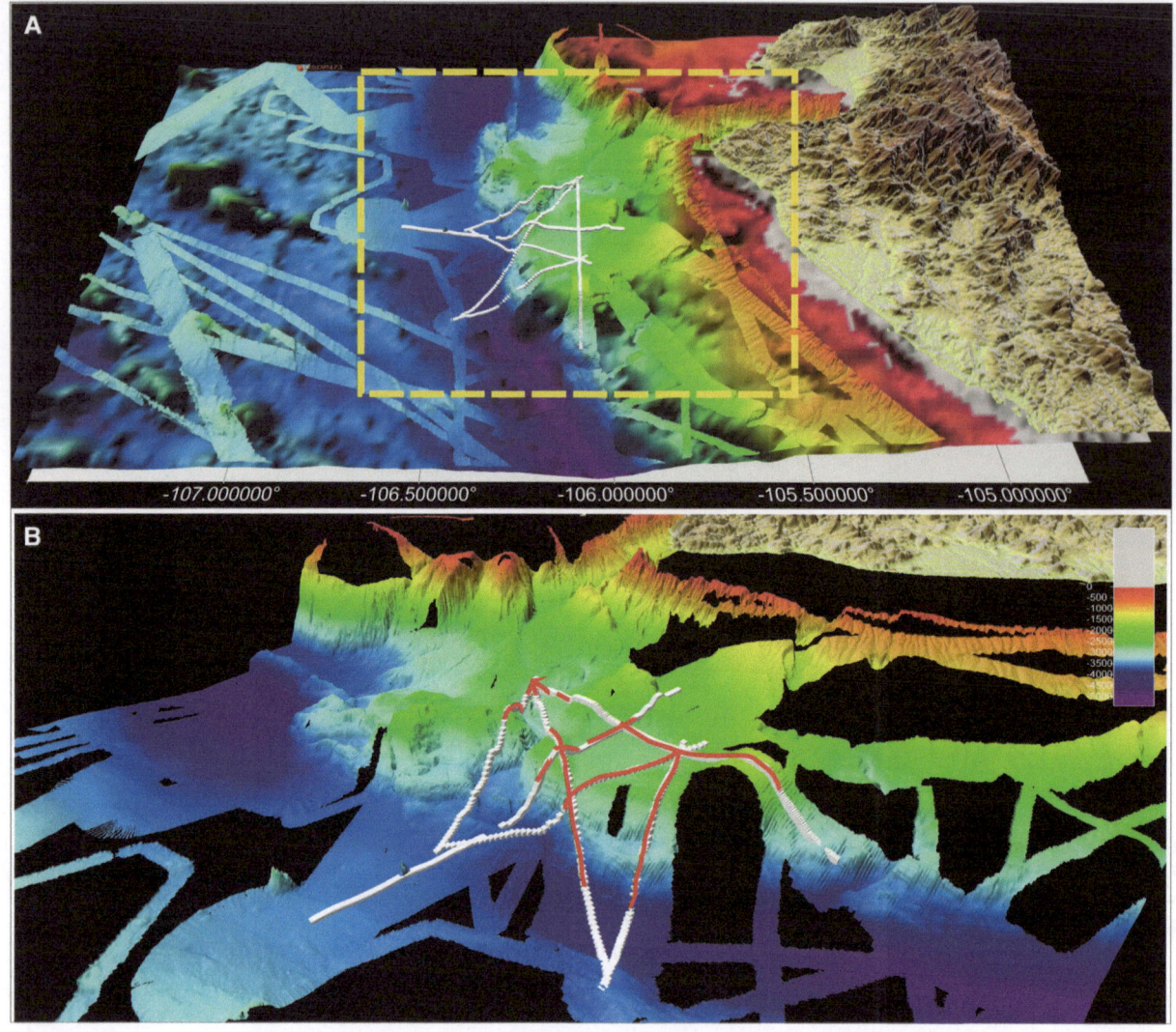

Figure 11
a Swath bathymetric data of TSUJAL and CORTES (BARTOLOME *et al.* 2011) surveys offshore the Jalisco Block. *Thin white lines* mark MCS profiles. **b** *Red lines* marks approximate location and distribution of BSR (and gas hydrates) offshore Puerto Vallarta

of water. In subduction zones, this process occurs along deep faults where oceanic plates enter, near the base of the mantle wedge at depths up to 150 m depth (MANEA and MANEA 2011) and always greater than 10 km. The décollement or basal thrust is identified in the interplate boundary, but the seismogenic zone is located in deeper areas of the subduction system not imaged by the TS06b. Subduction of serpentinized rocks enhances the generation of earthquakes and melting production in the mantle wedge, being

able to influence the position of the melting front (and volcanic arc).

6. Conclusions

The western margin of Mexico, in general, and the Jalisco Block, in particular, is a region where several tectonic plates interact, turning it into one of the most active seismic and volcanic zones in the

Americas. The investigation of subduction zones and associated seismogenic and volcanic areas becomes extremely important for the safety of people living in these areas. The reliability of a specific risk level regarding the future seismic activity in a region depends on the level of our understanding of the geodynamic processes, as well as the amount and quality of the available data. The structural and geometrical information provided by profile TS06b constitute an essential contribution to understand the geodynamic context of the area. Therefore, our work aimed to collect high quality geophysical data to make more reliable earthquake forecasts, which has important social implications on time-scales of months and years. The main conclusions are summarized as follows:

- Processing and interpretation of the line TS06b acquired during the TSUJAL survey has allowed to obtain, for the first time, a high quality seismic image of the entire internal structure of the Rivera Subduction Zone beneath North American plate. This paper shows the main structures in the Jalisco Block area at crustal scale, characterized by their geometry, morphology, dimension, deformation and fracturing of the internal structure, depth and their lateral continuity. In this way, accretionary prism, slope basins associated, forearc basin, crust–mantle boundary and interplate limit have been identified.
- Evidence of BSRs reflectors compiled using the available multichannel seismic data in the continental slope region of the entire Jalisco Subduction zone suggests that extensive gas hydrates accumulations may exist in this area.
- Normal faulting and the resulting horst and graben geometry are observed in the Rivera plate as a consequence of the bending of the plate as it begins to subduct beneath the NA plate. Fluid circulation of oceanic water can easily enter into the subduction system using these fault paths, helping the formation of gas hydrates and serpentinization.
- Subduction accretion is the dominant process at the Jalisco Block near Puerto Vallarta along the lower slope of the margin, in agreement with the occurrence of an accretionary prism and the young age of the Rivera plate

- The combination of a thick (1.6 km) sediment infill within the trench and an almost neutral strain in the Jalisco Block indicates that the area is prone to giant earthquake occurrence.

Acknowledgments

We acknowledge the Captain and crew of the RRS James Cook, the seismic UTM-CSIC technical team and Luis Ansorena for their professional work, and the officers and crew of the AMR Holzinger (Secretaría de Marina, Mexico) and CABO/UNAM J-GAP2013 Cruise (BO El Puma, Mexico) which made possible the success of the TSUJAL cruise. We gratefully acknowledge financial support from MINECO (Spain) through TSUJAL project (CGL2011-29474-C02-01) and "Ramon y Cajal" program (R. Bartolome), and from Mexico through CONACYT–FOMIXJAL 2008–96567 (2009); CONACYT–FOMIXJAL 2008–96539 (2009); CONACYT–FOMIXJAL 2010–149245 (2011); CONACyT-FOMIXJal (2012-08-189963) and also partial funding was provided by UNAM DGAPA Grant #IN115513-3. We also gratefully acknowledge the support of the Mexican Instituto Oceanográfico del Pacífico in Manzanillo, Secretaría de Defensa Nacional, Unidad Municipal de Proteccion Civil y Bomberos (Puerto Vallarta), Unidad Estatal de Proteccion Civil y Bomberos (Nayarit State), Unidad Municipal de Proteccion Civil y Bomberos (Jalisco State), Reserva de la Biosfera (Islas Marías) CONANP-SEMARNAT, Órgano Desconcentrado de Prevención y Readaptacion Social de la SEGOB, Secretaría de Relaciones Exteriores for their help during all the cruise. This work was carried out within the Grup de Recerca de la Generalitat de Catalunya B-CSI (2014 SGR 940).

REFERENCES

ALLAN, J.F. (1986). *Geology of the Northern Colima and Zacoalco Grabens, southwest Mexico: Late Cenozoic rifting in the Mexican Volcanic Belt.* Geol. Soc. Am. Bull., *97*, 473–485.

ALLAN, J.F., NELSON, S.A., LUHR, J.F., CHARMICHAEL, I.S.E., WOPAT, M., & WALLACE, P.J. (1991). Pliocene-Holocene rifting and associated volcanism in southwest Mexico: An exotic terrane in the making. In J. DAUPHIN & B. SIMONEIT (Eds.) The Gulf and Peninsular Provinces of the Californias, Am. Assoc. Pet. Geol. Bull., Memoir *47*, (pp. 425–445), Tulsa, AAPG.

Alvarez, R., & Yutsis, V. (2015). The elusive Rivera-Cocos plate boundary: not diffuse. In T. J. Wright, A. Ayele, D.J. Ferguson, T. Kidane, & C. Vye-Brown (Eds.), Magmatic Rifting and Active Volcanism, Geological Society, London, Special Publications, 420.

ANDREWS, V., STOCK, J., RAMIREZ-VAZQUEZ, C.A., & REYES-DAVILA, G. (2011). *Double-difference relocation of the aftershocks of the Tecoman, Colima, Mexico earthquake.* Pure and Applied Geophysics, *168*, 1331–1338.

ARGUS, D.F., GORDON, R.G., HEFLIN, M.B., CHOPO M.C., EANES, R.J., WILLIS, P., PELTIER, W.R., & OWEN, S.E. (2010). *The angular velocities of the plates and the velocity of Earth's centre from space geodesy.* Geophys. J. Int., *180*, 913–960.

BANDY, W.L. (1992). Geological and Geophysical Investigation of the Rivera-Cocos Plate Boundary: Implications for Plate Fragmentation, Ph.D. Dissertation, Texas A&M University, College Station.

BANDY W.L., & MORTERA GUTIÉRREZ, C.A. (2012). *Gas Hydrates in the southern Jalisco subduction zone as evidenced by bottom simulating reflectors in Multichannel Seismic Reflection Data of the 2002 BART/FAMEX campaign.* Geofísica Internacional, *51*(4), 393–400.

BANDY, W.L., & PARDO, M. (1994). *Statistical examination of the existence and relative motion of the Jalisco and Southern Mexico Blocks.* Tectonics, *13*(4), 755–768.

BANDY, W.L., MORTERA-GUTIERREZ, C., URRUTIA-FUCUGAUCHI, J., & HILDE, T.W.C. (1995). *The subducted Rivera-Cocos plate boundary: Where is it, what is it, and what is its relationship to the Colima rift?.* Geophysical Research Letters, *22*, 3075–3078.

BANDY, W.L., MICHAUD, F., BOURGOIS, J., CALMUS, T., DYMENT, J., MORTERA-GUTIÉRREZ, C.A., ORTEGA-RAMÍREZ, J., PONTOISE, B., ROYER, J.-Y., SICHLER, B., SOSSON, M., REBOLLEDO-VIEYRA, M., BIGOT-CORMIER, F., DÍAZ-MOLINA, O., HURTADO-ARTUNDUAGA, A.D., PARDO-CASTRO, G., & TROUILLARD-PERROT, C. (2005). *Subsidence and strike-slip tectonism of the upper continental slope off Manzanillo, Mexico.* Tectonophysics, *398*, 115–140.

BANDY, W.L., MICHAUD F., MORTERA-GUTIÉRREZ C.A., DYMENT, J., BOURGOIS, J., ROYER, J.-Y., CALMUS, T., SOSSON, M., & ORTEGA-RAMIREZ, J. (2011). *The Mid-Rivera-Transform Discordance: Morphology and tectonic development.* Pure and Applied Geophysics, *168*(8–9), 1391–1413. doi:10.1007/s00024-010-0208-8.

BANDY, W.L., KOSTOGLODOV, V., SINGH, S.K., PARDO, M., PACHECO, J., & URRUTIA-FUCUGAUCHI, J. (1997). *Implications of the October 1995 Colima-Jalisco Mexico earthquakes on the Rivera-North America Euler vector.* Geophys. Res. Lett., *24*(4), 485–488.

BARTOLOMÉ, R., DAÑOBEITIA, J.J., & CÓRDOBA, D. (2015). *Seismic research in western Mexico,* Sea Technology, *56*, 10, 25–29.

BARTOLOME, R., DAÑOBEITIA, J.J., MICHAUD, F., CÓRDOBA, D., & DELGADO-ARGOTE, L. (2011). *Imaging the seismic crustal*

structure of the western Mexican margin between 19°N and 21°N. Pure and Applied Geophysics, *168*, 1373–1389. doi:10.1007/s00024-010-0206-x.

BOURGOIS, J., & MICHAUD, F. (1991). *Active fragmentation of the North American plate at the Mexican Triple Junction Area off Manzanillo (Mexico).* Geomarine Letters, *11*, 59–65.

DAÑOBEITIA, J.J., CORDOBA, D., DELGADO-ARGOTE, L.A., MICHAUD, F., BARTOLOMÉ, R., FARRAN, M., CARBONEL, M., NUÑEZ-CORNÚ, F., & the CORTES-P96 WORKING GROUP (1997). *Expedition Gathers New Data on Crust Beneath Mexican West Coast.* EOS, Trans. Am. Geophys. Union, *78*(49), 565–572.

DAÑOBEITIA J.J., BARTOLOMÉ, R., NUÑEZ-CORNÚ, F., CORDOBA, D., BANDY, W., PRADA, M., NUÑEZ, D., CASTELLÓN, A., & ALONSO, J.L. (2015). *Crustal architecture at the collision zone between Rivera and North American plates at the Jalisco Block: TSUJAL project.* Pure and Applied Geophysics, This volume.

DEMETS, C., GORDON, R.G., ARGUS, D.F., & STEIN, S. (1994). *Effect of recent revisions to the geomagnetic reversal time scale on estimates of current plate motions.* Geophys. Res. Lett., *21*(20), 2191–2194.

DEMETS, C., GORDON, R.G., & ARGUS, D.F. (2010). *Geologically current plate motions.* Geophys. J. Int., *181*, 1–80.

DEMETS, C., & TRAYLEN, S. (2000). Motion of the Rivera plate since 10 Ma relative to the Pacific and North American plates and the mantle. In L. FERRARI, J. STOCK, & J. URRUTIA-FUCU-GAUCHI (Eds.), The Influence of plate interaction on post-Laramide magmatism and tectonics in Mexico (pp. 119–159), Tectonophysics, 31(1–4).

DEMETS, C., & WILSON, D. S. (1997). *Relative Motions of the Pacific, Rivera, North American, and Cocos Plates Since 0.78 Ma.* J. Geophys. Res, *102*(B2), 2789–2806.

FOSSEN, H. (2010). Structural geology. Cambridge University Press.

GEORGE, D.L., YUEN, D.A., MARUYAMA, S., & YANAI, S.C. (2011). What mechanisms produce the tsunami of 2011 Tohoku Earthquake: Elastic deformation or landslides?, paper presented at Asia Oceania Geosciences Society Fall Meeting, Taipei, Taiwan.

GRILLI, S.T., et al. (2012). *Numerical simulation of the 2011 Tohoku tsunami based on a new transient FEM co-seismic source: Comparison to far- and near-field observations.* Pure Appl. Geophys. doi:10.1007/s00024-012-0528-y.

HAACKE, R.R., WESTBROOK, G K., & HYNDMAN, R.D. (2007). *Gas hydrate, fluid flow and free gas: Formation of the bottom-simulating reflector.* Earth and Planetary Science Letters, *261*(3), 407–420.

HEURET, A., CONRAD, C., FUNICIELLO, F., LALLEMAND, S., & SANDRI, L. (2012). *Relation between subduction megathrust earthquakes, trench sediment thickness and upper plate strain.* Geophys. Res. Lett., *39*, L05304. doi:10.1029/2011GL050712.

JOHNSON, C.A., & HARRISON, C.G.A. (1989). *Tectonics and volcanism in central Mexico: A Landsat Thematic Mapper perspective.* Remote Sens. Envir., *28*, 273–286.

KANAMORI, H. (1977). Seismic and aseismic slip along subduction zones and their tectonic implications. In M. TALWANI & W.C. PITMAN III (Eds.), Island Arcs Deep Sea Trenches and Back-Arc Basins (pp. 163–174). Washington D.C., American Geophysical Union.

KANAMORI, H. (2006). *Lessons from the 2004 Sumatra-Andaman earthquake.* Philos. Trans. R. Soc. A., 364, 1927–1945.

KASTNER, M., KEENE, J.B., & GIESKES, J.M. (1977). *Diagenesis of siliceous oozes-I. Chemical controls on the rate of opal-A to opal-CT transformation an experimental study.* Geochimica et Cosmochimica Acta, *41*, 1041–1059.

KLITGORD, K., & MAMMERICKX, J. (1982). *Northern East Pacific rise: magnetic anomaly and bathymetric framework.* J. Geophys. Res., *100*, 24367–24392.

KOSTOGLODOV, V., & BANDY, W. (1995). *Seismotectonic Constraints on the Convergence Rate between the Rivera and North America Plates.* J. Geophys. Res., *100*(B9), 17977–17989.

KHUTORSKOY, M.D., DELGADO-ARGOTE, L.A., FERNÁNDEZ, R., KONONOV, V.I., & POLYAK, B.G. (1994). *Tectonics of the offshore Manzanillo and Tecpan basins, Mexican Pacific, from heat flow, bathymetric and seismic data.* Geofís. Intern., *33*, 161–185.

LEON SOTO, G., NI, J.F., GRAND, S.P., SANDOVOL, E., VALENZUELA, R.W., GUZMAN SPEZIALE, M., GÓMEZ GONZÁLEZ, J.M., & DOMÍNGUEZ REYES, T. (2009). *Mantle flow in the Rivera-Cocos subduction zone.* Geophy. J. Int., *179*, 1004–1012. doi:10.1111/j.1365-246X.2009.04352.x.

LONSDALE, P. (1995). *Segmentation and disruption of the East Pacific Rise in the mouth of the Gulf of California.* Marine Geophysical Researches, *17*, 323–359.

MANEA, M., MANEA, V.C., & KOSTOGLODOV, V. (2003). *Sediment fill in the Middle America Trench inferred from gravity anomalies.* Geofísica Internacional, *42*(4), 603–612.

MANEA, V.C., & MANEA, M. (2011). *Flat-slab thermal structure and evolution beneath central Mexico.* Pure and Applied Geophysics, *168*(8–9), 1475–1487.

McCLOY, C., INGLE, J. C., & BARRON, J. A. (1988). *Neogene stratigraphy, foraminifera, diatoms, and depositional history of María Madre Island Mexico: Evidence of Early Neogene marine conditions in the southern Gulf of California.* Marine Micropaleontology, *13*, 193–212.

MERCIER DE LEPINAY, B., MICHAUD, F., & the NAUTIMAT team (1997). *Large Neogene Subsidence along the Middle America trench off Mexico (18°-19°N): Evidence from Submersible Observations.* Geology, *25*(5), 387–390.

MICHAUD, F., DAÑOBEITIA, J., CARBONELL, R., BARTOLOMÉ, R., CORDOBA, D., DELGADO ARGOTE, L., NÚÑEZ CORNÚ, F., & MONFRET, T. (2000). *New insights into the subducting ocean crust in the Middle American Trench off western Mexico (17°-19°N).* Tectonophysics, *318*, 187–200.

MICHAUD, F., DAÑOBEITIA, J., BARTOLOMÉ, R., CARBONELL, R., DELGADO ARGOTE, L., CORDOBA, D., & MONFRET, T. (2001). *Did the East Pacific Rise subduct beneath the North American Plate (western Mexico)?.* Geo-Marine Letters, *20* (3), 168–173.

MICHAUD, F., MERCIER DE LEPINAY, B., BOURGOIS, J., & CALMUS, T. (1996). *Evidence for active extensional tectonic features within the Acapulco trench fill at the Rivera-North America plate boundary,* C.R. Acad. Sci., Paris, *321* série IIa, 521–528.

MINSHULL, T., BARTOLOMÉ, R., BYRNE, S., & DAÑOBEITIA, J.J. (2005). *Low heat flow from young oceanic lithosphere at the Middle America Trench off Mexico.* Earth and Planetary Science Letters, *239*, 33–41. doi:10.1016/j.epsl.2005.05.045.

MINSTER, J.B., & JORDAN, T.H. (1979). *Rotation vectors for the Philippine and Rivera plates (abstract).* Eos Trans. Am. Geophys. Union, *60*, 958.

NIXON, G.T. (1982). *The relationship between Quaternary volcanism in central Mexico and the seismicity and structure of subducted ocean lithosphere.* Geological Society of America Bulletin, *93*, 514–523.

NÚÑEZ-CORNÚ, F., CORDOBA, D., DAÑOBEITIA, J.J., BANDY, W., ORTIZ, M., BARTOLOME, R., NÚÑEZ, D., ZAMORA, A., ESPÍNDOLA, J.M.,

CASTELLON, A., ESCUDERO, C.R., TREJO, E., ESCALONA, F.J., SUÁREZ, C., MORTERA, C., & TSUJAL WORKING GROUP (2016). *Geophysical Studies across Rivera Plate and Jalisco Block, MEXICO: TsuJal Project.* Seism. Res. Lett, *87*(1), 59–72. doi:10.1785/0220150144.

NUÑEZ-CORNÚ, F.J., REYES-DÁVILA, G.A., RUTZ, M., TREJO-GÓMEZ, E., CAMARENA-GARCÍA, M.A., & RAMÍREZ-VAZQUEZ, C.A. (2004). *The 2003 Armería, México earthquake (Mw 7.4): Mainshock and early aftershocks.* Seism. Res. Lett, *75*, 734–743.

NUNEZ-CORNU, F.J., RUTZ, L.M., MARQUEZ-RAMIREZ, V., SUAREZ-PLASCENCIA, C., & TREJO-GOMEZ, E. (2011). *Using an enhanced data set for reassessing the source region of the 2003 Armeria, Mexico, earthquake.* Pure and Applied Geophysics, *168*, 1293–1302.

OCHOA-CHAVEZ, J.A., ESCUDERO, C.R., NÚÑEZ-CORNÚ, F.J., & BANDY, W.L. (2015). *P-wave Velocity Tomography from Local Earthquakes in Western Mexico,* Pure and Applied Geophysics, This volume.

OKAL, E.A., & BORRERO, J.C. (2011). *The "tsunami earthquake" of 22 June 1932 in Manzanillo, Mexico: Seismological study and tsunami simulations.* Geophys. J. Intl, *187*, 1443–1459.

PARDO, M., & SUÁREZ, G. (1995). *Shape of the subducted Rivera and Cocos plates in southern Mexico: Seismic and tectonic implications.* J. Geophys. Res, *100*, 12,357–12,373.

PELÁEZ GAVIRIA, J.R., MORTERA GUTIÉRREZ, C.A., BANDY, W.L., & MICHAUD, F. (2013). *Morphology and magnetic survey of the Rivera-Cocos plate boundary of Colima, Mexico.* Geofisica Internacional, *52*(1), 73–85.

QUINTANAR ROBLES, L., RODRIGUEZ-LOZOYA, H.E., ORTEGA, R., GOMEZ-GONZALEZ, J.M., DOMINGUEZ, T., JAVIER, C., ALCANTARA, L., & REBOLLAR, C.J. (2011). *Source characteristics of the 22 January 2003 M (w) = 7.5 Tecoman, Mexico, earthquake: New insights.* Pure and Applied Geophysics, *168*, 8–9, 1339–1353.

RAMÍREZ-HERRERA, M.T., Kostoglodov, V., & URRUTIA-FUCUGAUCHI, J. (2011). *Overview of recent coastal tectonic deformation in the Mexican subduction zone.* Pure Appl. Geophys., *168*, 1415–1433. doi:10.1007/s00024-010-0205-y.

RENARD, V., AUBOUIN, J., LONSDALE, P., & STEPHAN, J.F. (1980). *Premiers resultants d'une etude de la fose d'Amerique Central au sondeur multifaisceaux (Seabeam).* Geologie Marine, C.R.Acad. Sci., *291*, Sér. D, 137–142.

ROSS, D.A. (1971). *Sediments in the northern Middle America trench.* GSA Bulletin, *82*, 303–322.

RUFF, L.J., & KANAMORI, H. (1980). *Seismicity and the Subduction Process.* Phys. Earth. Planet. Inter., *23*, 240–252.

SANDWELL, D. T., & SMITH, W.H.F. (2009). *Global marine gravity from retracked Geosat and ERS-1 altimetry: Ridge Segmentation versus spreading rate.* J. Geophys. Res, *114*, B01411. doi:10.1029/2008JB006008.

SELVANS, M., STOCK, J.M., DEMETS, C., SANCHEZ, O., & MARQUEZ-AZUA, B. (2011). *Constraints on Jalisco Block Motion and Tectonics of the Guadalajara Triple Junction from 1998–2001 Campaign GPS Data.* Pure and Applied Geophysics, *168*(8–9), 1435–1447.

SCHOLL, D.W., KIRBY, S.H., & VON HUENE, R. (2011). Exploring a link between great and giant megathrust earthquakes and relative thickness of sediment and eroded debris in the subduction channel to roughness of subducted relief. Abstract TI4B-01 presented at 2011 Fall Meeting, AGU, San Francisco, Calif., 5–9 Dec.

SHEARER, P., & BÜRGMANN, R. (2010). *Lessons learned from the 2004 Sumatra-Andaman megathrust rupture.* Annu. Rev. Earth Planet. Sci., *38*, 103–131.

SINGH, S.K., PONCE, L., & NISHENKO, S. (1985). *The great Jalisco, Mexico earthquake of 1932: subduction of the Rivera plate.* Bull. Seismic. Soc. Amer., *75*, 1301–1313.

SUÁREZ, G., JARAMILLO, S.H., & BANDY, W.L. (2013). *Relative motion between the Rivera and North American plates determined from the slip directions of earthquakes.* Pure and Applied Geophysics, *170*, 2163–2172. doi:10.1007/s00024-013-0667-9.

UYEDA, S., & KANAMORI, H. (1979). *Back-arc opening and the mode of subduction.* J. Geophy. Res, *84*, 1049–1061.

YAGI, Y., MIKUMO, T., PACHECHO, J., & REYES, G. (2004). *Source Rupture Process of the Tecoman, Colima, Mexico Earthquake of 22 January 2003, Determined by Joint Inversion of Teleseismic Body-Wave and Near-Source Data.* Bull. Seism. Soc. Am., *94*, 1795–1807.

YANG, T., GRAND, S.P., WILSON, D., GUZMAN-SPEZIALE, M., GOMEZ-GONZALEZ, J.M., DOMINGUEZ-REYES, T., & NI, J. (2009). *Seismic structure beneath the Rivera subduction zone from finite-frequency seismic tomography.* J. Geophys. Res., *114*, B01302. doi:10.1029/2008JB005830.

(Received August 19, 2015, accepted June 10, 2016, Published online June 24, 2016)

Pure Appl. Geophys. 173 (2016), 3595–3614
© 2015 Springer International Publishing
DOI 10.1007/s00024-015-1221-8

Pure and Applied Geophysics

Geometric Aspects of the Full Moment Tensors in the Gulf of California and the Mexican East Pacific Rise

ROBERTO ORTEGA,[1] LUIS QUINTANAR,[2] and EDUARDO HUESCA-PÉREZ[3]

Abstract—The East Pacific Rise (EPR) and the Gulf of California (GC) have different tectonic histories. While the EPR has been present for 75 Ma, the GC started only 12.5 Myr. The region that links both systems is the Tamayo Fracture Zone, where a diffuse triple junction is located. A key question to be solved is whether the source mechanisms in this region reflect important variations from the GC to the EPR. Therefore, we analyzed the seismic moment tensors of the GC and the EPR using a full moment tensor inversion. This source model is useful in extensional regimes where isotropic components or complex faults are present. The full moment tensor is the best representation of the fault and slip direction in a rifting process because it resolves for six free parameters, including complex sources of pure shear dislocations. The analysis is similar to the deviatoric case, but the interpretation is different, because physical characteristics in the model allow for choosing a realistic style of rupture. Our results show that there are similarities between focal mechanisms determined by full moment tensors computed for the southern part of the GC and the EPR. We suggest that the EPR is tectonically linked to the GC not only at the diffuse triple junction region but also along the entire province. The rupture patterns of the GC and the EPR are slightly different: whereas the GC is partitioned by means of NW–SE faults, the EPR ruptures through a faulting system NE–SW. The geometrical relations of the extensional province of the GC and the EPR were present since the crustal thinning of the rifting process. Strain partitioning of faults explains easily the nature of the oblique divergence of the GC and the EPR. In addition, in our analysis, we observe clockwise rotation in the structures of the southern part of the GC, suggesting that there is a change in the spatial partitioning of this region.

Key words: Earthquake source observations, isotropic component, transform faults, Gulf of California, full moment tensor.

1. Introduction

The west coast of North America, at the latitude of the Baja California peninsula, was the site of a subduction zone in which the Farallon (Fig. 1) plate was subducting eastward beneath western North America (ATWATER 1970; STOCK and HODGES 1989; HAUSBACK 1984). As the Farallon plate continued to be subducted and consumed (~ 15 Ma), the Pacific plate, located west of the Farallon plate, came into contact with North America and began subducting beneath the North American plate, initiating the development of the Mendocino and Rivera triple junctions along southern California and northern Baja California (ATWATER 1970; STOCK and HODGES 1989). As subduction of the Pacific plate continued, the Mendocino triple junction moved northward, while the Rivera triple junction moved southward, lengthening the right-lateral transform plate boundary that had developed between the Pacific plate and North American plate (STOCK and HODGES 1989). This right-lateral system marked the early stages of the San Andreas fault system (ATWATER 1970).

The GC extends for ~ 1300 km and contains part of the plate boundary between the Pacific and North America plate intersecting to the north the San Andreas fault and to the south the East Pacific Rise (EPR). This margin undergoes a transition from continental extension to seafloor spreading that initiated ~ 6 Myr (ATWATER and STOCK 1998; AXEN et al. 2000) and can be geologically described as an *en echelon* array of right-lateral transform faults connected by short oceanic spreading centers. In the southernmost spreading center of the Gulf of California (GC), along the Alarcón Rise, new proto-oceanic crust started to form approximately

[1] Centro de Investigación Científica y de Educación Superior de Ensenada, Unidad La Paz. Miraflores 334. Col. Bellavista., 23050 La Paz, BCS, Mexico. E-mail: ortega@cicese.mx

[2] Instituto de Geofísica, Universidad Nacional Autónoma de México, Circuito de la Investigación Científica, 04510 Coyoacán, Mexico.

[3] CONACYT Research Fellow, Centro de Investigación Científica y de Educación Superior de Ensenada, Unidad La Paz. Miraflores 334. Col. Bellavista, 23050 La Paz, BCS, Mexico.

(a)

12.3 Ma
12.3-6 Ma
6-0 Ma
Baja California Microplate
Sierra Madre Occidental
Guadalupe microplate
Magdalena microplate
Jalisco Block
Tosco-Abreojos Fault
Rivera Plate
Pacific Plate
Cocos Plate
PACIFIC PLATE
NORTH AMERICAN PLATE
Alarcon Rise
East Pacific Rise
Rivera Plate
Rivera Transform
Cocos Plate
BAJA CALIFORNIA
TOSCO - ABREOJOS FAULT ZONE
LA PAZ FAULT
SAN JOSE DEL CABO TROUGH
CABO SAN LUCAS
CERRALVO TROUGH
TAMAYO FAULT
EAST PACIFIC RISE
CREST
MAGDALENA RIDGE
MARIA MAGDALENA RISE
ISLAS TRES MARIAS
MAINLAND MEXICO
MIDDLE AMERICA TRENCH
MATHEMATICIANS RIDGE
200 km
ACTIVE TRANSFORM OR SPREADING AXIS
TRENCH
FRACTURE ZONE OR FOSSIL TRANSFORM FAULT
INFERRED FROM MAGNETICS ALONE
FOSSIL SPREADING CENTER
WITH BATHYMETRIC EVIDENCE
OCEANIC CRUST FLOWLINE
DIRECTION OF LINEATION IN A REGION OF UNINTERPRETABLE MAGNETICS
112° W
110° W
108° W
113° W
111° W
109° W
107° W
25°N
24°N
23°N
22°N
21°N

(b)

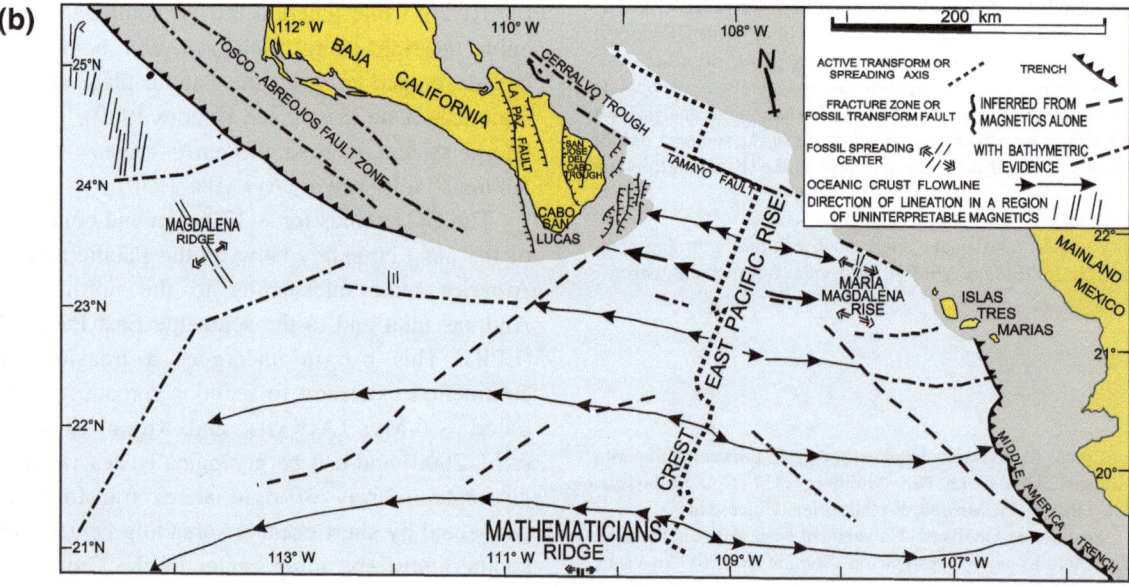

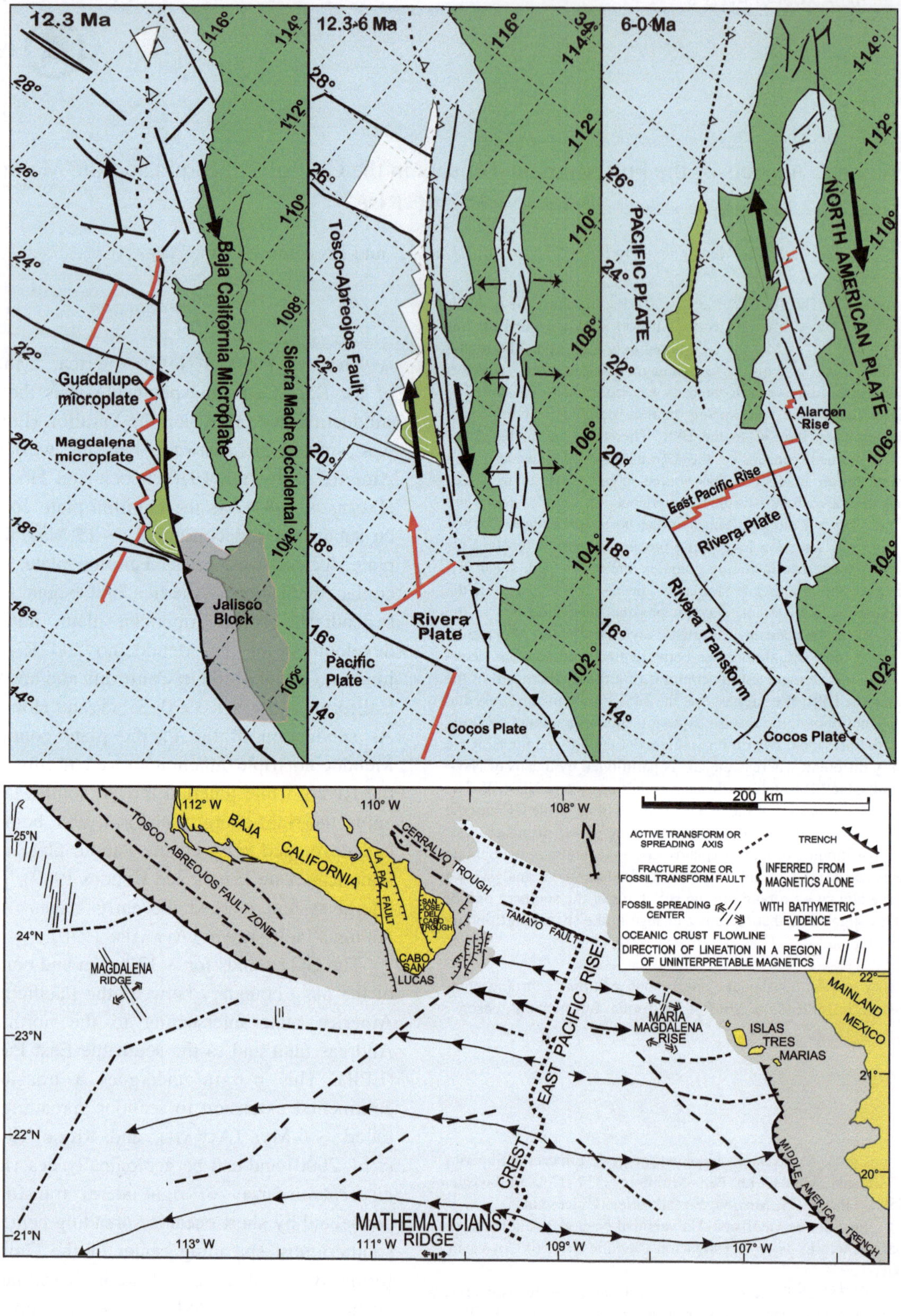

◀Figure 1
a Evolution of the Gulf of California (GC), where each *panel* represents different periods starting from 12.3 Ma and extending to 6.0 Ma. The last two stages of the *GC* are represented. **b** Current state of the southernmost part of the GC and the northern part of the East Pacific Rise (EPR). Figure modified from FLETCHER *et al.* (2007)

3–3.5 Myr (DeMETS 1995), and true seafloor spreading at the present observed rates started at 2.4 Ma (UMHOEFER *et al.* 2008). The EPR and Alarcón Rise both have magnetic anomalies on the oceanic crust (DeMETS 1995), while the Guaymas basin is a complex of sills and sediment layers that have an obscuring effect on the development of seafloor-spreading magnetic lineations (LONSDALE 1989; LIZARRALDE *et al.* 2007). The spreading centers of El Pescador, Farallon, and Carmen basins were formed around 2 Ma linking the Loreto basin (DORSEY and UMHOEFER 2000; MORTIMER *et al.* 2005) with the width of the low bathymetry basin surrounding the spreading centers. The initiation of seafloor spreading occurred approximately 12.5 Ma as an oblique-divergent plate boundary (STOCK and LEE 1994; ATWATER and STOCK 1998), while the time from onset of rifting to complete rupture of continental lithosphere in the southern gulf was only $\sim$6–10 Myr.

Deformation in this region is accommodated by both right-lateral strike-slip and normal-fault systems (e.g., ANGELIER *et al.* 1981; FLETCHER and MUNGUÍA 2000; UMHOEFER *et al.* 2002; PLATTNER *et al.* 2007). The boundaries of the GC are marked by long, right-lateral strike-slip faults separated by short spreading centers. The west side of the southern tip of the Baja California peninsula hosts an onshore to offshore fault array of roughly north-striking, left-stepping, mostly east-dipping normal faults known as the gulf-margin system (FLETCHER and MUNGUÍA 2000) (Fig. 1). At the southern part of the region, the EPR connects the GC through the Tamayo Fracture Zone Triple Junction (Fig. 2).

To the south of 25°N, the GC has a more recent tectonic history than the central and northern parts (Fig. 2). The onset of seafloor spreading in the southern part of the GC started between 6 and 2.4 Ma (LIZARRALDE *et al.* 2007; UMHOEFER *et al.* 2008) and took place with continental rifting (DeMETS 1995;

FLETCHER and MUNGUÍA 2000; PLATTNER *et al.* 2007; LIZARRALDE *et al.* 2007). This region, therefore, provides an opportunity to study a fault system development at a critical time during the rift-to-drift transition, when oceanic spreading has already begun and the continental rifting is still active. The spreading rate across the Alarcón Rise in the southern GC is $\sim$90 % that of the Pacific–North American plate boundary, indicating that a broad area of deformation takes up the deformation missing on the transform faults (DeMETS 1995; FLETCHER and MUNGUÍA 2000; PLATTNER *et al.* 2007).

Seismicity and geomorphic seafloor bathymetry (MUNGUÍA *et al.* 2006; FLETCHER and MUNGUÍA 2000) indicate that the array of normal faults rupturing the southern end of the Baja California peninsula is active and acts as a weak distributed shear zone that contributes to translating the Baja California peninsula blocks away from mainland Mexico (PLATTNER *et al.* 2007). The gulf margin is a zone of distributed extension as well as decreasing elevation and thinning crust (LIZARRALDE *et al.* 2007). Low shear velocities at depth, which are associated with mantle material rise, characterize most of the gulf (e.g., DI LUCCIO *et al.* 2014; PÉREZ-CAMPOS and CLAYTON 2013). Some authors have suggested a clockwise rotation of the maximum extension since the early Miocene (ANGELIER *et al.* 1981; MARTIN-BARAJAS *et al.* 2000), but this hypothesis lacks clear evidence and remains controversial. There are no Miocene faults exposed in the area, and the active normal faults in the peninsula have similar direction to the Gulf Extensional Province in this area (FLETCHER and MUNGUÍA 2000).

These normal faults control the topography of the southern tip of the Baja California peninsula and produce moderate-sized earthquakes (FLETCHER and MUNGUÍA 2000), whereas the gulf-margin fault array provides a relatively minor contribution to the overall plate divergence occurring in the region. The faults bound prominent Quaternary basins and offset Quaternary alluvial deposits (FLETCHER and MUNGUÍA 2000).

The seismicity of the GC basically consists of right-lateral strike-slip earthquakes located near transform faults and normal faulting that occur in fast-spreading centers (CASTRO *et al.* 2010; GOFF *et al.*

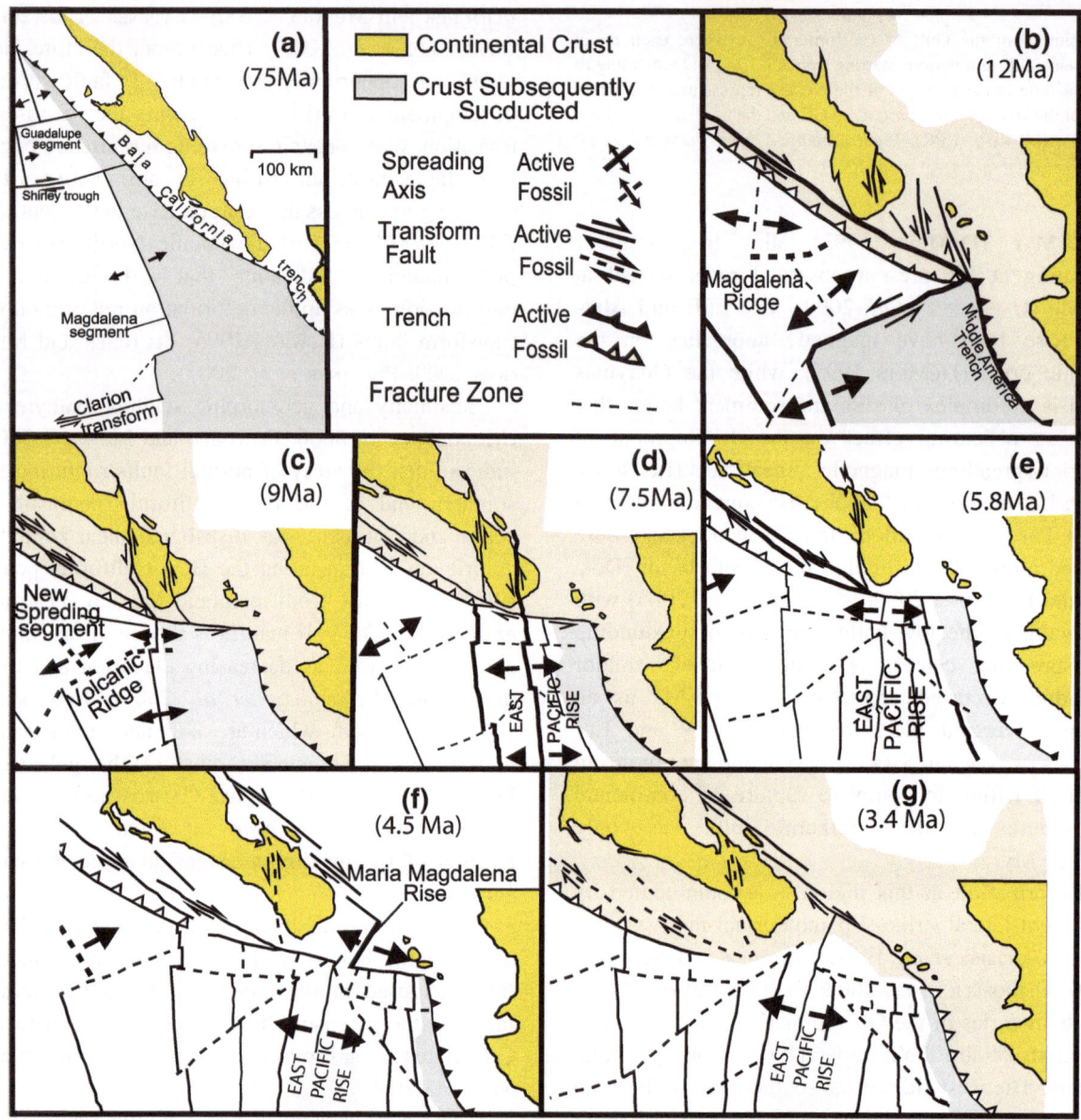

Figure 2
Tectonic history of the EPR. The formation of the GC is represented in **f** and **g** (figure modified from Lonsdale 1989)

1987). Previous studies have been done using tele-seismic data (e.g., SYKES 1968, 1970; MOLNAR 1973; GOFF *et al.* 1987) and local stations (e.g., LOMNITZ *et al.* 1970) and sonobuoys (e.g., THATCHER and BRUNE 1971; REID *et al.* 1973; REICHELE and REID 1977). Most of the seismicity distributes along the NW–SE direction, parallel to the Gulf axis, clustering in the main basins (Wagner, Delfin, Guaymas, Carmen, Farallon, Pescadero, and Alarcón) and spreading centers. In the north of the gulf, the earthquake locations scatter due to complexity in the fault system; in the southern locations, they are mostly confined to narrow zones near active faults. SUMY *et al.* (2013) found that much of the gulfs seismicity is accommodated via transform faults rather than the spreading centers, with a just a few moderate-magnitude earthquakes with extensional mechanisms consistent with observations along fast-spreading mid-ocean ridge systems (e.g., Fox *et al.* 2001; BOHNENSTIEHL *et al.* 2002; BOHNENSTIEHL and DZIAK 2009; DZIAK *et al.* 2009). Unusual focal mechanisms deviating away from the strike of the transform fault have been found by SUMY *et al.* (2013) and are observed in the intersection between ridge and transform faults (e.g., ENGELN *et al.* 1986; WOLFE *et al.* 1993). These anomalous events will be addressed with detail in this work.

2. Moment Tensor Analysis

In general, analyzing full moment tensors (FMT) from the seismic waveforms is challenging (HARA 1996). A FMT is a mathematical expression of six independent variables; however, on a routine basis, it is common practice to reduce the number of variables to five, assuming that the isotropic component is zero. This constraint is valid in most tectonic regimes where slip occurs entirely at the fault surface (e.g., subduction zones and strike-slip faults).

In this article, we present results in the GC and EPR using a FMT approach. In addition, the angle between the slip and the fault normal vector, referred to as angle θ, is obtained. Besides the angle θ, the proportion of volume change (k) and the constant volume (shear) component (T) are numerical indicators of the source rupture.

Moreover, ORTEGA *et al.* (2013) have been using these numbers as complexity indicators. Earthquakes are more complex as θ deviates from $\pi/2$ or as T and k deviate from zero. These parameters are obtained from the eigensolution of the FMT. In the GC, the earthquakes exhibit a clear isotropic component, and the constant volume parameter T is independent of scalar moments. The principal objective of this article is to find the first-order approximation of the normal surface of the FMT solution and to compare this with the deviatoric one.

There is enough evidence that many earthquakes rupture in a different way than a pure double couple dislocation. Some examples of FMT solutions are found in divergent margins, deep earthquakes, and volcanic fields at very different scales, ranging from negative magnitude (mining and human induced) to moderate to large earthquakes with tens of kilometers

in length (transform faults). In addition, the seismic instrumentation has increased and improved greatly, as have the modeling methods. There are numerous advantages of using FMT solutions over traditional moment tensors. The first, and more significant, is that it reflects the complexity of the source mechanism (ORTEGA *et al.* 2013). When studying the dislocation parameters of a single fault model, at least we need to know if there are no more faults involved. Another advantage of studying the FMT solutions is the possibility to obtain the elastic parameters (DUFUMIER and RIVERA 1997). Finally, probably the most challenging problem to be solved is to analyze the structural relationships in divergent or volcanic tectonic settings (MILLER *et al.* 1998).

The literature contains many examples of well-documented nondouble couple (NDC) earthquakes (AKI 1984; FOULGER and LONG 1984; JULIAN and SIPKIN 1985; DREGER *et al.* 2000; ORTEGA and QUINTANAR 2010) and laboratory observations (BRUNE *et al.* 1993) or changes in the material composition at the source (KNOPOFF and RANDALL 1970; VAVRYČUK 2004; BEN-ZION and AMPUERO 2009; SHI and BEN-ZION 2009). Recent studies have demonstrated that major earthquakes present NDC components, for example, (1) the 2000 Sumatra earthquake (ABERCROMBIE *et al.* 2003), (2) the 2010 Samoa Islands earthquake (LAY *et al.* 2010), and (3) the 2010 Cucapah–El Mayor, Mexicali, Mexico, earthquake (HAUKSSON *et al.* 2010). Some authors consider that the NDC component present in the FMT solutions is due to the errors induced by modeling a finite source as a point source (e.g., ADAMOVA and SILENY 2010).

We used some source parameters explained by HUDSON *et al.* (1989) that describe the moment tensor in terms of volume change, the constant volume parameter T, and the proportion of volume change parameter k (HUDSON *et al.* 1989). On the other hand, we used the angle between the resultant vector of slip and the fault system normal vector θ, $\hat{n}$, and $\hat{v}$.

3. Method

The definition of a moment tensor as a function of the isotropic fourth-order tensor c_{ijkl} and the vectors $n_k v_l$ represented in Fig. 3 is

(a)

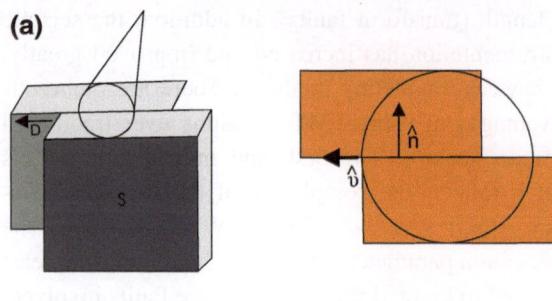

(b)

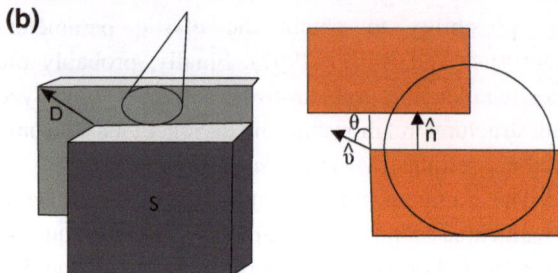

Table 1

Velocity model used in this study (from ORTEGA et al. 2013)

No	Depth (Km)	Velocity P (km/s)	Velocity S (km/s)	Density (g/cm^3)
0	2	4.00	2.28	2.30
1	5	6.00	3.42	2.40
2	8	6.40	3.64	2.67
3	55	7.50	4.20	3.10
4	22	7.52	4.21	3.20
5	23	7.60	4.25	3.20

$$\hat{n} = 1/2(\hat{s}_1 + \hat{s}_3) = 1/2((\hat{n} + \hat{v}) + (\hat{n} - \hat{v}))$$

$$\hat{v} = 1/2(\hat{s}_1 - \hat{s}_3) = 1/2((\hat{n} + \hat{v}) - (\hat{n} - \hat{v})). \quad (5)$$

The parameters θ, $\hat{n}$, and $\hat{v}$ are our unknowns, and they are useful for a tectonic interpretation of the geometry of the faults. We obtain them first from an inversion of the moment tensor, M_{ij}, as described in FORD *et al.* (2009); then we solve the characteristic equation; ($\hat{s}_n$ and e_n) and finally we obtain (4) and (5).

FMT inversions require a good velocity model (Table 1) to obtain robust results (KŘÍŽOVÁ *et al.* 2013). Furthermore, the difficulty in performing the moment tensor inversion of the six independent variables adds problems to the numerical stability of the solution. We used the variance reduction of the moment tensor inversion to represent the degree of confidence of the solution (FORD *et al.* 2009).

The characteristic equation of the moment tensor was performed via an iterative Jacobi method. The eigenvalue–eigenvector pairs correspond to Eq. (3). DUFUMIER and RIVERA (1997) noted that the elastic constant ratio λ/μ that is required to compute θ is better obtained by independent studies to avoid unrealistic rheological values. However, they represented a general model without physical constraints from the information of independent studies. In any case, nowadays, knowledge of the crustal structure is high, and we can estimate the elastic constants using different studies; in our case, we used the results of the CRUST2.0 project (BASSIN *et al.* 2000).

According to KAWAKATSU (1991), the moment tensor of transform systems may be the result of normal and strike-slip faults acting together. Moreover, ORTEGA *et al.* (2013) described a moment tensor model in which more than two faults with their

Figure 3
Representation of the **a** deviatoric model and **b** full moment tensor (FMT). The $\hat{n}$ and $\hat{v}$ directions and the scalar S (area) and D (slip) are represented

$$M_{ij} = c_{ijkl}n_k v_l SD$$
$$= [\lambda\delta_{ij}\delta_{kl} + \mu(\delta_{ik}\delta_{jl} + \delta_{il}\delta_{jk})]n_k v_l SD, \quad (1)$$

where n_k is the unit vector related to the fault and v_l is the unit vector related to the slip and where λ and μ are elastic constants. In the general case, v_l may not coincide with the fault plane. S is the scalar that defines the fault area, and D is the average slip.

Therefore, the FMT is (BEN-MENAHEN and SINGH 1981; AKI and RICHARDS 2002)

$$M_{ij} = SD[\delta_{ij}\lambda n_k v_k + \mu(v_i n_j + v_j n_i)]. \quad (2)$$

The characteristic solution is

$$\hat{s}_1 = \hat{n} + \hat{v}, \ e_1 = \lambda(\hat{n} \cdot \hat{v})SD + \mu(1 + (\hat{n} \cdot \hat{v})]SD,$$

$$\hat{s}_2 = \hat{n} \times \hat{v}, \ e_2 = \lambda(\hat{n} \cdot \hat{v})SD, \quad (3)$$

$$\hat{s}_3 = \hat{n} - \hat{v}, \ e_3 = \lambda(\hat{n} \cdot \hat{v})SD - \mu(1 - (\hat{n} \cdot \hat{v})]SD,$$

where $\hat{s}_n$ and e_n are the eigenvectors and eigenvalues, respectively.

Then, if $\hat{n} \cdot \hat{v} = \cos\theta$ (Fig. 3), it follows that

$$\theta = ar\cos[(\lambda/\mu)^{-1}(2e_2/(e_3 - e_1))]. \quad (4)$$

Therefore,

respective orientations represent a FMT. The moment tensor solution is the representation of the entire rupture process, and in some cases, the single dislocation model represented with strike, dip, and rake of the fault is not the best (Fig. 3). There are advantages in using the FMT model even in simple cases. If the solution is based on simple dislocation models, then the fault orientation and slip direction coincide with the dislocation parameters. In other words, in a pure shear double couple, $\hat{n}$ is directly related to the fault dislocation, and θ will be always perpendicular and pointing to the rake direction. If the model is complex, then the simple dislocation model cannot explain the data. One limitation shared by both full moment and simple dislocation models is the inversion itself. To obtain robust results, we need high-quality data. In some way, this limitation is virtually the same for both methods. But the regularization in the FMT has fewer constraints than the deviatoric case and in general is more difficult to be solved in the former case. This is the reason why the global moment tensor project solves the deviatoric case instead of the six independent variables, although the algorithm is capable of solving the FMT.

For clarity, when we refer to FMTs, we mean the solution of the tensor with the six free parameters of Mij. It is important to note that in this article, the NDC solutions are different to FMT solutions. However, in both cases, the double couple solution (DC) is always present. A NDC (deviatoric case) solution is composed of Compensated Linear Vector Dipole (CLVD) and DC, whereas a FMT comprises CLVD, explosion, and DC terms (isotropic case). What makes NDC and FMT "irregular" is that they cannot be represented by a single fault. NDC tensors are deviatoric and have all the properties of nonvolumetric change, including the fact that it is not possible to obtain the elastic parameters via moment tensor inversion (DUFUMIER and RIVERA 1997).

There are many ways of plotting a FMT with different types of parameters (CHAPMAN and LEANEY 2012; VAVRYČUK 2004; HUDSON et al. 1989). Probably the most popular is the source type plot (HUDSON et al. 1989), which is a simple representation of the source for general moment tensors that expresses the moment tensor as function of two basic parameters: constant volume (shear) component T and the proportion of volume change k (Fig. 4). This plot is capable of interpretation in terms of one or a combination of physical processes of failure resulting from an accumulated change in the stress field. The processes may take any form, such as shear or tensile fracture, explosion, or implosion.

On the basis of the general assumption that all the earthquakes have a maximum magnitude M, if we suppose that any moment tensor has the same probability of occurrence, we can plot in an equal area a source type plot. Some authors argue that the probability of occurrence of moment tensors is not uniform and that therefore the plot should not be restricted with a diamond shape that is symmetrical about the axis that corresponds to the intermediate eigenvalue $= 0$. At any rate, this plot is a good representation of the CLVD–DC–explosion components, and we used them as a graphic guide to illustrate if a moment tensor deviates from double couple.

In Fig. 4, all the possible eigenvalue combinations of FMTs in the $T - k$ space are obtained by fixing the eigenvectors and perturbing the eigenvalues to have different percentages of DC, CLVD, cracks, and explosions (Bowers and Hudson 1999), where

$$T = 2\frac{e'_1}{|e'_3|} \quad \text{and} \quad k = \frac{M_{\mathrm{ISO}}}{|M_{\mathrm{ISO}}| + |m'_3|}$$

and where e'_1, e'_2, and e'_3 are the deviatoric eigenvalues for the N, P, and T axes, respectively. They are defined as $e'_n = e_n - \frac{1}{3}\mathrm{trace}(M_{ij})$ In this case, the deviatoric eigenvalues are ordered in such a way that $|e'_1| \le |e'_2| \le |e'_3|$ and $M_{\mathrm{ISO}} = 1/3\mathrm{trace}(M_{ij})$. Then the value of θ is computed (Eq. 4), and we plot it in a source type plot.

4. Analysis

We analyzed 27 earthquakes that occurred in the GC and 8 earthquakes that were located in the northern part of the EPR (see Table 2). This catalog was extracted from the Global CMT database; waveform data were obtained from the National Seismic Service of Mexico (SSN) and the NARS-Baja array (TRAMPERT et al. 2003; CLAYTON et al. 2004). We calibrated all the SSN stations and prepared the response files in RESP format. All the

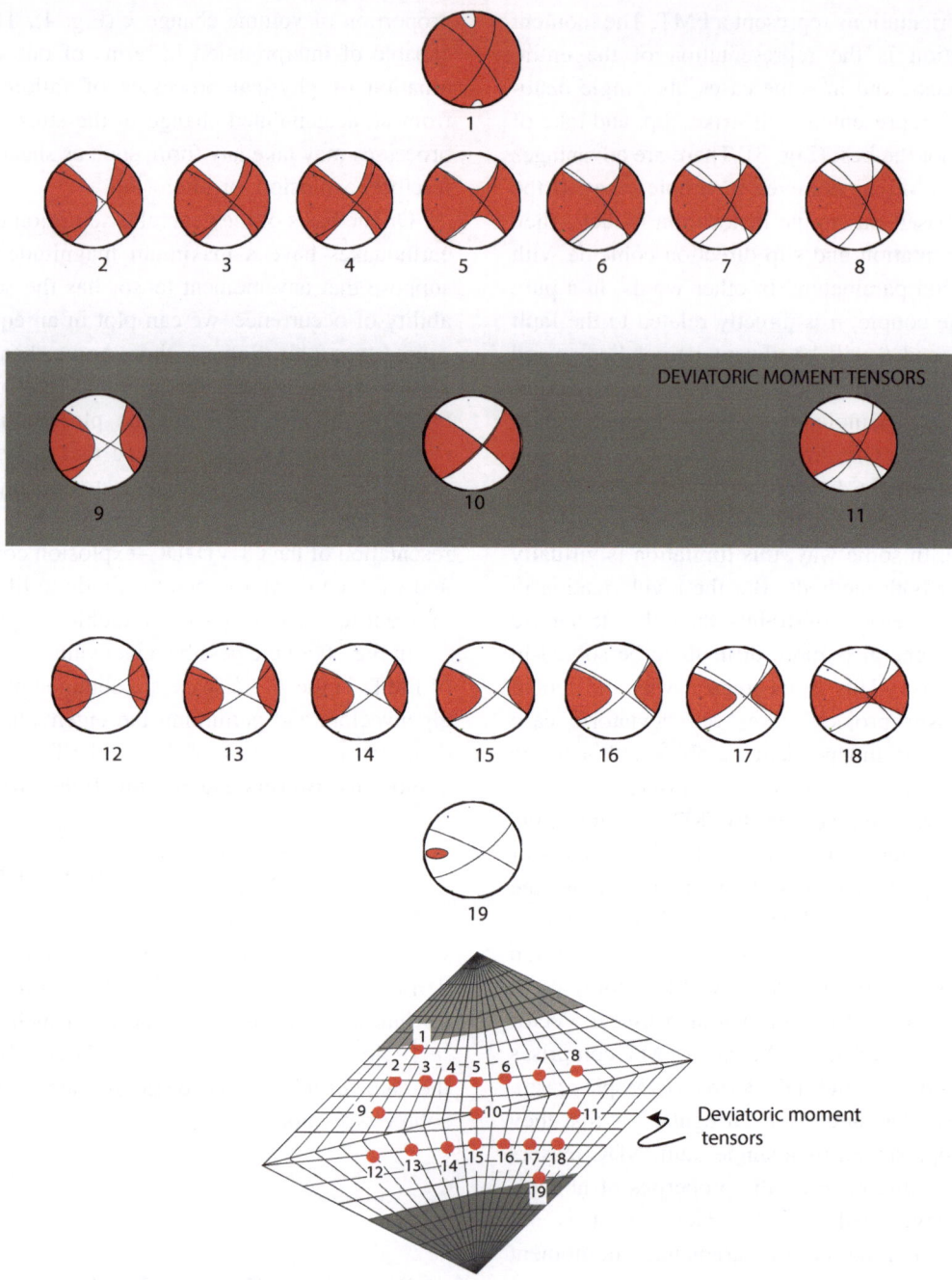

Figure 4
Focal mechanisms of FMTs in the source type plot (HUDSON *et al.* 1989). The central part is when $k = 0$ and corresponds to the deviatoric case

waveforms were visually inspected and preprocessed by applying a baseline correction and a cosine taper using SAC software, and the instrument response was removed. The synthetic seismograms were computed for 10 different depths ranging from 2 to 20 km using

the velocity model reported by ORTEGA *et al.* (2013). We used the scalar moment definition of Bowers to estimate the scalar moment (M_0). We filtered the synthetic and observed seismograms in the 25–50 s range and then computed all the parameters described

Table 2

List of events analyzed in this work

Ev. Id.	Date (dd/mm/yy)	Time (hh:mm:ss)	Latitude (°)	Longitude (°)	Depth (km)	mb	Ms	M_0 (N–m)	
1	03/10/02	16:08:30	23.32	−108.53	10.0	5.4	6.2	6.22e + 25	G
2	12/03/03	23:41:33	26.56	−110.59	10.0	5.5	6.5	3.90e + 25	G
3	12/11/03	04:54:56	28.97	−113.22	10.0	5.5	5.4	2.95e + 24	G
4	18/02/04	10:59:19	23.64	−108.82	10.0	5.4	5.6	6.66e + 24	G
5	24/09/04	14:43:11	28.57	−112.72	10.0	5.5	5.6	7.83e + 24	G
6	22/02/05	19:15:50	25.67	−109.97	10.0	5.1	–	2.36e + 24	G
7	05/06/05	08:28:51	23.67	−108.37	10.0	5.1	5.3	3.70e + 24	G
8	04/01/06	01:05:09	27.96	−112.14	10.0	5.0	5.1	9.29e + 23	G
9	04/01/06	08:32:32	28.16	−112.12	14.0	6.1	6.7	9.62e + 25	G
10	28/05/06	14:18:03	26.70	−111.23	10.0	4.7	–	6.60e + 23	G
11	06/12/06	14:09:59	27.59	−111.29	10.0	4.5	–	4.64e + 23	G
12	30/07/06	01:20:59	26.86	−111.21	10.0	5.3	5.9	8.93e + 24	G
13	23/02/07	04:16:15	25.97	−110.35	10.0	4.3	–	2.43e + 23	G
14	23/02/07	07:12:18	25.95	−110.52	10.0	4.6	–	7.19e + 23	G
15	23/02/07	07:30:12	25.99	−110.50	10.0	4.7	–	9.03e + 23	G
16	25/02/07	01:18:34	26.00	−110.46	10.0	4.7	–	6.75e + 23	G
17	25/02/07	15:00:41	26.15	−110.42	10.0	5.3	5.9	8.38e + 24	G
18	27/02/07	20:08:09	25.98	−110.46	10.0	4.8	4.6	4.21e + 23	G
19	13/03/07	02:59:04	26.26	−110.54	26.1	5.5	6.0	9.93e + 24	G
20	28/03/07	13:21:37	25.42	−109.71	10.0	5.0	5.1	6.95e + 23	G
21	28/03/07	14:28:55	25.50	−109.63	10.0	5.1	5.1	1.09e + 24	G
22	17/07/07	04:13:48	24.55	−109.09	10.0	4.8	–	1.23e + 24	G
23	01/09/07	19:14:23	24.91	−109.69	10.0	5.5	6.3	1.91e + 25	G
24	02/09/07	11:04:46	24.61	−109.82	10.0	4.9	–	7.69e + 23	G
25	05/09/07	03:07:37	24.35	−109.95	10.0	5.0	–	1.36e + 24	G
26	24/09/07	01:50:00	24.37	−109.91	10.0	4.7	4.8	3.23e + 23	G
27	05/05/08	16:28:06	24.80	−109.58	10.0	4.4	–	2.66e + 23	G
28	18/09/09	18:46:10	19.39	−108.43	10.0	5.2	5.6	3.49e + 24	E
29	20/07/09	10:44:55	19.20	−109.20	10.0	4.4	–	2.39e + 23	E
30	24/08/10	02:12:02	19.02	−107.17	10.0	5.5	6.1	2.34e + 25	E
31	03/05/12	04:59:54	21.21	−108.78	10.0	4.6	5.6	2.41e + 24	E
32	09/08/12	02:34:41	19.95	−109.33	10.0	5.1	5.6	4.11e + 24	E
33	01/10/12	17:32:32	18.61	−107.16	10.0	5.1	5.5	2.03e + 24	E
34	23/06/13	01:46:34	21.83	−108.00	10.0	4.8	4.7	5.52e + 23	E
35	07/12/13	21:11:15	19.46	−108.59	25.0	–	4.4	4.62e + 23	E

Extracted from Global CMT database. *Dashes* indicate that Ms was not reported by GCMT

in the previous section. It is worth noting that both the frequency band used and the stations analyzed (epicentral distance) are in agreement to the criteria suggested by FUKUYAMA and DREGER (2000). The FMT inversion (Table 3) was carried out using the methodology described by FORD et al. (2009) and ORTEGA et al. (2013). As a measure of fit quality, the variance reduction, VR, given by

$$VR = \left[1 - \sum_i \sqrt{(data_i - synth_i)^2} \Big/ \sqrt{data_i^2} \right] \times 100$$

where *data* and *synth* are the data and Green's function time series, respectively, was calculated for

each fit; the summation is performed for all stations and components. The VR is maximized when the error is low. The regularization matrix is based on a least squares approach that minimizes the observed and predicted waveforms, but it is not focused on depth phases or higher frequencies inversions. However, we only used moderate earthquakes, and not low magnitude events that may increase high uncertainty.

The graphical representation of these solutions is displayed in Fig. 5. In Table 2, we present the FMT parameters (θ, T, $k\hat{n}$, and $\hat{v}$) as described in the method section. The values of λ and μ were obtained using the Vp and Vs values from the CRUST2.0

Table 3

Full moment tensor solution

I.D.	Depth (km)	M_0 (N–m)	m11	m12	m13	m22	m23	m33	Reg
1	18	2.40E + 18	1.44E + 25	−9.44E + 24	−4.71E + 24	−1.37E + 25	−3.01E + 24	9.47E + 24	G
2	10	3.57E + 18	−2.66E + 25	1.16E + 25	−1.32E + 24	3.15E + 25	1.58E + 24	5.21E + 24	G
3	8	4.74E + 17	−1.93E + 24	−1.37E + 23	2.03E + 23	2.22E + 24	2.64E + 23	−4.05E + 24	G
4	6	8.97E + 17	−5.18E + 24	6.05E + 23	1.59E + 24	1.67E + 24	−1.92E + 24	−6.81E + 24	G
5	12	9.15E + 17	−6.09E + 24	4.19E + 24	−3.51E + 23	7.51E + 24	−1.38E + 24	1.40E + 24	G
6	16	4.64E + 17	−3.29E + 24	7.15E + 23	−4.22E + 23	2.15E + 24	−7.41E + 22	−2.44E + 24	G
7	8	4.94E + 17	−3.22E + 24	1.06E + 24	4.87E + 23	1.94E + 24	−2.53E + 23	−2.89E + 24	G
8	8	1.26E + 17	−4.13E + 23	−5.61E + 23	7.35E + 22	7.88E + 23	−2.01E + 23	−7.16E + 23	G
9	8	1.06E + 19	−7.48E + 25	2.90E + 25	−2.73E + 24	7.89E + 25	−1.23E + 25	3.52E + 25	G
10	18	8.38E + 16	−4.09E + 23	6.83E + 22	−1.77E + 23	7.97E + 23	1.18E + 23	1.68E + 23	G
11	6	4.06E + 16	−3.24E + 23	1.29E + 23	−1.34E + 22	3.13E + 23	−1.13E + 23	−4.40E + 22	G
12	10	7.88E + 17	−5.40E + 24	1.10E + 24	−9.99E + 23	7.64E + 24	6.84E + 23	1.09E + 24	G
13	10	1.58E + 16	−1.28E + 23	2.71E + 22	2.81E + 22	1.79E + 20	−2.64E + 22	−9.44E + 22	G
14	10	8.35E + 16	−3.80E + 23	5.75E + 22	1.30E + 23	1.72E + 23	−1.17E + 23	−7.60E + 23	G
15	10	1.06E + 17	−6.02E + 23	9.70E + 22	1.62E + 23	7.84E + 22	−1.20E + 23	−9.12E + 23	G
16	10	7.14E + 16	−3.65E + 23	1.54E + 23	1.15E + 23	−6.90E + 22	−5.39E + 22	−6.42E + 23	G
17	6	6.21E + 17	−4.05E + 24	1.40E + 24	2.00E + 24	4.98E + 24	−1.99E + 24	−1.86E + 24	G
18	10	4.28E + 16	−3.17E + 23	9.08E + 22	9.20E + 22	1.12E + 23	−7.23E + 22	−2.30E + 23	G
19	6	1.06E + 18	−6.70E + 24	−7.89E + 23	4.76E + 24	6.34E + 24	−2.51E + 24	−4.65E + 24	G
20	14	9.32E + 16	−3.73E + 23	2.08E + 23	1.82E + 23	7.07E + 23	−2.62E + 23	3.47E + 23	G
21	16	1.75E + 17	−2.05E + 23	2.89E + 23	2.02E + 23	1.36E + 24	−6.43E + 23	6.30E + 23	G
22	16	1.15E + 17	−1.01E + 24	−2.72E + 23	3.03E + 23	6.88E + 23	3.47E + 23	2.60E + 23	G
23	8	1.79E + 18	−9.22E + 24	6.88E + 24	−1.28E + 24	9.54E + 24	−2.80E + 24	−8.83E + 24	G
23	8	6.56E + 16	−5.27E + 23	1.91E + 23	1.33E + 23	4.02E + 23	1.32E + 23	−1.39E + 23	G
25	18	2.04E + 17	2.35E + 23	1.08E + 24	−1.96E + 22	−4.48E + 23	5.17E + 23	−1.25E + 24	G
26	10	3.30E + 16	−1.11E + 23	2.09E + 23	2.95E + 22	−9.63E + 21	−2.16E + 21	−1.42E + 23	E
27	10	3.88E + 16	−3.56E + 23	−1.15E + 23	8.05E + 21	6.15E + 22	6.55E + 22	−6.91E + 22	E
28	10	7.88E + 17	−5.40E + 24	1.10E + 24	−9.99E + 23	7.64E + 24	6.84E + 23	1.09E + 24	E
29	6	8.97E + 17	−5.18E + 24	6.05E + 23	1.59E + 24	1.67E + 24	−1.92E + 24	−6.81E + 24	E
30	10	6.21E + 17	−4.05E + 24	1.40E + 24	2.00E + 24	4.98E + 24	−1.99E + 24	−1.86E + 24	E
31	11	8.97E + 17	−5.18E + 24	6.05E + 23	1.59E + 24	1.67E + 24	−1.92E + 24	−6.81E + 24	E
32	10	7.14E + 16	−3.65E + 23	1.54E + 23	1.15E + 23	−6.90E + 22	−5.39E + 22	−6.42E + 23	E
33	8	1.06E + 19	−7.48E + 25	2.90E + 25	−2.73E + 24	7.89E + 25	−1.23E + 25	3.52E + 25	E
34	10	7.14E + 16	−3.65E + 23	1.54E + 23	1.15E + 23	−6.90E + 22	−5.39E + 22	−6.42E + 23	E
35	8	1.06E + 19	−7.48E + 25	2.90E + 25	−2.73E + 24	7.89E + 25	−1.23E + 25	3.52E + 25	E

Columns from *left* to *right*: event identifier, depth of the best solution in the inversion, scalar moment tensor, the six components of the moment tensor (m11, m12, m13, m22, m23, m33, respectively), geographical region of the event (*G* Gulf of California; *E* East Pacific Rise)

(BASSIN *et al.* 2000) project, and the depth was estimated by choosing the maximum variance reduction from systematic moment tensor inversions with depths ranging from 2 to 26 km at a 2 km step.

We emphasize that not all our solutions have the same reliability; despite we tried to analyze the most stations possible, considering its epicentral distance (band-frequency used) and azimuthal coverage, several facts should be considered: there are some differences in the quality of the data. While the GC has a dense network along the Gulf of California and the Northwestern part of Mexico, the EPR is far from the seismic network and only high-magnitude

earthquakes can be analyzed. In addition, the EPR has low seismicity rate and earthquake location is difficult to perform due to the remoteness of the region.

The seismicity patterns of both regions (EPR and GC) are different, but there are some differences in the quality of the data. While the GC has a dense network along the peninsula and the northwestern part of Mexico, the EPR is far from the seismic network, and only high-magnitude earthquakes can be analyzed. In addition, the EPR has a low seismicity rate, and earthquake locations are difficult to determine due to the remoteness of the region.

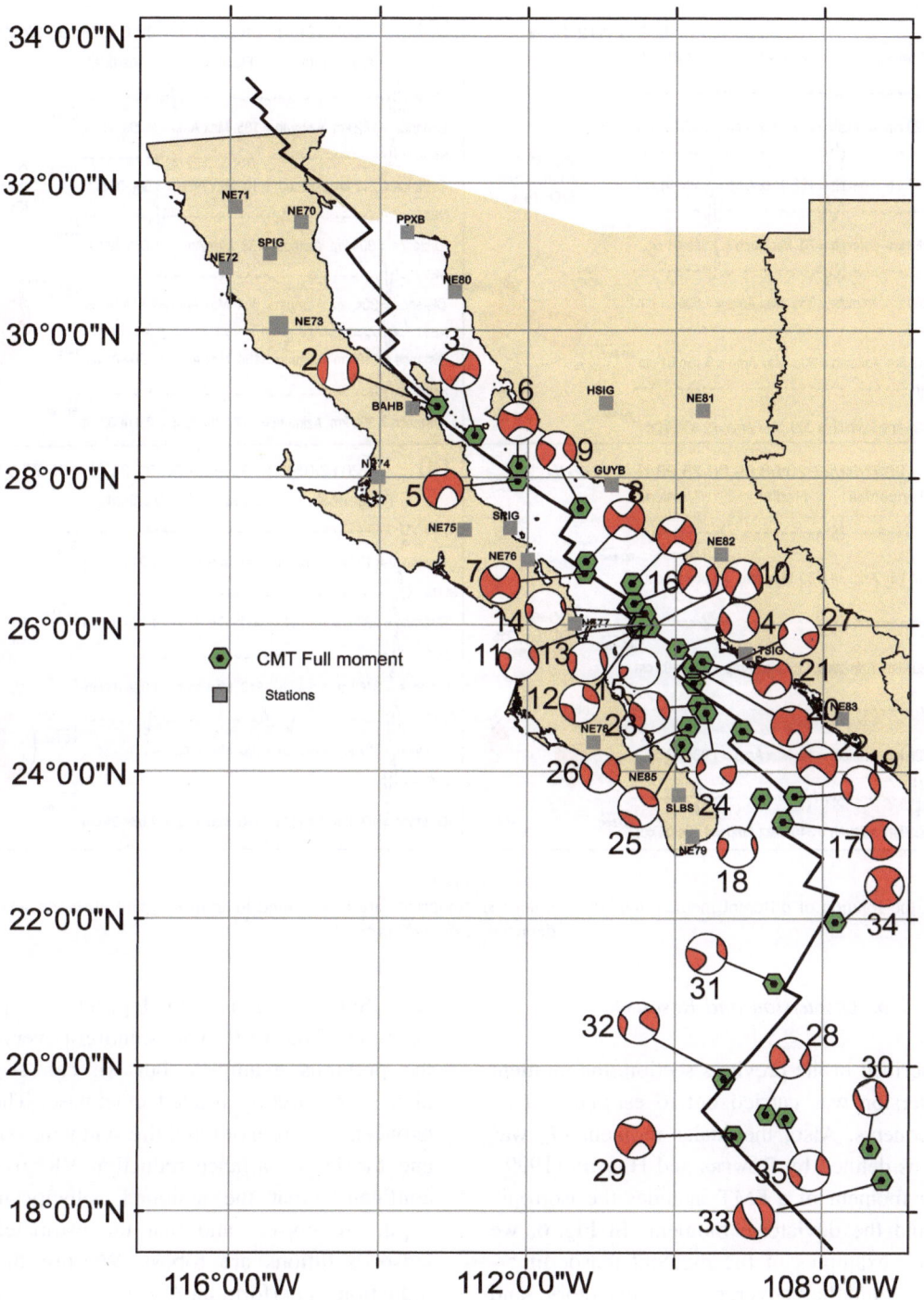

Figure 5
Results of the FMT analysis for all the earthquakes represented in Table 2

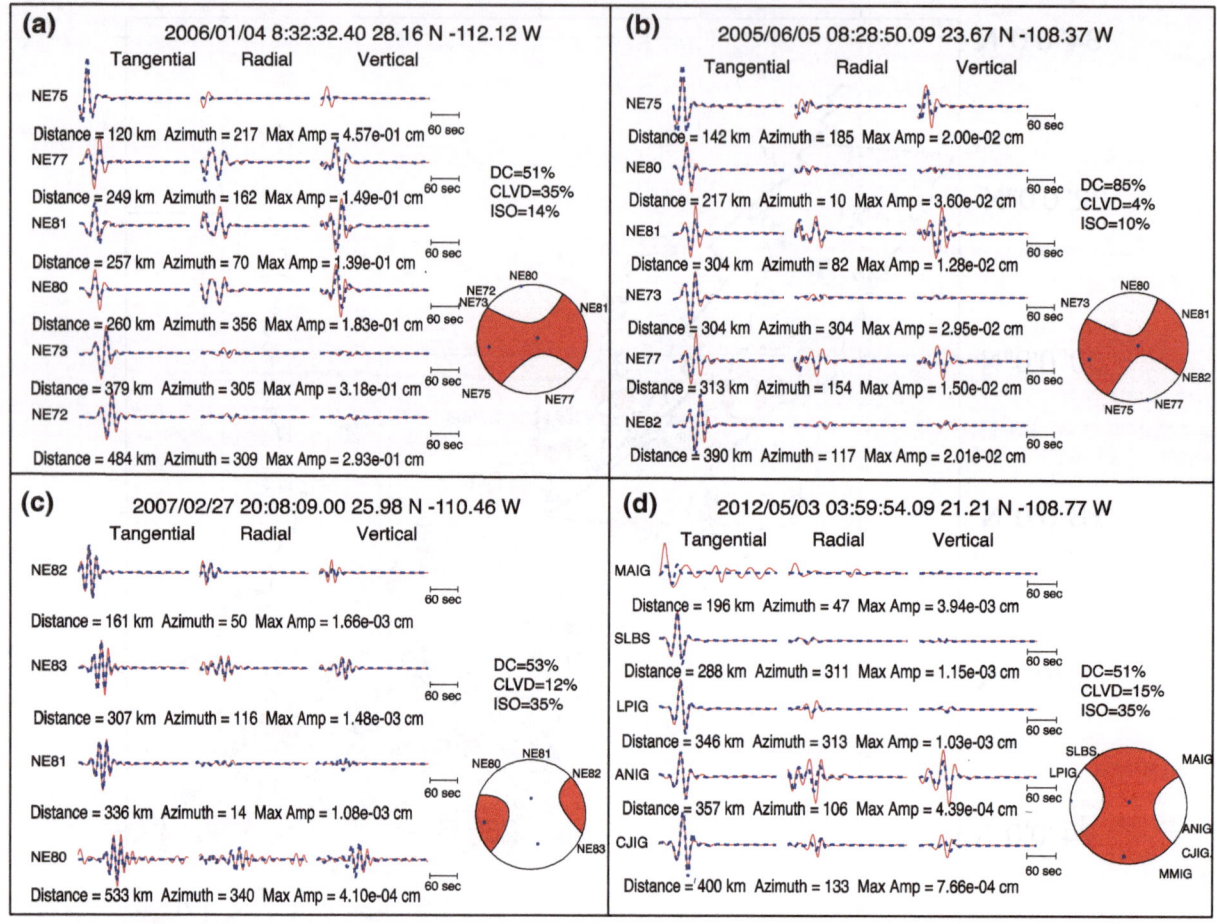

Figure 6
Examples of the solution of different moment tensors. Synthetic seismograms are represented in *solid red*, and observed seismograms are depicted in *dotted blue*

5. Discussion and Results

As described in the previous section, the moment tensor inversion was carried out to estimate all its free components. Also, the scalar moment M_0 was estimated as defined by BOWERS and HUDSON (1999). The scalar moment of a FMT includes the isotropic moment and the deviatoric moment. In Fig. 6, we present four examples of the moment tensor inversions. The azimuthal coverage is satisfactory, and the moment tensor estimation appears to be good. In Fig. 6a, we present a moment tensor that is closer to a double couple (DC = 85 %, CLVD = 4 %, ISO = 10 %); in contrast, in the examples presented in Fig. 6b, c, the double couple component is approximately 50 % and the contributions of CLVD

and ISO are higher. In Fig. 6d, we present an example of the EPR. This solution is very similar to the previous examples, but we can see that the planes are slightly rotated clockwise. The good fit between the observed and the synthetic seismograms and the large variance reduction VR give us good confidence that the assumed velocity model and depth are correct and that the estimated moment tensor solutions are robust. We use the variance reduction to perform an F test statistics (Fig. 7). The F test shows that only 10 events (1, 3, 6, 8, 18, 19, 21, 22, 23 and 26) improve the variance reduction using the full moment tensor. Because the probabilities for those cases are too small, improvement is considered significant if the p value is lower than 0.001 under a 0.05 level; the rest are not considered

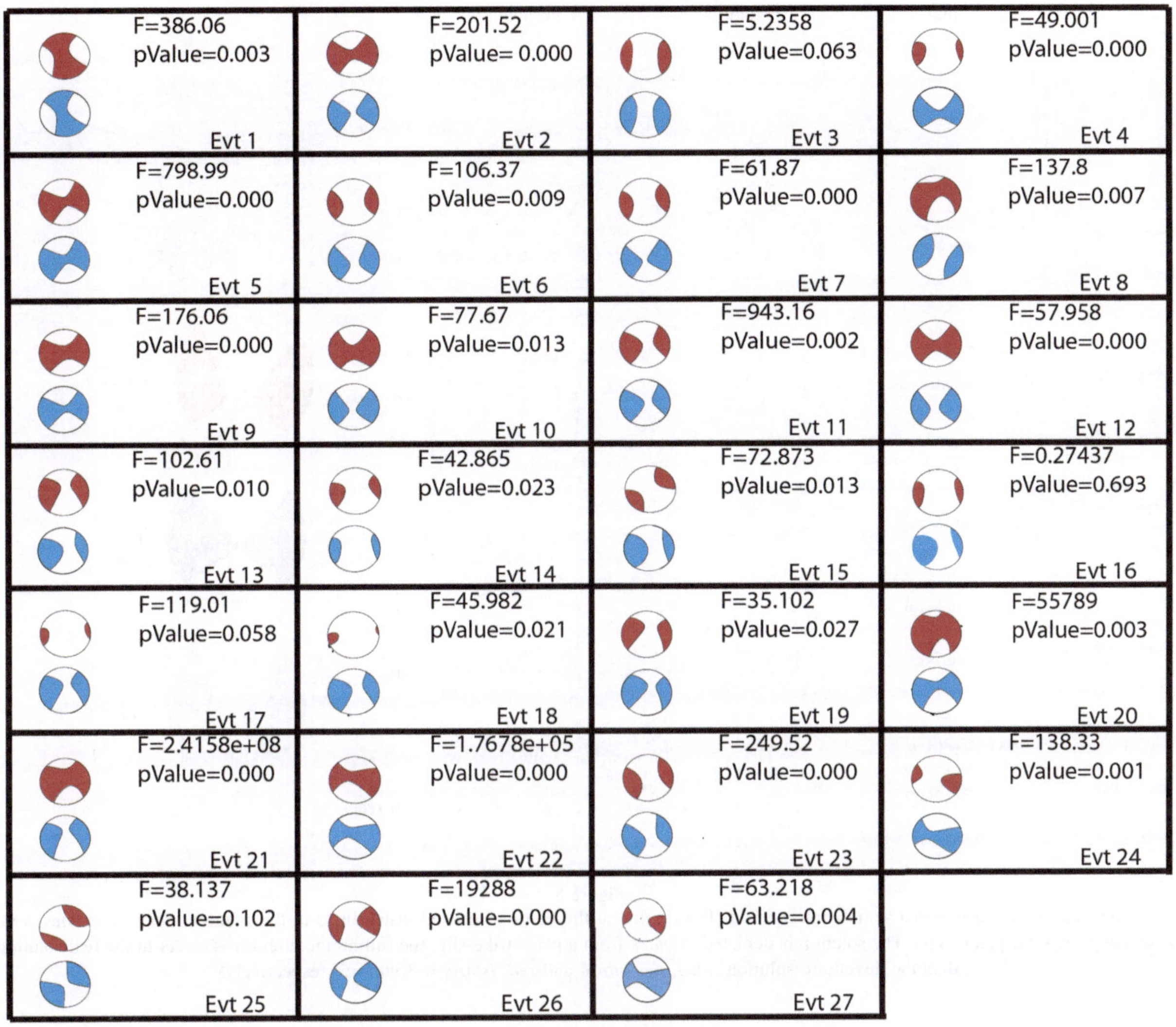

Figure 7

*F test statistics of variance reduction for the 27 events analyzed in the Gulf of California. The panels for each event show the full moment tensor (*red*) and deviatoric focal mechanisms along with the F value and probability (p) obtained in the statistics. In this analysis, the test shows that only 10 events (1, 3, 6, 8, 18, 19, 21, 22, 23 and 26) improve their VR using the full moment tensor (p < 0.001)*

to improve. Comparing the full moment tensor analysis with the deviatoric approach (Fig. 8) we observe that there is no substantial improvement of the solution. We computed synthetic seismograms with a single shear slip fault. The synthetic seismograms were calculated using a modal summation technique (HERRMANN 2013), whereas the inversion was carried out with the reflectivity method described by FORD *et al.* (2009). In this case, the error is due to numerical problems of the reflectivity method. However, for complex faults composed of

two shear slip mechanisms with different magnitudes, the synthetic seismograms show that the full moment tensor solution is closer to the strike and dip of the main even (Fig. 9), although both inversions have similar variance reductions. In Fig. 9, we compared both inversions using another synthetic example. In this case, the new synthetic was computed with two shear slip mechanisms in different directions. These directions are consistent with transform and normal faults in the Gulf of California. We can see that the full moment tensor is closer

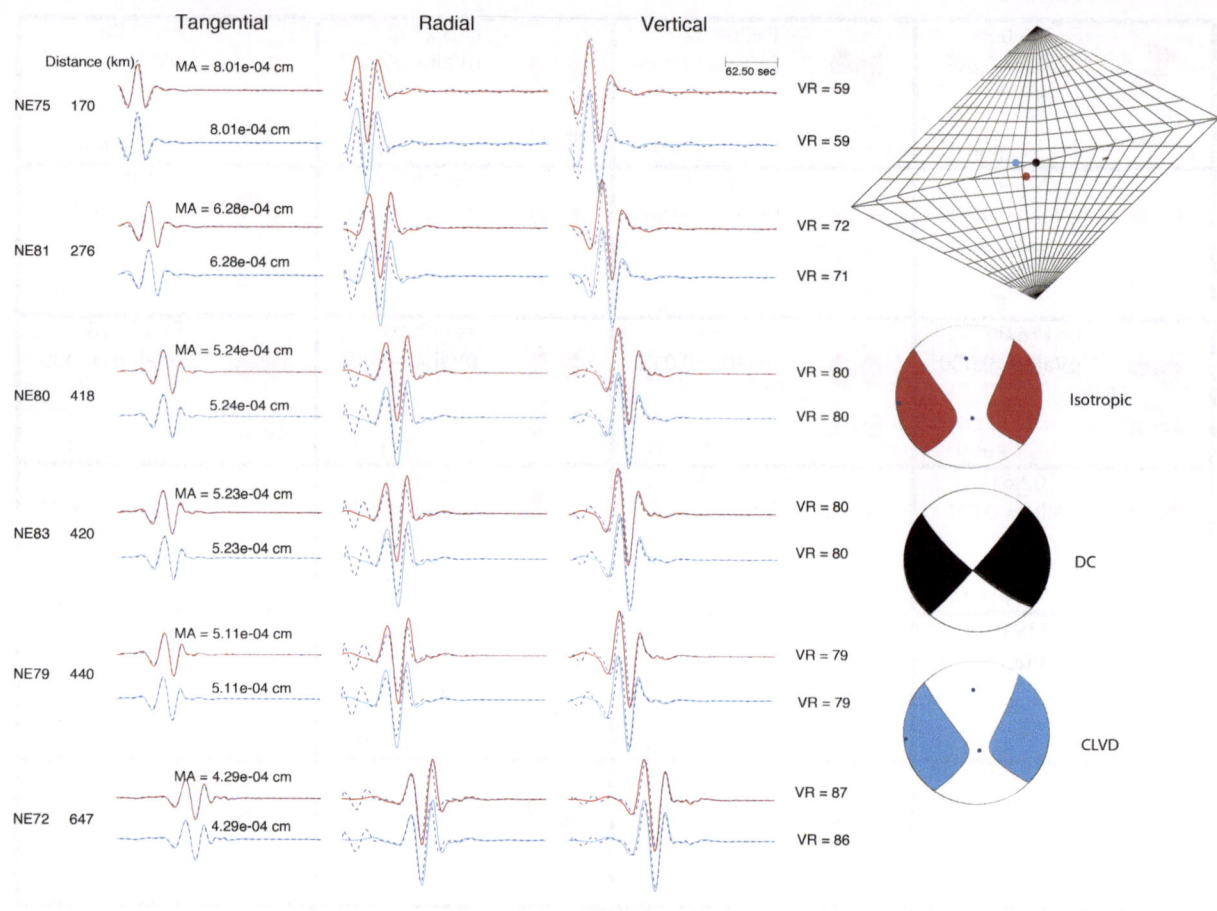

Figure 8
Synthetic example using a modal summation method. The ringing at the first part caused instabilities in the solution. Both inversions were done using the same parameters. The solution is depicted in *black* from a pure strike-slip, the full moment tensor is closer to the real solution than the deviatoric solution (*blue, deviatoric* and *red, isotropic solutions*, respectively)

to the origin (0, 0) in the source time plot, suggesting that this solution is best represented as double couple.

This model has been proposed as a possible rupture for many years (KAWAKATSU 1991); most recently, the Mayor-Cucapah earthquake in Baja California has been explained as a complex rupture (WEI *et al.* 2011). The reason is that estimating fault parameters of complex sources has six free variables and not five. In a single double couple, we need to estimate the Euler angles, namely, strike, dip, and rake, whereas in the complex source, we need six (three for each fault). Therefore, the numerical solution of five free variables assuming that the trace is zero will map in the deviatoric case. In the FMT, we need to estimate two directions $\hat{n}$ and $\hat{v}$; both are

independent and do not have to be orthogonal. The k factor that is supposed to account for a volumetric change is only an effective value that reflects the difference of mapping the space solution of two double couples into one double couple and a free slip. In this case, the free slip is acting as the numerical stability that allows the system to solve freely the six parameters. The isotropic component is part of this free slip model and is just mapping the solution of two double couples into two normal directions ($\hat{n}$ and $\hat{v}$).

The nodal planes of the moment tensor solutions of the GC and EPR are similar; however, we observe that the general trends of the focal mechanisms have minor differences. For example, in the region close to the city of La Paz, the earthquakes

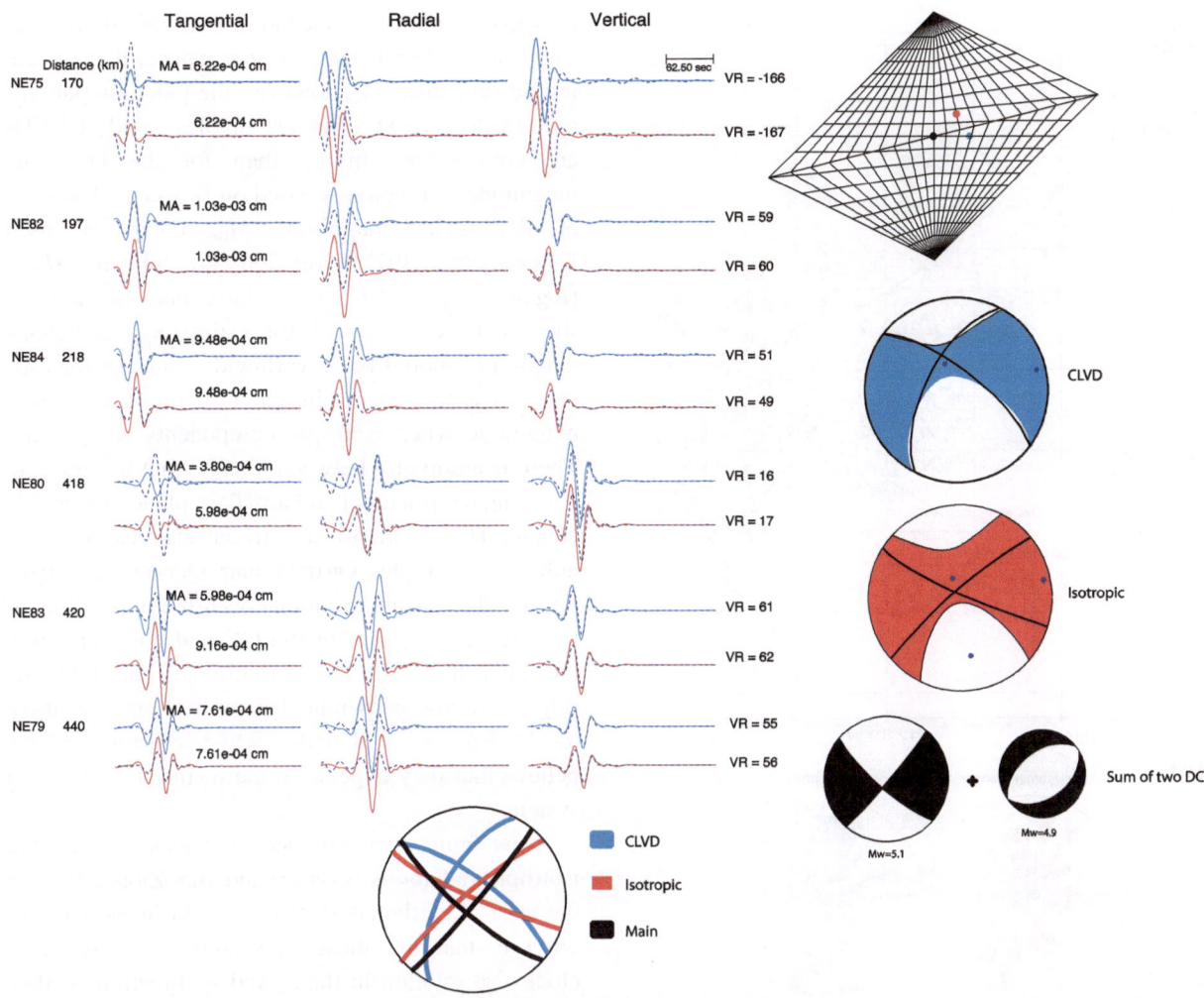

Figure 9

Synthetic example using a modal summation method of a complex source. The two sources are chosen to simulate a typical transform system in the GC; both solutions are depicted in *black*, and the size is proportional to the scalar moment. The FMT solution has a fault plane closer to the real solution if it is compared to the deviatoric moment tensor

of the Cerralvo fault show more isotropic components than those located in the principal transform system (Fig. 9). When we compare the solution of the classical deviatoric case with the FMT solution (Fig. 10), we observe that the general trends of compressional and extensional regions are similar, but in the case when an earthquake has a high CLVD, the FMT is best resolved with six free variables. For the GC, we represented all the normal plane surfaces of $\hat{n}$ and $\hat{v}$ (Table 4) in a rose diagram with steps of $10°$ and plotted them with the general trend of the Transform Fault System of the GC (Fig. 10b), and we also computed $\hat{n}$ and $\hat{v}$ for

the deviatoric case. We did not include the EPR mechanisms because some were rotated and the comparison would be inadequate. The trend was estimated using a least squares regression of the Cartesian coordinates of the Transform Fault System of the GC. In Fig. 10, we observe that the general trend of the fault system (dotted line) follows one of the principal directions of the rose diagram; on the other side, when we compare the trend of the GC with the deviatoric solution, the bin that should match with the population of faults does not coincide. In this article, we are interested only in studying the first-order characteristics of the fault

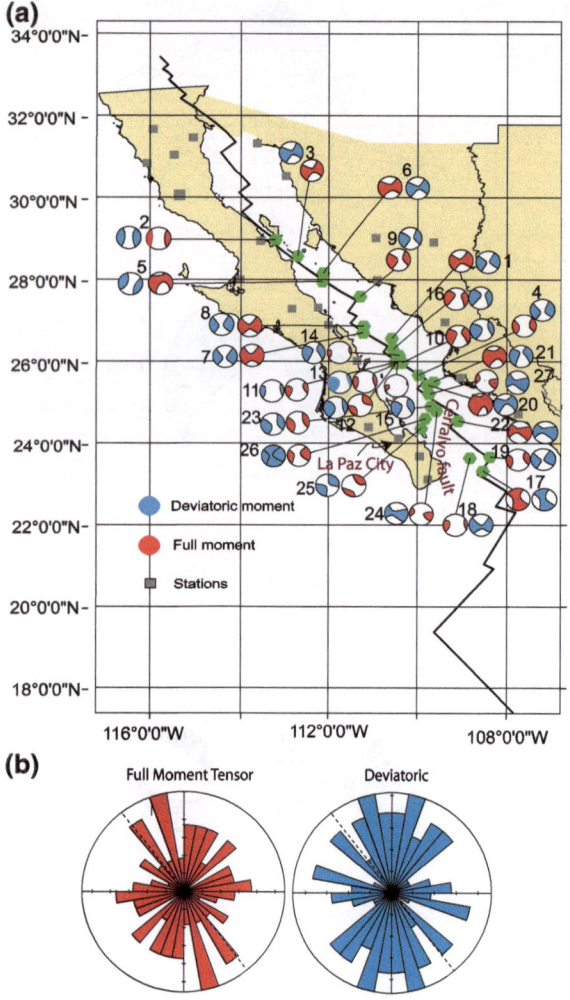

(a)

(b)

Full Moment Tensor   Deviatoric

Figure 10

(*top*) Comparison of the FMT and deviatoric solutions. The *red* solution was computed using the full moment inversion; at its side is depicted the deviatoric solution. (*bottom*) Rose diagram of the trend of all data of $\hat{n}$ and $\hat{v}$. The general trend of the transform fault is *plotted* with a *discontinued line*. Only the GC earthquakes are compared

trend and the angle of slip (for the six free variables) of the two planes (principal and auxiliary, for the DC), so we conclude that the FMT analysis is better than the deviatoric one.

The deviatoric DC case has a different interpretation if compared with the FMT. First, we find that the fault trend is bearing at slightly different angles; this is not surprising, because there is a trade-off between the slip angle and the trend when we solve for the double couple or for the FMT that includes the

isotropic component. Second, the magnitude is measured in a different way. At first glance, they should not be very different; however, the radiation patterns of tensile cracks, vector dipoles, and CLVDs are considerably higher than for the DC. The magnitude estimation is based on Bowers's definition of the scalar magnitude M_0. Later we used KANAMORI's (1977) relation to obtain $M_w = (\log M_0/1.5) - 10.73$. We believe that the standard shear-dislocation model for estimating earthquake magnitude should be a worldwide standard for estimating earthquake size instead of defining a different magnitude when isotropic components are present. There is ambiguity between $\hat{n}$ and $\hat{v}$ that is similar to choosing the principal and auxiliary plane of a double couple. This seems to be a trivial exercise, but it is not. For example, ORTEGA and QUINTANAR (2010) discussed a seismic event that occurred in one limb of the transform system of the GC and concluded that the active strike-slip fault was oriented parallel to the ridge and not perpendicular to it. Some of these mechanisms are still under active research, and we believe that they depend on the maturity of the fault system.

New faults may produce earthquakes with high isotropic radiations (CASTRO and BEN-ZION 2013). In the case of earthquakes that occurred in the GC, we observe that in those mechanisms, there is a clockwise rotation in the s_1 and s_3 directions if they are compared to the EPR solutions (Fig. 10). Some authors have argued that this change in direction occurred in the Miocene (ANGELIER *et al.* 1981; LONSDALE 1989), while others have argued that there is not enough evidence (FLETCHER and MUNGUÍA 2000). We believe that this is a spatial partition of fault kinematics in the entire region and it is not a temporal variation effect of the principal extension direction since the Miocene.

Our final conclusions are the following:

1. Earthquake moment tensors using a FMT analysis show that GC and EPR are similar and represent a homogeneous state of stress. However, there is a small rotation at the EPR.

2. Isotropic components are present in the region, but they are more evident in incipient faults. In those cases, FMT represents this type of source.

Table 4

Parameters obtained from the full moment tensor inversion

I.D.	Θ (°)	T	k	n_{trend} (°)	n_{plunge} (°)	v_{trend} (°)	v_{plunge} (°)	VR (%)	%ISO	%DC	%CLVD
1	64.1	0.51	0.14	275.3	56.8	251.6	−12.8	82.5	16	41	43
2	81.3	0.11	0.09	76.1	4.7	98.5	0.4	76.1	10	80	10
3	123.4	−0.38	−0.26	121.5	−2.6	249.7	0.6	73.5	28	45	27
4	138.7	−0.29	−0.38	43.6	−11	16	−12	81.6	40	43	17
5	82.7	0.05	0.1	120.2	−17	61.8	−1.6	81.5	10	85	4
6	136.1	−0.64	−0.26	188.2	−5.8	311.1	7.3	70.1	29	26	45
7	142.3	−0.67	−0.28	59.8	8.9	183.8	−10.6	80.6	32	23	45
8	130.4	−0.91	−0.09	316.5	9.4	40.3	−19.8	75.8	11	8	80
9	69.5	0.41	0.12	102.5	−15.5	77.3	−11.1	79.2	14	51	35
10	73.3	0.05	0.22	193.1	−6.4	313.3	−3.4	82.8	22	73	4
11	100.9	−0.29	−0.05	283.4	17.2	260.4	13	74.7	5	67	28
12	80.7	0.03	0.14	51.5	6.7	110.4	2.4	80	14	83	2
13	152.5	−0.13	−0.47	351.6	14.3	252.7	1.3	72.6	48	46	7
14	128	−0.07	−0.39	83.4	1.3	311.6	−6.5	75.9	39	57	4
15	151.9	−0.2	−0.45	57	−1.2	351.5	−8.3	68	46	43	11
16	165.1	−0.07	−0.5	73.7	17.3	336.3	−17.9	62.4	54	40	5
17	97.6	−0.17	−0.05	219.7	−4	29.4	6.5	88.3	5	78	16
18	125.5	−0.19	−0.34	296.7	7.6	308.5	9.6	79.4	35	53	12
19	99.2	0.03	−0.16	90.8	−17.1	170.2	−6.4	86.2	16	82	2
20	65.4	0.22	0.24	5.4	11.9	215.1	2.1	84.2	25	58	16
21	69.1	−0.31	0.34	114.4	−29.2	113.9	−29.1	75.8	36	44	20
22	79.9	0.39	−0.02	233.2	−14.5	341.9	−19.5	74.5	2	60	38
23	134.9	−0.85	−0.16	167.9	−18.8	0.1	17.1	81.4	19	12	68
24	105.1	−0.23	−0.13	52	18.8	152.6	−0.2	78.7	14	66	20
25	123.6	−0.45	−0.24	171.5	−47.9	265	37.8	54.2	26	41	33
26	125.3	−0.42	−0.26	272.9	−8.8	287.3	2.5	73	29	41	30
27	108.7	0.25	−0.31	333.3	−21.5	222.1	−0.4	65.3	33	51	17

3. We observe a clockwise rotation as we approach the southern faults of the GC and EPR, implying that there is a spatial partition of fault kinematics in the region.

4. The full moment tensor inversion resolves best the complex faults, composed mainly of two double couple; this kind of rupture is present in many tectonic settings worldwide, like transform systems.

Acknowledgments

This work was funded by CONACYT Mexico under grant 133910. We acknowledge the field assistance of Sergio Mayer and Alfredo Aguirre for maintaining the seismic network of Baja California. Also, we thank Raul Ochoa for helping to prepare the figures. The authors acknowledge the comments from two anonymous reviewers.

REFERENCES

ABERCROMBIE, R.E., ANTOLIK, M., and EKSTROM, G. (2003), *The June 2000 Mw 7.9 Earthquakes South of Sumatra: Deformation in the India–Australia Plate*, J. Geophys. Res. *108*(B1), 1–16. doi:10.1029/2001JB000674.

ADAMOVA, P., and SILENY, J. (2010), *Non-Double-Couple Earthquake Mechanism as an Artifact of the Point-Source Approach Applied to a Finite-Extent Focus*, Bull. Seismol. Soc. Am. *100*, 447–457. doi:10.1785/0120090097.

AKI, K. (1984), *Evidence for Magma Intrusion during the Mammoth Lakes Earthquakes of May 1980 and Implications of the Absence of Volcanic (Harmonic) Tremor*, J. Geophys. Res. 89(B9), 7689–7696.

AKI, K., and RICHARDS, P., *Quantitative Seismology* (University Science Books, Sausalito 2002).

ANGELIER, J., COLLETTA, B., CHOROWICZ, J., ORTLIEB, L., and RANGIN, C. (1981), *Fault Tectonics of the Baja California Peninsula and the Opening of the Sea of Cortez, Mexico*, J. Struct. Geol. *3*(4), 347–357. doi:10.1016/0191-8141(81)90035-3.

ATWATER, T. (1970), *Implications of Plate Tectonics for the Cenozoic Tectonic Evolution of Western North America*, Geol. Soc. Am. Bull. *81*, 3513–3536. doi:10.1130/0016-7606(1970)81 [3513:IOPTFT]2.0.

ATWATER, T., and STOCK, J. (1998), *Pacific–North America Plate Tectonics of the Neogene Southwestern United States: An update*, Int. Geol. Rev. *40*, 375–402.

AXEN, G.J., GROVE, M., STOCKLI, D., LOVERA, O.M., ROTHSTIEN, D.A., FLETCHER, J.M., FARLEY, K., and ABBOTT, P.L. (2000), *Thermal Evolution of Monte Blanco Dome: Low-Angle Normal Faulting during Gulf of California Rifting and Late Eocene Denudation of the Eastern Peninsular Ranges*, Tectonics *19*, 197–212.

BASSIN, C., LASKE, G., and MASTERS, G. (2000), *The Current Limits of Resolution for Surface Wave Tomography in North America*, Eos Trans. AGU *81*:F897

BEN-MENAHEN, A., and SINGH, S.J., *Seismic Waves and Sources* (Springer, New York 1981).

BEN-ZION, Y., and AMPUERO, J.P. (2009), *Seismic Radiation from Regions Sustaining Material Damage*, Geophys. J. Int. *178*, 1351–1356, doi:10.1111/j.1365-246X.2009.04285.x.

BOHNENSTIEHL, D.R., and DZIAK, R.P., *Mid-ocean Ridge Seismicity, In Encyclopedia of Ocean Sciences*, 2nd ed. (ed. STEELE, J.H., TUREKIAN, K.K., and THORPE, S.A.) (Elsevier, Boston 2009) pp. 837–851.

BOHNENSTIEHL, D.R., TOLSTOY, M., DZIAK, R.P., FOX, C.G., and SMITH, D.K. (2002), *Aftershock Sequences in the Mid-ocean Ridge Environment: An Analysis Using Hydroacoustic Data*, Tectonophysics *354*(1–2), 49–70. doi:10.1016/S0040-1951(02)00289-5.

BOWERS, D., and HUDSON, J.A. (1999), *Defining the Scalar Moment of a Seismic Source with a General Moment Tensor*, Bull. Seismol. Soc. Am. *89*(5), 1390–1394.

BRUNE, J.N., BROWN, N.S., and JOHNSON, P.A. (1993), *Rupture Mechanism and Interface Separation in Foam Rubber Models of Earthquakes: A Possible Solution to the Heat Flow Paradox and the Paradox of Large Overthrusts*, Tectonophysics *218*, 59–67.

CASTRO, R. R., A. PEREZ-VERTTI, I. MENDEZ, A. MENDOZA, and L. INZUNZA (2010). *Location of moderate-sized earthquakes recorded by the NARS-Baja Array in the Gulf of California region between 2002 and 2006*, Pure Appl. Geophys. *168*, no. 8–9, 1279–1292, doi:10.1007/s00024-010-0177-y.

CASTRO, R.R., and BEN-ZION, Y. (2013), *Potential Signatures of Damage-Related Radiation from Aftershocks of the 4 April 2010 (Mw 7.2) El Mayor–Cucapah Earthquake, Baja California*, Bull. Seismol. Soc. Am. *103*, 1130–1140. doi:10.1785/0120120163.

CHAPMAN, C.H., and LEANEY, W.S. (2012), *A New Moment Tensor Decomposition for Seismic Events in Anisotropic Media*, Geophys. J. Int. *188*, 343–370. doi:10.1111/j1365-246X.2011.05265.x

CLAYTON, R.W., TRAMPERT, J., REBOLLAR, C.J., RITSEMA, J., PERSAUD, P., PAULSSEN, H., PÉREZ-CAMPOS, X., van WETTUM, A., PÉREZ-VERTTI, A. and F. DI LUCCIO (2004). *THE NARS-BAJA ARRAY IN THE GULF OF CALIFORNIA RIFT ZONE*, Margins Newsl. *13*, 1–4.

DEMETS, C. (1995), *A Reappraisal of Seafloor Spreading Lineations in the Gulf of California: Implications for the Transfer of Baja California to the Pacific Plate and Estimates of Pacific–North America Motion*, Geophys. Res. Lett. *22*(24), 3545–3548. doi:10.1029/95GL03323.

DI LUCCIO, F., PERSAUD, P., and CLAYTON, R.W. (2014), *Seismic Structure beneath the Gulf of California: A Contribution from Group Velocity Measurements*, Geophys. J. Int. *199*(3), 1861–1877.

DORSEY, R.J., and UMHOEFER, P.J. (2000), *Tectonic and Eustatic Controls on Sequence Stratigraphy of the Pliocene Loreto Basin, Baja California Sur, Mexico*, Geol. Soc. Am. Bull. *112*, 177–199.

DREGER, D.S., TKALCIC, H., and JOHNSTON, P.A. (2000), *Dilatational Processes Accompanying Earthquakes in the Long Valley Caldera*, Science *288*, 122–125.

DUFUMIER, H., and RIVERA, L. (1997), *On the Resolution of the Isotropic Component in Moment Tensor Inversion*, Geophys. J. Int. *131*, 595–606. doi:10.1111/j.1365-246X.1997.tb06601.x.

DZIAK, R.P., BOHNENSTIEHL, D.R., MATSUMOTO, H., FOWLER, M.J., HAXEL, J.H., TOLSTOY, M., and WALDHAUSER, F. (2009), *January 2006 Seafloor-Spreading Event at 9°50'N, East Pacific Rise: Ridge Dike Intrusion and Transform Fault Interactions from Regional Hydroacoustic Data*, Geochem. Geophys. Geosyst. *10*. doi:10.1029/2009GC002388.

ENGELN, J.F., WIENS, D.A., and STEIN, S. (1986), *Mechanisms and Depths of Atlantic Transform Earthquakes*, J. Geophys. Res. *91*, 548–577.

FLETCHER, J., and MUNGUÍA, L. (2000), *Active Continental Rifting in Southern Baja California, Mexico: Implications for Plate Motion Partitioning and the Transition to Seafloor Spreading in the Gulf of California*, Tectonics *19*, 1107–1123.

FLETCHER, J.M., GROVE, M., KIMBROUGH, D., LOVERA, O., and GEHRELS, G.E. (2007), *Ridge-Trench Interactions and the Neogene Tectonic Evolution of the Magdalena Shelf and Southern Gulf of California: Insights from Detrital Zircon U-Pb Ages from the Magdalena Fan and Adjacent Areas*, Geol. Soc. Am. Bull. *119*, 1313–1336. doi:10.1130/B26067.1.

FORD, S.R., DREGER, D.S., and WALTER, W.R. (2009), *Identifying Isotropic Events Using a Regional Moment Tensor Inversion*, J. Geophys. Res. *114*, B01306, 1–12, doi:10.1029/2008JB005743.

FOULGER, G., and LONG, R.E. (1984), *Anomalous Focal Mechanisms: Tensile Crack Formation on an Accreting Plate Boundary*, Nature *310*, 43–45.

FOX, C.G., MATSUMOTO, H., and LAUC, T.-K.A. (2001), *Monitoring Pacific Ocean Seismicity from an Autonomous Hydrophone Array*, J. Geophys. Res. *106*(B3), 4183–4206. doi:10.1029/2000JB900404.

FUKUYAMA, E., and DREGER, D. (2000), *Performance Test of an Automated Moment Tensor Determination System for the Future "Tokai" Earthquake*, Earth Planets Space *52*, 383–392.

GOFF, J.A., BERGMAN, E.A., and SOLOMON, S.C. (1987), *Earthquake Source Mechanism and Transform Fault Tectonics in the Gulf of California*, J. Geophys. Res. *92*, 10485–10510.

HARA, T. (1996), *Determination of the Isotropic Component of Deep Focus Earthquakes by Inversion of Normal-Mode Data*, Geophys. J. Int. *127*(2), 515–528.

HAUKSSON, E., STOCK, J., HUTTON, K., YANG, W., VIDAL-VILLEGAS, J.A., and KANAMORI, H. (2010), *The 2010 Mw 7.2 El Mayor-Cucapah Earthquake Sequence, Baja California, Mexico and Southernmost California, USA: Active Seismotectonics along the Mexican Pacific Margin*, Pure Appl. Geophys. *168*(8–9), 1255–1277.

HAUSBACK, B.P., *Cenozoic Volcanic and Tectonic Evolution of Baja California Sur, Mexico*, PhD thesis (University of California, Berkeley 1984).

HERRMANN, R. B. (2013), *Computer programs in seismology: An evolving tool for instruction and research*, Seism. Res. Lettr. *84*, 1081-1088, doi:10.1785/022011009

HUDSON, J.A., PEARCE, R.G., and ROGERS, R.M. (1989), *Source Type Plot for Inversion of the Moment Tensor*, J. Geophys. Res. *94*(B1), 765–774.

JULIAN, B.R., and SIPKIN, S.A. (1985), *Earthquake Processes in the Long Valley Caldera Area, California*, J. Geophys. Res. *90*(B13), 11155–11169.

KANAMORI, H. (1977), *The Energy Release in Great Earthquakes*, J. Geophys. Res. *82*, 2981–2987. doi:10.1029/JB082i020p02981.

KAWAKATSU, H. (1991), *Enigma of Earthquakes at Ridge-Trans-form-Fault Plate Boundaries: Distribution of Non-double Couple Parameter of Harvard CMT Solutions*, Geophys. Res. Lett. *18*(6), 1103–1106.

KNOPOFF, L., and RANDALL, M.J. (1970), *The Compensated Linear Vector Dipole: A Possible Mechanism for Deep Earthquakes*, J. Geophys. Res. *75*, 4957–4963.

KŘÍŽOVÁ, D., ZAHRADNÍK, J., and KRATZI, A. (2013), *Resolvability of Isotropic Component in Regional Seismic Moment Tensor Inversion*, Bull. Seismol. Soc. Am. *103*, 2460–2473. doi:10.1785/0120120097.

LAY, T., AMMON, C.J., KANAMORI, H., RIVERA, L., KOPER, K.D., and HUTKO, A.R. (2010), *The 2009 Samoa-Tonga Great Earthquake Triggered Doublet*, Nature *466*, 964–968. doi:10.1038/nature09214.

LIZARRALDE, D., AXEN, G.J., BROWN, H.E., FLETCHER, J.M., GONZÁ-LEZ-FERNÁNDEZ, A., HARDING, A.J., HOLBROOK, W.S., KENT, G.M., PARAMO, P., SUTHERLAND, F., and UMHOEFER, P.J. (2007), *Variation in Styles of Rifting in the Gulf of California*, Nature *448*, 466–469. doi:10.1038/nature06035.

LOMNITZ, C., MOOSER, F., ALLEN, C.R., BRUNE, J.N., and THATCHER, W. (1970), *Seismicity and Tectonics of the Northern Gulf of California Region, Mexico: Preliminary Results*, Geof. Int. *10*, 37–48.

LONSDALE, P., (1989), *Geology and tectonic history of the Gulf of California, in Winterer, E.L., et al., eds., The eastern Pacific Ocean and Hawaii*: Boulder, Colorado, Geological Society of America, Geology of North America, v. N, p. 499–521.

MARTIN-BARAJAS, A., FLETCHER, J.M., LOPEZ-MARTINEZ, M., and MENDOZA BORNDA, R. (2000), *Waning Subduction and Arc Volcanism in Baja California: The San Luis Gonzaga Volcanic Field*, Tectonophysics *318*, 27–51.

MILLER, A.D., JULIAN, B.R., and FOULGER, G.R. (1998), *Three Dimensional Seismic Structure and Moment Tensors of Non-double Couple Earthquakes at the Hengill-Gresdallur Volcanic Complex, Iceland*, Geophys. J. Int. *133*, 309–325.

MOLNAR, P. (1973), *Fault Plane Solutions of Earthquakes and Direction of Motion in the Gulf of California and on the Rivera Fracture Zone*, Geol. Soc. Am. Bull. *84*, 1651–1658.

MORTIMER, E., GUPTA, S., and COWIE, P. (2005), *Clinoform Nucleation and Growth in Coarse-Grained Deltas, Loreto Basin, Baja California Sur, Mexico: A Response to Episodic Accelerations in Fault Displacement*, Basin Res. *17*, 337–359. doi:10.1111/j.1365-2117.2005.00273.x.

MUNGUÍA, L.M., GONZÁLEZ, M., MAYER, S., and AGUIRRE, A. (2006), *Seismicity and State of Stress in the La Paz–Los Cabos Region, Baja California Sur, Mexico*, Bull. Seismol. Soc. Am. *96*(2), 624–636. doi:10.1785/0120050114.

ORTEGA, R., and QUINTANAR, L. (2010), *Seismic Evidence of a Ridge-Parallel Strike Slip Fault off the Transform System in the Gulf of California*, Geophys. Res. Lett. *37*, L06301. doi:10.1029/2009GL042208.

ORTEGA, R., QUINTANAR, L., and RIVERA, L. (2013), *Full Moment Tensor Variations and Isotropic Characteristics of Earthquakes in the Gulf of California Transform Fault System*, Pure. Appl. Geophys. *171*, 2805–2817. doi:10.1007/s00024-013-0758-7.

PÉREZ-CAMPOS, X., and CLAYTON, R.W. (2013), *Evidence of Upper-Mantle Processes Related to Continental Rifting versus Oceanic Crust in the Gulf of California*, Geophys. J. Int. *194*(2), 952–960.

PLATTNER, C., MALSERVISI, R., DIXON, T.H., LaFEMINA, P., SELLA, G.F., FLETCHER, J., and SUAREZ-VIDAL, F. (2007), *New Constraints on Relative Motion between the Pacific Plate and Baja California Microplate (Mexico) from GPS Measurements*, Geophys. J. Int. *170*(3), 1373–1380. doi:10.1111/j.1365-246X.2007.03494.x.

REICHELE, M., and REID, I. (1977), *Detailed Study of Earthquake Swarms from the Gulf of California*, Bull. Seismol. Soc. Am. *67*, 159–171.

REID, I., REICHELE, M., BRUNE, J., and BRADNER, H. (1973), *Micro-earthquake Studies Using Sonobuoys: Preliminary Results from the Gulf of California*, J. R. Astron. Soc. *34*, 365–379.

SHI, Z., and BEN-ZION, Y. (2009), *Seismic Radiation from Tensile and Shear Point Dislocations between Similar and Dissimilar Solids*, Geophys. J. Int. *179*, 444–458.

STOCK, J.M., and HODGES, K.V. (1989), *Pre-Pliocene Extension around the Gulf of California and the Transfer of Baja California to the Pacific Plate*, Tectonics *8*(1), 99–115. doi:10.1029/TC008i001p00099.

STOCK, J.M., and LEE, J. (1994), *Do Microplates in Subduction Zones Leave a Geological Record?*, Tectonics *13*, 1472–1487.

SUMY, D.F., GAHERTY, J.B., WON-YOUNG, K., DIEHL, T., and COLLINS, J.A. (2013), *The Mechanisms of Earthquakes and Faulting in the Southern Gulf of California*, Bull. Seismol. Soc. Am. *103*, 487–506. doi:10.1785/0120120080.

SYKES, L.R., *Seismological Evidence for Transform Faults, Seafloor Spreading and Continental Drift, In The History of the Earth's Crust (ed. Phinney, X.)* (Princeton University Press, Princeton 1968) pp. 120–150.

SYKES, L. R., (1970), *Focal mechanism solutions for earthquakes along the world rift system*, Bull Seismol Soc Am *60*, 1749–1752

THATCHER, W., and BRUNE, J. (1971), *Seismic Study of an Oceanic Ridge Earthquake Swarm in the Gulf of California*, Geophys. J. R. Astron. Soc. *22*, 473–489.

TRAMPERT, J., PAULSSEN, H., vanWETTUM, A., RITSEMA, J., CLAYTON, R., CASTRO, R., REBOLLAR, C., and PEREZ-VERTTI, A. (2003), *New Array Monitors Seismic Activity Near the Gulf of California in Mexico*, Eos Trans. AGU *84*, 29–32.

UMHOEFER, P.J., MAYER, L., and DORSEY, R.J. (2002), *Evolution of the Margin of the Gulf of California Near Loreto, Baja California Peninsula, Mexico*, Geol. Soc. Am. Bull. *114*, 849–868. doi:10.1130/0016-7606(2002)114<0849:EOTMOT>2.0.CO;2.

UMHOEFER, P.J., SUTHERLAND, F., KENT, G., HARDING, A., LIZAR-RALDE, D., SCHWENNICKE, T., FLETCHER, J., HOLBROOK, W.S., and AXEN, G. (2008), *Synchronous Changes in Rift-Margin Basins and Initiation of the Alarcón Spreading Ridge and Related Transform Fault, Southwestern Gulf of California*, Geol. Soc. Am. Abstr. Prog. *40*(6), 151.

VAVRYČUK, V. (2004), *Inversion for Anisotropy from Non-Double-Couple Components of Moment Tensors*, J. Geophys. Res. *109*, B07306, doi:10.1029/2003JB002926.

WEI, S., FIELDING, E., LEPRINCE, S., SLADEN, A., AVOUAC, J.P., HELMBERGER, D., HAUKSSON, E., CHU, R., SIMONS, M., HUDNUT, K., HERRING, T., and BRIGGS, R. (2011), *Superficial Simplicity of the 2010 El Mayor–Cucapah Earthquake of Baja California in Mexico*, Nat. Geosci. *4*, 615–618. doi:10.1038/ngeo1213.

WOLFE, C., BERGMAN, E., and SOLOMON, S. (1993), *Oceanic Transform Earthquakes with Unusual Mechanisms or Locations: Relation to Fault Geometry and State of Stress in Adjacent Lithosphere*, J. Geophys. Res. *98*(B9), 16187–16211.

(Received June 30, 2015, revised November 11, 2015, accepted December 1, 2015, Published online December 22, 2015)

Pure Appl. Geophys. 173 (2016), 3615–3629
© 2015 Springer Basel
DOI 10.1007/s00024-015-1184-9

Pure and Applied Geophysics

The 2006 Bahía Asunción Earthquake Swarm: Seismic Evidence of Active Deformation Along the Western Margin of Baja California Sur, Mexico

Luis Munguía,[1] ⓘ Sergio Mayer,[2] Alfredo Aguirre,[2] Ignacio Méndez,[1] Mario González-Escobar,[1] and Manuel Luna[1]

Abstract—The study of the Bahía Asunción earthquake swarm is important for two reasons. First, the earthquakes are clear evidence of present activity along the zone of deformation on the Pacific margin of Baja California. The swarm, with earthquakes of magnitude M_w of up to 5.0, occurred on the coastline of the peninsula, showing that the Tosco-Abreojos zone of deformation is wider than previously thought. Second, the larger earthquakes in the swarm caused some damage and much concern in Bahía Asunción, a small town located in the zone of epicenters. We relocated the larger earthquakes with regional and/or local seismic data. Our results put the earthquake sources below the urban area of Bahía Asunción, at 40–50 km to the north of the teleseismically determined epicenters. In addition, these new locations are in the area of epicenters of many smaller events that were located with data from local temporary stations. This area trends in an E–W direction and has dimensions of approximately 15 km by 10 km. Most earthquakes had sources at depths that are between 4 and 9 km. A composite focal mechanism for the smaller earthquakes indicated right-lateral strike–slip motion and pure-normal faulting occurred during this swarm. Interestingly, the ANSS earthquake catalog of the United States Geological Survey (USGS) reported each one of these faulting styles for two large events of the swarm, with one of these earthquakes occurring 2 days before the other one. We associate the earthquake with strike–slip mechanism with the San Roque Fault, and the earthquake with the normal faulting style with the Asunción Fault. However, there is need of further study to verify this possible relation between the faults and the earthquakes. In addition, we recorded peak accelerations of up to 0.63g with an accelerometer installed in Bahía Asunción. At this site, an earthquake of M_w 4.9 produced those high values at a distance of 4.1 km. We also used the acceleration dataset from this site to estimate the linear response of sediments lying beneath the station. The resulting average amplification function has a fundamental resonance frequency of about 5 Hz and shows amplification factors of 2–4 for motions at frequencies in the range 2–8 Hz. A comparison of this amplification function with the response of soils to shaking from larger events shows that above 20 Hz the amplification of motion in the larger events decreases relative to the weak-motion response.

Key words: Pacific margin of Baja California Sur, earthquake swarm in Bahía Asunción, transtensional deformation, high ground accelerations from moderate magnitude earthquakes.

1. Introduction

Figure 1 shows a map of the seismicity of the northwest region of Mexico between N22° and N29°. This seismicity includes earthquakes with magnitudes equal to or larger than 4.0 that occurred in the period 1960–2014. We notice that although the higher density of epicenters is associated with the Gulf of California fault systems, which accommodates most of the relative motion between the Pacific and North American plates, a small number of earthquakes with magnitudes of up to ~5.5 occurred in or near the region of this study (see Fig. 1; Table 1).

In that period, other events of similar magnitudes occurred to the south and offshore Baja California Sur. Although transform motion along the Tosco-Abreojos fault zone supposedly lasted until the opening of the Gulf of California at 5–3.6 Ma (Spencer and Normark 1979), the occurrence of earthquakes along the western margin of Baja California Sur is evidence of continuing fault activity (Fletcher and Munguía 2000; Michaud *et al.* 2004, 2007, 2011). Further, it is believed that faults causing those earthquakes isolate Baja California as one microplate that is not yet rigidly coupled to the Pacific plate (e.g., Ortlieb 1991; DeMets and Dixon 1999; Fletcher and Munguía 2000). Michaud *et al.* (2004)

[1] División de Ciencias de la Tierra, Centro de Investigación Científica y de Educación Superior de Ensenada, B. C., Carretera Ensenada-Tijuana No. 3918, Zona Playitas, C. P. 22860 Ensenada, Baja California, Mexico. E-mail: lmunguia@cicese.mx
[2] Centro de Investigación Científica y de Educación Superior de Ensenada, B. C., Unidad La Paz, Baja California Sur, Miraflores 334, Fraccionamiento Bellavista, C. P. 23050 La Paz, Baja California Sur, Mexico.

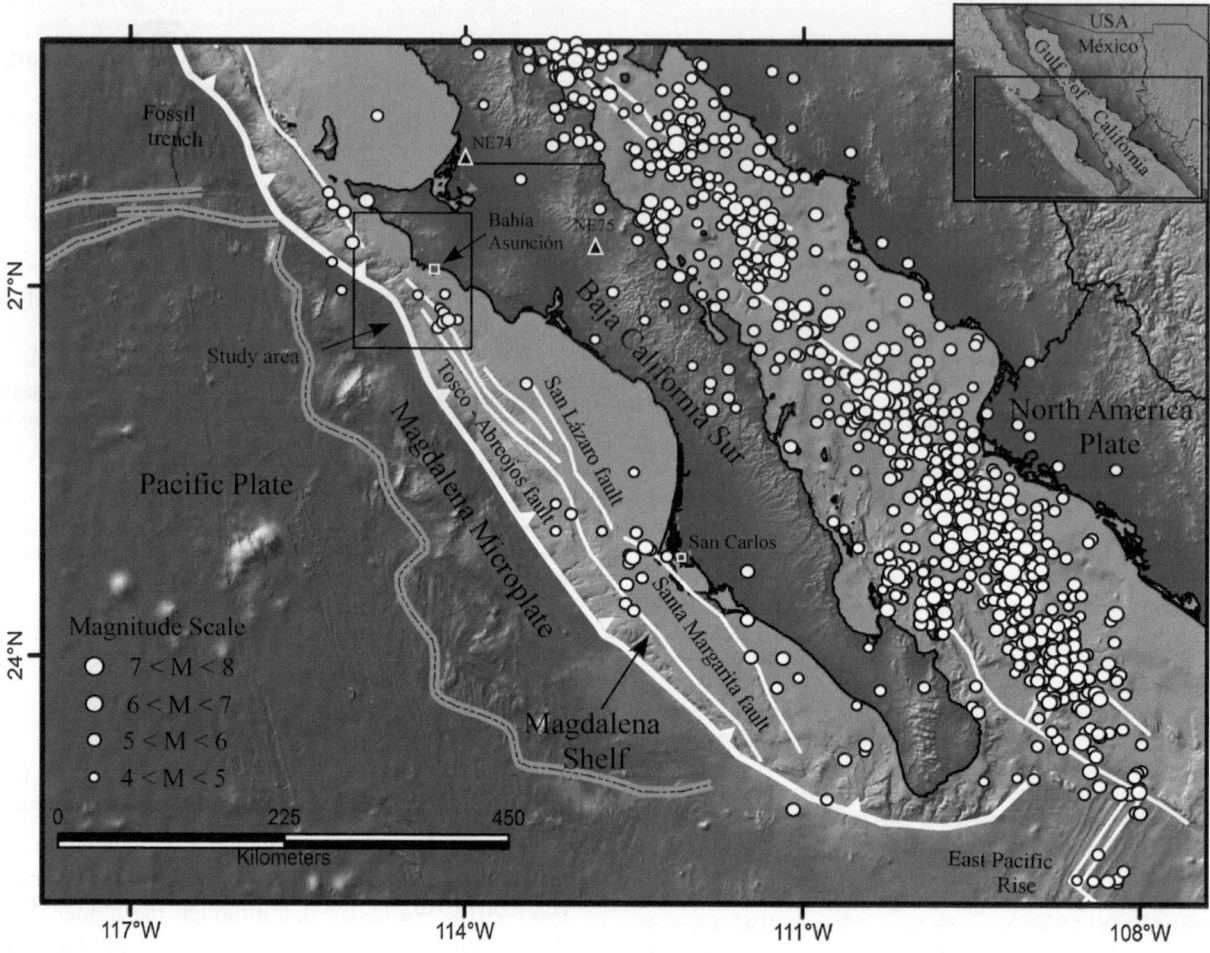

Figure 1
Seismicity map of Baja California Sur and the Gulf of California regions for the period 1960–2014. The figure also shows the major tectonic features outlined in *white* and the area of this study that includes the town of Bahía Asunción. The fault traces are from NORMARK *et al.* (1987), as taken from the work of FLETCHER *et al.* (2007). *Triangles* drawn to the north and east of the study area mark the sites of two closer NARS Baja stations that recorded the Bahía Asunción earthquakes. In the Fig. 2, we show more details of the area outlined by the *black rectangle* in this figure

estimated relative velocities between the Pacific plate and the Baja California block. Their results are compatible with the right-lateral transtensional motion that is indicated by focal mechanisms for earthquakes along the Tosco-Abreojos fault system (see also MICHAUD *et al.* 2011; among others).

The 2006 earthquake swarm that we study here occurred in the area of the town Bahía Asunción, at the northwestern margin of Baja California Sur. The shaking produced by the larger earthquakes was so intense that people evacuated the town on two occasions. Fortunately, there were no fatalities and only minor damage resulted from those earthquakes.

The North California Earthquake Data Center (NCEDC) catalog reported preliminary determinations of epicenters for the larger events in the swarm. The accuracy of those epicenters, however, was poor since most recording stations were located in the United States. Due to poor station coverage, the teleseismic locations provided only good insight into the region of the earthquakes, but specifying which fault produced them remained uncertain. In this study, we use regional and/or locally recorded data to relocate the larger events. Experience shows that when the station coverage for an earthquake is poor its preliminary location will have large errors. This

Table 1

Earthquakes with magnitude 4.0 or greater that occurred between 1963 and 2014 in the study region

No.	Year	Month	Day	Hour	Minute	Longitude	Latitude	Mag
1	1963	03	19	14	13	−115.20	27.20	4.1
2	1966	03	09	14	02	−114.90	27.70	5.5
3	1975	09	28	04	42	−113.45	26.22	4.4
4	1990	01	13	20	05	−115.09	27.60	5.0^a
5	1992	05	22	14	09	−115.02	27.36	5.1^a
6	1992	05	22	16	02	−115.12	26.98	4.0
7	2006	11	26	23	07	−114.25	26.66	4.3
8	2006	11	27	07	39	−114.22	26.98	4.1
9	2006	11	29	14	34	−114.16	26.73	5.0^a
10	2006	11	29	16	15	−114.20	26.79	4.5
11	2006	12	01	19	33	−114.20	26.72	4.9^a
12	2006	12	08	21	36	−114.44	26.94	4.1
13	2007	01	29	12	02	−114.23	26.82	4.1
14	2007	01	29	12	52	−114.20	26.95	4.0
15	2010	05	09	10	42	−115.19	27.67	4.8
16	2010	05	11	23	08	−115.23	27.76	4.5
17	2012	04	12	07	49	−113.51	27.88	4.4
Coordinates of relocated epicenters								
7	2006	11	26	23	07	−114.387	26.915	4.3
8	2006	11	27	07	39	−114.359	26.938	4.1
9	2006	11	29	14	34	−114.273	27.119	5.0^a
10	2006	11	29	16	15	−114.333	27.130	4.5
11	2006	12	01	19	33	−114.256	27.121	4.9^a
12	2006	12	08	21	36	−114.310	27.130	4.1

The epicenter coordinates are those taken from the NCEDC catalog. We relocated the earthquakes with numbers 7–12 in this study; their new epicenter coordinates are given in the last six rows of this table. Superscript letter in the last column of the table indicates moment magnitude, M_w. For all other events the magnitude is body wave magnitude, m_b

has occurred for several earthquakes of the region of Fig. 1 (REICHLE *et al.* 1976; MUNGUÍA *et al.* 1977; GOFF *et al.* 1987; FLETCHER and MUNGUÍA 2000; CASTRO *et al.* 2011). Thus, to find good correlations between the earthquakes and their causative faults, the use of near-source recorded data is mandatory.

In this study, we analyze the near-source strong-motion and weak-motion data collected in the area of the Bahía Asunción earthquake swarm with ten portable seismic stations. Although we missed four of the initial larger events of the swarm, the deployed stations still recorded many smaller events and one M_w 4.9 earthquake. We present and discuss our results in terms of epicenter determinations for the larger and smaller events, fault-plane solutions for two of the larger events and a comparison of these solutions with a composite focal mechanism for the smaller earthquakes. In addition, we use the strong-motion data recorded at one site located in downtown Bahía Asunción to investigate the response of the soils beneath the station.

To our knowledge, this is the first study of earthquakes that occurred, and were locally recorded, in the Bahía Asunción area. Thus, our results should contribute to the understanding of both the regional deformation models and the potential hazard from future earthquakes in this locality.

2. Seismotectonic Background

From the 1960–2014 seismicity map of the Baja California Sur and the Gulf of California regions, which includes only earthquakes with magnitudes that are greater than or equal to 4.0, it is evident that most seismicity of the western part of Mexico occurs in association with the Gulf of California–Salton

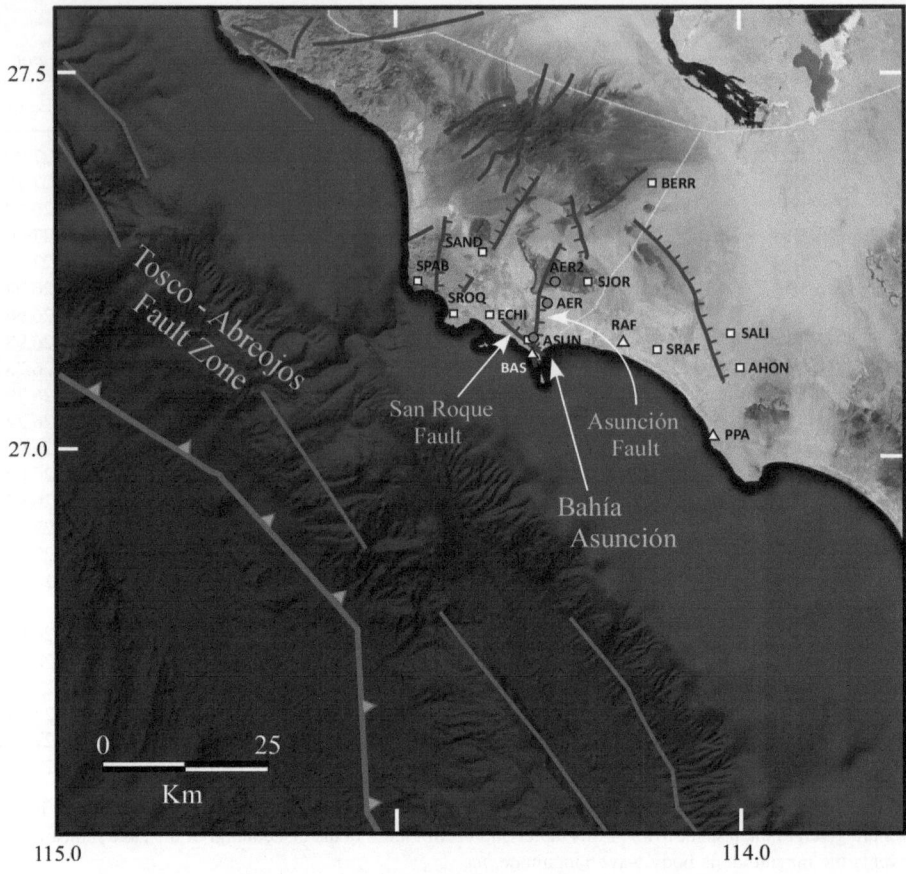

Figure 2
Detailed map of the study area (*black rectangle* in Fig. 1) showing some faults of the Tosco-Abreojos fault system (*gray lines*) and the distribution of temporary stations used to record the Bahía Asunción earthquake swarm. The faults located on land are from Carta Geológico-Minera de Baja California Sur (Servicio Geológico Mexicano). Instruments sites: *triangles* mark the K2 accelerometers; *gray circles* mark the Guralp seismometers; *squares* mark the smoked-paper seismographs

Trough fault system. This active system constitutes an important segment of the main boundary between the Pacific and North America plates.

The major tectonic features that cut through or are located close to the study area (black rectangle in Fig. 1) are a fossil trench, which is a remnant of the subduction of the Farallon plate, and the Tosco-Abreojos and San Lázaro-Santa Margarita fault systems. These fault systems trend approximately N–NW, dip to east and control the position and geometry of asymmetric transtensional basins (e.g., NORMARK et al. 1987; SPENCER and NORMARK 1979; MICHAUD et al. 2005; FLETCHER et al. 2007; BROTHERS et al. 2012; SALAZAR 2014).

The earthquake swarm studied here occurred 250 km west of the main plate boundary, on the northwestern margin of Baja California Sur, Mexico.

According to NCEDC catalog, 17 earthquakes in the 4.0- to 5.5-magnitude range occurred in or near the area of study between 1963 and 2012 (Table 1). MICHAUD et al. (2005) and MICHAUD et al. (2011) show the focal mechanisms for the earthquakes in Table 1 and for other events occurring on the western margin. The events in Table 1 that occurred during November 2006 and January 2007 are part of the earthquake swarm that we will analyze here. Although the earthquakes occurring along the Pacific margin of Baja California are low in number, they are still seismological evidence of stress transfer from the Gulf of California fault system to the west coast of Baja. Based on this, some researchers suggested that the capture of Baja California by the Pacific plate is still an ongoing process (NORMARK et al. 1987; DEMETS 1995; FLETCHER and MUNGUÍA 2000;

MUNGUÍA *et al.* 2006; MICHAUD *et al.* 2005; FLETCHER *et al.* 2007; among many others).

Figure 2 is a zoom of the study area showing the regional offshore geologic faults. The fault traces in this figure are from NORMARK *et al.* (1987), as taken from the work of FLETCHER *et al.* (2007). The fault traces drawn on land are from Carta Geológico-Minera de Baja California Sur, published by Servicio Geológico Mexicano. In our study area, the faults of most interest are the San Roque and Asunción Faults, features that trend in NW–SE and N–S directions, respectively. Both faults cut across the town of Bahía Asunción, and as we will show later, the earthquake swarm investigated here was associated with one or perhaps the two of these faults.

3. Post-earthquake Field Observations

The larger events of the swarm produced motions of strong intensity at Bahía Asunción. Some town residents reported that a little before and during the stronger earthquakes, a roaring noise was always present. However, post-earthquake field surveys revealed only slight damage to most structures in the town (Fig. 3). Small ground fissures across the landscape, cracking in plaster covering masonry, and books and other objects thrown from shelves are examples of the earthquake effects in town. There was also evidence of toppling of electric power line poles and water tanks during the more intense shaking. According to the testimony of a few residents, outside of 40 km from downtown the ground motions were not felt anymore. This was an indication of proximity of the earthquake sources to Bahía Asunción.

Another indication of the close location of the earthquakes to town is the phenomenon of soil liquefaction induced in a zone known as El Salitral. This place is located 1.5 km northeast of downtown Bahía Asunción and ~400 m from the ocean. The unconsolidated and saturated shallow material (~3 m thickness) at this site consists of fine sediments, clays, and medium-grade sandy material. We will show later that in spite of their moderate sizes, the

Figure 3

Photographs showing four different evidences of damage and earthquake effects. **a** Small ground fissures observed across the landscape (a pen was included for scale); **b** cracking in plaster covering masonry; **c** evidence of slight toppling of industrial tanks during the more intense shaking; **d** sand boils resulting from the soil liquefaction induced in El Salitral, 1.5 km east of downtown Bahía Asunción (a 12-in. *ruler* is shown for scale)

earthquakes produced high peak ground accelerations at short distances from the sources. Even though the duration of the more intense ground shaking in the earthquakes was too short (about 1 s), the water pressure reached the required level to produce liquefaction-induced features like the sand boils in Fig. 3. The earthquake-induced soil liquefaction and surface ground features observed emphasize the need to characterize the soil response of the epicentral area.

4. Seismic Instruments and Data

Figure 2 shows the distribution of temporary stations deployed in the neighborhood of Bahía Asunción. Two stations were equipped with K2 strong-motion accelerographs that had internal FBA-23 sensors, both from Kinemetrics. These devices were 2gs full-scale instruments that recorded free-field accelerations at a rate of 200 samples per second. The remaining stations were equipped with 1-s period vertical seismometers (SS-1, from Kinemetrics) connected to smoked-paper recorders (MEQ from Sprengnether). Throughout the entire recording period, we removed some stations and reinstalled them at new sites to improve the azimuthal coverage. The network operated from November 27, 2006 to March 24, 2007, and recorded many small events. However, most of the earthquakes recorded occurred during the month of December 2006 (see Fig. 4).

5. Data Analysis and Results

5.1. Epicenter Locations

The earthquakes studied here were located using a crustal velocity model composed of four flat homogeneous layers over a half space. This model had the following speeds and layer thicknesses from top to bottom: 4.0 km/s, 2 km; 6.0 km/s, 5 km; 6.4 km/s,

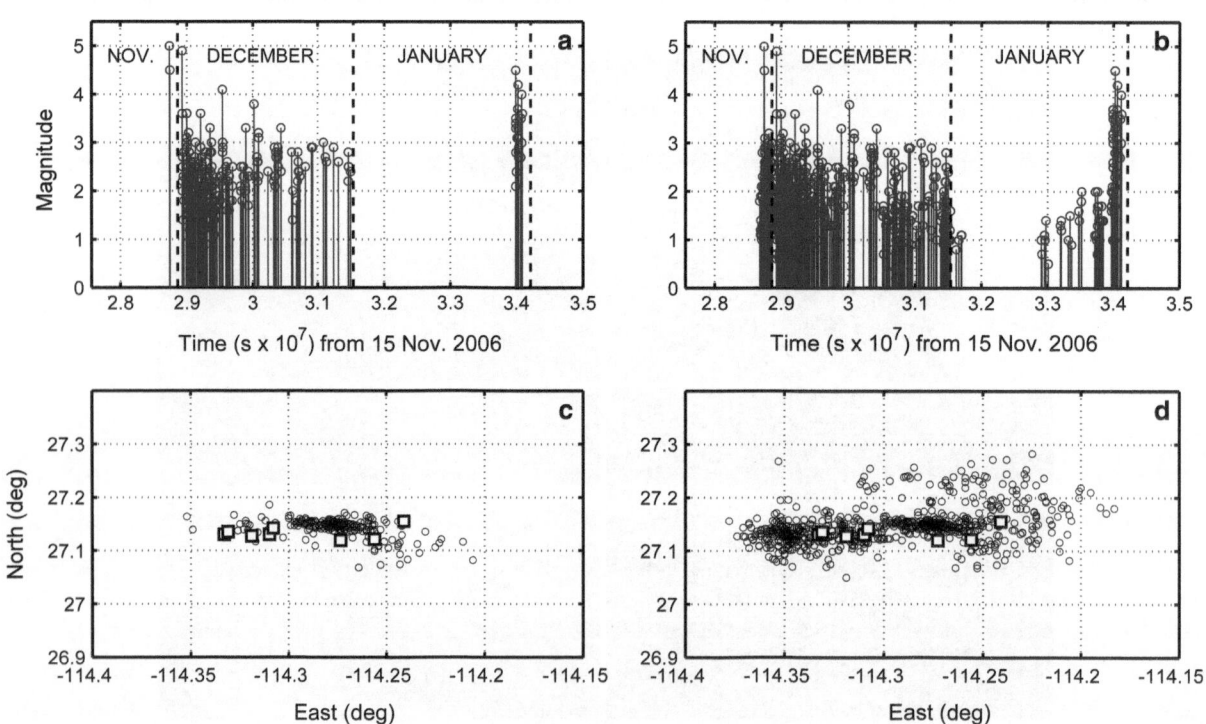

Figure 4

Temporal distribution of well-located earthquakes in the Bahía Asunción swarm (**a**) and a map with the correspondent epicenters (**c**). Parts **b**, **d** show the same time distribution of events in **a** and **b** plus an additional number of events for which the epicenter locations were less reliable. The *squares* in the maps of this figure mark the epicenters of earthquakes with magnitudes 4.0 or larger that occurred in November–December 2006 and at the end of January 2007

7 km; and 6.9 km/s, 10 km. The lower infinite half space had a P-wave velocity of 7.6 km/s (FLETCHER and MUNGUÍA 2000). We used this model in combination with the program Hypocenter (LIENERT and HAVSKOV 1995), a program that is part of the Seisan software package (HAVSKOV and OTTEMÖLLER 2001).

5.2. Relocation of the Larger Events of the Swarm

The epicenters of six stronger earthquakes of the swarm were teleseismically determined and reported in the NCEDC catalog (events with numbers 7–12 in Table 1). These NCEDC epicenters gave us insight into the region of the swarm. However, experience tells us that earthquakes of the region that are located only with data from seismic stations in North America (US) are usually mislocated by much as 50 km (REICHLE *et al.* 1976; MUNGUÍA *et al.* 1977; GOFF *et al.* 1987; FLETCHER and MUNGUÍA 2000; CASTRO *et al.* 2011; MUNGUÍA *et al.* 2015). Because of this, we took advantage of data recorded by some regional stations of the NARS Baja Array (CLAYTON *et al.* 2004) to improve the teleseismically determined epicenters. At the time of occurrence of the earthquakes under study, the NARS stations covered the Gulf of California region. A shorter distance range and better azimuthal coverage provided by this array allowed us to obtain improved locations for some of the larger earthquakes.

On the above basis, we relocated the epicenters of six stronger earthquakes of the swarm (events with numbers 7–12 in Table 1). The first two earthquakes (events 7 and 8 in the Table 1 and in the Fig. 5), however, occurred before the installation of temporary stations in the region. To relocate these earthquakes, we used only the P-wave data provided by four NARS Baja stations located at distances of 100–550 km from the sources. Since no local data were available for these particular events, and the gap in the coverage of the NARS stations was big (∼295°), their re-located epicenters are not reliable. Therefore, we do not base any conclusion on the location of these particular events. Regarding the other four events, their epicenters were located with data from local stations and one or more NARS stations. The estimated focal depths for those events

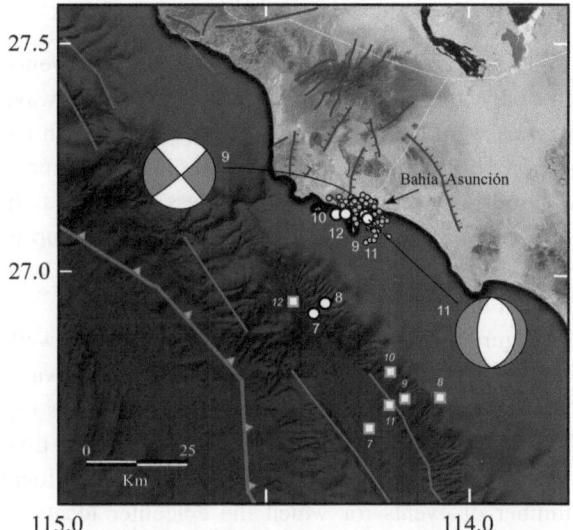

Figure 5

Map showing the faults in the study area (*gray lines*), the epicenters of the larger magnitude events determined teleseismically (*white squares and smaller numbers*), the relocated larger magnitude events (*white filled larger circles and larger-sized numbers*), and 256 epicenters determined for smaller earthquakes of the swarm (*smaller gray circles*). The focal mechanisms for the M_w 5.0 and 4.9 earthquakes of the swarm (events 9 and 11 in Table 1) are shown. These mechanisms are from the ANSS catalog of USGS

were of about 5–7 km, with average location errors of up to 3 km.

In the map of Fig. 5, we used white squares and small numbers to indicate the NCDEC epicenters of the six larger events of the swarm. In the same map, the relocated epicenters are marked with white circles and larger numbers. One to eight local stations provided data for the relocation of the epicenters. In particular, the epicenter with the number 11 was determined with data from eight local stations plus data from two closer NARS stations. We also located the earthquake 12 with data from six local stations and from data of one nearby station of the NARS Baja network. As we will show in the next section, these epicenter relocations are highly consistent with the distribution of epicenters of a large number of smaller events that we obtained with data from our local stations. This fact confirms that the earthquakes originated just below downtown Bahía Asunción, and not to the south in the sea, as the NCEDC preliminary epicenters indicated.

5.3. *Hypocenter Location of the Smaller Events*

The local network recorded many small events, but only those events with clear P- and S-wave arrivals were located. The earthquakes for which the station coverage was good had associated root mean squared location errors (rms) of less than 0.3 s. In such cases, the spatial location errors were of up to 3 km in the horizontal and vertical directions, with median errors of 1.3 km in both directions.

Figure 4 shows the temporal distribution of well-located earthquakes in the Bahía Asunción swarm (a) and a map with the correspondent epicenters (c). Parts (b) and (d) of the figure show the same time distribution of events in (a) and (b) plus an additional number of events for which the epicenter locations were less reliable. We recorded many other smaller events, but since they were not located, they are not included in the figure. The epicenters of earthquakes with magnitudes 4.0 or larger that occurred in November–December 2006 and at the end of January 2007 are marked with squares in the maps of Fig. 4. These maps show that the earthquake distribution pattern determined by using the well-located epicenters does not change when we add the less accurate epicenters.

The smaller gray circles in Fig. 5 are the epicenters determined for 256 earthquakes of the swarm. Most of those events occurred during December 2006. Since all of these events were located only with data from the local stations, their location errors are small. As observed in the swarms of other seismic regions, the well-located earthquakes of Bahía Asunción occurred also tightly clustered in space. The distribution of the epicenter defines an approximate band of 15 km by 10 km with an E–W apparent trend. Vertically, the clustering was also evident, with the great majority of the sources occurring at depths varying from 4.0 to 9.0 km. This hypocenter distribution, however, does not show a direction of focal depth increase, if any.

Figure 5 shows the surface traces of two normal faults named San Roque and Asunción (From Carta Geológico-Minera de Baja California Sur, Servicio Geológico Mexicano). The San Roque Fault extends parallel to the Tosco-Abreojos fault system, but the Asunción Fault is near perpendicular to it. Since these sub-perpendicular faults cut through the area of epicenters, it is likely that the earthquakes occurred in association with those faults. We will come back to this in the section of focal mechanisms.

The accurate location of the 2006 Bahía Asunción earthquake swarm on the coastline of Baja California Sur has one important implication. That is, earthquakes of moderate magnitudes that occur along the Pacific margin not necessarily are the result of motions along known major faults. The fact that the earthquake swarm occurred to the east of the Tosco-Abreojos fault zone indicates that a wider zone of the western margin accommodates part of the relative motion between the Pacific and North America plates. Munguía *et al.* (2015, submitted) found a similar result from earthquakes recorded in the area of San Carlos, Baja California Sur. In that region, the seismicity also occurred to east of the Tosco-Abreojos fault system. The seismic activity of Bahía Asunción and San Carlos are thus evidences of wider zones of deformation along the western margin of Baja California Sur, at least in those areas.

5.4. *Focal Mechanisms*

The ANSS Comprehensive Catalog of the United States Geological Survey (USGS) reported the focal mechanisms for the two larger events of the earthquake swarm (M_w 5.0 and 4.9). These mechanisms correspond to the events with the numbers 9 and 11 in the Fig. 5 and in Table 1. It is worth noting that although these earthquakes occurred close to each other, their fault-plane solutions are quite different. In order of occurrence, earthquake 9 occurred 2 days before earthquake 11.

Geographically, both epicenters are located at the convergence of the San Roque and Asunción Faults. These features that trend in NW–SE and near N–S directions, respectively, seem to agree well with the fault planes in the reported mechanisms. We suggest, then, the possibility that earthquake 9 occurred in association with the San Roque Fault, while earthquake 11 resulted from slip on the Asunción Fault. If true, then the San Roque Fault should be more of the strike–slip type than of a normal style. Consequently, it would seem natural to expect small earthquakes of the swarm with one or the other style of faulting. For some events, the mechanism would be predominantly strike–slip, and for others of pure-normal faulting style. However, because of the small size of the earthquakes and the limited number of first-motion polarities per event, there is no a simple way to classify the small earthquakes by their style of faulting.

With the two possible styles of earthquake faulting in mind, we obtained a composite fault-plane solution by taking advantage of the spatial proximity of earthquakes. For this, we analyzed the polarities of P-wave first vertical motions produced by 42 of the better-located earthquakes. The earthquakes chosen had magnitudes above 2.5 and sources located inside a volumetric zone of about 2-km radius, centered at a depth of 7 km. Further, the earthquakes were selected without restrictions on their times of occurrence, so all chosen earthquakes occurred during the entire period of recording.

Figure 6 shows the composite fault-plane solution obtained. Based on this figure, we first recognize that some compressions or dilatations from the 42 events considered plot in the wrong quadrants. Nonetheless, the much larger number of compressions or dilatations

that fell consistently in the correct quadrants allows us to make some inferences. We noted first that the polarity data plotted seem to be consistent with two possible mechanisms. One is a right-lateral, strike–slip faulting mechanism, as shown in Fig. 6 with the white and shaded quadrants. The other one is a pure-normal faulting mechanism, as drawn in the same figure with the dashed lines. We found that with small discrepancies these mechanisms are very similar to those of the M_w 5.0 and 4.9 events (events 9 and 11 in Fig. 5). The consistency of the P-wave first motions for these two fault-plane solutions is good, and supports our aforementioned conjecture of having earthquakes in the swarm with either strike–slip or normal styles of faulting. We interpret these earthquakes with distinct styles of faulting as being the result of interaction between the San Roque and Asunción Faults. We thus speculate that the strike–slip earthquake that occurred 2 days earlier acted like a trigger mechanism for the pure-normal event.

Scientific knowledge about the San Roque and Asunción Faults and the way they interact is poor in the literature. For the northern segment of the Pacific margin, where the Bahía Asunción earthquakes occurred, MICHAUD et al. (2011) reported the existence of only transcurrent faults. However, MICHAUD et al. (2005) documented the existence of transcurrent and normal faults at the southern end of the Tosco-Abreojos fault system, similar to what we are reporting for the area of Bahía Asunción. A geology map of the Vizcaino region, published by Servicio Geológico Mexicano, show several on-land normal faults that are perpendicular or near perpendicular to the shoreline of Baja California Sur (see Fig. 2). Particularly the area of the Bahía Asunción earthquakes looks complex in that map, with some faults dipping east and others dipping west. Apparently, the presence of such faults control vertical displacements and/or tilting of some blocks of rock in the area (e.g., ORTLIEB 1991). Although the Bahía Asunción earthquake data indicate fault motions of the strike–slip and normal styles in the area, they are insufficient to draw solid conclusions about the kinematics of faults in the region. What is clear is a need of further research to improve the present knowledge about the way in which faults along the Pacific margin interact between each other.

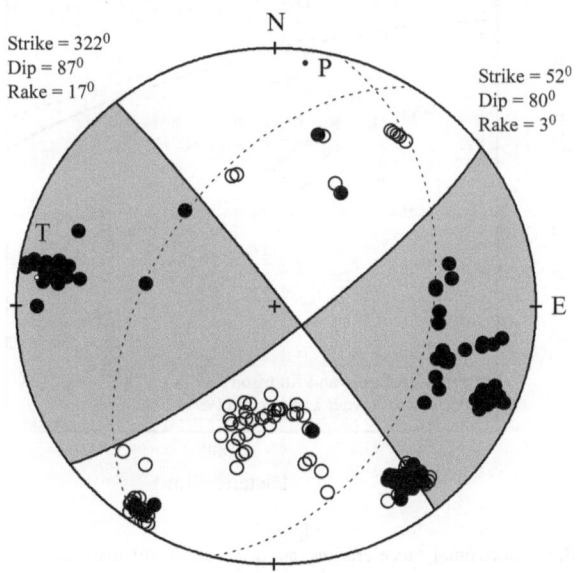

Figure 6
Composite fault-plane solution prepared with data from 42 of the better-located earthquakes with magnitudes above 2.5. All these earthquakes had hypocenters inside a volume of about 2-km radius centered at a depth of 7 km. The fault planes shown as *dashed lines* also satisfy the P-wave polarity data (see text for a discussion)

Table 2

Absolute values of peak motion parameters of the four larger earthquakes recorded at the BAS station

Date (d m y)	Time (h min)	M_w	Dist. (km)	Peak Accel. (g)			Peak Vel. (cm/s)			Peak Disp. (cm)		
				N–S	Z	E–W	N–S	Z	E–W	N–S	Z	E–W
01 12 06	19 33	4.9	4.1	0.37	0.36	0.63	5.48	3.80	13.35	0.24	0.21	0.57
01 12 06	19 46	3.8	4.9	0.13	0.07	0.12	1.73	0.64	2.81	0.06	0.02	0.11
02 12 06	14 43	3.7	3.8	0.10	0.07	0.11	2.02	0.85	1.59	0.08	0.02	0.04
02 12 06	15 35	3.6	5.4	0.06	0.03	0.10	1.24	0.31	1.00	0.06	0.01	0.03

5.5. Acceleration Data

Two accelerometers recorded at the BAS and PPA sites. BAS was located in the southern part of Bahía Asunción, at ~1.5 km from downtown. Since the sources of the earthquakes in this swarm occurred beneath the town, BAS recorded a large number of earthquakes at distances from 2 to 8 km. The second accelerometer recorded initially at Punta Prieta (PPA), which was located ~28 km from the sources. Because of such large distance, the accelerations recorded at this site were of significantly lower amplitude than those recorded at BAS. Thus, after some time, we moved PPA to a site marked as RAF in Fig. 2, but due to technical problems, this station recorded only a few small events.

The strongest earthquake recorded by the local array was an M_w 4.9 earthquake that occurred on 1 December 2006 at 19:33 h (UTC). This event is marked with the number 11 in Fig. 5. The peak ground accelerations recorded at BAS from this earthquake were 0.37, 0.36 and 0.63g on the North–South, Vertical, and East–West components, respectively (see Table 2). At the time of occurrence of this event, however, BAS was the only strong-motion station in operation. Two days before this event another M_w 5.0 earthquake occurred at nearly the same place (relocated event number 9 in Fig. 5). This other event occurred on 29 November 2006, at 14:34 h (UTC). The fact that both earthquakes were nearly of the same size explains why residents of Bahía Asunción felt the motions with similar intensities. In Table 2, we see that other events with magnitudes about 4.0 produced ground accelerations of around 0.10g at BAS. The last two events of this table were the only events recorded at both strong-motion stations. PPA, located at ~28 km from those

events, recorded peak accelerations of 0.002 and 0.001g on its E–W component for the third and fourth events of Table 2, respectively. It was also of interest to note that for most earthquakes both strong-motion stations recorded the larger peak acceleration on the E–W component of motion. The median value of ratios of the N–S to E–W peak accelerations is 0.4 (Fig. 7).

Figure 7 shows a plot of peak horizontal accelerations versus distance. Only data from earthquakes with magnitudes equal to 2.5 or greater are included

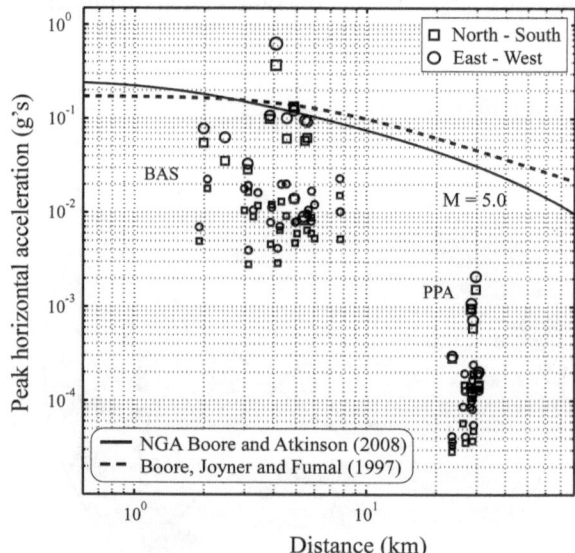

Figure 7

Peak horizontal acceleration as a function of distance from earthquakes with magnitudes between 2.5 and 4.9. Data at distances from 2 to 8 km were recorded by an accelerometer installed in Bahía Asunción (BAS in Fig. 2). BAS recorded the larger peak horizontal acceleration values (0.63 and 0.37g) for the M_w 4.9 earthquake of 1 December 2006, at 19:33 h (UTC). Curves from two attenuation models ($V_{S30} = 450$ m/s) for earthquakes of magnitude 5 are included for comparison purposes in the figure. To compare with our data, the figure includes the attenuation curves from two widely used models. In the calculation of these curves, we assumed an average S-wave velocity in the top 30 m (V_{S30}) of 450 m/s. From Fig. 7, it is evident that for earthquakes of magnitude 5 and distances greater than 20 km both attenuation curves overestimate the ground motions recorded at PPA. This fact implies that the ground motions from earthquakes of the swarm attenuate much faster with distance than predicted by the models.

We obtained the ground velocity and displacement time series through one and two integrations of the acceleration waveforms with respect to time. As part of the integration process, the time series were band-pass filtered between 0.7 and 50 Hz. Table 2 includes some results of the process of record integration. In particular, for the M_w 4.9 earthquake that produced the higher peak accelerations at BAS, the corresponding peak velocity and peak displacement were 13.5 cm/s and 0.57 cm, respectively. Figure 8 shows the three components of acceleration recorded for this earthquake, as well as the response spectra calculated from the horizontal components of motion. These spectra constitute another convenient way to analyze the frequency characteristics of the earthquake motions.

To our knowledge, this is the first study of earthquakes that occurred in the Bahía Asunción area. The strongly felt earthquakes caused common effects on the ground and slight damage to some structures in town. Because of this, the acceleration response spectra of Fig. 8 might be of interest to the engineering community. These data are representative of ground motions that could result from earthquake sources located in the subsurface of downtown Bahia Asunción.

5.6. Soil Response at BAS

Soft soils amplify or attenuate the earthquake motions in accordance with the intensity and period of the seismic waves. Also recognized is the fact that sediments amplify linearly the weak motions from small earthquakes (relative to rock) (e.g., BORCHERDT 1994; SEED et al. 1997). However, under the larger strains induced by stronger earthquakes the linearity property breaks down, with a consequent reduction of the shear modulus and a de-amplification of the high-frequency motions.

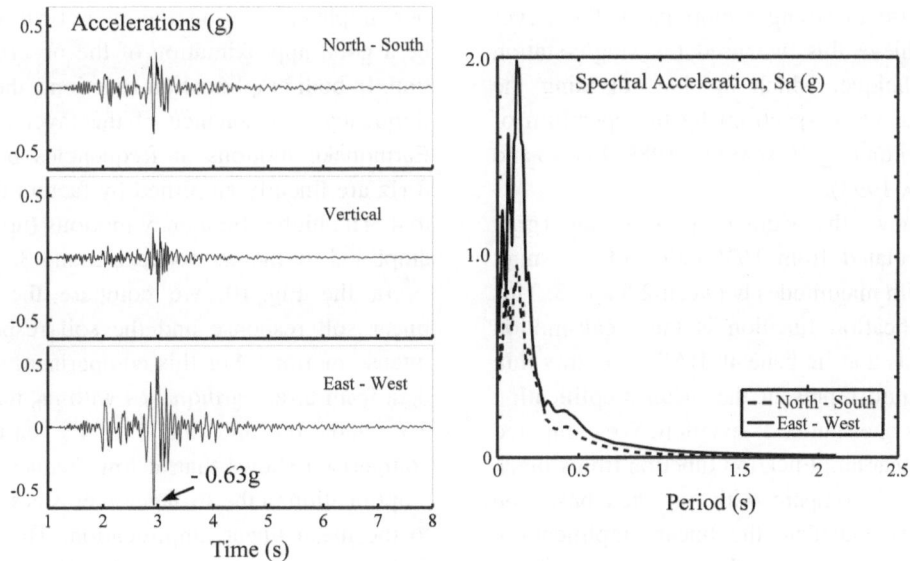

Figure 8

Three components of acceleration recorded for the M_w 4.9 earthquake of 1 December 2006 at 19:33 h (UTC). The 0.63g peak acceleration, observed on the East–West component of motion, was the highest acceleration instrumentally recorded. The *right panel* of the figure shows the acceleration response spectra calculated from the horizontal components of motion

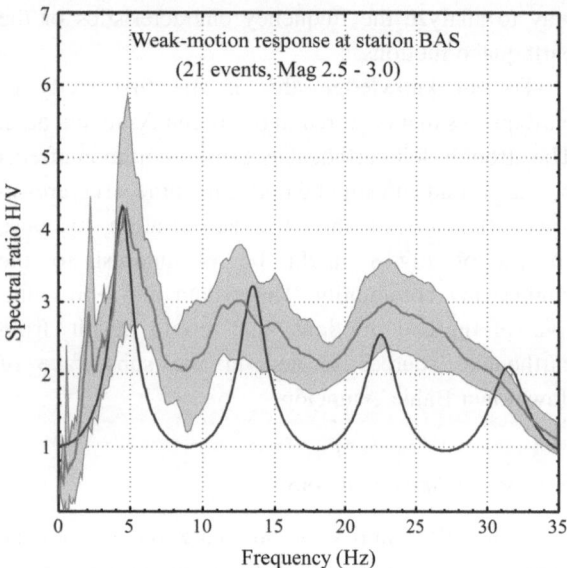

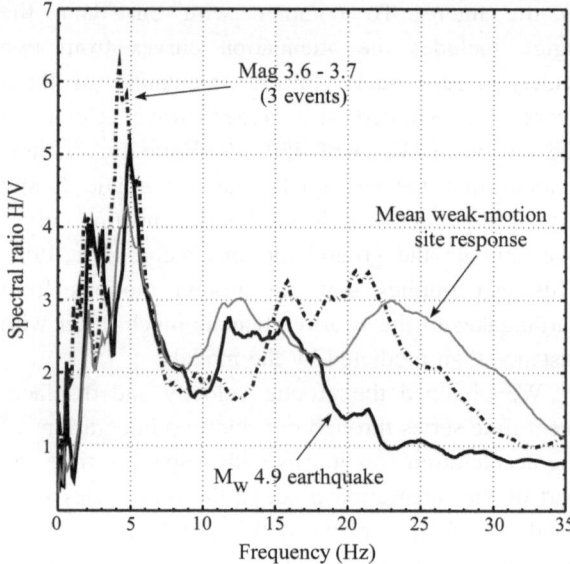

Figure 9

Mean site amplification of ground motions at the station BAS (*gray thicker line*). *Lines* bordering the *gray area* correspond to the mean site amplification plus and minus one standard deviation. A theoretical amplification function is included for comparison purposes

Figure 10

Comparison of the weak-motion site response to the soil response to stronger earthquakes. The weak-motion response of soils beneath BAS was calculated from H/V ratios of 21 small earthquakes with magnitudes between 2.5 and 3. This average amplification function is compared here with the response of soils to earthquakes with m_b magnitudes 3.6, 3.7, and 3.8 and one M_w 4.9 event (see text for a discussion)

Here, we investigate the soil response beneath BAS. We first estimate the weak-motion response at the site using data from the smaller earthquakes. Then, we compare the resulting linear response and the soil response to stronger motions of the larger events. To achieve this, we used the single-station H/V ratio technique, which involves dividing the horizontal shear-wave spectrum by the spectrum of the vertical motion (e.g., NAKAMURA 1989; LERMO and CHÁVEZ-GARCÍA 1993).

Figure 9 shows the average spectral ratio (gray tick line) calculated from H/V ratios of 21 small earthquakes with magnitudes between 2.5 and 3. This average amplification function is the weak-motion response of soils that lie beneath BAS. The gray thin lines in the figure represent the mean amplification plus and minus one standard deviation. We computed a theoretical linear amplification function (thick black line in Fig. 8) to compare with our data based on observations. To calculate the linear amplification, we made the assumption of a single soil layer overlying an infinite half space. The thickness, shear-wave velocity, and density assumed for the layer were 25 m, 450 m/s, and 1.8 g/cm³, respectively. For

the infinite half space, we used a shear-wave velocity of 1500 m/s and a density of 2.8 g/cm³. In spite of our simple calculation, the theoretical result seems to be a good approximation of the observed amplification. In both amplification functions, the fundamental frequency of resonance of the layer is about 5 Hz. Earthquake motions at frequencies between 2 and 8 Hz are linearly amplified by factors that go from 2 to 4. The higher frequency motions (up to 30 Hz) are amplified by factors between 2 and 3.

In the Fig. 10, we compare the weak-motion linear soil response and the soil response to more intense motions. For this comparison, we considered data from three earthquakes with m_b magnitudes 3.6, 3.7, and 3.8 and an M_w 4.9 earthquake. Our comparison shows that at low frequencies the mean amplification in the stronger events is similar in shape to the mean linear amplification. The peak amplifications, however, are somewhat higher in comparison with those of the weak-motion soil response. It seems likely, then, that our estimated weak-motion response at BAS is only a lower limit of the true amplification.

Another feature to observe from Fig. 10 is that the soil response to stronger events stays within ± one standard deviation of the weak-motion amplification up to ∼20 Hz. Above that frequency, the amplification of motion in the larger events decreases relative to the weak-motion linear response. This reduction starts occurring first for the M_w 4.9 earthquake, since at higher levels of strain the soils respond more nonlinearly.

The above observations provide insights about the possible nonlinear behavior of local soils in response to intense ground shaking. Yet, the soil liquefaction induced in the epicentral area is perhaps the best indicator of such phenomenon. Liquefaction of saturated sandy materials was observed at El Salitral, which is located east of Bahía Asunción. In this zone, sand boils developed and water spouted into the air (Fig. 3). It appears that large soil softening, due to high excess pore water pressure during the more intense shaking, resulted in strong nonlinear response of the soil at this location. Making a quantitative statement about the degree of soil nonlinearity, however, is difficult and beyond the scope of this study.

6. Conclusions

We analyzed data from an earthquake swarm that occurred in the neighborhood of Bahía Asunción. The relocation of some initial larger earthquakes (M_w 4–5) of this swarm puts the sources below the town of Bahía Asunción, at 40–50 km to the north of the teleseismically determined epicenters. With temporary seismic stations deployed in the epicenter area, we recorded many subsequent smaller events. These events were located around the epicenters of the larger initial earthquakes, within a zone of approximately 15 km by 10 km. All events, small and large, occurred at depths mostly between 4 and 9 km.

A composite focal mechanism prepared with P-wave polarity data from the smaller earthquakes could be interpreted as indicating right-lateral strike–slip motion and pure-normal faulting occurred during the swarm. This interpretation is supported by the fact that for two larger events of the swarm the ANSS earthquake catalog (USGS) reported these two styles of faulting, with one of the earthquakes occurring 2 days before the other one. We suggest that the earthquake with strike–slip mechanism could be associated with the San Roque Fault and the earthquake with normal faulting style likely occurred on the Asunción Fault. However, there is need of further local studies to verify this possible relation between these faults and the earthquakes.

The earthquakes of the swarm occurred along on-land active faults that extend parallel and near perpendicular to the shoreline. This fact makes the swarm more relevant, because its association with these faults indicates that active crustal deformation takes place to the east of the Tosco-Abreojos fault system. The focal mechanisms of the earthquakes are also indicative of a wider transtensional zone of deformation that accommodates part of the relative slip between the Pacific and North America plates.

We also analyzed the near-source acceleration data recorded in a site located in downtown Bahía Asunción. Of much interest was to note that an M_w 4.9 earthquake produced horizontal peak accelerations of up to $0.63g$ at a distance of 4.1 km. On this base, it is likely that other events of the swarm with about the same magnitudes, but not recorded instrumentally, had produced similar strong motions in Bahía Asunción. Such high values of ground accelerations explain the concern among the residents, as well as some minor damage to houses and the liquefaction of soils in the epicentral area.

From the analysis of spectral ratios of horizontal to vertical motions, we found that the soils beneath Bahía Asunción amplify the ground motion by factors of 2–4 at frequencies between 2 and 8 Hz. Above 20 Hz, the amplification of motion in the larger events showed a decrease relative to the weak-motion linear response of the soil. Finally, we make emphasis in that all of these aspects of the acceleration dataset are of importance in assessments of the seismic hazard for Baja California Sur.

Acknowledgments

Financial support for this research was provided by Centro de Investigación Científica y de Educación Superior de Ensenada, B. C. (CICESE). We also want to express our gratitude to Guillermo Gutiérrez de Velasco, of CICESE at La Paz, Baja California Sur, Pedro G. Osuna López of H. XII Ayuntamiento de Mulegé, Luis Martínez Pozo, Subdelegate of Bahía Asunción, Mario Ramade of Federación de Cooperativas de Baja California, Managing personnel of Cooperativa Leyes de Reforma, Cesar Redona Camacho of Sociedad Cooperativa California en San Ignacio, Ramón Villa Villegas, Armando Naranjo Rivera of XIII District of Mulegé, David A. Maraver Romero of Servicio Geológico Mexicano (SGM), Antonio de Jesús Ceseña Moyrón of Seguridad Pública y Tránsito Municipal of Bahía Asunción, and to personnel of Civil Protección of Baja California Sur. We acknowledge Carlos Gutiérrez and technicians of Centro Nacional de Prevención de Desastres for providing some seismic data they collected in the epicentral area with an analog seismograph and two weak-motion digital stations. We appreciate the field service provided by Martín Romero to our analog stations. Many other people supported our study in different forms (technical support, economic funds, transportation vehicle, materials and accessories, etc.). Our deepest thanks to all. The authors would also like to acknowledge the work of Dr. William Bandy and two anonymous reviewers. Their in-depth reviews and constructive comments and suggestions were helpful in improving the manuscript.

References

BORCHERDT, R. D. (1994). Estimates of Site-Dependent Response Spectra for Design (Methodology and Justification). Earthquake Spectra, vol. 10(4), pp. 617–653.

BROTHERS, D., A. HARDING, A. GONZÁLEZ-FERNÁNDEZ, W. S. HOLBROOK, G. KENT, N. DRISCOLL, J. FLETCHER, D. LIZARRALDE, P. UMHOEFER and G. AXEN (2012). Farallon slab detachment and deformation of the Magdalena Shelf, southern Baja California, Geophys. Res. Lett., 39, L09307, doi:10.1029/2011GL050828.

CASTRO, R. R., C. VALDÉS-GONZÁLEZ, P. SHEARER, V. WONG, L. ASTIZ, F. VERNON, A. PÉREZ-VERTTI, and A. MENDOZA (2011). The 3 August 2009 Mw 6.9 Canal de Ballenas Region, Gulf of California, Earthquake and Its Aftershocks. Bull. Seism. Soc. Am., 101, No. 3, pp. 929–939, doi:10.1785/0120100154.

CLAYTON, R.W., J. TRAMPERT, C. REBOLLAR, J. RITSEMA, P. PERSAUD, H. PAULSSEN, X. PÉREZ-CAMPOS, A. WETTUM, A. PÉREZ-VERTTI, F. DILUCCIO (2004). The NARS-Baja Seismic Array in the Gulf of California Rift Zone. Margins Newsletter, No. 13, 1–4.

DEMETS, C. A. (1995). Reappraisal of seafloor spreading lineations in the Gulf of California: implications for the transfer of Baja California to the Pacific plate and estimates of Pacific-North America motion, Geophys. Res. Lett., 22, 3545–3548.

DEMETS, C. A. and DIXON, T. H. (1999). New kinematic models for Pacific-North America motion from 3 Ma to present, I: Evidence for steady motion and biases in the NUVEL-1A model, Geophysical Research Letters, 26, 1921–1924.

FLETCHER, J. M., and L. MUNGUÍA (2000). Active continental rifting in southern Baja California, Mexico: implications for plate motion partitioning and the transition to sea floor spreading in the Gulf of California, Tectonics, 19, No. 6, 1107–1123.

FLETCHER, J.M., M. GROVE, D. KIMBROUGH, O. LOVERA, and G. E. GEHRELS (2007). Ridge-trench interactions and the Neogene tectonic evolution of the Magdalena Shelf and southern Gulf of California: Insights from detrital zircon U-Pb ages from the Magdalena fan and adjacent areas. Geol. Soc. Am. Bull., 119; no. 11/12; pp. 1313–1336; doi:10.1130/B26067.1.

GOFF, J. A., E. A. BERGMAN, and S. C. SOLOMON (1987). Earthquake source mechanisms and transform fault tectonics in the Gulf of California, J. Geophys. Res., 92, 10,485–10,510.

HAVSKOV, J., and L. OTTEMÖLLER (Editors) (2001). SEISAN: the earthquake analysis software for Windows, SOLARIS, and LINUX, Version 7.2. Manual, Institute of Solid Earth Physics, University of Bergen, Norway.

LERMO, J., and F. J. CHAVEZ-GARCIA (1993). Site effects evaluation using spectral ratios with only one station, Bull. Seism. Soc. Am. 83, no. 5, 1574–1594.

LIENERT, B. R. E. and J. HAVSKOV (1995). A computer program for locating earthquakes both locally and globally. Seismol. Res. Lett., 66:26–36.

MICHAUD, F., SOSSON, M., ROYER, J.-Y., CHABERT, A., BOURGOIS, J., CALMUS, T., MORTERA, C., BIGOT-CORMIER, F., BANDY, W., DYMENT, J., PONTOISE, B., and SICHLER, B. (2004), Motion partitioning between the Pacific plate, Baja California and the North America plate: The Tosco-Abreojos fault revisited: Geophysical Research Letters, v. 31, p. L08604, doi:10.1029/2004GL019665.

MICHAUD F, T. CALMUS, M. SOSSON, J-Y. ROYER, J. BOURGOIS, A. CHABERT, F. BIGOT-CORMIER, B. BANDY, C. MORTERA-GUTIÉRREZ (2005). La zona de falla Tosco-Abreojos: un sistema lateral derecho activo entre la placa Pacífico y la península de Baja California. Boletín de la Sociedad Geológica Mexicana. Tomo LVII, Núm. 1., pp. 53–63.

MICHAUD, F., CALMUS, T., ROYER, J.-Y., SOSSON, M., BANDY, B., MORTERA-GUTIÉRREZ, C., DYMENT, J., BIGOT-CORMIER, F., CHABERT, A., and BOURGOIS, J. (2007), Right lateral active faulting between southern Baja California and the Pacific plate: The Tosco-Abreojos fault, In Alanis-Alvarez, S. A. and Nieto-Samaniego, A. F., eds., Celebrating the Centenary of the Geological Society of Mexico: Geological Society of America, special paper 422, pp. 287–300, doi:10.1130/2007.2422(09).

MICHAUD, F., CALMUS, T., RATZOV, G., ROYER, J.-Y., SOSSON, M., BIGOT-CORMIER, F., BANDY, W. and MORTERA-GUTIÉRREZ, C. (2011). Active deformation along the southern end of the Tosco-Abreojos fault system: New insights from multibeam swath bathymetry. Pure Appl. Geophys., 168(8–9), 1363–1372, doi:10.1007/s00024-010-0193-y.

MUNGUÍA, L., M. REICHLE, A. REYES, R. SIMONS, and J. N. BRUNE (1977). Aftershocks of the 8 July, 1975 Canal de las Ballenas, Gulf of California, earthquake, Geophys. Res. Lett., 4, 507–509.

MUNGUÍA, L., M. GONZÁLEZ, S. MAYER, and A. AGUIRRE (2006). Seismicity and State of Stress in the La Paz–Los Cabos Region,

Baja California Sur, Mexico. Bull. Seismol. Soc. Am., *96*, No. 2, pp. 624–636, April 2006, doi:10.1785/0120050114.

MUNGUÍA, L., M. GONZÁLEZ, M. NAVARRO, T. VALDEZ, S. MAYER, A. AGUIRRE, V. WONG, and M. LUNA (Submitted to PAGEOPH, 2015). A local seismic study at the central part of the Magdalena Shelf, in the Pacific margin of Baja California Sur, Mexico.

NAKAMURA, Y. (1989). A method for dynamic characteristics estimation of subsurface using microtremor on the ground surface, Quarterly Report of the Railway Technical Research Institute *30*, no. 1, pp. 25–33.

NORMARK, W.R., J.E. SPENCER, and J. INGLE (1987). Geology and Neogene history of the Pacific c continental margin of Baja California Sur, Mexico, in SCHOLL, D.W., *et al.*, eds., Geology and resource potential of the continental margin of western North America and adjacent ocean basins—Beaufort Sea to Baja California: Houston, Texas, Circum-Pacific Counsel for Energy and Mineral Resources, Earth Science Series, *6*, p. 449–472.

ORTLIEB, L. (1991). Quaternary Vertical Movements along the Coasts of Baja California and Sonora. In: J. P. DAUPHIN and B. T. SIMONEIT (Eds.), The Gulf and Peninsular province of the Californias, Amer. Asoc. Petrol. Geol. Mem., *47*, pp. 447–480.

REICHLE, M. S., G. F. SHARMAN, and J. N. BRUNE (1976). Sonobuoy and teleseismic study of Gulf of California transform fault earthquake sequences. Bull. Seismol. Soc. Am., *66*, No.5, pp. 1623 1641.

SALAZAR, C. R. M. (2014). Evidencias estructurales y sismo estratigráficas en la parte central de la plataforma Magdalena, margen occidental de Baja California a partir de sísmica de reflexión. MsC. thesis, 89 pp., Centro de Investigación Científica y de Educación Superior de Ensenada, B. C., Ensenada, B. C., México.

SEED, R. B., CHANG, S. W., DICKENSON, S. E., and BRAY, J. D. (1997) "Site-Dependent Seismic Response Including Recent Strong Motion Data." Proc., Special Session on Earthquake Geotechnical Engineering, XIV International Conf. On Soil Mechanics and Foundation Engineering, Hamburg, Germany, A. A. Balkema Publ., Sept. 6–12, pp. 125–134.

SPENCER, J. E. and NORMARK, W. R. (1979). Tosco-Abreojos fault zone: A Neogene transform plate boundary within the Pacific margin of southern Baja California, Mexico. Geology, V. *7*, pp. 554–557.

(Received July 1, 2015, revised September 25, 2015, accepted September 26, 2015, Published online October 26, 2015)

Pure Appl. Geophys. 173 (2016), 3631–3644
© 2015 Springer International Publishing
DOI 10.1007/s00024-015-1217-4

▌Pure and Applied Geophysics

Active Crustal Deformation in the Area of San Carlos, Baja California Sur, Mexico as Shown by Data of Local Earthquake Sequences

LUIS MUNGUÍA,[1] [iD] MARIO GONZÁLEZ-ESCOBAR,[1] MIGUEL NAVARRO,[1] TITO VALDEZ,[1] SERGIO MAYER,[2] ALFREDO AGUIRRE,[2] VICTOR WONG,[1] and MANUEL LUNA[1]

Abstract—We analyzed earthquakes of sequences that occurred at different times near San Carlos, a town of approximately 5000 inhabitants. The seismic sequences happened during March–April 1989, October 2000–June 2001, and 5–15 February 2004 at about 200 km west of the Pacific-North America plate boundary. The strong shaking from initial earthquakes of the first two sequences prompted the installation of temporary seismic stations in the area. With data recorded by these stations, we found an earthquake distribution that is consistent with the northwest segment of the Santa Margarita fault. Both the focal depth, that seemed to increase in E–NE direction, and a composite fault-plane solution, obtained from polarity data of the small earthquakes, were also consistent with the main characteristics of that fault. We also found that our normal-faulting mechanism (east side down) was quite similar to centroid moment tensor solutions for earthquakes with M_w 5.4 and 5.3 that occurred in the area in February 2004. It is likely, then, that these larger earthquakes also occurred along the Santa Margarita Fault. To get some insight into the regional stress pattern, we compared the above mechanisms with mechanisms reported for other earthquakes of the Pacific margin of Baja California Sur and the Gulf of California regions. We observed that focal mechanisms of the two regions have T axes of stress that plunge sub horizontally in E–NE average direction. The corresponding P axes have N–NW average trend, but for the Pacific earthquakes these axes plunge at angles that are ∼35° larger than those for the Gulf earthquakes. These more vertically inclined P axes of compressive stress mean substantial oblique fault motions. The mixture of oblique and strike-slip components of fault motions, as the focal mechanisms show, confirms a transtensional stress regime for the region. Before this research, we knew little about the seismicity and styles of faulting in the area. Now we know that earthquakes can occur along the coastline of Baja California, at 60 km east of the Tosco-Abreojos fault system. We conclude that transtensional deformation is taking place across a wide zone of the Pacific margin of Baja California. Finally, we

point out that although the studied earthquakes were of small magnitude, they might serve as a reminder of the danger that future larger events pose to San Carlos.

Key words: Pacific margin of Baja California, seismic activity, transtensional stress regime.

1. Introduction

The region of this study is located along the Pacific coast of Baja California Sur, at the central part of the Magdalena Shelf (Fig. 1). This zone includes San Carlos, a town of approximately 5000 inhabitants. In the past 25 years, several earthquakes rocked this small town and caused minor damage to houses and alarm among residents. The inflicted damage consisted of ground fissures, cracking in plaster covering masonry, and books and other objects thrown down from shelves. The earthquake effects and distribution of damage were only noticeable at short distances from the sources.

In this paper, we present the results of analyses of small-magnitude earthquake sequences recorded instrumentally for the first time in the neighborhood of San Carlos. These sequences occurred during March–April 1989, October 2000–June 2001, and 5–15 February 2004. Before this study, the seismicity of the area was unknown mostly due to a lack of local seismic stations. Some earthquakes of the first two sequences produced intense ground motions at San Carlos, and that prompted the installation of small temporary seismic networks in the area. The 2004 events, on the other hand, occurred far from San Carlos. In spite that two of those events were of M_w 5.4 and 5.3, no one in town felt their motions.

---
[1] División de Ciencias de la Tierra, Centro de Investigación Científica y de Educación Superior de Ensenada, B. C., Carretera Ensenada-Tijuana No. 3918, Zona Playitas, C. P. 22860, Ensenada, Baja California, Mexico. E-mail: lmunguia@cicese.mx
[2] Centro de Investigación Científica y de Educación Superior de Ensenada, B. C., Unidad La Paz, Baja California Sur, Miraflores 334, Fraccionamiento Bellavista, C. P. 23050, La Paz, Baja California Sur, Mexico.

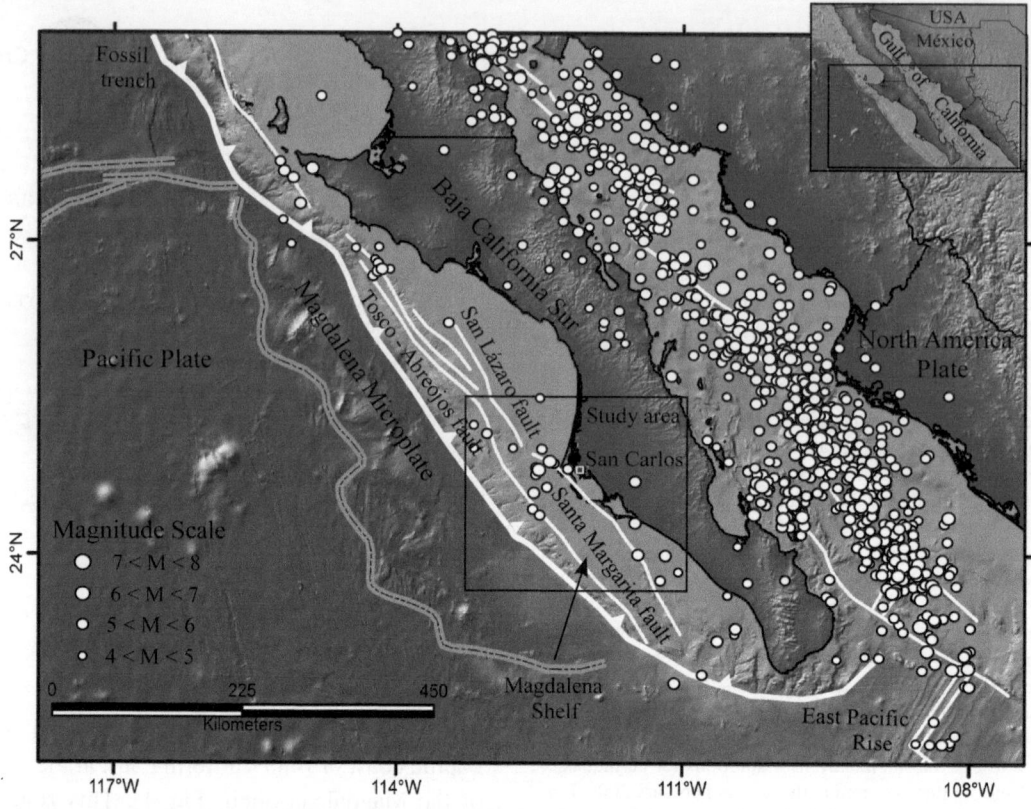

Figure 1
Seismicity of the Baja California Sur-Gulf of California region for the period from 1960 to 2014, as reported in the Northern California Earthquake Data Center (NCEDC)

Because of this, we did not install temporary stations in the area. Nonetheless, stations of the NARS-Baja seismic network (CLAYTON *et al.* 2004) recorded the two larger events of the sequence at distances between 150 and 600 km. With data from those stations, we relocated the epicenters of the 2004 earthquakes.

We present the results of our study in terms of earthquake locations and their correlation with a known fault of the study area. In addition, we describe how a composite focal mechanism obtained for the recorded small earthquakes and two mechanisms of previous larger events of the area are indicative of a transtensional stress regime for the region. We investigated the condition of regional stress by comparing these focal mechanisms with mechanisms for earthquakes of the Pacific margin and the Gulf of California region. The results also

provide insights about a zone of concentration/release of stresses near San Carlos, with potential seismic hazard of unknown level to this town.

2. *Tectonic Setting and Regional Seismicity*

Prior to ~29 Ma, the Farallon and several microplates were subducting along the western margin of the American continent (ATWATER 1970). About 12 million years ago, the western North America subduction process ceased and the plate-margin slip concentrated mostly in the Gulf of California (LONSDALE 1991). At that time, the Pacific plate began the capture of the Baja California microplate (ATWATER 1989; BOHANNON and PARSONS 1995). Since then, most of the relative slip between the Pacific and North American plates has taken place along the transform fault system that runs through the Gulf of California. There is, however, the hypothesis that faults of the coastal and offshore zones west of Baja California accommodate a fraction of the relative plate motion (e.g., HUMPHREYS and WELDON 1991; DEMETS et al. 1995; DIXON et al. 2000; FLETCHER and MUNGUÍA 2000; MICHAUD et al. 2005; MUNGUÍA et al. 2006; FLETCHER et al. 2007). If we accept this hypothesis, then the assumption that the Baja California peninsula transferred completely to the Pacific plate by 3.6 Ma might not be valid (DEMETS 1995; STOCK and HODGES 1989; MICHAUD et al. 2004).

The major tectonic features that cut through the study area are the Tosco-Abreojos and San Lázaro-Santa Margarita fault systems (Fig. 1). These faults trend approximately north-northwest, dip to the east, and control the position and geometry of two asymmetric transtensional basins. Detailed reviews of the tectonics and fault systems of the region can be found in papers by SPENCER and NORMARK (1979), NORMARK et al. (1987), MICHAUD et al. (2005, 2011); FLETCHER et al. (2007), BROTHERS et al. (2012) and SALAZAR (2014) among others.

Figure 1 also shows the seismicity of Baja California Sur and the Gulf of California regions for the period from 1960 to 2014. The data plotted include earthquakes in the 3–7 magnitude range obtained from the Northern California Earthquake Data Center (NCEDC) catalog. As the figure shows, the faults in the Gulf of California produce the highest rate of seismicity. Yet, some events reaching magnitudes of up to 5.4 had occurred off the west coast of Baja California. Though these earthquakes are rather small in number, they give clear evidence of tectonic activity along the western margin of the Baja California peninsula.

In the 54-year period of seismicity considered in Fig. 1, 24 events with magnitudes between 3.5 and 5.4 occurred along the offshore faults of our study area (inner rectangle in the figure). All of these events were located with data from stations in North America and it thus seems natural to expect some bias in the epicenters. Results of previous studies have demonstrated that this has been the case for other earthquakes of the region. When near-source data were available for an earthquake location, the locally and teleseismically determined epicenters

differed by up to ~ 50 km (REICHLE et al. 1976; MUNGUÍA et al. 1977; GOFF et al. 1987; FLETCHER and MUNGUÍA 2000; CASTRO et al. 2011).

Figure 2 is a plan view of the regional faults of the area and the epicenters of the larger earthquakes that occurred there in the 1960–2014 period (NCEDC catalog). In that period, the more relevant earthquakes of the region had magnitudes 5.3, 5.4, and 5.3. The numerals 3, 13, and 16 identify the locations of these events on the map of Fig. 2. The first of those events occurred on 26 January 1968; the other two occurred on 9 February 2004. Figure 3 shows graphically the chronological order of occurrence of the 24 events of the study area. From that figure, we note that nine earthquakes occurred in 2004, seven of which had magnitudes 4.0 or larger.

3. Micro Earthquake Data Collection

3.1. 1989 Fieldwork Procedure

On 30 March and 4 April 1989, people in San Carlos felt two earthquakes of moderate intensity. Although these earthquakes produced only light damage in the town, they were the cause of much concern among the population. Due to the lack of seismic stations, there were no local instrumental recordings of those earthquakes. In addition, none of these 1989 earthquakes appeared in the listings of global earthquake catalogs. Based only on the information provided by some local inhabitants, we speculate here that those events had magnitudes between 3 and 3.5. Our reasoning for this conjecture is as follows. First, we noted that global catalogs of distant seismic stations usually report earthquakes of the region with $M > 3.5$. However, none of the 1989 earthquakes under study appeared in the listings of such catalogs. Second, we propose the lower magnitude bound on the basis that people start feeling local earthquakes with magnitudes 3.0, or slightly lower, depending upon distance.

After the first 15 days of seismicity, we installed six temporary stations in the area of the epicenters. The stations were equipped with 1-s natural period seismometers (Kinemetrics, model Ranger SS-1) connected to smoked-paper seismographs

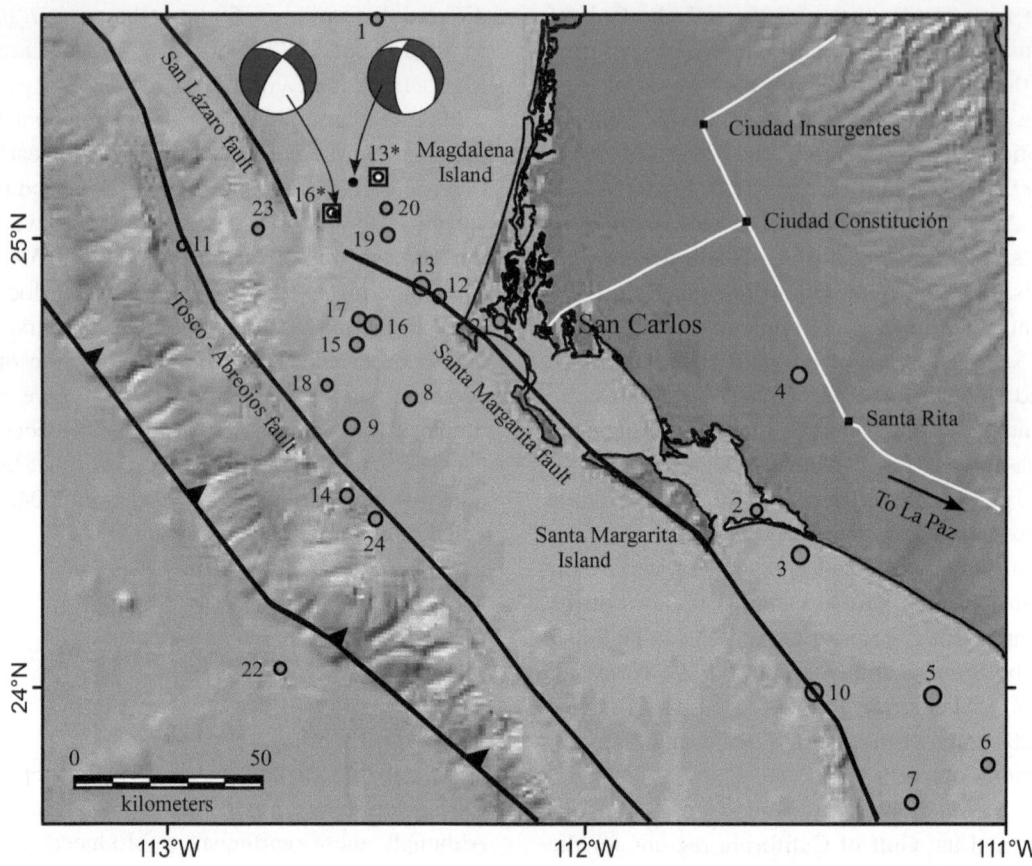

Figure 2

Zoom of the study area to show the epicenters of earthquakes that occurred in the 1960–2014 period (*circles*), as reported in the NCEDC catalog. All of these events had magnitudes between 3.5 and 5.4. The three earthquakes with the larger magnitudes (5.3, 5.4 and 5.3) have the numbers 3, 13 and 16, respectively. The first of these three events occurred on 26 January 1968; the other two occurred on 9 February 2004 (see Fig. 3). *White circles* inside *squares* indicate relocated epicenters for the 2004 events. *Black dots* and beach balls represent the epicenters and double-couple fault plane solutions reported in the CMT catalog for these events

(Sprengnether, model MEQ 800). The triangles in Fig. 4 mark the sites occupied by these analog stations. In a 29-day period of operation, the network recorded 21 earthquakes with magnitudes of up to 3.2. We describe the epicenter location process in a later section of this paper.

3.2. 2000–2001 Fieldwork Procedure

The second period of earthquake activity started on October 2000 and ceased by June 2001. On the 27 and 28 of October 2000, two earthquakes produced intense ground shaking at San Carlos. On that occasion, a seismic station located at a distance of

190 km south of San Carlos recorded those initial events. Based on data from that station and on the magnitude relationship of LEE *et al.* (1972), we calculated duration magnitudes of 3.5 for both earthquakes. As in the 1989 earthquake series, those initial earthquakes produced only small cracks and fissures on some local constructions.

Residents of the cities of Constitución and Insurgentes, located at 50- and 60-km distances from downtown San Carlos (see Fig. 2), respectively, did not feel the earthquakes. Due to scarceness of people living between San Carlos and those cities, there was no information on the earthquake effects at intermediate distances. However, a few people living or

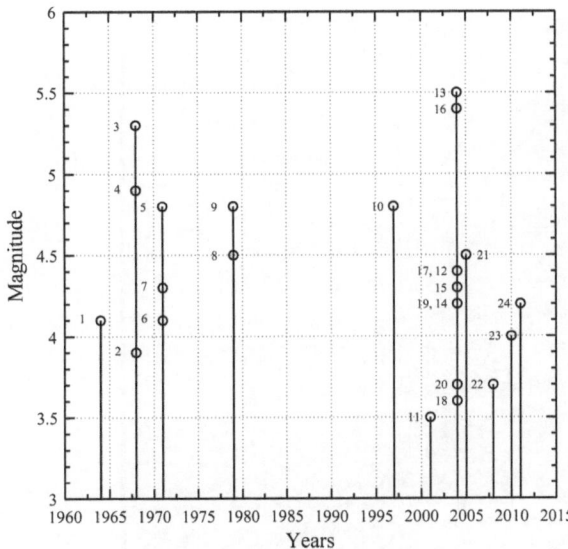

Figure 3

Chronological order of occurrence of the earthquakes plotted in Fig. 2. Earthquakes with the numbers 13 and 16 occurred on 9 February 2004

working north and southeast of San Carlos, but at distances larger than 15 km from downtown, did not perceive the motions from those events. Our conclusion was that the earthquakes caused alarm and minor damage only at distances of less than 15 km from downtown San Carlos.

At 2 weeks from the first two events, we had installed eight digital seismic recorders (white outlined pentagons in Fig. 4) and four analog stations (white squares in Fig. 4) in the area. The digital instruments consisted of three-axial K2 accelerographs or K2 recorders with external Episensor accelerometers (all from Kinemetrics). These digital stations recorded the seismic signals at a rate of 200 samples per second. The four analog stations used were of the same type as those in the 1989 fieldwork. This combined network recorded for a period of 8 months.

4. Data Analysis and Results

4.1. Hypocenter Locations

The earthquake hypocenters were determined with the program Hypocenter (LIENERT and HAVSKOV

1995), a program that is part of the Seisan software package (HAVSKOV and OTTEMÖLLER 2001). We used this location program with the crustal velocity model of FLETCHER and MUNGUÍA (2000), which consists of four layers over a half space with the following P-wave velocity and thickness characteristics, respectively: 4.0 km/s, 2 km; 6.0 km/s, 5 km; 6.4 km/s, 7 km; and 6.9 km/s, 10 km. The lower infinite half-space had a P-wave velocity of 7.6 km/s.

In Fig. 4, circles with a plus sign inside represent the locations of fourteen of the 1989 earthquakes. The epicenters of those earthquakes do not correlate with any of the borderland faults that lie to the west of Baja California. Rather, they spread out perpendicular to the trend of those regional faults. However, we must be cautious on this, because it could be, in part, a result of the sparseness of our seismic stations. Only the epicenters that fall closer to downtown San Carlos had good station coverage and so they were more reliably located. For those events, the estimated focal depths are between 1 and 8 km. Micro earthquakes located at larger distances from San Carlos have depths that increase eastward from 10 to 16 km. In short, not all of these 1989 events were located with good precision, but still they are seismic evidence of tectonic activity in the area of our study.

In the 2000–2001 fieldwork, we recorded many earthquakes in digital and analog format. This time, we took advantage of the high-quality digital recordings to make accurate readings of the P- and S-wave arrival times. These times, combined with time readings from recordings of the analog stations, were the basis for our hypocenter location process. Figure 4 shows the epicenters of 31 earthquakes (white circles) that could be well located. These locations resulted from the use of 5–14 arrival times (with median of eight readings) from well-distributed stations around the epicenters. It is worth noting here that all digital stations recorded within 25 km of the epicenters. Because of this, the horizontal and vertical location errors were quite small. Median values of the location errors are rms of 0.07 s, horizontal error (erH) of 0.6 km, and depth error (erZ) of 0.6 km. Table 1 contains the results of the location process for the earthquakes of this sequence.

An interesting feature of these 2000–2001 earthquakes is that, even though they occurred evenly

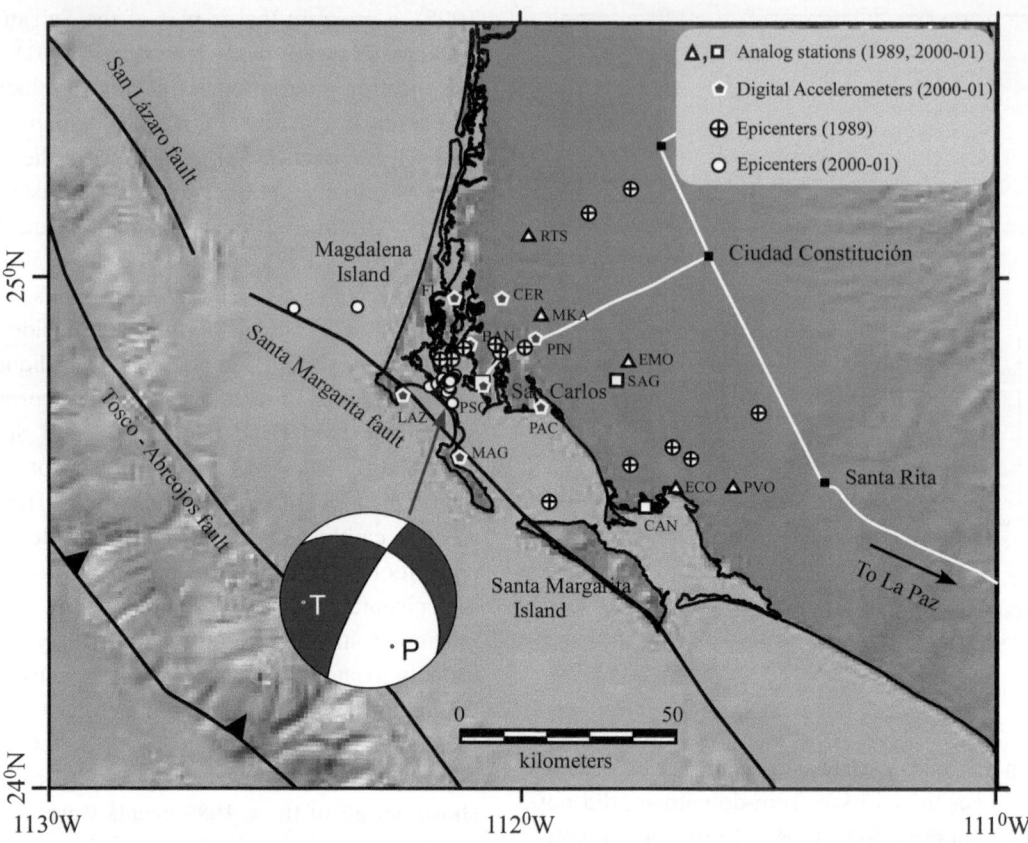

Figure 4
Map that shows the sites of the recording seismic stations together with the epicenters of earthquakes recorded in the 1989 and 2000–2001 field surveys

distributed in time along a 7-month period, their sources occurred tightly clustered in a zone of 5-km radius. Only two micro earthquakes of M_w 2.3 and 2.5 occurred offshore of the Magdalena Island, at distances of 35 and 45 km from downtown San Carlos. Earthquakes of this zone of concentrated activity occurred at 5–10 km from San Carlos.

4.2. Relocation of the M_w 5.4 and 5.3 Earthquakes of February 2004

Although the 2004 earthquake sequence developed during the period 5–15 February, the two stronger events of the series occurred on 9 February. The NCEDC teleseismic locations for these particular events are marked with the numbers 13 and 16 in Fig. 2. In San Carlos, people did not feel these earthquakes, and that is why we did not install

temporary instruments on that occasion. Nevertheless, seismic stations of the NARS-Baja array recorded the two larger events at epicentral distances of 150 to 600 km. These distances are rather large, but the azimuthal coverage provided by these stations was better than that of the teleseismic stations. Using the data from NARS stations we attempted to refine the global epicenter determinations of the NCEDC listings. For the M_w 5.4 and 5.3 earthquakes, the newly located epicenters are shown in Fig. 2 with white circles inside squares (numbers 13* and 16*). We note that the relocated epicenters are ~30 km NNW of the epicenters reported in the NCEDC catalog and that the distances from these earthquakes to San Carlos are larger than originally reported. This explains why the residents of San Carlos did not feel these earthquakes. It is also worth noting from Fig. 2 that the new epicenters lie close to the global

Table 1

Epicenter locations of 31 events of the 2000–2001 earthquake series

Date yr mo day	Time hr min	Lat. (°)	Long. (°)	Depth (km)	M_w	NR	Az (°)	CD (km)	rms (s)	erH (km)	erZ (km)
2000 11 24	22 41	24.78	−112.16	5.71	1.3	6	303	7.0	0.02	0.30	0.40
2000 11 28	20 21	24.77	−112.16	7.86	1.5	7	283	7.3	0.07	0.90	0.90
2000 12 04	15 13	24.78	−112.19	6.64	1.8	7	297	10.4	0.08	0.90	0.60
2000 12 04	19 34	24.78	−112.18	6.84	1.6	7	293	9.3	0.07	0.80	0.40
2000 12 08	08 22	24.78	−112.18	6.95	1.8	7	295	9.6	0.08	0.90	0.50
2000 12 09	16 29	24.79	−112.17	9.62	1.9	6	288	10.9	0.02	0.30	0.40
2000 12 11	20 37	24.80	−112.16	10.14	1.8	5	284	9.2	0.01	0.20	0.20
2000 12 31	23 59	24.77	−112.17	6.90	1.5	7	287	8.2	0.07	0.80	0.50
2001 01 02	22 07	24.77	−112.16	6.91	1.5	5	285	7.7	0.03	0.50	0.60
2001 01 18	01 27	24.94	−112.35	5.63	2.5	6	337	23.7	0.09	0.50	0.30
2001 01 26	04 48	24.80	−112.13	9.49	1.8	7	263	4.7	0.08	0.90	0.80
2001 03 01	01 05	24.78	−112.15	6.43	1.6	14	117	6.4	0.11	0.30	0.90
2001 03 16	05 05	24.94	−112.48	6.00	2.2	8	299	15.2	0.12	1.90	1.50
2001 03 18	17 50	24.78	−112.15	7.05	1.9	10	206	6.7	0.07	0.60	0.60
2001 03 23	02 39	24.77	−112.17	5.56	1.4	7	288	8.3	0.04	0.60	0.90
2001 03 23	03 13	24.78	−112.15	7.00	1.7	14	108	6.1	0.08	0.20	0.30
2001 04 22	09 28	24.78	−112.18	5.88	1.5	6	291	8.9	0.05	0.90	1.40
2001 04 29	07 02	24.80	−112.14	6.44	2.3	8	200	8.7	0.07	0.70	1.20
2001 05 12	21 42	24.79	−112.14	6.76	1.9	11	114	5.6	0.08	0.30	0.70
2001 05 13	00 36	24.78	−112.14	7.83	2.2	12	107	5.6	0.07	0.20	0.50
2001 05 25	14 35	24.78	−112.16	6.40	1.7	8	303	7.0	0.07	0.80	0.90
2001 05 25	18 17	24.77	−112.16	6.96	1.5	8	307	7.4	0.09	1.00	0.40
2001 05 31	11 07	24.78	−112.15	7.11	1.7	11	109	6.3	0.09	0.30	0.60
2001 06 03	20 54	24.77	−112.15	6.63	1.7	10	111	6.3	0.08	0.30	0.70
2001 06 04	06 37	24.78	−112.15	8.15	1.3	6	188	6.0	0.02	0.30	0.40
2001 06 07	23 59	24.79	−112.14	8.47	2.1	10	168	5.2	0.08	0.50	0.60
2001 06 08	00 01	24.79	−112.14	8.45	1.7	10	172	5.3	0.06	0.40	0.50
2001 06 08	02 53	24.80	−112.15	6.55	1.4	8	289	6.4	0.07	0.80	0.80
2001 06 08	03 03	24.79	−112.14	7.57	1.8	10	165	4.7	0.06	0.40	0.50
2001 06 11	21 49	24.79	−112.15	7.87	1.7	8	291	6.2	0.07	0.70	0.60
2001 07 01	01 31	24.79	−112.14	5.46	1.4	10	114	5.6	0.12	0.50	1.10

NR is the number of arrival times used in the earthquake locations; CD is distance to the closest station, rms is the root mean square value and erH and erZ are horizontal and vertical location errors, respectively

Centroid Moment Tensor locations (CMT) (black dots) (DZIEWONSKI *et al.* 1981; EKSTRÖM *et al.* 2012).

4.3. Moment Magnitude of the 2000–2001 Micro Earthquakes

The ground accelerations recorded from the 2000–2001 earthquakes were of small amplitude and of little interest from the engineering point of view. Such low accelerations did not warrant a detailed analysis in this study. We only used the acceleration data to estimate the size of the micro earthquakes. For that, we first calculated the seismic moment, Mo, of the earthquakes first, via standard spectral analysis of the recorded shear waves (BRUNE 1970). We then used the moments to obtain the moment magnitude, M_w, by using the formula of HANKS and KANAMORI (1979). The magnitudes estimated for the recorded micro earthquakes were in the 1.3–2.5 range. For such small events, the recorded peak ground accelerations were up to 8.4 cm/s^2 at distances of 5–10 km from the sources.

4.4. Focal Depth and Composite Fault Plane Solution for the 2000–2001 Earthquakes

Figure 5a shows an enlarged view of the zone of earthquakes that occurred to the west of San Carlos

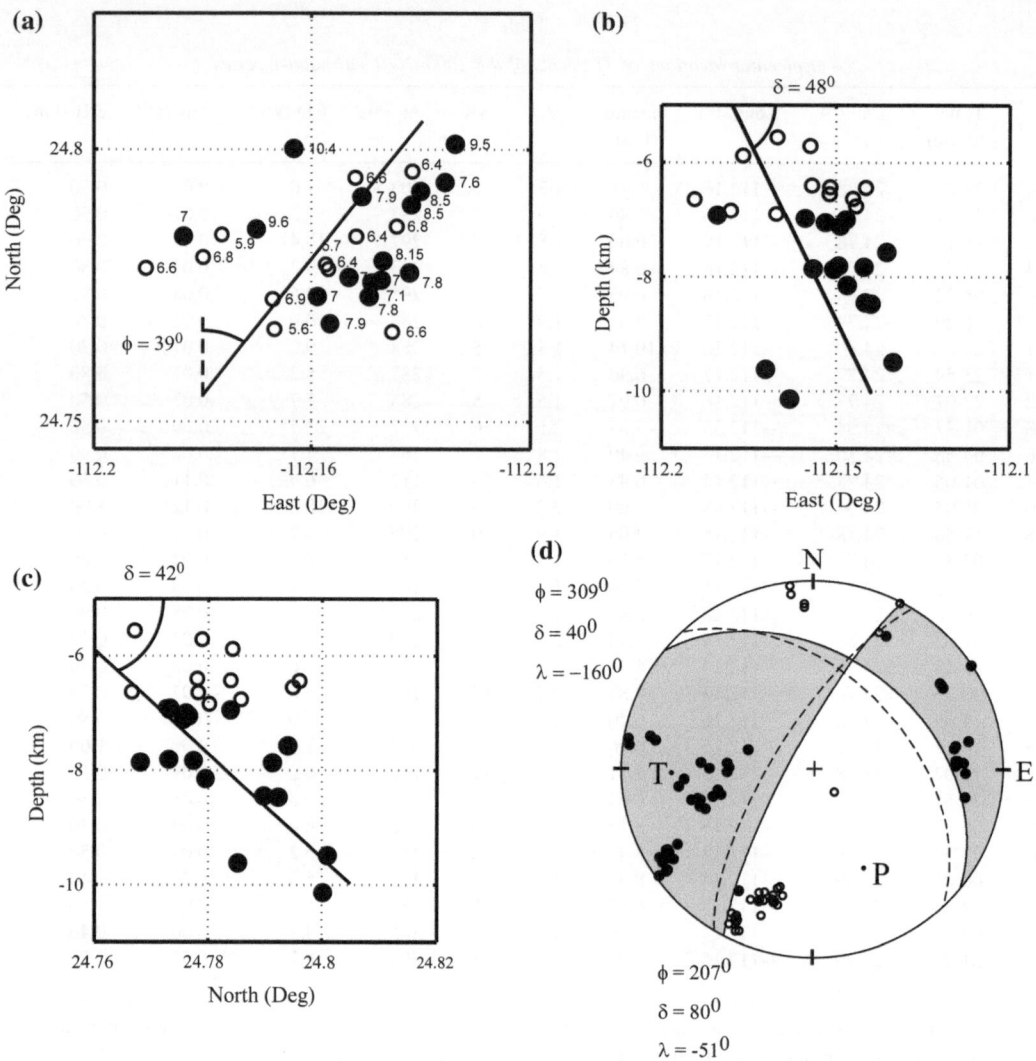

Figure 5
a Enlarged view of the area of clustered earthquakes that occurred during the 2000–2001 interval. The *number* located to one side of each epicenter indicates the focal depth of the correspondent event. *Solid circles* stand for earthquakes with depths equal to 7.0 or larger. The line plotted with azimuth of 39° shows an apparent trend of the epicenters. In **b** and **c**, we show plots of focal depth versus the east and north geographic coordinates, respectively. The *straight line* on these plots corresponds to the line drawn in the map of Fig. 5a projected from 5 to 10 km depth. **d** Composite fault plane solution determined with P-wave polarity data from the same group of earthquakes. The *dashed lines* in **d** represent the CMT double-couple solution for the M_w 5.4 event of 9 February 2004, at 01:24 h (UTC)

during the 2000–2001 interval. Numbers located to one side of the epicenters indicate the focal depth of the earthquake sources, which vary from 5.6 to 10.4 km. Those earthquakes for which the focal depths were equal to 7.0 km or larger are marked with black circles in the figure. Although the epicenters were located within a small area, it

appears that they follow NE trend, as indicated by the straight line plotted at an azimuth of 39°.

Figure 5b and c shows plots of focal depth versus the east and north geographic coordinates, respectively. The straight line on these plots corresponds to the line drawn in the map of Fig. 5a, projected from 5 to 10 km depth. These two figures show that the focal depth increased in E–NE direction, at an angle from the surface that is between 42° and 48°. This might not be strongly convincing, due to the limited number of events and the small volume that contains the hypocenters. Nevertheless, if this trend in the focal depth is true, then the data would be consistent with the composite fault plane solution of Fig. 5d, as is described next.

The fault plane solution of an earthquake is a useful tool to describe the style of faulting in a given region. In this study, however, due to the small size of the analyzed earthquakes we could not obtain fault plane solutions for individual events. Nonetheless, using the P-wave polarity data from 21 of the closely spaced events, we determined the composite fault plane solution of Fig. 5d. The assumption here was that all of these events occurred on the same fault plane and with the same focal mechanism, which seemed to be a valid assumption given the closely spaced sources.

It is worth noting that our composite fault plane solution is quite similar to the global CMT double-couple solutions for the M_w 5.4 and 5.3 events of 9 February 2004 (see Fig. 2; Table 2). For comparison, we drew the CMT mechanism of the M_w 5.4 event of 9 February 2004 (at 01:24 h) (dashed lines) on top of the fault plane solution of Fig. 5d. One may see that the two mechanisms are nearly equal, even though the stronger earthquake occurred 3 years later and at ~60 km northwest of the San Carlos events (event 13* in Fig. 2). Such strong similarity between the focal mechanisms tells us that the two earthquake sequences occurred along the same active fault, but at different times.

Our fault plane solution has a plane that strikes in SE–NW direction and dips to NE, consistent with the San Lázaro-Santa Margarita and the Tosco-Abreojos faults systems that cut through the area of study (e.g., NORMARK et al. 1987; SPENCER and NORMARK 1979; FLETCHER et al. 2007; BROTHERS et al. 2012; SALAZAR 2014). The seismicity analyzed here, however, was located 5 km east of the northwest segment of the Santa Margarita Fault and ~50 km east of the Tosco-Abreojos Fault. It is more likely, then, that the earthquakes had occurred in association with the Santa Margarita Fault. The apparent depth-increasing E-NE trend of the earthquakes plus features of the focal mechanisms seems to support this conclusion.

Based on the spatial location of the analyzed earthquakes, we confirm the existence of active faults east of the Tosco-Abreojos fault system. As shown here, such faults have the potential to generate earthquakes of low to intermediate magnitudes. This also implies that transtensional deformation is taking place in a wide zone along the Pacific margin of Baja California Sur, and it is not limited to the Tosco-Abreojos fault system. More seismic data are required to improve the present understanding of the geometry and kinematics of the active faults of the region.

4.5. Examination of the Current State of Stress in the Region

In this section, our goal is to get some insights into the state of the tectonic stress that prevails in the region. For this, we will compare the focal mechanisms considered in this study with the focal

Table 2

Global CMT fault plane solution data for two earthquakes that occurred on February 2004 and for the composite fault plane solution determined in this study

Date yr mo day	Time hr mn	Latitude (°)	Longitude (°)	Depth (km)	M_w	Strike (°)	Dip (°)	Rake (°)
CMT fault plane solutions								
2004 02 09	01 24	25.06	−112.60	12	5.4	317	49	−157
						211	73	−43
2004/02/09	09 03	25.13	−112.56	12	5.3	312	37	−138
						186	66	−60
Composite fault plane solution for the 2000–2001 earthquakes								
Fault plane 1						309	40	−160
Fault plane 2						207	80	−51

mechanisms for other earthquakes of the region. To begin that, Table 3 summarizes the azimuth and plunge angles of P and T stress axes from the mechanisms of this study and from other regional earthquakes. The first earthquakes to consider are two M 5.3 events reported by MOLNAR (1973). These earthquakes occurred in August 1969 southwest of Todos Santos, at 240 km southeast of San Carlos; Figure 6a shows their fault-plane solutions. Figure 6b shows projections, onto the equatorial plane, of the P axes (continuous lines) and the T axes (discontinuous lines) from Molnar's mechanisms and from the mechanism of this study. Concentric circles were drawn to serve as scale for the plunging angles. With this scale, the longer the projection lines, the more horizontal the P or T axes are.

Figure 6b also includes global average orientations of the P and T axes of stress from mechanisms for earthquakes of La Paz–Los Cabos and of the Gulf of California regions (MUNGUÍA et al. 2006). We indicate those regional stress orientations with the arrows labeled P_{LPC}, P_{GC}, T_{LPC}, and T_{GC}. MUNGUÍA et al. (2006) determined the P_{GC} and T_{GC} stress orientations from highly consistent directions of P and T axes on mechanisms for 13 representative events of the Gulf of California fault system studied by GOFF et al. (1987). These earthquakes had their epicenters distributed along the Gulf fault system, from the Delfin Basin zone (29°N) to the Tamayo Fracture Zone (23°N). Thus, the calculated trends for

P_{GC} and T_{GC} are close approximations to the orientations of the regional stresses that drive the Gulf of California fault system. For these Gulf earthquakes, MUNGUÍA et al. (2006) estimated average plunge angles of $\sim 9°$ and $\sim 12°$ for the P_{GC} and T_{GC} axes, respectively. Such near horizontal axes of stress are clear indication of predominant strike-slip faulting within the Gulf region.

The P_{LPC} and T_{LPC}, on the other side, are average trends of the stress axes calculated from earthquakes of La Paz–Los Cabos region (MUNGUÍA et al. 2006). At the southern part of that region, the analyzed earthquakes had mechanisms that showed predominant strike-slip motions. The mechanisms reported by MOLNAR (1973) (Fig. 6a, b) are additional examples of earthquakes with predominant strike-slip faulting at the southern part of La Paz–Los Cabos region. At the northern part of this region (24°–24.5°N), including the eastern margin of Baja California Sur, the earthquakes were of the normal-fault type. In this case, the P axes had a mean plunge angle of 55° at an azimuth of 168°, whereas the T axes were nearly horizontal, with a 61° average trend (see Table 3).

According to information from Table 3, the T axes on focal mechanisms for earthquakes of the Gulf of California and of the northern La Paz–Los Cabos region are nearly horizontal. In addition, we see that the orientation of those axes varies only within 36°. These variations in the orientation of the axes of stress are probably due to motions of Baja

Table 3

Azimuth and plunge angles of P and T axes determined from focal mechanisms for the earthquakes studied herein and from earthquakes of previous studies

Events	P axes		T axes	
	Azimuth (°)	Plunge (°)	Azimuth (°)	Plunge (°)
2004	Event 1: 166	44	88	14
	Event 2: 136	58	76	12
2006	154	40	88	26
Average of 2004 and 2006 data:	152	44	84	17
Northern La Paz–Los Cabos region	169	57	65	8
Entire La Paz–Los Cabos region	158	41	61	7
Gulf of California	170	9	84	12
MOLNAR (1973)	Event 1: 156	40	55	12
	Event 2: 142	32	51	0

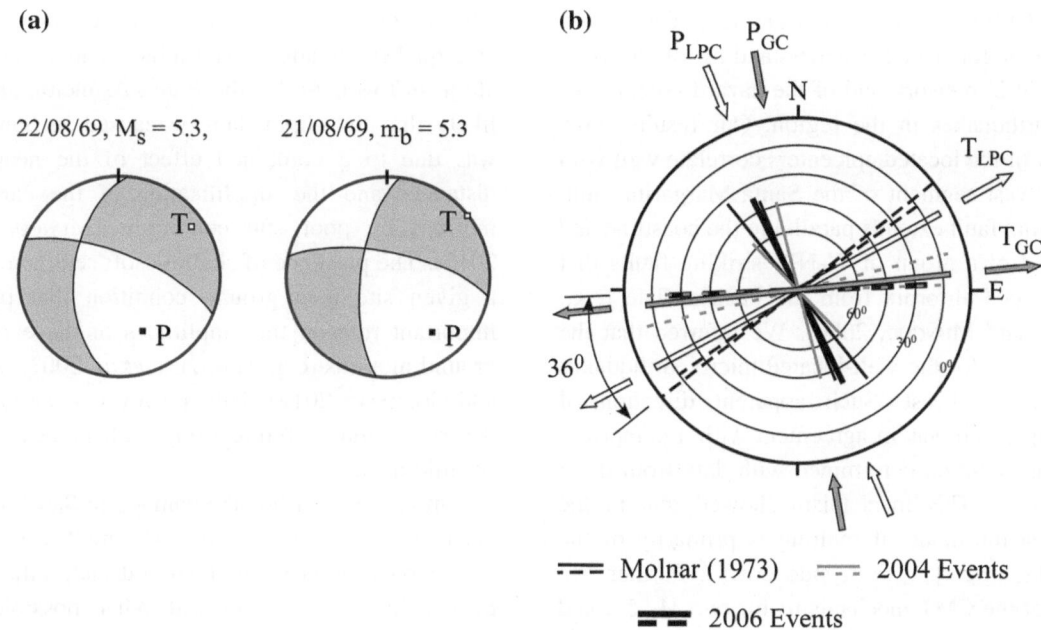

(a)

22/08/69, M_s = 5.3,   21/08/69, m_b = 5.3

(b)

=== Molnar (1973)   === 2004 Events

=== 2006 Events

Figure 6

a Fault plane solutions for earthquakes of the Pacific margin of Baja California Sur, as reported in the study of MOLNAR (1973). **b** P and T axes from the focal mechanisms in **a** and from the mechanisms obtained in this study projected onto the equatorial plane. The concentric circles are a scale for the plunge angles. With this scale, the longer the lines of projection, the more horizontal the P or T axes are. *Arrows* with the P_{LPC}, P_{GC}, T_{LPC}, and T_{GC} labels are the global average stress orientations for the La Paz–Los Cabos (LPC) and for the Gulf of California (GC) regions, as taken from MUNGUÍA *et al.* (2006)

California as a micro plate. We envisage a complex tectonic situation for the region since Baja California is loosely coupled with the North America plate and its transfer to the Pacific plate is not yet complete (e.g., FLETCHER and MUNGUÍA 2000; MICHAUD *et al.* 2004; PLATTNER *et al.* 2009; among others). Under this situation, the seismicity occurs in response to a combination of local and regional tectonic stresses.

In contrast to the T axis, the P axes from mechanisms of the La Paz–Los Cabos region earthquakes plunge at an average angle that is higher (55°) than that for the Gulf earthquakes (9°). In this study, we found a similar result for the case of the 2004 and 2006 earthquakes. In this case, since the mechanisms of these events and our composite mechanism were all similar, we averaged the plunge and azimuth of the P and T axes. For the P axis, the results showed an average plunge angle that is about 35° larger than the plunge angle of mechanisms of the Gulf earthquakes. Such higher plunge angle reflects an important component of oblique fault motion. Thus, the mixture of oblique- and strike-slip faulting is clear evidence of a transtensional stress regime in the

region of study. In a study of the 2006 earthquake swarm of Bahía Asunción, MUNGUÍA *et al.* (2015) reported closely spaced hypocenters for earthquakes with normal and strike-slip fault mechanisms. This swarm occurred 350 km northwest of San Carlos and ~35 m to the east of the Tosco-Abreojos fault. The seismicity of Bahía Asunción and San Carlos areas is thus evidence of wide zones of transtensional deformation along the western margin of Baja California Sur. According to this, the western margin of Baja California Sur is a region characterized by active tectonism, with faults that accommodate part of the relative plate motion, as proposed in many other studies (e.g., HUMPHREYS and WELDON 1991; DIXON *et al.* 2000; FLETCHER and MUNGUÍA 2000; MICHAUD *et al.* 2005; MUNGUÍA *et al.* 2006; FLETCHER *et al.* 2007).

5. Summary and Conclusions

We studied earthquakes that occurred along the western margin of Baja California Sur, at 200 km from the Gulf of California fault system. Their study is important for a better understanding of the seismotectonic framework and of the hazard potential of future earthquakes in the region. Our results show that most of the located epicenters correlate well with the northwest segment of the Santa Margarita fault. This normal fault extends parallel to the coastline and forms part of a group of W-NW striking faults that isolate Baja California from the Pacific plate (e.g., FLETCHER and MUNGUÍA 2000). We inferred that the source depths for the well-located micro earthquakes deepen to northeast. Such apparent direction of increasing depth was in agreement with a composite fault plane solution determined with data from these small events. This mechanism showed that in the study area the mode of faulting is primarily of the normal faulting type (east side down). Further, we found that the CMT mechanisms for two M_w 5.4 and 5.3 earthquakes that occurred in the area on 9 February 2004 had nearly the same mechanism. This fact suggested to us that the larger size earthquakes occurred also along the Santa Margarita Fault. It seems then that the Santa Margarita fault is one of the most active faults to the west of Baja California Sur. At least in the period of historic seismicity considered here, this fault, and possibly other faults of the Tosco-Abreojos fault zone, accommodated some of the total Pacific–North American slip rate. Other earthquakes of intermediate magnitude that have occurred to the south and north of our study area also had focal mechanisms showing substantial components of normal faulting (MOLNAR 1973; FLETCHER and MUNGUÍA 2000; MICHAUD et al. 2005, 2011; MUNGUÍA et al. 2006, MUNGUÍA 2015). Then, these overall results support the hypothesis of plate motion partitioning between the Gulf of California and the Pacific margin of Baja California (e.g., HUMPHREYS and WELDON 1991; DEMETS 1995; DIXON et al. 2000; FLETCHER and MUNGUÍA 2000; MICHAUD et al. 2005, MUNGUÍA et al. 2006; FLETCHER et al. 2007; MUNGUÍA et al. 2015).

Another important feature of this study deals with the size of the analyzed earthquakes. The earthquakes seem to be of little relevance due to their low magnitudes. However, accurate location of their sources revealed a zone in which stresses accumulate and release more often in the region. The earthquakes of that zone occur at 6- to 10-km distance of San Carlos.

With magnitudes of slightly over 3.0, some of those earthquakes already caused minor damage and some alarm in town. As for the Bahía Asunción area, it is likely that the slight damage caused in San Carlos was due to a combined effect of the near-source distances and the amplification of the earthquake motions by poor soil conditions (MUNGUÍA et al. 2015). The presence of shallow soft sediments below a given site is a ground condition that plays an important role on the amplitudes of the earthquake ground motions (e.g., HARTZELL et al. 2003; MUNGUÍA and GONZÁLEZ 2012). This, however, is an issue that deserves detailed future study with more and better seismic data.

Up to now, earthquake damage in San Carlos has been minimal because of the small size of the occurred earthquakes. In the past decades, the activity of the Santa Margarita and other possible faults resulted only in earthquakes of low-to-intermediate magnitudes. However, the possible occurrence of future stronger earthquakes near San Carlos is something that we cannot discard. This is a possibility that poses a seismic hazard of unknown level to the town. Hence, the earthquakes analyzed in this study should serve as a remainder to San Carlos residents of the danger upon occurrence of stronger events.

Acknowledgments

The authors are grateful to numerous local authorities of Comondú and San Carlos for the economic and/or logistic support they provided to us during the field surveys. They also thank Javier Gaitán from Universidad Autónoma de Baja California Sur for logistic support provided during the 1989 field survey. The main financial support for this study was provided by Centro de Investigación Científica y de Educación Superior de Ensenada, B. C. (CICESE). The authors deeply appreciate the administrative support provided during the project by personnel of this institution. The authors would also like to acknowledge the work of Dr. William Bandy and two anonymous reviewers. Their reviews and constructive comments and suggestions were helpful in improving the manuscript.

REFERENCES

ATWATER, T. (1970). *Implications of plate tectonics for the Cenozoic tectonic evolution of western North America*, Geol. Soc. Am. Bull., *81*, 3513–3536.

ATWATER, T. (1989). Plate tectonic history of the northeast Pacific and western North America, in The Eastern Pacific Ocean and Hawaii, E. L. Winterer, D. M. Hussong, and R. W. Decker (Editors), Geol. Soc. Am., Boulder, Colorado, 21–72.

BOHANNON, R. G., and T. PARSONS (1995). *Tectonic implications of post-30 Ma Pacific and North American relative plate motions*, Geol. Soc. Am. Bull., *107*, 937–959.

BROTHERS, D., A. HARDING, A. GONZÁLEZ-FERNÁNDEZ, W. S. HOLBROOK, G. KENT, N. DRISCOLL, J. FLETCHER, D. LIZARRALDE, P. UMHOEFER, and G. AXEN (2012), *Farallon slab detachment and deformation of the Magdalena Shelf, southern Baja California*, Geophys. Res. Lett., *39*, L09307, doi:10.1029/2011GL050828.

BRUNE, J. N. (1970). *Tectonic stress and the spectra of seismic shear waves from earthquakes*, J. Geophys. Res., *75*, 4997–5009.

CASTRO, R. R., C. VALDÉS-GONZÁLEZ, P. SHEARER, V. WONG, L. ASTIZ, F. VERNON, A. PÉREZ-VERTTI, and A. MENDOZA (2011). *The 3 August 2009 M_w 6.9 Canal de Ballenas Region, Gulf of California, Earthquake and Its Aftershocks*. Bull. Seismol. Soc. Am., *101*, No. 3, pp. 929–939, doi: 10.1785/0120100154.

CLAYTON, R.W., J. TRAMPERT, C. REBOLLAR, J. RITSEMA, P. PERSAUD, H. PAULSSEN, X. PÉREZ-CAMPOS, A. WETTUM, A. PÉREZ-VERTTI, F. DiLuccio (2004). *The NARS-Baja Seismic Array in the Gulf of California Rift Zone*. Margins Newsletter, No. 13, 1-4.

DeMETS, C. A. (1995). *Reappraisal of seafloor spreading lineations in the Gulf of California: implications for the transfer of Baja California to the Pacific plate and estimates of Pacific-North America motion*, Geophys. Res. Lett., *22*, 3545–3548.

DIXON, T., F. FARINA, C. DeMETS, F. SUÁREZ-VIDAL, J. M. FLETCHER, B. MÁRQUEZ-AZUA, M. MILLER, O. SÁNCHEZ, and P. UMHOEFER (2000). *New kinematic models for Pacific-North America motion from 3 Ma to Present. II Evidence for Baja California shear zone:* Geophys. Res. Lett., *27*, p. 3961–3964.

DZIEWONSKI, A. M., T. A. CHOU and J. H. WOODHOUSE (1981). *Determination of earthquake source parameters from waveform data for studies of global and regional seismicity*, J. Geophys. Res., *86*, 2825–2852, doi: 10.1029/JB086iB04p02825.

EKSTRÖM, G., M. NETTLES, and A. M. DZIEWONSKI (2012). *The global CMT project 2004-2010: Centroid-moment tensors for 13,017 earthquakes*, Phys. Earth. Planet. In., *200–201*, 1–9, doi:10.1016/j.pepi.2012.04.002.

FLETCHER, J. M., and L. MUNGUÍA (2000). *Active continental rifting in southern Baja California, México: implications for plate motion partitioning and the transition to seafloor spreading in the Gulf of California*, Tectonics, *19*, No. 6, 1107–1123.

FLETCHER, J.M., M. GROVE, D. KIMBROUGH, O. LOVERA, and G. E. GEHRELS (2007). *Ridge-trench interactions and the Neogene tectonic evolution of the Magdalena Shelf and southern Gulf of California: Insights from detrital zircon U-Pb ages from the Magdalena fan and adjacent areas*. Geol. Soc. Am. Bull., *119*; no. 11/12; p. 1313–1336; doi: 10.1130/B26067.1.

GOFF, J. A., E. A. BERGMAN, and S. C. SOLOMON (1987). *Earthquake source mechanisms and transform fault tectonics in the Gulf of California*, J. Geophys. Res., *92*, 10,485–10,510.

HANKS, T.C. and H. KANAMORI (1979). *Moment magnitude scale.* J. Geophys. Res., *84* (B5): pp. 2348–2350. doi: 10.1029/JB084iB05p02348.

HARTZELL, S., DAVID CARVER, ROBERT A. WILLIAMS, STEPHEN HARMSEN, and ASPASIA ZERVA (2003). *Site Response, Shallow Shear-Wave Velocity, and Wave Propagation at the San Jose, California, Dense Seismic Array*. Bull. Seism. Soc. Am., Vol. *93*, No. 1, pp. 443–464.

HAVSKOV, J., and L. OTTEMÖLLER (Editors) (2001). SEISAN: the earthquake analysis software for Windows, SOLARIS, and LINUX, Version 7.2. Manual, Institute of Solid Earth Physics, University of Bergen, Norway.

HUMPHREYS, E. D., and R. J. WELDON, II (1991). Kinematic constraints on the rifting of Baja California, in The Gulf and Peninsular Province of the Californias, J. P. Dauphin and B. R. T. Simoneit (Editors), American Association of Petroleum Geologists Memoir *47*, 217–229.

LEE, W. H. K., R. E. BENNETT, and K. L. MEAGHER, (1972). A method of estimating magnitude of local earthquakes from signal duration. U.S. Geological Survey Open File Report, 28 pp.

LIENERT, B. R. E. and J. HAVSKOV (1995). *A computer program for locating earthquakes both locally and globally*. Seismol. Res. Lett., *66*:26-36.

LONSDALE, P. F. (1991). Structural patterns of the Pacific floor offshore of peninsular California, in The Gulf and Peninsular Province of the Californias, edited by J. P. Dauphin and B. R. T. Simoneit, AAPG Mem., *47*, 87–125.

MICHAUD, F., SOSSON, M., ROYER, J.-Y., CHABERT, A., BOURGOIS, J., CALMUS, T., MORTERA, C., BIGOT-CORMIER, F., BANDY, W., DYMENT, J., PONTOISE, B., and SICHLER, B. (2004), *Motion partitioning between the Pacific plate, Baja California and the North America plate: The Tosco-Abreojos fault revisited*: Geophysical Research Letters, v. *31*, p. L08604, doi: 10.1029/2004GL019665.

MICHAUD F, T. CALMUS, M. SOSSON, J-Y. ROYER, J. BOURGOIS, A. CHABERT, F. BIGOT-CORMIER, B. BANDY, C. MORTERA-GUTIÉRREZ (2005). *La zona de falla Tosco-Abreojos: un sistema lateral derecho activo entre la placa Pacífico y la península de Baja California*. Boletín de la Sociedad Geológica Mexicana. Tomo LVII, Núm. 1., p. 53-63.

MICHAUD, F., CALMUS, T., RATZOV, G., ROYER, J. Y., SOSSON, M., BIGOT-CORMIER, F., BANDY, W. and MORTERA-GUTIÉRREZ, C. (2011). *Active deformation along the southern end of the Tosco-Abreojos fault system: New insights from multibeam swath bathymetry*. Pure Appl. Geophys., *168*(8–9), 1363–1372, doi:10.1007/s000240100193y.

MOLNAR, P. (1973). *Fault plane solutions of earthquakes and direction of motion in the Gulf of California and in the Rivera fracture zone*, Geol. Soc. Am. Bull., *84*, 1651–1658.

MUNGUÍA, L., M. REICHLE, A. REYES, R. SIMONS, and J. N. BRUNE (1977). *Aftershocks of the 8 July 1975 Canal de las Ballenas, Gulf of California earthquake*, Geophys. Res. Lett., *4*, 507–509.

MUNGUÍA, L., M. GONZÁLEZ, S. MAYER, and A. AGUIRRE (2006). *Seismicity and State of Stress in the La Paz–Los Cabos Region, Baja California Sur, México*. Bull. Seismol. Soc. Am., *96*, No. 2, pp. 624–636, April 2006, doi: 10.1785/0120050114.

MUNGUÍA, L., S. MAYER, A. AGUIRRE, I. MÉNDEZ, M. GONZÁLEZ-ESCOBAR, and M. LUNA (2015). *The 2006 Bahía Asunción earthquake swarm: Seismic evidence of active deformation along the western margin of Baja California Sur, Mexico*. Pure Appl. Geophys. Doi 10.1007/s00024-015-1184-9.

MUNGUÍA, L. and M. GONZÁLEZ (2012). Linear and nonlinear soil response at the Mexicali Valley, Baja California, México during the El Mayor-Cucapah earthquake of 4 April 2010 (M_w 7.2) and other past earthquakes of the region. Abstract of SSA Annual Meeting, April 2012, San Diego, California, U. S. (ID: 15559).

NORMARK, W.R., J.E. SPENCER, and J. INGLE (1987). Geology and Neogene history of the Pacific continental margin of Baja California Sur, Mexico, in Scholl, D.W., *et al.* eds., Geology and resource potential of the continental margin of western North America and adjacent ocean basins—Beaufort Sea to Baja California: Houston, Texas, Circum-Pacific Counsel for Energy and Mineral Resources, Earth Science Series, 6, p. 449–472.

PLATTNER, C., R. MALSERVISI, R. GOVERS (2009). *On the plate boundary forces that drive and resist Baja California motion.* Geology, *37*, no. 4, p. 359–362; doi: 10.1130/G25360A.1.

REICHLE, M. S., G. F. SHARMAN, and J. N. BRUNE (1976). *Sonobuoy and teleseismic study of Gulf of California transform fault earthquake sequences.* Bull. Seismol. Soc. Am., *66*, No. 5, pp. 1623 1641.

SALAZAR, C. R. M. (2014). Evidencias estructurales y sismo estratigráficas en la parte central de la plataforma Magdalena, margen occidental de Baja California a partir de sísmica de reflexión. MsC thesis, 89 pp., Centro de Investigación Científica y de Educación Superior de Ensenada, B. C., Ensenada, B. C., México.

SPENCER, J. E., and W. R. NORMARK (1979). *Tosco-Abreojos fault zone: A Neogene transform plate boundary within the Pacific margin of southern Baja California, Mexico*: Geology, *7*, p. 554–557, doi: 10.1130/0091-613.

STOCK, J. M. and K. V. HODGES (1989). *Pre-Pliocene extension around the Gulf of California and the transfer of Baja California to the Pacific plate.* Tectonics, *8*, No. 1, p. 99–115.

(Received June 19, 2015, revised November 24, 2015, accepted November 25, 2015, Published online December 15, 2015)

Pure Appl. Geophys. 173 (2016), 3645–3661
© 2016 Springer International Publishing
DOI 10.1007/s00024-016-1396-7

Pure and Applied Geophysics

Structural and Seismic Stratigrapic study in the Center of the Magdalena Shelf in the Western Margin of Baja California Based on Seismic Reflection Data

MARIO GONZÁLEZ-ESCOBAR,[1] ROSA M. SALAZAR-CÁRDENAS,[1] LUIS MUNGUÍA,[1] ARTURO MARTÍN,[1] and FRANCISCO SUÁREZ-VIDAL[1]

Abstract—The Magdalena Shelf is a shallow, low-relief surface located along the Baja California Pacific margin. As part of a forearc basin, the shelf was a convergent margin setting before the oblique divergent plate boundary formed in the Gulf of California at 12 Ma. It is thought that since 12–8 Ma, this basin has been a transtensional or strike–slip basin. To constrain the geometry, structural characteristics and some stratigraphic relationships, an active-source, seismic-reflection study was carried out in the central part of the shelf. As a result, the analyzed data show faults, basins and unconformities. Two out of four observed basins are clearly controlled by the Santa Margarita and San Lázaro faults that dip ~40° NE; a third basin is controlled by the Tosco-Abreojos fault. These three basins are part of the deformation zone that is associated with the Tosco-Abreojos fault system. The Iray-Margarita basin, on the other hand, is a fourth basin located at the northeast sector of the study area. An additional feature observed is a stepover lying between the overlapping ends of the Santa Margarita and San Lázaro faults. Small faults oriented sub-parallel to the above major faults are present, mainly throughout the western sector of the study area. Some of those minor faults cut through the seafloor indicating recent tectonic activity. Santa Margarita, San Lázaro and Tosco-Abreojos are also the names given to half-grabens controlled by the active faults that have the same names. The first two basins are affected by many more small faults in comparison with what we see in the third basin. Tectonically, this means that those two basins are the more active in the area of study. In all four basins, the upper seismic sequence consists of sediments controlled by faults of Neogene age. We found that the Iray-Santa Margarita basin is the deepest of all four basins (beyond the resolution of the data, >5 km), and lack of minor faults there indicates that the basin is not tectonically active. Two unconformities are present in the region, unconformity-1 of Miocene age and unconformity-2 of Paleocene-Eocene and Cretaceous ages. Unconformity-1 is present in the entire region, while Unconformity-2 is present mainly in the Iray-Magdalena and San Lázaro basins. In the Iray-Magdalena basin the sedimentary sequences show an uplift that took place in late Jurassic time along its western portion. We also see that moderate compressive deformation increases gradually from west to east. A chaotic-reflector oriented

NW–SE is observed with an irregular shape in all regions. It shows very superficial in some places, sub-parallel to the major faults and in association with the subduction complex. Seismic activity reported for the region showed a strong correlation with the fault plane of the Santa Margarita fault, indicating that the Tosco-Abreojos deformation zone is ~90 km wide in the area. Finally, the Santa Margarita fault should be considered in future hazard-risk assessments because of their proximity to Puerto San Carlos. As it has been reported, local seismic events ($M < 4$) have generated alarm in the local populace, as well as soil liquefaction and minor damage to structures in the region.

Key words: Magdalena Shelf, forearc basin, Tosco-Abreojos fault, Santa Margarita-San Lázaro fault.

1. Introduction

The Baja California peninsula is a fragment of continental crust rifted from the continental region of Mexico and captured by the Pacific plate at 10–15 Ma (Normark et al. 1987; Michaud et al. 2004, 2005, 2006, 2007, 2011; Fletcher et al. 2007; Plattner et al. 2007; Brothers et al. 2012, amongst others). The formation and subsequent capture of continental microplates in subduction margins are controlled somehow by the underlying oceanic slab, possibly through basal tensile or shear distributed and transmitted through the interface of the continental crust and underlying relic slab. It is also controlled by the interaction between the subduction zone and oceanic ridge, as well as the subsequent evolution of triple junctions (Nicholson et al. 1994; Bohannon and Parsons 1995; Brothers et al. 2012). Convergent margins are initiated due to the ageing of the lithosphere, the cooling and increased density thereof, and decoupling in the ocean-continent boundary (Keary and Vine 1990). In this transformation of passive

[1] División de Ciencias de la Tierra, Centro de Investigación Científica y de Educación Superior de Ensenada (CICESE), C.P. 22860 Ensenada, Baja California, Mexico. E-mail: mgonzale@cicese.mx

margin (divergent) to active margin (convergent), significant changes start to occur in the ancient margin, especially in the neighborhood of the trench and the subduction zone.

The Magdalena Shelf, at the west of the Baja California microplate, is considered a forearc basin (Figs. 1, 2a), product of the subduction of the Farallon plate under the North American plate (Mammerickx and Klitgord 1982). This basin was an area of subsidence in front of the volcanic arc, the Comondú Group (Umhoefer et al. 2001) (Fig. 1), and

Figure 2
Schematic section of the forearc basin along the Pacific margin of Baja California Sur. **a** Volcanoes, plutons and the subduction complex (accretionary wedge) along the inner wall of the trench (from Dickinson 1995). Deep Sea Drilling Program (DSDP) site 471 (Yeats et al. 1981). **b** Section II-II' across the province Iray-Magdalena, as taken from García-Domínguez (1976). The locations of three exploratory wells are also shown. **c** Seismic profile OW reported by Brothers et al. (2012)

was a forearc basin in a convergent margin setting before the oblique divergent plate boundary formed in the Gulf of California at 12 Ma. There was a

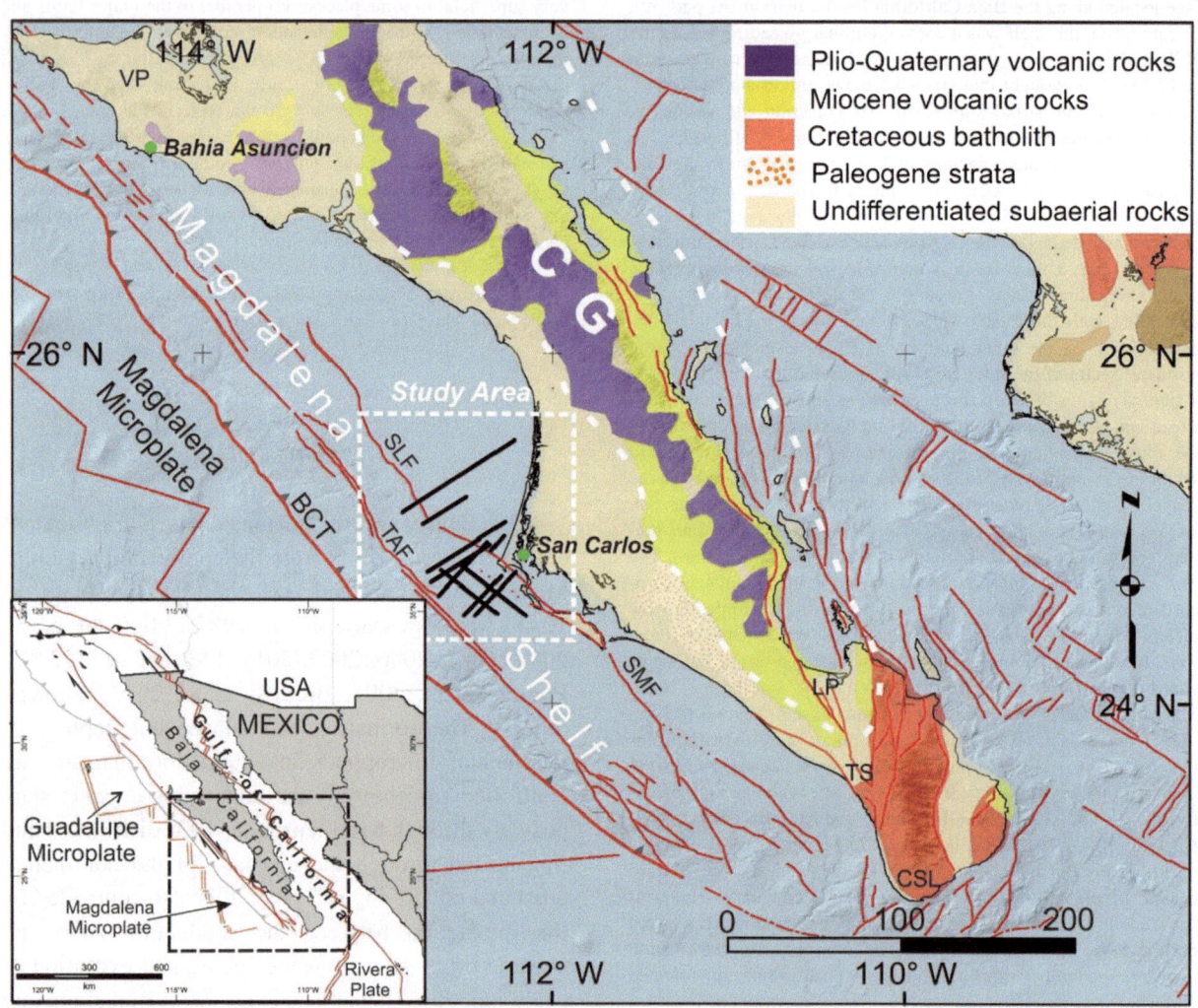

Figure 1
Inset Study region (*black box*) and major fault systems in the Baja California-Gulf of California region. CG: Comondú Group (from Umhoefer et al. 2001). *Black lines* seismic reflection profiles used in this work. *BCT* Baja California trench, *TAF* Tosco-Abreojos fault, *SMF* Santa Margarita fault, *SLF* San Lázaro fault, *VP* Vizcaino Peninsula, *LP* La Paz, *TS* Todos Santos, *CSL* Cabo San Lucas. Map generated with information of Brothers et al. (2012) and Fletcher et al. (2007)

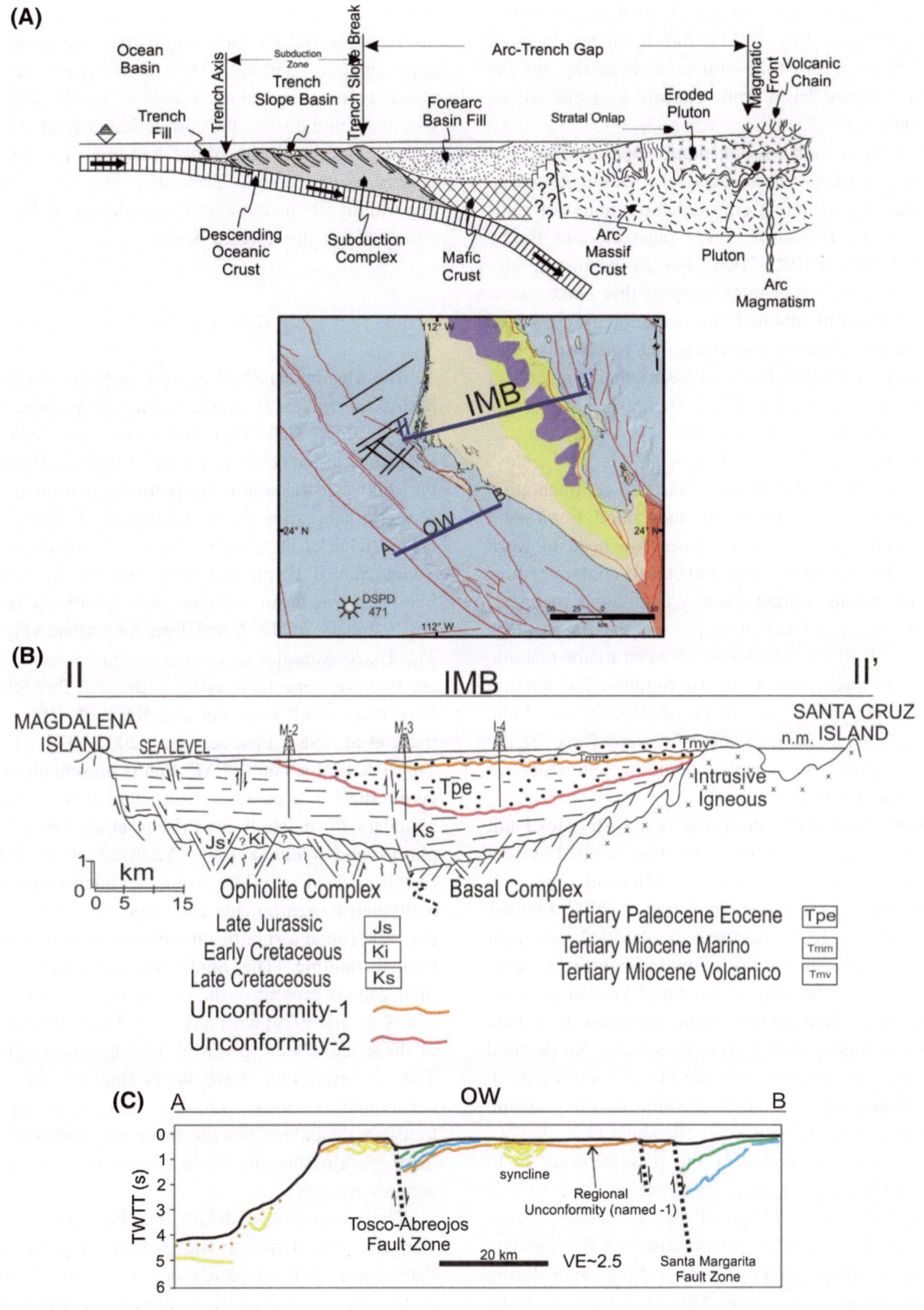

probable transition period from 12 to 8 Ma, and since 8 Ma (and possibly 12 Ma) this basin has been of transtensional or strike–slip type, breaking up the former forearc basin from the mid-Miocene to the present.

Due to abundant sedimentation and tectonic subsidence, a forearc basin acts as a sedimentary trap that can accumulate a considerable amount of sediments (e.g., Dickinson 1995; Ingersoll and Busby 1995; Ingersoll 1982, 2000). Forearc basins are elements of convergent plate margins that make major contributions in volume to the accretionary growth of continents. Modern convergent-margin basins, particularly in the Pacific, have been studied in recent years (e.g., Ingersoll 1979; Dickinson and Seely 1979; Dickinson et al. 1987; Michaud et al. 2004, 2005, 2007, 2010; Plattner et al. 2007). Nevertheless, the evolution and control of sedimentation in these basins are still poorly understood. For a good understanding of a forearc basin, we need to learn about its geometry and structural characteristics. Forearc basins contain a nearly continuous record of the marginal sedimentation process that takes place as a result of the interaction between major tectonic plates. In forearc basins, the stratigraphic architecture reflects changes in the shape of deposition, which, due to interaction with the downward force of the oceanic plate, causes variations in the sediment load (Dickinson 1995).

In the area of this study, the Tosco-Abreojos Fault System (TABFS) has been the most studied system (Normark et al. 1987; Michaud et al. 2004, 2005, 2007, 2011; Fletcher et al. 2007; Plattner et al. 2007, 2009; Brothers et al. 2012, amongst others). Although the evolution and control on sedimentation in the Magdalena Shelf are, in general, understood, there are few studies that show the details of the sediment buildup in some sectors. No detailed images have yet been presented of the subsurface in the central part the shelf, except for one seismic profile in its southeast sector (Brothers et al. 2012).

The main objective of the present study is to update the state of knowledge about the structure and stratigraphy in a sector of the Magdalena Shelf (Figs. 1, 2, 3). For this, we use seismic reflection data gathered in the region by Petróleos Mexicanos during the 1970s and early 1980s. This work focuses on the processing and interpretation of 11 seismic reflection profiles of ∼400 km total length that cover an area of approximately 350 km². We define their seismic characteristics, including structural basins, acoustic basement, and discuss the correlation of fault systems and stratigraphy with those reported for another sectors of the region. These data provide a unique opportunity to increase our knowledge of the tectonics and of this forearc basin.

2. Geological Setting

The Magdalena Shelf is constituted by a series of basins and ridges. It extends along the western margin of Baja California Sur, from the Vizcaino Peninsula, at the north, to the tip of Baja California at the south. At the south, the platform is limited by a series of submarine basins southwest of Todos Santos. Fletcher et al. (2007) refer to this feature as the Todos Santos Basin and note that the Magdalena Shelf and this basin coincide closely with the southern boundary of the Magdalena microplate (Fig. 1). The Tosco-Abreojos and Santa Margarita-San Lázaro are two Neogene fault systems that cut through the Magdalena Shelf (Spencer and Normark 1979; Normark et al. 1987; Fletcher et al. 2007) (Fig. 1). The Tosco-Abreojos Fault (TAF) that is present along the entire shelf is a geological structure that marks the boundary of the study area (Legg et al. 1991; Humphreys and Weldon 1991; Michaud et al. 2004). Fletcher et al. (2007), from an interpretation of bathymetric escarpments and structures of basins in the area, found a significant component of east-down normal faulting. The Santa Margarita-San Lázaro fault zone is present to the east of the TAF (Spencer and Norwark 1979; Normark et al. 1987). The names of these faults were given by Fletcher et al. (2007). They inferred that these faults that cut the shelf accommodate some of the integrated dextral transtension and not just the strike–slip component of plate margin shearing as is commonly assumed in tectonic models.

The presence of Adakites has been attributed to slab detachment and to the melting of young (less than 5 Ma old) subducted oceanic crust (Brothers et al. 2012 and references therein) in the region.

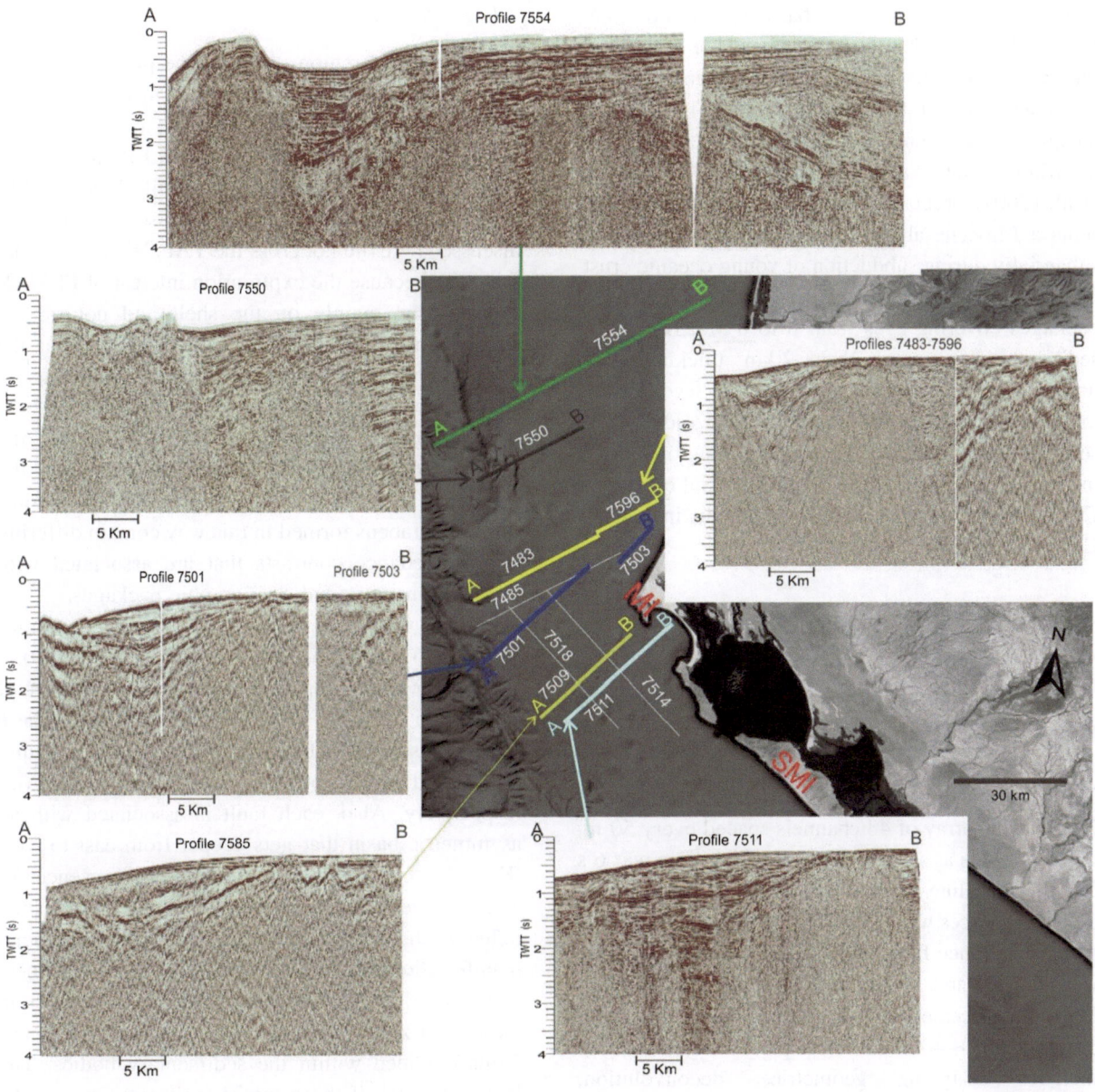

Figure 3
Some of the seismic reflection profiles examined in this study. *SMI* Santa Margarita Island, *MI* Magdalena Island. The *numbers* represent the names of the profiles

These rocks, dated between 5 and 6 Ma, have been sampled in Santa Margarita Island (Bonini and Baldwin 1998; Figs. 1, 3) and in different places in Baja California Sur (Calmus et al. 2011). Their emplacement began at roughly the same time as the onset of dextral transtension along the Magdalena Shelf (Michaud et al. 2006; Brothers et al. 2012). To east of the study region, García-Domínguez (1976),

using seismic reflection and wells data from the Iray-Magdalena basin (see profile II-II', Fig. 2), reported two unconformities, one of Miocene age and one of Paleocene–Eocene and Cretaceous ages. Brothers et al. (2012), using a seismic reflection profile located 70 km south of the study area, correlated their interpretation with the lithostratigraphy sampled at DSDP Site 471 (Yeats et al. 1981), showing a generalized structure of the basin (see profile OW, Fig. 2). They say that the sediments in the Magdalena fan are turbidities (13–14.5 Ma) overlaying the irregular layer of the middle unit (8–13 Ma) and terrigenous sediments of 8–0 Ma; an unconformity of the Miocene age was also reported (Fig. 2). Fletcher et al. (2007) proposed that this unconformity represents a Miocene abrasion surface that was exposed subaerially during subduction of young oceanic crust. Upon the unconformity there are three sedimentary packages covering ages from 8 to 0 Ma. Below are sediments of greater than 2 km thickness and unknown age (Brothers et al. 2012). The recognizable major tectonic events affecting the Mesozoic-Tertiary sedimentation in Baja California are a regional uplift in late Jurassic time, a prolonged subaerial erosion in Cretaceous times, and Batholith domain in Albian-Cenomanian time.

3. Seismic Reflection Data

The Magdalena Shelf was subject to a 2-D seismic reflection exploration program during the 1970s and early 1980s by Petróleos Mexicanos (PEMEX). The geometry to acquire the seismic profiles (Figs. 1, 2, 3) was an array of 48 channels spaced every 50 m, with 7 airguns as sources. The recording time was 6 s and the sampling interval was 0.002 s. The distance between sources was 25 m. The shot pattern indicates that the distance from the source to the first receiver was 307 m, and to the last receiver was 2657 m. Here, we processed and interpreted seismic reflection profiles of ~400 km. The processing sequence includes assigning geometries, deconvolution, grouping by CDP, velocity analysis, NMO, stacking, and migration (Kirchhoff) (Yilmaz 2001).

4. Results

The major elements that are seen in the data are the San Lázaro, Santa Margarita and Tosco-Abreojos fault systems, the Iray-Magdalena, San Lázaro, Santa Margarita and Tosco-Abreojos basins, the acoustic basement and some unconformities. These are described in the following sections.

4.1. Fault Systems

The studied region presents one prominent trend that corresponds to three major fault systems oriented in the NW–SE direction. These are, the Santa Margarita (SMF), San Lázaro (SLF) and Tosco-Abreojos (partially observed in profile 7554) (TAF) (Figs. 3, 4, 5, 6, 7, 8, 9). The seismic profiles discussed here did not cross the TAF. We believe that this is so because the exploration interest of PEMEX was focused mainly on the shelf and not on the deformation zone. It has been recognized that the Tosco-Abreojos basin is controlled by TAF (Figs. 4, 5, 6), which dips to the east. Each one of these three basins, forming a syn-rift divergent wedge of the sedimentary sequence. Major faults displace the strata and control the morphology of this basins. The half-grabens formed in this way contain differing large impedance contrasts that are associated with tilted sedimentary and stratigraphic packages.

From the interpretation of seismic profiles 7483–7596 and 7485–7503 (Figs. 4, 5, 6, 7, 8, 9), a stepover or relay ramp region was seen at the place of connection between the SMF and the SLF. These faults, considered as being part of the same structure, are located to the north and south of such stepover, respectively. Also, each fault is associated with an asymmetric basin that gets deeper from east to west (Figs. 4, 5, 9). In this sector the presence of compressive stresses are inferred by the way seismic reflectors bend. The SLF and SMF have superficial manifestations that show clear control of the local bathymetry, as reported in previous studies. Also, in these fault zones we noted a large number of minor faults confined within the sedimentary bodies. The fact that some of those minor faults are seen on the ocean bottom is evidence of local recent activity (e.g., Figs. 3, 4, 5, 6, 7, 8). Based on the ocean floor morphology, Fletcher et al. (2007) depicted some of those faults. The TAF zone is also associated with an asymmetric basin (Figs. 4, 5, 6), which shows a series of sub-parallel faults between the SMF-SLF and the TAF systems. In the TAF sector, those structures, previously unknown, dip to northeast and southwest, indicating the presence of a basin. The TAF, SLF and SMF major systems cut recent sediments and develop active depocenters. This means that they are still

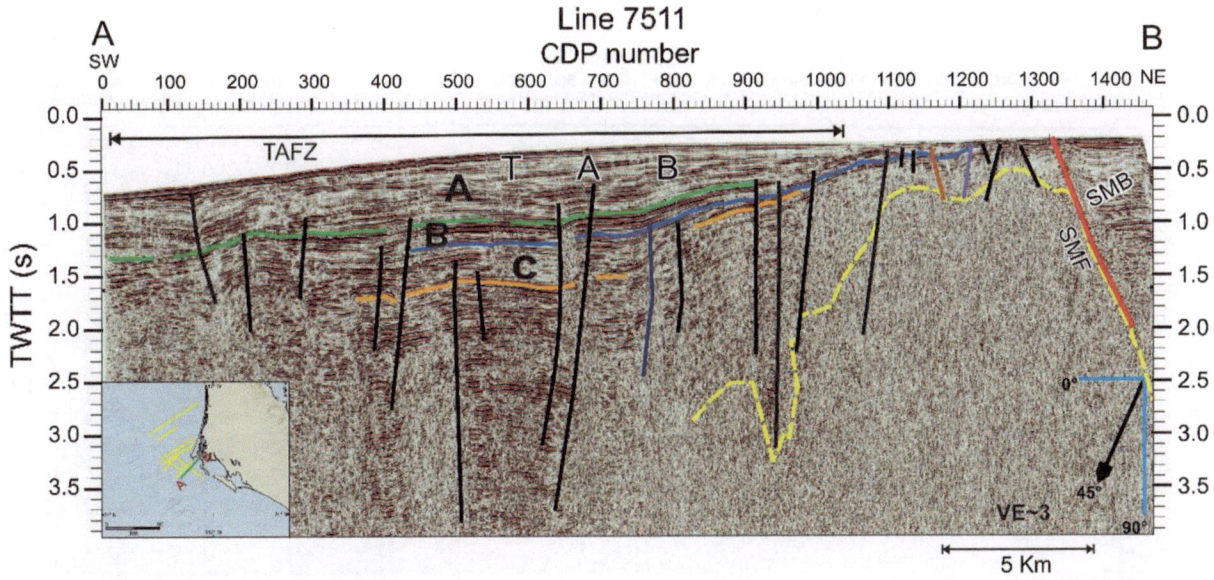

Figure 4
Seismic profile 7511 (see *inset map*). The figure shows two major asymmetric transtensional basins: Tosco-Abreojos (TAB) and Santa Margarita (SMB) basins. This second basin does not show clearly as a semi-graben but we believe it is the same body reported by Brothers et al. (2012), to southeast of the study area. TAB is cut by a series of secondary faults and reach sedimentary thicknesses of up to ~3.5 s (TWTT). The acoustic basement reflector is diffuse (*yellow line*). The horizons A, B and C are discussed in the text. *Orange line* represents unconformity-1. TAFZ: Tosco-Abreojos Fault Zone. The *color code* is the same used in all profiles

active and accommodate part of the deformation that is induced by the interaction of the North American and Pacific plates. The traces of the SLF and SMF interpreted here are slightly different in some sectors compared to those reported in the literature.

4.2. Basins

Four basins are present on the seismic profiles studied here. The Iray-Magdalena basin (IMB) lies on the eastern side of the study area (Figs. 7, 8). The other three asymmetric fault-bounded basins lie on the western side of the study area. These three basins are the Santa Margarita basin (SMB), bounded by the SMF (Fig. 4), the San Lázaro basin (SLB), bounded by SLF (Figs. 5, 6), and the Tosco-Abreojos basin (TAB), bounded by the TAF (Figs. 4, 5).

4.2.1 Iray-Magdalena Basin (IMB)

Profile 7554 shows a cross-section in the marine region of the IMB (Figs. 7, 8). This basin is also partially imaged on the seismic line 7550. Profile

7554 has an orientation of N60°E and a length of 80 km. The most prominent feature in this profile is the presence of two unconformities: an upper one named Unconformity-1 and a lower one named Unconformity-2. These unconformities are delineated in our figures with the orange and pink lines, respectively. Unconformity-1, which is the shallowest, is present in all our seismic profiles (Figs. 4, 5, 6, 7, 8). This unconformity is also present 90 km southwest of the study area (Brothers et al. 2012). As profile 7554 shows, below Unconformity-1 there is a thick package of seismic reflectors, and further down is the Unconformity-2. Unconformity-2 shows in Profile 7554 (Fig. 8) as a strong seismic reflector. Seismic reflector packages cut by this unconformity are clearly seen in the profile 7554. This unconformity, which is present only in this basin, represents a subaerial erosion surface. According to García-Domínguez (1976), the ages of these unconformities are Miocene, at ~8 Ma, (Unconformity-1) and Upper and Lower Cretaceous and occasionally Upper Jurassic (Unconformity-2). Only a little number of minor faults are seen in this basin, in comparison with

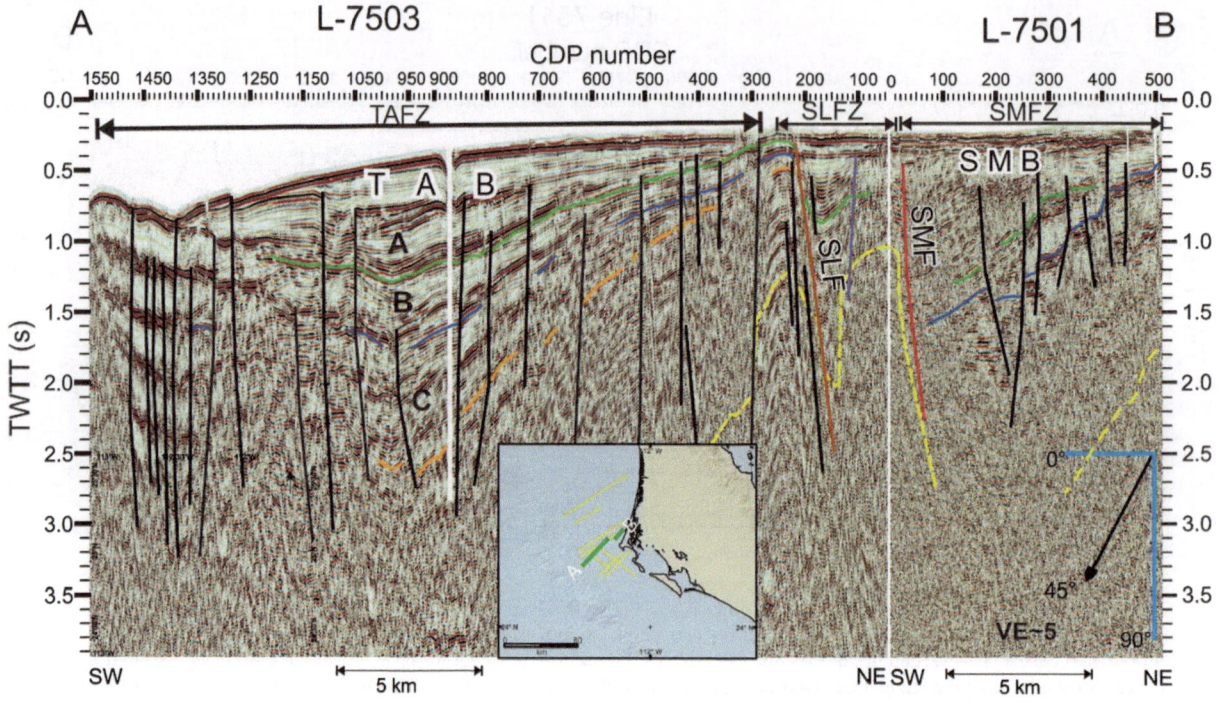

Figure 5

Seismic profile 7503–7501 (see *inset map*). The two major asymmetric basins, TAB and SMB, are shown. These basins are cut by a series of secondary faults and reach sedimentary thicknesses of up to ∼3.5 s in TAB, and ∼2.5 s in SMB. The acoustic basement reflector is diffuse (*yellow line*). The horizon A, B and C, are discussed in the text. *Orange line* represents unconformity-1. Basins are separated by a high structure that is associated with the subduction complex. *TAFZ* Tosco-Abreojos Fault Zone; *SLFZ* San Lázaro Fault Zone; *SMFZ* Santa Margarita Fault Zone

the large number of small faults present in the other three basins located to southwest (TAB, SLB and SMB). A larger impedance contrast is associated with the unconformities that progressively increase in amplitude eastward. Another feature of interest is the conservation of thicknesses between the horizons of the seismic reflectors below Unconformity-1. Above this unconformity, the sediment package thickens to the southwest, in a pattern of semi-parallel wavy reflectors (rolling) of continuous strong amplitude (Fig. 8). At the northeast sector of the profile of Fig. 8 (near CDP 700) a seismic reflector is cut by Unconformity-1, but this could be a case of over-migration. Below the Unconformity-1, we may see again a pattern of interleaved reflectors that define a package of semi-parallel reflectors with continuous strong amplitude. This package, which deepens to the southwest, overlies the Unconformity-2. This latter unconformity is considered as an erosional unconformity of the IMB.

Two sedimentary packages (Unit I and II) are cut by the Unconformity-2 (Fig. 8). Unit I is present from ∼1.5 s (TWTT) and apparently it is deeper than the resolution of the data. Unit II is present from ∼1.0 S (TWTT) to 2.5 s. Unit II is defined by a package of parallel and lateral reflectors of continuous strong amplitude that dips to northeast with a large thickness. The bottom half of this package is characterized by reflectors ending in downlap to the east on continuous reflectors of the base and the top half is characterized by parallel seismofacies, very continuous reflectors of low amplitude, that terminate at Unconformity-2. The seismic reflectors of this package dip to the east due to a lifting of the structural high located to the west. Sediment depths of up to ∼4.5 s (two-way travel time, TWTT) are observed in the area off the coast, so we may infer that the basin depths range from 5 to 6 km. This basin is separated from the SMB and SLB by a compressive and chaotic/diffuse structural high, defined by a diffractor

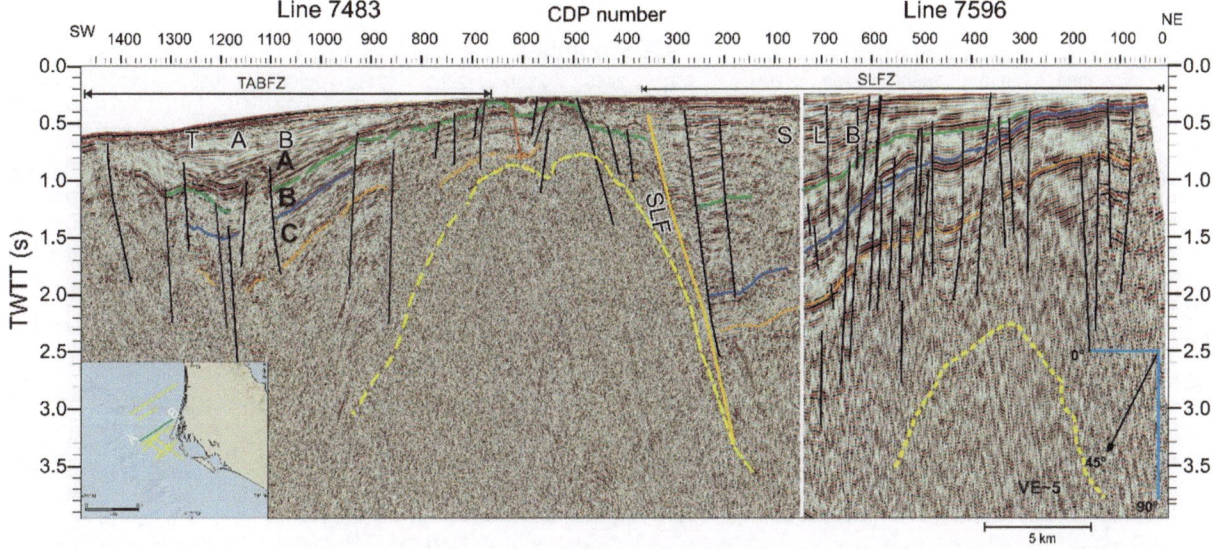

Figure 6
Seismic profile 7483–7596 (see *inset map*). The two major asymmetric basins, TAB and SLB are shown. The basins are cut by a series of secondary faults and reach sedimentary thicknesses of up to ~3.0 s in TAB, and ~3.5 s in SMB. The acoustic basement reflector is diffuse (*yellow line*). The horizons, A, B and, C are discussed in the text. *Orange line* represents unconformity-1. The surface of the seabed has relief and is cut by some faults. Basins are separated by a high structure that is associated with the subduction complex. *TAFZ* Tosco-Abreojos Fault Zone; *SLFZ* San Lázaro Fault Zone

which is associated with a subduction complex (Figs. 2, 9) and is in some places is very shallow. This diffracting structure is interpreted as the acoustic basement in study area.

4.2.2 Santa Margarita-San Lázaro Basin (SMB-SLB)

Santa Margarita and San Lázaro are asymmetric, transtensional basins bounded by the SMF and SLF, respectively (Figs. 4, 5, 6, 7, 8). We have considered that both basins are the same structure. To the south of the stepover is the SMB as shown by seismic profiles 7483–7596 and 7485–7503, while to the north is the SLB. A series of seismic reflections are observed in the two depocenters; the deeper one is seen in the northernmost seismic profile which reaches a depth of 3.5 s (TWTT). The divergent strata suggest that basin subsidence was syn-depositional. The syn-rift sedimentary sequence was limited by the thickness and divergence of this sequence in the direction of the SMF and SLF zones in the respective basin. Sediments within each of these semi-grabens are controlled by the SMF and the SLF to the west, and become thicker as one approaches

these faults. On top of the basement high, there is a relief that gives shape to two depocenters spaced 2 km apart. The depocenter that is seen in profile L-7554 (Fig. 8) has a depth of ~3.5 s (TWTT) and corresponds to the SLB. This depocenter is limited by the east-dipping listric SLF. The profile L-7511 (Fig. 4) shows the depocenter that corresponds to the SMF. Brothers et al. (2012) also reported this basin, but at 70 km southeast of this region.

4.2.3 Tosco-Abreojos Basin (TAB)

Southwest of the area of the seismic profiles, there is another depocenter (TAB) that is limited to the west by the TAF, as reported in previous studies (e.g., Normark et al. 1987; Michaud et al. 2004). The position of the TAF was reported by these authors. They reported bathymetric features that are very close to those seen in the profiles of this study (e.g., Fig. 5). A structural high that extends semi-parallel to the SMF-SLF system (Figs. 4, 5, 6, 7, 8, 9) separates the TAB from SMB-SLB. We noted that TAB has fewer faults than SMB and SLB, indicating that these latter basins are the more active basins in the study

441

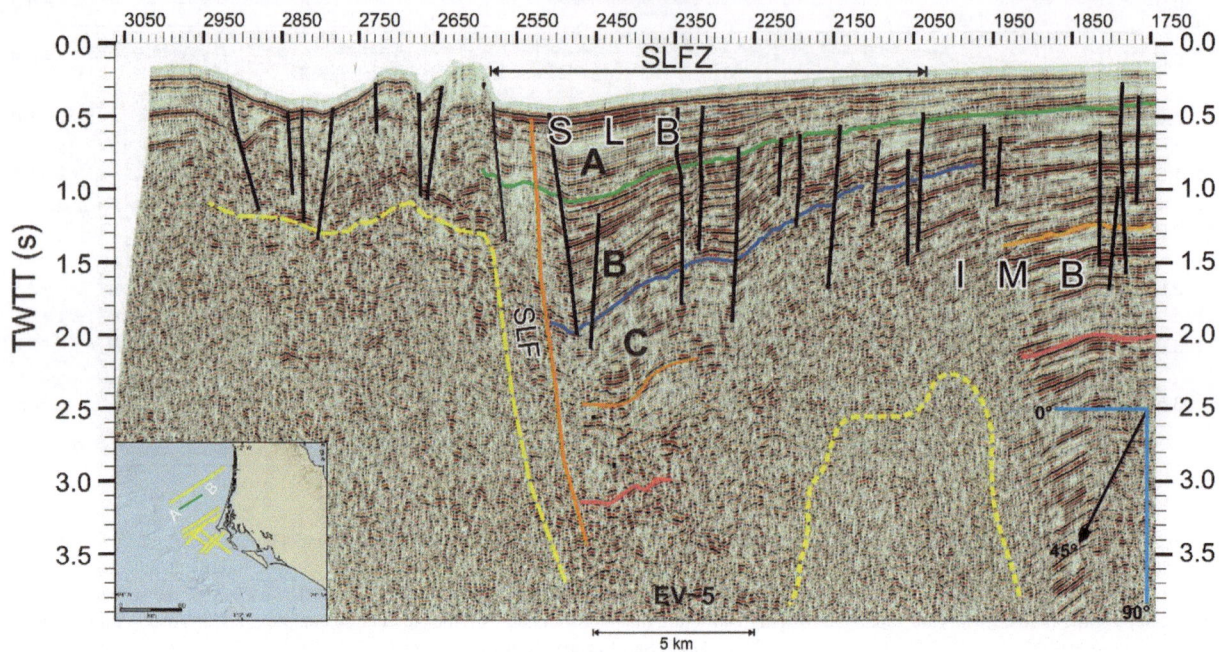

Figure 7

Seismic profile 7550. The figure shows the San Lázaro Basin (SLB). This profile shows partially the Iray-Magdalena Basin (IMB). The SLB has more of 3.5 s of thickness. The IMB clearly shows that the depths are over 4.0 s. At the top of this section a relationship of the onlap and downlap of seismic reflectors is observed, possibly due to transgressions and regressions of the sea level. *Orange* and *pink lines* represent Unconformity-1 and Unconformity-2, respectively. *SLFZ* San Lázaro Fault Zone

area. In TAB, a larger impedance contrast is associated with Unconformity-1, which progressively increases in amplitude and dips to the west. In this basin, the young and less consolidated sediments are separated from the older and more consolidated sediments by this unconformity. A series of seismic reflectors are observed too.

4.3. Nature of the Acoustic Basement and Basement Highs Between Depocenters

In all seismic profiles no clear acoustic basement is observed. A classical strong seismic reflector showing the change in acoustic impedance from sediment to basement is not observed. The absence of such a clear reflector could be due to sediments being intruded in deep basins to form a new generation of recycled crust due to the process and cessation of subduction. To the east of the structural high, there is a package of forearc basin sediments. The acoustic

basement is interpreted over the structural high as an elongated body extending in an NW direction and ~ 8–10 km wide (Figs. 4, 5, 6, 7, 8, 9). This reflector is attributed to the oceanic crust that crops out as an ophiolite complex in the Magdalena and Santa Margarita Islands (Fig. 3), consisting of blue schist and basaltic rocks (Hagstrum and Sedlock 1998); also the presence of amphibolites, quartz-mica schists and Adakites has been reported (Bonini and Baldwin 1998). It is likely that exhumation of this body of continental crust, in the Santa Margarita and Magdalena Islands, was due to activity of SMF and SLF.

4.4. Seismic Reflectors

Here we make an attempt to find a correlation between our seismic profiles by considering some similarities observed in the characteristics of the seismic reflectors. Our results are compared with those of Brothers et al. (2012), who used a seismic

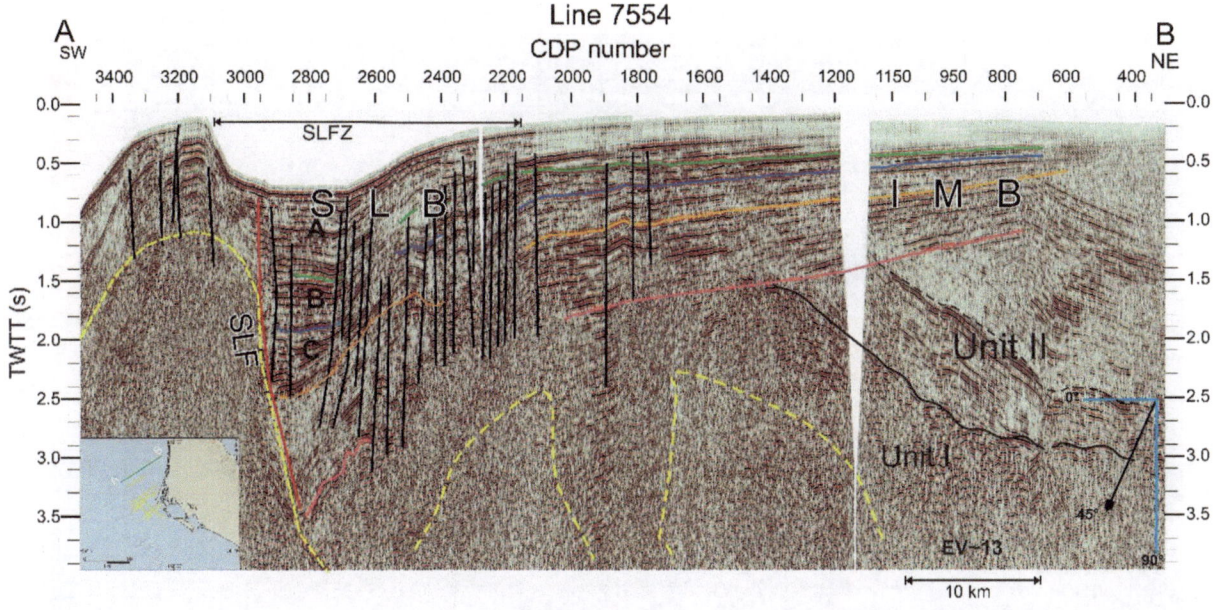

Figure 8
Seismic profile 7554. SLB and IMB basins are shown. The SLB has more of 3.5 s of thickness. The IMB clearly shows that the depths are over 4.0 s. The IMB shows the Unconformities-1 and -2 (see text). The acoustic basement reflector is more diffuse. Towards the northeast sector there is a sediment thickness of more than 4 km and the presence of faults is not observed. The Units I and II, to the right of this profile, are cut by the Unconformity-2. *SLFZ* San Lázaro Fault Zone

profile and data of a well of the Deep Sea Drilling Project (DSDP-Site 471) (Yeats et al. 1981) (Fig. 2). For this, we use their nomenclature: sediment packages A, B, C and the two unconformities that we identified (Figs. 4, 5, 6, 7, 8). The thickness of packages A, B and C increases westward through the area of the TAF-TAB, then displays a pronounced syncline. The same behavior is observed for the regions SMF-SMB and SLF-SLB. As the Figs. 5, 6 and 7 show, the sediment package A is the thickest of all packages. It has typical *onlap* and *downlap* stratigraphic boundaries, which indicate transgressions and regressions of the sea level. The layers in package B do not diverge and thicken along the syncline. A few small faults present in the sediments of this package are indicative of a slight deformation. The C package contains pre-rift sediments (i.e., sediments deposited before the opening of the basin). Such package is limited at its base by a seismic reflector that is considered here as Unconformity-1, and below this unconformity a thick package of sediments older than ∼8 Ma exists (Brothers et al. 2012). Unconformity-1 is the shallowest of all

unconformities observed in this study. It is present in all seismic profiles (Figs. 4, 5, 6, 7, 8) and it was also seen at 70 km to the southeast of the study area by Brothers et al. (2012). A second unconformity (Unconformity-2) is present in profile 7554 as a strong seismic reflector that cuts other seismic reflector packages (Fig. 8). That unconformity, which is only present in the IMB, has the appearance of a subaerial erosion surface. The profile 7554 of Fig. 8 shows clear positions of the unconformities 1 and 2. In the Fig. 7, the profile 7550 shows also two unconformities that seem to correlate with those seen in the other profile. Given the time window that exists between each pair of unconformities in the profiles, there is no doubt that they are continuation of each other.

The seismic reflection profile 7554 is the most significant of all (Fig. 8). The main feature of such profile is the constant thicknesses between the horizons of seismic reflectors observed to the northeast. At the top of this profile there is a bundle of parallel reflectors with continuous strong amplitude. Here, the shelf is wider, has a sedimentary package

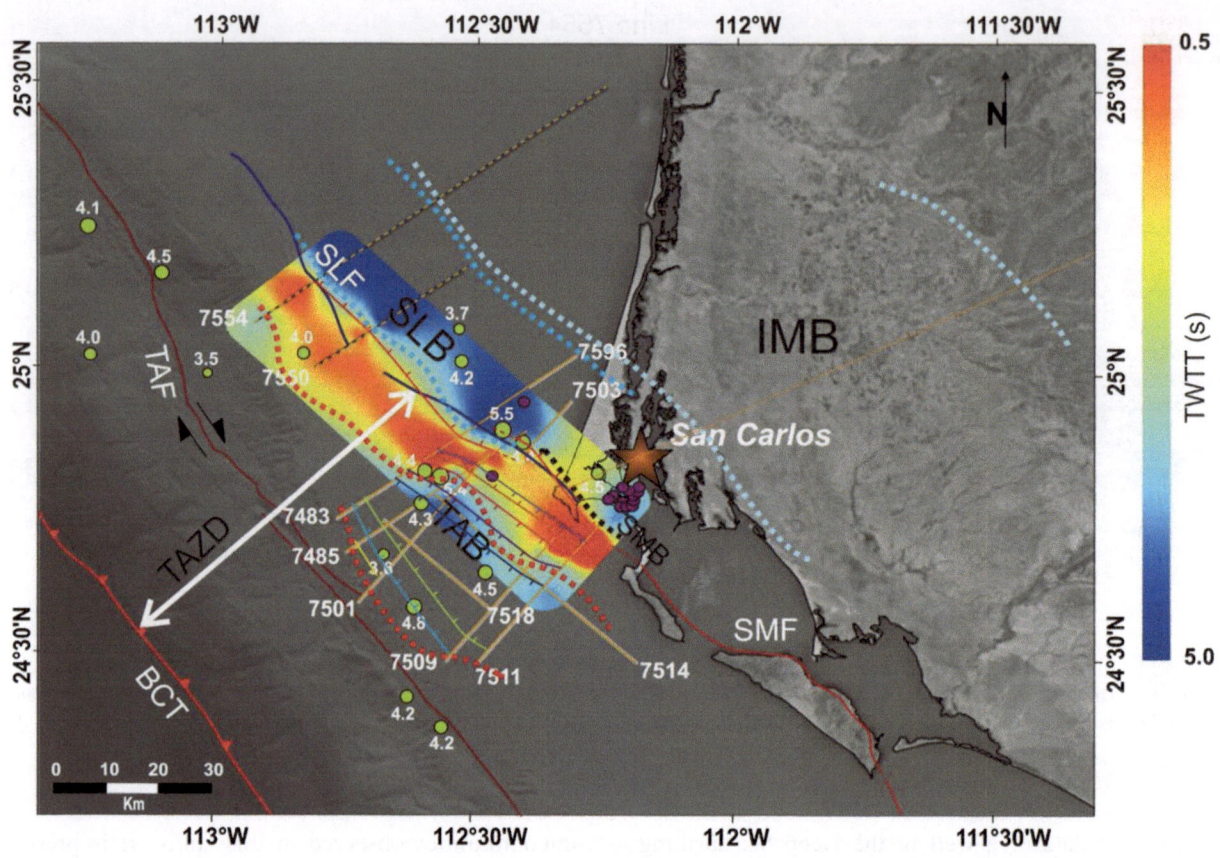

Figure 9

Structures correlated based on different seismic lines and seismic activity reported by NEIC and Munguía et al. (2015b). Between the Tosco-Abreojos Fault zone and the Santa Margarita and San Lázaro Faults, the continental shelf is cut by minor faults that follow the structural pattern of major faults. *Colors* represent the reflective acoustic basement obtained in this work, showing the structural high (in *red*). Two Basins are separated, by the San Lázaro Fault (SLF) and Santa Margarita Fault (SMF). This last fault delimits the block of basement of the Santa Margarita and Magdalena islands (Sedlock 1993; Fletcher et al. 2007). *IMB* Iray-Magdalena Basin; *SMB* Santa Margarita Basin; *SLB* San Lázaro Basin, *TAB* Tosco-Abreojos Basin; *TAZD* Tosco-Abreojos zone of deformation

that is thicker than sediment thicknesses seen in the other profiles. Further, the profile does not show small faults that would be evidence of tectonic activity. In this zone, the deepest and oldest sediments are strongly inclined at the two unconformities, reaching a sediment thickness of more than 3 km. To the east of SLB, the seismic profiles show sedimentary packages with thicknesses that decrease to east, but we do not know what happens beyond the end of those profiles. The seismic profile 7554 shows that the basin thins and then widens to achieve thicknesses greater than ∼4 s (TWTT). In addition, the younger and less deformed sediments at the eastern part of the study area are separated by two identified unconformities. No evidence of the presence of active small

faults is noted. The top of Unit II, of the seismic reflector present clinoforms that dip to northeast. Usually the top is defined by truncated reflectors (toplap). Unconformity-2 is the only unconformity present in the northwest part of the region of study, including the San Lázaro basin.

5. Discussion

5.1. Seismicity and Active Faults

The tectonic activity in the area of study is evidenced by the seismicity reported by the National Earthquake Information Center (NEIC), in their catalogs from 1960 to present, and more recently by Munguía et al. (2015a, b). Munguía et al. reported seismic activity in the region of Bahía Asunción between November 2006 and March 2007 (M_w up to 5.0), and in San Carlos, Baja California Sur, in 2000 and 2001, with earthquakes with magnitudes ranging from 1.2 to 3.2 (Fig. 9). Geophysical and geodetic studies indicate that the Tosco-Abreojos and San Benito fault zones accommodate deformation between the peninsula of Baja California and the Pacific plate (Michaud et al. 2004, 2005, 2007, 2011). Our seismic profiles show faults that cut through the seafloor, indicating that those faults are tectonically active (see Figs. 3, 4, 5, 6, 7, 8). These observations are consistent with other studies that suggested that the region is tectonically active (e.g., Michaud et al. 2011). SMF and SLF cut recent sediments, creating depocenters and accommodating an unknown fraction of the relative slip between the Pacific and North American plates. Thus, both faults accommodate some transtensional shear and not just the strike–slip component of plate margin shearing, as is commonly assumed in tectonic models.

The seismic activity reported by Munguía et al. (2015b) occurred practically below Puerto San Carlos. The earthquakes had sources at an average focal depth of around 5 km, and a distribution of epicenters that showed a strong correlation with the fault plane of the SMF (Figs. 9, 10). The fault plane solution for those earthquakes has a plane that strikes in a SE–NW direction and dips 40° to NE, in consistency with the San Lázaro-Santa Margarita fault (Fig. 9). Figure 9 shows the seismic lines that extend in front of Puerto San Carlos, two observed basins and the seismogenic zone associated to SMF. The SMF must be considered in future risk studies because, as Munguía et al. (2015a, b) reported, seismic events occurred along the Magdalena Shelf with magnitudes of up to 5.0 have generated much alarm within the populace, as well as soil liquefaction and small cracks in structures of San Carlos and Bahía Asunción.

On the other hand, more faults are observed in SLB than in TAB (Fig. 6), evidencing that the San Lázaro fault zone is more active than the Tosco-Abreojos fault zone, at least in the sector studied here. We conclude that the Tosco-Abreojos zone of deformation (Fig. 9) in this region is ~90 km wide,

extending eastward from TAF to SLF-SMF. A similar result was published by Munguía et al. (2015a) for the San Carlos and Bahía Asunción regions. Our evidence of active transtensional deformation in the study area are some identified small faults that cut recent sediments. Transtensional deformation occurs by means of vertical and horizontal relative slip motions. Both type of fault motions are difficult to interpret in our seismic reflection profiles. However, Munguía et al. (2015a, b) reported active transtensional deformation along the western margin of Baja California Sur, based on focal mechanisms of recent earthquakes occurred in the areas of San Carlos and Bahía Asunción. Fletcher et al. (2000) proposed that the SMF and SLF were responsible for the uplift of the Santa Margarita and Magdalena Islands in Pliocene-Quaternary time.

5.2. The Subduction Complex

The subduction complex is close to the surface of the seabed and emerging in some places. The sedimentary volumes and characteristics of their sequences in the forearc basin indicate that their respective deposits are thin to the east and thicker to the west. The lenticular shape of the reflectors and the chaotic nature of the sediments, along the TAF, SMF and SLF regions, suggest sediment transport from east to west.

5.3. Seismic Stratigraphy

The seismic stratigraphy in the Magdalena Shelf shows various sedimentary bodies with the presence of two unconformities. The upper package in these bodies is formed by recent sediments deposited in three basins (half-grabens). These basins are controlled by Neogene faults that cut the continental shelf. According to Brothers et al. (2012), if the extension associated with the TAF and SMF started after the dextral transcurrent movement initiation, then the sequences A and B were deposited before the development of these faults, and therefore can be used as piercing points to calculate the dextral slip displacement. Brothers et al. (2012), interpreted the unconformity-1 that shows in the seismic profile as being of Miocene age; the lower unit is undated. We

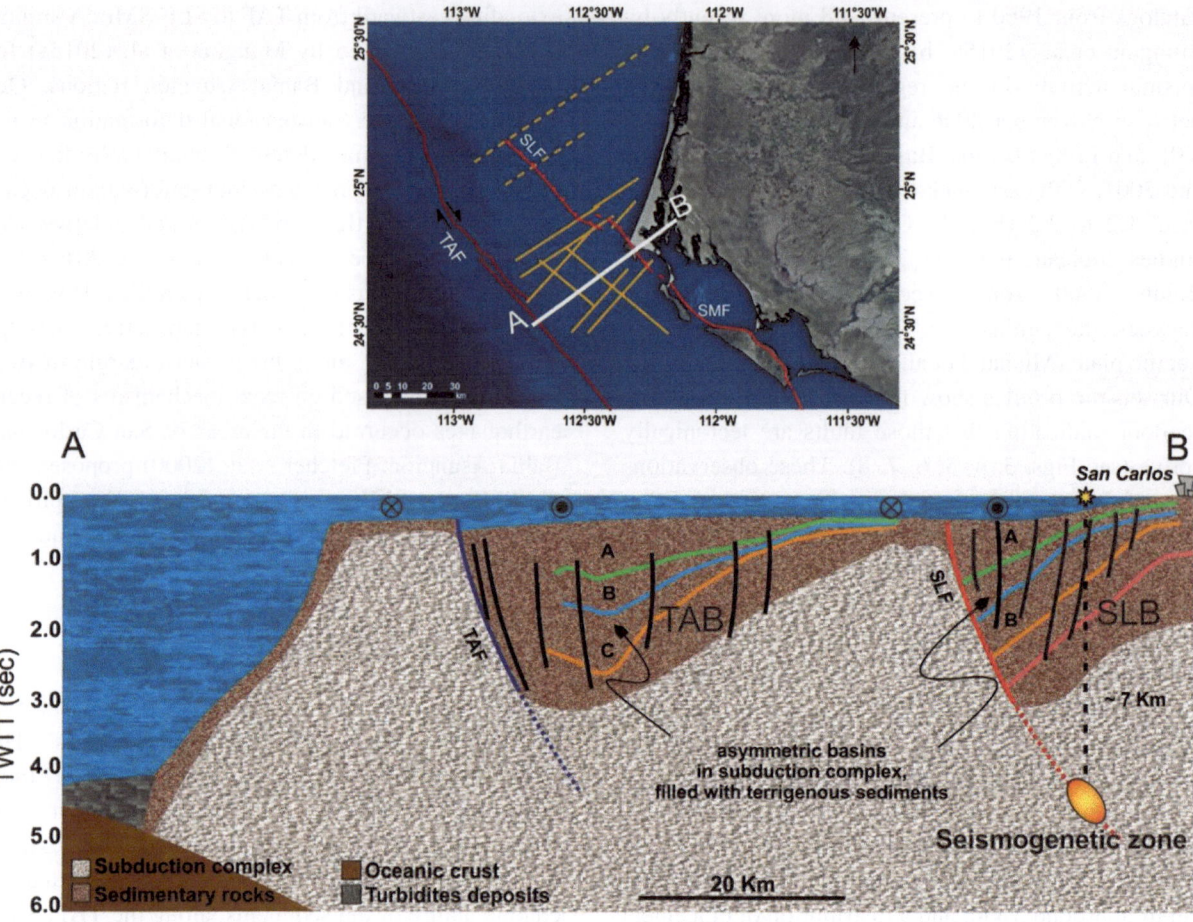

Figure 10

a Location of the schematic profile shown in part, **b** across the Magdalena Shelf. **b** Profile based on the seismic lines analyzed. This scheme is very similar to that proposed by Brothers et al. (2012) 50 km south of the study area. Two asymmetric basins are shown: the Tosco-Abreojos and the Santa Margarita-San Lázaro, controlled by major faults. The seismogenic zone proposed is based on the seismicity reported by Munguía et al. (2015b). *TAB* Tosco-Abreojos Basin; *SLB* San Lázaro Basin

believe that their Miocene-aged Unconformity-1 is the same as the one we observed in our profiles (Fig. 8). If true, then that unconformity extends regionally and is present in the IMB, TAB, SLB and SMB. On the other hand, Unconformity-2 is observed only in the profile 7554 (Fig. 8). This unconformity seems to be also present in the profile 7550, but it is difficult to correlate. The Unconformities 1 and 2 that can be correlated agree with the profile II-II' (Fig. 2) reported by García-Domínguez (1976). This information is reliable in the sense that the wells that were studied are closer to the study area and do not shows the existence of any structure (i.e., fault or unconformity). Figures 8 and 2 show that the

sediments below the Miocene Unconformity-1, of the Paleocene-Eocene ages, can represent filling of the forearc basin.

Units A, B and C cover the horizons of the Paleocene with horizontal reflectors, laterally continuous and with strong amplitude. This is indicative of an alternating low and high-energy depositional environment that is typical of a slowly subsiding continental shelf (Sangree and Widmier 1977). The sedimentological interpretation of Unit II is a unit that may include shelf environments and break the continental slope (margin deltas platform?). Alternatively, it may be proximally lenticular forms of submarine fan facies. The base in those observations is interpreted as lenticular bodies formed on the continental shelf and resulting from greater compaction of shaly facies. The inclination of the Jurassic sediments shows the major tectonic event that produced a regional uplift in late Jurassic times that affected the Mesozoic-Tertiary sedimentation in Baja California. Unconformity-2 is the result of prolonged subaerial erosion in Cretaceous times. The regional uplift in the western portion of the basin during Lower Tertiary times may have modified the limits of the uplift. Moderate compressive deformation increases gradually from west to east through lower Eocene times. Therefore, the basin has greater sedimentary thicknesses to the northwest of Magdalena Island, where the Magdalena Shelf becomes wider and runs to the southeast, skirting to the east of the Magdalena-Margarita Islands and towards Baja California.

6. Conclusions

Based on the processing and interpretation of seismic reflection profiles, we defined the structure and seismic stratigraphy of a sector of the Magdalena Shelf located off the west coast of Baja California Sur, Mexico.

Two major faults were identified: Santa Margarita Fault (SMF) and San Lázaro Fault (SLF), which are part of the Tosco-Abreojos zone of deformation. These faults have been reported previously in the literature, but we found that their geographic locations are slightly different in some places than previously reported, mainly in the area of the reoriented bedding between these two normal faults (SMF and SLF) that overlap in map view and have the same dip direction (relay ramp). This sector has the presence of compressive stress.

We examined four basins, namely the Iray-Magdalena, Santa Margarita, San Lázaro and Tosco-Abreojos. Sediment evolution of the Santa Margarita and San Lázaro half-grabens are controlled by the Santa Margarita and San Lázaro faults, respectively. The Tosco-Abreojos basin is controlled by the Tosco-Abreojos Fault, although that zone was not covered by our seismic profiles.

All four basins show sedimentary packages with strong amplitude reflectors that correlated well with those that Brothers et al. (2012) reported to be 70 km to the southeast of the study region. The uppermost package is formed by recent sediments deposited in two basins that are controlled by Neogene-aged faults that cut the continental shelf. Under this sedimentary package there is an unconformity (Unconformity-1) of the Miocene. We think that this Unconformity-1 extends regionally to the north of our study area and to the south, to the area studied by Brothers et al. (2012).

The Iray-Magdalena basin is a wider basin located in the northeast sector of the area of study. A significant lack of faults in this basin indicates that such sector has not been tectonically active since the time during which the lower layers of Unit II were tilted due to uplift of the ridge to the west. In this basin, and at ~500 m below Unconformity-1, the seismic profiles show a strong unconformity that is not present or is not easy to see in the other basins (Fig. 8). The Iray-Magdalena basin has another unconformity (Unconformity-2) that is deeper and more meaningful from the tectonic point of view. The age of this lower unit is Paleocene–Eocene and Cretaceous, according to García-Domínguez (1976).

One sedimentary sequence shows a regional uplift in late Jurassic times that affected the Mesozoic-Tertiary sedimentation in the Baja California. Regional uplift in the western portion of the basin during Lower Tertiary times may have modified the limits of the uplift. Moderate compressive deformation increases gradually from west to east through lower Eocene times. A chaotic reflector oriented NW–SE is observed, which is quite irregular in its base along the study area, very superficial in some places, is sub-parallel to the Tosco-Abreojos fault, and is associated with a subduction complex.

Seismic profiles show faults that cut through the seafloor, which indicates that the region is tectonically active. Seismic activity in the region has a strong correlation with the fault plane of SMF. This fault should be considered in future risk assessments since it is located just below Puerto San Carlos, and as reported in other studies, even small seismic events (of magnitude up of 5.0) generate alarm, soil liquefaction and minor damage to houses and other structures in the region (Munguía et al. 2015a, b).

Acknowledgments

We thank PEMEX and Ing. Antonio Escalera for permission to publish these results (PEP-DE-35-2015). This research was funded by a PEMEX contract to CICESE. The Landmark University Grant Program (2008-UGP-008005 to CICESE), OpendTect V5.0 provided licenses for the use of software and Google Earth Pro grant Educators. Discussions with Ramón Mendoza and Clemente Gallardo are acknowledged. We thank Sergio Arregui and Martin Pacheco-Romero for technical support. Constructive comments and suggestions by Dr. William Bandy and two anonymous reviewers were helpful in improving the manuscript.

References

Bohannon, R. G., & Parsons, T. (1995). Tectonic implications of post-30 Ma Pacific and North American relative plate motions. *Geological Society of America Bulletin, 107*(8), 937–959.

Bonini, J. A., & Baldwin, S. L. (1998). Mesozoic metamorphic and middle to late Tertiary magmatic events on Magdalena and Santa Margarita Islands, Baja California Sur, Mexico: Implications for the tectonic evolution of the Baja California continental borderland. *Geological Society of America Bulletin, 110*(8), 1094–1104.

Brothers, D., Harding, A., González-Fernández, A., Holbrook, W. S., Kent, G., Driscoll, N., et al. (2012). Farallon slab detachment and deformation of the Magdalena Shelf, southern Baja California. *Geophysical Research Letters, 39*(9), 1–7. doi:10.1029/2011GL050828.

Calmus, T., Pallares, C., Maury, R. C., Agillón-Robles, A., Bellon, H. M., Benoit, M., et al. (2011). Volcanic markers of the post-subduction evolution of Baja California and Sonora, Mexico: Slab tearing versus lithospheric rupture of the Gulf of California. *Pure and Applied Geophysics, 168*(8–9), 1303–1330. doi:10.1007/s00024-010-0204-z.

Dickinson, W.R. (1995). Forearc basins. Tectonic and sedimentary basins. In Busby, C. & Antonio Azor, A., (Eds), Wiley-Blackwell Books, 221–261.

Dickinson, W. R., Armin, R. A., Beckvar, N., Goodlin, T. C., Janecke, S. U., Mark, R. A., et al. (1987). Geohistory analysis of rates of sediment accumulation and subsidence for selected California basins. In R. V. Ingersoll & W. G. Ernst (Eds.), *Cenozoic basin development of coastal California* (pp. 1–23). Prentice-Hall, Rubey, VI: Inglewood Cliffs.

Dickinson, W. R., & Seely, D. R. (1979). Structure and stratigraphy of forearc regions. *American Association of Petroleum Geologists Bulletin, 63*, 2–31.

Fletcher, J.M., Eakins, B.A., Sedlock, R.L., Mendoza-Borunda, R., Walter, R.C., Edwards, R.L., & Dixon, T.H. (2000). Quaternary and Neogene slip history of the Baja-Pacific plate margin: Bahia Magdalena and the southwestern borderland of Baja California: Eos (Transactions, American Geophysical Union), 81, F1232.

Fletcher, J. M., Grove, M., Kimbrough, D., Lovera, O., & Gehrels, G. E. (2007). Ridge-trench interactions and the Neogene tectonic evolution of the Magdalena shelf and southern Gulf of California: insights from detrital zircon U-Pb ages from the Magdalena fan and adjacent areas. *Geological Society of America Bulletin, 119*(11–12), 1313–1336.

García-Domínguez, G. (1976). Prospección Geológica en Baja California. II Simposium de geología de Subsuelo. Superintendencia General de Exploración-Pemex.

Hagstrum, J. T., & Sedlock, R. L. (1998). Remagnetization of Cretaceous forearc strata on Santa Margarita and Magdalena Islands, Baja California Sur: implications for northward transport along the California margin. *Tectonics, 17*(6), 872–882.

Humphreys, E. D., & Weldon II, R. J. (1991). Kinematic constraints on the rifting of Baja California. In J. P. Dauphin & B. R. T. Simoneit (eds.) The Gulf and Peninsular Province of the Californias, American Association of Petroleum Geologists, Memoir 47, 217–229.

Ingersoll, R. V. (1979). Evolution of the Late Cretaceous forearc basin, northern and central California. *Geological Society of America Bulletin, 90*, 813–826.

Ingersoll, R.V. (1982). Initiation and evolution of the Great Valley forearc basin of northern and central California, U.S.A. In J. K. Leggett (ed.) Trench-forearc geology: sedimentation and tectonics on modern and active plate margins. Geological Society of London, Special Publication, 10, 459–467.

Ingersoll, R.V. (2000). Models for origin and emplacement of Jurassic ophiolites of northern California. In Y. Dilek, E. M. Moores, D. Elthon, & A. Nicolas (eds.) Ophiolites and oceanic crust: new insights from field studies and the Ocean Drilling Program. Geological.

Ingersoll, R.V., & Busby, C. (1995). Tectonics of sedimentary basins. In C. J. Busby, & R. V. Ingersoll (eds.) Tectonics of sedimentary basins. Oxford, Blackwell Science, pp. 1–51.

Keary and Vine, 1990, Global Tectonics. Blackwell Oxford, 302.

Legg, M.R., Wong, O., & Suarez, V. (1991). Geologic structure and Tectonics of the Inner Continental Borderland of the Northern Baja California. In J. P. Dauphin, & B. Simoneit (eds.) The Gulf and Peninsular Province of the Californias, AAPG Memoir, 47, 145–196.

Mammerickx, J., & Klitgord, K. D. (1982). Northern East Pacific Rise: evolution from 25 m.y. to the present. *Journal of Geophysical Research, 87*, 6751–6759.

Michaud, F., Calmus, T., Ratzov, G., Royer, J.-Y., Sosson, M., Bigot-Cormier, F., et al. (2011). Active deformation along the southern end of the Tosco-Abreojos fault system: new insights from multibeam swath bathymetry. *Pure and Applied Geophysics, 168*(8–9), 1363–1372. doi:10.1007/s00024-010-0193-y.

Michaud, F., Calmus, T., Royer, J.-Y., Sosson, M., Bandy, B., Mortera-Gutierrez, C., Dyment, J., Bigot-Cormier, F., Chabert, A., & Bourgois, J. (2007). Right lateral active faulting between southern Baja California and the Pacific plate: the Tosco-Abreojos fault, In S. A. Alanis-Alvarez & A. F. Nieto-Samaniego (eds.), Celebrating the centenary of the geological society of Mexico: Geological Society of America, special paper 422, 287–300. doi:10.1130/2007.2422(09).

Michaud, F., Calmus, T., Sosson, M., Royer, J.-Y., Bourgois, J., Chabert, A., et al. (2005). La zona de falla Tosco-Abreojos: un sistema lateral derecho activo entre la placa Pacifico y la península de Baja California. *Boletín de la Sociedad Geológica Mexicana. Tomo LVII, 1*, 53–63.

Michaud, F., Royes, J. Y., Bourgois, J., Dyment, J., Calmus, T., Bandy, M., et al. (2006). Oceanic-ridge subduction vs. slab break off: plate tectonic evolution along the Baja California Sur continental margin since 15 Ma. *Geology, 34*(1), 13–16. doi:10.1130/g22050.1.

Michaud, F., Sosson, M., Royer, J.-Y., Chabert, A., Bourgois, J., Calmus, T., et al. (2004). Motion partitioning between the Pacific plate, Baja California and the North America plate: the Tosco-Abreojos fault revisited. *Geophysical Research Letters, 31*, L08604. doi:10.1029/2004GL019665.

Munguia, L., González-Escobar, M., Navarro, M., Valdez, T., Mayer, S., Aguirre, A., Wong, V., Luna, M. (2015b). Active crustal deformation in the area of San Carlos, Baja California Sur, Mexico as shown by data of Local Earthquake Sequences. Pure and Applied Geophysics. doi:10.1007/s00024-015-1217-4. Published online: 15 December 2015.

Munguía, L., Mayer, S., Aguirre, A., Méndez, I., González-Escobar and M., Luna, M. (2015a). The 2006 Bahía Asunción earthquake swarm: Seismic evidence of active deformation along the western margin of Baja California Sur, Mexico. Pure and Applied Geophysics. doi:10.1007/s00024-015-1184-9. Published online: 26 October 2015.

Nicholson, C., Sorlien, C. H. C., Atwater, T., Crowell, J. C., & Luyendyk, B. P. (1994). Microplate capture, rotation of the western Transverse Ranges, and initiation of the San Andreas transforms as a low-angle fault system. *Geology, 22*, 491–495.

Normark, W.R., Spencer, J.E., Ingle, J. (1987). Geology and Neogene history of the Pacific C continental margin of Baja California Sur, Mexico, In D. W. Scholl (ed.), Geology and resource potential of the continental margin of western North America and adjacent ocean basins—Beaufort Sea to Baja California: Houston, Texas, Circum-Pacific Counsel for Energy and Mineral Resources, Earth Science Series, 6, pp. 449–472.

Plattner, C., Malservisi, R., Dixon, T., Lafemina, P., Sella, G., Fletcher, J. M., et al. (2007). New constraints on relative motion between the Pacific plate and Baja California microplate (Mexico) from GPS measurements. *Geophysical Journal International, 170*(3), 1373–1380.

Plattner, C., Malservisi, R., & Govers, R. (2009). On the plate boundary forces that drive and resist Baja California motion. *Geology, 37*(4), 359–362.

Sangree, J. B., & Widmier, J. M. (1977). Seismic stratigraphy and global changes of sea level, Part 9: seismic interpretation of clastic depositional facies. In Charles E. Payton (Ed.), *Seismic Stratigraphy–applications to hydrocarbon exploration, AAPG Memoir 26*. Tulsa: AAPG.

Sedlock, R.L. (1993). Mesozoic geology and tectonics of blueschist and associated oceanic terranes in the Cedros-Vizcaino-San Benito and Magdalena-Santa Margarita regions, Baja California, Mexico, in Mesozoic Paleogeography of the Western Unitd States, Vol. II. In G.C. Dunne & K.A. McDugall (Eds.), *Society of Economic Paleontologists and Mineralogists, Pacific Section*, (pp.113–125). Los Angeles, CA.

Spencer, J. E., & Norwark, W. (1979). Tosco-Abreojos fault zone: a Neogene transform plate boundary within the Pacific margin of southern Baja California, Mexico. *Geology, 7*, 554–557.

Umhoefer, P. J., Dorsey, R. J., Willsey, S., Mayer, L., & Renne, P. (2001). Stratigraphy and geochronology of the Comondú Group near Loreto, Baja California sur, Mexico. *Sedimentary Geology, 144*, 125–147.

Yeats, R.S., Haq, B.U., Barron, J.A., Couch, J., Denham, C., Douglas, A.G., Grechin, V.I., Leinnen, M., Niem, A., Palverma, S., Pisciotto, K.A., Poore, R.Z., Shibata, T., & Wolfart, R. (1981). Initial reports of the Deep Sea Drilling Project: Washington, D.C., U.S. Government Printing Office, p. 967.

Yilmaz, O. (2001). Seismic Data Analysis. Investigations in Geophysics, Society of Exploration Geophysicists. Second Edition. Tulsa, p. 1000.

(Received November 26, 2015, revised September 6, 2016, accepted September 8, 2016, Published online September 22, 2016)

Reprinted from the journal

William L. Bandy et al., Geodynamics of the Latin American Pacific Margin (2017), 3663
© 2017 Springer International Publishing
DOI 10.1007/978-3-319-51529-8_24

Pure and Applied Geophysics

CrossMark

Erratum to: Geodynamics of the Latin American Pacific Margin

EDITED BY WILLIAM L. BANDY,[1] JUANJO DAÑOBEITIA,[2] CARLOS MORTERA GUTIÉRREZ,[3] YURI TARAN,[4] and RAFAEL BARTOLOMÉ[5]

Erratum to: William L. Bandy et al., Geodynamics
of the Latin American Pacific Margin,
DOI 10.1007/978-3-319-51529-8

On the title page and copyright page iv the spelling of the 3rd editor's name was incorrect.

On copyright page iv the affiliation of the 2nd editor was incorrect.

The correct author name and affiliation has been updated in book.

The updated online version of the original book can be found under DOI 10.1007/978-3-319-51529-8.

[1] Instituto de Geofísica, Universidad Nacional Autónoma de México, Ciudad Universitaria, Coyoacán, 04510 Mexico DF, Mexico.

[2] Unidad de Tecnología Marina, CSIC; Centro Mediterráneo de Investigaciones Marinas y Ambientales Paseo; Marítimo de la Barceloneta; 37-49, 08003 Barcelona, Spain.

[3] Instituto de Geofísica, Universidad Nacional Autónoma de México, 04510 Mexico, Mexico.

[4] Volcanology Department, Institute of Geophysics, UNAM, 3000 Av. Universidad, 04510 Mexico D.F., Mexico.

[5] Instituto de Ciencias del Mar-CSIC, P. Marítimo de la Barceloneta, 37-49, 08003 Barcelona, Spain.